Intelligent Systems Reference Library

Volume 201

The aim of this series is to publish a Reference Library, including novel advances and developments in all aspects of Intelligent Systems in an easily accessible and well structured form. The series includes reference works, handbooks, compendia, textbooks, well-structured monographs, dictionaries, and encyclopedias. It contains well integrated knowledge and current information in the field of Intelligent Systems. The series covers the theory, applications, and design methods of Intelligent Systems. Virtually all disciplines such as engineering, computer science, avionics, business, e-commerce, environment, healthcare, physics and life science are included. The list of topics spans all the areas of modern intelligent systems such as: Ambient intelligence, Computational intelligence, Social intelligence, Computational neuroscience, Artificial life, Virtual society, Cognitive systems, DNA and immunity-based systems, e-Learning and teaching, Human-centred computing and Machine ethics, Intelligent control, Intelligent data analysis, Knowledge-based paradigms, Knowledge management, Intelligent agents, Intelligent decision making, Intelligent network security, Interactive entertainment, Learning paradigms, Recommender systems, Robotics and Mechatronics including human-machine teaming, Self-organizing and adaptive systems, Soft computing including Neural systems, Fuzzy systems, Evolutionary computing and the Fusion of these paradigms, Perception and Vision, Web intelligence and Multimedia.

Indexed by SCOPUS, DBLP, zbMATH, SCImago.

All books published in the series are submitted for consideration in Web of Science.

More information about this subseries at http://www.springer.com/series/8578

Antonio Lloris Ruiz ·
Encarnación Castillo Morales · Luis Parrilla Roure ·
Antonio García Ríos · María José Lloris Meseguer

Arithmetic and Algebraic Circuits

Springer

Antonio Lloris Ruiz
Departamento de Electrónica y
Tecnología de Computadores
Campus Universitario Fuentenueva
Universidad de Granada
Granada, Spain

Encarnación Castillo Morales
Departamento de Electrónica y
Tecnología de Computadores
Campus Universitario Fuentenueva
Universidad de Granada
Granada, Spain

Luis Parrilla Roure
Departamento de Electrónica y
Tecnología de Computadores
Campus Universitario Fuentenueva
Universidad de Granada
Granada, Spain

Antonio García Ríos
Departamento de Electrónica y
Tecnología de Computadores
Campus Universitario Fuentenueva
Universidad de Granada
Granada, Spain

María José Lloris Meseguer
Oficina Española de Patentes y Marcas
O.A. (OEPM), Madrid, Spain

ISSN 1868-4394 ISSN 1868-4408 (electronic)
Intelligent Systems Reference Library
ISBN 978-3-030-67265-2 ISBN 978-3-030-67266-9 (eBook)
https://doi.org/10.1007/978-3-030-67266-9

This Springer imprint is published by the registered company Springer Nature Switzerland AG
The registered company address is: Gewerbestrasse 11, 6330 Cham, Switzerland

To our children and grandchildren

Julio

Lucía, Adriana and Pablo

José Luis and Sofía

Marina

Ana, Carmen and Jaime

who are the future

Prologue

Arithmetic Circuits are those digital circuits with inputs interpreted as numbers and whose outputs provide the results of some arithmetic operation over the inputs (addition, subtraction, multiplication, or division). These initial objectives (the elemental arithmetic operations) have been expanded so any mathematical function (trigonometrics, exponentials, logarithmics, etc.) is included as the purpose of the arithmetic circuits.

As a first definition, *Algebraic Circuits* are those digital circuits whose behaviour can be associated with any algebraic structure. Specifically, a polynomial is associated to each circuit, so that the evolution of the circuit will correspond to the algebraic properties of the polynomial. **LFSR**s (**L**inear **F**eedback **S**hift **R**egisters) and **CA**s (**C**ellular **A**utomata), included in this first denomination of algebraic circuits, are grouped under the name of *basic algebraic circuits*.

As a second definition, *Algebraic Circuits* are those digital circuits implementing the different operations within some algebraic structure. Specifically, in this book, this definition references to finite or Galois fields. The implementation of this *algebraic circuits* requires **LFSR**s and some basic arithmetic circuits.

This book is an expansion of our previous book *Algebraic Circuits*, including now arithmetic circuits, as both, arithmetic and algebraic, have many in common. Besides the addition of new material, each chapter includes a collection of exercises for didactic purposes.

The reader mainly interested in algebraic circuits will find the corresponding materials in Chaps. 1, 2, 3, 9, 10, 11, and 12; those interested only in arithmetic circuits may obviate Chaps. 3, 9, 10, 11. and 12.

Each chapter has been written as autonomous as possible from the rest of the book, thus avoiding back-consultation; this has the drawback of some redundancy, particularly of Chaps. 1 and 2 with those devoted to arithmetic circuits.

Chapter 1 is devoted to number systems, and a complete revision of the different representations of integer numbers is made, including redundant systems. The main procedures for the implementation of the fundamental arithmetic operations (addition, subtraction, multiplication, division, and square root) are also presented.

The implementation of those arithmetic circuits used for the construction of algebraic circuits is the purpose of Chap. 2. Addition, subtraction, multiplication, division (with special attention to modular reduction), and square root are implemented. Also, comparators and shifters, which can be considered to actually perform arithmetic operations but usually not considered as such, are described.

Chapter 3 deals with residue number systems, which are systems for numerical representation with interesting applications under the appropriate circumstances. Also, the Galois fields GF(p) are introduced in this chapter, since modular operations for prime values of p have to be implemented in GF(p).

Chapter 4 is mainly dedicated to the floating-point representation of real numbers, used profusely in arithmetic circuits. Rounding schemes and the IEEE 754 standard are presented, as well as circuit design to implement the main floating-point arithmetic operations. Also, the logarithmic system for real number representation is described.

Chapters 5–8 expand the basic arithmetic circuits presented in Chap. 2. As mentioned above, and in order to make each chapter autonomous, some redundance is allowed in these chapters.

Addition and subtraction are explored in detail in Chap. 5, specially all the questions associated to carry propagation for the construction of fast adders. Multioperand adders are described, to be used in the design of multipliers.

Multiplication is approached in Chap. 6, studying both combinational and sequential multipliers, as well as some special multipliers.

Division, the most complex of the basic arithmetic operations, is the object of Chap. 7. Division algorithms and their hardware implementations are sufficiently covered.

The computation of the most commonly used mathematical functions (logarithms, exponentials, trigonometrics, etc.) is the purpose of Chap. 8. The **CORDIC** algorithm is used as a general introduction to the procedures described in this chapter.

The basic algebraic circuits are the objective of Chap. 9. Regarding **LFSR**s, classic circuits, those storing a single bit in each cell, are introduced first. Then, they are generalized defining circuits with cells storing more than one bit each. The **CA**s studied in this chapter are mainly one-dimensional and linear, although two-dimensional **CA**s are also defined.

Chapter 10 is devoted to the Galois fields GF(2^n), presenting circuits to implement sums, products, divisions, squares, square roots, exponentiations, and inversions using power representation and the standard, normal, and dual basis. Also, the operations in the composite Galois fields GF($(2^n)^m$) are detailed.

Chapter 11 is parallel to Chap. 10, but refers to Galois fields GF(p^n) and GF($(p^n)^m$).

Chapter 12 presents two very simple cryptographic applications of Galois fields: the first is based on the use of discrete logarithms, and as a real example, the Galois field GF(2^{233}) is used. The second is devoted to elliptic curves, and as a real example, the Galois field GF(2^{192}—2^{64}—1) is used.

All related mathematical fundamentals concerning Galois fields are divided into three appendices structuring everything that is used in the corresponding chapters, without any demonstration of most of the theorems and algorithms. The objective of these appendices is to provide an immediate source and to unify the nomenclature. Readers interested in in-depth details may use the indicated references. In Appendix A, the postulates and theorems about Galois fields are provided. Appendix B is devoted to the algebra of polynomials, paying particular attention to the different forms of representation. Appendix C includes all matters relating to elliptic curves used in the application examples developed in Chap. 12. Appendix D elaborates on errors, an important question when dealing with arithmetic circuits. Finally, Appendix E describes some important algorithms for function implementation, while Chebyshev and Legendre sequences of orthogonal polynomials are also presented.

Written as a self-contained text, this book is mainly intended as a practical reference for designers of hardware applications, but also may be used as textbook for courses on arithmetic and/or algebraic circuits. The exercises at the end of each chapter facilitate the practice of the corresponding concepts.

Contents

Chapter 1
Number Systems

This chapter is devoted primarily to study the various representations used for integers, both binary and decimal, but also mentioned the fixed point representation of fractional numbers. Next, we analyze the basic operations (addition, subtraction, multiplication and division) with integers, without going into implementation details. Chapter finishes studying the representation of integers using signed digits. The following chapter discusses in more detail these elementary operations, presenting different implementations.

1.1 Introduction

The use of numbering systems for keeping records of cereals or livestock, and enabling the transactions or exchanges between different groups, are part of the first signs of civilization in the various cultures that emerged at the dawn of mankind, when humans went from nomadic to sedentary and agriculture and livestock raise.

In the most primitive number systems, each entity to account (sheep, amount of wheat, soldier, etc.) is represented by a more or less idealized image. For example, to verify that the shepherd who is entrusted with a flock is loss free, a pebble can be placed in a bowl for each head of cattle. Thus, when the flock returns it can be easily checked that at least there are so many head of cattle as pebbles in the bowl. Something similar was done in certain cultures when a war expedition was initiated. At the departure, each man deposited a stone (or other representative object) in a receptacle, and at the return of the expedition, each soldier picked up a stone from the receptacle. The stones not collected corresponded to casualties in battle. Note that the meaning of *Calculus* in Latin is precisely *pebble*, referring to the use of these objects in elementary arithmetic.

Either way, in all civilizations the need of using specific representations for groups of entities arose. For example, a herd of 147 sheep can be represented by 147 clay balls. Now if you want to reduce the number of balls, it is possible to use balls of

A. Lloris Ruiz et al., *Arithmetic and Algebraic Circuits*, Intelligent Systems Reference Library 201, https://doi.org/10.1007/978-3-030-67266-9_1

different sizes or shapes, representing the same herd with a large ball, four medium and seven small. Thus arises, by accounting needs, numbering bases, which certainly were not introduced for their better or worse suitability to the needs of calculation (i.e., they were more or less suitable for different arithmetic operations), but only by convenience in representation.

Due to anthropomorphic reasons is logical to assume (as indeed happened) that the first bases used were 5 and 10, because always have been counted with the fingers, and in one hand we have 5 fingers and between the two we have 10. The same explanation seems to be valid for the use of the base 20 in widely separated civilizations if in this case are used as reference both hands and feet. Even the use of the base 12, currently held in certain contexts, is explained by anthropologists because the phalanges of the four fingers opposing the thumb are twelve; jerking a thumb each of the twelve phalanges, with one hand it is easy to count up to 12. What it lacks is a clear explanation from the Sumerian use of the base 60, use that we continue for measuring time and angles. The base 60 seems to correspond to the joint use of bases 5 and 12; with one hand we can count dozens, and with the other we can count up to five dozen; with two hands we can easily count to sixty.

1.1.1 Additional Notation

Currently the base used in complete generality is 10, sometimes coexisting, in certain contexts, with the bases 60 and 12. To represent the different quantities are no longer used pebbles or clay pellets, but the figures from 0 to 9, whose origin is in India and came to the western world by the Arabs, which is why it is known among us as Arabic numerals. Sometimes (for example, to number the front pages of some books) Roman numerals are used, in which numeric values are represented using letters of the alphabet: basically the symbols I (one), V (five), X (ten), L (fifty), C (hundred), D (five hundred) and M (thousand), with some modifications in order to represent large numbers.

The Roman numeral system, in a similar way than other civilizations, is based on the additional principle for representing numbers. As known, when representing a given value using Roman numbers, it is decomposed in additions or differences of a prefixed values (one, five, ten, ...). In this sense, the Roman numeration is an **additional notation**. Although the values to be added are ordered from the highest to lowest, the value represented by each symbol not depends on the position occupied in the number being represented. As an example, in the number MMCMXLVIII, each M takes the value 1000, and C = 100, but because it is at the left of the third M, this value must be subtracted. The same occurs with the X with relation with the L, and the three I have the value 1 each one, resulting the final value of 2948.

As it is well known, Roman numerals result very cumbersome for arithmetic operations which explains the underdevelopment of the calculus in this culture.

1.1.2 *Positional Notation*

The usual decimal number system employs the important concept of positional or relative value: it is a **positional notation**, where each figure represents different values depending on the position occupied. As an example, in number 6362′65, the first 6 has the value 6000, the second 60, and the third 0’6, according to:

$$6362'65 = 6 \times 10^3 + 3 \times 10^2 + 6 \times 10^1 + 2 \times 10^0 + 6 \times 10^{-1} + 5 \times 10^{-2}$$

This form of representation makes arithmetic operations easier, improving the representation of large numbers.

While not always aware of it, in our daily life we use numbering systems handling simultaneously different bases. For example, the time taken by our computer for performing a complex calculation can be measured in days, hours, minutes and seconds. For the minutes and seconds, base 60 is used, while base 24 for hours. In any case, the addition and subtraction of time intervals are easily performed. Thus, adding 2 days, 15 h, 36 min, 18 s to 3 days, 12 h, 25 min and 52 s we have:

2	15		36	18
3	12		25	53
5	27		61	71

resulting after appropriate reductions, 6 days, 4 h, 2 min, 11 s.

Obviously, during a long time, only natural number were used, then the concepts of negative numbers, rational number, irrational numbers and imaginary number were appearing. In this chapter are considered essentially different representations of integers.

1.2 Positional Notation Using One Base

In this section, the representations of unsigned numbers with positional notation are considered; in Sect. 1.4, signed number representation will be contemplated.

In general, any unsigned value N, can be represented as a weighted sum of powers of another value, b, called base or radix, as follows:

$$N_b = a_n b^n + a_{n-1} b^{n-1} + \cdots + a_1 b^1 + a_0 b^0 + a_{-1} b^{-1} + \cdots + a_{-m} b^{-m} + \ldots \quad (1.1)$$

being N_b the representation of N in base b. If the base b is preset, N_b is given by the figures $a_n a_{n-1} \ldots a_1 a_0 a_{-1} \ldots a_{-m} \ldots$, and consists of an integer part $(a_n a_{n-1} \ldots a_1 a_0)$ and a fractional part $(a_{-1} \ldots a_{-m} \ldots)$, which can contains infinite figures. Until now, each figure in base b can take values in 0, 1, ..., $b - 1$ (b different values), in which case the expression of N in base b is unique; that is, if $a_i \in \{0, 1, \ldots, b - 1\}$, there is a unique expansion of N as a sum of powers of b.

Later, representations using positional notation in which each figure can take integer values in ranges different from $\{0, 1, \ldots, b - 1\}$ will be considered. However, always will be assumed that 0 is among the values that each figure can take, and that the integer values that each figure can take are consecutives.

Independently of b, the integer part of N_b represents the part of N being greater or equal than unity, while the fractional part corresponds to the part of N being less than unity. Therefore, if the same number N is represented in two different bases, b_1 and b_2, and N_1, N_2 are the respective representations, the integer part of N_1 will be equal to the integer part of N_2, and the fractional part of N_1 will be equal to the fractional part of N_2.

When using positional notation without limiting the number of digits, any number can be represented. In fact, given an integer number N to be represented using radix b, n digits are required, being n the value resulting from:

$$b^{n-1} < N + 1 \leq b^n$$

Conversely, if the number of digits, n, is set, the maximum integer value, N, that can be represented is given by:

$$N = b^n - 1$$

In real systems for data storing and/or processing, always will be a limit in the number of available digits, thus, there will be a limited range for values representation.

The radix b can take any value: it can be positive, negative, integer, fractional, rational, irrational or imaginary. Nevertheless, the more reasonable selection consists on using natural numerals (the other options, in general, are not advantageous), then being the minimal value $b = 2$ ($b = 1$ obviously has no sense). Radix 2 is widely used in the computers world, as known.

1.2.1 *Most Efficient Radix*

Thinking about the values of b, it results logical the raising of the question if, under certain assumptions, there is a recommended value for b, i.e. if there is a most efficient basis, regardless of anthropological reasons.

The objective in the different chapters of this book is the designing of circuits implementing arithmetic and algebraic operations. These circuits, when computing a given operation between two digits, will be more complex as the value of b raises. The same is true for the memory elements needed for storing each digit. In other words, the cost c_d per digit, for both processing and storing information, is a function of b, $c_d = f(b)$. Assuming that $f(b)$ is lineal [6]:

$$c_d = k_d b$$

Thus, for a number with n digits, the total cost will be:

$$C = k_d bn$$

If a range of M values is required for data processing, using radix b, n digits will be required, and:

$$M = b^n - 1 \quad \Rightarrow \quad n = \frac{\ln M}{\ln b}$$

thus:

$$C = k_d b \frac{\ln M}{\ln b} = K \frac{b}{\ln b} \quad (K = k_d \ln M)$$

The best radix b will be the one minimizing $\boldsymbol{C}$:

$$\frac{dC}{db} = k \frac{\ln b - 1}{(\ln b)^2} = 0 \quad \Rightarrow \quad \ln b = 1 \quad \Rightarrow \quad b = e$$

Then, the most efficient radix, assuming linearity for $\boldsymbol{f}(b)$, will be $\boldsymbol{e} = 2'718\ldots$ If an integer radix is desired, the recommended value is $b = 3$, because is the nearest integer to $\boldsymbol{e}$ number. Radix $b = 2$ results slightly less efficient, but has the advantage that electronic implementations in this base are the most reliable (it is easier to distinguish between two states instead of three states) and is the base used in computers.

1.2.2 Base Conversion

When using positional notation with only one radix, often there is a need of making a conversion from one base to another. Given the representation $\boldsymbol{N}_1$, of a number in base b_1, the goal is to obtain the representation $\boldsymbol{N}_2$ corresponding to the same number, but in terms of base b_2. For completing this conversion, the base b_1 or b_2, where the arithmetic operations for the conversion will be computed, must be chosen. Usually, one of the bases is 10, and it will be the preferred option. Thus, without entailing loss of generality, in what follows it is assumed that one of the bases implied in the conversion process is 10. Consequently, the problem of base conversion will be decomposed in two: the conversion from base 10 to any other base b, and from base b to base 10. First, the conversion from base 10 to base b is considered.

Given a number $\boldsymbol{N}_{10}$ represented in base 10, the issue consists on finding its representation $\boldsymbol{N}_b$ in another base b. In the base conversion, the integer part and the fractional part are processed separately. Starting with the integer part, if N_{10} is an integer number [i.e., the integer part of (1.1)], represented by the digits $a_n a_{n-1} \ldots a_1 a_0$, in base b, applying the Horner scheme it results:

$$N_{10} = N_b = a_n b^n + a_{n-1} b^{n-1} + \ldots + a_1 b^1 + a_0 b^0 =$$
$$= (\ldots (a_2 b + a_{n-1}) b + a_{n-2}) b + \ldots + a_1) b + a_0 \tag{1.2}$$

As known, when dividing a dividend D by a divisor d, a quotient C and a remainder r, are generated:

$$D = C \cdot d + r \tag{1.3}$$

Comparing (1.2) and (1.3), it can be concluded that a_0 is the remainder resulting from dividing N_{10} por b (the division is performed in base 10). If the quotient of this division is named C_0, then:

$$C_0 = (\ldots ((a_n b + a_{n-1}) b + a_{n-2}) b + \ldots + a_2) b + a_1,$$

resulting a_1 as the remainder of dividing C_0 by b, and so on for the quotients C_1, C_2, etc., until the last possible division, for which the remainder is a_{n-1}, and the quotient will be a_n.

In other words, for converting the representation of an integer N in base 10 to base b, the initial value N is divided by b, and all the successive quotients. Then, the digits of N represented in base b are, from the most significant to the less significant, the last quotient followed by all the remainders, from the last one to the fist one. As an example, the decimal number 947 can be expressed using radix 6 as follows:

$$947 = 157 \times 6 + 5$$
$$157 = 26 \times 6 + 1$$
$$26 = 4 \times 6 + 2$$

resulting:

$$947_{10} = 4215_6$$

and, converting 947_{10} to binary representation, it results on:

$$947_{10} = 1110110011_2$$

We will consider now the fractional part. Given a fractional number N_{10}, its polynomial expression in base b is the fractional part of (1.1):

$$N_{10} = N_b = a_{-1} b^{-1} + a_{-2} b^{-2} + \cdots$$

Multiplying by b (in base 10) the two members of the equality, it results:

$$N_{10}b = a_{-1} + a_{-2}b^{-1} + \cdots$$

Thus, a_{-1} is the integer part resulting from the multiplication of N_{10} by b. Subtracting a_{-1} from both of the members and multiplying again by b can be obtained a_{-2}:

$$(N_{10}b - a_{-1})b = a_{-2} + a_{-3}b^{-1} + \cdots$$

and so on. While the result of the multiplication by b is nonzero, new digits are added to the fractional part. This process may no terminate, resulting infinite digits in the fractional part. Some examples of base conversion are the following:

$$0.743_{10} = 0.512564\ldots_7$$
$$0.6_{10} = 0.100100100\ldots_2$$

For the inverse conversion, from base b to base 10, whether if converting the integer part and the fractional part, results enough to make the operations implied by the polynomial development (1.1) of N_b. As an example,

$$\begin{aligned} 4,601.73_8 &= 4\times 8^3 + 6\times 8^2 + 1\times 8^0 + 7\times 8^{-1} + 3\times 8^{-2} = 4\times 512 \\ &+ 6\times 64 + 1 + 7\times 0.125 \\ &+ 3\times 0.015625 = 2433.921875_{10} \end{aligned}$$

$$101101.0101_2 = 2^5 + 2^3 + 2^2 + 2^0 + 2^{-2} + 2^{-4} = 45.3125_{10}$$

A radix conversion especially simple occurs when one of the radices is a power of the other, i.e., the conversion is performed between radix B and b being:

$$B = b^m$$

In this situation, each digit in base B corresponds to m digits in base b. Thus, the conversion may be completed digit by digit in B or by m-digit blocks in b. Equally simple results the case when the two radices are power of the same number. As an example, when converting between bases 4 and 8, both of them power of 2, base 2 may be used as intermediate stage:

$$2301.201_4 = 10\ 11\ 00\ 01.10\ 00\ 01_2 = 010\ 110\ 001.100\ 001_2 = 261.41_8$$

Some bases powers of two are considered in Sect. 1.2.3.

When considering fractional numbers (less than unity), like $3/7_{10}$, for converting to another base, (2 in this example), dividing by 7 is required, resulting a periodic decimal number, and later multiplying iteratively by 2. Nevertheless it results more convenient the interleaving of both operations, thus multiplying by 2 before dividing by 7, as follows:

$$\left\lfloor \frac{3}{7} \right\rfloor = 0, \quad \left\lfloor \frac{6}{7} \right\rfloor = 0, \quad \left\lfloor \frac{12}{7} \right\rfloor = 1$$

where $\lfloor x \rfloor$ is the greatest integer less than or equal to x. Taking into account that 12/7 − 7/7 = 5/7, this is the remainder after the first 1. Continuing with the remainder of the last division, multiplied by 2, it results:

$$\left\lfloor \frac{10}{7} \right\rfloor = 1, \quad \frac{10}{7} - \frac{7}{7} = \frac{3}{7}$$

Since 3/7 has appeared before, a periodic sequence is obtained. Finally, we have:

$$3/7_{10} = 0.011011\ldots_2$$

It is clear that the binary representation of any decimal fraction will be periodic. In fact, if the denominator is N, there will be $N - 1$ different remainders, and a repetition will be outlined in a maximum of $N - 1$ iterations.

1.2.3 *Bases Power of Two*

Computers are built using binary circuits, thus radix 2 is the natural option for operating in them. The **binary number system** (or binary numbers) is a positional system, where the different positions are power of 2:

$$\ldots .32\ 16\ 8\ 4\ 2\ 1.\ 0.5\ 0.25\ldots$$

Each digit of a binary number can take the 0 and 1 values. Thus, it can be represented by a binary variable (as example, the state of a flip-flop). Also, it could be an option the use of radix −2, resulting the **negabinary number system**, advantageous in some situations, as will be outlined later. With this radix, the different positions take the value:

$$\ldots . -32\ 16\ -8\ 4\ -2\ 1.-0.5\ 0.25\ldots$$

Arithmetic operations (addition, subtraction, multiplication and division) can be performed in binary using the corresponding tables, as shown in Table 1.1.

Table 1.1 (a) Addition table, (b) subtraction table, (c) multiplication table

A	Carry	*S*	Borrow	Multiplication
$0+0=0$	0	$0-0=0$	0	$0\times 0=0$
$0+1=1$	0	$0-1=1$	1	$0\times 1=0$
$1+0=1$	0	$1-0=1$	0	$1\times 0=0$
$1+1=0$	1	$1-1=0$	0	$1\times 1=1$
(a)		(b)		(c)

The following examples show the application of these tables when using positive operands, and considering positive results.

Addition example
The addition of decimal number 19 y 53 in binary is:

$$
\begin{array}{rcccccccccl}
 & & & 1 & 0 & 1 & 1 & 1 & & \leftarrow \text{Carries} \\
19 & \rightarrow & & & 1 & 0 & 0 & 1 & 1 & \\
+53 & \rightarrow & + & 1 & 1 & 0 & 1 & 0 & 1 & \\
\hline
72 & \rightarrow & 1 & 0 & 0 & 1 & 0 & 0 & 0 &
\end{array}
$$

Subtraction example
Subtracting 34 from 85 it results 51:

$$
\begin{array}{rcccccccccl}
 & & 1 & 0 & 0 & 0 & 1 & 0 & & \leftarrow \text{Borrows} \\
85 & \rightarrow & 1 & 0 & 1 & 0 & 1 & 0 & 1 & \\
-34 & \rightarrow & - & 1 & 0 & 0 & 0 & 1 & 0 & \\
\hline
51 & \rightarrow & 0 & 1 & 1 & 0 & 0 & 1 & 1 &
\end{array}
$$

Multiplication example
Multiplying 25 by 13 it results 325:

$$
\begin{array}{rccccccccccc}
25 & \rightarrow & & & & & 1 & 1 & 0 & 0 & 1 \\
\times 13 & \rightarrow & & & & \times & & 1 & 1 & 0 & 1 \\
\hline
 & & & & & & 1 & 1 & 0 & 0 & 1 \\
 & & & & & 0 & 0 & 0 & 0 & 0 & \\
 & & & & 1 & 1 & 0 & 0 & 1 & & \\
 & & & 1 & 1 & 0 & 0 & 1 & & & \\
\hline
325 & \rightarrow & 1 & 0 & 1 & 0 & 0 & 0 & 1 & 0 & 1
\end{array}
$$

Division example
When dividing 437 by 38 it results a quotient of 11 and a remainder of 19:

437 →	1	1	0	1	1	0	1	0	1	1	0	0	1	1	0	← 38
	−1	0	0	1	1	0				1	0	1	1			← 11
		1	0	0	0	0	1	0								
			−1	0	0	1	1	0								
				1	1	1	0	0	1							
				−1	0	0	1	1	0							
19 →					1	0	0	1	1							

Of these four arithmetic operations, the only essential is the sum, meaning that the other operations may be computed using algorithms based on sums. Another elemental operation implemented frequently in digital circuits is the comparison. Given two binary numbers of the same length, (in general, two characters A and B), this operation has to decide on the relative value of the binary representation of both: if $A > B$, or $A = B$, or $A < B$. In the next chapter the synthesis of these arithmetic units is studied.

When representing a magnitude in a computer, a limited number of bits is available, regardless of the chosen radix representation. The most extended and simple solution is the use of a fixed number n of bits for numeral representation. The finiteness in the number of bits involves limits on the numeric values that can be represented. When trying to represent a number that is outside these limits (i.e., a very large or very small) and the computer is not programmed to anticipate this contingency, an error known as **overflow** (very large) or **underflow** (very small) happens. Thus, when performing calculations on a computer it is important to consider the possible overflows or underflows, for any errors that may entail.

The binary expression of a numeral may get a large number of bits, resulting disadvantageous for handling. The octal and hexadecimal bases are used for obtaining more compact representations.

When using octal base, the symbols 0, 1, 2, 3, 4, 5, 6 and 7 are used. As $8 = 2^3$, one octal digit corresponds to three binary digits, and vice versa. Thus, converting from octal to binary (or binary to octal) results easy: only expanding (or compressing) the digits from the decimal point is required, adding zeros to the right if needed. As an example:

46057.213_8	=	100	110	000	101	111'	010	001	011_2
		↔	↔	↔	↔	↔	↔	↔	↔
		4	6	0	5	7	2	1	3

11	000	100	111	001.	101	110	1_2	=	30471.564_8
↔	↔	↔	↔	↔	↔	↔	↔		
3	0	4	7	1	5	6	4		

In the last line, note that two zeros have been added implicitly to the right of the fractional part, resulting $100_2 = 4_8$.

When using radix 16 or hexadecimal (usually indicated by H sub index), the 0, 1, 2, 3, 4, 5, 6, 7, 8, 9, A, B, C, D, E, F symbols are used. Because $16 = 2^4$, one hexadecimal digit corresponds to four binary digits, and vice versa. Thus, for converting from hexadecimal to binary (or binary to hexadecimal) only expanding (or compressing, adding zeros to the left or the right, if needed) the digits, after the decimal point is required, as follows:

$3A57.2C3_H$	=	0011	1010	0101	0111'	0010	1100	0011_2
		↔	↔	↔	↔	↔	↔	↔
		3	A	5	7	2	C	3

1	0000	1000	1111.	1001	1101	1_2	=	$108H.9D8_H$
↔	↔	↔	↔	↔	↔	↔		
1	0	8	H	9	D	8		

Again, in this example, three zeros have been added to the fractional part of the binary number, resulting $1000_2 = 8_H$.

1.2.4 Modular Arithmetic

Sometimes, the operations with integer number are restricted to a limited range. In this situation the modular arithmetic may be interesting. Given a base or modulo M, $M > 0$, in the modular arithmetic, both the operands and the operations results are into the range $0 \leq X < M$. Given an integer number N out of this range, its modular representation is derived by means of the modular reduction operation. The modular reduction consists on assigning to each integer N, the positive remainder resulting from dividing N by M. The result of the modular reduction is represented as N **mod** M (or also as $|N|_M$ or **mod** (N, M)). As an example, for 17, being reduced modulo 7 is 17 **mod** 7 = 3, and for −3 modulo 5 is $-3\textbf{mod}5 = 2$.

Given a modulo, M, and two integers, A and B, with modular representations r and s, respectively, we have:

$$r = A \textbf{ mod } M \quad \Rightarrow \quad \frac{A}{M} = C + \frac{r}{M}$$
$$s = B \textbf{ mod } M \quad \Rightarrow \quad \frac{B}{M} = D + \frac{s}{M}$$

For the modular representation of the sum of A and B, $(A + B)$ **mod** M, we have:

$$\frac{A + B}{M} = (C + D) + \frac{r + s}{M}$$

Taking into account that $r + s$ may be greater than M, from the last equation it results:

$$(A + B)\,\mathbf{mod}\,M = (A\,\mathbf{mod}\,M + B\,\mathbf{mod}\,M)\,\mathbf{mod}\,M$$

For the modular representation of the product aA it results:

$$\frac{aA}{M} = aC + \frac{ar}{M}$$

Again, because ar may be greater than M, from this equality it results:

$$aA\,\mathbf{mod}\,M = \{a(A\,\mathbf{mod}\,M)\}\,\mathbf{mod}\,M$$

Applying all of these expressions to the development of an integer number N in positional notation as a sum of powers of the radix,

$$N = a_n b^n + a_{n-1} b^{n-1} + \cdots + a_1 b^1 + a_0 b^0$$

we have:

$$N\,\mathbf{mod}\,M = \left\{a_n\left(b^n\,\mathbf{mod}\,M\right) + a_{n-1}\left(b^{n-1}\,\mathbf{mod}\,M\right) + \cdots \right.$$
$$\left. + a_1\left(b^1\,\mathbf{mod}\,M\right) + a_0\left(b^0\,\mathbf{mod}\,M\right)\right\}\,\mathbf{mod}\,M$$

The modular reductions of the different powers of the radix can be pre-computed, being possible the simplification of the $N\mathbf{mod}M$ computing applying the following expression:

$$\left|a_n b^n + \cdots + a_0 b^0\right|_m = \left|a_n\left|b^n\right|_m + \cdots + a_0\left|b^0\right|_m\right|_m$$

Obviously, if M is a power of the base, $M = b^k$, the modular reduction of $N = a_n b^n + a_{n-1} b^{n-1} + \cdots + a_1 b^1 + a_0 b^0$ is trivial:

$$N\mathbf{A}\mathbf{mod}\,M = a_{k-1} b^{k-1} + a_{k-2} b^{k-2} + \cdots + a_1 b^1 + a_0 b^0$$

Other procedure for performing the modular reduction (named **multiplicative modular reduction**) is based on state M as the difference: $M = b^k - a$, being $1 \le a < b^{k-1}$. In this case, $N\mathbf{mod}M$ can be easily computed by means of successive multiplications by a, and modular reductions b^k, as it is shown in the following. Starting from:

$$N = c_0\, b^k + r_0$$

Multiplying c_0 (and the successive quotients) by a, and reducing modulo b^k, we obtain:

$$ac_0 = c_1\, b^k + r_1$$
$$ac_1 = c_2\, b^k + r_2$$
$$\dots$$
$$ac_{i-1} = c_i\, b^k + r_i$$

In each iteration, the result is multiplied by a and divided by b^k. Because $a < b^{k-1}$, after p iterations a zero quotient is obtained:

$$ac_{p-1} = 0\, b^k + r_p$$

Rearranging these equalities:

$$N = c_0\, b^k + r_0$$

$$0 = -ac_0 + c_1\, b^k + r_1$$
$$0 = -ac_1 + c_2\, b^k + r_2$$
$$0 = -ac_{i-1} + c_i\, b^k + r_j$$
$$\dots$$
$$0 = -ac_{p-1} + r_p$$

Memberwise adding the equalities:

$$N = c_0\left(b^k - a\right) + c_1\left(b^k - a\right) + \cdots + c_i\left(b^k - a\right) + \cdots + c_{p-1}\left(b^k - a\right) + r_0 + r_1 + \cdots + r_p$$

And:

$$N \mathbf{\ mod\ } \left(b^k - a\right) = \left(r_0 + r_1 + \cdots + r_p\right) \mathbf{\ mod\ } \left(b^k - a\right)$$

Thus, the remainder modulo $b^k - a$ of N may be computed by multiplying iteratively by a, being N the first product. Each product is decomposed in two fragments, $A_i b^k + B_i$. Each $A_i \neq 0$ is the new multiplicand, and the calculus ends when $A_i = 0$. The remainder is ΣB_i**mod**$(b^k - a)$.

In modular arithmetic, the addition, subtraction and multiplication (each of these three operations will be represented from now on with the symbol $\diamond$) of two integers is defined as follows. The division has sense only in some cases.

Given A, B and M ($0 \leq A < M$, $0 \leq B < M$), $A \diamond B$**mod**M is obtained by calculating $A \diamond B$ using normal arithmetic, and making a modular reduction to the result of $A \diamond B$.

Later, in Chap. 3, the modular arithmetic will be deeply studied when introducing the Residue Number System (**RNS**). Concretely, will be shown the properties derived when M is prime, being the possibility of defining division one of them.

1.2.5 Fractional Numbers: Fixed Point Representation

A fractional number consists of an integer part and a fractional part, both of them with a limited number of digits and separated by the named decimal mark (a point or a comma, or an apostrophe, depending on the country): 23.456, 0.0027 and 378.42196 are examples of fractional numbers. The real number can have a decimal development with infinite decimal digits (all of the irrational numbers and some rational). When a limited number of digits is used, like in digital systems, any real number will be represented as a fractional number.

For representing a fractional number N, besides the representation of the different digits, the representation (or indication) of the decimal mark position is required. In general, the number of the fractional digits can vary from a number N to another. Otherwise, if premised that the number of fractional digits is fixed for any number N, there is no need to indicate the decimal mark position. As an example, using 5 digits for the fractional part, the numbers 23.456, 0.0027 and 378.42196 will be represented as 2345600, 270 and 37842196. This idea of assigning always a fixed number of fractional digits, is the one used in the named **fixed point representation.**

If using positional notation with fixed point representation, each digit represents a different power of the radix and the decimal mark position can be indicated giving the power of the radix representing any digit. Specifically, the less significant digit is the one used for these task, defining the known as **unit in the last position** (*ulp*), i.e., which is the power of the radix corresponding to the less significant digit. If k digits are used for the fractional part, then $ulp = b^{-k}$. Obviously, *ulp* is also the difference between two consecutive representable values using k fractional digits. Thus, it can be used for measuring the **precision** that can be achieved with this representation. As a consequence, the difference between two consecutive values cannot be less than *ulp*.

When using a total of n digits (k for the fractional part and y $n - k$ for the integer part) for representing positive fractional number in positional notation in base b, the decimal value M of any represented number $d_{n-k-1}\ldots d_0 d_{-1}\ldots d_{-k}$ is:

$$M = \sum_{i=-k}^{n-k-1} d_i b^i$$

With k digits for the fractional part and $n - k$ for the integer part, the less positive representable value nonzero is b^{-k} (in this case, all of the digits are zero except the last, taking the 1 value), the maximum representable value is $b^{n-k} - b^{-k}$ (now, the n digits take the value $b - 1$); i.e.:

$$b^{-k} \leq M \leq b^{n-k} - b^{-k}$$

The various arithmetic operations using fractional numbers with fixed point can be performed, except minimal corrections, as if the number where integers. Specifically, the addition or subtraction of two fractional numbers is achieved by adding or subtracting as integer numbers, resulting a fractional number with the same number of fractional digits. However, in the case of multiplication, for achieving a result with the same number of fractional digits as the operands, after multiplying like integers, the result must be right shifted as many positions as fractional digits are being used. When dividing, the quotient will results an integer number, and the remainder will has the same number of fractional digits like the operands.

In that follows, when considering arithmetic operations with integer numbers, the various algorithms can be translated to fractional numbers represented with fixed point, introducing the adequate modifications, consisting on shifting the result when necessary.

When the representation of very large or very small numbers is required, the preferred solution is the scientific notation, in which real numbers are represented as the product of a fractional number by a power of the base: $0.000126 = 1.26 \times 10^{-4}$. Based on the scientific notation, the floating point representation has been developed, being the representation for real number used normally in computers.

Chapter 4 will be devoted to the floating point representation of binary and decimal numbers.

1.3 Multiple Radix Representations

The idea of representing a numeral as the addition of a fixed radix powers can be extended allowing the use of more than one radix. In that follows, two options for using multiple radixes are described.

1.3.1 Double Radix

In [3] the use of a double radix is proposed (specifically the base consisted of the numbers 2 and 3) so that an integer number N, is expressed in this system as:

$$N = \sum d_{ij} 2^i 3^j, \quad d_{ij} = 0, 1 \tag{1.4}$$

Thus, N is decomposed as the sum of products of powers of 2 and 3. Note that binary and ternary numbers results as particular cases of (1.4) making $j = 0$ and $i = 0$, respectively. The main advantage of this representation using two radices resides on the reduced number of summands appearing in the development (1.4), which could be translated onto an arithmetic operations simplification. Other radices different from 2 and 3 can be used, and also more than two radices for performing the decomposition of (1.4).

1.3.2 Mixed Radix

Another generalization of the positional notation consists on selecting the weights of the several digits by means of criteria different from using successive powers of one or more radices, as mentioned until now. This generalization opens multiple possibilities, being the most notable the known as **mixed radix representation**, as described in the following. Given n integers, $\{b_n, ..., b_1\}$, known as radices, the following weights are defined:

$$p_1 = 1, p_2 = p_1 b_1, \ldots, p_i = p_{i-1} b_{i-1}, \ldots, p_n = p_{n-1} b_{n-1}, \qquad (1.5)$$

where each weight p_i is associated to the radix with the same sub index, b_i.

Thus, any integer X, $0 \leq X < \Pi b_i = M$, can be represented in a mixed radix system $\{b_n, ..., b_1\}$ as follows:

$$X = a_n p_n + a_{n-1} p_{n-1} + \cdots + a_2 p_2 + a_1 p_1 \qquad (1.6)$$

being $0 \leq a_i < b_i$. Given a mixed radix system $\{b_n, ..., b_1\}$, X is represented by the digits $a_n a_{n-1} \ldots a_2 a_1$, and any integer in the range $[0, M - 1]$ has an unique representation. As an example, in the Table 1.2 the representations of the integer from 0 to 23 in the mixed radix system $\{4, 3, 2\}$ are shown. In this case, it results:

$$p_1 = 1, p_2 = p_1 b_1 = 1 \times 2 = 2, p_3 = p_2 b_2 == 2 \times 3 = 6$$

With respect to the coefficients, as $b_1 = 2$, a_1 can take values from the set $\{0, 1\}$; with $b_2 = 3$, it results $a_2 \in \{0, 1, 2\}$, $b_3 = 4 \Rightarrow a_3 \in \{0, 1, 2, 3\}$. Thus, any integer N in the range (0, 23) can be uniquely decomposed as $= a_3 \times 6 + a_2 \times 2 + a_1$, and

Table 1.2 Mixed radix system {4, 3, 2}

No	$a_3\ a_2\ a_1$	No	$a_3\ a_2\ a_1$	No	$a_3\ a_2\ a_1$	No	$a_3\ a_2\ a_1$
0	0 0 0	**6**	1 0 0	**12**	2 0 0	**18**	3 0 0
1	0 0 1	**7**	1 0 1	**13**	2 0 1	**19**	3 0 1
2	0 1 0	**8**	1 1 0	**14**	2 1 0	**20**	3 1 0
3	0 1 1	**9**	1 1 1	**15**	2 1 1	**21**	3 1 1
4	0 2 0	**10**	1 2 0	**16**	2 2 0	**22**	3 2 0
5	0 2 1	**11**	1 2 1	**17**	2 2 1	**23**	3 2 1

Table 1.3 Mixed radix system {2, 3, 4}

No	$a_3\ a_2\ a_1$	No	$a_3\ a_2\ a_1$	No	$a_3\ a_2\ a_1$	No	$a_3\ a_2\ a_1$
0	0 0 0	**6**	0 1 2	**12**	1 0 0	**18**	1 1 2
1	0 0 1	**7**	0 1 3	**13**	1 0 1	**19**	1 1 3
2	0 0 2	**8**	0 2 0	**14**	1 0 2	**20**	1 2 0
3	0 0 3	**9**	0 2 1	**15**	1 0 3	**21**	1 2 1
4	0 1 0	**10**	0 2 2	**16**	1 1 0	**22**	1 2 2
5	0 1 1	**11**	0 2 3	**17**	1 1 1	**23**	1 2 3

N will be written as $N = a_3a_2a_1$. As an example, $11_{10} = 1 \times 6 + 2 \times 2 + 1 = 121$, and $21_{10} = 3 \times 6 + 1 \times 2 + 1 = 311$.

When using the mixed radix system {2, 3, 4}, the same range of values can be represented, from 0 to 23, but the representations of numbers are, in general, different from those obtained with the system {4, 3, 2}. Now, $p_1 = 1$, $p_2 = 4$, $p_3 = 12$; and a_1 can take values from the set {0, 1, 2, 3}, a_2 from the set {0, 1, 2}, a_3 from the set {0, 1}. Table 1.3 shows the representation of the integer from 0 to 23 in the mixed radix system {2, 3, 4}.

When all radices are equal, $b_n = \ldots = b_1 = b$, the mixed radix representation is reduced to the positional representation with one radix, b. Thus, the positional representation with one radix is a particular case of mixed radix representation.

From (1.6) it is clear that a_n is the quotient resulting from de division of X by p_n, being the remainder R_1:

$$R_1 = a_{n-1}p_{n-1} + \cdots + a_2p_2 + a_1p_1 \tag{1.7}$$

This division by p_n allows the calculus of a_n, which will be performed using a positional representation system (binary or decimal, for example). Again, from (1.7) it results a_{n-1} as the quotient from dividing R_1 by p_{n-1}. Repeating this process, all the digits of the development with mixed radix can be computed. In the last division, the divider will be p_2, the quotient a_2, and the remainder a_1, because $p_1 = 1$. In this way, a procedure for converting from a positional system with one radix to a mixed radix system is provided.

For converting from the mixed radix system to a positional system with one radix, it results enough evaluating the development (1.6).

Example 1.1 Find the representation of 27435 in the mixed radix system {2, 3, 5, 7, 11, 13}.

First, 27,435 lies into the representation range. In fact, in this system, any integer less than $2 \times 3 \times 5 \times 7 \times 11 \times 13 = 30{,}030$ can be presented.

The weights are:

$$p_1 = 1,\ p_2 = 13,\ p_3 = 13 \times 11 = 143,\ p_4 = 143 \times 7 = 1001,$$
$$p_5 = 1001 \times 5 = 5005,\ p_6 = 5005 \times 3 = 15015.$$

The quotient resulting from dividing 27,435 by 15015 is a_6. Thus:

$$27{,}435 = 15{,}015 \times 1 + 12{,}420 \Rightarrow a_6 = 1$$

In a similar way, a_5, a_4, a_3 and a_2 are obtained, and a_1 is the last remainder:

$$\begin{aligned} 12{,}420 &= 5005 \times 2 + 2410 \Rightarrow a_5 = 2 \\ 2410 &= 1001 \times 2 + 408 \Rightarrow a_4 = 2 \\ 408 &= 143 \times 2 + 122 \Rightarrow a_3 = 2 \\ 122 &= 13 \times 9 + 5 \Rightarrow a_2 = 9, a_1 = 5 \end{aligned}$$

Resulting:

$$27{,}435 = 1 \times 15{,}015 + 2 \times 5005 + 2 \times 1001 + 2 \times 143 + 9 \times 13 + 5 = 122{,}295$$

■

Example 1.2 Find the decimal value, N, corresponding to the number 5(10)6(13) represented in the mixed radix system {7, 11, 13, 15}.

First, the weights in this mixed radix system are:

$$p_1 = 1,\ p_2 = 15,\ p_3 = 13 \times 15 = 195,\ p_4 = 195 \times 11 = 2145$$

Thus:

$$N = 2145 \times 5 + 195 \times 10 + 15 \times 6 + 13 = 12{,}778$$

■

The mixed radix system results of special interest for the Residue Number System, as will be outlined in Chap. 3.

1.4 Negative Integer Numbers

So far, only natural number (i.e., positive integers) representations have been considered. The next step will consist on include the representation of negative integer numbers. As known, independently of the base of the number, two symbols (+ and −) are used for numbers sign representation. This corresponds to a binary situation, and a bit (or a digit) could be used for sign representation. The convention is to use 0 for representing the + sign, and 1 for the − sign. Thus, for representing any number A in base b, $n + 1$ digits are used, being one of them the digit corresponding to the sign, specifically the most significant digit. The others n digits corresponds to the magnitude or module of the number, assuming that n is greater or equal than the minimum number of digits required for representing A in base b.

Table 1.4 **SM**, two's complement (C2), one's complement (C1) and biased (−16 and −15) values of the five bit numbers

$a_4a_3a_2a_1a_0$	**SM**	C2	C1	−16	−15	$a_4a_3a_2a_1a_0$	**SM**	C2	C1	−16	−15
00000	+0	+0	+0	−16	−15	10000	−0	−16	−15	+0	+1
00001	+1	+1	+1	−15	−14	10001	−1	−15	−14	+1	+2
00010	+2	+2	+2	−14	−13	10010	−2	−14	−13	+2	+3
00011	+3	+3	+3	−13	−12	10011	−3	−13	−12	+3	+4
00100	+4	+4	+4	−12	−11	10100	−4	−12	−11	+4	+5
00101	+5	+5	+5	−11	−10	10101	−5	−11	−10	+5	+6
00110	+6	+6	+6	−10	−9	10110	−6	−10	−9	+6	+7
00111	+7	+7	+7	−9	−8	10111	−7	−9	−8	+7	+8
01000	+8	+8	+8	−8	−7	11000	−8	−8	−7	+8	+9
01001	+9	+9	+9	−7	−6	11001	−9	−7	−6	+9	+10
01010	+10	+10	+10	−6	−5	11010	−10	−6	−5	+10	+11
01011	+11	+11	+11	−5	−4	11011	−11	−5	−4	+11	+12
01100	+12	+12	+12	−4	−3	11100	−12	−4	−3	+12	+13
01101	+13	+13	+13	−3	−2	11101	−13	−3	−2	+13	+14
01110	+14	+14	+14	−2	−1	11110	−14	−2	−1	+14	+15
01111	+15	+15	+15	−1	−0	11111	−15	−1	−0	+15	+16

Three conventions for representing signed number will be considered: Sign-Magnitude (**SM**), base complement (two's complement in binary) and base-1 complement (one's complement in binary). In all three cases the representation of positive numbers is the same: the sign bit is 0, and the magnitude is given like in natural numbers. The conventions differ in the negative numbers representation, as will be remarked in that follows, describing the addition and subtraction operations. Later, multiplication and division will be considered. Also, biased representations will be described, allowing the representation of positive and negative numbers without using a sign bit. In Table 1.4, the equivalences among the different representations for five bits binary numbers including the sign are shown.

1.4.1 SM Representation

Sign-Magnitude representation (**SM**) is the one used daily in the decimal system. Negative numbers are represented by placing the − symbol before the magnitude: 97 and −97 represent the same magnitude, the first one positive, and the second one negative. In binary, negative numbers are represented as follows: the sign bit is 1, and the magnitude or module is given in binary. Using n digits for the magnitude, it is possible to represent number from $+(b^n - 1)$ to $-(b^{n-1})$. Any number out of this range will produce an overflow.

Arithmetic operations with **SM** representation are performed in the same way as when using positive numbers, as described before. In addition and subtraction (only two operands), taking into account the operands signs and their relative value, the result sign is decided, and the type of operation to complete for obtaining the final result. Specifically, the addition of two numbers with the same sign will produce a result taking the same sign than the operands, and with a magnitude being the sum of the operands magnitudes. If the numbers have different sign, the magnitudes have to be compared, and the result has the sign of the operand with greater magnitude, being the magnitude of the result the difference between the two operands magnitudes. Similar rules are applied for subtraction. Therefore, a circuit for adding and subtracting signed number using **SM** representation, must include a comparator, an adder and a subtractor. Examples of addition and subtraction of **SM** binary numbers using one bit for sign and four bits for magnitude are:

$$\begin{array}{rcr} +8 & \rightarrow & 01000 \\ +(-4) & \rightarrow & +10100 \\ \hline +4 & \rightarrow & 00100 \end{array} \qquad \begin{array}{rcr} +4 & \rightarrow & 00100 \\ +(-8) & \rightarrow & +11000 \\ \hline -4 & \rightarrow & 10100 \end{array}$$

$$\begin{array}{rcr} +8 & \rightarrow & 01000 \\ -(-4) & \rightarrow & -10100 \\ \hline +12 & \rightarrow & 01100 \end{array} \qquad \begin{array}{rcr} -4 & \rightarrow & 10100 \\ -(-8) & \rightarrow & -11000 \\ \hline +4 & \rightarrow & 00100 \end{array}$$

Note that the subtractor, when needed, can be changed by a combination of an adder and a inverter. In fact, if a carry bit is added (i.e., adding 2^n if n bits are used for magnitude), that will be discarded once the operation is accomplished:

$$A - B + 2^n = A + \left(2^n - 1 - B\right) + 1$$

But $\overline{B} = (2^n - 1 - B)$ can be computed from B complementing each bit. Thus, $A - B$ can be calculated adding A and $\overline{B}$ and adding 1 to the result. As an example, consider the addition (+5) + (−8): comparing the magnitudes the result will be negative, with a magnitude of:

$$8 - 5 = 8 + \overline{5} + 1 = 1000 + 1010 + 0001 = 10011$$

that results 0011 if the carry is discarded. Joining the sign bit and the magnitude, the final result is 10011. Later will be shown this is equivalent to use one's complement. The presented subtraction way only produces correct results if the minuend is greater than the subtrahend. In other words, for adding/subtracting with **SM** representation, a comparator is needed in order to decide which operand is the minuend and which the subtrahend when a subtraction is required.

In addition and subtraction operations, overflow of the result can be produced. Specifically, when adding, overflow can appear when the two summands have the same sign (never if the summands have different sign) and in subtraction when the two operands have different sign (never if they have the same sign). Overflow can be detected testing if a carry is produced from the most significant digit of the magnitude to the sign bit. As example, when adding 9 and 12 with a sign bit and four bits for the module, we have:

Sign ↓	1 ← Carry	
0	1001	← 9
+0	1100	← 12
0	0101	← 5

A carry is generated from the most significant digit of the magnitude, which is equivalent to an overflow. The same occurs when adding -9 y -12, as can be easily verified.

In **SM** representation, the sign bit and the magnitude bits are separately processed. For the magnitude, a fixed number of bits are reserved, and, if an expansion in the number of bits is required (when performing a format conversion, or during a calculus process), the addition of zeros to the left of the most significant digit and to the right of the less significant digit (if fractional numbers are used) is sufficient. The following series of bits represents the same number in **SM** representation: 1001.011, 0100.0110, 0001001.011, ... Previously to the magnitude is the sign bit, thus, if the previous bit series will correspond to negative numbers, the complete representation including the sign bit will be: 11001.011, 101001.0110, 10001001.011.

As shown in Table 1.4, when using **SM**, there are two representations for 0 (+0 and -0), with different sign bit. This redundancy can be a disadvantage in practice, when the zero value should be detected.

1.4.2 Complement Representations

When using complement representation in positional notation over base b, both positive and negative numbers are represented using positive values. For a specific representation, a complementation constant $\boldsymbol{C}$ is chosen and any negative number $-N$ is represented as $\boldsymbol{C} - N \geq \mathbf{0}$. Positive numbers remain unmodified.

For the complement representation to be useful, the value range for negative numbers should not overlap with the range reserved for the positive numbers. Assuming that n digits are used for representing positive numbers, a positive number P will be in the range $0 \leq P < b^n$ and $ulp = 1$. When using fractional numbers, with k digits for representing the integer part, and f digits for the fractional part, the range of possible values is $b^{-f} \leq P < b^k$, and $ulp = b^{-f}$. Nevertheless the modifications regarding the ranges, the following ideas for integer numbers are easily translated to fractional numbers.

Thus, considering only integer numbers, the range for negative numbers should be chosen satisfying that $C - N \geq b^n$. No other restriction limits the range, but it is reasonable to impose that the negative numbers range to be similar to the positive numbers range. So, if $0 \leq N < b^n$, the greater value for N is $b^n - 1$, and:

$$C \geq b^n + b^n - 1 = b^{n+1} - 1$$

(If using fractional numbers, it is easy to verify that $C \geq b^{k+1} - ulp$, being k the number of integer digits).

Note that the condition $C \geq b^{n+1} - 1$ makes equivalent the complement representation to a modular representation, modulo C.

In practice, two values are used for C: $C = b^{n+1}$ and $C = b^{n+1} - 1$, resulting the base complement representation, and the base-1 complement representation, respectively.

1.4.2.1 Base Complement Representations

Given an n bits positive number in base b, $N = a_{n-1} \ldots a_0$, the base b complement representation of $-N$ is defined as $b^{n+1} - \sum_{i=0}^{n-1} a_i b^i$, that can be rewritten as follows:

$$b^{n+1} - \sum_{i=0}^{n-1} a_i b^i = b^{n+1} - b^n + \sum_{i=0}^{n-1} \left(b^{i+1} - b^i\right) + 1 - \sum_{i=0}^{n-1} a_i b^i =$$

$$= (b-1)b^n + \sum_{i=0}^{n-1} (b - 1 - a_i) b^i + 1$$

Thus, the base b complement representation of $-N$ can be obtained changing each N digit a_i by $(b - 1 - a_i)$ an adding 1 to the result. The $(b - 1 - a_i)$ value is known as the a_i **digit complement**. Consequently, in the base complement representation, given a positive number, $+N$, the complement, $-N$, can be calculated complementing each digit and adding 1 to the result. In this complementing operation, it appears a new digit, a_n, that will be the new most significant digit, being $a_n = 0$ for positive numbers, and $a_n = (b - 1)$ for negative numbers. This digit will be known as the **sign digit**.

As the sign digit can take only two values, it can be reduced to one bit, the **sign bit**, s_n, defined as:

$$s_n = \frac{a_n}{b-1}$$

resulting $s_n = 0$ for positive numbers and $s_n = 1$ for negative.

The range of values when using n digits for magnitude in the base complement representation is $-b^{n+1} \leq R \leq b^{n+1} - 1$.

Given a negative number $-N$ ($a_n a_{n-1} \ldots a_0$, with $a_n = b - 1$) in base complement, the corresponding positive value $+N$ results from performing the complementing operation. In fact:

$$b^{n+1} - \sum_{i=0}^{n} a_i b^i = \sum_{i=0}^{n} \left(b^{i+1} - b^i\right) + 1 - \sum_{i=0}^{n} a_i b^i = \sum_{i=0}^{n-1} (b - 1 - a_i) b^i + 1$$

Thus, each digit a_i is changed with its complement, $b - 1 - a_i$, and the result is increased by 1, leading to a sign digit equal to 0. If a sign bit is used instead of the sign digit, the complementation of the magnitude digits remains unchanged, but the sign bit must be complemented in binary.

The zero value is represented in base complement as a positive number and has a unique representation, because $b^{n+1} - 0$ results 0 when the carry (b^{n+1}) is discarded.

Note that the decimal value X (independently if the number is a positive or a negative one) corresponding to the number $s_n a_{n-1} \ldots a_0$, represented using base b complement (with sign bit instead sign digit), is given by:

$$X = -s_n b^n + \sum_{i=0}^{n} a_i b^i \tag{1.8}$$

In fact, if $s_n = 0$ (positive number), it results from (1.8) that $X = \sum_{i=0}^{n-1} a_i b^i$, as expected. If $s_n = 1$ (negative number), then the value resulting from (1.8) is $X = -b^n + \sum_{i=0}^{n-1} a_i b^i$. By the other hand, the represented value is:

$$-\left\{1 + \sum_{i=0}^{n-1} (b - 1 - a_i) b^i\right\} = -1 - \sum_{i=0}^{n-1} b^{i+1} + \sum_{i=0}^{n-1} b^i + \sum_{i=0}^{n-1} a_i b^i = -b^n + \sum_{i=0}^{n-1} a_i b^i$$

being equal to X.

1.4.2.2 Base −1 Complement Representations

Given a positive number with n bits in base b, $N = a_{n-1} \ldots a_0$, the base-1 complement representation of $-N$ is defined as $b^{n+1} - 1 - \sum_{i=0}^{n-1} a_i b^i$, that can be developed as follows:

$$b^{n+1} - 1 - \sum_{i=0}^{n-1} a_i b^i = b^{n+1} - b^n + \sum_{i=0}^{n-1} \left(b^{i+1} - b^i\right) - \sum_{i=0}^{n-1} a_i b^i = (b - 1) b^n$$

$$+\sum_{i=0}^{n-1}(b-1-a_i)b^i$$

Thus, each digit a_i of N is changed by its complement $(b - 1 - a_i)$. Again, in this complementing operation appears one additional digit, (that will be the most significant digit), being 0 for positive numbers, and $(b - 1)$ for negative ones. This digit will be the **sign digit**, which can be substituted by a **sign bit**.

Using n digits for the magnitude, the values range available for representing number using base-1 complement representation is $-b^{n+1} + 1 \leq R \leq b^{n+1} - 1$. So, base-1 complement representation presents a lower range than base complement.

Given a negative number $-N$ ($a_n a_{n-1} \ldots a_0$, with $a_n = b - 1$) in base-1 complement, the corresponding positive number $+N$ can be obtained applying the complementing operation. In fact:

$$b^{n+1} - 1 - \sum_{i=0}^{n-1} a_i b^i = \sum_{i=0}^{n}\left(b^{i+1} - b^i\right) - \sum_{i=0}^{n} a_i b^i = \sum_{i=0}^{n-1}(b-1-a_i)b^i$$

Thus, each digit a_i is substituted by its complement, $b - 1 - a_i$, and the sign digit is converted to 0. Again, if a bit sign is used instead of a digit sign, the magnitude digits remain unchanged, but the sign bit must be complemented in binary.

About the zero value in base-1 complement representation, there are two representations, one positive and other negative: $+\,0$ and $b^{n+1} - 1$. In fact, $b^{n+1} - 1 - 0 = b^{n+1} - 1$, is a negative number, and different from $+\,0$. This double representation of zero introduces a redundancy in the base-1 complement representation.

The decimal value X (independently if the number is positive or negative) corresponding to the number $s_n a_{n-1} \ldots a_0$, represented using base-1 complement (with sign bit instead sign digit), is given by:

$$X = -s_n\left(b^n - 1\right) + \sum_{i=0}^{n-1} a_i b^i$$

In fact, if $s_n = 0$ (positive number), it results $X = \sum_{i=0}^{n-1} a_i b^i$, as expected. If $s_n = 1$ (negative number), then the value resulting for X is $X = 1 - b^n + \sum_{i=0}^{n-1} a_i b^i$. By the other hand, the value represented is:

$$-\left\{\sum_{i=0}^{n-1}(b-1-a_i)b^i\right\} = -\sum_{i=0}^{n-1} b^{i+1} + \sum_{i=0}^{n-1} b^i + \sum_{i=0}^{n-1} a_i b^i = 1 - b^n + \sum_{i=0}^{n-1} a_i b^i$$

being equal to X.

1.4.2.3 Base Complement Addition and Subtraction

The main advantage of complement representations is the reduction of addition and subtraction to only one operation, as detailed in that follow. As usual, only two operands will be assumed, with $n + 1$ digits each one, being the most significant digit the sign.

When the summands are represented using base complement, and both of them are positive, the sum $A + B$ can be computed directly, including the sign digit, and resulting the correct result, except if $A + B \geq b^n$, then producing an overflow that will be studied in Sect. 1.4.2.5.

If one of the summands is positive and the other negative (i.e., we are calculating $A + (b^{n+1} - B)$) with $A \geq B$, the result must be positive, $A - B$. In this situation, the sum produces the correct result and a carry digit b^{n+1}:

$$A + \left(b^{n+1} - B\right) = b^{n+1} + (A - B)$$

Thus, the correct result is obtained discarding the carry.

If one of the summands is positive and the other negative (again, we are calculating $A + (b^{n+1} - B)$), but $A < B$, the result of the sum $A + (b^{n+1} - B)$ must be negative. In fact:

$$A + \left(b^{n+1} - B\right) = b^{n+1} - (B - A)$$

If the two summands are negative, i.e. $(b^{n+1} - A) + (b^{n+1} - B)$, with $A + B < b^n$ (if $A + B \geq b^n$, an overflow will be produced, as will be seen in Sect. 1.4.2.5), the result must be negative, $b^{n+1} - (A + B)$. The direct sum of both of the representations also provides the correct result. In fact:

$$\left(b^{n+1} - A\right) + \left(b^{n+1} - B\right) = b^{n+1} + \left\{b^{n+1} - (A + B)\right\}$$

Because $A + B < b^n$, $\{b^{n+1} - (A + B)\}$ is a negative number. The direct sum generates the correct result plus a carry digit, b^{n+1}. Thus, again, the carry must be discarded for obtaining the final result.

Resuming, the direct sum when using base complement representations, always generates the correct result (except overflow situations). In this operation, the carry is discarded.

If a sign bit is used instead of the sign digit, all described before is valid, but the sign bit must be added in binary.

Subtraction is converted in addition complementing de subtrahend, that is, $A - B = A + (-B)$. In this way, addition and subtraction can be joined in only one operation.

1.4.2.4 Addition and Subtraction Using Base-1 Complement

Considering base-1 complement representation, the sum $A + B$ when the two summands are positive, can be computed by means of the direct sum (including the sign digit or bit), obtaining the correct result (in a similar way than **SM**), except if an overflow is produced, situation analyzed in the next section.

If one of the summands is positive and the other is negative (i.e., $A + (b^{n+1} - 1 - B)$ is calculated) being $A \geq B$, the result, $A - B$, must be positive. The direct addition does not provide the correct result in this situation:

$$A + \left(b^{n+1} - 1 - B\right) = b^{n+1} + (A - B) - 1$$

Thus, a carry digit, b^{n+1}, is generated, and the correct result appears diminished in one. For the correct result to be obtained, adding 1 to $(A - B) - 1$ when a carry is generated, is needed. This correction is known as **end-around carry**.

If one of the summands is positive and the other negative $\{A + (b^{n+1} - 1 - B)\}$ with $A < B$, the sum $A + (b^{n+1} - 1 - B)$ provides the correct result, which is negative:

$$A + \left(b^{n+1} - 1 - B\right) = b^{n+1} - 1 - (B - A)$$

If the two summands are negatives, $(b^{n+1} - 1 - A) + (b^{n+1} - 1 - B)$, with $A + B < b^n$ (if $A + B \geq b^n$ an overflow is generated, as will be seen in the next section), a negative result must be generated, $b^{n+1} - 1 - (A + B)$. The direct sum does not provide the correct result. In fact:

$$\left(b^{n+1} - 1 - A\right) + \left(b^{n+1} - 1 - B\right) = b^{n+1} + \left\{b^{n+1} - 1 - (A + B)\right\} - 1$$

Because $A + B < b^n$, $\{b^{n+1} - 1 - (A + B)\}$ is a negative number. Also, a carry digit, b^{n+1}, is produced. Again, if the end-around carry correction is applied, the correct result is obtained.

As a conclusion, performing the direct sum when using base-1 complement representation generates the correct result if the end-around carry correction is applied.

When a sign bit is used instead the sign digit, the sign bit must be added in binary.

Again, subtraction can be performed using addition by means of subtrahend complementation. Thus, $A - B = A + (-B)$, joining addition and subtraction in only one operation.

Comparing the two complement representations, complementation results easier in base-1 complement (there is no need of adding 1), but the addition operation results more simple in base complement (there is no need of end-around carry correction). Moreover, the base-1 complement results redundant, introducing same disadvantages for zero detection. As a consequence, the most used representation is the base complement one.

1.4.2.5 Overflows

Overflow in addition or subtraction, whatever of the representation used, only can be produced when the final operation computed consists on the addition of two numbers with the same sign. When using **SM** representation, the sign bit is treated in a different way than the magnitude bits, and overflow is detected because a carry from the most significant digit to the sign digit (or sign bit) is generated.

When using complement representations, overflow is detected because the addition of two numbers with the same sign produces a number with the other sign. In fact, consider sum $A + B$ both in base or base-1 complement representation, and $A + B \geq b^n$. When the two summands are positive, the sign digits of A and B are 0, and adding the carry generated in $A + B$ to the sign, the resulting sign digit will be non zero.

If the two summands are negative, in base complement $(b^{n+1} - A)$ and $(b^{n+1} - B)$, with $A + B \geq b^n$, the direct sum of both representations results on:

$$\left(b^{n+1} - A\right) + \left(b^{n+1} - B\right) = b^{n+1} + \left\{b^{n+1} - (A + B)\right\}$$

Thus, a carry digit, b^{n+1}, is generated, which in base complement is discarded. By the other hand, $A + B \geq b^n$, $\{b^{n+1} - (A + B)\}$ resulting a non negative number.

In base-1 complement, if the two summands are negative, $(b^{n+1} - 1 - A)$ and $(b^{n+1} - 1 - B)$, with $A + B \geq b^n$, it results:

$$\left(b^{n+1} - 1 - A\right) + \left(b^{n+1} - 1 - B\right) = b^{n+1} + \left\{b^{n+1} - 1 - (A + B)\right\} - 1$$

Now, a carry digit, b^{n+1}, is generated, which must be taken into account for the end-around carry. By the other hand, $A + B \geq b^n$, $\{b^{n+1} - 1 - (A + B)\}$ resulting a non-negative number.

Next, two examples of additions and subtractions of decimal numbers in 10 complement, and 9 complement are presented, respectively.

Example 1.3 Additions (and subtractions) of decimal numbers using base complement representation.

Let us use three digits for the magnitude. There will be a fourth digit for sign representation, which will be 0 for positive numbers, and 9 for negative ones. Thus, the value range goes from -1000 to $+999$. Some negative numbers are:

$$(-327) \rightarrow 9673; \; (-548) \rightarrow 9452; \; (-732) \rightarrow 9268$$

When adding a positive number and a negative one, we have:

$$548 + (-327) \rightarrow 0548 + 9673 = (1)0221;$$
$$548 + (-732) \rightarrow 0548 + 9268 = 9816 \rightarrow -184$$

When adding summands with different sign, no overflow can be produced. The carry generated in the first sum is discarded. The first result is a positive number, and the second, a negative number. In both cases, the result is correct.

When adding two negative numbers, the following occurs:

$$(-548) + (-327) \to 9452 + 9673 = (1)9125;$$
$$(-548) + (-732) \to 9452 + 9268 = (1)8720$$

In the first sum, a correct result is obtained, while in the second one, an overflow is produced. The overflow is detected by means of the sign digit which takes a value of 8, resulting different from the two operands sign. Other overflow situation can be produced when adding two positive numbers, as follows:

$$0548 + 0732 = 1280$$

Again, the overflow is detected from the sign digit, taking a 1 value, and thus, different from the two operands sign.

If a sign bit is used, some examples of negative numbers are:

$$(-327) \to 1673; (-548) \to 1452; (-732) \to 1268$$

When adding a positive number to other negative, we have:

$$548 + (-327) \to 0548 + 1673 = (1)0221;$$
$$548 + (-732) \to 0548 + 1268 = 1816$$

In these sums, overflow is not possible. The carry generated in the first sum is discarded. The firs result is a positive number, and the second one a negative number. In both cases, the result is correct.

When adding two negative numbers, it results the following:

$$(-548) + (-327) \to 1452 + 1673 = (1)1125;$$
$$(-548) + (-732) \to 1452 + 1268 = (1)0720$$

In the first sum, a correct result is obtained while in the second one, an overflow is produced. The overflow is detected by the sign bit, being different from the sign bit of the two operands.

Another overflow situation can be produced when adding two positive numbers, as follows:

$$0548 + 0732 = 1280$$

Again the overflow is detected by means of the sign bit, being different from the sign bit of the two operands. ■

Example 1.4 Additions (and subtractions) of decimal numbers using base-1 complement representation.

Now, we will consider decimal numbers with three digits for magnitude, and a fourth digit for sign, which will be 0 for positive numbers and 9 for negative ones. The values range will go from −999 to +999. Some negative numbers are:

$$(-327) \rightarrow 9672; (-548) \rightarrow 9451; (-732) \rightarrow 9267$$

Adding a positive number to other negative, and applying the end-around carry correction, it results the following:

$$548 + (-327) \rightarrow 0548 + 9672 = (1)0220 = 0220 + 1 = 0221$$
$$548 + (-732) \rightarrow 0548 + 9267 = 9815$$

In this additions overflow is not possible because the sign of the summands are different. The first result is a positive number, and the second one a negative number. In both cases, the results are correct.

When adding two negative numbers, it result the following:

$$(-548) + (-327) \rightarrow 9451 + 9672 = (1)9123 + 1 = 9124$$
$$(-548) + (-732) \rightarrow 9451 + 9267 = (1)8718 + 1 = 8719$$

In the first sum, a correct result is obtained, while in the second one, an overflow is produced. The overflow is detected because the sign digit takes the value 8, being different from the sign digit of the two operands.

Other overflow situation can be produced when adding two positive numbers, as follows:

$$0548 + 0732 = 1280$$

Again, the overflow is detected by means of the digit sign, which takes the value 1, being different from the two operands sign.

If a sign bit is used on instead of a sign digit, some negative numbers are:

$$(-327) \rightarrow 1672; (-548) \rightarrow 1451; (-732) \rightarrow 1267$$

Adding a positive number to other negative, and applying the end-around carry correction, the following is obtained:

$$548 + (-327) \rightarrow 0548 + 1672 = (1)0220 = 0220 + 1 = 0221$$
$$548 + (-732) \rightarrow 0548 + 1267 = 1815$$

In this sums, overflow cannot be produced. The first result is a positive number, and the second one a negative number. In both cases, the result is correct.

When adding two negative numbers, we have:

$$(-548) + (-327) \rightarrow 1451 + 1672 = (1)1123 + 1 = 1124$$
$$(-548) + (-732) \rightarrow 1451 + 1267 = (1)0718 + 1 = 0719$$

The first sum provides the correct result. In the second one, an overflow is generated, being detected by the sign bit, which results different from the sign bit of the two operands.

Other overflow situation is given when adding two positive numbers, as follows:

$$0548 + 0732 = 1280$$

Again, overflow is detected by means of the sign bit, which results different from the sign bit of the two operands. ■

The next sections are devoted to the detailed study of the complement representations for binary numbers. Later, the complement representations will be used for computing decimal additions and subtractions.

1.4.2.6 Two's Complement Representation

Considering the absolute value A, the negative number $-A$ can be represented in two's complement using $n + 1$ bits, $a_n \ldots a_0$, as the binary value of $2^{n+1} - A$. Then, the sign bit is $a_n = 1$. An example can be outlined considering 8 bits for representing positive and negative numbers in two's complement. In this context, the representation of some numbers is:

$$+45 \rightarrow 00101101; -45 \rightarrow 11010011$$
$$+90 \rightarrow 01011010; -90 \rightarrow 10100110$$

Given $+45 \rightarrow 00101101$, for deriving -45 the following subtraction must be calculated:

$$\begin{array}{r} 100000000 \\ -\,00101101 \\ \hline 11010011 \end{array}$$

resulting $-45 \rightarrow 11100111$.

In a similar way, given a negative number, the positive value is derived subtracting from 2^{n+1}. As an example, given $-90 \rightarrow 10100110$

$$
\begin{array}{r}
100000000 \\
-\ 10100110 \\
\hline
01011010
\end{array}
$$

resulting $+90 \rightarrow 01011010$.

From the examples before, the negation operation (i.e., obtaining $-A$ from $+A$, or $+A$ from $-A$) in two's complement can be derived as follows: *subtracting from 2^{n+1} the complete representation (i.e., including the sign bit) of $+A$ or $-A$ as an unsigned number*. By the other way, simple rules can be applied for performing complementation. Specifically, subtraction from 2^{n+1} can be complete as follows: *complementing all of the bits, and adding 1 to the result as unsigned binary number*. Other equivalent procedure is: *searching for the 1 less significant, remaining this 1 and all the 0's to the right, while complementing all of the bits to the left*. The probe of this rules are left as an exercise for the reader.

The decimal value X (both for positive and negative numbers) corresponding to the binary number $a_n \ldots a_0$ represented using two's complement is given by:

$$X = -a_n 2^n + \sum_{i=0}^{n-1} a_i 2^i$$

Thus, the most significant bit (the sign bit), contributes to the decimal value with a negative input and the rest of the bits with a positive input.

The main advantage provided by the two's complement representation is that adding two numbers with different signs should not take into account the sign: just adding the numbers, considering the sign bit as another summand, always resulting a correct sum (except overflows, that will be studied later). See some examples with eight bit numbers:

$$
\begin{array}{rcl}
(+49) & \rightarrow & 00110001 \\
+(+54) & \rightarrow & 00110110 \\
\hline
(+103) & \rightarrow & 01100111
\end{array}
\qquad
\begin{array}{rcl}
(+49) & \rightarrow & 00110001 \\
+(-54) & \rightarrow & 11001010 \\
\hline
(-5) & \rightarrow & 11111011
\end{array}
$$

$$
\begin{array}{rcl}
(-49) & \rightarrow & 11001111 \\
+(+54) & \rightarrow & 00110110 \\
\hline
(+5) & \rightarrow & 00000101 \\
\text{Carry} & 1 &
\end{array}
\qquad
\begin{array}{rcl}
(-49) & \rightarrow & 11001111 \\
+(-54) & \rightarrow & 11001010 \\
\hline
(-103) & \rightarrow & 10011001 \\
\text{Carry} & 1 &
\end{array}
$$

In these examples, obtaining the correct result considering the sign bit like any other bit for adding, and discarding the output carry is checked. Thus, for subtracting two numbers, just the subtrahend should be two's complemented, and added to the minuend, being subtraction reduced to a sum operation. As a consequence, a circuit for adding/subtracting numbers represented in two's complement, can be implemented using an adder and a two's complement block (the latter is not really necessary as will be outlined later).

When adding a positive number and a negative one with a positive result, or when adding two negative numbers, a carry in the sign bits is generated. This carry is discarded.

In two's complement representation, the sign bit and the modulo bits are processed simultaneously. If the number of bits needs to be extended, the extension to the left must be performed by using the sign bit. So, if the number is positive the extension will be made with 0's, and if the number is negative, extension will be made using 1's. When operating with fractional numbers, the extension to the right of the less significant bit must be completed using 0's. As an example, the serial of bits in two's complement 01101001.011, 0001101001.01100, 000001101001.011, …,, corresponds to the same positive number. In a similar way, 11101001.011, 1111101001.01100, 111111101001.011, …, are representing the same negative number.

In order to show the correctness of the sign bit extension, the previous sums are repeated in that follows, but extending from eight to ten the number of bits.

(+49)	→	0000110001	(+49)	→	0000110001
+(+54)	→	0000110110	+(−54)	→	1111001010
(+103)	→	0001100111	(−5)	→	1111111011

(−49)	→	1111001111	(−49)	→	1111001111
+(+54)	→	0000110110	+(−54)	→	1111001010
(+5)	→	0000000101	(−103)	→	1110011001
Carry	1		Carry	1	

When adding, overflow can be produced with two positive numbers or two negative ones. Overflow situations are detected by checking the sign bits of the operands and the result. Specifically an overflow will be produced when adding two positive numbers the result presents negative sign, and when adding two negative numbers the result is positive, as shown in the following examples:

(+93)	→	01011101	(−93)	→	10100001
+(+54)	→	00110110	+(−54)	→	11001010
(+147)	≠	10010011	(−147)	≠	01101011

Thus, a logic circuit for overflow detection can be derived from the following function:

$$F = a_n b_n \bar{r}_n + \bar{a}_n \bar{b}_n r_n$$

being a_n, b_n and r_n the sign bits of the two operands, and the result, respectively.

Multiplication and division in two's complement are more complex than using **SM**, as will be shown.

When using two's complement, there is only one representation for the 0 (+0), as shown in Table 1.4, which constitutes an advantage for zero detection issues. In the general case, the values range for representation using n bits for the magnitude goes from -2^n to $+(2^n - 1)$.

1.4.2.7 One's Complement Representation

Given the absolute value A, for representing the negative number $-A$ using one's complement representation with $n + 1$ bits, $a_n \ldots a_0$, proceed as follows: the sign bit is $a_n = 1$, and the remaining n bits are the ones corresponding to the binary value of $(2^n - 1) - A$. As an example, consider using 8 bits for representing positive and negative numbers in one's complement. In this context, the representation of some numbers is the following:

$$+45 \rightarrow 00101101; -45 \rightarrow 11010010$$
$$+90 \rightarrow 01011010; -90 \rightarrow 10100101$$

Given $+45 \rightarrow 00101101$, for obtaining -45 the following subtraction is performed

$$\begin{array}{r} 11111111 \\ -\,00101101 \\ \hline 11010010 \end{array}$$

and then $-45 \rightarrow 11010010$.

In a similar way, given a negative number represented using one's complement, and subtracting it from $(2^{n+1}-1)$ the positive number is obtained. As an example, given $-90 \rightarrow 10100110$:

$$\begin{array}{r} 11111111 \\ -\,10100110 \\ \hline 01011001 \end{array}$$

and then $+90 \rightarrow 01011001$.

From the examples presented can be outlined that the negation operation using one's complement can be performed as follows: considering the complete representation of the number (i.e., including the sign bit) $+A$ or $-A$, as an unsigned number, and subtracting it from (2^{n+1-1}). By the other hand, subtraction from (2^{n+1-1}) can be calculated complementing all of the number bits. Thus, the negation is simpler in one's complement representation than in two's complement one.

The decimal value X (both for positive and negative numbers) of the binary number $a_n \ldots a_0$ represented using one's complement is:

$$X = -a_n(2^n - 1) + \sum_{i=0}^{n-1} a_i 2^i$$

In this situation, the most significant bit (the sign bit) contributes to the decimal value with a corrected negative input (decreased in 1), and the rest of the bits with a positive input.

The sum using one's complement representation requires the end-around carry correction. See some examples with eight bit numbers:

(+49)	→	00110001	(+49)	→	00110001
+(+54)	→	00110110	+(−54)	→	11001001
(+103)	→	01100111	(−5)	→	11111010

In these examples the carry from the sign bit is 0, being the initial result the correct one. In the following two examples, the carry will be 1, and the end-around carry correction will be requires in order to correct the initial result.

(−49)	→	11001110	(−49)	→	11001110
+(+54)	→	00110110	+(−54)	→	11001001
		00000100			10010111
Carry	1	+1	Carry	1	+1
Sum (+5)	→	00000101	Sum (−103)	→	10011000

When using one's complement representation, the sign bit and the modulo bits are processed together. If the number of bits has to be extended, the extension to the left must be made with the sign bit itself (with 0's if the number is positive and 1's if the number is negative). If the number is fractional, the extension to the right of the less significant bit must be complete also with the sign bit, due to the end-around carry correction. As an example, the bits serial 01101001.011, 0001101001.01100, 000001101001.011, ..., represents to the same positive number, and 11101001.011, 1111101001.01111, 111111101001.0111, ..., are representing the same negative number.

Overflow situations are detected in the same way than in the case of two's complement representation: if the resulting sign bit is different when adding two numbers with the same sign bit.

Again, multiplication and division using one's complement representation results more complex than using **SM**, as will be shown later.

One's complement representation has two assignations for the 0 (+0 and −0), as shown in Table 1.4. This fact can be a disadvantage when zero detection is needed. In general, the values range with n bits for the modulo, goes from $-(2^n - 1)$ to $+(2^n - 1)$.

1.4.3 Biased Representation

Given m bits for representing positive and negative numbers, when using biased representation, all of the bits are treated as modulo bits (there is no sign bit in this case). Then, the represented number, N, is the binary value of the m bits, B, minus a fix bias, D:

$$N = B - D$$

When using biased representation, there is only one assignment for zero, being a non redundant representation.

Usually, D is $D = 2^{m-1}$ or $D = 2^{m-1} - 1$. Making $D = 2^{m-1}$, numbers from -2^{m-1} to $2^m - 1 - 2^{m-1} = 2^{m-1} - 1$ can be represented, being $2^m < \texttt{Pos} < 2^{m-1}$ the range reserved for positive numbers, and $2^{m-1} \geq \text{Neg} \geq 0$ the range for negative numbers. Making $D = 2^{m-1} - 1$, numbers from $-2^{m-1} + 1$ to $2^m - 1 - 2^{m-1} + 1 = 2^{m-1}$, can be represented with the same ranges for positive numbers ($2^m < \texttt{Pos} < 2^{m-1}$), and negative ones ($2^{m-1} \geq \text{Neg} \geq 0$). Both for $D = 2^{m-1}$ and for $D = 2^{m-1} - 1$, the most significant bit for positive numbers is 1, and for negative ones is 0. Attending to this most significant bit, if it is interpreted as a sign bit, with $D = 2^{m-1}$ the zero value is positive, and with $D = 2^{m-1} - 1$, the zero is negative.

As an example, when using 4 bits and $D = 8$, 1111 will correspond to +7 (i.e., 15 − 8), 1000 will represent to 0 (8 − 8), and 0000 will be −8 (0 − 8). Making $D = 7$, 1111 will correspond to +8 (i.e., 15 − 7), 0111 will represent to 0 (7 − 7), and 0000 will be −7 (0 − 7).

This biased representation is also known as **excess** representation.

Given two numbers, $N1$ and $N2$, with biased representations $B1$ and $B2$, respectively:

$$N1 = B1 - D \qquad N2 = B2 - D$$

then,

$$B1 = N1 + D \qquad B2 = N2 + D$$

Thus, the biased representation of the $N1$ and $N2$ addition will be $B3 = N1 + N2 + D$. In order to obtain the sum representation from the representation of $N1$ and $N2$, both representations must be added, and then subtract the biased D, as shown in the following:

$$B3 = N1 + N2 + D = (N1 + D) + (N2 + D) - D = B1 + B2 - D$$

In a similar way, the biased representation of the difference between $N1$ and $N2$ ($B4 = N1 - N2 + D$) can be derived from the $N1$ and $N2$ representations subtracting them, and then adding the biased D. In fact:

$$B4 = N1 - N2 + D = (N1 + D) - (N2 + D) + D = B1 - B2 + D$$

Thus, additions and subtraction must be implemented as different operations, and the biased is always involved, subtracting or adding it to the previous sum or subtraction. Making $D = 2^{m-1}$, adding or subtracting D is equivalent to complementing the most significant bit. In the following examples with five bit numbers (see Table 1.4), this issues are shown:

8	→	11000	8	→	11000
+(+2)	→	+10010	+(−2)	→	+01110
		01010			00110
+10	→	11010	+6	→	10110

Thus, the addition/subtraction when using biased representation with $D = 2^{m-1}$ can be implemented by means of a binary adder/subtractor of m bits (in **SM**), and complementing the most significant bit of the result.

When $D = 2^{m-1} - 1$, additionaly to the most significant bit complementation, the result must be increased in 1 if adding, and decreased in 1 if subtracting. Thus, the addition/subtraction in this case, can be implemented by using a binary adder/subtractor (in **SM**) with a carry/borrowing initialized to 1, followed by a complementation of the most significant bit of the result.

With respect to overflows, they can appear only when adding two numbers with the same sign, or when subtracting two numbers with different signs. Making $D = 2^{m-1}$ or $D = 2^{m-1} - 1$ overflows are easily detected attending to the most significant bits (which can be interpreted as sign bits) of the operands and the result. The following examples with five bit numbers (see Table 1.4) using biased representation with $D = 16$ show these issues:

+12	→	11100	−12	→	00011
+(+9)	→	+11001	+(−9)	→	+00110
		10101			01001
Correction		00101	Correction		11001

In these two examples, overflow is produced, and detection can be performed analyzing the most significant bits of the operands, a_n y b_n, and of the result (once corrected), r_n. Again, the overflow detector can be implemented synthesizing the following three variables logic function:

$$F = a_n b_n \bar{r}_n + \bar{a}_n \bar{b}_n r_n$$

Biased representation present multiple difficulties for multiplication and division, and these operations are not usually implemented when using this representation.

Table 1.4 show the different five bit values which can be represented using biases of $2^4 = 16$ and $2^4 - 1 = 15$.

1.4.4 Advantages and Disadvantages of the Different Representations

None of the several representations described in this chapter presents absolute advantages over the other. Previously, comparisons related with addition and subtraction operations have been made, and using this criterion, complement representations result advantageous. Among the complement representation, base complement is recommended because of the end-around carry correction is not needed. In the next section, that **SM** is the most adequate representation for multiplication will be shown.

Comparison is other frequently implemented operation. For this operation, it is preferred that the number representation comply the same order relation than the represented values. This is not accomplished by **SM** or complement representations (see Table 1.4), and only is verified by biased representations. Thus, if the system to be designed is based on the comparison operation, the biased representation is the recommended one.

1.5 Binary Numbers Multiplication

In previous sections, when presenting the different representations of binary numbers, has been introduced the methods for adding and subtracting integers. Now, the multiplication of two signed integer binary numbers A and B, ($A = a_{n-1} \ldots a_0, B = b_{n-1} \ldots b_0$) will be considered, depending on using **SM** or complement representations.

1.5.1 SM Representation

SM representation is recommended for multiplication implementation. Independently of the A and B signs, the bit sign is calculated separately from the result magnitude, $R = A \cdot B$. The R sign (whatever the number of operands) is computed by performing the XOR function of the operands sign bits. The product magnitude is calculated from the operands magnitudes.

With respect to the wide of the processed numbers, we will consider A and B to have the same size, n bits, $A = a_{n-1}a_{n-2}...a_0$, $B = b_{n-1}b_{n-2}...b_0$, where $n - 1$ bits are reserved for magnitude, and 1 for sign. Then, the magnitude of R, will be $2n - 2$ bits wide. In fact, if $\text{MA} = a_{n-2}...a_0$ and $\text{MB} = b_{n-2}...b_0$ are the A and B magnitudes, respectively, then $\text{MA} < 2^{n-1}$, $\text{MB} < 2^{n-1}$; so $(A \cdot B) < 2^{2n-2}$. Thus, for representing R, including the sign bit, $2n - 1$ bits are required. The sign bit will be equal to $\text{XOR}(a_{n-1}, b_{n-1})$, and the rest of the bits, $r_{2n-3} ...r_0$, will be obtained multiplying $a_{n-2}...a_0$ by $b_{n-2}...b_0$ as unsigned binary numbers. For operative reasons, usually the size of R is fixed to $2n$, the double of each one of the operands, reserving the $2n - 2$ less significant bits ($r_{2n-3}...r_0$) to the magnitude, the sign to r_{2n-1}, and r_{2n-2} will be always zero. Then, $R = r_{2n-1}r_{2n-2}...r_0$, $r_{2n-1} = \text{XOR}(a_{n-1}, b_{n-1})$, and $r_{2n-2} = 0$. The magnitude of the result, $\text{MR} = (\text{MA})\cdot(\text{MB})$, can be calculated by means of the following expression:

$$\text{MR} = \sum_{i=0}^{n-2} (\text{MA})b_i 2^i$$

Example 1.5 Being $A = 1011\ 1001$ and $B = 0101\ 0011$ two binary numbers represented in SM, for obtaining $A \times B$ it results:

$$\text{Sign}(R) = \text{XOR}(1, 0) = 1$$
$$\text{Magnitude}(R) = 0111001 \times 1010011 = 01001001111011$$

which, applying the previous expression for MR, is calculated adding seven partial products, each one being shifted one position to the left with respect to the previous one, as detailed in Sect. 1.2.3.

Thus, using 16 bits for R, assigning the most significant bit to the sign, and the next to zero, it results $R = 1001001001111011$. ■

1.5.2 Complement Representations

When both of the operands are positive, multiplication is implemented as in SM representation. In that follows, the different situations in which some of the operands are negative are considered separately.

Of course, if some of the operands are negative, multiplication can be completed as if both of the operands are positive. For this, the negative operands are complemented, next the corresponding positive values are multiplied, and finally, if the result is negative, its complement is generated.

In the next, a procedure for multiplying directly the operands, without transforming them into positive values by complementation, will be shown. A size or n bits for each operand and $2n$ for the result are assumed.

1.5.2.1 One Operand Negative and the Other Positive

In that follows, A will be assumed to be a negative operand and B a positive one. First, two's complement representation will be considered, and next, one's complement one.

Being $\overline{A} = 2^n - A$ the A two's complement, the negative result of multiplying A, and B, will be:

$$R = 2^{2n} - \overline{A} \cdot B$$

where $\overline{A} \cdot B$ is the product of two positive numbers. Multiplying directly A by B it results:

$$R' = A \cdot B = \left(2^n - \overline{A}\right)B = 2^n B - \overline{A} \cdot B$$

Thus, in order to obtain R from R', the difference $R - R'$ must be added:

$$R - R' = 2^{2n} - 2^n B = \left(2^n - B\right) \cdot 2^n$$

where $(2^n - B) \cdot 2^n$ is the two's complement of B, shifted n bits.

As a consequence, direct multiplication using two's complement with a positive operand and other negative, must be followed by a correction. This correction consists on adding the complement of the positive operand shifted n bits, to the result.

Now we are to consider one's complement representation. If $\overline{A} = 2^n - 1 - A$ is the one's complement of A, the negative result obtained from multiplying A and B, being A negative and B positive, must be:

$$R = 2^{2n} - 1 - \overline{A} \cdot B$$

where $\overline{A} \cdot B$ is the product of two positive numbers. If A and B are multiplied directly, it results:

$$R' = A \cdot B = \left(2^n - 1 - \overline{A}\right)B = 2^n \cdot B - B - \overline{A} \cdot B$$

For deriving R from R', the difference $R - R'$ must be added:

$$R - R' = 2^{2n} - 1 - 2^n B + B = \left(2^{2n} - 1 - 2^n B\right) + B$$

where $(2^{2n} - 1 - 2^n B)$ is the one's complement of B previously shifted n bits.

Thus, direct multiplication in one's complement of a positive operand and a negative one, must be followed by a correction consisting on adding the positive operand and the complement of the positive operand (previously shifted n bits).

Example 1.6 $A = 1101\ 0101\ (-43)$ and $B = 0010\ 1101\ (+45)$ are numbers represented in two's complement. Multiplying them directly, it results:

$$R' = 1101\ 0101 \times 0010\ 1101 = 0010\ 0101\ 0111\ 0001$$

The two's complement of B, shifted 8 bits is 1101 0011 0000 0000. Adding this correction to R' it results:

$$R = 0010\ 0101\ 0111\ 0001 + 1101\ 0011\ 0000\ 0000 =$$
$$= 1111\ 1000\ 0111\ 0001(-1935)$$

which is the correct result.

If A = 1101 0100 (−43) and $B = 0010\ 1101$ (+45) are numbers represented in one's complement, the direct multiplication of A and B it results:

$$R' = 1101\ 0100 \times 0010\ 1101 = 0010\ 0101\ 0100\ 0100$$

The B one's complement previously shifted 8 bits, is 1101 0010 1111 1111. Adding B, the total correction to add is:

$$1101\ 0010\ 1111\ 1111 + 0010\ 1101 = 1101\ 0011\ 0010\ 1100$$

Finally, adding this correction to R' it results:

$$R = 0010\ 0101\ 0100\ 0100 + 1101\ 0011\ 0010\ 1100 =$$
$$= 1111\ 1000\ 0111\ 0000(-1935)$$

which is the correct result. ■

1.5.2.2 Both Operands Negative

Starting with the two's complement representation, $\overline{A} = 2^n - A$ and $\overline{B} = 2^n - B$ are the two's complement of A and B. The positive number resulting from multiplying A and B, both of them negative, must be:

$$R = \overline{A} \cdot \overline{B}$$

where $\overline{A} \cdot \overline{B}$ is the product of two positive numbers. The direct multiplication of A and B results:

$$R' = A \cdot B = \left(2^n - \overline{A}\right) \cdot \left(2^n - \overline{B}\right) = 2^{2n} - 2^n\overline{B} - 2^n\overline{A} + \overline{A} \cdot \overline{B}$$

where $2^n\overline{A}$ and $2^n\overline{B}$ are the complements of A and B, respectively, shifted n bits.

If $2^n\overline{A}$ and $2^n\overline{B}$ are added to R', it results:

$$R' + 2^n\overline{A} + 2^n\overline{B} = 2^{2n} + \overline{A} \cdot \overline{B}$$

that it is equal to R except the carry bit corresponding to 2^{2n}. Thus, for deriving the correct result, the carry bit must be discarded once applied the corrections described. As a resume, when using two's complement representation, if the two operands are negative, the direct multiplication must be followed by a correction. This correction consist on adding the two operands complements shifted n bits, and discarding the final carry bit. Note that other possibility can be performing the complement of the operands, and then multiplying these complements.

Now, we will consider the one's complement representation. If $\overline{A} = 2^n - 1 - A$ and $\overline{B} = 2^n - 1 - B$ are the one's complement of A and B, the positive result obtained from multiplying A and B, both of them negative, is:

$$R = \overline{A} \cdot \overline{B}$$

where $\overline{A} \cdot \overline{B}$ is the product of two positive numbers. The direct multiplication of A and B results:

$$R = A \cdot B = \left(2^n - 1 - \overline{A}\right) \cdot \left(2^n - 1 - \overline{B}\right) = 2^{2n} - 2^{n+1} - 2^n\overline{B} - 2^n\overline{A} + \overline{A} \cdot \overline{B} + \overline{A} + \overline{B} + 1$$

In this case, the correction is quite complex, being recommended performing the complementation of the operands, and then multiplying them as positive numbers.

1.6 Division and Square Root of Binary Integer Numbers

In this section, simple procedures for division and square root implementation for integer binary numbers are approached. Both of these operations are grouped because the procedures presented are based in successive subtractions.

The following implementations are devoted to integer numbers, but they can be easily extended to fractional numbers.

1.6.1 Division

Division is a more complex operation than sum and multiplication. Given two operands, the dividend D and the divisor d, two results have to be generated, the quotient c and the remainder r:

$$D = d \cdot c + r, \quad |r| < |d|$$

The division is not defined for $d = 0$, and in that follows, will be assumed that $d \neq 0$. When D and d are positive, also c and r are positive numbers. In this situation, there are several algorithms for computing the division of positive numbers, obtaining c and r. In these algorithms, subtraction and shifting are used, in a similar way than the school division algorithm (see Sect. 1.2.3). In this section, two integer numbers will be processed (D and d), and two integer will be obtained as a result (c and r).

When some of the operands are negative, r has the same sign than D by convention, and the sign bit of c is calculated as the **XOR** function of the two operands sign bits, as made in the multiplication. Table 1.5 shows the sign of the results in the different situations.

When using **SM**, c and r are derived from D and d by dividing the magnitudes of D and d as positive numbers. In this way, the c and r magnitudes are obtained, and the sign bits corresponding to c and r are derived from:

$$\text{sign}(r) = \text{sign}(D), \quad \text{sign}(c) = \text{sign}(D) \oplus \text{sign}(d)$$

When using complement representations, modifications of positive numbers division algorithms can be used. Nevertheless, these modifications are so complex, that the most recommendable is to use **SM** representation, converting to it the negative numbers (dividend, divisor, or both) represented in complement, and later converting the negative results (quotient, remainder, or both) from **SM** to the complement representation.

In the following, the division of positive numbers without sign bit will be considered. The usual in all division cases is to assume that the dividend has the double of bits than the divisor, while assigning to the quotient and the remainder the same number of bits than the divisor. Specifically, it will be assumed that the dividend is $2n$ bits wide, and that divisor, quotient and remainder are n bits wide each one. With this restriction, some issues can appear related to the number of bits assigned to the quotient. In fact, if $D \geq d \cdot 2^n$, then $c \geq 2^n$ and c cannot be represented using n bits:

Table 1.5 Sign in division

Dividend	+	+	−	−
Divisor	+	−	+	−
Quotient	+	−	−	+
Remainder	+	+	−	−

Table 1.6 Division example

$D = 10011101$, $d = 1101$, $n = 4$. It is $D < d \cdot 2^4$

Step	i	D	$d \cdot 2^{i-1}$	G	c_{i-1}
1	4	10011101	1101000	110101	$c_3 = 1$
2	3	110101	110100	01	$c_2 = 1$
3	2	0001	11010	$G < 0$	$c_1 = 0$
4	1	0001	1101	$G < 0$	$c_0 = 0$

Result: $c = 1100$, $r = 0001$

an overflow is generated. For avoiding this overflow situation, we will assume $D < d \cdot 2^n$. If D and d do not verify $D < d \cdot 2^n$, d will be shifted to the left (or D to the right, or both) so many positions as needed for verifying the condition. The D and d shiftings will be conveniently transferred to c and r.

Division of two positive numbers without sign bit by successive differences, can be carried out in a similar way than manual division. The algorithm starts calculating the most significant bit, and computing one bit of the quotient in each of the iterations. After n iterations, the quotient is calculated and the remainder is stored in r, as follow.

Algorithm 1.1

Algorithm for division

```
r ← D, i ← n
while i > 0 do
  G ← r − d·2^(i−1), i ← i −1;
  if G < 0, then c_(i−1) ← 0, else r ← G, c_(i−1) ← 1
```

End algorithm

Table 1.6 shows an application example of this algorithm. In this example, an 8 bit number ($D = 10011101$) is divided by a 4 bit number ($d = 1101$). Thus, in this case $n = 4$ and $D < d \cdot 2^4$, resulting a quotient and a remainder of 4 bits each. The calculus is completed in four iterations, as indicated in Table 1.6.

1.6.2 Square Root

If $A = a_{2n-1} \ldots a_0$, is a positive integer number without sign bit with $2n$ bits (a zero in the most significant position can be added if needed), its integer square root, $R = A^{1/2}$, will has n bits ($R = r_{n-1} \ldots r_0$). The calculus of R can be carried out bit to bit starting by the most significant one, r_{n-1}. In the calculus process, after obtaining the bits from r_{n-1} to r_{i+1}, it can be written for the r_i bit the following:

$$R = R_1 \times 2^{i+1} + r_i \times 2^i + R_0$$

being $R_1 = r_{n-1} \ldots r_{i+1}$ the part of R calculated until this step, and $R_0 = r_{i-1} \ldots r_0$ the part of R which will be calculated in the next iterations.

For designing the calculus algorithm, the following difference is defined:

$$D = A - \left(R_1 \times 2^{i+1}\right)^2$$

Thus, in the next iteration, once included r_i, this difference will take the value:

$$D^+ = A - \left(R_1 \times 2^{i+1} + r_i \times 2^i\right)^2 = A - \left(R_1 \times 2^{i+1}\right)^2 - 2R_1 \times r_i \times 2^{2i+1} \\ - \left(r_i \times 2^i\right)^2 = D - (4R_1 + 1)r_i \times 2^{2i}$$

In this development, $(r_i)^2 = r_i$, because r_i is 0 or 1. The difference D always must be positive, and then, the algorithm consists on computing

$$D^+ = D - (4R_1 + 1)2^{2i}$$

If $D^+ \geq 0$, then $r_i = 1$ and D is updated with D^+; if $D^+ < 0$, then $r_i = 0$ and D remains unchanged.

The square root of a number A of $2n$ bits can be carried out using successive differences, as in the following algorithm.

Algorithm 1.2 *Algorithm for square root*

```
D ← A, R ← 0, i ← n
while i > 0 do
  B ← D − (4R + 1) 2^(2i), i ← i − 1;
  if B < 0, then r_(i−1) ← 0, else D ← B, r_(i−1) ← 1
End algorithm
```

As shown, one of the bits of the square root is calculated in each iteration, starting with the most significant one. After n iterations the square root is completed, and the remainder is stored in D_f (the final value in D in each iteration).

Table 1.7 shows an application example of this algorithm. The square root of an 8 bit number ($A = 10101110$) is computed. In this case, $n = 4$, and a 4 bits square root, $r_3r_2r_1r_0$, is obtained with a 3 bits remainder $b_2b_1b_0$.

Table 1.7 Square root example

Step	i	D_i	D_f	$(4R + 1)2^{2i}$	B	r_{i-1}
1	4	10101110	1101110	1000000	1101110	$r_3 = 1$
2	3	1101110	11110	1010000	11110	$r_2 = 1$
3	2	11110	11110	110100	$B < 0$	$r_1 = 0$
4	1	11110	101	11001	101	$r_0 = 1$

Result: $R = 1101$, remainder $= 101$

1.7 Decimal Numbers

Almost for all digital systems where arithmetic operations are performed, the numbers are internally represented using radix 2. Nevertheless, in some cases decimal representations are used, operating in radix 10.

There are several representations for decimal digits. The most used is the **BCD** (**B**inary-**C**oded **D**ecimal), which each digit is represented by its binary value, as shown in Table 1.8. When using this representation, a decimal number is coded using four bits for each decimal digit, performing a direct translation, as presented in Table 1.8. As an example:

$$483 = 0100\ 1000\ 0011$$

BCD is an example of weighted codification, because the bit of each position contributes with a given weight to the final value. Specifically, the weights used in **BCD** code are 8, 4, 2, 1, corresponding to the binary representation of the decimal digit. There are several weight combinations, even using positive and negative weights. In some situations, one of these codes can be interesting for performing arithmetic operations instead of **BCD**. As an example, in Table 1.8 codes with weights (2, 4, 2, 1), (7, 4, 2, 1), (8, 4, −2, −1) and (4, 4, 1, −2) are provided.

In **BCD** code, four bits are used for representing one decimal digit. In some situations, for complying with more general standards including other characters codifications (including alphanumeric ones), more than the minimum four bits are used. This is the case of the **ASCII** (**A**merican **S**tandard **C**ode for **I**nformation **I**nterchange) code, which uses eight bits (one byte) for each symbol codification. In **ASCII**, the first four bits of the ten digits are the same (0011), and can be used for distinguishing if a character is numeric; the four less significant bits match with the corresponding to **BCD**.

Table 1.8 Different codifications for decimal digits

	BCD	2421	7421	84-2-1	441-2	Excess-3	Gray	2 in 5	Johnson
0	0000	0000	0000	0000	0000	0011	0010	00011	00000
1	0001	0001	0001	0111	0010	0100	0110	00101	00001
2	0010	0010	0010	0110	0101	0101	0111	00110	00011
3	0011	0011	0011	0101	0111	0110	0101	01001	00111
4	0100	0100	0100	0100	0100	0111	0100	01010	01111
5	0101	1011	0101	1011	1010	1000	1100	01100	11111
6	0110	1100	0110	1010	1101	1001	1101	10001	11110
7	0111	1101	1000	1001	1111	1010	1111	10010	11100
8	1000	1110	1001	1000	1100	1011	1110	10100	11000
9	1001	1111	1010	1111	1110	1100	1010	11000	10000

One code used in some situations because of presenting some advantages on certain operations is the **excess-3** one, which is an unweighted code. This code is derived from **BCD** adding 3 to each character. Other codes without weights presented in Table 1.8 are the **Gray** code, the **2-in-5** code, and the **Johnson** code. Gray code is characterized because two contiguous characters differ in only one bit, avoiding the generation of erroneous characters in a transition between two contiguous correct characters. In the 2-in-5 code, each character has two 1's and three 0's, resulting easy the detection of errors in one bit (errors in two bits cannot be detected). The Johnson code is generated in a natural way from the Johnson counter.

1.7.1 BCD Sum

BCD codification is the most used for decimal digits representation. Moreover, the use of decimal arithmetic is increasing in digital systems. When using binary as internal representation, digital systems have to make radix conversions from decimal to binary and vice versa when reading data or when transferring them to the outside. If the computational cost of these conversions is not relevant, it results interesting the use of binary as internal representation, because of the simplification of the circuits for processing and storing. But, if the computational cost is significant, like in a pocket calculator, then internal **BCD** representation can be the best option.

Additionally to time resources consumed by the radix conversion, errors can be produced in these conversions. When representing in binary some decimal fractions like 0.1_{10}, they cannot be exactly represented in binary (in this example, we have a periodic fraction), and some error is produced when expressing them in binary.

Nowadays these arguments are encouraged by the increasing number of financial transactions through internet. These are decimal data, and the results must be provided in decimal radix. Significant rounding errors can appear when converting from a radix to another if the internal operations are performed in binary, being then preferred to make all of the processing in decimal.

In that follows, we will consider the sum of **BCD** positive decimal number (without sign).

When using **BCD**, decimal numbers are represented using radix 16, but using only the first 10 combinations. Then, the addition of two **BCD** characters can be performed using radix 16 (i.e. using binary arithmetic), followed by a correction. In order to apply the correction, two cases must be considered. When the sum is lower or equal than 9, the addition of two **BCD** digits, M y N, is calculated by adding the four bits of the two numbers in binary:

$$M + N < 10 \Rightarrow (M + N)\,\mathbf{mod}10 = (M + N)\,\mathbf{mod}16$$

As an example:

$$
\begin{array}{rcl}
1 & \rightarrow & 0001 \\
+4 & \rightarrow & 0100 \\
\hline
5 & \rightarrow & 0101
\end{array}
\qquad
\begin{array}{rcl}
7 & \rightarrow & 0111 \\
+2 & \rightarrow & 0010 \\
\hline
9 & \rightarrow & 1001
\end{array}
\qquad
\begin{array}{rcl}
5 & \rightarrow & 0101 \\
+3 & \rightarrow & 0011 \\
\hline
8 & \rightarrow & 1000
\end{array}
$$

When the result of adding the two digits is greater than 9, then a carry for the next digit must be generated, and the result of the present digit will be $S = M + N - 10$. Thus, in this case:

$$M + N > 9 \Rightarrow (M + N)\,\mathbf{mod}10 = (M + N - 10)$$

But,

$$(M + N - 10) = \{(M + N - 10) + 16\}\mathbf{mod}16 = (M + N + 6)\,\mathbf{mod}16$$

As a consequence, the correction to be introduced when the result is greater than 9 consists on adding 6 to the binary result. In fact, if $A + B > 9$, the binary sum does not provide the correct result, as shown in:

$$
\begin{array}{rcl}
6 & \rightarrow & 0110 \\
+7 & \rightarrow & 0111 \\
\hline
 & & 1101
\end{array}
\qquad
\begin{array}{rcl}
9 & \rightarrow & 0111 \\
+2 & \rightarrow & 0010 \\
\hline
 & & 1011
\end{array}
\qquad
\begin{array}{rcl}
9 & \rightarrow & 1001 \\
+8 & \rightarrow & 1000 \\
\hline
 & 1 & 0001
\end{array}
$$

In the first two cases ($6 + 7$ y $9 + 2$), the obtained results are not **BCD** characters, and the carry for the next digit is not generated. In the third case ($9 + 8$), the result is a **BCD** character, but not the correct one, and the carry is generated correctly. As mentioned before, the correction to be introduced for continuing operating like in binary numbers when the result of the sum is greater than 9 consist on adding 6 (0110), also in binary, to the result.

Applying the correction to the three previous examples, we have:

$$
\begin{array}{rr}
 & 1101 \\
 & 0110 \\
\hline
1 & 0011
\end{array}
\qquad
\begin{array}{rr}
 & 1011 \\
 & 0110 \\
\hline
1 & 0001
\end{array}
\qquad
\begin{array}{rr}
 & 0001 \\
 & 0110 \\
\hline
 & 0111
\end{array}
$$

Note that 6 must be added if the initial sum result is greater than 9, i.e.:

1010
1011
1100
1101
1110
1111

Also, if the initial sum is generating a carry, adding 6 is needed. Naming r_3, r_2, r_1, r_0 to the bits resulting from the initial binary sum, and a_+ to the carry generated in it, the condition for adding 6 (i.e. when the correction is required) is given by the 1 value of the following logic function $\boldsymbol{F}$:

$$\boldsymbol{F} = r_3r_2 + r_3r_1 + a_+ \tag{1.8}$$

Other expression for the correction function $\boldsymbol{F}$ can be used. Note that the initial sum $A + B$ generates a carry (a_+) if this sum provides a result greater than 15. If 6 is added to $A + B$, $A + B + 6$ generates a carry (that we will name a_{++}) if $A + B > 9$. Thus, adding 6 is required when $a_+ = 1$ or when $a_{++} = 1$, and the function $\boldsymbol{F}$ can also be expressed as:

$$\boldsymbol{F} = a_+ + a_{++} \tag{1.9}$$

Using expression (1.3) for $\boldsymbol{F}$, adding two **BCD** characters can be completed as follows: generating the two possible results ($A + B$ and $A + B + 6$) and selecting one of them depending on the $\boldsymbol{F}$ value.

For adding positive **BCD** numerals with any number of digits, the corresponding **BCD** digits must be added with the correction introduced when required, and taking into account the carry from the previous digit. So, as an example:

572	→	0101	0111	0010	
+365	→	0011	0110	0101	
		1000	1101	0111	← Binary sum
		1	0110		← Correction and carry
937	→	1001	0011	0111	← Decimal sum

1.7.2 *Negative Decimal Numbers*

When representing negative decimal numbers, codifications introduced before (**BCD** and **ASCII**, as an example) can be used, adding all required for including the sign. Using an additional bit (0 for positive numbers, and 1 for negative ones), sign can be indicated. Nevertheless, due to format issues, the usual it is using the same number of bits for the sign than for a digit. Then, the positive sign for a **BCD** number is represented by 0000, and the negative sign by 1111 (or any value different than 0000).

Once decided the representation of the sign of decimal numbers, any of **SM**, base complement (10's complement), of base-1 complement (9's complement) representations can be used.

When using **SM** representation of **BCD** numbers including four bits for the sign we have:

$$+483 = 0000\ 0100\ 1000\ 0011$$
$$-483 = 1111\ 0100\ 1000\ 0011$$

Addition and subtraction operations using **SM** are calculated in a similar way than in the binary case: taking into account the relative values of the operands and their signs, the sign of the result is carried out, and the magnitude is computed adding or subtracting the operands magnitudes, as in the following examples:

+572	→	0000	0101	0111	0010	
−365	→	1111	0011	0110	0101	
+207	→	0000	0010	0000	0111	← Subtraction

−572	→	0000	0101	0111	0010	
+365	→	1111	0011	0110	0101	
−207	→	1111	0010	0000	0111	← Subtraction

−572	→	1111	0101	0111	0010	
−365	→	1111	0011	0110	0101	
		1111	1000	1101	0111	← Binary sum
			1	0110		← Correction and carry
−937	→	1111	1001	0011	0111	← Decimal sum

When using base complement representation for decimal numbers, the sign digit appears. As seen en Sect. 1.4.2, the sign digit is 0000 for positive numbers, and $b - 1 = 1001$ for the negative ones. In this case, when adding and subtracting, this digit is operated as any other one. Nevertheless, it results better using 1111 as sign digit for negative numbers, because the complementation of this digit is easier (only complementing each bit is needed), and all works like with the $b - 1$ value. Thus, in the following 1111 will be assumed as sign digit for negative numbers, while the magnitude will be the 10's complement of the positive number. As an example, the 10's complement of 572 is $1000 - 572 = 428$, and the one of 365 is 635. The numbers -572 and -365 are represented in 10's complement as:

$$-572 = 1111\ 0100\ 0010\ 1000;\ -365 = 1111\ 0110\ 0011\ 0101$$

When using 10's complement, subtraction is converted in addition like in the binary case. The addition of two positive numbers is performed like in **SM**.

When adding, the carry generated in the sign digits is discarded. As an example:

+572	→	0000	0101	0111	0010	
−365	→	1111	0110	0011	0101	
		1111	1011	1010	0111	← Binary sum
			0110	0110		← Correction
		1	1			← Carry
+207	→	0000	0010	0000	0111	← Decimal sum

−572	→	1111	0100	0010	1000	
+365	→	0000	0011	0110	0101	
		1111	0111	1000	1101	← Binary sum
				1	0110	← Correction and carry
−207	→	1111	0111	1001	0011	← Decimal sum

−572	→	1111	0100	0010	1000	
−365	→	1111	0110	0011	0101	
		0000	1010	0101	1101	← Binary sum
		1111	0110	1	0110	← Correction and carry
−937	→	1111	0000	0110	0011	← Decimal sum

When using 9's complement for representing **BCD** digits, the positive numbers are represented as in **SM** and 10's complement representations. In the case of negative numbers, for the same reasons than with 10's complement, the 1111 is assumed as sign digit, and the magnitude is the 9's complement of the positive number. As an example for three digits numbers, the 9's complement of 572 is $999 - 572 = 427$, and the one of 365 is 634. Thus, the representation for −572 and −365 will be:

$$-572 = 1111\ 0100\ 0010\ 0111;\ -365 = 1111\ 0110\ 0011\ 0100$$

When using 9's complement, subtraction is converted in addition, like in the binary case. The processing of additions and subtraction in 9's complement representation is performed in a similar way than the one's complement binary case, being necessary the end-around carry correction. The addition of two positive numbers is completed as in the 10's complement situation. The following examples illustrate the other cases:

+572	→	0000	0101	0111	0010	
−365	→	1111	0110	0011	0100	
		1111	1011	1010	0110	←Binary sum
			0110	0110		←Correction
		1	1		1	←Carry and end-around carry
+207	→	0000	0010	0000	0111	←Decimal sum

−572	→	1111	0100	0010	0111	
+365	→	0000	0011	0110	0101	
		1111	0111	1000	1100	←Binary sum
				1	0110	←Correction and carry
−207	→	1111	0111	1001	0010	←Decimal sum

−572	→	1111	0100	0010	0111	
−365	→	1111	0110	0011	0100	
		0000	1010	0101	1011	←Binary sum
		1111	0110		0110	←Correction and carry
				1	1	←Carry and end-around carry
−937	→	1111	0000	0110	0010	←Decimal sum

Overflow detection in **BCD** adders for the described representations is proposed as an exercise.

1.7.3 Packed BCD Codification (CHC)

BCD uses four bits for the ten character codification. Thus, from the 16 possible combinations, only 10 are used: it is a low efficient codification. In order to save bits in data transmission and storing, a more compact codification is desirable.

BCD efficiency can be improved codifying simultaneously several digits. The best results are obtained when grouping **BCD** characters three in three. With this technique three **BCD** characters can be coded by using 10 bits, and only 24 from the 1024 possible combinations are wasted.

Example 1.7 Efficiency of an m characters code which is using n bits is defined as the quotient $m/2^n$ (characters really represented/characters that can be represented with n bits). In Table 1.9 the efficiency values corresponding to coding jointly several **BCD** characters, are shown. In order to improve the efficiency obtained when grouping characters three in three, grouping 59 in 59 is required, which present practical problems. Each 59 **BCD** characters group can be coded using 196 bits, instead of the

Table 1.9 Several **BCD** digits codifications efficiency

A	*B*	*E*
1	4	0.625
2	7	0.781
3	10	0.977
4	14	0.610
5	17	0.763
6	20	0.954
…	…	…
59	196	0.996

A: number of **BCD** characters
B: bits to use
E: coding efficiency

197 required by the **CHC** method. Thus, the reduction in the number of bits achieved using 59 characters groups in comparison with those required by the 3 characters groups used in **CHC** is not significant. ■

1.7.3.1 CHC Coding

If $A = abcd$ is a **BCD** digit, A will be considered small when $a = 0$, and A will be considered large when $a = 1$. Thus, A will be small if it is between 0 and 7, and the eight possibilities are coded with *bcd* bits. A will be large if its value is 8 or 9, and these two possibilities are coded with the d bit (b and c are 0). With this convention, when a **BCD** digit is small, three bits are required for specifying it, while when it is large, only one bit is needed.

The **Chen-Ho** [1] proposal for coding together three **BCD** digits, improved by Cowlishaw [2] (kwown as **CHC** method), is based on using a Huffman code. Given three **BCD** digits without packing $ABC = abcd\ efgh\ ijkm$, 12 bits are required. The **CHC** method allows coding all de ABC values using only 10 bits, *pqrstuvwxy*.

With *aei* bits, the eight possible combinations of large and small digits, are detected, as the Table 1.10 shows. The most probable combination (with a 51.2% of probability) is the one corresponding to three small digits. In this situation, 9 bits are required (3 for each of the digits): *bcdfghjkm*. Because of it, when packing, one bit is dedicated to point out if the three digits are small or not. This bit is named v in the Table 1.10 taking the value $v = 0$ if the digits are small, (first row in this table) and $v = 1$ if not (seven remaining rows of this table). When two digits are small and one large (38.4% of probability), 7 bits are required (second, third and fourth rows of Table 1.10), and 3 bits can be used for indicating these situations (specifically, $v = 1$, and wx are used for distinguishing among the three rows). When one digit is small and two large (9.6% of probability), five bits are required (fifth, sixth, and seventh rows of Table 1.10), and 5 bits are available for indicating this combination

Table 1.10 **CHC** codification (from unpacked **BCD** to packed **BCD**)

aei	ABC	Group	Probability	Codification *pqrstuvwxy*
000	SSS	All small	51.2%	bcdfgh0jkm
001	SSL			bcdfgh100m
010	SLS	One large	38.4%	bcdjkh101m
100	LSS			jkdfgh110m
011	SLL			bcd10h111m
101	LSL	One small	9.6%	fgd01h111m
110	LLS			jkd00h111m
111	LLL	All large	0.8%	00d11h111m

(specifically, *vwx* = 111, and *st* are used for distinguishing among the three rows). When the three digits are large (0.8% of probability), only three bits are required (*d*, *h* and *m*, eigth row of Table 1.10), remaining 7 bits for indicating this situation (*pqstvwx* = 0011111).

With assignations described in Table 1.10, 24 combinations remains unused: those corresponding to *stvwx* = 11111 (last row of Table 1.10) and *pq* = 01, 10, 11, with the eight possible *ruy* combinations.

Thus, for coding a three **BCD** characters group, the last column of Table 1.10 is used, leading to the following ten bits $p\ q\ s\ r\ t\ u\ v\ w\ x\ y$ expressions as sum of products:

$$
\begin{aligned}
p &= \bar{a}b + a\bar{i}j + a\bar{e}fi \\
q &= \bar{a}c + a\bar{i}k + a\bar{e}gi \\
r &= d \\
s &= \bar{a}\bar{e}f + \bar{e}\bar{i}f + \bar{a}e\bar{i}j + ei \\
t &= \bar{a}\bar{e}g + \bar{e}\bar{i}g + \bar{a}e\bar{i}k + ai \\
u &= h \\
v &= a + e + i \\
w &= a + ei + \bar{e}\bar{i}j \\
x &= e + ai + \bar{a}\bar{i}k \\
y &= m
\end{aligned}
$$

Example 1.8 For coding 483 = 0100 1000 0011 = *abcd efgh ijkm* with ten bits using **CHC** codification, the procedure will be the following.

In this case *aei* = 010, corresponding to the third row of Table 1.10. Thus, the codifications is *pqrstuvwxy* = *bcdjkh*101*m* = 1000111011.

Note that using previous expressions for $p\ q\ s\ r\ t\ u\ v\ w\ x\ y$, the same coding is carried out. ■

Given a serial of **BCD** characters, the packing process consists on grouping them three in three, and coding each of the groups using the ten bits *pqsrtuvwxy*, as described in the expressions before. If the last **BCD** characters group has less than three characters, two possible situations can occur: the last group has only one **BCD** character, and can be coded with only four bits, or the last group has two **BCD** characters, and then seven bits are required.

For coding only one digit (we will assume that it is *C*) with **CHC**, $pqrstu = 000000$, and $vwxy = ijkm$ are assigned (two first rows of Table 1.10).

For coding two digits (we will assume that they are *B* and *C*), $pqr = 000$ (rows 1, 2, 3 and 7 of Table 1.10), and *stuvwx* are used for coding the two characters. The six or three bits unused in the described cases (which are assigned to zero), can be discarded when storing or transmitting the information.

1.7.3.2 CHC Decoding

For unpacking a three digit **BCD** group coded by using **CHC** (i.e., for obtaining *abcd efgh ijkm* from *pqrstuvwxy*), Table 1.11 is used. This table is derived from Table 1.10, and leads to the following combinational expressions as sum of products:

$$
\begin{aligned}
a &= vw\bar{x} + vwx\bar{s} + vwxst\\
b &= p\bar{v} + p\bar{w} + pxst\\
c &= q\bar{v} + q\bar{w} + qxs\bar{t}\\
d &= r\\
e &= v\bar{w}x + vwx\bar{t} + vwxs\\
f &= s\bar{v} + sv\bar{x} + pvwx\bar{s}t\\
g &= t\bar{v} + tv\bar{x} + qvwx\bar{s}t\\
h &= u\\
i &= v\bar{w}\bar{x} + vwxs + vwxt\\
j &= \bar{v}w + sv\bar{w}x + pvw\bar{x} + pvwx\bar{s}\bar{t}
\end{aligned}
$$

Table 1.11 **CHC** codification (from unpacked **BCD** to packed **BCD**)

vwxst	*abcd efgh ijkm*
0–	0pqr 0stu 0wxy
100–	0pqr 0stu 100y
101–	0pqr 100u 0sty
110–	100r 0stu 0pqy
11100	100r 100u 0pqy
11101	100r 0pqu 100y
11110	0pqr 100u 100y
11111	100r 100u 100y

$$k = \bar{v}x + tv\bar{w}x + qvw\bar{x} + qvwx\bar{s}\bar{t}$$
$$m = y$$

Example 1.9 Given the ten bits *pqrstuvwxy* = 10010001110, obtain the three BCD digits corresponding to this codification.

In this case, *vwxst* = 11110, corresponding to the seventh row of Table 1.11. Thus, the three coded characters are *ABC* = 0*pqr* 100*u* 100*y* = 0100 1000 1000.

Using previous expressions for *a b c d e f g h i j k m* the same result is obtained. ■

1.8 Signed Digits

As commented in previous sections, when assigning decimal values in complement representations, all digits are assigned to positive values except the most significant one, which is negative. Nevertheless, in the negabinary system (Sect. 1.2.3), the even digits contribute with a positive value to the represented number, while the odd digit contribute with a negative one. This idea about a positive or negative contribution to the represented value of the different digits can be generalized, defining a sign for each one of the digits.

1.8.1 Negative Digits

When using positional notation, a different sign can be assigned to each of the digits, contributing to the total value of the represented number with a positive or negative input. If each digit has its own sign, the global sign of the number can be eliminated. As an example, with four decimal signed digits, values from −9999 (will be (−9)(−9)(−9)(−9), with all digits negative) to +9999 (all digits positive) can be represented, resulting a symmetric range, with a total of 19999 different values. Using these ideas the majority of the values in the representation range (in this case from −9999 to +9999), like −3281, can be represented using several ways, resulting a **redundant representation system**:

$$-3281 = (-3)(-2)(-8)(-1) = -3000 - 200 - 80 - 1$$
$$-3281 = (-4)(+7)(+1)(+9) = -4000 + 700 + 10 + 9$$
$$-3281 = (-3)(-3)(+1)(+9) = -3000 - 300 + 10 + 9$$
$$-3281 = (-4)(+8)(-8)(-1) = -4000 + 800 - 80 - 1$$
$$\ldots$$

Note that when numbers does not have a unique sign, determining if a number is positive or negative is not immediate. In fact, when using signed digits, it is the non-zero most significant digit sign which determines if a number is positive or negative.

In a similar way, comparing two number results more complex when using signed digits than using unsigned ones.

In a general way, when using signed digits, the resulting number systems present redundancy. This fact can result advantageous for some addition algorithms, as will be shown in Sect. 1.9. In the previous example about signed decimal digits, it has been assumed that each digit can take 19 different values, (from −9 to +9, including 0). Note that for each of the 19999 values, will be a mean of:

$$(19^4/19999) \approx 6.5$$

different representations.

In order to make more compact the representation, in the following, the + sign will be discarded in the positive digits, and the negative ones will be represented by means of putting a dash over the digit. Thus, −3281 can be written as:

$$-3281 = \bar{3}\bar{2}\bar{8}\bar{1} = \bar{4}719 = \bar{3}\bar{3}19 = \bar{4}8\bar{8}\bar{1} = \ldots$$

The idea of assigning a different sign to each of the digits can be implemented defining all the different values that each digit can take. Always will be assumed that 0 is in the possible values which each digit can take, and that these possible values will be consecutive integers (i.e., there are no steps in the possible values ranges). Thus, in a number system using positional notation with radix b and signed digits, will be assumed that each of the digits can take values from the set $C = [-a, c]$, and that it can take at least b different values, i.e. $a + c + 1 \geq b$. In this way, a continuous range of values is obtained. Note that if each digit can take exactly b different values the system is not redundant. The range can be or not symmetrical, like it is shown in the following example:

Example 1.10 If $b = 5$, $C = \{\bar{1}, 0, 1, 2, 3\}$, the representation system is no-redundant and asymmetrical. The smallest representable number with n digits will result from making all digits $\bar{1}$, i.e.:

$$A = -5^{n-1} - 5^{n-2} - \cdots - 1 = -(5^{n-1}5 - 1)/(5 - 1) = -(5^n - 1)/4$$

The greatest representable number is obtained when making all digits equal to + 3, i.e.:

$$B = 3(5^{n-1} + 5^{n-2} + \cdots + 1) = 3(5^n - 1)/4$$

When $b = 5$ and $C = \{\bar{2}, \bar{1}, 0, 1, 2\}$, a non-redundant but symmetrical system is obtained. The smallest representable number with n digits is:

$$C = -2(5^{n-1} + 5^{n-2} + \cdots + 1) = -2(5^{n-1}5 - 1)/(5 - 1) = -(5^n - 1)/2$$

and the greatest representable number is

$$D = 2(5^{n-1} + 5^{n-2} + \cdots + 1) = (5^n - 1)/2$$

■

1.8.2 Conversion Between Representations

For converting from an unsigned digit representation to other with signed digits, each of the original digits non belonging to the final set C is converted like illustrated in the following example:

Example 1.11 Let's consider the unsigned digits integer number 324014 with $b = 5$. For converting it to another representation with $C = \{\bar{1},\ 0,\ 1,\ 2,\ 3\}$ $b = 5$, the two digits with value 4 must be converted. Using that $4 = 5 - 1 = 1\,\bar{1}$ (5 is the unit in the following digit), we have:

$$324014 = 32\,(5-1)\,01(5-1) = 3\,2\,\bar{1}\,0\,1\,\bar{1} + 010010 = 3\,3\,\bar{1}\,0\,2\,\bar{1}$$

The conversion process can be arranged like an addition:

$$\begin{array}{cccccccccccccc} 3 & 2 & 4 & 0 & 1 & 4 & \rightarrow & 3 & 2 & \bar{1} & 0 & 1 & \bar{1} \\ & & & & & & & 0 & 1 & 0 & 0 & 1 & 0 \\ \hline & & & & & & & 3 & 3 & \bar{1} & 0 & 2 & \bar{1} \end{array}$$

In this process, several stages can be required if in each iteration new digits not belonging to C appear. As an example, for converting 324014 to $C = \{\bar{2},\ \bar{1},\ 0,\ 1,\ 2\}$, $b = 5$, the digits 3 (equivalent to $5 - 2$) and 4 (equivalent to a $5 - 1$) must be converted. Thus

$$\begin{array}{cccccccccccccc} 3 & 2 & 4 & 0 & 1 & 4 & \rightarrow & & \bar{2} & 2 & \bar{1} & 0 & 1 & \bar{1} \\ & & & & & & & 1 & 0 & 1 & 0 & 0 & 1 & 0 \\ \hline & & & & & & & 1 & \bar{2} & 3 & \bar{1} & 0 & 2 & \bar{1} \end{array}$$

$$\begin{array}{cccccccc} \rightarrow & 1 & \bar{2} & \bar{2} & \bar{1} & 0 & 2 & \bar{1} \\ & 0 & 1 & 0 & 0 & 0 & 0 & 0 \\ \hline & 1 & \bar{1} & \bar{2} & 1 & 0 & 2 & \bar{1} \end{array}$$

For converting a negative number, the process starts transferring the sign to each of the digits, and then, continuing like in the previous cases. As an example, with b

$= 5$, -324014 leads to convert $\bar{3}\ \bar{2}\ \bar{4}\ 0\ \bar{1}\ \bar{4}$. For converting it to other representation with $C = \{\bar{1}, 0, 1, 2, 3\}$, must be used that $\bar{4} = \bar{1}\ 1$, $\bar{3} = \bar{1}\ 2$ and $\bar{2} = \bar{1}\ 3$. Appling the corresponding conversions we have:

$$\bar{3}\ \ \bar{2}\ \ \bar{4}\ \ 0\ \ \bar{1}\ \ \bar{4}\ \ \rightarrow\ \ \begin{array}{ccccccc} & 2 & 3 & 1 & 0 & \bar{1} & 1 \\ \bar{1} & \bar{1} & \bar{1} & 0 & 0 & \bar{1} & 0 \\ \hline \bar{1} & 1 & 2 & 1 & 0 & \bar{2} & 1 \end{array}\ \rightarrow$$

$$\rightarrow\ \ \begin{array}{ccccccc} \bar{1} & 1 & 2 & 1 & 0 & 3 & 1 \\ 0 & 0 & 0 & 0 & \bar{1} & 0 & 0 \\ \hline \bar{1} & 1 & 2 & 1 & \bar{1} & 3 & 1 \end{array}$$

■

For the inverse conversion, i.e. for converting from a signed digit representation to other with unsigned digits, only subtracting the negative digits from the positive ones is required, as presented in the following example.

Example 1.12 Let's consider the signed digits number $2\ \bar{1}\ 1\ \bar{2}\ 0\ 1\ \bar{2}\ 1\ \bar{1}$ with $b = 4$. For converting it to another representation with $C = [0, 1, 2, 3]$, only the following subtraction has to be completed:

$$\begin{array}{r} 201001010 \\ -010200201 \\ \hline 130200203 \end{array}$$

■

In practical implementations, symmetric C sets are considered, being then the range of representable values also symmetric.

1.8.3 Binary Signed Digits (BSD)

The representations of binary integer numbers with signed digits are known as **BSD** representations (**B**inary **S**igned **D**igit). In this case, there is only a possible set of values, which is $C = \{\bar{1}, 0, 1\}$.

Thus, with n signed digits, the range for values N which can be represented is:

$$-(2^n - 1) \leq N \leq (2^n - 1)$$

BSD gives redundant representations. As an example, when using five bits, the number 15 can be represented in the following ways:

$$15 = 01111 = 1\,\bar{1}\,1\,11 = 10\bar{1}\,11 = 100\bar{1}\,1 = 1000\,\bar{1}$$

Among the different radix 2 with signed digits representations of a number, the **minimal representation** is defined as the one including the lowest number of non-zero digits. For the number 15, the minimal representation is $1000\,\bar{1}$.

The global sign of a **BSD** number (i.e. if the number is positive or negative) is the one of the non-zero most significant digit, as in the general case.

For converting an integer N represented in base b ($b \neq 2$) to **BSD**, the procedure is the one defined in Sect. 1.2.2 (i.e., successively dividing by 2), but using as possible remainder $+1, 0$ and -1, and selecting $+1$ or -1 for making even the quotient. Thus, if C_n is the actual quotient we have two cases. If C_n is even, then the remainder r_n is zero. If C_n is odd the remainder must be $r_n = 2 - C_n\textbf{mod}4$, for $C_{n+1} \leftarrow (C_n - r_n)/2$ resulting even. The algorithm can be written as:

Algorithm 1.3

Algorithm for BSD representation

while $C_n > 0$ **do**

  1) if C_n **is** even, **then** $r_n \leftarrow 0$, **else** $r_n \leftarrow 2 - C_n\textbf{mod}4$;

  2) $C_{n+1} \leftarrow (C_n - r_n)/2$

End algorithm

Table 1.12. shows the different iterations for converting the decimal number 477 to radix 2 with signed digits. The result is $477_{10} = 1000\,\bar{1}\,00\,\bar{1}\,01_2$.

Two's complement representation of a number can be easily converted to the set $C = \{\bar{1}, 0, 1\}$. In fact, remembering that the sign bit in two's complement has a negative weight:

$$X = -a_n 2^n + \sum_{i=0}^{n-1} a_i 2^i$$

Table 1.12 Expression of 477 in **BSD**

C_n	N**mod**4	r_n	$(C_n - r_n)/2$
477	Odd, 1	+1	238
238	Even	0	119
119	Odd, 3	−1	60
60	Even	0	30
30	Even	0	15
15	Odd, 3	−1	8
8	Even	0	4
4	Even	0	2
2	Even	0	1
1	Odd, 1	+1	0

replacing the sign bit when is 1 by $\overline{1}$ (if 0, it remains unchanged) the conversion is completed. Thus, a positive number represented in two's complement is converted to **BSD** without changing any bit, and a negative number is converted changing the sign bit from 1 to $\overline{1}$. As an example, the binary number represented in two's complemented as 10011101 will be $\overline{1}0011101$ in **BSD**. Note that this number can have other **BSD** representations:

$$\overline{1}0011101 = \overline{1}0100\,\overline{1}\,1\,\overline{1} = \overline{1}\,1\,\overline{1}\,0\,0\,\overline{1}\,1\,\overline{1} = \ldots$$

Converting a one's complement number to $C = \{\overline{1},\ 0,\ 1\}$ results a bit more complex. In this case, if the sign bit is 1, it is changed to $\overline{1}$, and 1 must be added to the less significant bit, because:

$$X = -a_n(2^n - 1) + \sum_{i=0}^{n-1} a_i 2^i = -a_n 2^n + \sum_{i=0}^{n-1} a_i 2^i + a_n$$

Now, the conversion to **BSD** of an n bits binary number, $N_x = x_{n-1}\ldots x_0$, is approached, resulting the representation $N_y = y_{n-1}\ldots y_0$ in **BSD**. Usually this conversion is oriented to carry out a minimal or quasi-minimal **BSD** representation (in general, **BSD** representations with the largest possible number of zeros are desired). Thus, the next paragraphs are devoted to achieve these representations depending on the initial binary representation: unsigned, **SM**, or complement.

When N_x is represented in complement, the first step consists on obtaining an initial **BSD** representation by using the conversion techniques previously described (if N_x is represented in two's complement, this initial conversion is not required, as will be seen later).

When N_x is negative, and using **SM** representation, conversion can be completed as in the case of a positive number, applying the algorithms being detailed next, and later changing 1 by $\overline{1}$ and vice versa in all digits for actualizing the global sign.

1.8.3.1 Zeros and Ones Chains

When converting from N_x to N_y, any 1's chain included in N_x can be replaced by a 1 followed by a 0's chain and followed by a $\overline{1}$ in N_y, as will be detailed. In fact, if we have a chain of m 1's, ...01...10.... and the less significant 1 has the value 2^p, the most significant one will be 2^{p+m-1}, and the value V of the m 1's will be:

$$V = \sum_{i=0}^{m-1} 2^{p+i} = 2^{p+m} - 2^p$$

Thus the m 1's chain can be replaced by one 1 instead of the 0 preceding to the 1's chain, followed by $m - 1$ 0's, and one $\overline{1}$, instead of the less significant 1. Thus, if

Table 1.13 Booth codification in radix 2

x_ix_{i-1}	y_i	Comment
00	0	0's chain
01	1	End of 1's chain
10	$\overline{1}$	Starting of a 1's chain
11	0	1's chain

N_x is a natural number (i.e., the most significant bit is not interpreted as a sign bit), and $x_{n-1} = 1$ is part of a 1's chain to be replaced, then N_y must be extended with the bit $y_n = 1$.

Replacing 1's chains in N_x can be made in a systematic way by exploring the consecutive bit pairs x_ix_{i-1}, trying to detect the beginning and the end of 1's chains. Exploration can be carried out from the left to the right or vice versa. If exploration is made from the less significant bit to the most significant one, the starting of a 1's chain will be marked by $x_ix_{i-1} = 10$, and the end by $x_ix_{i-1} = 01$. If we are in a 0's chains will be $x_ix_{i-1} = 00$, and if we are in a 1's chain, $x_ix_{i-1} = 11$. From each pair x_ix_{i-1} (i = 0, ..., $n - 1$), the corresponding digit y_i is obtained, as detailed in Table 1.13. Because we are searching for 1's chains, should be assumed that $x_{-1} = 0$. If the exploration is performed from the most significant bit to the less significant one, the table to use is the same, Table 1.13, and $x_n = 0$ should be assumed. This procedure is known as Booth algorithm in radix 2, and can be applied both in series (sequentially) as in parallel.

For converting a negative number N_x expressed in two's complement ($x_{n-1} = 1$), to **BSD** performing the minimum number of changes possible, $y_{n-1,1} = \overline{1}$ must be made, remaining the rest of the bits unchanged, as seen before. Otherwise, when 1's chains of the **BSD** representation are removed, and $x_{n-2} = 1$, $y_{n-1,2} = 1$ must be made. Adding these two inputs, it results $y_{n-1} = y_{n-1,1} + y_{n-1,2} = 0$. Summarizing, if N_x is a number expressed in two's complement, the conversion to N_y is performed by applying Table 1.13, without adding the bit $y_n = 1$ when $x_{n-1} = 1$. Except for the adding of 1 to the less significant bit, all of the above is applicable to one's complement negative numbers.

When $x_{n-1}...x_0$ contains isolated 1's, application of the described algorithm can produce bad results in the sense of $y_{n-1}...y_0$ containing more non-zero bits than $x_{n-1}...x_0$. As an example, applying Table 1.13 to 01010101, it results $1\,\overline{1}\,1\,\overline{1}\,1\,\overline{1}\,1\,\overline{1}$, Thus, it is better leaving isolated 1's unchanged than replace them by $1\,\overline{1}$.

Isolated 0's in $x_{n-1}...x_0$ lead to similar issues when applying the presented algorithm. For illustrating them, consider the following sequence:

$$...011...1011...10...$$

with an isolated 0 between two 1's chains. Replacing the two 1's chains, the sequence leads to:

$$\ldots 100 \ldots \bar{1}100 \ldots \bar{1}0 \ldots$$

However, using $0 = 1 + \bar{1}$, the original sequence can be written as one 1's chain plus an isolated $\bar{1}$, as follows:

$$\ldots 011 \ldots 1011 \ldots 10 \ldots = \ldots 011 \ldots 1111 \ldots 10 .. + \ldots 000 \ldots 0\bar{1}00 \ldots 00 \ldots$$

Replacing the 1's chain, we have:

$$\ldots 011 \ldots 1111 \ldots 10 \ldots + \ldots 000 \ldots 0\bar{1}00 \ldots 00 \ldots = \ldots 100 \ldots 0000 \ldots \bar{1}0 \ldots$$
$$+ \ldots 000 \ldots 0\bar{1}00 \ldots 00 \ldots = \ldots 100 \ldots 0\bar{1}00 \ldots 10 \ldots$$

containing more 0's than the first transformation, $\ldots 100 \ldots \bar{1}100 \ldots \bar{1}0 \ldots$

1.8.3.2 Canonical Codification

From the above, emerges the need of an specific processing for isolated 0's and 1's. First, note that for detecting them examining at least three bits of N_x is required (or have information about more than two bits). In the algorithm known as **canonical codification**, two consecutive bits are examined, $x_{i+1}x_i$ ($i = 0, \ldots, n - 1$), from the less significant bit to the most significant one, starting with x_0, and generating the corresponding y_i bit in each iteration. Additionally, a c_i bit, named **carry** bit, for signaling if there is a 0's chain ($c_i = 0$) or a 1's chain ($c_i = 1$) on the right of x_i, is used. It is clear that must be $c_0 = 0$, because the objective is detecting 1's chains. In a similar way, must be $x_n = 0$ for obtaining y_{n-1}. Performing the appropriate modifications, this codification could also be computed from the left to the right.

In order to apply the canonical codification (from the right to the left), Table 1.14 can be used, where the above ideas for replacing isolated 0's or 1's have been included. Specifically, the isolated 1's remain unchanged, and the isolated 0's are replaced by $\bar{1}$, signaling to the next iteration that a 1's chain is on the right $\left(0 = 1 + \bar{1}\right)$ by making

Table 1.14 Canonical codification

$x_{i+1}x_ic_i$	y_ic_{i+1}	Comment
000	00	0's chain
001	10	End of 0's chain
010	10	Isolated 1
011	01	1's chain
100	00	0's chain
101	$\bar{1}1$	Isolated 0
110	$\bar{1}1$	Starting of 1's chain
111	01	1's chain

$c_{i+1} = 1$. Again, if N_x is a natural number (the most significant bit is not a sign bit), and $x_{n-1} = 1$, then N_y must be extended with the bit $y_n = 1$. If N_x is a two's complement number, all mentioned previously is applied.

When using canonical codification on 01010101, all of the 0's and 1's are identified as isolated ones, and remain unchanged, as seen in the following example.

Example 1.13 If $x_7x_6x_5x_4x_3x_2x_1x_0 = 01010101$, converting it by using canonical codification, we have:

$$x_1x_0c_0 = 010 \rightarrow y_0c_1 = 10;\ x_2x_1c_1 = 100 \rightarrow y_1c_2 = 00$$
$$x_3x_2c_2 = 010 \rightarrow y_2c_3 = 10;\ x_4x_3c_3 = 100 \rightarrow y_3c_4 = 00$$
$$x_5x_4c_4 = 010 \rightarrow y_4c_5 = 10;\ x_6x_5c_5 = 100 \rightarrow y_5c_6 = 00$$
$$x_7x_6c_6 = 010 \rightarrow y_6c_7 = 10;\ x_8x_7c_7 = 000 \rightarrow y_7 = 0$$

Thus, $y_7y_6y_5y_4y_3y_2y_1y_0 = 01010101$. In a similar way, can be checked that 00101010 remains unchanged when applying the canonical codification. ■

Canonical codification provides the greatest number of zero digits: with n bits, there are a mean of $(2n/3)$ 0's when using canonical codification. Thus, it is a minimal representation, and because it cannot include two non-zero adjacent digits, it is known also as a **NAF** (**N**on-**A**djacent **F**orm) representation. In [4] other procedures for obtaining **NAF** representations are presented. Note that the algorithm for canonical codification presented here is computed sequentially, while in [5], a parallel one is described, introducing some additional computing cost.

1.8.3.3 Booth Codifications

Although less efficient than canonical codification, because the minimal representation is not guaranteed, Booth algorithm (previously introduced in radix 2) provides simpler procedure for **BSD** codification. Some improvements are described in what follows.

In the Booth algorithm in radix 4, two bits of N_y, $y_{i+1}y_i$, are generated simultaneously, by examining three bits of N_x, $x_{i+1}x_ix_{i-1}$ (i = 0, 2, 4, ...), with $x_{-1} = 0$, and shifting two bits in each iteration. Thus, for converting an n bits character, $n/2$ iterations are required. About detecting isolated 0's and 1's, Table 1.15 can be used, as also for the conversion process.

If N_x is a natural number containing an odd number of bits, for making it to have an even number of bits, it is extended with $x_n = 0$. If N_x is represented in complement, containing an odd number of bits, then it is extended by using the sign bit, $x_n = x_{n-1}$.

If N_x is a natural number and $x_{n-1} = 1$, then N_y must be extended with the $y_n = 1$ bit. If N_x is a two's complement number, the conversion to N_y is calculated by applying Table 1.15, without adding the bit $y_n = 1$ when $x_{n-1} = 1$.

Example 1.14 If $x_7x_6x_5x_4x_3x_2x_1x_0 = 01010101$, conversion by using the Booth algorithm in radix 4 will be:

Table 1.15 Booth codification in radix 4

$x_{i+1}x_ix_{i-1}$	$y_{i+1}y_i$	Comment
000	00	0's chain
001	01	End of 1's chain in x_{i-1}
010	01	Isolated 1
011	10	End of 1's chain in x_i
100	$\bar{1}$ 0	Starting of a 1's chain in x_{i+1}
101	0 $\bar{1}$	Isolated 0
110	0 $\bar{1}$	Starting of 1's chain in x_i
111	00	1's chain

$$x_7x_6x_5 = 010 \rightarrow y_7y_6 = 01;\ x_5x_4x_3 = 010 \rightarrow y_5y_4 = 01$$
$$x_3x_2x_1 = 010 \rightarrow y_3y_2 = 01;\ x_1x_0x_{-1} = 010 \rightarrow y_1y_0 = 01$$

Thus, $y_7y_6y_5y_4y_3y_2y_1y_0 = 01010101$, as when applying the canonical codification.

If $x_7x_6x_5x_4x_3x_2x_1x_0 = 00101010$, conversion by using the Booth algorithm in radix 4 results in:

$$x_7x_6x_5 = 001 \rightarrow y_7y_6 = 01;\ x_5x_4x_3 = 101 \rightarrow y_5y_4 = 0\bar{1}$$
$$x_3x_2x_1 = 101 \rightarrow y_3y_2 = 0\bar{1};\ x_1x_0x_{-1} = 101 \rightarrow y_1y_0 = 0\bar{1}$$

Thus, $y_7y_6y_5y_4y_3y_2y_1y_0 = 0\,1\,0\,\bar{1}\,0\,\bar{1}\,0\,\bar{1}$, that is not a minimal representation. ■

Extending the radix 4 Booth algorithm, the radix 8 Booth algorithm can be carried out. Now, three bits of N_y, $y_{i+2}y_{i+1}y_i$, are generated simultaneously, by examining four bits from N_x, $x_{i+2}x_{i+1}x_ix_{i-1}$ (i = 0, 3, 6, …), with $x_{-1} = 0$, and shifting three bits in each iteration. Thus, for converting an *n* bits character, *n*/3 iterations are required. For this conversion, Table 1.16 can be used.

When N_x has a number of bits not being multiple of three, it must be extended with one or two bits, like in the Booth algorithm in radix 4 (with 0 or the sign bit).

Example 1.15 If $x_7x_6x_5x_4x_3x_2x_1x_0 = 00101010$, it is extended making:

$$(x_8)x_7x_6x_5x_4x_3x_2x_1x_0 = (0)00101010$$

Converting it by applying the radix 8 Booth algorithm, we have:

$$x_8x_7x_6x_5 = 0001 \rightarrow (y_8)y_7y_6 = (0)01;\ x_5x_4x_3x_2 = 1010 \rightarrow y_5y_4y_3 = \bar{1}01$$
$$x_2x_1x_0x_{-1} = 0100 \rightarrow y_2y_1y_0 = 010$$

Thus, $y_7y_6y_5y_4y_3y_2y_1y_0 = 01\,\bar{1}\,01010$, which is different than those obtained in Example 1.14, and it is not a minimal representation. ■

Applying the same idea, Booth codification ban be extended to greater bases (16, 32, …), exploring in each case a greater number of N_x bits.

Table 1.16 Booth codification in radix 8

$x_{i+2}x_{i+1}x_ix_{i-1}$	$y_{i+2}y_{i+1}y_i$	Comment
0000	000	0's chain
0001	001	End of 1's chain in x_{i-1}
0010	001	Isolated 1 in x_i
0011	010	End of 1's chain in x_i
0100	010	Isolated 1 in x_{i+1}
0101	011	Isolated 1 in x_{i+1} and end of 1's chain in x_{i-1}
0110	011	Two isolated 1's in $x_{i+1}x_i$
0111	100	End of 1's chain in x_{i+1}
1000	$\overline{1}00$	1's chain starting in x_{i+2}
1001	$\overline{1}01$	1's chain starting in x_{i+2} and 1's chain ending in x_{i-1}
1010	$\overline{1}01$	1's chain starting in x_{i+2} and an isolated 1 in x_i
1011	$0\overline{1}0$	Isolated 0 in x_{i+1}
1100	$0\overline{1}0$	1's chain starting in x_{i+1}
1101	$00\overline{1}$	Isolated 0 in x_i
1110	$00\overline{1}$	1's chain starting in x_i
1111	000	1's chain

1.8.3.4 Codification of $\overline{1}$, 0, 1

The codification of the three values $\overline{1}$, 0, 1, requires using, at least, two bits. There are 24 possibilities, although only 9 are strictly different if permutations and complementations are considered the same codification. In any case, there are codifications better than other for some applications. As an example, sign detection (i.e. if N_y is positive or negative) is simpler in some codifications than other. The more frequent codifications are Negative-Positive (*n*, *p*), and the Sign-Value (*s*,*v*) ones, and are detailed in the following.

The (*s*, *v*) codification uses one bit for the sign, and other for the value. Assuming that zero is positive, the 10 combination remains as don´t care, as indicated in Table 1.17.

When using (*n*, *p*) codification, the two bits are named x^+ and x^-, and the **BSD** value of the *x* digits is:

$$x = x^+ - x^-$$

Table 1.17 Codification of $\overline{1}$, 0, 1

	sv	x^+x^-
$\overline{1}$	11	01
0	00	00 or 11
1	01	10

Table 1.17 shows (n, p) codification. In this case, the zero can be coded as 00 and 11, being possible remaining one of these codes as don´t care.

In practice, for some applications like multiplication, the values $\overline{1}$, 0, 1, are really not coded. The different values combinations are translated to actions, and the corresponding control signals for activating these actions are generated.

1.8.3.5 BSD Decodification

For converting a **BSD** number to another representation with unsigned digits (i.e. for converting from $C_1 = \{\overline{1}, 0, 1\}$ to $C_2 = \{0, 1\}$), in principle only subtracting negative digits from positive ones is needed. But, depending on the target representation system, actions can be different.

If N_{BSD} is the number to convert, $N_{BSD}(+)$ will be its positive part, and $N_{BSD}(-)$ will be its negative part:

$$N_{BSD} = N_{BSD}(+) - N_{BSD}(-)$$

For carrying out $N_{BSD}(+)$ and $N_{BSD}(-)$, (n, p) codification is preferred, because of the two bits of every **BSD** digit, the first one is part of $N_{BSD}(+)$ and the second one is part of $N_{BSD}(-)$. The previously mentioned subtraction must be correctly performed depending on the target representation, such as seen in Sect. 1.4. In any case, always a parallel subtractor with the same size of N_{BSD} will be required.

For converting an n digits **BSD** number to **SM**, the new representation will has $n + 1$ bits: a bit sign must be added, which value will be obtained by comparing $N_{BSD}(+)$ and $N_{BSD}(-)$.

For converting N_{BSD} (n digits) to two's complement, N_{C2} ($n + 1$ bits), an algorithm based on replacing each of the N_{BSD} digits for the corresponding N_{C2} bit starting from the less significant one, can be used. The objective is avoiding the value $\overline{1}$ to appear in N_{C2}. For describing the algorithm, first N_{BSD} will be assumed to be positive and some obvious replacements will be shown. It is clear that $1\,\overline{1} = 01$ and, in general, $10\ldots0\overline{1} = 01\ldots11$. Thus, starting from the less significant digit of N_{BSD}, the algorithm is:

(1) Exploring N_{BSD} digit to digit, leaving unchanged the 0's or 1's, until a $\overline{1}$ is reached.
(2) Replacing the first $\overline{1}$ found by 1, and the rest of $\overline{1}$ by 0, and replacing all the 0's by 1 until the first 1 is reached.
(3) Replacing the first 1 found by 0. Go to 1).

These steps are repeated until removing all $\overline{1}$ value. If N_{BSD} is positive (as it has been assumed), the most significant bit non-zero has the 1 value, and the algorithm presented is sufficient for completing the conversion. For positive numbers, this algorithm can be used for converting from **BSD** to **SM** (the sign bit is 0).

When N_{BSD} is negative, the most significant digit non-zero takes the $\overline{1}$ value, and the most significant digit will be 0 or $\overline{1}$. In this case, the expression $\overline{1} = \overline{1}1$, or, in general $0 \ldots 0\overline{1} = \overline{1}1 \ldots 11$ must be applied. The most significant digit must be processed separately, and the previous algorithm can be rewritten as follows:

(1) Exploring N_{BSD} digit to digit, leaving unchanged the 0's or 1's, until a $\overline{1}$ is reached.
(2) Replacing the first $\overline{1}$ found (not being the most significant digit) by 1, and the rest of $\overline{1}$ (not being the most significant digit) by 0 until the first 1 is reached. Replacing all the 0's (not being the most significant bit) by 1 until the first 1 is reached.
(3) Replacing the first 1 found by 0. If the most significant bit has not been reached, go to 1).
(4) Replacing the most significant digit (0 or $\overline{1}$) by 11.

The result provided by this algorithm is expressed in two's complement. For converting from **BSD** to one's complement, this algorithm can be used, adding 1 at the end.

1.8.3.6 BSD Applications

BSD codifications appear in different applications, being highlights multiplication, division and exponentiation. In that follows, multiplication will be considered (later, exponentiation will be analyzed).

The $A \times B$ multiplication using sums and shifts, also known as school algorithm, consists on adding A several times, with the corresponding shifts. The number of sums is equal to the number of non-zero digits of the B multiplier minus 1. Thus, if a minimal (or quasi-minimal) representation is used for B, the multiplication will be carried out with a minimal number of operations, which, in this case, will be additions and subtractions.

Usually, Booth codifications with bases greater than two, exploring simultaneously several multiplier bits, are used. In this situation, adding or subtracting different multiples of A is required. Tables 1.13, 1.14, 1.15 and 1.16 must be rewritten for multiplication, indicating in each case the A multiple to be added.

1.9 Redundant Number Systems

The arithmetic operation most frequently used in digital systems is addition, thus resulting important optimizing its implementation in terms of performance and resources. Parallel implementation, as shown previously, presents a problem: carry propagation produces a slowing of the circuits, and makes the delay dependent on the size of the sum (i.e. on the number of bits of the summands).

Thus, the use of carry free procedures, or procedures with limited carry propagation is desirable. Later will be shown how the Residue Number System avoids carry propagation. Also, the problem can be reduced by means of a fast carry propagation, or detecting when carry propagation has been finished, without always waiting the worst case time. These options will be evaluated later, when studying the sum with more detail.

1.9.1 Carry Propagation

Using redundant number systems with positional notation, carry propagation can be limited, taking advantage of redundancy, as shown in the following example.

Example 1.16 Let's consider the sum of decimal numbers with $C_1 = [0, 9]$, but using $C_2 = [0, 19]$ for intermediate results. Obviously, in the first sum no carry is generated, but in the following carry can appear. Because of this, after each sum, the range must be adjusted, generating the corresponding carry if needed. In this way, carry propagation is limited to adjacent positions. As an example, we will show the sum of several numbers as 5408, 6287, 8710 and 9598. The process is the following: after each sum, the result obtained in each position is decomposed in a carry (0 or 1) and a digit being into the C_1 range.

	5	4	0	8	
	6	2	8	7	
....	....	....	....	....	....
	11	6	8	15	First sum
1	0	0	1	0	Carry
	1	6	8	5	Digit in C_1 range
....	....	....	....	....	....
1	1	6	9	5	Adjusting result
	8	7	1	0	Second summand
....	....	....	....	....	....
1	9	13	10	5	Second sum
	1	1	0	0	Carry
1	9	3	0	5	Digit in C_1 range
....	....	....	....	....	....
1	10	4	0	5	Adjusting result
	9	5	9	8	Third summand
....	....	....	....	....	....
1	19	9	9	13	Third sum
1	0	0	1	0	Carry
1	9	9	9	3	Digit in C_1 range
....	....	....	....	....	....
2	9	9	10	3	Adjusting result(Final result)

As can be checked, for the intermediate results, after the corresponding adjusts, the set $C_3 = [0, 10]$ is used. When processing the adjustment, carry can take the values 0 or 1, and it is propagated only to the following digit in the successive sums. Although each sum is made in two stages, in practice, it can be completed in only one by using for the sum in each digit the carry from the previous digit, which only depends of the digits to be summed in that position. In any case, only anticipation of the carry modifying the sum in the following sum is required, but no more.

Each digit in the final result is into the range $C_3 = [0, 10]$. If we want transforming them to the set $C_1 = [0, 9]$, the possible carries must be propagated, and this propagation can affect to all the positions. But this possible propagation only has to be completed one time, at the end of all sums. Thus, this sum procedure with different value ranges for the digits to process saves carry propagations when several numbers have to be added, but not if only two numbers have to be summed. ■

Now, the general case of a positional number systems with signed digits, in base b, being $C_b = [-m, n]$, will be approached. The target is analyzing the conditions making the carry propagating only to the next digit, when adding two numbers, and being all the digits in the intermediate sums into the range $C_b = [-m, n]$. The carry will take its values from the set $C_c = [-c_1, c_2]$, being c_1 and c_2 such as when two digits in a position are taking the extreme values of C_b, after subtracting the carry, the result in that digits is in C_b. In other words, when:

$$-2m = -c_1 b + A \quad \Rightarrow \quad A = -2m + c_1 b$$

A must be in C_b. In the same way, when

$$+2n = c_2 b + B \quad \Rightarrow \quad B = 2n - c_2 b$$

B must be in C_b.

The previous conditions, without taking into account the previous carry, can be written as inequalities as:

$$-m \leq -2m + c_1 b \leq n$$
$$-m \leq 2n - c_2 b \leq n$$

When adding the possible carry generated in the previous digit, for avoiding generating new carries, and so, avoiding its propagation, the result being into C_b is required. The extreme cases which can appear, and allowing to narrow c_1 and c_2 values, are the following two:

(a) when the two digits to be summed are $-m$ (a carry $-c_1$ will be generated) and also, the previous carry takes one of the extreme values $-c_1$ or c_2; and
(b) when the two digits to be summed are n (a carry c_2 will be generated) and also, the previous carry takes one of the extreme values $-c_1$ or c_2.

Taking into account the previous carry, from all the above it results the following four inequalities:

$$
\begin{aligned}
-2m - c_1 + c_1 b &\geq -m \\
-2m + c_2 + c_1 b &\leq n \\
2n - c_1 - c_2 b &\geq -m \\
2n + c_2 - c_2 b &\leq n
\end{aligned}
$$

From this inequalities, the following shirts result for c_1 and c_2:

$$
\frac{m}{b-1} \leq c_1 \leq \frac{2m(b-1) + n(b-2)}{b(b-1)}
$$
$$
\frac{n}{b-1} \leq c_2 \leq \frac{2m(b-1) + n(b-2)}{b(b-1)}
$$

These conditions are necessary for avoiding carry propagation, but are not sufficient, because of the correct carry selection issue when there are several possibilities, as shown in the following.

When the carry range is $C_c = [-c_1, c_2]$, it can take $c_1 + c_2 + 1$ different values. One issue to take into account and which has not been approached until now, consists on determining in what cases each one of the possible carry values must be applied. When possible, this carry selection will be made in terms of the digits being summed, without taking into account the previous carry. It is clear that when adding two digits taking the extreme values of the range, the extreme carry values must be applied, as has been carried out when obtaining inequalities relative to c_1 and c_2. Thus, when the two digits of each summand are $-m$, then the carry must be $-c_1$, and when both of them are n, carry must be c_2. But, in the rest of situations, what values from partial sums are transferred from one value to other of the carry must be selected. The following example illustrates these issues.

Example 1.17 Let's consider $b = 10$ and $C_b = [-9, 9]$. From the previous inequalities for determining the c_1 and c_2 ranges, it results that c_1 as c_2 can take the values 1 or 2. The two digits sum, without taking into account the preceding carry, will be into the range $[-18, 18]$.

We will start considering the case $c_1 = c_2 = 1$. In this situation, when the two digits partial sum is into the range $[-18, -9]$, carry must be -1. In this way, for every precedent carry, -1 or $+1$, the final value will be into the correct C_b range. When the two digits partial sum is into the range $[-8, -2]$, due to the same reason, carry can be -1 or 0. When the two digits partial sum is into the range $[-2, 2]$, carry will be 0. When the two digits partial sum is into the range $[2, 8]$, carry can be 1 or 0. Finally, when the two digits partial sum is into the range $[9, 18]$, carry must be 1.

In the case of $c_1 = c_2 = 2$, when the two digits partial sum is -18, carry must be -2. When the sum is into the range $[-17, -13]$, carry can be -1 or -2. When it is into the range $[-12, -8]$, carry must be -1. When it is into the range $[-7, -3]$, carry can be -1 or 0. When it is into the range $[-2, 2]$, carry must be 0. When it is into the range [3, 7], carry can be 0 or 1. When into the range [8, 12], carry must be 1. When it is into the range [13, 17], carry can be 1 or 2. Finally when the partial sum is 18, carry must be 2. Thus, there are several value ranges in partial results which can use different carry values.

As an exercise, the cases $c_1 = 1$, $c_2 = 2$; and $c_1 = 2$, $c_2 = 1$ are proposed. ■

In each of the situations in the previous example, carry propagation can be limited to only adjacent positions. Thus, the carry generated in a given position only depends on the partial sum in this position, and only can modify the next position digit. However, in some situations, wider propagations must be allowed, or previous digits must be taken in consideration when selecting the carry in a given digit.

Removing carry propagation is essential for fast completion of addition (or subtraction) operation, moreover taking int0o account that the rest of arithmetic operations such as multiplication and division, are based on repeated additions and subtractions.

1.9.2 Binary Case

When $b = 2$, with $C_b = [-1, 1]$, it results $C_c = [-1, 1]$, thus being $c_1 = 1$, $c_2 = 1$. Table 1.18 provides the values for the sum and carry bits for each possible partial sum. When the partial sum takes the values -1 or 1, two options with respect the sum and carry bits (s, c) are possible. If one of them is selected without taking into account the precedent carry, none of the two guarantees the final sum being in C_b, when the precedent carry will be summed. Thus, for the final sum being in C_b, one of the two options must be chosen depending on the previous digits. Specifically, in both rows of Table 1.18, the sum bit can be -1 and 1. The value -1 must be selected when the precedent carry cannot be 1. As shown in Table 1.18, the carry in a given stage cannot be -1 if none of the digits takes the value -1, and cannot be 1 if one of the digits takes the value -1. With all of these considerations, Table 1.19 results for the binary addition of signed digits. Although for obtaining sum and carry bits in a given position i, taking into consideration two positions digits (i and $i-1$) is needed, carry propagation is avoided, because it is limited to the next position only.

When the two summands are given in conventional representation (i.e., with $C_b = [0, 1]$), and using $C_b = [-1, 1]$ and $C_c = [-1, 1]$ for the result, then none of the

Table 1.18 Signed digits binary sum

Partial sum	Sum and carry (s, c)
-2	$(0, -1)$
-1	$(-1, 0)$ or $(1, -1)$
0	$(0, 0)$
1	$(1, 0)$ or $(-1, 1)$
2	$(0, 1)$

Table 1.19 Sum and carry taking into account precedent digits

Partial sum i	Bits $i-1$	Sum and carry (s, c)
-2	Regardless	$(0, -1)$
-1	None -1	$(-1, 0)$
-1	Any -1	$(1, -1)$
0	Regardless	$(0, 0)$
1	None -1	$(-1, 1)$
1	Any -1	$(1, 0)$
2	Regardless	$(0, 1)$

Table 1.20 Conventional summands

Partial sum i	Sum and carry (s, c)
0	$(0, 0)$
1	$(-1, 1)$
2	$(0, 1)$

bits $i-1$ can take the value -1, and the partial sum can take the values 0, 1 or 2. Thus, Table 1.19 can be reduced to Table 1.20.

1.10 Conclusion

This Chapter has been devoted to all relative to the different numeral representations, and to the most frequent arithmetic operations fundamentals. With respect to the contents that will be developed in the following chapters, in the second one, some basic arithmetic circuits will be described, using the ideas of this first chapter.

1.11 Exercises

1.1 Calculate the most efficient radix when the cost c_d per digit is $c_d = k_d b^2$. Idem when $c_d = k_d b^{1/2}$.

1.2 Convert to base 2 the following decimal numbers: 76.0436, 48.122, 0.0025.

1.3 Convert to base 3 the following numbers given in base 7: 5023.042, 304.52, 4.332.

1.4 Convert to base 7 the following fractional decimal numbers: 2/5, 3/8, 1/3.

1.5 Convert to binary the following numbers given in hexadecimal: 36A04, 76B5, 43E0D1, 1807C, F20A.54.

1.6 Using the double radix (2 and 3) representation obtain the simplest development of the following natural decimal numbers: 49, 532, 608.

1.7 Find the representation of the decimal value 13.821 in the mixed radix system $\{2, 3, 7, 17, 23\}$.

1.8 Find the decimal value, N, corresponding to the number 69(13)(18) represented in the mixed radix system $\{7, 11, 17, 23\}$.

1.9 In base 7 do the followings arithmetic operations using the SM representation: $236 + 543$, $603 - 306$, $-532 + 125$, $-411 - 244$.

1.10 In base 7 do the followings arithmetic operations using the 6-complement representation: $236 + 543$, $603 - 306$, $-532 + 125$, $-411 - 244$.

1.11 In base 7 do the followings arithmetic operations using the 7-complement representation: $236 + 543$, $603 - 306$, v $532 + 125$, $-411 - 244$.

1.12 In base 7 design a biased representation with four digits. In this biased representation system, do the followings arithmetic operations: $236 + 543$, $603 - 306$, $-532 + 125$, $-411 - 244$.

1.13 Using the **SM** representation do the followings products of 8-bits binary numbers:

$$00101101 \times 01110101,$$
$$01101101 \times 10100101,$$
$$11010010 \times 00101101,$$
$$10110011 \times 11010110.$$

1.14 Using the 1-complement representation do the followings products of 8-bits binary numbers:

$$00101101 \times 01110101,$$
$$01101101 \times 10100101,$$
$$11010010 \times 00101101,$$
$$10110011 \times 11010110.$$

1.15 Using the 2-complement representation do the followings products of 8-bits binary numbers:

$$00101101 \times 01110101,$$
$$01101101 \times 10100101,$$
$$11010010 \times 00101101,$$
$$10110011 \times 11010110.$$

1.16 Do the followings divisions of two positive binary numbers without sign bit:

$$00101101 \div 0101,$$
$$01101101 \div 1010,$$
$$11010010 \div 0101,$$
$$10110011 \div 1001.$$

1.17 Compute the square root of the followings positive binary numbers without sign bit:

$$\begin{gathered} 00101101, \\ 01101101, \\ 11010010, \\ 10110011. \end{gathered}$$

1.18 Indicate the necessary corrections to add three **BCD** characters using binary adders.

1.19 Code in **BCD SM** the followings decimal numbers and do the indicated operations: $678 + 293$, $678 - 293$, $-678 + 293$, $-678 - 293$.

1.20 Code in **BCD** 9-complement the followings decimal numbers and do the indicated operations: $678 + 293$, $678 - 293$, $-678 + 293$, $-678 - 293$.

1.21 Code in **BCD** 10-complement the followings decimal numbers and do the indicated operations: $678 + 293$, $678 - 293$, $-678 + 293$, $-678 - 293$.

1.22 Design a procedure for the overflow detection in **BCD** adders when using **SM**, 10' complement and 9'complement representations.

1.23 Using the packed **BCD** codification **CHC**, code the followings decimal numbers: 145, 2347, 87023.

1.24 Given the three groups of 10 bits 0110011010, 1101001001, 1011001110, obtain the three **BCD** digits corresponding to each of these CHC codifications.

1.25 Obtain the range of possible values with $b = 7$ and $C = \{\bar{2}, \bar{1}, 0, 1, 2, 3, 4, 5\}$. In this redundant system, obtain the different representations of the followings decimal numbers: 4987, 3045 y 682.

1.26 Code in the BSD representation the followings decimal numbers: 4987, 3045 y 682.

1.27 Recode the followings bit-chains using the canonical codification: 101100110101, 110100100100 y 101100111001.

1.28 Recode the followings bit-chains using the Booth in radix 4 codification: 101100110101, 110100100100 y 101100111001.

1.29 Recode the followings bit-chains using the Booth in radix 8 codification: 101100110101, 110100100100 y 101100111001.

1.30 Convert to 2-complement the followings binary numbers coded in BSD: $\bar{1}011\bar{1}011010\bar{1}$, $1\bar{1}010\bar{1}100\bar{1}00$ y $10\bar{1}100\bar{1}1100\bar{1}$.

1.31 Design a system to add several base 7 numbers in which the carries in the intermediate additions will propagate only to the next position.

1.32 Solve the cases $c_1 = 1$, $c_2 = 2$, and $c_1 = 2$, $c_2 = 1$ proposed in Example 1.17.

1.33 Let $b = 7$ and $C_b = [-6, 6]$. Determine the possible carries depending of the result when adding two digits.

References

1. Chen, T.C., Ho, T.: Storage-efficient representation of decimal data. CACM (1), 49–52 (1975)
2. Cowlishaw, M.: Densely packed decimal encoding. In: IEE Proceedings of Computers and Digital Techniques, vol. 149, issue no. 3, pp. 102–104 (2002, May)
3. Dimitrov, V.S., Jullien, G.A., Miller, W.C.: Theory and applications of the double-base number system. IEEE Trans. Comput. **48**(10), 1098–1106 (1999 October)
4. Ebeid, N., Hasan, M. A.: On binary signed digit representations of integres. Des Codes Crypt **42**, 43–65 (2007). (Springer)
5. Koç, Ç.K.: Parallel canonical recoding. Electron. Lett. **32**(22), 2063–2065 (1996)
6. Miller, R.E.: Switching Theory. Wiley (1965)

Chapter 2
Basic Arithmetic Circuits

This chapter is devoted to the description of simple circuits for the implementation of some of the arithmetic operations presented in the previous chapter. Specifically, the design of adders, subtracters, multipliers, dividers, comparators and shifters are studied, with the objective of providing the design guidelines for these specific application circuits. The arithmetic circuits presented will be used in the next chapters for the implementation of algebraic circuits.

This introductory chapter on arithmetic circuits is the only necessary for the building of algebraic circuits and will be extended with specific chapters devoted to the different operations.

2.1 Introduction

This section presents the previous aspects related to the arithmetic circuits: differences between serial and parallel information, pipelining, or circuits multiplicity for increasing performance. Although these concepts will be probably known by the reader, they are included in order to provide an immediate reference.

2.1.1 Serial and Parallel Information

When transmitting or processing information, two extreme structures can be considered: serial and parallel information. Briefly, we have serial information when the bits integrating each of the information blocks are transmitted or processed at different times. On the contrary, we have parallel information when the bits composing each information block are transmitted or processed simultaneously.

The clearest example for discriminating between serial and parallel information resides on information transmission. Assuming the design of a system for performing

A. Lloris Ruiz et al., *Arithmetic and Algebraic Circuits*, Intelligent Systems Reference Library 201, https://doi.org/10.1007/978-3-030-67266-9_2

some numerical calculations formed by several subsystems, and being each data 8 bits wide (i.e., 8-bit words must be processed), the information among the different subsystems can be transmitted using 8 wires. In this case, the 8 bits are transmitted simultaneously, at the same time, thus being parallel information. But this information can be also transmitted using only 1 wire, sending the 8-bit of a data block bit by bit, with a predetermined order and at 8 different times, constituting serial information. When using serial information, usually the first bit being transmitted and/or processed it the less significant one, but it could be also the most significant one.

Intermediate situation between serial and parallel structures can be considered. Each word can be divided into blocks (known as digits), being processed in parallel the bits corresponding to each digit, but being the different digits transmitted or processed in a serial way. As an example, a 64-bit word can be processed or transmitted in serial (taking 64 cycles), in parallel (taking only one cycle by using a 64-wire bus), in 16-digit of 4 bits (taking 16 cycles by using a 4-wire bus), in 8-digit of 8 bits (taking 8 cycles by using a 8-wire bus), etc.

2.1.2 *Circuit Multiplicity and Pipelining*

Every digital circuit **C** (Fig. 2.1a) establishes a correspondence between the inputs, E, and the outputs, S, $S = \mathrm{F}(E)$. Given an input at a given time, the most efficient circuit in terms of temporal response will be the combinational circuit capable of generating the output in the same cycle when the input has been arrived. The complexity of this circuit depends mainly on the number of the input bits (input size). If the output is not needed in the same cycle, probably a simpler sequential circuit can be built generating the output S some clock cycles later from the input E arrival. However, in the case of a continuous input data flow, and being necessary generating a result in each cycle, the complexity of a circuit can produce large delays, preventing the output being generated in the same cycle than the input arrival. For maintaining a continuous data flow at the output, two alternatives can be considered: circuit multiplicity and pipelining, as detailed in the following.

Circuit multiplicity (Fig. 2.1b) consists on using m identical circuits (as many as the delay introduced by each one of the circuits), working in parallel. The inputs of the m circuits are connected to the outputs of a 1-to-m demultiplexer which input is E. The function of the demultiplexer consists on driving the data to the circuit C_i being available in each time for starting the calculation. The outputs of the m circuits are connected with the inputs of an m-to-1 multiplexer, with output S. The function of the multiplexer consists on selecting at each time the output of the circuit being finished the calculation. In this way, during the first m cycles no result is generated, and from this moment, a calculation result will be generated at each cycle. Note that the result collected at the output in a given time corresponds to the inputs introduced m cycles before. This delay between input and output sequences is known as the **latency** of the system. Circuit multiplicity presents the advantage of simplicity because the design

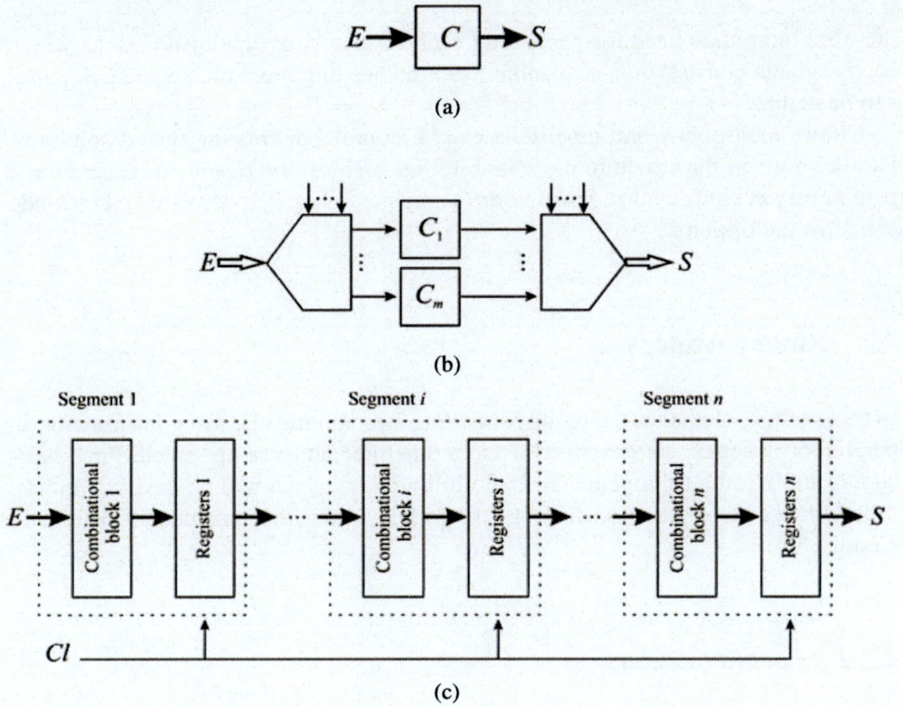

Fig. 2.1 **a** Circuit. **b** Multiplicity. **c** Pipelining

is reduced to place as many circuits in parallel as indicated by the latency. As a drawback, the cost of the system can result excessive.

The **pipelining** of a combinational circuit (Fig. 2.1c), in their simplest version, consists on modifying the original circuit dividing it into n segments, each one completing the corresponding processing in one clock cycle. Each of the segments includes a register storing its output, making it available for the next segment. The registers at the different segments are controlled by the same clock signal. The pipelined circuit allows a continuous data flow at the input E, and after the initial n cycles delay due to the different segments (the latency of the pipelined circuit), a continuous data flow at the output S, is obtained. Thus, the output at a time corresponds with the input introduced into the pipelined circuit n clock cycles before. Each segment executes one part of the complete calculation, being then n data sets computed in parallel, each one with a different phase and at a different segment. Note that this structure is similar to the assembly line of a factory, where the global task is decomposed into simpler tasks, in such a way that each assembly machine (with the corresponding workers) performs only one of these elementary tasks.

When using pipelined circuits with the structure presented in Fig. 2.1c, each segment generates correctly its output in one cycle, and is used only once for generating each result. More complex circuits can be used where some or all of the segments are used more than once for generating each result. Also, pipelined circuits which each segment consists on a sequential machine needing more than one clock pulse, can be defined.

Circuit multiplicity and pipelining can be combined, creating mixed solutions. Parallel units in the structure presented in Fig. 2.1b can be pipelined, generating a result every m clock cycles. Some of the segments in Fig. 2.1c structure can include elements multiplicity.

2.2 Binary Adders

In this section, elementary circuits for adding two summands, using information in parallel or in series, are described. Also a pipelined adder is presented. First, half-adders are introduced, together with the full-adders, which will be the basic blocks for building adder circuits, and also will be widely used in the remainder of arithmetic circuits.

2.2.1 *Parallel Adders*

In the following, simplest binary adders are described. Thus, it will be assumed the situation of adding two positive numbers without sign bit. As an example, let consider the addition:

$$A + B = 1011 + 0011$$

Arranging these two summands as usual, one below the other, as it is done in Fig. 2.2a, first the two bits corresponding to the position 2^0 are added, obtaining the bit of the result at the same position. For obtaining the bit at position 2^i ($i = 1, \ldots, n$) of the result, the two bits at this position are added together with the precedent carry. Partial sums and the carry for the next stage are obtained from addition tables (Table 1.1a), which are repeated in Fig. 2.2b and c using other arrangement. In our example, the result is: $1011 + 0011 = 1110$.

$$A + B = 1011 + 0011$$

1001
+0101
1110

(a)

xy	0	1
0	0	1
1	1	0

(b)

xy	0	1
0	0	0
1	0	1

(c)

Fig. 2.2 **a** Addition examples. Two bits addition tables: **b** Sum. **c** Carry

The functions corresponding to the partial sum, *s*, and to the carry, *a*, are:

$$s = \overline{x}y + x\overline{y} = x \oplus y; \qquad a = x \cdot y$$

Synthesizing these two functions as a combinational block (using two **AND-OR** gate levels or using **XOR** gates), in the way represented in Fig. 2.3a, it results the circuit known as half-adder. This block, represented in Fig. 2.3b, allows the less significant bit of a sum to be obtained, while the remainder of the bits require two half-adders to be calculated. Connecting in cascade several half-adders in the way represented in Fig. 2.3c, binary numbers with an arbitrary number of bits can be added. For obtaining the carry in a given stage, the **OR** operation must be performed over the carries generated by the two half-adders, because the two half-adders in a same stage cannot produce simultaneously carry '1', as can be easily proved.

The calculation of sum and carry at each position can be also performed by means of a combinational block known as full-adde r, with three inputs (the two summand bits at this position, *x* and *y*, plus the previous carry, a_-) and two outputs, *S* and a_+. From the truth table of the two functions (Fig. 2.4a) to be synthesized by this block, it results:

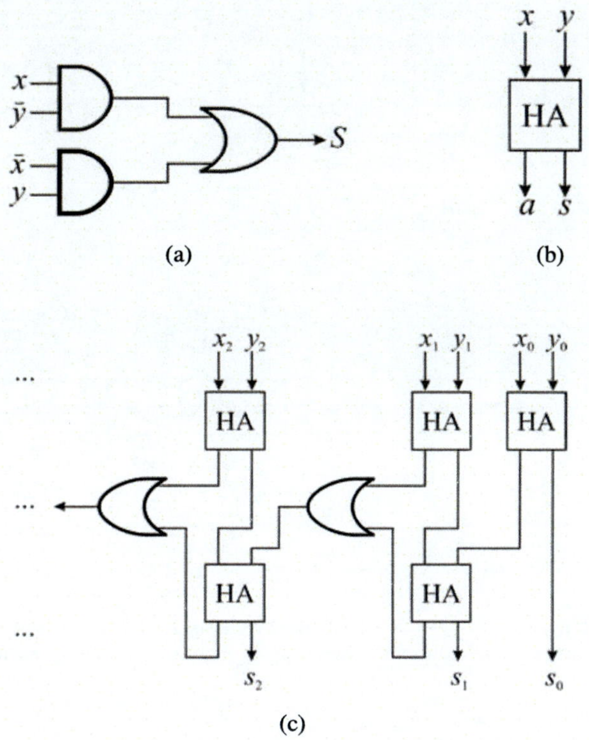

Fig. 2.3 Half-adder: **a** Circuit. **b** Representation. **c** Cascading

xya_-	Sa_+
000	00
001	10
010	10
011	01
100	10
101	01
110	01
111	11

(a)

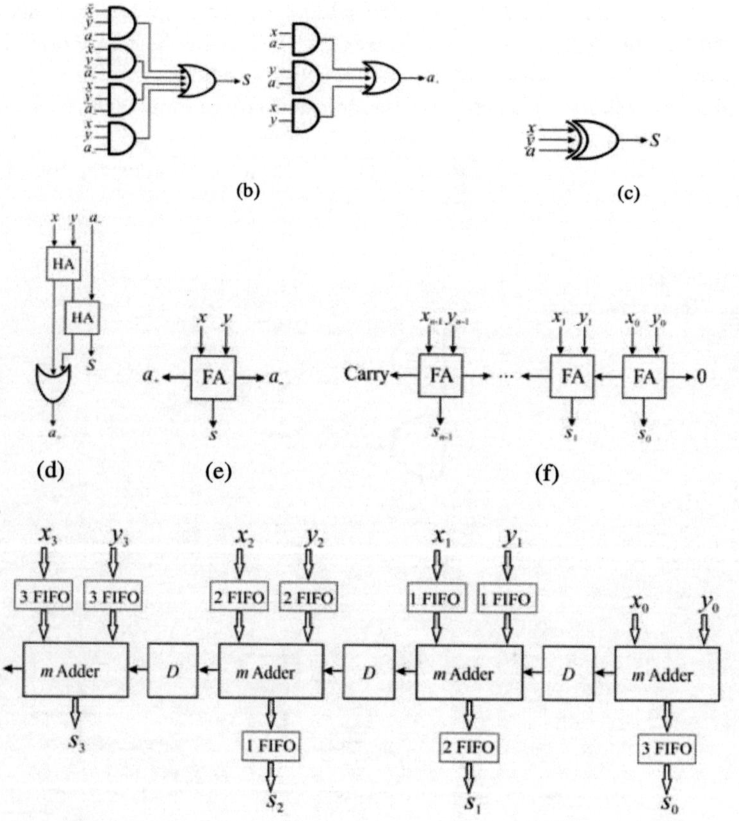

(g)

Fig. 2.4 Full adder: **a** Truth table. Synthesis: **b** AND-OR. **c** With an XOR gate. **d** With half-adders cascading. **e** Representation. **f** Ripple carry adder. **g** Pipelined ripple carry adder

$$S = \overline{x}\,\overline{y}a_- + \overline{x}y\overline{a}_- + x\overline{y}\,\overline{a}_- + xya_- = x \oplus y \oplus a_-; \; a_+ = xa_- + ya_- + xy$$

The full-adder block can be implemented using **AND-OR** synthesis (Fig. 2.4b) or using a **XOR** gate for the sum *S* (Fig. 2.4c), a_+ can be synthesized as shown in Fig. 2.4b or concatenating two adders plus an **OR** gate (Fig. 2.4d), and it is represented in Fig. 2.4e.

For adding *n*-bit numbers with parallel information, simply connect *n* full adders in cascade (Fig. 2.4f). The parallel *n*-bit adder resulting is known as **ripple carry adder**. This adder presents the drawback of the delay introduced by the carry propagation through the successive stages. In fact, the result at the carry output of the most significant bit of the sum must wait for any change at the carry input of the less significant bit being propagated. When the size of the summands (number of bits) is not excessive (from 4 to 16), or the circuit's performance is not relevant, this drawback has no impact. However, when the size of the operands is large or a high operation speed is required, may be the result of the addition can not be generated correctly in one cycle. In this situation, alternative solutions accelerating carry propagation should be used, leading to carry look-ahead adders, or special procedures for adding. Pipelining of circuit detailed in Fig. 2.4f, the addition of more than one bit in each stage, and the addition of more than two summands at a time, are among the options for speeding up the adders operation.

When using biased representation, as shown in Sect. 1.4.3, and making $D = 2^{m-1}$, the same adders presented here can be used appending an inverter for the most significant bit. Similarly, if $D = 2^{m-1-1}$, in addition to complementing the most significant bit, the initial carry must be 1.

2.2.2 *Pipelined Adders*

In several applications, like those involved in digital signal processing, a continuous data flow with multiple additions must be made. In this situation, the ripple carry adder results unsuitable because of their excessive delay, but it can be easily pipelined introducing registers in the appropriate locations. Assuming *r* is the delay corresponding to a full adder, and *f* is the clock frequency, then the maximum length *m* of the adder providing the result in each of the cycles will be:

$$m \leq \frac{1}{r \cdot f}$$

For building an *n*-bits adder, it must be divided into *s* segments, being:

$$s \geq \frac{n}{m}$$

Obviously, if $n \neq m \cdot s$, then one of the segments (usually the first one or the last one) can be shorter than the rest.

Each segment will be separated from the following by a **D** flip-flop in order to store the carry between stages. The inputs and outputs will be separated, in general, by means of register stacks (**FIFO** registers). All of the registers will have so many bits as the corresponding segment length (in the previous example, m bits), and the size or depth of each one of the stacks (i.e., the number of registers stacked) will depend on the segment position, with the objective of properly synchronizing inputs and outputs, as represented in Fig. 2.4g for an adder composed by four m-bits segments. The depth of each **FIFO** is indicated in Fig. 2.4g by the first digit in their name. The latency of these adders is 4.

2.2.3 *Serial Adders*

When the summands (X and Y) are serially available bit by bit (being the first bit the less significant one), they can be added using a full adder and a **D** flip-flop in order to store the carry generated for the next stage, as shown in Fig. 2.5a. For a correct operation, the **D** flip-flop must be initialized to 0. At the output S, the addition is obtained serially. The final carry will remain at D, but it can be transferred to S introducing one '0' into each input after the most significant bits of both summands.

For serial operands digit by digit (the first digit is the less significant one, again) a parallel digit adder and a **D** flip-flop (initialized to '0') are required, as shown in Fig. 2.5b. The digit adder can be built using as many full adders as the size of the digit. Again, the final carry remains in the **D** flip-flop, but it can be transferred to S introducing one digit with all zeros into each input after the most significant digits of both summands.

Comparing serial processing with parallel processing, it is clear that the series circuits are simpler than the parallel, both in number of gates (less full adders in this case) and the number of inputs and outputs. With regard to the processing time, with serial structures as many computation cycles as blocks forming each word are required, whereas with parallel information only one cycle is sufficient. However,

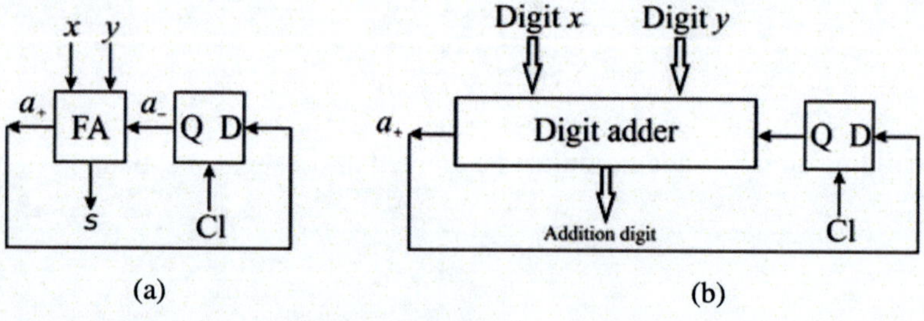

Fig. 2.5 Serial adder: **a** bit by bit. **b** Digit by digit

the serial adder, because it is simpler than the parallel, withstand higher speeds than parallel, i.e. the serial adder will require more cycles, but each cycle can be of shorter duration.

2.3 Binary Subtractors

Subtraction tables (Table 1.1b, and repeated in Fig. 2.6a) implementing the functions corresponding to the partial difference *r*, and the borrow, *d*, are:

$$r = \overline{x}y + x\overline{y} = x \oplus y$$
$$d = \overline{x}y$$

Synthesizing *r* and *d* functions (*r* fits with the partial sum *S* from the half-adder), half-subtractors are obtained, which can be cascaded in a similar way than shown in Fig. 2.3c for half-adders, allowing the subtraction of binary numbers with any number of bits, as shown in Fig. 2.6b.

Also full-subtractors can be designed for 1-bit characters. In this case, the truth table corresponding to *x–y*, including the previous borrow, is given in Fig. 2.6c, resulting the following functions:

$$R = \overline{x}\,\overline{y}\,d_- + \overline{x}\,y\,\overline{d}_- + x\,\overline{y}\,\overline{d}_- + x\,y\,d_- = x \oplus y \;\oplus\; d_-$$

$$d_+ = \overline{x}\,d_- + \overline{x}\,y + y\,d_-$$

For subtracting unsigned binary *n*-bit numbers, *X–Y*, the *ripple-borrow subtracter* of Fig. 2.6d can be used. When $X \geq Y$, this subtractor generates the correct result being the final borrow 0, as can be easily checked by the reader. When $X < Y$, the final borrow is 1, and the result is not correct. Thus, the result generated is correct only when it is positive. For taking into account negative results with this subtractor, a comparator must be included for detecting which operand is the greatest one, as was shown when introducing the **SM** representation. Other alternative is the use of complement adders/subtractors, as detailed in the following.

Because of the common part with the full-adder and the full-subtractor, are often built as a single adder/subtractor block, with a control input for selecting between the two operations.

As shown in Sect. 1.4.3, when using biased representations and making $D = 2^{m-1}$, the same subtractors described for **SM** can be used, adding an inverter for the most significant bit. In a similar way, if $D = 2^{m-1-1}$, an inverter must be added, with an initial borrow of 1.

About the subtraction using complement representations, when using two's complement, subtraction consists on adding to the minuend, the two's complement of the subtrahend. By the other hand, the complementation is performed by

Difference

xy	0	1
0	0	1
1	1	0

Borrow

xy	0	1
0	0	1
1	0	0

(a)

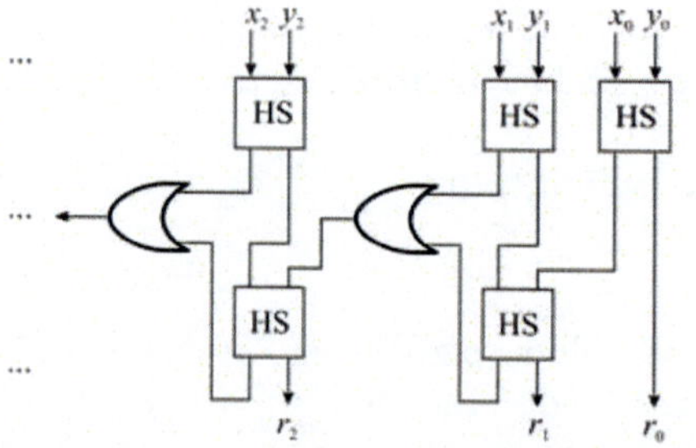

(b)

xyd_-	Sd_+
000	00
001	11
010	11
011	01
100	10
101	00
110	00
111	11

(c)

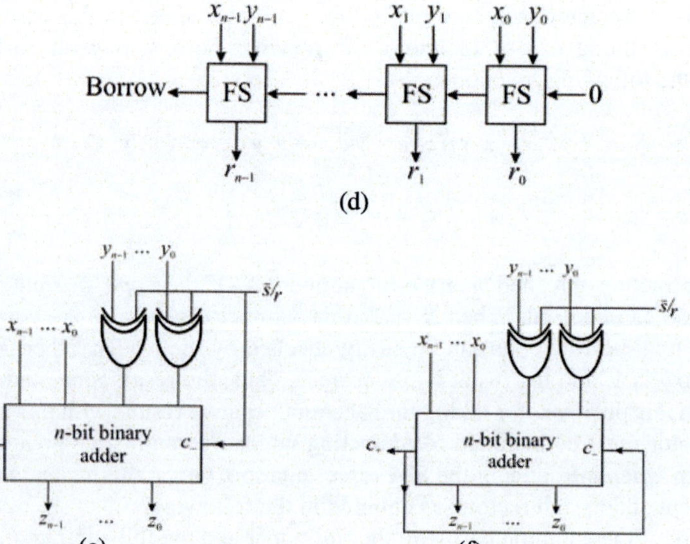

(d) (e) (f)

Fig. 2.6 **a** Subtraction table. **b** Half-subtractors cascading. **c** Full-subtractor table. **d** Full-subtractors cascading. Adder/subtractor: **e** two's complement. **f** One's complement

complementing all bits and adding 1 to the result. Joining these ideas, the circuit of Fig. 2.6e can be carried out for implementing a two's complement adder/subtractor. The control signal $\overline{s}/r$ must be 0 for adding, and 1 for subtracting (the detailed analysis of the circuit is left as an exercise for the reader). In this circuit, making $X = x_{n-1} \cdots x_0 = 0 \cdots 0$, and $\overline{s}/r = 1$, the two's complement of Y is obtained. With a similar idea, Fig. 2.6f shows a one's complement adder/subtractor, as can be easily checked. In this situation, the end-around carry must be included, using the carry out as input carry. This end-around condition makes the two's complement advantageous with respect to one's complement representation, as seen comparing Fig. 2.6e and f.

2.4 Multipliers

In the following, some simple circuits for multiplication, both combinational and sequential, for integer binary numbers will be described. Also, the design of circuits for multiplying by a constant and for raising to an integer power will be approached.

2.4.1 *Combinational Multipliers*

To give an idea of how to build these circuits, without trying too much detail, we will first consider multipliers for binary coded positive integers (without sign bit). Such multipliers are widely used in signal processing applications and can be the core of multipliers when using signed binary numbers.

When multiplying an m-bit number A by an n-bit number B (both unsigned positive numbers), the product P will take $m + n$ bits. In fact:

$$A \leq 2^m - 1$$

$$B \leq 2^n - 1$$

thus $P = A \cdot B \leq (2^m - 1)(2^n - 1) = 2^{m+n} - 2^m - 2^n + 1$. Then, except in the situations $m = 1$ or $n = 1$, $m + n$ bits are required for representing P.

The most elemental multiplier is the one for one bit characters, which table is presented in Table 1.1c (and repeated in Fig. 2.7a). In this case, the operation is the **AND** function, and the result is represented by using only one bit.

When multiplying 2-bit integer positive numbers, $X = x_1x_0$ e $Y = y_1y_0$, 4 bits are required for representing the product $M = X \cdot Y$. This multiplier can be designed as a combinational circuit with four inputs and four outputs, which truth table and circuit are presented in Fig. 2.7b and c. This circuit can be also interpreted as a base-4 multiplier of two 1-digit characters, and synthesized by using elemental multipliers and adders. In fact, Fig. 2.7d details the X by Y multiplication, and Fig. 2.7e presents the circuit with this design strategy, using four 1-bit multipliers and two half-adders. With independence of the design used, a multiplier of two 2-bit characters (or two 1-digit base-4 characters) is represented as in Fig. 2.7f.

For building circuits enabling the multiplication of characters with any number of bits, the same techniques used for 2-bit numbers can be used. As an example, for

x	y	P
0	0	0
0	1	0
1	0	0
1	1	1

(a)

x_1	x_0	y_1	y_0	m_3	m_2	m_1	m_0
0	0	0	0	0	0	0	0
		0	1	0	0	0	0
		1	0	0	0	0	0
		1	1	0	0	0	0
0	1	0	0	0	0	0	0
		0	1	0	0	0	1
		1	0	0	0	1	0
		1	1	0	0	1	1
1	0	0	0	0	0	0	0
		0	1	0	0	1	0
		1	0	0	1	0	0
		1	1	0	1	1	0
1	1	0	0	0	0	0	0
		0	1	0	0	1	1
		1	0	0	1	1	0
		1	1	1	0	0	1

$$m_3 = x_1 x_0 y_1 y_0$$

$$m_2 = x_1 y_1 (\overline{x}_0 + \overline{y}_0)$$

$$m_1 = x_0 y_1 (\overline{x}_1 + \overline{x}_0) + x_1 y_0 (\overline{x}_0 + \overline{y}_1)$$

$$m_0 = x_0 y_0$$

(b)

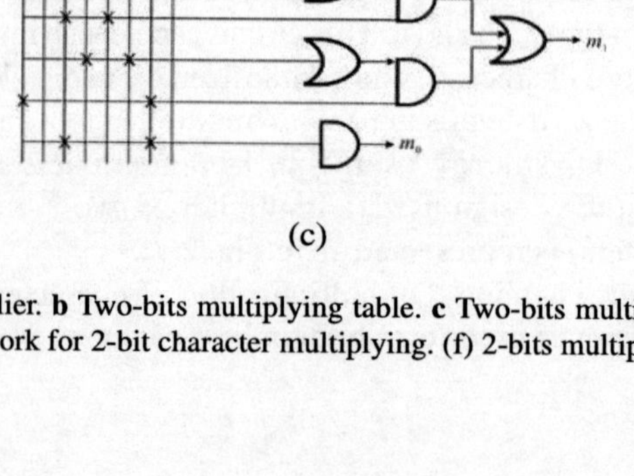

(c)

Fig. 2.7 **a** 1-bit multiplier. **b** Two-bits multiplying table. **c** Two-bits multiplier circuit. **d** *X* by *Y* multiplication. (e) Network for 2-bit character multiplying. (f) 2-bits multiplier

Fig. 2.7 (continued)

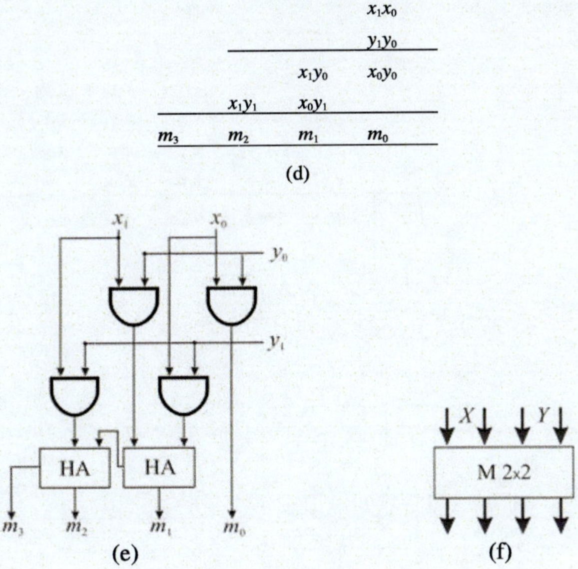

multiplying 4-bit charactersa base-16 elemental multiplier), a combinational circuit with eight inputs and eight outputs can be synthesized and implemented using a programmable device (a **ROM** for example) or in any other way. Nevertheless, when the size of the characters to be multiplied increases, this synthesis technique leads to bulky and difficult to manage circuits. In this situation the multipliers are synthesized by combining elemental multipliers and adders. Figure 2.8a shows the method of operation for multiplying two 4-bit numbers. The circuit in Fig. 2.8b implements mimetically this method using 1-bit multipliers, half-adders and full-adders. Also, 2-bit multipliers and adders could be used as building blocks. In this case, being $X = X_1X_0$ ($X_1 = x_3x_2$, $X_0 = x_1x_0$), $Y = Y_1Y_0$ ($Y_1 = y_3y_2$, $Y_0 = y_1y_0$), multiplication can be performed as described in Fig. 2.8c, and the circuit of Fig. 2.8d is also a 4-bit characters multiplier.

Building methods based on the use of elementary multipliers and adders allow the design of combinational multipliers of any size.

In general, given two base b numbers (P and Q) to be multiplied, they can be decomposed into two or more pieces, and then processing these pieces using less complex resources. As an example, decomposing P and Q into two pieces it results:

$$P = p_{n-1}b^{n-1} + p_{n-2}b^{n-2} + \cdots + p_1b + p_0 = P_Hb^s + P_L$$
$$Q = q_{n-1}b^{n-1} + q_{n-2}b^{n-2} + \cdots + q_1b + q_0 = Q_Hb^s + Q_L$$
$$P \cdot Q = (P_Hb^s + P_L) \cdot (Q_Hb^s + Q_L) = P_HQ_Hb^{2s} + (P_HQ_L + P_LQ_H)b^s + P_LQ_L$$

Note that s must be chosen close to $(n - 1)/2$ in order to make the circuits simpler.

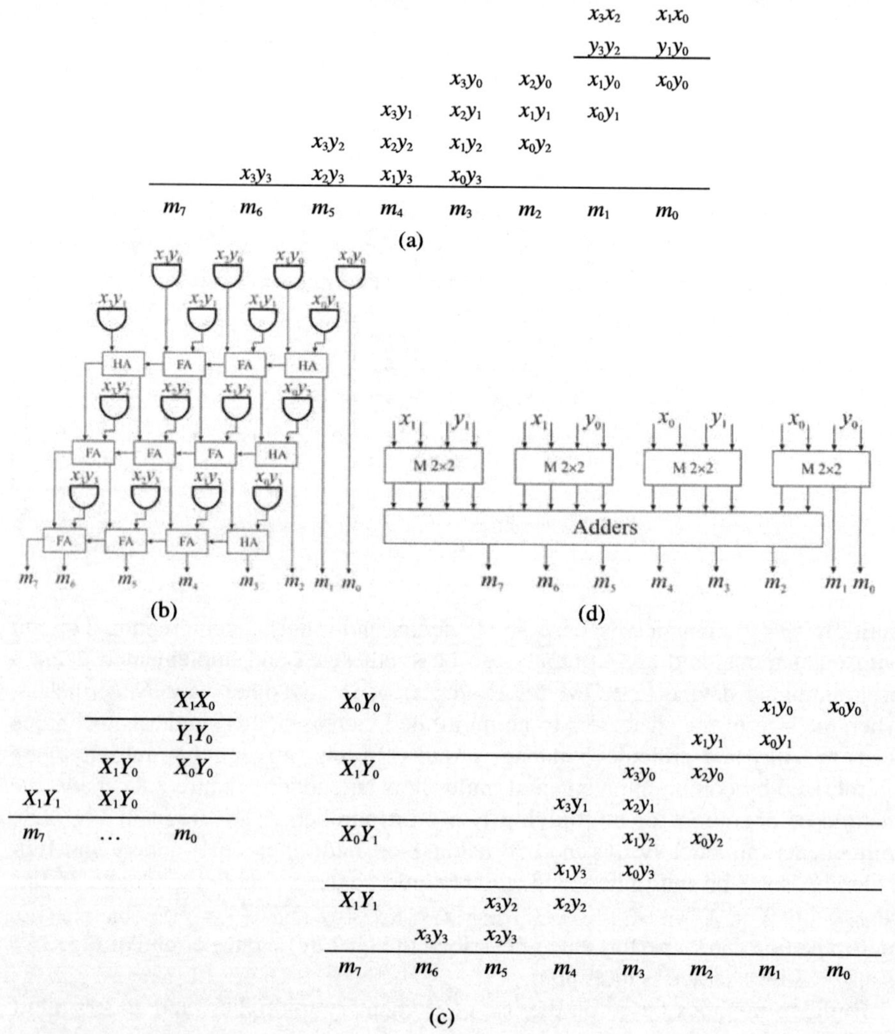

Fig. 2.8 **a** Four-bit characters multiplication. **b** Circuit for multiplication. **c** X by Y multiplication. **d** Network for multiplying X by Y

With the expressions above, four multipliers and the corresponding adders are required for obtaining $P_H Q_H, P_H Q_L, P_L Q_H$ and $P_L Q_L$. In order to reduce the number of multipliers, although at expenses of increasing the number of adders, the product $P \cdot Q$ can be expressed as:

$$P \cdot Q = P_H Q_H b^{2s} + ((P_H + P_L) \cdot (Q_H + Q_L) - P_H Q_H - P_L Q_L) b^s + P_L Q_L$$

In this way, only three multipliers are required for getting $P_H Q_H$, $(P_H + P_L) \cdot (Q_H + Q_L)$ y $P_L Q_L$. Obviously, each one of the partial products can be computed using iteratively the same procedure for decomposing each operand into two or more chunks.

2.4.2 *Sequential Multipliers*

The designs above allow the multiplication of two unsigned binary numbers in only one clock cycle. It is possible the resulting circuits being so much complex for the designer convenience or introducing excessive delay for responding into the clock cycle required for the general system performance. In this situation, pipelining of the circuits above or the design of the sequential circuit must be approached for providing simpler circuits, at the expense of more iterations for completing the multiplication operation. Let's consider the construction of a multiplier for *n*-bits unsigned binary numbers. If $X = x_{n-1}...x_0$ is the multiplicand, being available in a parallel output register, $Y = y_{n-1}...y_0$ is the multiplier, which is available in a shift register with serial output, and $R = r_{2n-1}...r_0$, is the output which will be available in a 2*n*-bit register (initialized to zero), we have the structure presented in Fig. 2.9a. With this circuit, the multiplication can be completed in *n* clock cycles (as many as bits have the multiplier operand), so that in each cycle, the partial sum corresponding to each multiplier bit is added to the previous result properly shifted. If the partial bit is '0', the corresponding partial sum will be zero, and when the multiplier bit is '1', the partial sum will be equal to the multiplicand. The operation can start from the most significant bit or from the less significant one, and in each case the previous result must be shifted in a different direction: to the left when starting from the most significant bit, and to the right when starting from the less significant one. As an example, when starting from the most significant bit of the multiplier, the multiplication algorithm will be:

Algorithm 2.1

First algorithm for sequential integer multiplication

```
R ← 0, i ← 0
while i < n, do
      R ← R⃖ + y_{n-1-i} X, i ← i + 1;
end
```

End algorithm

where $\overleftarrow{R}$ is the previous content of *R* shifted one position to the left, and $y_{n-1-i}X$ is the current partial product.

This algorithm can be implemented using the circuit of Fig. 2.9a. A latch, *X*, with parallel output, a shift register, *Y*, activated by falling edge, and a register *R*, with parallel output, parallel input and activated by rising edge, are used. In addition to these registers, *n* **AND** gates are used to generated the partial products. As the least significant bit of each partial product is directly the bit r_0 of the corresponding

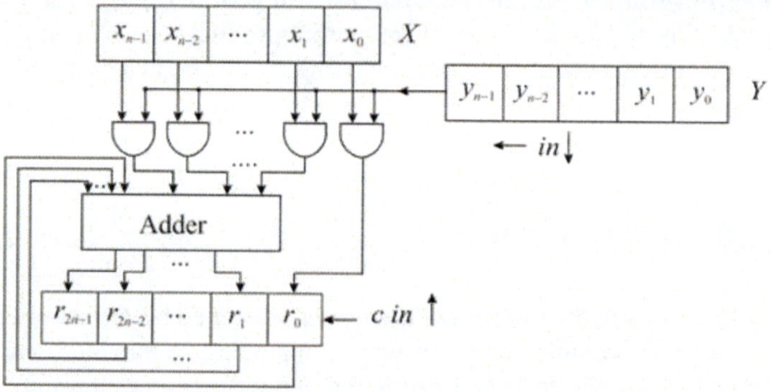

(a)

1) $y_3 = 1$	$\overleftarrow{R} \rightarrow$	00000000	
	$y_3x \rightarrow$	1011	
		00001011	$\rightarrow R$
2) $y_2 = 1$	$\overleftarrow{R} \rightarrow$	00010110	
	$y_2x \rightarrow$	1011	
		00100001	$\rightarrow R$
3) $y_1 = 0$	$\overleftarrow{R} \rightarrow$	01000010	
	$y_1x \rightarrow$	0000	
		01000010	$\rightarrow R$
4) $y_0 = 1$	$\overleftarrow{R} \rightarrow$	10000100	
	$y_0x \rightarrow$	1011	
		10001111	$\rightarrow R$

(b)

Fig. 2.9 First serial-parallel multiplier: **a** Circuit. **b** Example ($X = 1011$, $Y = 1101$)

partial sum, this bit is stored directly in R and it is not an input of the adder. Thus, just an $(2n-1)$-bit adder (usually one less than to be laid down for the final result), whose entries are, first, the n-products **AND** $x_{n-1}y_{n-1-i}, x_{n-2}y_{n-1-i}, \ldots, x_1y_{n-1-i}$ (see Fig. 2.9a), and secondly, the bits $r_{2n-2} \ldots r_0$ of the previous result. The $(2n-1)$-bits of the adder output are stored in $r_{2n-1} \ldots r_1$. This allows the displacement to the left of the previous results. A counter modulo n would suffice to control the operation of this multiplier. As an example, the results that are generated in the four iterations that have to be executed by multiplying the 4-bit numbers $X = 1011$ by $Y = 1101$ are given in Fig. 2.9b.

If the multiplication starts by the least significant bit of the multiplier, the algorithm is as follows:

Algorithm 2.2
Second algorithm for sequential integer multiplication

```
R ← 0, i ← 0
while i < n, do
        R ← R⃗ + yi X, i ← i + 1;
end
```

End algorithm

where $\vec{R}$ is the previous value of R shifted to the right, and y_iX is the present partial product.

This algorithm can be implemented using the circuit of Fig. 2.10a. A latch, X, with parallel output is used again for the multiplicand. However, the multiplier can be stored in the lower half of the register R, such that the most significant half of R (n bits) form a register with parallel output and parallel load, and the n-bit least significant of R form a shift register, Y. The register R is loaded or displaced in the falling edge of each clock pulse. In order to generate the partial products, n **AND** gates are used and an n-bits (as many as bits in the multiplicand) adder, whose inputs are, first, the bits corresponding to the partial product in each iteration, $x_{n-1}y_i, \ldots, x_0y_i$, and otherwise, the $r_{2n-1} \ldots r_n$ bits of the previous result (see Fig. 2.10a). The $n + 1$ output bits of the adder are stored in $r_{2n-1} \ldots r_{n-1}$ (recall that r_{n-1} is the serial input of the shift register and, in each iteration there is a shift to the right of Y). With all this, the shift of the previous results is achieved. Again, to control the operation of the multiplier a module n counter is sufficed. As an example, the results generated in the four iterations by multiplying the 4-bit numbers, $X = 1011$ by $Y = 1101$, are given in Fig. 2.10b.

The circuits with the structures of Figs. 2.9a and 2.10a can be called **serial-parallel multiplier** due to the multiplier is serial data and the multiplicand is parallel data. A simpler but more expensive solution in terms of calculation time, would be the **serial-serial multiplier**, where in each iteration one bit of the multiplicand and one of the multiplier would be multiplied; it is left as an exercise.

In each iteration of the serial-parallel multiplier, a multiplier bit and the multiplicand, M, are multiplied. This circuit can be transformed into another allowing that M could be multiplied by more than one bit of the multiplier in each iteration. For example, multiplying in each iteration by two bits of the multiplier, for an n-bit multiplier, the multiplication would be available in $n/2$ iterations. Again, the design of these circuits is left as an exercise.

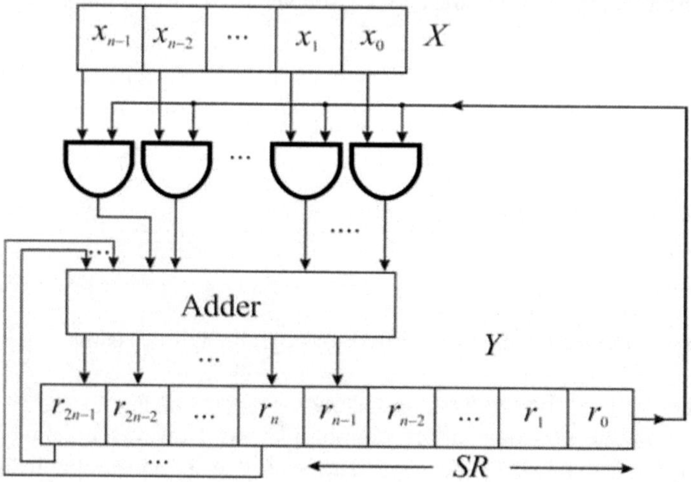

(a)

1) $y_0 = 1$	$R \rightarrow$	00001101	
	$y_0x \rightarrow$	1011	
		01011110	$\rightarrow \vec{R}$
2) $y_1 = 1$	$R \rightarrow$	01011110	
	$y_1x \rightarrow$	0000	
		00101111	$\rightarrow \vec{R}$
3) $y_2 = 0$	$R \rightarrow$	00101111	
	$y_2x \rightarrow$	1011	
		01101111	$\rightarrow \vec{R}$
4) $y_3 = 1$	$R \rightarrow$	01101111	
	$y_3x \rightarrow$	1011	
		10001111	$\rightarrow \vec{R}$

Fig. 2.10 Second serial-parallel multiplier: **a** Circuit. **b** Example ($X = 1011$, $Y = 1101$)

2.4.3 Multiplying by a Constant

The multiplication of a set of data for one or more constants is an operation that must be performed frequently. Of course, any multiplier can be used for this purpose, as described previously. However, in this case, when one of the operands is constant, simpler circuits can be designed for this specific purpose. For example, let suppose the case of a circuit for multiplying any unsigned 8-bit binary number, $X = x_7 \ldots x_0$, by 25. Given that $25_{10} = 11001_2$, to multiply by 25 is equivalent to adding the three terms given in Fig. 2.11a. Thus, using two 8-bit parallel adders, this multiplication can be implemented, as shown in Fig. 2.11a, generating a 13-bit result, $R = r_{12} \ldots r_0$. Compared to a generic multiplier circuit, the reduction to be achieved with this specific circuit is evident. This idea of using parallel adders will be called **solution 1** for multiplying by a constant.

In general, both adders and subtractors can be used for the decomposition of the multiplier M. This is equivalent to use signed digits in the decomposition of M, and from the minimal representation of M a simple multiplier circuit may be obtained.

If full-adders and half-adders are used as building blocks, the circuit for multiplying by 25 can be reduced more. Specifically just 11 full adders and 5 half adders are required, as shown in Fig. 2.11b. This is the **solution 2** for multiplying by a constant. Solution 2, when considering the design at a lower level, usually produces simpler circuits than **solution 1**.

Another way to build specific multiplier when the multiplier is a constant, M, using adders digits (of adequate size in each situation), consists in decomposing the constant M in factors, which in turn are decomposed into sums of powers of two. For example, for $M = 25$, it results:

$$25 \cdot X = (4 + 1) \cdot (4 + 1) \cdot X = (4 + 1) \cdot (4 \cdot X + X) = 4 \cdot Y + Y$$

where $Y = 4 \cdot X + X$. Therefore, the multiplication is performed using two adders of appropriate length, as shown in Fig. 2.11c. Multiplication by a power of two is reduced to a displacement, which does not require circuitry. If X is of n bits, the adder 1 of Fig. 2.11c must be an n-bit adder, and the adder 2 an $(2n + 1)$-bit adder. This solution to multiply by a constant is called **solution 3**.

It is also possible to use that $25 = 3 \times 8 + 1$, and again this multiplication can be implemented using two adders, resulting in the circuit of Fig. 2.11d.

Developments that can be used to multiply by a constant, up to 100, are given in Table 2.1. The powers of 2 do not need one adder (obviously not included in the table); in this table, one adder/subtractor is enough for 31 constants; for 54 constants two adders/subtractors are needed; only for 8 constants three adder/subtractors are required.

Two factors products are only used in Table 2.1, since only reaches 100. Obviously, products with more than two factors can be used, which may make sense to constants greater than those shown in Table 2.1. For example, $5049_{10} = 9 \times 17 \times 33$ and, according with Table 2.1, it can be implemented with three adders, since each

					x_7	x_6	x_5	x_4	x_3	x_2	x_1	x_0
		x_7	x_6	x_5	x_4	x_3	x_2	x_1	x_0			
	x_7	x_6	x_5	x_4	x_3	x_2	x_1	x_0				
r_{12}	r_{11}	r_{10}	r_9	r_8	r_7	r_6	r_5	r_4	r_3	r_2	r_1	r_0

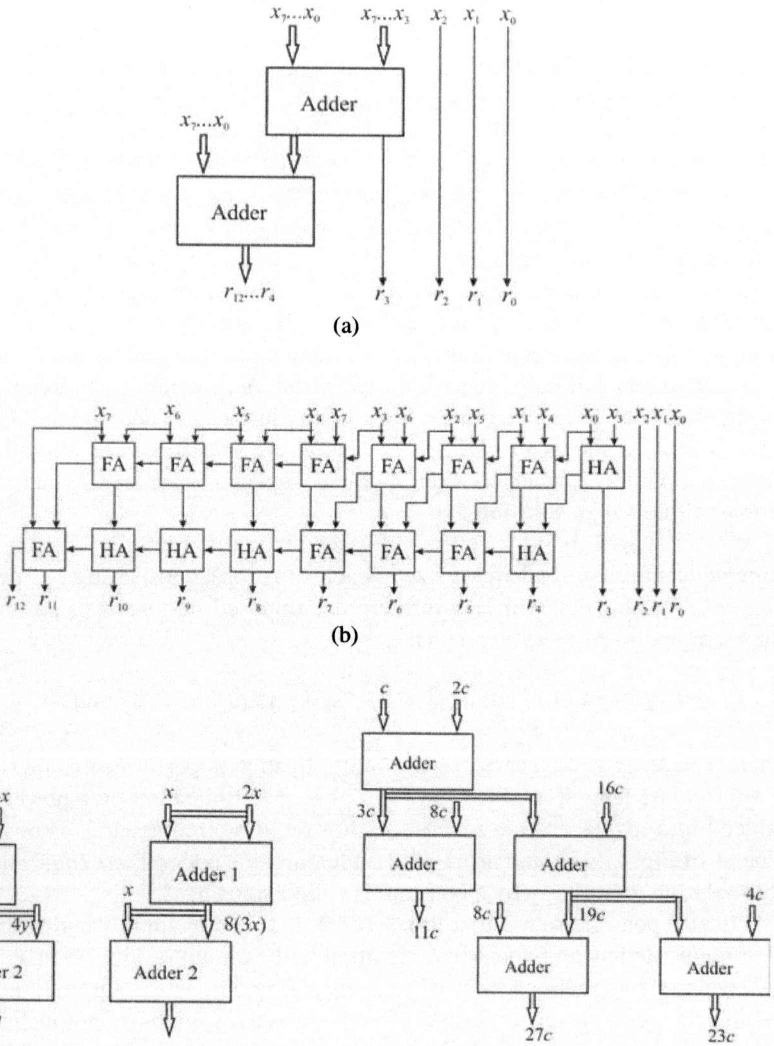

Fig. 2.11 Multiplying by 25: **a** First solution. **b** Second solution. **c** Third solution. **d** Other implementation for the third solution. **e** Multiplying by 11, 19, 23 and 27

Table 2.1 Multipliers 1–100

No.	No. A/S	Develop	No.	No. A/S	Develop
3	1	2 + 1	53	3	32 + 16 + 4 + 1
5	1	4 + 1	54	2	6 × 9; 64 − 8 − 2
6	1	4 + 2	55	2	64 − 8 − 1
7	1	8 − 1	56	1	64 − 8
9	1	8 + 1	57	2	64 − 8 + 1
10	1	8 + 2	58	2	64 − 8 + 2
11	2	8 + 2 + 1	59	2	64 − 4 − 1
12	1	8 + 4	60	1	64 − 4
13	2	8 + 4 + 1	61	2	64 − 4 + 1
14	1	16 − 2	62	1	64 − 2
15	1	16 − 1	63	1	64 − 1
17	1	16 + 1	65	1	64 + 1
18	1	16 + 2	66	1	64 + 2
19	2	16 + 2 + 1	67	2	64 + 2 + 1
20	1	16 + 4	68	1	64 + 4
21	2	16 + 4 + 1	69	2	64 + 4 + 1
22	2	16 + 4 + 2	70	2	64 + 4 + 2
23	2	16 + 8 − 1	71	2	64 + 8 − 1
24	1	16 + 8	72	1	64 + 8
25	2	16 + 8 + 1; 5 × 5	73	2	64 + 8 + 1
26	2	16 + 8 + 2	74	2	64 + 8 + 2
27	2	3 × 9; 32 − 4 − 1	75	2	15 × 5
28	1	32 − 4	76	2	64 + 8 + 4
29	2	32 − 4 + 1	77	3	64 + 8+4 + 1
30	1	32 − 2	78	2	5 × 16 − 2
31	1	32 − 1	79	2	5 × 16 − 1
33	1	32 + 1	80	1	5 × 16
34	1	32 + 2	81	2	5 × 16 + 1; 9 × 9
35	2	5 × 7; 32 + 2+1	82	2	5 × 16 + 2
36	1	32 + 4	83	3	64 + 16 + 2 + 1
37	2	32 + 4 + 1	84	2	5 × 16 + 4
38	2	32 + 4 + 2	85	2	17 × 5
39	2	32 + 8 − 1	86	3	64 + 16 + 4 + 2
40	1	5 × 8; 32 + 8	87	3	64 + 16 + 8 − 1; 3 × 29
41	2	32 + 8+1	88	2	8 × 11; 5 × 16 + 8; 3 × 32 − 8
42	2	32 + 8+2	89	3	64 + 16 + 8+1

(continued)

Table 2.1 (continued)

No.	No. A/S	Develop	No.	No. A/S	Develop
43	3	32 + 8+2 + 1	90	2	3 × 30; 5 × 18
44	2	32 + 8+4	91	3	64 + 32 − 4 − 1
45	2	5 × 9	92	2	3 × 32 − 4
46	2	32 + 16 − 2	93	2	3 × 31
47	2	32 + 16 − 1	94	2	3 × 32 − 2
48	1	32 + 16	95	2	3 × 32 − 1
49	2	32 + 16 + 1	96	1	3 × 32
50	2	32 + 16 + 2	97	2	3 × 32 + 1
51	2	3 × 17	98	2	3 × 32 + 2
52	2	32 + 16 + 4	99	2	3 × 33
			100	2	3 × 32 + 4; 5 × 20

factor only needs one adder; using signed digit $5049_{10} = 101000\bar{1}00\bar{1}001_2$, and four adders/subtractors are required to build the multiplier.

Another decomposition of multiplier M of interest to explore consists on finding dividers on the form $2^i \pm 2^j$, which in some cases can lead to simpler circuits. For example, the case of multiply $17 \times 41 = 697 = (16 + 1)\ (1 + 8 + 32) = (16 + 1) + 8\ (16 + 1) + 32(16 + 1)$. With this decomposition, the multiplication can be done with three adders, while starting form the development $697 = 1011 00\bar{1}001$ four adders/subtractors are required.

When the same data have to be multiplied by multiple constants, it is possible to organize the process so that different partial products can be shared in the different calculations, as can be seen in the following example.

Example 2.1 **Let suppose the case of multiply simultaneously by 11, 19, 23 and 27.**

Developing these constants as follows $11 = 8 + 3$, $19 = 16 + 3$, $23 = 19 + 4$, $27 = 19 + 8$ or $27 = 11 + 16$, the multiplier can be made using five adders, sharing intermediate results, as it is depicted in Fig. 2.11e. ■

2.5 Exponentiation

To raise a number N to a power P (P integer) consists on multiplying the number N by itself P times. Therefore, with an appropriate multiplier, any integer number N can be raise to any power. First the calculation of the square of N is considered, where N is an unsigned integer in base 2. As illustrative example, let suppose a 8-bit number, $N = x_7x_6x_5x_4x_3x_2x_1x_0$. The $N \times N$ multiplication is shown in Fig. 2.12a, where it has been applied that $x_ix_i = x_i$. Moreover, when a column has $x_ix_j + x_jx_i =$

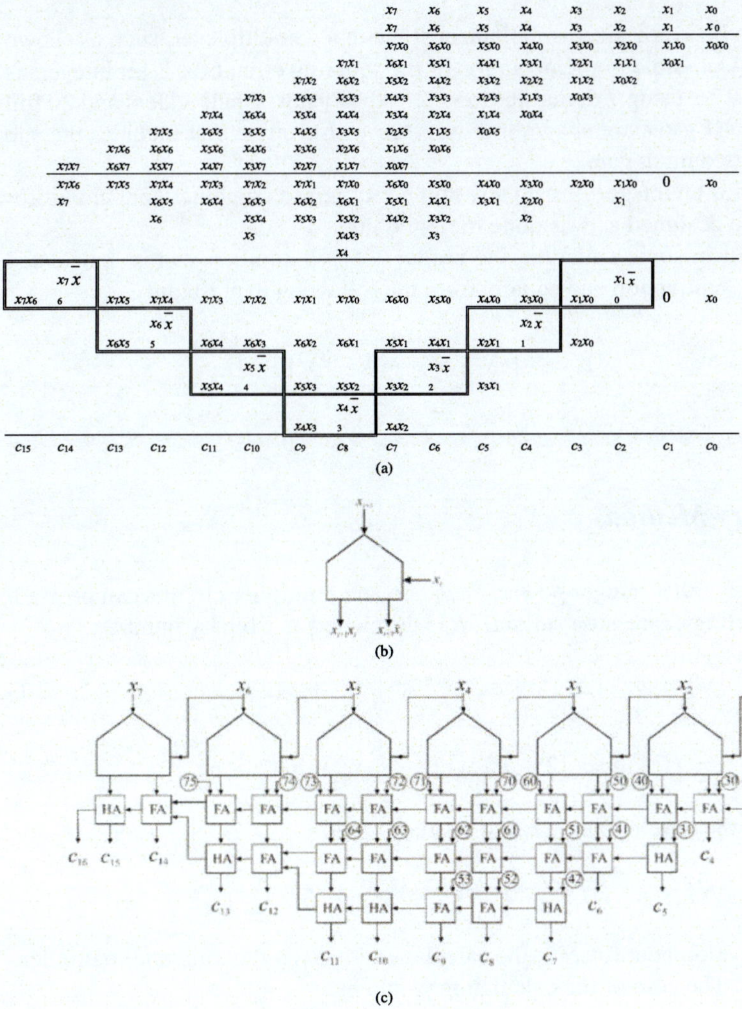

Fig. 2.12 **a** Square. **b** Demultiplexer. **c** Circuit for squaring

$2x_ix_j$ obviously it can be moved to the next column as x_ix_j. Also, when in a column is $x_i + x_ix_j$:

$$x_i + x_ix_j = x_i\left(x_j + \overline{x}_j\right) + x_ix_j = 2x_ix_j + x_i\overline{x}_j$$

and $2x_ix_j$ can be moved to the next column as x_ix_j. Considering all these replacements, the summands to be used for calculating the square can remain as in Fig. 2.12a.

With respect to the implementation of the different products, the products expressed as $x_{j+1}\,\overline{x}_j$ and $x_{j+1}x_j$ that appear in adjacent columns (highlighted in

Fig. 2.12a) can be synthesized simultaneously with a demultiplexer, such as shown in Fig. 2.12b. A possible implementation of the squaring circuit for 8-bit integers is given in Fig. 2.12c, using 7 demultiplexers, 21 **AND** gates, 7 half adders and 20 full adders. The **AND** gates are shown in Fig. 2.12c with a circle that includes the sub indexes of the two input gate.

Obviously, for an integer N with any number of bits, a combinational circuit for squaring can be designed as was done for eight bits.

If it is useful in some situation, the product of two numbers can be calculated using addition, subtraction and square, from the following expression:

$$XY = \frac{1}{4}\{(X+Y)^2 - (X-Y)^2\}$$

2.5.1 *Binary Methods*

To raise N to any other integer power, P, square and multiplier circuits can be used. To obtain a starting expression suitable, P is developed as a binary number:

$$P = p_{m-1}2^{m-1} + p_{m-2}2^{m-2} + \cdots + p_1 2 + p_0 \tag{2.1}$$

$$= ((\ldots(p_{m-1}2 + p_{m-2})2 + \cdots)2 + p_1)2 + p_0 \tag{2.2}$$

Thus, using the development (2.2) it results:

$$N^P = (\ldots((N^{p_{m-1}})^2 \cdot N^{p_{m-2}})^2 \cdot \ldots \cdot N^{p_1})^2 \cdot N^{p_0}$$

With this development for N^P, the calculation involves squaring and multiplication, iteratively. The core of the calculation would be:

```
1) R ← R²;
2) if pi=1, R ← RN;
```

The result remains in the register R, which initially must be $R \leftarrow 1$. The complete algorithm could be as follows:

Algorithm 2.3

First algorithm for exponentiation

```
R ← 1, i ← m-1;
while i > 0, do
  begin
     1) if pi=1 then R ← RN;
     2) R ← R², i ← i - 1;
```

```
    end
  if p0=1 then R ← RN;
```

End algorithm

The bits p_i of the binary development of P are processed in this algorithm starting with the most significant one, hence, this method is generally known as *binary method from left to right*.

A possible processing unit for exponentiation using the above algorithm (the control signals are not included) is represented in Fig. 2.13a. This circuit includes a register *R*, a multiplier and a squarer.

Other development of N^P using (2.1) is the following:

$$N^P = (N^{2^0})^{p_0} \cdot (N^{2^1})^{p_1} \cdot \ldots \cdot (N^{2^{m-2}})^{p_{m-2}} \cdot (N^{2^{m-1}})^{p_{m-1}}$$

From this development another algorithm for the exponentiation can be designed. Again, the calculation consists on to square and to multiply, according to the following core:

```
1) N ← N²;
2) if pm-1-i=1, R ← RN;
```

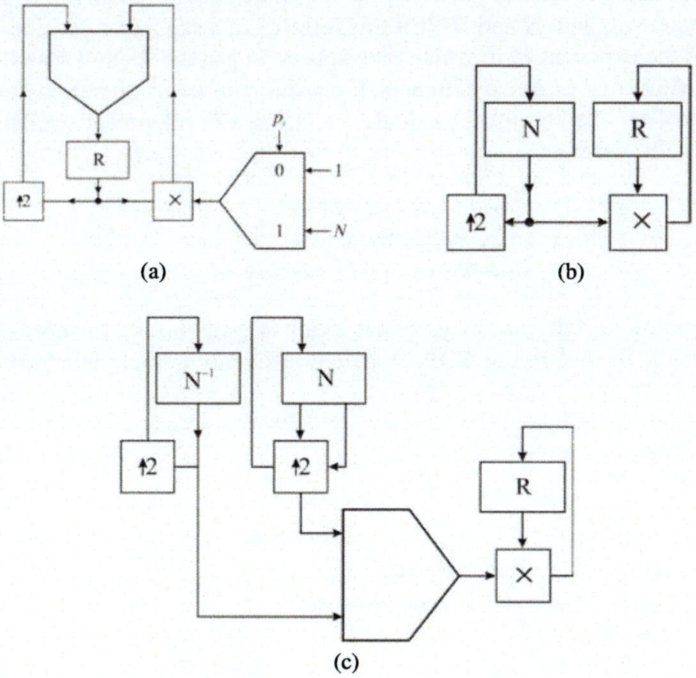

Fig. 2.13 Exponentiation: **a** First solution. **b** Second solution. **c** Exponentiation using the canonic development

Initially $R \leftarrow 1$. The result remains in R. The algorithm for this case could be the following one:

Algorithm 2.4
Second algorithm for exponentiation

```
R ← 1, i ← m-1;
while i > 0, do
  begin
    1) if p_{m-1-i}=1 then R ← RN;
    2) N ← N², i ← i - 1;
  end
```

End algorithm

The bits p_i of the binary development of P are processed in this algorithm starting with the least significant one, hence, this method is generally known as *binary method from right to left*.

A possible circuit for exponentiation using the above algorithm (the control signals are not included) is represented in Fig. 2.13b. This circuit includes two registers, a multiplier and a squarer.

When operating in the context of certain algebraic structures can be arranged and easily operate with both N and N^{-1}. If this is the case, a canonical development can be used for the exponent P, in which they appear, in general, both +1 and −1 (that is, both positive and negative exponents), but the number of operations will be in average, smaller. The core of the calculation, using a development similar to (2.3) would be in this case:

```
1) N ← N²; N⁻¹ ← N⁻²;
2) if p_{m-1-i}=1, then R ← RN;
   else if p_{m-1-i}=-1, then R ← RN⁻¹;
```

The algorithm in this case may be that which is given below, the corresponding circuit could be the one in Fig. 2.13c (the control signals are not included).

Algorithm 2.5
Third algorithm for exponentiation

```
R ← 1, i ← m-1;
while i > 0, do
  begin
    1) if p_{m-1-i}=1 then R ← RN;
       else if p_{m-1-i}=-1, then R ← RN⁻¹;
    2) N ← N², N⁻¹ ← N⁻², i ← i - 1;
  end
```

End algorithm

If the exponent P can be factorized, $P = Q \bullet R$, then the exponentiation can be decomposed into two phases:

$$N^P = N^{Q \cdot R} = (N^Q)^R$$

If the exponent P is developed using any base, b:

$$P = p_{m-1}b^{m-1} + p_{m-2}b^{m-2} + \cdots + p_1 b + p_0$$
$$= ((\ldots(p_{m-1}b + p_{m-2})b + \cdots)b + p_1)b + p_0$$

the binary method, both from left to right and from right to left, can be extended to the base b, with appropriate modifications. In the algorithm in base b from left to right it must be calculated:

$$N^P = (\ldots((N^{p_{m-1}})^b \cdot N^{p_{m-2}})^b \cdot \ldots \cdot N^{p_1})^b \cdot N^{p_0}$$

Since the coefficients p_j are not only 0 or 1, it is necessary to multiply by N^j ($j =$ 1, 2, ..., $b - 1$) and to raise to the power b.

In the algorithm in base b from right to left it must be calculated:

$$N^P = (N^{b^0})^{p_0} \cdot (N^{b^1})^{p_1} \cdot \ldots \cdot (N^{b^{m-2}})^{p_{m-2}} \cdot (N^{b^{m-1}})^{p_{m-1}}$$

Thus, it is required to raise to the powers 1, 2, ..., b.

2.5.2 Additive Chains

The developments (2.1) and (2.2) transform the exponent P into an addition and, applying that the exponents are additive, the binary developments emerge. This same idea is used for additive chains.

Given P a positive integer, an additive chain for P is a sequence of integers, p_0, $p_1, \ldots, p_n$, such that $p_0 = 1, p_n = P, p_i = p_j + p_k, i > j \geq k, p_i \neq p_j$ for $i \neq j$. Then, the two first elements of each additive chain are always 1 and 2; the third element can be only 3 or 4, and the remainder elements are obtained adding two previous elements (that can be a previous repeated element).

A particular case of additive chains are known as Brauer chains [2]; for these additive chains $p_i = p_{i-1} + p_k, i - 1 \geq k$. Thus, using a Brauer chain, to obtain the next element of the chain, the present element is used in the involved addition. For implementation purposes, it is obvious that using always the previous result is very interesting. A procedure for constructing this type of additive chains, which are the most used, is described in [2]. This procedure is not the only possibility. To apply this method an integer e is chosen and P is developed in base $b = 2^e$:

$$P = p_{m-1}b^{m-1} + p_{m-2}b^{m-2} + \cdots + p_1 b + p_0$$

The following Brauer additive chain can be constructed for P, being composed by m sections that have to be adequately linked. The first section is $\{1, 2, 3, \ldots, 2^e - 1\}$; the second section is $\{2p_{m-1}, 4p_{m-1}, 8p_{m-1}, \ldots, bp_{m-1}, (bp_{m-1} + p_{m-2})\}$; the third section is $\{2(bp_{m-1} + p_{m-2}), 4(bp_{m-1} + p_{m-2}), 8(bp_{m-1} + p_{m-2}), \ldots, b(bp_{m-1} + p_{m-2}), b(bp_{m-1} + p_{m-2}) + p_{m-3}\}$; ...; the last section is $\{2(b(\ldots(bp_{m-1} + p_{m-2}) \ldots + p_1), 4(b(\ldots(bp_{m-1} + p_{m-2}) \ldots + p_1), 8(b(\ldots(bp_{m-1} + p_{m-2}) \ldots + p_1), \ldots, b(b(\ldots(bp_{m-1} + p_{m-2}) \ldots + p_1), b(b(\ldots(bp_{m-1} + p_{m-2}) \ldots + p_1) + p_0\}$, as it is done in the following example.

Example 2.2 **To obtain Brauer additive chains for 26,221, with $e = 1, 2, 3, 4$ and 5.**

Choosing $e = 1$ ($b = 2^e = 2$) it is:

$$26{,}221 = 2^{14} + 2^{13} + 2^{10} + 2^9 + 2^6 + 2^5 + 2^3 + 2^2 + 1$$

Thus, $p_{14} = 1, p_{13} = 1, p_{12} = 0, p_{11} = 0, p_{10} = 1, p_9 = 1, p_8 = 0, p_7 = 0, p_6 = 1, p_5 = 1, p_4 = 0, p_3 = 1, p_2 = 1, p_1 = 0, p_0 = 1$. The first section of the Brauer chain is 1; the second section is 2 and 3; the third section is 6; the fourth section is 12; the fifth section is 24, 25; the sixth section is 50, 51; the seventh section is 102, the eighth section is 204; the ninth section is 408, 409; the tenth section is 818, 819; the eleventh section is 1638, the twelfth section is 3276, 3277; the thirteenth section is 6554, 6555; the fourteenth section is 13110; and the fifteenth section is 26,220, 26,221. Thus the additive chain is formed by 23 elements {1, 2, 3, 6, 12, 24, 25, 50, 51, 102, 204, 408, 409, 818, 819, 1638, 3276, 3277, 6554, 6555, 13,110, 26,220, 26,221}.

Choosing $e = 2$ ($b = 2^e = 4$) it results:

$$26{,}221 = 1 \times 4^7 + 2 \times 4^6 + 1 \times 4^5 + 2 \times 4^4 + 1 \times 4^3 + 2 \times 4^2 + 3 \times 4 + 1$$

Therefore $p_7 = 1, p_6 = 2, p_5 = 1, p_4 = 2, p_3 = 1, p_2 = 2, p_1 = 3, p_0 = 1$. The first section of the chain Bauer is 1, 2, 3; the second section is 2, 4, 6; the third section is 12, 24, 25; the fourth section is 50, 100, 102; the fifth section is 204, 408, 409; the sixth section is 818, 1636, 1638; the seventh section is 3276, 6552, 6555; and the eighth section is 13110, 26,220, 26,221. In the second section the 2 should be removed, which is already in the first section; the 4 can also be removed, since it is not needed to build the following elements. Thus the additive chain has 22 elements: {1, 2, 3, 6, 12, 24, 25, 50, 100, 102, 204, 408, 409, 818, 1636, 1638, 3276, 6552, 6555, 13,110, 26,220, 26,221}.

Choosing $e = 3$ ($b = 2^e = 8$) it is:

$$26{,}221 = 6 \times 8^4 + 3 \times 8^3 + 8^2 + 5 \times 8 + 5$$

Therefore $p_4 = 6, p_3 = 3, p_2 = 1, p_1 = 5, p_0 = 5$. The first section of the Bauer chain is 1, 2, 3, 4, 5, 6, 7; the second section is 12, 24, 48, 51; the third section is 102, 204, 408, 409; the fourth section is 818, 1636, 3272, 3277; the fifth section is 6554, 13,108, 26,216, 26,221. From the first sections 4 and 7 can be suppressed, which are not used lately. Thus the additive chain is 21 elements {1, 2, 3, 5, 6, 12, 24, 48, 51, 102, 204, 408, 409, 818, 1636, 3272, 3277, 6554, 13,108, 26,216, 26,221}.

Choosing $e = 4$ ($b = 2e = 16$) it results:

$$26{,}221 = 6 \times 16^3 + 6 \times 16^2 + 6 \times 16 + 13$$

Thus, $p_3 = 6$, $p_2 = 6$, $p_1 = 6$, $p_0 = 13$. The first section of the Bauer chain is 1, 2, 3, 4, 5, 6, 7, 8, 9, 10, 11, 12, 13, 14, 15; the second section is 12, 24, 48, 96, 102; the third section is 204, 408, 816, 1632, 1638; and the fourth section is 3276, 6552, 13104, 26,208, 26,221. From the first sections 3, 5, 7, 8, 9, 10, 11, 14 and 15 can be suppressed, which are not subsequently used, although 13 is used later, can also be delete building it like 12 + 1, thereby facilitating the link of the first section to the second; making this, the fourth section would be 3276, 6552, 13,104, 26,208, 26,220, 26,,221. In this way the additive chain is 20 elements {1, 2, 4, 6, 12, 24, 48, 96, 102, 204, 408, 816, 1632, 1638, 3276, 6552, 13,104, 26,208, 26,220, 26,221}.

Choosing $e = 5$ ($b = 2^e = 32$) it results:

$$26{,}221 = 25 \times 32^2 + 19 \times 32 + 13$$

Therefore $p_2 = 25$, $p_1 = 19$, $p_0 = 13$. The first section of the Bauer chain is 1, 2, 3, 4, 5, 6, 7, 8, 9, 10, 11, 12, 13, 14, 15, 16, 17, 18, 19, 20, 21, 22, 23, 24, 25, 26, 27, 28, 29, 30, 31; the second section is 50, 100, 200, 400, 800, 819, and the third section is 1638, 3276, 6552, 13,104, 26,208, 26,221. From the first sections 3, 5, 7, 8, 9, 10, 11, 14, 15, 16, 17, 18, 20, 21, 22, 23, 24, 26, 27, 28, 29, 30 and 31 can be suppressed, since they are not subsequently used. In this way the additive chain is 20 elements string {1, 2, 4, 6, 12, 13, 19, 25, 50, 100, 200, 400, 800, 819, 1638, 3276, 6552, 13,104, 26,208, 26,221}.

Using an additive chain with P elements, P-1 operations must be performed (multiplications and squares) to calculate the corresponding power, since the element 1 does not involve any operation. Thus, with a chain of 20 elements, 19 operations are required.

Considering $26{,}221 = 2017 \times 13$, 26,221 can be expressed as the concatenation of two additive chains, corresponding to 2017 and to 13. For 2017 the additive Brauer chain with 15 elements can be constructed {1, 2, 3, 6, 12, 24, 31, 62, 124, 248, 252, 504, 1008, 2016, 2017}, for 13 the Brauer additive chain with 6 elements can be constructed {1, 2, 3, 6, 12, 13}. Using these two additive chains, 19 operations are required again to calculate the corresponding power, as the initial 1 of each chain does not involve any operation. ■

From Example 2.2 it is clear that a Brauer chain with $e = 1$ corresponds to the binary method from left to right described above. It can be considered that a Brauer chain with $e > 1$ is only a generalization of the binary method from left to right; in fact the exponentiation method using a Bauer chain is also known as *the method* 2^e *from left to right*.

There is not available an algorithm that guarantees the shortest chain.

The same circuit proposed for the binary method from left to right (Fig. 2.13), with minor modifications, can be used to calculate any power using additive chains.

2.6 Division and Square Root

This section will consider simple circuits to divide integers and to extract the integer square root of integer numbers.

2.6.1 *Combinational Divisors*

Let consider two unsigned binary numbers: D (m-bit dividend), and d (n-bit divisor). In this case, the division consists on finding two unsigned binary numbers, C (quotient) and r (remainder), $r < d$, such that:

$$D = C \cdot d + r$$

The division is defined for $d \neq 0$. Therefore, in what follows it is assumed that this condition is met, i.e., before dividing, it is checked if $d \neq 0$, and only in this affirmative case, the division is performed.

With the condition $r < d$, C and r are unique, and to calculate them, an immediate combinational solution might be thought obviously. To implement the division it is sufficed, for example, a **ROM** of $m + n$ address bits and m outputs (2^{m+n} words of m bits), in which the quotient and the remainder corresponding to every possible (D, d) are written. For real cases this is not a feasible combinational solution since $m + n$ will result almost always too large to directly synthesize the corresponding functions (for example, to include all possible outputs in a **ROM**).

Another combinational solution is possible that attempts to mimic the division algorithm as a serie of subtractions and shifts. Before addressing this alternative, more aspects about the operands have to be established. Specifically it is assumed that, as usual, the relationship between the lengths of the dividend and the divisor is $m = 2n - 1$, and that the most significant bit of the divisor, d, is 1. Neither of these assumptions implies restriction, because on the one hand, the size of the operand can always be adjusted by adding zeros, and, second, by shifting the divisor it is possible to make 1 the most significant bit; after division, the shifts made in the divisor must be properly transferred to the quotient and the remainder to obtain correct results. With these assumptions, with n-bits for both the quotient and to the remainder, all possible results may be represent, and the division is made in n-steps, in each of which a bit of the quotient is obtained.

In what follows, an example will be used to reach a combinational divider circuit: let $n = 4$, $D = 0110101$, $d = 1011$. The four stages of calculation for this case are detailed in Fig. 2.14a. In the first stage d is subtracted from the four most significant bits of D ($D_6D_5D_4D_3$); if the result is positive (and therefore no output borrow), the quotient bit is 1, and the difference $D_6D_5D_4D_3 - d$ passes to the next stage as the most significant bits of the modified dividend. If the result is negative (i.e., there is output borrow), the quotient bit is 0, and the dividend unchanged passes to the

1) $D_6D_5D_4D_3 \rightarrow$	0110	
	-1011	
	Negative	$\rightarrow c_3 = 0$
2) $D_6D_5D_4D_3 \rightarrow$	1101	
	-1011	
	0010	$\rightarrow c_2 = 1$
3) $D_6D_5D_4D_3 \rightarrow$	0100	
	-1011	
	Negative	$\rightarrow c_1 = 0$
4) $D_6D_5D_4D_3 \rightarrow$	1001	
	-1011	
	Negative	$\rightarrow c_0 = 0$
	Remainder	$\rightarrow 1001$

(a)

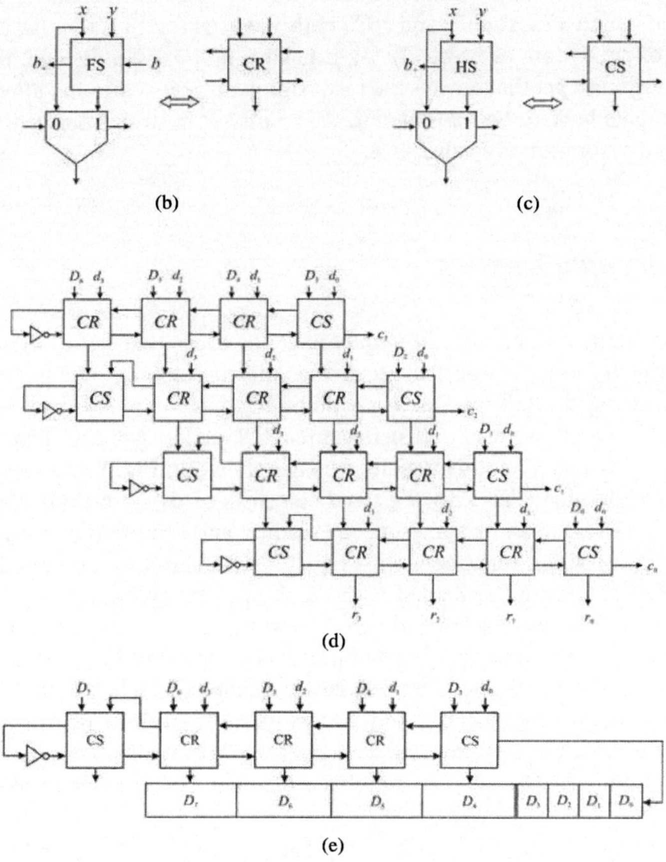

Fig. 2.14 **a** Division example; **b** CR cell; **c** CS cell; **d** combinational divisor of 7 by 4 unsigned bits. **e** Sequential divisor of 7 by 4 unsigned bits

next stage. In other words, the quotient bit is the complement of the borrow of the subtractor output, and if the quotient bit is 0, D without changing is selected for the next stage, while if the quotient bit is 1, the most significant bits of the dividend bit must be modified selecting $D_6D_5D_4D_3 - d$. Therefore, with a full subtractor, **FS**, to take into account the possible borrow of the previous bit, plus one 2-to-1 multiplexer, the circuit necessary for processing each bit can be constructed, as shown with cell **CR** of Fig. 2.14b. If for a given bit (as with the least significant bit) no input borrows are to be considered, the full subtractor **FS** can be replaced by a half subtractor, **HS**, resulting in the **CS** cell, simpler than the **CR** of Fig. 2.14c.

The second and subsequent iterations consist on repeating the same as the first iteration, using in each case the unmodified or modified dividend which has resulted in the previous iteration. Then, by subtracting, the divisor is shifted one position to the right in each iteration. The remainder, $r_3 \ldots r_0$, is obtained in the fourth iteration.

The circuit for dividing a number of seven bits by other of four bits is detailed in Fig. 2.14d, in which 12 **CR** cells and 7 **CS** cells are used (or 19 **CR** cells, if only one single type of cells want to be used). As it has been already indicated, the divisor has to be adjusted to get that always the most significant bit is a 1, and after division, these movements have to be translated to the results. It is straightforward to extend these design divisors for any value of n.

2.6.2 *Sequential Divisors*

The most common divisors are the sequential. The ideas that led to the divisor of Fig. 2.14d can be used to construct a divisor that divides D, of $2n - 1$ bits, by d, of n bits, using n clock pulses. As a particular case it is still assumed that $n = 4$. Figure 2.14e shows a circuit using three **CR** cells, two **CS** cells, one 4-bit latch to store the divisor, d (this register it is not shown in Fig. 2.14e), and an 8-bit register for the dividend, D. This register D consists of two 4-bit register: the first register ($D_7D_6D_5D_4$) must be simultaneous reading and writing (i.e., master-slave), the second ($D_3D_2D_1D_0$) must be a shift register with serial input and output. The registers d and D have to be loaded with the data to be processed before starting operation. Obviously always $D_7 = 0$ before starting to divide, and the divisor will have been shifted so the most significant bit of d is 1. The shift register ($D_3D_2D_1D_0$) is used to store the bits of the quotient. It is easily verified that this circuit in Fig. 2.14e does the same as the one in Fig. 2.14d, unless using four clock pulses. Therefore, after four iterations, the quotient is stored in $D_3D_2D_1D_0$ and the remainder of the division is sorted in $D_7D_6D_5D_4$. Again, this circuit can be extended immediately to any value of n.

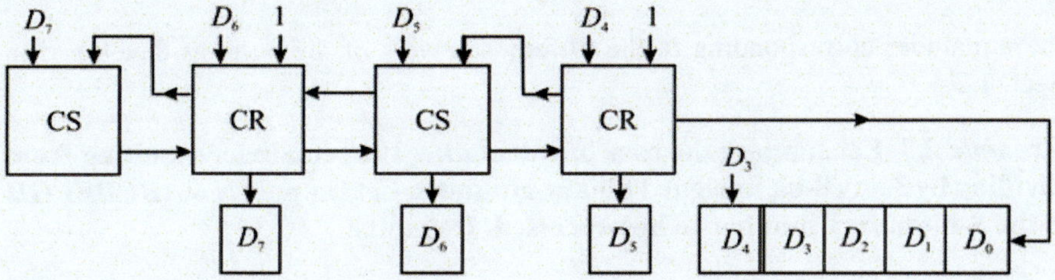

Fig. 2.15 Sequential divisor by 10_{10}

2.6.3 *Dividing by a Constant*

In some applications, as in the scaling or in the change of base or in the modular reduction, it is necessary to divide a data set by the same constant. For this purpose different specific circuits can be used. In what follows it is assumed that integer data are going to be divided by an integer constant.

It is easy to see that, strictly, just dividers to divide by odd numbers have to be designed. Indeed, the division of a number of m bits by 2^n is simplified to n shifts: the n least significant bits are the remainder of the division, and the m–n most significant bits are the quotient. Moreover, any even number can be decomposed into the product of an odd number by a power of two:

$$C = I \times 2^n \Rightarrow \frac{N}{C} = \frac{N}{I}\frac{1}{2^n}$$

A first solution to design a divider by a constant, even or odd, consists on to particularize the generic circuits of Fig. 2.14d and e for the divisor that wants to be used. For example, the sequential divisor of Fig. 2.14e is shown in Fig. 2.15; this sequential divisor is particularized to divide by 10 ($10_{10} = 1010_2$) any unsigned 7-bit integer data. Obviously, the same result is obtained by dividing by 5, and then by 2. In any case, these particularized circuits provide both the quotient and the remainder of the division.

The next three sections are also devoted to the implementation of the division by a constant, but considering those cases in which only one of the results is of interest: the quotient or the remainder.

2.6.4 *Modular Reduction*

In some cases only one of the two results of the division is of interest. If only the remainder is of interest, it is a modular reduction, as shown in Sect. 1.2.4. After obtaining the remainder, $R = N\mathbf{mod}C$, the difference N-R is a multiple of C. The following example shows a case study of modular reduction based on calculating the remainder corresponding to the different powers of the base, as developed in Sect. 1.2.4.

Example 2.3 **Let suppose the case of calculating the remainder resulting from dividing by 5 any 8-bit unsigned binary numbers. Let suppose $N = ABCDEFGH$ is the 8-bit binary number to be processed. It results:**

$$\begin{aligned} N\mathbf{mod5} &= (A2^7 + B2^6 + C2^5 + D2^4 + E2^3 + F2^2 + G2^1 + H)\mathbf{mod5} \\ &= \{A(2^7\mathbf{mod5}) + B(2^6\mathbf{mod5}) + C(2^5\mathbf{mod5}) + D(2^4\mathbf{mod5}) \\ &\quad + E(2^3\mathbf{mod5}) + F(2^2\mathbf{mod5}) + G(2^1\mathbf{mod5}) + H\}\mathbf{mod5} \end{aligned}$$

Calculating the remainders of the different powers it results:

$$\begin{aligned} &2^7\mathbf{mod5} = 3; \quad 2^6\mathbf{mod5} = 4; \quad 2^5\mathbf{mod5} = 2; \quad 2^4\mathbf{mod5} = 1; \\ &2^3\mathbf{mod5} = 3; \quad 2^2\mathbf{mod5} = 4; \quad 2^1\mathbf{mod5} = 1 \end{aligned}$$

Thus:

$$N\mathbf{mod5} = (3A + 4B + 2C + D + 3E + 4F + 2G + H)\mathbf{mod5}$$

Applying this expression, the modular reduction can be made using three blocks:

$$\begin{aligned} L &= (3A + 4B + 2C + D)\mathbf{mod5}; \\ M &= (3E + 4F + 2G + H)\mathbf{mod5}; \\ N &= (L + M)\mathbf{mod5} \end{aligned}$$

The calculations for L and M are identical, and in what follows reference to L will be made. The sum ($\Sigma = 3A + 4B + 2C + D$) and the remainder (**mod5**) for each combination of the inputs are shown in the table of Fig. 2.16a. It is immediate that the value of Σ can be obtained with the circuit of Fig. 2.16b. It can be probed that to get the remainder it is just necessary adding 3 to Σ when Σ is equal to 5, 6 or 7, or the carry is $c = 1$, i.e., it must add 3 when the function $F = s_2s_0 + s_2s_1 + c$ is equal to 1. This correction, when $\Sigma = 10_{10}$, gives 5 as remainder, when it should be 0. Therefore the result should be correct in this situation, to be 0 instead of 5. Since $\Sigma = 10_{10}$ only for $ABCD = 1111$, with a **NAND** gate this exceptional situation can be controlled. Joining the two successive corrections it results the circuit of Fig. 2.16c.

It is easy to see that to calculate N the same circuit for L or M can be used, although in the case of N it is not necessary to correct the value 10, since it can not appear.

In short, with three blocks as that in Fig. 2.16b the remainder from dividing by 5 any 8-bit binary number can be calculated. ■

Fig. 2.16 a Table with additions and remainders. **b** Circuit for Σ. **c** Circuit for **mod5**

ABCD	Σ	**mod5**
0000	0	0
0001	1	1
0010	2	2
0011	3	3
0100	4	4
0101	5	0
0110	6	1
0111	7	2
1000	3	3
1001	4	4
1010	5	0
1011	6	1
1100	7	2
1101	8	3
1110	9	4
1111	10	0

(a)

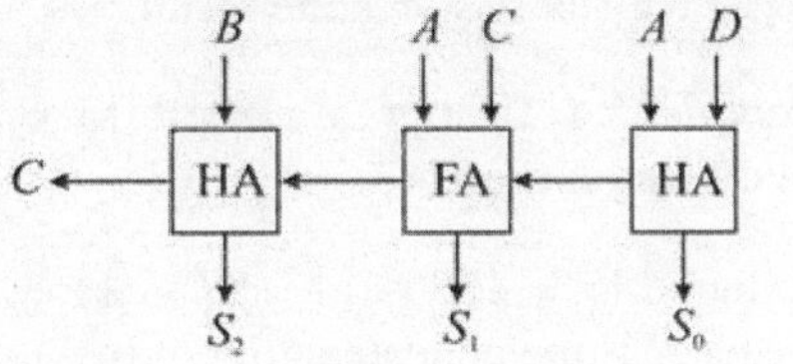

(b)

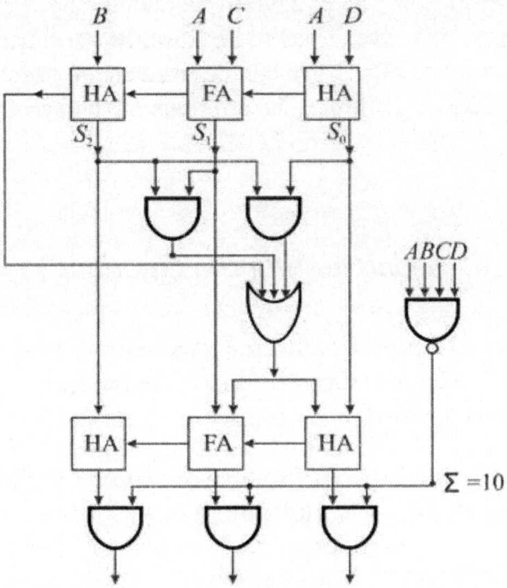

(c)

The modular reduction algorithm based on successive modular multiplications (modular multiplicative reduction, see Sect. 1.2.4) can also be used. Let $M = 2^k - a$, $1 \leq a < 2^k - 1$. To calculate $N\textbf{mod}M$, being N an n-bit integer number, an n-bit register N can be used, which is the concatenation of P, of n–k bits, and Q, of k bits, by applying the following algorithm:

Algorithm 2.6

First algorithm for modular reduction

```
R ← 0;
while P ≠ 0, do
      R ← R+Q, N ← P·a;
      R ← R+Q;
while R ≥ M, do
      R ← R–M;
```

End algorithm

The register R stores the result.

Example 2.4 **Design a circuit to operate with 16-bit binary numbers for calculating Nmod245.**

Since $245 = 256 - 11 = 2^8 - 11 = 2^k - a$ (it results $k = 8$, $a = 11$), the 16-bit register N is the concatenation of two 8-bit registers, P and Q. An auxiliary register, R, is used to store intermediate results, and which is initially set to zero. In each iteration R and Q have to be added, and P has to be multiplied by 11. The number N to be reduced has to be initially introduced into the register N. With all this, the circuit of Fig. 2.17a can be used as the processing unit for this calculation.

As an example, the contents of the various registers when N = 1101 0011 1000 0010 are shown in Fig. 2.17b. ■

2.6.5 *Calculating the Quotient by Undoing the Multiplication*

When it is known that the remainder is zero (i.e., it is an exact division) then iterative procedures that try to "undo" the multiplication [4] can be used, such as dividing by 3, as detailed in the following example. For division by 5, see [3].

Example 2.5 **Let suppose the case of the division of any 8-bit integer number N = *QRSTUVWZ* multiple of 3, by 3 (zero remainder).**

The 6-bit quotient is $C = abcdef$. The task consists on obtaining *abcdef* from *QRSTUVWZ*. The bits *QRSTUVWZ* are related to the bits *abcdef*. In fact, multiplying *abcdef* by $3_{10} = 11_2$, as follows, it results:

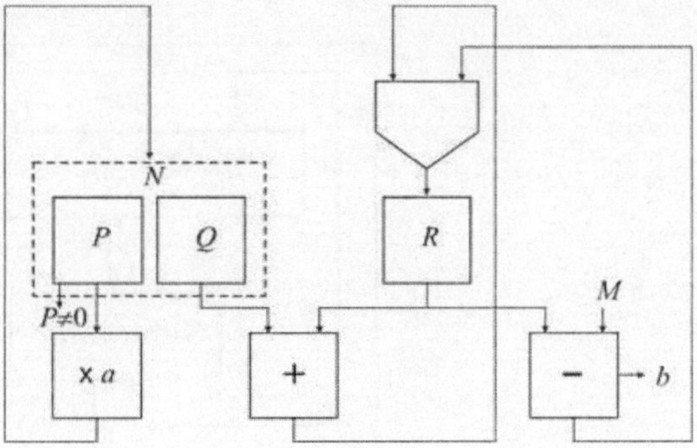

(a)

Step	Operation	P	Q	R
0	$R \leftarrow 0$	1101 0011	1000 0010	0000 0000
1	$R \leftarrow R + Q, N \leftarrow P{\cdot}a$	0000 1001	0001 0001	1000 0010
2	$R \leftarrow R + Q, N \leftarrow P{\cdot}a$	0000 0000	0110 0011	1001 0011
3	$R \leftarrow R + Q$	0000 0000	0110 0011	1111 0110
4	$R \leftarrow R - M$	0000 0000	0110 0011	0000 0001

(b)

Fig. 2.17 Multiplicative modular reduction. **a** Processing unit. **b** Example N = 1101 0011 1000 0010

		a	b	c	d	e	f
					x	1	1
		a	b	c	d	e	f
	a	b	c	d	e	f	
Q	R	S	T	U	V	W	Z

It is clear that $f = Z$. As $W = e + f$, e is obtained by subtracting f to $QRSTUVW$; specifically it is the least significant bit of $M = QRSTUVW - f$. After deletion of the least significant bit of M with a right shift, d is obtained by subtracting e, and so on for the remaining bits of the quotient. In short, by shifts and subtractions the quotient bits are obtained, one each time, from the least significant to the most significant, with an algorithm, which core may be obtained as follows:

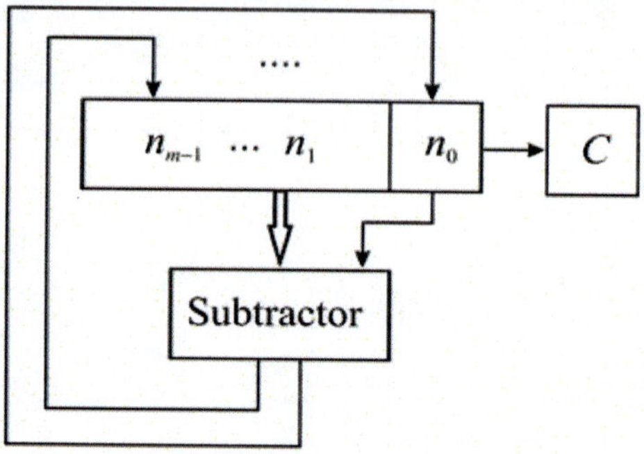

Fig. 2.18 Divider by 3

$$C \leftarrow \overset{n_0 \rightarrow}{C}, \quad N \leftarrow \overrightarrow{N - n_0}$$

where C (where it will be the quotient) is a shift register in which, at each iteration, the least significant bit in register N (n_0 bit) is entered. This division can be implemented with the circuit of Fig. 2.18. ■

The procedure for dividing by a constant when the remainder is zero applied in Example 2.5 can be easily extended to any divisor, with appropriate modifications. For example, repeating this procedure for each case, to divide by $5_{10} = 101_2$, two bits of the quotient can be obtained in each iteration, and to divide by $9_{10} = 1001_2$, three bits of the quotient can be obtained in each iteration.

When the binary divider development has more than two ones, the intermediate operations can be more complex, even the minimal signed digit development can be used. For example, for $7 = 111 = 100\overline{1}$ a procedure involving additions instead of subtractions can be designed. In general, both sums and subtractions may appear.

2.6.6 *Calculating the Quotient by Multiplying by the Inverse of the Divisor*

Another method for dividing by a constant, particularly when only the quotient is of interest, consists on multiplying by its inverse. The inverse of the first odd integers are given in Table 2.2; the inverse of an even number is simply obtained by shifts of the inverse of the greater odd which it is a multiple.

When the task consists on dividing by a constant whose inverse is a periodic fraction, to obtain the quotient a process in which simply multiplying the dividend by the first period or by the first two periods (usually more than two periods are not necessary) of the inverse of the divisor can be designed, and then adding 1 to the least significant position, as it is demonstrated in the following example.

Example 2.6 **Design a circuit to divide any unsigned 8-bit integer number N, multiple of 5 (it means, with remainder 0), by $5_{10} = 101_2$.**

Table 2.2 Inverses of the first integers

$1/3 = 0.\overline{01}$
$1/5 = 0.\overline{0011}$
$1/7 = 0.\overline{001}$
$1/9 = 0.\overline{000111}$
$1/11 = 0.\overline{0001011101}$
$1/13 = 0.\overline{000100111011}$

The largest multiple of 5 with 8 bits is $255_{10} = 11111111_2$. The quotient in this case is $51_{10} = 110011_2$. According to Table 2.2, the inverse of 5 is $1/5 = 0.\overline{0011}$. Multiplying 11111111 by 0.0011 it results the following integer:

$$11111111 \times 0.0011 = 101111 = 47_{10}$$

which differs in three of 51. Therefore it is not suffice to multiply by the first period to generate a result that differs in one from the correct. Using two periods, it results the following integer:

$$11111111 \times 0.00110011 = 110010 = 50_{10}$$

which differs in one of 51. Therefore, to calculate the quotient of dividing by 5 any 8-bit multiple of 5 (non-zero), it is enough to multiply by 0'00110011, and to add 1 to the integer part of the result. Moreover, the multiplication by 0'00110011 can be done in two iterations, with a single adder, or in one iteration, with two adders.

A circuit for dividing by 5, in one iteration, including the correction of adding 1 to the result of the multiplication by 0'00110011, is shown in Fig. 2.19a. A first adder of 8 bits, A1, whose inputs are $E_1 = N$ and $E_2 = \overrightarrow{N}$ (that is, N shifted right one position) and whose 10-bit output is S1, is used. S1 (unshifted and shifted four positions) is the input to a second adder A2; only the most significant 6 bits of the output of A2 are used; to add these 6 bits and 1 the adder A3 is used, consisting of six half adders. ■

The procedure for obtaining the quotient in an exact division using the multiplication by the inverse consists in analyzing the behavior of the largest possible multiple of the divisor, and to decide how many periods of the inverse should be used. This particular procedure outlined in Example 2.6 can be easily refined and extended to obtain the integer quotient when dividing any number by 5, whether be or not a multiple of 5.

With the multiplication by the inverse there is a procedure for scaling values within a predetermined range. When it is applied without refinements to any value, not necessarily a multiple of the constant scaling, the maximum error that can be committed is 1.

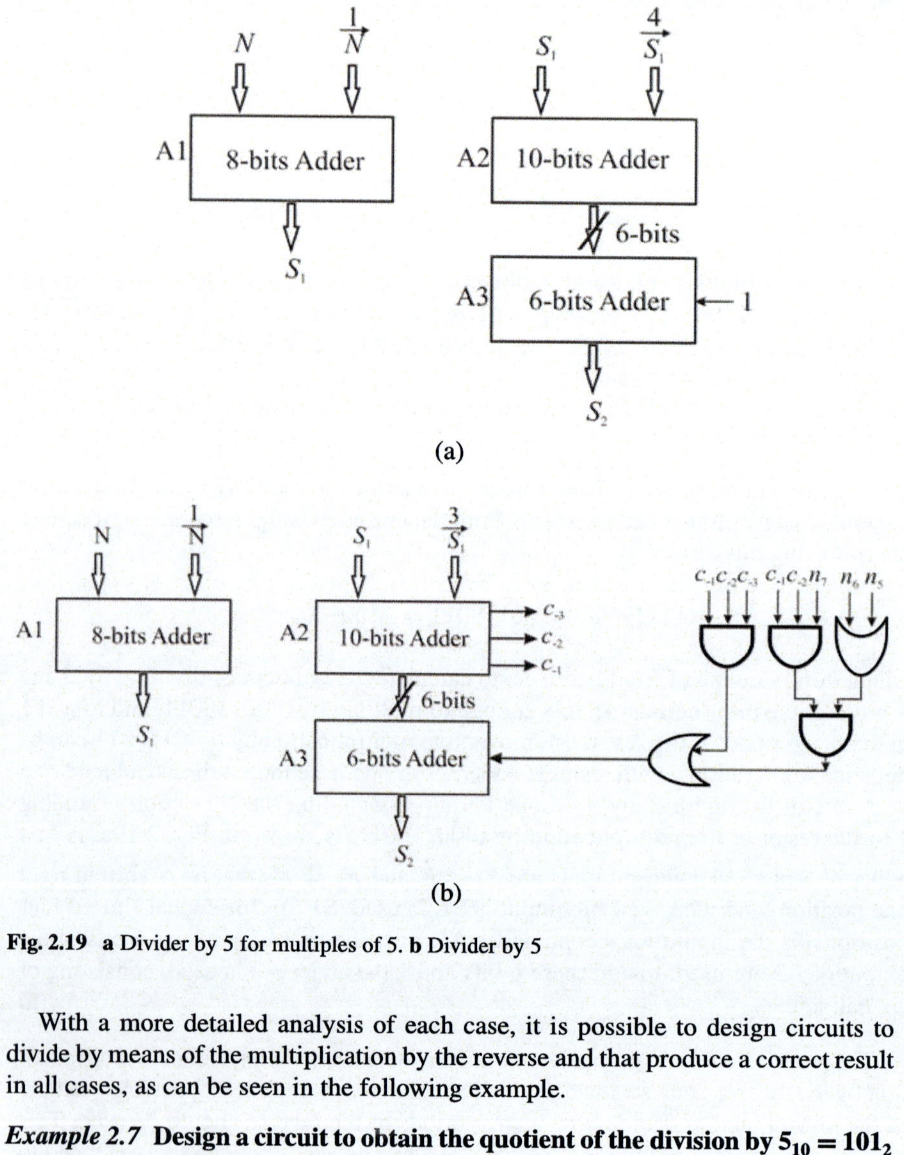

Fig. 2.19 **a** Divider by 5 for multiples of 5. **b** Divider by 5

With a more detailed analysis of each case, it is possible to design circuits to divide by means of the multiplication by the reverse and that produce a correct result in all cases, as can be seen in the following example.

Example 2.7 **Design a circuit to obtain the quotient of the division by $5_{10} = 101_2$ of any unsigned 8-bit integer number N ($n_7 \ldots n_0$).**

According to Example 2.6, if N is a multiple of 5, it is necessary to add 1 to the result of multiplying N by 0.00110011. It is easy to see that if N is not a multiple of 5, the integer part of $N \times 0.00110011$ is the correct quotient. Therefore, if for this case is intended to construct a similar circuit to that in Fig. 2.19a, it is necessary to separate the multiples of 5 from the other values. This can be done by analyzing the fractional part of $N \times 0.00110011$. With a detailed analysis of the different cases, it is concluded that N is a multiple of 5 if the three most significant bits of the fractional part $c_{-1}c_{-2}c_{-3}$ are equal to 111 (this is true for $N \leq 160$), or if the two most significant bits $c_{-1}c_{-2}$ are equal to 11, and $N > 160$. As $160_{10} = 10100000_2$,

$$N \geq 160 \rightarrow n_7(n_6 + n_5) = 1$$

Therefore, the condition for adding 1 is that the following function F be equal to 1:

$$F = c_{-1}c_{-2}c_{-3} + c_{-1}c_{-2}n_7(n_6 + n_5)$$

From all this, the circuit of Fig. 2.19b will generate the correct quotient of any 8-bit unsigned integer, when it is divided by 5. ■

Other procedures for dividing by an integer are based on the following expressions, which can be easily verified making the corresponding divisions:

$$\left(1 - 2^{-n}\right)^{-1} = 1 + 2^{-n} + 2^{-2n} + 2^{-3n} + \cdots$$
$$\left(1 + 2^{-n}\right)^{-1} = 1 - 2^{-n} + 2^{-2n} - 2^{-3n} + \cdots$$

Given an integer p, it is possible to find two integers q and n, such that:

$$p \times q = 2^n - 1 \rightarrow \frac{1}{p} = \frac{q}{2^n - 1}$$

But

$$\left(2^n - 1\right)^{-1} = 2^{-n}\left(1 - 2^{-n}\right)^{-1} = 2^{-n}\left(1 + 2^{-n} + 2^{-2n} + 2^{-3n} + \cdots\right)$$

From this equation:

$$\frac{1}{p} = q \times 2^{-n}\left(1 + 2^{-n} + 2^{-2n} + 2^{-3n} + \cdots\right)$$

Another option is:

$$p \times q = 2^n + 1 \rightarrow \frac{1}{p} = \frac{q}{2^n + 1}$$
$$\left(2^n + 1\right)^{-1} = 2^{-n}\left(1 + 2^{-n}\right)^{-1} = 2^{-n}\left(1 - 2^{-n} + 2^{-2n} - 2^{-3n} + \cdots\right)$$
$$\frac{1}{p} = q \times 2^{-n}\left(1 - 2^{-n} + 2^{-2n} - 2^{-3n} + \cdots\right)$$

Therefore, for dividing by p it is enough to multiply by q, shifted n places, and for the corresponding sum $A = 1 + 2^{-n} + 2^{-2n} + 2^{-3n} + \cdots$ or $B = 1 - 2^{-n} + 2^{-2n} - 2^{-3n} + \cdots$. Only the first summands of each addition are used, as can be seen in the following example. Of these two last possibilities, the integer q leading to a simpler procedure is chosen in each case. Possible products applicable to the first integers are shown in Table 2.3.

Example 2.8 **Obtain the expression corresponding to the division by 5 and by 7 using Table 2.3.**

The development of $(2^n + 1)^{-1}$ is used for the division by 5, where $q = 1, n = 2$. It results:

$$\begin{aligned}\frac{1}{5} &= 1 \times 2^{-2}(1 - 2^{-2} + 2^{-4} - 2^{-6} + \cdots) \\ &= 0.01(1 - 0.01 - 0.0001 - 0.0000001 + \cdots) \\ &= 0.01(0.11 + 0.000011 + \cdots) = 0.\overline{0011}\end{aligned}$$

Table 2.3 Applied products to the first integers

p	$p \times q$	$2^n \pm 1$	n
3	3 × 1	3	2
5	5 × 1	5	2
7	7 × 1	7	3
9	9 × 1	9	3
11	11 × 3	33	5
13	13 × 5	65	6
15	15 × 1	15	4
17	17 × 1	17	4
19	19 × 27	513	9
21	21 × 3	63	6
23	23 × 89	2047	11
25	25 × 41	1025	10
27	27 × 19	513	9
29	29 × 565	16,385	14
31	31 × 33	1023	10
33	33 × 31	1023	10
35	35 × 117	4095	12
37	37 × 7085	262,145	18
39	39 × 105	4095	12
41	41 × 25	1025	10
43	43 × 3	129	7
45	45 × 91	4095	12
47	47 × 178481	8,388,607	23
49	49 × 42799	2,097,151	21
51	51 × 5	255	8

The development of $(2^n - 1)^{-1}$ is used for the division by 7, where $q = 1, n = 3$. It results:

$$\frac{1}{7} = 1 \times 2^{-3}(1 + 2^{-3} + 2^{-6} - 2^{-9} + \cdots)$$
$$= 0.001(1 + 0.01 + 0.00001 + 0.000000001 + \cdots) = 0.\overline{001}$$

Of course, the obtained expressions are identical to those given in Table 2.2. ■

2.6.7 Modular Reduction (Again)

The idea developed in the previous section to obtain the quotient multiplying by the inverse of the divisor can be used to implement the modular reduction, $N\mathbf{mod}m$. It involves using a good approximation for the value of the quotient of N divided by m. As seen in Sect. 2.6.6, using an appropriate value for $M = 1/m$ and multiplying by N, the correct value of the quotient is obtained, or a sufficiently approximate value, c_a, so that the following algorithm can calculate $R = N\mathbf{mod}m$:

Algorithm 2.7

Second algorithm for modular reduction

```
R ← N-N·M·m;
while R ≥ m, do
    R ← R-m;
```

End algorithm

If n digits are used for operating in the base b, the $N \bullet M$ product can be expressed as:

$$c_a = \left\lfloor \frac{N}{b^k} \frac{b^n}{m} \frac{1}{b^{n-k}} \right\rfloor$$

so that, pre calculating $M = \frac{b^n}{m}$, it is enough to multiply M by the n–k most significant digits of N to obtain c_a, as it is probed in the following example. The Barrett modular reduction method [1] basically consists on this.

Example 2.9 **Design a procedure to obtain Nmod13, with N of 8 bits.**

With these data $2^8/13 \approx 19_{10} = 10011_2$ can be used. Let consider the extreme case $N = 1111\ 1111$. It is straightforward to check that $c = 10011$.

- For $k = 4$: $c_a =$ five most significant bits of $1111 \times 10011 = 10001$. Therefore $c - c_a = 2$ (should subtract twice).
- For $k = 3$: $c_a =$ five most significant bits of $11111 \times 10011 = 10010$. Therefore $c - c_a = 1$ (should subtract once).
- For $k = 2$, 1 and 0, the same value is obtained for c_a (10010).

Using $2^8/13 \approx 19.5_{10} = 10011.1_2$ again for the extreme case $N = 1111\ 1111$ (will remain $c = 10011$), it results:

- For $k = 4$: c_a = five most significant bits of $1111 \times 100011.1 = 10010$. Therefore $c - c_a = 1$ (should subtract once).
- For $k = 3$ it results the same value for c_a (10010).
- For $k = 2$: c_a = five most significant bits of $1111 \times 100011.1 = 10010$. Therefore $c = c_a$ (subtraction is not necessary).

In conclusion, for this application works well using $2^8/13 \approx 19.5_{10} = 10011.1_2$ and do $k = 2$. ■

2.6.8 *Square Root*

The square root can be extracted by successive subtractions, such as seen in Sect. 1.6.1. The obtained circuits are very similar to those implementing the division. For example, the circuit of Fig. 2.20a, which uses the same **CR** and **CS** cells from the divider (Fig. 2.14b and c) extracts the integer binary square root of any 8-bit integers, $a_7 \ldots a_0$. In this case, the combinational circuit for calculating the square root has four stages or rows, each one calculating $D^+ = D - (4R_1 + 1)2^{2i}$. If $D^+ \geq 0$, then $r_i = 1$ and D is substituted by D^+; if $D^+ < 0$, then $r_i = 0$ and D is unchanged. The result is a square root of four bits, $r_3r_2r_1r_0$, and a remainder of five bits, $b_4b_3b_2b_1b_0$.

The integer square root of a binary number of 8 bits, $A = a_7 \ldots a_0$, may be calculated with the sequential circuit of Fig. 2.20b using four iterations. A shift register is used to store A, called SR1, with double shift at each iteration, so that in the a_i and a_{i-1} outputs are successively obtained a_7 and a_6, a_5 and a_4, a_3 and a_2, a_1 and a_0. The successive bits of the result are written to a normal shift register called SR2. The results of successive subtractions are written to a read-write parallel register, R3. Initially R3 must be zero. After four iterations, the root is stored in SR2 and the remainder is stored in R3. It is easy to verify that, with the specified operating conditions, the circuit of Fig. 2.20b performs in each iteration the same action as the corresponding row of the circuit of Fig. 2.20a.

2.7 BCD Adder/Substracter

From Sect. 1.7.1 it results that a circuit to add two **BCD** characters, $X = x_3x_2x_1x_0$ and $Y = y_3y_2y_1y_0$, can be constructed with four binary adders plus the correction circuit for adding 6 when appropriate. Calling $R = r_3r_2r_1r_0$ to the partial result generated by the four binary adders, and calling a_+ to the partial carry, 6 must be added when $F = 1$, for which two expressions are given:

$$F = r_3r_2 + r_3r_1 + a_+ = a_+ + a_{++}$$

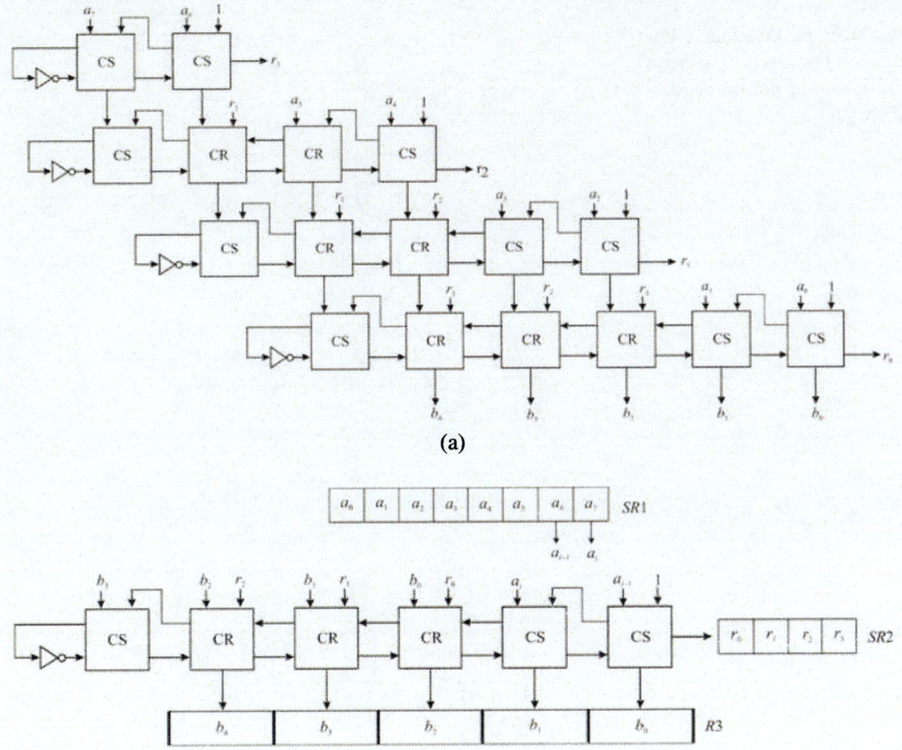

(a)

(b)

Fig. 2.20 Square root **a** Combinational circuit. **b** Sequential circuit

This *F* function also gives the final carry. Therefore, the circuit of Fig. 2.21a or Fig. 2.21b is an adder for **BCD** digits. Unsigned decimal numbers of *n* digits can be added by cascading *n* adder circuits of **BCD** digits as depicted in Fig. 2.21c.

The sign digit has to be included if the subtraction has to be implemented. The **SM** representation is not recommended for subtraction, since prior to the operation itself, the two operands should be compared. However, if the 9's complement representation is used, basically the same structure in Fig. 2.21c can be used to add and subtract. Just it is necessary to change the sign of the subtrahend and 9's complement each of its digits. The truth table for the 9's complement generation of each digit is shown in Fig. 2.22a, and the corresponding circuit is shown in Fig. 2.22b; to change the sign it is enough to invert the bits with which it is encoded. Using the control signal $\overline{s}/r$, to be 0 for the sum and to 1 for the subtraction, in Fig. 2.22c has an adder/subtractor for two **BCD** numbers of $n - 1$ digits plus a sign digit represented in 9's complement; in this circuit it is taken into account the end-around carry.

Regarding the 10's complement representation, it is important to remember that the 10's complement of a **BCD** number can be obtained from the 9's complement, by adding 1. Using this idea, and considering that in this case there is no end-around

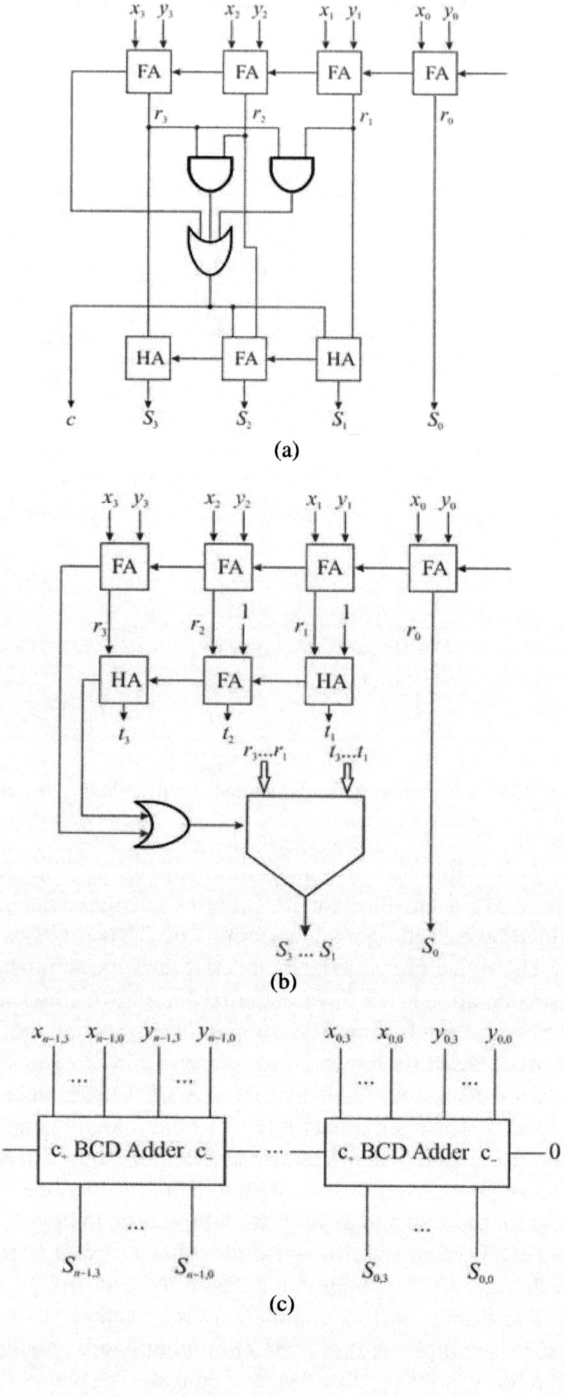

Fig. 2.21 **BCD** Adder. **a** For digits. **b** For digits by using a multiplexer. **c** For numbers of length n

$x_3x_2x_1x_0$	$z_3z_2z_1z_0$
0000	1001
0001	1000
0010	0111
0011	0110
0100	0101
0101	0100
0110	0011
0111	0010
1000	0001
1001	0000
1010	----
1011	----
1100	----
1101	----
1110	----
1111	----

$$z_3 = \overline{x_3 + x_2 + x_1}$$
$$z_2 = x_2 \oplus x_1$$
$$z_1 = x_1$$
$$z_0 = \bar{x}_0$$

(a)

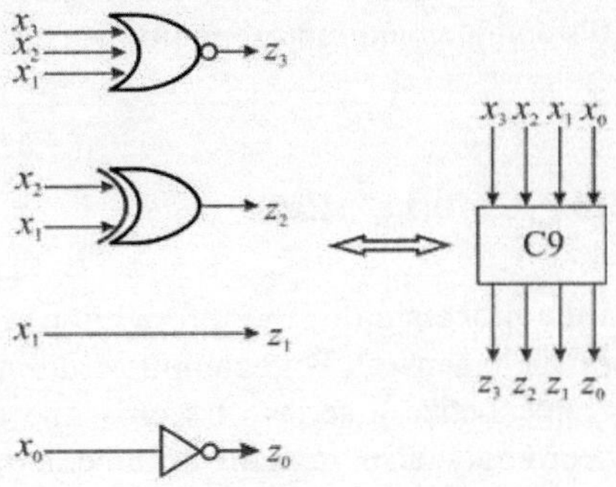

(b)

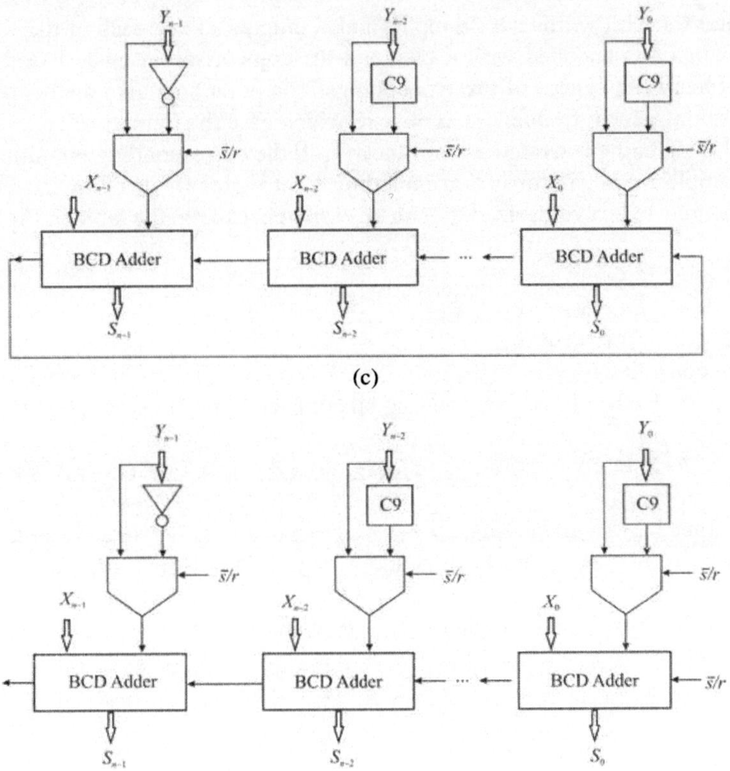

(c)

(d)

Fig. 2.22 **a** Truth table for the 9's complement. **b** Circuit to calculate 9's complement. **c** 9's complement adder/subtractor. **d** 10's complement adder/subtractor

carry, the circuit of Fig. 2.22d is an adder/subtractor for **BCD** numbers represented in 10's complement.

Comparing the circuits of Fig. 2.22c and d, it is obvious that it is preferable the 10's complement representation versus the 9's complement representation.

2.8 Comparators

In the processing of the information it is common to have to compare two words or data in general. For example, ordering from lowest to highest a table of numbers or alphabetize a series of words, the elements are compared in pairs and sorted accordingly; also, in many arithmetic operations different numerical results have to be compared.

Let X and Y be two elements to sort; the comparators can be used for this task. A comparator for n-bit words has $2n$ inputs and m outputs so that each of the elements to be compared is encoded with n bits, and the comparison is made based on the value in binary (unsigned) of these encodings. The m outputs give the result of the comparison; the most frequent is $m = 3$, in which case the outputs are $X > Y$, $X = Y$, $X < Y$, each being activated as well they fulfill the corresponding condition.

The simplest comparator is that including 1-bit words ($n = 1$). In this case the three functions to be synthesized, as can be easily checked in the table in Fig. 2.23a, are:

For $X > Y$: $f_2(x, y) = x\overline{y}$
For $X = Y$: $f_9(x, y) = xy + \overline{x}\overline{y}$
For $X < Y$: $f_4(x, y) = \overline{x}y$

A common value for n is 4 ($X = x_3 \ldots x_0$, $Y = y_3 \ldots y_0$). For this case, the output $X = Y$ will be 1 when the corresponding bits of each input are equal; this means:

$$F(X = Y) = f_9(x_3, y_3) \cdot f_9(x_2, y_2) \cdot f_9(x_1, y_1) \cdot f_9(x_0, y_0)$$

The output $X > Y$ will be 1 if $x_3 > y_3$, or $x_3 = y_3$ and $x_2 > y_2$, or $x_3 = y_3$ and $x_2 = y_2$ and $x_1 > y_1$, or $x_3 = y_3$ and $x_2 = y_2$ and $x_1 = y_1$ and $x_0 > y_0$; it means:

$$\begin{aligned} F(X > Y) = {} & f_2(x_3, y_3) + f_9(x_3, y_3) \cdot f_2(x_2, y_2) \\ & + f_9(x_3, y_3) \cdot f_9(x_2, y_2) \cdot f_2(x_1, y_1) + f_9(x_3, y_3) \cdot f_9(x_2, y_2) \\ & \cdot f_9(x_1, y_1) \cdot f_2(x_0, y_0) = f_2(x_3, y_3) + f_9(x_3, y_3) \cdot (f_2(x_2, y_2) \\ & + f_9(x_2, y_2) \cdot (f_2(x_1, y_1) + f_9(x_1, y_1) \cdot f_2(x_0, y_0))) \end{aligned}$$

The expression for the function corresponding to the output $X < Y$ is parallel to the one for $X > Y$, substituting > by < (it means, f_2 by f_4). It is obvious that once that two of the comparator outputs are known, the third can be obtained from these two. Concretely,

xy	x>y	x=y	x<y
00	0	1	0
01	0	0	1
10	1	0	0
11	0	1	0

(a)

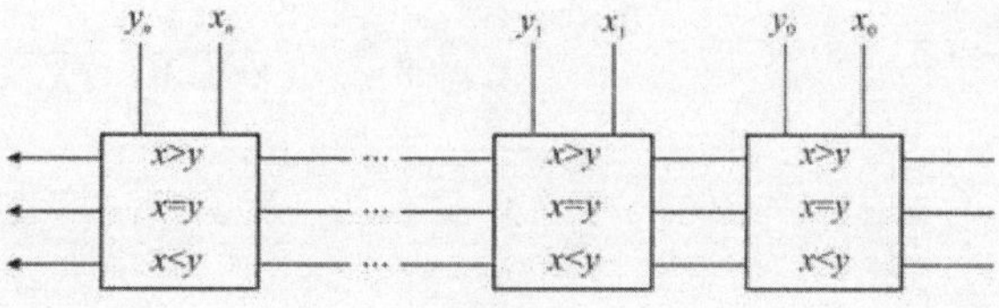

(b)

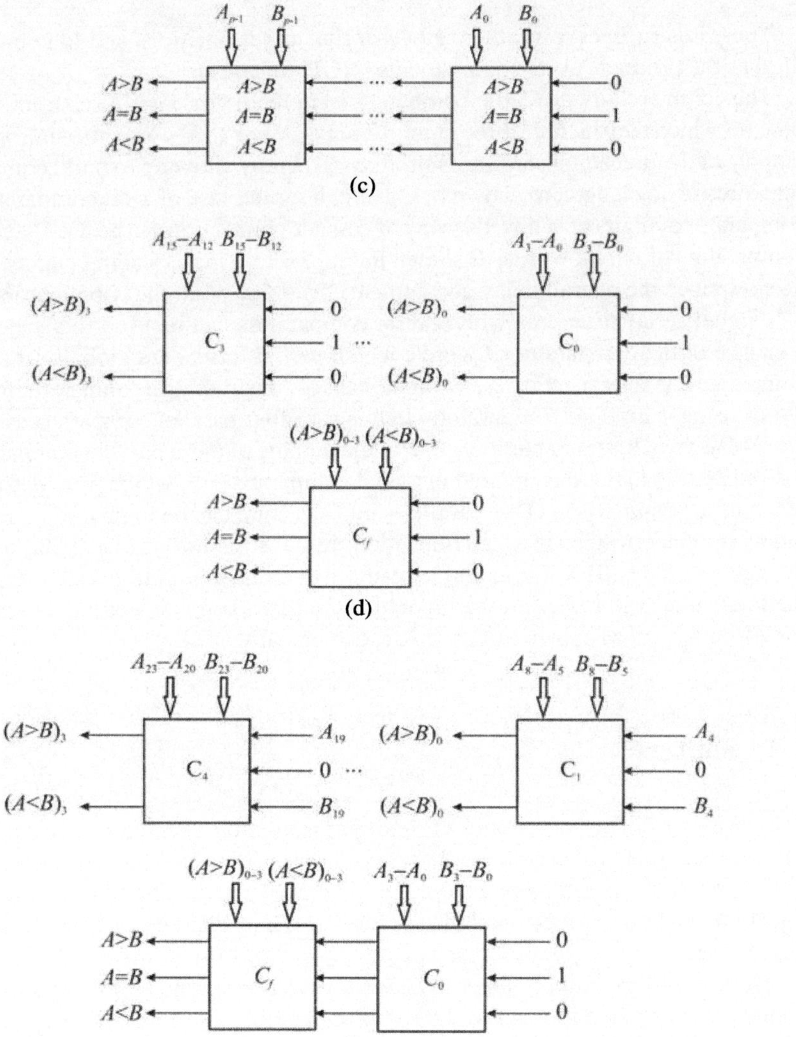

(e)

Fig. 2.23 Comparators. **a** Table for 1-bit comparator. **b** Cascade connection. **c** Cascade connection of digit comparator. **d** Parallel-serial connection of digit comparators. **e** Comparator of 24-bit words with parallel-serial connection

$$X = Y \Leftrightarrow \overline{(X > Y)} \cdot \overline{(X < Y)}$$

$$X > Y \Leftrightarrow \overline{(X = Y)} \cdot \overline{(X < Y)}$$

$$X > Y \Leftrightarrow \overline{(X = Y)} \cdot \overline{(X > Y)}$$

Thus, it is suffices to synthetize two of the output functions and to construct the third as the products of its complements (**NOR** function).

The commercially available comparators are ready for possible cascade connection, for which they include three inputs ($X > Y_{in}, X = Y_{\text{in}}, X < Y_{in}$), through which the outputs of the preceding stage are introduced, thereby allowing to build comparators for words of any length, as shown in Fig. 2.23b for the case of 1-bit comparators. For comparator of digits of m bits, this way of cascade connection can be used to construct comparators of pm bit words, as shown in Fig. 2.23c. This cascading connection can be slow since the overall delay accumulates the delay of all the comparators.

Parallel-serial structures with several comparators can be used to accelerate the response of the comparator of words, which partial results are globalized at a final comparator, as shown in Fig. 2.23d for the case of constructing a comparator for 16-bit words using 4-bit digit comparators. In this case four parallel comparators are used, $C_3 \ldots C_0$. Two digits are composed with the outputs of these parallel comparators, $A > B$ and $A < B$, that are compared in a final comparator, C_f, which provides the final result of the comparison. This parallel-serial structure can be improved by using the inputs provided to the cascade connection; in the structure of Fig. 2.23d are set to the neutral values 010. For example, using five comparators in parallel, $C_4 \ldots C_0$, the final comparator C_f, and the inputs for the cascading connection, a comparator for 24-bit words, as shown in Fig. 2.23e, can be built.

2.9 Shifters

A k-positions shifter is a circuit whose input is an n-bit character, $E = e_{n-1} \ldots e_0$, and whose output is also an n-bit character, $S = s_{n-1} \ldots s_0$, which, when the shifting have to be made, it is obtained from the input E by means of k-shifts, either to the right or to the left, as stated, as shown in Fig. 2.24; if no displacement has to be made, then $S = E$.

In a shift of k positions there will be k bits of S to which no bit of E is applied: the k most left bits when moving to the right, or the k most right bits on the left shifts. For these k bits of S the values to be assigned has to be established, usually using one of the following two options:

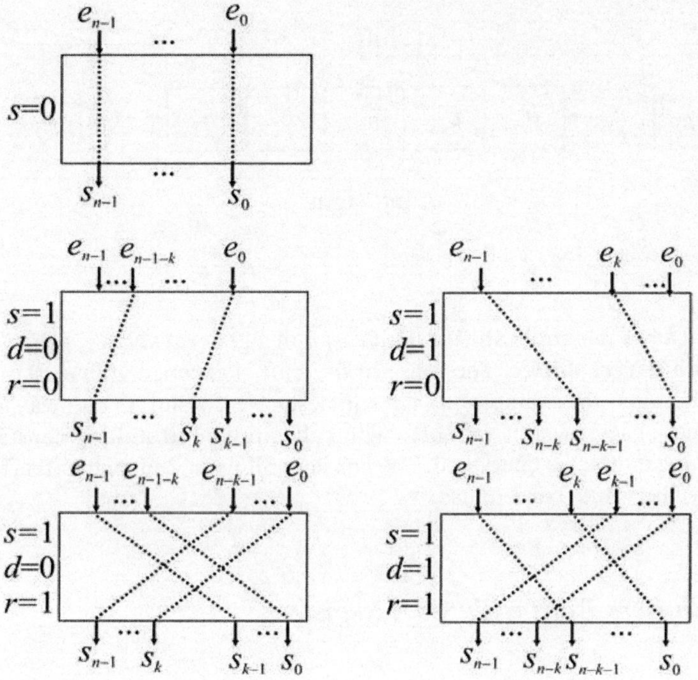

Fig. 2.24 Actions of the shifters

a. filled with constant (all zero or all one, although other combinations are possible);
b. filled with the k bits of E that would be unmatched (i.e., the most right on the right shifts, or the most left on the left shifts; in both cases it consists on rotating the input in the sense that apply). For example, in a shifter with two positions ($k = 2$) for 8-bit characters ($n = 8$), with zero padding, if $E = 10011101$, a shift to the right will result $S = 00100111$, and a shift to the left be $S = 01110100$; if the padding were with remaining input bits (i.e., a rotation) in a shift to the right will be $S = 01100111$, and in a shift to the left $S = 01110110$.

Therefore, to define the action of a shifter it must be specify:

- the size n of the characters to be shifted,
- whether it has to perform or not the shift, with the variable s,
- the magnitude of the shift k,
- if the shift is to the right or to the left, with the variable d,
- if the padding is a constant value or by rotation, with the variable r,
- and finally, the value of the constant filling, where appropriate, with the c variable.

The size n of the characters are supposed to be pre-established, usually equal to the size of the characters being processed. The decision to perform the shift is described, for example, with $\mathbf{s = 1}$ ($\mathbf{s = 0}$, no shift). The magnitude of the shift k can

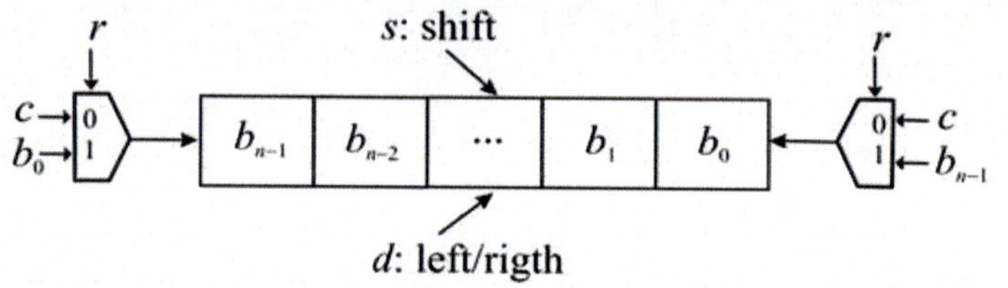

Fig. 2.25 Shifter built using a shift register

range from ***k*= 1** in **simple shifter** until ***k* ≤ *n*** in a **general shifter**, called by some authors as a **barrel shifter**. The sense of the shift is encoded with ***d*** (0 on the left and 1 to the right, for example); in the simplest case the shift is one-way, in which case it is not necessary that variable. With ***r*** the fill type (0 stuffing constant, 1 for rotation, for example) is encrypted. The constant filling, ***c***, can match (this would be the easiest) the value 0 or 1 to use.

2.9.1 Shifters Built with Shift Registers

The most obvious and simplest solution to construct a shifter is to use a shift register. The register depends on the features desired for the shifter. Using a bidirectional universal shift register, a shifter with all possible benefits can be built, as shown in Fig. 2.25. A standard shift register can shift a position (to the left or to the right) on each clock pulse.

The drawback of this solution is the time it can take to make a shift. In effect, a shift of *k*-position takes *k* clock pulses (each position takes a pulse), and, sometimes, it is an unbearable delay due to the performance degradation involved. Therefore a strictly combinational solution is the option, without using memory elements, as shown below.

2.9.2 Combinational Shifters

Using multiplexers as building blocks it is very easy to get a shifter with any performance. Let consider first the design of a shifter for *k* fixed and a default value of *n*. In this case the control variables are: *s*, *d*, *r* and *c*. It is straightforward to check that the circuit of Fig. 2.26a acts as a fixed shifter of *k*-positions; it is sufficient to obtain the outputs $s_{n-1}\ldots s_0$ from the multiplexers for all combinations of *s*, *d*, and *r*, and comparing them to the outputs generated in Fig. 2.24.

The *k*-position shifter of Fig. 2.26a consists of three levels of multiplexing (with 2-to-1 multiplexers), which can be replaced by a single level of multiplexing using

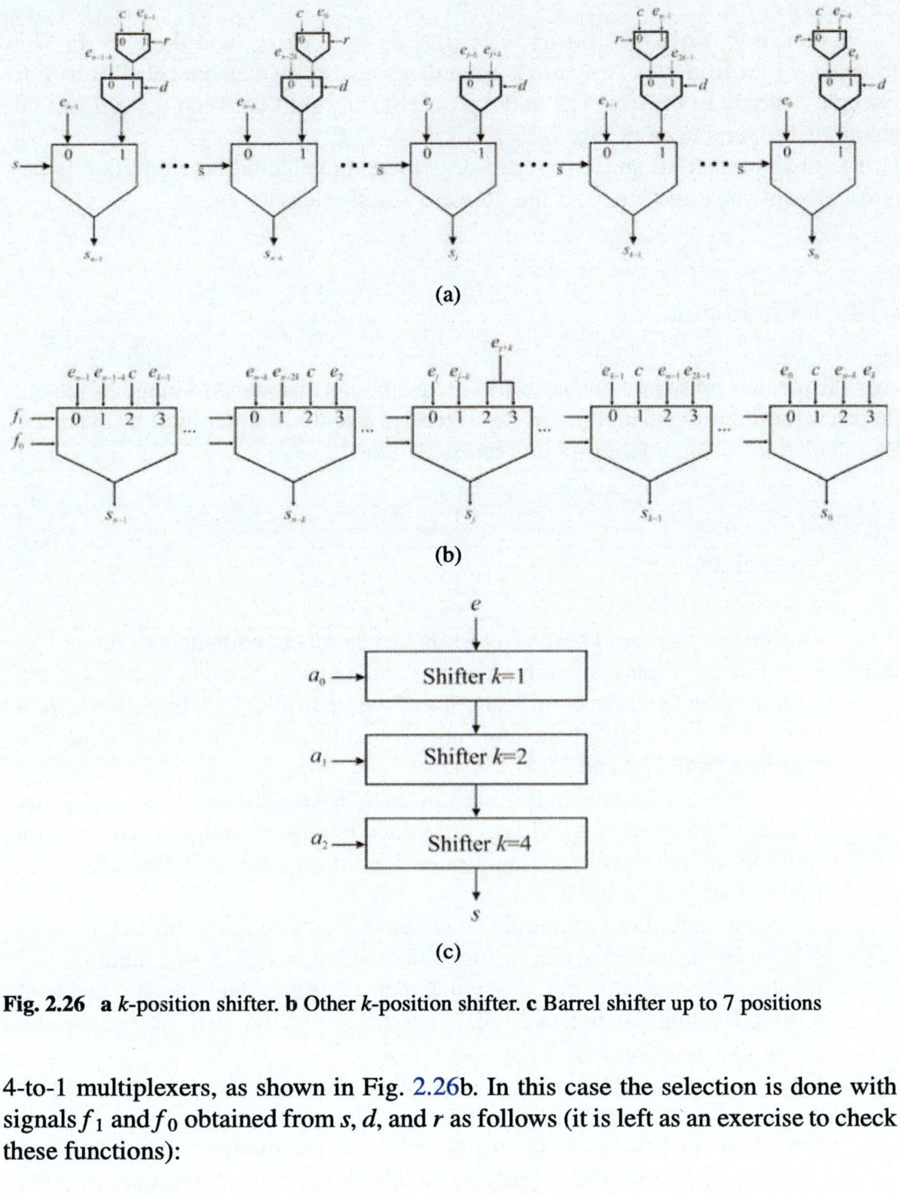

Fig. 2.26 **a** *k*-position shifter. **b** Other *k*-position shifter. **c** Barrel shifter up to 7 positions

4-to-1 multiplexers, as shown in Fig. 2.26b. In this case the selection is done with signals f_1 and f_0 obtained from *s*, *d*, and *r* as follows (it is left as an exercise to check these functions):

$$f_1 = sd \quad f_0 = s(\bar{d}\,\bar{r} + d\,r)$$

Let suppose that *p* fixed shifters with different values of *k* are used, such that $k = 2^n$, $n = 0, \ldots, p - 1$. Each of these shifters has its own control input s_i, for deciding whether or not made the corresponding shift; all other control inputs (*d*, *r* and *c*) are common to all shifters. It is easy to check that with these *p* shifters, acting in cascade, each one over the output of the previous one, any shift $k < 2^p$ can be get.

It is sufficient to write k in binary, $k = a_{p-1}\ a_{p-2}\ \ldots\ a_1\ a_0$, and do $s_i = a_i$. The obtained shifter from this structure is sometimes also known as **barrel shifter**. For example, with three shifters (1, 2 and 4 positions) any shift between 0 and 7 can be accomplished, as shown in Fig. 2.26c.

It is obvious that all shifter circuits described above could be simplified if the shifts were in one direction, or if the fill were of a single type, etc.

2.10 Conclusion

This Chapter has presented the basic arithmetic circuits that are used in the following chapters, for the implementation of the algebraic circuits and that will be extended for calculations using floating-point representation.

2.11 Exercises

2.1. Design a serial-serial multiplier for binary numbers without sign bit.

2.2. Design a serial-parallel multiplier similar to that described in Sect. 2.4.2, but starting with the most significant bits of the multiplier and, in each iteration multiply by two bits of the multiplier (i.e., in each iteration the circuit can multiply by 0, 1, 2 and 3).

2.3. Design a serial-parallel multiplier similar to that described in Sect. 2.4.2, but starting with the less significant bits of the multiplier and, in each iteration multiply by two bits of the multiplier (i.e., in each iteration the circuit can multiply by 0, 1, 2 and 3).

2.4. Design a multiplier to multiply simultaneously by 5, 7, 11 and 13.

2.5. Obtain Brauer additive chains for 34781, with $e = 1, 2, 3, 4, 5$, and 6.

2.6. Obtain the quotient and the remainder corresponding to the dividend 110100101 and the divisor 11101.

2.7. Design a combinational divisor for $n = 6$.

2.8. Design a combinational circuit to calculate the remainder from dividing by 13 any 8-bit unsigned binary number (see Example 2.3).

2.9. Design a combinational circuit to calculate the quotient obtained when dividing by 5 any 8-bits binary number without sign bit, multiple of 5, by undoing the multiplication.

2.10. Design a combinational circuit to calculate the quotient obtained when dividing by 7 any 8-bits binary number without sign bit, multiple of 7, as it is done in Example 2.6 for 5.

2.11. Design a combinational circuit to calculate the square root of any 10-bit unsigned binary number.

2.12. Design a sequential circuit to calculate the square root of any 10-bit unsigned binary number.

2.13. Design a combinational circuit to calculate directly the 10's complement (not adding 1 to the 9's complement).

2.14. Design a comparator of 32-bits words with parallel-serial connection.

2.15. Design a combinational shifter to rotate any 8-bits word 2 positions to the right or 4 positions to the left, using one control signal.

2.16. Obtain functions f_1 and f_2 used in Fig. 2.26b.

References

1. Barrett, P.: Implementing the Rivest, Shamir and Adleman public-key encryption algorithm on a standard digital signal processor. In: Odlyzko, A.M. (ed.) Advances in Cryptology—CRYPTO'86 Proceedings. LNCS, vol. 263, pp. 311–323. Springer (1987)
2. Brauer, A.: On addition chain. Bull. Am. Math. Soc. **45**, 736–739 (1939)
3. Sites, R.L.: Serial binary division by ten. IEEE Trans. Comput. **23**, 1299–1301 (1974)
4. Srinivasan, P., Petra, F.E.: Constant-division algorithms. IEE Proc. Comput. Technol. **141**(6), 334–340 (1994)

Chapter 3
Residue Number Systems

Residue Number Systems have probed their potential for computation-intensive applications, especially those related to signal processing. Their main advantage is the absence of carry propagation between channels in addition, subtraction and multiplication. Thus, high-performance systems may be built for applications involving only these operations using Residue Number Systems. On the other hand, modular operation associated to these Residue Number Systems are those to be implemented for Galois Fields GF(p), which are the objective of the following chapters related with algebraic circuits.

Theoretical basis for Residue Number Systems will be presented in this chapter, starting from the residue algebra, as well as some of the fundamentals circuits for implementing the main modular operations that will be used in following chapters.

3.1 Introduction

A Residue Number System, RNS in the following, represents integer numbers within a predefined range [4, 5]. Residue Number Systems make use of multiple radices and represent each number through the residues it generates over each base. Given an integer N and a positive integer base b, the residue r (or positive remainder resulting from dividing N over b) corresponds to:

$$r = N - bxc, \quad r \text{ and } c \text{ integers}, b > r \geq 0 \tag{3.1}$$

In this context, each base b is also known as modulus. It is not strictly necessary for the moduli to be positive, but negative moduli have no special interest, so in the following all moduli will be supposed to be positive.

The residue r in (3.1) will be represented as either $R(N, b)$ or $|N|_b$. Quotient c in (3.1) is the integer part of $\frac{N}{b}$ and will be represented as $\left\lfloor \frac{N}{b} \right\rfloor$. Thus, (3.1) can be also written as:

A. Lloris Ruiz et al., *Arithmetic and Algebraic Circuits*, Intelligent Systems Reference Library 201, https://doi.org/10.1007/978-3-030-67266-9_3

$$N = \left\lfloor \frac{N}{b} \right\rfloor b + R(N, b) \tag{3.2}$$

As an example, $R(5, 7) = 5$. In the same way $R(5, 2) = 1$, $R(5, 3) = 2$, $R(5, 5) = 0$. It also happens that $R(12, 7) = R(19, 7) = \cdots = 5$; generally, two integers whose difference is a multiple of base b have the same residue, i.e.:

$$R(N \pm kb, b) = R(N, b) \tag{3.3}$$

It is evident that the definition in (3.1) also holds for negative integers. As an example, $R(-2, 5) = 3$ (the residue is always defined as positive). The following theorem, related to negative integers, is easy to prove.

Theorem 3.1 *Given $R(N, b) = r$, $r \neq 0$, then $R(-N, b) = b - r$.*

Proof It is supposed that $N = bc + r$; thus:

$$-N = -bc - r = -bc - b + b - r = -b(c + 1) + (b - r)$$

It is obvious that $(b - r)$ is positive and satisfies (3.1). Thus $R(-N, b) = b - r$. q.e.d.

The value $b - r$ is called the complement of r with respect to b; and will be represented as $\bar{r}$. In this way, if the residue of a positive number, N, is known, the residue of $-N$ is simply the complement of the residue of N.

It is immediate that if $R(N, b) = 0$, then $R(-N, b) = 0$.

3.2 Residue Algebra

Given a modulus m, integers can be classified in equivalence classes using the residue they generate for this modulus. Evidently, for a modulus m there will be m equivalence classes, and each one of them can be represented (as it will be done in the following) by the smallest of its elements. As an example, there are 7 classes for $m = 7$, which are (0, 1, 2, 3, 4, 5, 6).

Between modulus m classes (which will be designated as CMm in the following) different operations may be defined as follows:

Modulo m addition: Give c_1 and c_2 (c_1, $c_2 \in$ CMm), their addition $c_3 = c_1 + c_2$ ($c_3 \in$ CMm) is defined as the class where the addition of the representatives of c_1 and c_2 belongs to.

As an example, Table 3.1 corresponds to the addition in CM7. It is obvious that the neutral element for addition in CMm, whatever m is, is 0.

Given any two integers, x and y, and a modulus m, the following expression is easily proved:

Table 3.1 Addition in CM7

+	**0**	**1**	**2**	**3**	**4**	**5**	**6**
0	0	1	2	3	4	5	6
1	1	2	3	4	5	6	0
2	2	3	4	5	6	0	1
3	3	4	5	6	0	1	2
4	4	5	6	0	1	2	3
5	5	6	0	1	2	3	4
6	6	0	1	2	3	4	5

$$R(x \pm y, m) = R[R(x, m) \pm R(y, m), m] \tag{3.4}$$

In fact, from (3.2):

$$x = \left\lfloor \frac{x}{m} \right\rfloor m + R(x, m)$$
$$y = \left\lfloor \frac{y}{m} \right\rfloor m + R(y, m)$$

Thus,

$$\begin{aligned} R(x \pm y, m) &= R\left[\left\lfloor \frac{x}{m} \right\rfloor m + R(x, m) \pm \left\{\left\lfloor \frac{y}{m} \right\rfloor m + R(y, m)\right\}, m\right] \\ &= R\left[\left\{\left\lfloor \frac{x}{m} \right\rfloor \pm \left\lfloor \frac{y}{m} \right\rfloor\right\} m + R(x, m) \pm R(y, m), m\right] \\ &= R[R(x, m) \pm R(y, m), m] \end{aligned}$$

since $\left\{\left\lfloor \frac{x}{m} \right\rfloor \pm \left\lfloor \frac{y}{m} \right\rfloor\right\} m$ is a multiple of m and (3.3) can be applied.

Equality (3.4) can be summarized as *the residue of addition is equal to the addition of residues.*

It is clear that a correspondence exists between integer addition and addition in CMm. As an example, $16 + 25 = 41$ would be translated to CM7 as $2 + 4 = 6$; as $R(41, 7) = 6$.

The negative element (also known as additive inverse or opposite) of a given element $c_1 \in$ CMm is defined as other element $c_2 \in$ CMm satisfying that $c_1 + c_2 = 0$. It is immediate that all elements in CMm, for any m, have a negative, which is unique. As an example, Table 3.2 shows the negative of each element in CM7. The negative of element a will be represented as $(-a)$.

The existence of this negative element allows defining the subtraction operation:

$$a - b = a + (-b)$$

Table 3.2 Opposites in CM7

	Negative
0	0
1	6
2	5
3	4
4	3
5	2
6	1

Table 3.3 Subtraction in CM7

–	0	1	2	3	4	5	6
0	0	6	5	4	3	2	1
1	1	0	6	5	4	3	2
2	2	1	0	6	5	4	3
3	3	2	1	0	6	5	4
4	4	3	2	1	0	6	5
5	5	4	3	2	1	0	6
6	6	5	4	3	2	1	0

It is clear that the complement of an element coincides with the negative of this element. Table 3.3 illustrates subtraction in CM7; each cell includes the value of the difference between row and column coordinates. It is clear from this table that subtraction is not commutative.

Once again it is easy to check the correspondence between integer subtraction and subtraction in CMm. As an example, $16 - 25 = -9$ would be expressed in CM7 as $2 - 4 = 2 + (-4) = 2 + 3 = 5$; and $R(-9, 7) = 5$.

It is immediate that, for any value of m, the set CMm, with modulo m addition as defined above, is an additive finite group [1].

Modulo m product: Given c_1 and c_2 ($c_1, c_2 \in$ CMm), their product $c_3 = c_1 \times c_2$ ($c_3 \in$ CMm) is defined as the class which the product of the representatives of c_1 and c_2 belongs to.

As example, Table 3.4 illustrates multiplication in CM7; from this table it is obvious that multiplication is commutative and that the neutral element for the product in CMm, whatever m is, is 1.

Given any two integers, x and y, and a modulus m, it is easy to check that the following expression holds:

$$R(xy, m) = R[R(x, m)R(y, m), m] \tag{3.5}$$

As a fact,

Table 3.4 Multiplication in CM7

×	0	1	2	3	4	5	6
0	0	0	0	0	0	0	0
1	0	1	2	3	4	5	6
2	0	2	4	6	1	3	5
3	0	3	6	2	5	1	4
4	0	4	1	5	2	6	3
5	0	5	3	1	6	4	2
6	0	6	5	4	3	2	1

$$\begin{aligned} R(xy, m) &= R\left[\left\{\left\lfloor\frac{x}{m}\right\rfloor m + R(x, m)\right\}\left\{\left\lfloor\frac{y}{m}\right\rfloor m + R(y, m)\right\}, m\right] \\ &= R\left[\left\{\left\{\left\lfloor\frac{x}{m}\right\rfloor\left\lfloor\frac{y}{m}\right\rfloor m + \left\lfloor\frac{x}{m}\right\rfloor R(y, m) + \left\lfloor\frac{y}{m}\right\rfloor R(x, m)\right\}m\right.\right. \\ &\quad \left.\left. + R(x, m)R(y, m)\right\}, m\right] \\ &= R[R(x, m)R(y, m), m] \end{aligned}$$

since $\left\{\left\lfloor\frac{x}{m}\right\rfloor\left\lfloor\frac{y}{m}\right\rfloor m + \left\lfloor\frac{x}{m}\right\rfloor R(y, m) + \left\lfloor\frac{y}{m}\right\rfloor R(x, m)\right\}m$ is a multiple of m and (3.3) may be applied.

Expression (3.5) may be summarized as *the residue of the product is the product of the residues.*

It is again clear that a correspondence exists between integer multiplication and multiplication in CMm. As an example, $16 \times 25 = 400$ would be translated into CM7 as $2 \times 4 = 1$; with R(400, 7) = 1.

Given $c \in$ CMm, $c \neq 0$, for the successive powers of c it will be held that $R(c^p, m)$, $\forall p$, can only have, as a maximum, m different values; the remainders of the successive powers will appear again with periodicity less than m, except for a given exponent e, $R(c^e, m) = 0$, which will make $R(c^{e+i}, m) = 0$, $\forall i > 0$. On the other hand, if c and m are relatively prime, it is obvious that $R(c^p, m) \neq 0$, $\forall p$.

The inverse (also known as multiplicative inverse) of a given element, $c_1 \in$ CMm, is other element, $c_2 \in$ CMm, satisfying $c_1 \times c_2 = 1$. The inverse of c in CMm will be represented as $\left|\frac{1}{c}\right|_m$ or c^{-1}.

It is immediate that the 0 element has no inverse. It is also immediate that if an element has an inverse, this is unique. In the same way, it is easy to check the symmetry property (if a is the inverse of b, then b is the inverse of a). For any $c \neq 0$, the following theorem holds:

Theorem 3.2 *Given $c \neq 0$ ($c \in$ CMm), if m and c are relatively prime,* $\exists \left|\frac{1}{c}\right|_m$.

Proof Since m and c are relatively prime, for every p there is a value q, $q > p$, satisfying:

$$R\left(c^p, m\right) = R\left(c^q, m\right)$$

which can always be expressed as:

$$R\left(c^{p}, m\right) = R\left(c^{q-p}c^{p}, m\right) = R\left(c^{q-p}, m\right)R\left(c^{p}, m\right)$$

If c and m are relatively prime, $R(c^p, m) \neq 0$, and, using the cancellation law, the following will hold:

$$R\left(c^{q-p}, m\right) = 1$$

which can also be expressed as:

$$R\left(c^{q-p-1}c, m\right) = R\left(c^{q-p-1}, m\right)R(c, m) = 1$$

or equivalently:

$$\left|\frac{1}{c}\right|_{m} = R\left(c^{q-p-1}, m\right)$$

q.e.d.

It is also easily proved that $\exists \left|\frac{1}{c}\right|_m$ only if c and m are relatively prime. Obviously, if m is prime, for any $c \neq 0$ ($c \in CMm$), $\exists \left|\frac{1}{c}\right|_m$. If c has an inverse, it holds that:

$$R\left(c^{-1}, m\right)R(c, m) = 1 \tag{3.6}$$

As an example, Table 3.5 includes the inverses of every element in CM7 having one.

The division of a by c in CMm may be defined as:

$$\frac{a}{c} = a \times \left|\frac{1}{c}\right| m$$

Thus, the quotient will be defined only when c and m are relatively prime. Table 3.6 shows division in CM7; each cell includes the value of the quotient of the row and column coordinates. It is clear from this Table that division is not commutative.

Table 3.5 Inverses in CM7

	Inverse
1	1
2	4
3	5
4	2
5	3
6	6

Table 3.6 Division in CM7

:	1	2	3	4	5	6
1	1	4	5	2	3	6
2	2	1	3	4	6	5
3	3	5	1	6	2	4
4	4	2	6	1	5	3
5	5	6	4	3	1	2
6	6	3	2	5	4	1

While for addition, subtraction and multiplication there is a correspondence between integer operation and the corresponding operation in CMm, this is not generally true for division. For integers, division of D by d results in a quotient, c, and a remainder, r. Division in CMm, when defined, results only in the quotient, not a remainder, so a correspondence between integer division and division in CMm can only be established when the first is exact (null remainder); as an example, 24:6 = 4 has a null remainder, and would correspond in CM7 to 3:6 = 4. When the remainder is not null, integer division and division in CMm are not related; as an example, 5:2 = 6 in CM7, while using integers 5:2 results in quotient 2 and remainder 1.

For the computation of the inverse of an element it is useful to apply the following theorem, from Fermat (it is known as the Fermat Minor Theorem, in order to be distinguished from the famous Fermat Last Theorem).

Theorem 3.3 *If **p** is a prime number and **a** is a positive integer, with **p** and **a** relatively prime, then $R(a^p, p) = R(a, p)$.*

Proof Using induction, it is immediate to verify the theorem for $a = 0$ y $a = 1$; it is also easily proofed (expanding the binomial $(1 + 1)^p$) for $a = 2$. It is admitted that it holds for a generic a, and it is proofed for $a + 1$. In fact, expanding $(a + 1)^p$ yields to:

$$(a+1)^p = a^p + \binom{p}{1}a^{p-1} + \binom{p}{2}a^{p-2} + \cdots + \binom{p}{p-2}a^2 + \binom{p}{p-1}a + 1 \tag{3.7}$$

From this expansion, $R((a+1)^p, p)$ is equal to the addition of the residues generated by the n +1 terms of the second member of (3.7). All of the n – 1 central terms in expansion (3.7) are multiples of p and will result in a modulo p null remainder. Thus, for obtaining $R((a+1)^p, p)$ would only remain the first and last terms in (3.7). Having in mind that $R(a^p, p) = R(a, p)$ is supposed to hold, then:

$$R\big((a+1)^p, p\big) = R\big(a^p, p\big) + R(1, p) = R(a, p) + R(1, p) = R(a+1, p)$$

q.e.d.

The previous theorem can also be expressed as:

$$R\left(a^{p-1}, p\right) = 1 \tag{3.8}$$

In fact, under the condition of Theorem 3.3, $\exists a^{-1} \neq 0$ and (3.6) holds. Multiplying by $R(a^{-1}, m)$ both members of $R(a^p, p)= R(a, p)$ yields to:

$$R\left(a^{-1}, p\right)R\left(a^{p}, p\right) = R\left(a^{-1}, p\right)R(a, p)$$

from which (3.8) is derived.

As a Corollary of (3.8), if the proper conditions are verified, then:

$$\frac{1}{a} = a^{-1} = a^{p-2} \tag{3.9}$$

In fact, $R(a^{p-1}, p)= R(aa^{p-2}, p) = 1$, from which (3.9) is derived. Thus, under the conditions of Theorem 3.3, (3.9) establishes a procedure for obtaining the inverse of a given element.

From the definitions and properties of the four operations (addition, subtraction, multiplication and division) it is clear that, if m is a prime number, then these operations are defined in CMm for all cases (but division by 0).

The set CMm (m prime), with modulo m addition and modulo m product, forms finite field or Galois field (see Appendix A). The number of elements in the finite field is known as the order of a finite field; in this case, the order is m.

It is easy to check that, when m is prime, the set of non-null elements in CMm (i.e., CMm* = CMm − {0}), with the modulo m product, as defined above, forms a multiplicative group [2]. As an example, for $m = 7$, CM7* = {0, 1, 2, 3, 4, 5, 6} − {0} = {1, 2, 3, 4, 5, 6}, whose product is shown in Table 3.4 suppressing the first row and first column, and it is immediate that it forms a multiplicative group.

Moreover, all elements in CMm* may be expressed as powers of some other element in CMm*. These elements, from which all other elements can be generated by successive exponentiation, are known as primitive elements or generating elements. In CM7* the primitive elements are 3 y and 5; and from Table 3.4 it is derived:

$$3^0 = 1; 3^1 = 3; 3^2 = 2; 3^3 = 6; 3^4 = 4; 3^5 = 5$$
$$5^0 = 1; 5^1 = 5; 5^2 = 4; 5^3 = 6; 5^4 = 2; 5^5 = 3$$

Successive powers of these elements generate the same elements previously generated. The exponents to use, in this example, are the elements of CM6; in general, the exponents will always be the elements of CM(m − 1).

Given a primitive element, p, in CMm*, for an element $a \in$ CMm* it holds $a = p^e$. Applying the definition of logarithm, the exponent e is also known as radix-p discrete logarithm of a, $\lg_p a$. It is evident that if $e <$ ord(p), the radix-p discrete logarithm of a is unique.

Using primitive elements, multiplication in CMm* can be transformed into an addition in CM($m-1$), in the same way than in ordinary arithmetic the logarithm of the product is the addition of logarithms. Thus, given any two elements in CMm*(a, $b \in$ CMm*) and a primitive element in CMm*, $p \in$ CMm*, then:

$$a = p^e, b = p^i;$$

so the product of these two elements will be

$$a \times b = p^e \times p^i = p^{e+i}$$

Thus, the product of a and b can be obtained by adding $e + i$.

For computing discrete logarithms in a multiplicative group with not too many elements, a table may be constructed, such as Table 3.7 for $m = 31$ y $p = 3$.

Using adequate logarithm tables, other operations can also carried out, as it is the case of the Zech logarithm tables, built as described in the following. Given a multiplicative group and a primitive element p, each value a is associated to b, such as $1 + p^a = p^b$. As an example, Table 3.8 shows the Zech logarithm table for $m =$

Table 3.7 Logarithms for $m = 31$ y $p = 3$

a	lg3***a***	***a***	lg3***a***	***a***	lg3***a***	***a***	lg3***a***
0	$-\infty$	8	12	16	6	24	13
1	0	9	2	17	7	25	10
2	24	10	14	18	26	26	5
3	1	11	23	19	4	27	3
4	18	12	19	20	8	28	16
5	20	13	11	21	29	29	9
6	25	14	22	22	17	30	15
7	28	15	21	23	27		

Table 3.8 Zech logarithms for $m = 31$ and $p = 3$

a	***b***	***a***	***b***	***a***	***b***	***a***	***b***
0	24	8	29	16	9	24	1
1	18	9	15	17	27	25	28
2	14	10	5	18	20	26	4
3	16	11	22	19	11	27	13
4	8	12	2	20	25	28	12
5	3	13	10	21	6	29	17
6	7	14	23	22	21	30	24
7	26	15	$-\infty$	23	19	$-\infty$	0

31 and $p = 3$.

Using the Zech logarithm table, the addition of two elements, p^m and p^n, with $m < n$, is:

$$p^m + p^n = p^m\left(1 + p^{n-m}\right) = p^{m+k}$$

here $1 + p^{n-m} = p^k$ is looked for in the Zech table.

When the number of elements is too high, a table is not viable, so the computation of the discrete logarithm of *a* would require the computation of the successive powers of the radix *b* and the comparison of each result with *a* until a match is found. If a large enough group is selected, none of these procedures for the computation of discrete logarithms is a feasible solution. As an example, there are more than 10^{75} elements in GF(2^{256}), and a computer able to carry on 10^{10} computations per second (each computation would comprise exponentiation and comparison) would require an average of more than 10^{58} years for computing a discrete logarithm. This situation, where given a value the computation of its discrete logarithm is almost impossible, is known as the problem of the discrete logarithm.

3.3 Integer Representation Using Residues

Given a set of *n* moduli, $\{m_1,\ldots, m_n\}$, any integer *N* can be represented by the *n* residues that it generates over the moduli set. As an example, given the moduli {2, 3, 5}; the integer 17 is represented as (1, 2, 2), since *R*(17, 2) = 1, *R*(17, 3) = 2, *R*(17, 5) = 2. It is evident that the representation for 17 is unique, but this same representation may also correspond to several numbers; specifically, 47, 77, 107, ..., hold the same representation than 17, as can be easily proved. Thus, given a moduli set, each representation corresponds to an infinite number of integers. The conditions under which this representation is of interest are specified in the following theorem:

Theorem 3.4 *Given a set of relatively prime moduli, $\{m_1, \ldots, m_n\}$, whose product is D (known as* ***dynamic range****):*

$$D = \Pi\, m_i, i = 1, \ldots, n$$

any two different integers, A and B (A < D, B < D), have different residue representations.

Proof The representations of *A* and *B* over the moduli set $\{m_1, \ldots, m_n\}$ will be [$R(A, m_1), \ldots, R(A, m_n)$] and [$R(B, m_1), \ldots, R(B, m_n)$], so from (3.2):

$$R(A, m_1) = A - p_1 \times m_1, \ldots, R(A, m_n) = A - p_n \times m_n$$

$$R(B, m_1) = B - q_1 \times m_1, \ldots, R(B, m_n) = B - q_n \times m_n$$

If both representations are equal, then

$$A - p_1 \times m_1 = B - q_1 \times m_1, \ldots, A - p_n \times m_n = B - q_n \times m_n$$

or

$$A - B = (p_1 - q_1) \times m_1, \ldots, A - B = (p_n - q_n) \times m_n$$

so $A - B$ is zero or a multiple of M, as opposed to the initial hypothesis, and both representations have to be different.

q.e.d.

From all above, each moduli set defines a different RNS, with its own dynamic range. As an example, the moduli set {2, 3, 5}, whose product is 30, allows representing uniquely any set of 30 consecutive integers, which may be from 0 to 29, as shown in Table 3.9.

Given the RNS with moduli $\{m_1, \ldots, m_n\}$, the representation of any integer X in this RNS will be noted as $x_1, \ldots, x_n$. It is evident that, from Theorem 3.1, the representation of $-X$ will be $\bar{x}_1, \ldots, \bar{x}_n$. As an example, Table 3.9 shows the representation of negative integers, and this can be checked. The RNS {2, 3, 5} can also be used for representing the 30 integers between −15 and +14, or between −14 and +15 (or between any other integers that bound 30 consecutive integers). Thus, it has to be clear for each RNS which range of values is in use. In the same way, if the limits of the range might be surpassed as the result of any operation, the error would not be detectable, since the resulting representation would keep corresponding to a value in the range.

In the RNS $\{m_1, \ldots, m_n\}$, the range of values for the different residues is $[0, m_1 - 1], \ldots, [0, m_n - 1]$. As it has been done for other numeric systems, redundancy

Table 3.9 RNS {2, 3, 5}

		Moduli					Moduli					Moduli		
+*N*	−*N*	2	3	5	+*N*	−*N*	2	3	5	+*N*	−*N*	2	3	5
0	−30	0	0	0	10	−20	0	1	0	20	−10	0	2	0
1	−29	1	1	1	11	−19	1	2	1	21	−9	1	0	1
2	−28	0	2	2	12	−18	0	0	2	22	−8	0	1	2
3	−27	1	0	3	13	−17	1	1	3	23	−7	1	2	3
4	−26	0	1	4	14	−16	0	2	4	24	−6	0	0	4
5	−25	1	2	0	15	−15	1	0	0	25	−5	1	1	0
6	−24	0	0	1	16	−14	0	1	1	26	−4	0	2	1
7	−23	1	1	2	17	−13	1	2	2	27	−3	1	0	2
8	−22	0	2	3	18	−12	0	0	3	28	−2	0	1	3
9	−21	1	0	4	19	−11	1	1	4	29	−1	1	2	4

can be used in some of the residues; for the modulus m_i the range [0, e_i] can be used, with $e_i \geq m_i - 1$. As an example, for the modulus 5, the range [0, 7] might be used, so two representations would be available for 0 (0 and 5), for 1 (1 and 6) and for 2 (2 and 7). Even though this double representation may seem inconvenient, it will enhance the implementation of some operations, as it will be shown later.

3.4 Arithmetic Operations Using Residues

In this section it will be shown how to add, subtract and multiply integers represented by their residues, as well as the results. It is supposed that the RNS is defined by the set of relatively prime moduli $\{m_1, \ldots, m_n\}$, with dynamic range $D = \Pi m_i$, $i = 1, \ldots, n$, and that any two integers X and Y are represented in this RNS as $(x_1, \ldots, x_n)$, $(y_1, \ldots, y_n)$, respectively. For an operation $X \square Y$ (where $\square$ is either addition, subtraction or multiplication), the representation in this RNS of $X \square Y$, represented as $R(X \square Y, D)$, is $\{R(X \square Y, m_1), \ldots, R(X \square Y, m_n)\}$, and it will be proofed below that $R(X \square Y, m_i)) = R(x_i \square y_i, m_i)$, $i = 1, \ldots, n$. Thus, the global computation is reduced to the computation over each individual modulus, which requires the following theorem.

Theorem 3.5 $R(X \square Y, D) = R(x_1 \square y_1, m_1), \ldots, R(x_n \square y_n, m_n)$

Proof The proof of this theorem just requires equalities (3.4) and (3.5). By definition:

$$R(X \square Y, D) = \{R(X \square Y, m_1), \ldots, R(X \square Y, m_n)\}$$

Applying (3.4) and (3.5) to each reminder:

$$R(X \square Y, m_i) = R(R(X, m_i) \square R(Y, m_i), m_i) = R(x_i \square y_i, m_i)$$

q.e.d.

As an example, in the RNS {2, 3, 5}

$$\begin{array}{rcc} 11 & \rightarrow & 1, 2, 1 \\ +17 & \rightarrow & 1, 2, 2 \\ \hline 28 & \rightarrow & 0, 1, 3 \end{array} \qquad \begin{array}{rcc} 7 & \rightarrow & 1, 1, 2 \\ \text{x}4 & \rightarrow & 0, 1, 4 \\ \hline 28 & \rightarrow & 0, 1, 3 \end{array}$$

Addition (or subtraction or multiplication) is carried out in this RNS with three parallel additions (or subtraction or multiplications), corresponding to each modulus. In general, there will be as many parallel operations as moduli in the RNS. The main advantage of all this is that the parallel operations are mutually independent, and there is no carry propagation between moduli. Therefore, the global speed of the operation will only depend on the speed of the operation over each one of the moduli, and will be dictated by the slowest one; it will be thus advisable to the operation speeds over the different moduli to be as close as possible. Obviously, there may be carry propagating within each modulus operation.

In case the result of the operation exceeds the dynamic range *D*, this result obtained from this form of operation will be incorrect. Thus, it is necessary to guarantee that the results from the operations are going to be maintained within the dynamic range; one way of achieving this is to use a large enough dynamic range, analyzing in detail the most unfavorable cases.

3.5 Mixed Radix System Associated to Each RNS

Given the RNS with moduli set $\{m_1, \ldots, m_n\}$, the mixed radix system $\{m_1, \ldots, m_n\}$ can be associated to it (see Sect. 1.3.2), with the following weights in the mixed radix system:

$$p_1 = 1, \quad p_2 = p_1 m_1, \ldots, p_i = p_{i-1} m_{i-1}, \ldots, p_n = p_{n-1} m_{n-1}$$

i.e.,

$$p_1 = 1, \quad p_2 = m_1, \quad p_3 = m_1 m_2, \ldots, p_n = \sum_{i=1}^{n-1} m_i$$

So

$$\begin{aligned} X &= a_n p_n + \cdots + a_2 p_2 + a_1 p_1 \\ &= a_n \sum_{i=1}^{n-1} m_i + a_{n-1} \sum_{i=1}^{n-2} m_i + \cdots + a_2 m_1 + a_1 \end{aligned} \tag{3.10}$$

This association of the mixed radix system to each RNS makes sense if the change from one representation to another is easy. In the following, both transformations will be shown.

Given the representation of *X*, $(x_1, \ldots, x_n)$, in the RNS with moduli $\{m_1, \ldots, m_n\}$, it is necessary to find its representation in the mixed radix system $\{m_1, \ldots, m_n\}$; i.e., to find the coefficients $a_1, \ldots, a_n$ as a function of $x_1, \ldots, x_n$. Since all the weights p_i, but the first, are multiples of m_1, then $R(X, m_1) = a_1$. But, by definition, $R(X, m_1) = x_1$ and $a_1 = x_1$.

For computing a_2, expression (3.10) can be transformed by subtracting a_1 from both members of the equality, and dividing them by m_1. It is clear that:

$$a_2 = R\left(\frac{X - a_1}{m_1}, m_2\right) \tag{3.11}$$

If the transformations in (3.11) are carried out with the representation of X in the associated RNS, a_2 will be the transformed remainder corresponding to m_2.

The rest of coefficients can be computed by iteratively repeating this process. It is obvious that if

$$X_{i+1} = \frac{X_i - a_i}{m_i}$$

is defined, with $X_1 = X$, then

$$a_i = R(X_i, m_i)$$

Thus, the algorithm for transforming from the RNS with moduli $\{m_1, ..., m_n\}$ to the mixed radix system $\{m_1, ..., m_n\}$ requires $n - 1$ iterations, each one with a subtraction and a multiplication by the multiplicative inverse of the corresponding modulus m_i; both subtraction and multiplication in each iteration are computed over the RNS $\{m_1, ..., m_n\}$. As an example, 23 in the RNS {2, 3, 5} is represented as (1, 2, 3). For obtaining its representation in the mixed radix system {2, 3, 5}, $a_1 = 1$. Subtracting (1, 1, 1) from (1, 2, 3) in the RNS {2, 3, 5}, the result is (0, 1, 2). It is now required to divide by 2, i.e., multiply by the multiplicative inverse of 2 in each modulus. In particular:

$$\left|\frac{1}{2}\right|_3 = 2, \qquad \left|\frac{1}{2}\right|_5 = 3$$

and thus it results (0, 2, 1) for the representation in the RNS {2, 3, 5} of the transformed X. In this way, $a_2 = 2$. Subtracting this value 2 in the corresponding positions yields to (0, 2, 1) – (0, 2, 2) = (0, 0, 4). Multipliying (0, 0, 4) by $\left|\frac{1}{3}\right|_5 = 2$ leads to (0, 0, 3) and $a_3 = 3$. The representation of 23 in the mixed radix system is (a_3, a_2, a_1) = (3, 2, 1).

For the opposite transformation, given the representation of X, $(a_1, ..., a_n)$, in the mixed radix system $\{m_1, ..., m_n\}$, it is necessary to find its representation in the RNS with moduli $\{m_1, ..., m_n\}$, i.e., to find the coefficients $x_1, ..., x_n$ as a function of $a_1, ..., a_n$. For this, let's remind the expression as products of the weights p_i. All the terms in (3.10) but the least significant are multiples of m_1. Thus, $R(X, m_1) = a_1$. But, by definition, $R(X, m_1) = x_1$, i.e., $x_1 = a_1$. In the same way, all the terms but the least significant two are multiples of m_2, so $x_2 = R(a_2 m_1 + a_1, m_2)$. In general:

$$x_i = R\left(\sum_{j=1}^{i} a_j p_j,\ m_i\right)$$

As an example, in the mixed radix system {2, 3, 5}, we have the following weights: $p_1 = 1$, $p_2 = m_1 = 2$, $p_3 = m_1 m_2 = 2 \times 3 = 6$; as it was obtained above, in this system $23 = 3 \times 6 + 2 \times 2 + 1$. For transforming into the RNS {2, 3, 5} we have: $x_1 = a_1 = 1$; $x_2 = \text{R}(2 \times 2 + 1, 3) = 2$; $x_3 = \text{R}(3 \times 6 + 2 \times 2 + 1, 5) = 3$.

3.6 Moduli Selection

The first issue to solve when a RNS is going to be used is to decide which range of values has to be covered. This dynamic range D will evidently depend on the application in which the RNS will be used, and has to make that no incorrect results are generated by some partial result going beyond D (it will be noted as overflow); this can be guaranteed choosing a large enough D or by scaling the partial results when overflow is possible.

After fixing the value of D, it is necessary to decide how many moduli m_i are going to be used, and which specific values they are going to have, so $D \leq \Pi m_i$, i =1, ..., n.

Regarding the number of moduli, if only a few are chosen and D is large, the moduli will have large values. As a consequence, the time required for arithmetic operations may be long and no noticeable advantage is derived from the use of the RNS. If many moduli are used, there will be also an excessive number of channels that can delay the response of the system or require more hardware; specially, conversion processes are slower as more moduli are used, as it will be shown later. Thus, it is necessary to find a compromise that provides the maximum advantage in the use of the RNS.

A very important factor, which influences both the number of moduli to use and their values, is the convenience for the moduli having similar values; in particular, it is desirable that the different moduli make use of the same (or close) number of bits for representing their reminders. This will presumably make the computation times very similar for all channels, which will contribute to achieve the best system throughput. In order to get the maximum profit from the coding possibilities of the k bits assigned to a given modulus, m_a, it is convenient to make $2^k - m_a$ as small as possible.

As an example, if the selected dynamic range requires 40 bits (i.e., $D \leq 2^{40}$) and it is intended to use 6-bit moduli, then 7 moduli have to be used, provided that there are 7 relatively prime 6-bit integers covering this dynamic range.

Once the number of bits for each modulus has been fixed, it is necessary to select the moduli themselves. If n bits are going to be used for representing residues, the largest modulus that can be selected is 2^n, which is always and advisable choice since all arithmetic operations on this modulus will be easily implemented. Another recommended modulus is $2^n - 1$, which is always relatively prime to 2^n and leads again to an easy implementation of arithmetic operations. Modulus $2^n + 1$ is also often used, since arithmetic operations may be easily implemented on it. The rest of moduli to use will be chosen as the largest possible relatively prime to the previously selected, until the desired range is covered. As an example, continuing with the selection of seven moduli whose residues can be represented in six bits, it may start selecting $2^6 = 64$ and $2^6 - 1 = 63 = 3^2 \times 7$. If 65 is discarded as 7 bits would be required for codification, the next selectable moduli are 61 and 59, which are prime; $57 = 3 \times 19$ cannot be selected, since it is not relatively prime to 63; $55 = 5 \times 11$ can be selected, as well as 53 and 47, which are prime; thus, the seven moduli may be $\{47, 53, 55, 59, 61, 63, 64\}$. It is easily proved that $\Pi m_i \geq 2^{40}$, so the selection is valid. $2^5 - 1 = 31$ can also be used, instead of 47, since arithmetic operation implementation is expected to be simpler for 31 than for 47; in this case, the seven moduli may be $\{31, 53, 55, 59, 61, 63, 64\}$. It is easily proved that $\Pi m_i \geq 2^{40}$ once more, so this selection is also valid.

3.7 Conversions

It will usually be required to convert data from positional systems with radix b (2 or 10, generally) to RNS, and the other way around. In the following, conversions in both ways will be considered.

3.7.1 *From Positional Notation to RNS*

Given the RNS defined by the set $\{m_1, ..., m_n\}$ of relatively prime moduli, with dynamic range $D = \Pi m_i$, $i = 1, ..., n$, and an integer E expressed in a positional system with base b, $0 \leq E < D$, for obtaining the representation of E in this RNS it is just required to obtain the remainders $r_i = E \bmod m_i$, $i = 1, ..., n$. Thus, n modular reductions are necessary, which may be achieved with a division with only its remainder being used, or applying any of the specific procedures detailed in Sects. 1.2.4 (precomputation of the remainders of the different powers of b), 2.6.4 (multiplicative modular reduction) and 2.6.7 (Barret modular reduction).

As an example, given the RNS {2, 3, 5} and numbers in the binary system, the dynamic range is [0, 29]; numbers in this range require 5 bits for their binay expression. Assuming $N = a_4a_3a_2a_1a_0$, it is immediate that:

$$\mathrm{R}(N, 2) = a_0$$

since the remaining positions corresponds to multiples of 2.

$$\mathrm{R}(N, 3) = \mathrm{R}(a_4 + 2a_3 + a_2 + 2a_1 + a_0, 3)$$

since R(16, 3) = 1, R(8, 3) = 2, R(4, 3) = 1, R(2, 3) = 2, R(1, 3) = 1.

$$\mathrm{R}(N, 5) = R(a_4 + 3a_3 + 4a_2 + 2a_1 + a_0, 5)$$

since R(16, 5) = 1, R(8, 5) = 3, R(4, 5) = 4, R(2, 5) = 2, R(1, 5) = 1.

It is obvious that, for power-of-2 moduli, $M = 2^n$, modular reduction of binary numbers is direct: it just consists of the n less significant bits of the number to be reduced.

For $2^{n-1} \leq M < 2^n$, when two binary integers lesser than M are added modulo M, their addition is represented with $n + 1$ bits. In this case, a reducer from $n + 1$ bits to n bits should be use. In the same way, when they are multiplied, the reducer has to be from $2n$ to n bits. In the following these two types of reducers will be studied.

For reducing from $n + 1$ to n bits it is enough to implement the reduction of 2^n, as it is done in the following example for $M = 251$, which is the largest prime that can be represented with 8 bits.

Example 3.1 Design a modular arithmetic circuit for reducing from 9 to 8 bits, with $M = 251$.

Let it be $X = x_8 2^8 + \cdots + x_1 2 + x_0$ the integer to be reduced, so $R = r_7 2^7 + \cdots + r_1 2 + r_0 = X \text{mod} 251$ has to be obtained. Since $2^8 \text{mod} 251 = 5 = 4 + 1$, the reducer may be implemented with an adder. In fact:

$$\begin{aligned}&\left(x_8 2^8 + x_7 2^7 + x_6 2^6 + x_5 2^5 + x_4 2^4 + x_3 2^3 + x_2 2^2 + x_1 2 + x_0\right) \mathbf{mod}\, 251\\&= x_7 2^7 + x_6 2^6 + x_5 2^5 + x_4 2^4 + x_3 2^3 + (x_2 + x_8) 2^2 + x_1 2 + x_0 + x_8\end{aligned}$$

Thus, with an adder network as this depicted in Fig. 3.1a, the modular reduction is achieved. It is easy to check that multiplicative modular reduction yields the same solution.■

Reduction from $2n$ to n bits can be implemented with any of the procedures cited above (see Sect. 2.5.4). The following is an example.

Example 3.2 Design a modular arithmetic circuit for reducing from 16 to 8 bits, with $M = 251$.

Let it be $X = x_{15} 2^{15} + \cdots + x_1 2 + x_0$ the integer to be reduced, so $R = r_7 2^7 + \cdots + r_1 2 + r_0 = X \text{mod} 251$ has to be obtained. Since $2^8 \text{mod} 251 = 5 = 4 + 1$ and $2^9 \text{mod} 251 = 10 = 8 + 2$, the reducer may be implemented using an adequate adder. In fact:

$$\begin{aligned}x_{15} 2^{15} + \cdots + x_{12} + x_0 &= \left(x_{15} 2^7 + \cdots + x_9 2 + x_8\right) 2^8\\&+ x7 2^7 + \cdots + x_1 2 + x_0 = A 2^8 + B\left(A 2^8 + B\right) \mathbf{mod}\, 251\\&= \{A(4 + 1) + B\} \mathbf{mod}\, 251\\&= \left\{(A + B) \mathbf{mod}\, 251 + A 2^2 \mathbf{mod}\, 251\right\} \mathbf{mod}\, 251\end{aligned}$$

If the number to be reduced is the result of a multiplication, the most extreme case that may arise for $M = 251$ is $250 \times 250 = 62{,}500$. In that case, it is easily checked that the last modular reduction is not required, leading to:

$$\left(A 2^8 + B\right) \mathbf{mod}\, 251 = \{A(4 + 1) + B\} \mathbf{mod}\, 251$$

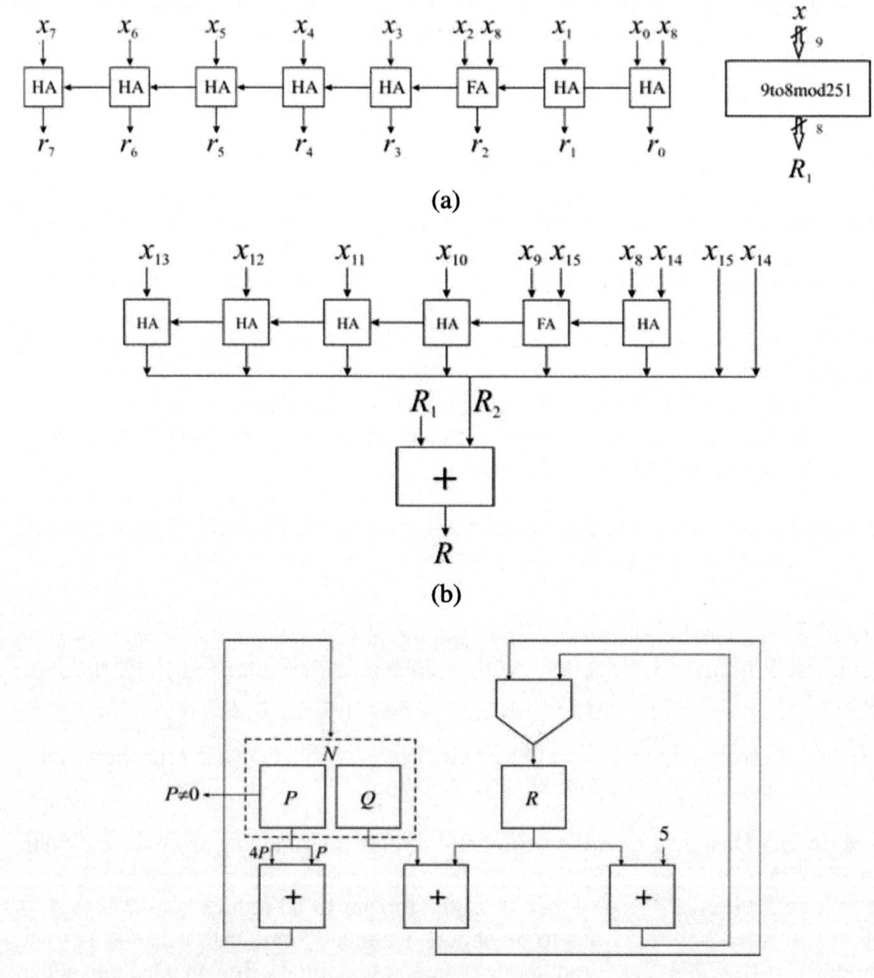

Step	Operation	P	Q	R
0	R ← 0	1111 0100	0010 0100	0000 0000
1	R ← R + Q, N ← 5P	0000 0100	1100 0100	0010 0100
2	R ← R + Q, N ← 5P	0000 0000	0001 0100	1110 1000
3	R ← R + Q	0000 0000	0001 0100	1111 1100
4	R ← R − M	0000 0000	0001 0100	0000 0001

(d)

Fig. 3.1 Reducer circuits: a From 9 to 8 bits. b From 16 to 8 bits using adders. c From 16 to 8 bits using multiplicative reduction. d Multiplicative reduction example

$$= (A + B)\,\mathbf{mod}\,251 + A2^2\,\mathbf{mod}\,251$$

After adding A and B, $R_1 = (A + B)\text{mod}251$ can be computed with the circuit of Fig. 3.1a (Example 3.1). On the other hand:

$$R_2 = A2^2\,\mathbf{mod}\,251 = \left(x_{15}2^9 + x_{14}2^8 + \cdots + x_9 2^3 + x_8 2^2\right)\mathbf{mod}\,251$$
$$= x_{13}2^7 + x_{12}2^6 + x_{11}2^5 + x_{10}2^4 + (x_9 + x_{15})2^3$$
$$+ (x_8 + x_{14})2^2 \cdots + x_{15}2 + x_{14}$$

Thus, a circuit as this in Fig. 3.1b can achieve modular reduction from 16 to 8 bits.

A second implementation may use multiplicative modular reduction, which can be carried out as follows. Since $251 = 256 - 5 = 2^8 - 5 = 2^{\,k} - a$ ($k = 8$, $a = 5$), the 16-bit register N will be the concatenation of two 8-bit registers, P and Q. An auxiliary register R is used for storing intermediate results and that initially has to be set to zero. R and Q have to be added in each iteration, and P has to be multiplied by 5 (since $5 = 4 + 1$, multiplying by 5 is achieved adding P and 4P). The number N to be reduced is initially introduced in the register N. With all this, the circuit in Fig. 3.1c, similar to that in Fig. 2.17a, can be used as processing unit for this computation. For computing $R - M$, an adder calculating $R + 2^n - M = R + 2^8 - 251 = R + 5$ can be used; if $R > M$, this adder generates a carry.

As an example, Fig. 3.1d shows the contents of the different registers for $N = 62{,}500 = 1111\ 0100\ 0010\ 0100$.

Using that:

$$\frac{1}{251} = 0'00000001000001010001\ldots_2$$

the desired remainder may be computed in a third way as:

$$N\,\mathbf{mod}\,251 = N - \left\lfloor N\frac{1}{251}\right\rfloor 251$$

Nonetheless, it is easy to conclude that the resulting circuit will be more complex than the preceding ones. ■

Modular reduction for $M = 2^n - 1$ can be easily implemented, since $2^n\text{mod}(2^n - 1) = 1$. As an example, for reducing a word of $2n$ bits, $A = A_H 2^n + A_L$, it is enough just to use that $A\text{mod}(2^n - 1) = (A_H + A_L)\text{mod}(2^n - 1)$. Thus, $S = A_H + A_L$ is computed, and in case $S \geq 2^n - 1$, $2^n - 1$ is subtracted from S. It will be shown later (Sect. 3.8.1) that this reduction can be carried out with a 1's complement adder.

Modular reduction for $M = 2^n + 1$ can also be easily implemented, since $2^n\text{mod}(2^n + 1) = -1$. As an example, reducing a word of 2^n bits, $A = A_H 2^n + A_L$, has to take into account that $A\text{mod}(2^n + 1) = (A_L - A_H)\text{mod}(2^n + 1)$. Thus, $R = A_L - A_H$ is computed and, in case it is negative, $2^n + 1$ is added.

3.7.2 From RNS to Positional Notation

The inverse conversion, from RNS to a positional system with base *b*, can be carried out making use of the following theorem, known as the Chinese remainder theorem [5].

Theorem 3.6 *Given the representation* $(x_1, \ldots, x_n)$ *of an integer* X, $0 \leq X < D$, *in the* ***RNS*** *defined by the set* $\{m_1, \ldots, m_n\}$ *of relatively prime, with dynamic range* $D = \Pi m_i$, $i = 1, \ldots, n$, *the value of* X *can be obtained as:*

$$X = R\left(\Sigma x_i M_i I_i, D\right),$$

where $M_i = D/m_i$, *and* I_i *is the multiplicative inverse of* M_i *in* $\boldsymbol{CMm_i}$.

Proof Prior to the proof itself, given any modulus m and one of its multiples, *km*, it is immediate that, for any integer *A* the following expression holds:

$$R(R(A; km), m) = R(A, m) \tag{3.12}$$

Proof of $X = R(\Sigma x_i M_i I_i, D)$ parts from the premise that the representation of X in the RNS $\{m_1, \ldots, m_n\}$ is unique. Thus, if $R(R(\Sigma x_i M_i I_i, D), m_j) = x_j, j = 1, \ldots, n$, then $X = R(\Sigma x_i M_i I_i, D)$. Since $D = k_i m_i$, $i = 1, \ldots, n$, applying (3.12) leads to:

$$R\left(R\left(\Sigma x_i M_i I_i, D\right), m_j\right) = R\left(\Sigma x_i M_i I_i, m_j\right)$$

All terms in $\Sigma x_i M_i I_i$ but the *j*-th are multiples of m_j. In this way:

$$R\left(\Sigma x_i M_i I_i, m_j\right) = R\left(x_j M_j I_j, m_j\right) \quad j = 1, \ldots, n$$

By definition $M_j I_j = 1$, since I_j is the multiplicative inverse of M_j in CMm_j. Thus,

$$R\left(x_j M_j I_j, m_j\right) = x_j$$

q.e.d.

As an example, the positive integer represented by (1, 1, 3) in the RNS {2, 3, 5} is computed as:

$$D = 30, M_1 = 15, M_2 = 10, M_3 = 6, I_l = 1, I_2 = 1, I_3 = 1;$$

$$\begin{aligned} X &= R(\Sigma x_i M_i I_i, M) \\ &= \mathrm{R}(1 \times 15 \times 1 + 1 \times 10 \times 1 + 3 \times 6 \times 1, 30) \\ &= \mathrm{R}(43, 30) = 13 \end{aligned}$$

Using the Chinese Remainder Theorem requires a modulo D addition, which is usually a large value requiring a lot of hardware. Another form of conversion from RNS to positional notation, which skips this disadvantage, consists on using the mixed-radix system as an intermediate stage. Thus, given a number represented in an RNS, it is transformed to its representation in the mixed-radix systems associated to the RNS; this transformation just requires modulo m_i additions. After that, the mixed-radix representation is transformed into the positional notation for the desired base.

3.8 Modular Circuits

As it has been proved above, the different operation in an RNS are translated to the different moduli. Thus, modular circuits are of interest for RNS implementations. The main circuits for operating modulo M are described in this section. The operands are supposed to be in the adequate range, i.e., in the set $C_M = \{0, 1, 2, \ldots, M - 1\}$.

If M is prime, the set $C_M = \{0, 1, 2, \ldots, M- 1\}$ with addition and product operations defines a Galois Field GF(M), as it is detailed in Appendix A. In this way, the circuits below will be also used for operating with Galois Fields in the next chapters.

3.8.1 Addition and Subtraction

Given $x, y \in C_M$, a combinational solution may be used for the modular addition $(x + y)\mathrm{mod}M$, as in the following example. This solution is recommendable or practical for small values of M. The corresponding circuits may be implemented with logic gates or a ROM. Obviously, the simplest case is for $M = 2^q$; in this case, binary adders do suffice.

Example 3.3 Given x and y, as 3-bit binary numbers, design a circuit for computing $(x + y)\mathrm{mod}5$, assuming $x, y < 5$.

Table 3.10 Modulo 5 addition

+	**000**	**001**	**010**	**011**	**100**
000	000	001	010	011	100
001	001	010	011	100	000
010	010	011	100	000	001
011	011	100	000	001	010
100	100	000	001	010	011

Table 3.10 shows the addition $s = (x + y)\text{mod}5$. Assuming $s = s_2s_1s_0$, the corresponding combinational functions, which include a lot of don't cares, are

$$s_2 = \sum \text{m}(4, 11, 18, 25, 32) + \text{D}$$
$$s_1 = \sum \text{m}(2, 3, 9, 10, 16, 17, 24, 28, 35, 36) + \text{D}$$
$$s_0 = \sum \text{m}(3, 8, 10, 17, 20, 24, 27, 34, 36) + \text{D}$$

$$\text{D} = \sum \text{m}(5, 6, 7, 13, 14, 15, 21, 22, 23, 29, 30, 31, 37, 38, 39, 40, 41, 42, 43, 4445, 46, 47, 48, 49, 50, 51, 52, 53, 54, 55, 56, 57, 58, 59, 60, 61, 62, 63)$$

The AND-OR synthesis may be:

$$s_2 = \bar{x}_2\bar{x}_1\bar{x}_0y_2 + x_2\bar{y}_2\bar{y}_1\bar{y}_0 + x_1\bar{x}_0y_1\bar{y}_0 + x_1x_0\bar{y}_1y_0 + \bar{x}_1x_0y_1y_0$$
$$s_1 = x_2y_2 + \bar{x}_1\bar{x}_0y_1y_0 + x_1x_0\bar{y}_1\bar{y}_0 + x_1\bar{x}_0\bar{y}_2\bar{y}_1 + \bar{x}_1x_0\bar{y}_1y_0 + \bar{x}_2\bar{x}_1y_1\bar{y}_0$$
$$s_0 = x_2y_2 + x_2y_1\bar{y}_0 + x_1\bar{x}_0y_2 + x_0\bar{y}_2\bar{y}_1\bar{y}_0 + \bar{x}_2\bar{x}_1\bar{x}_0y_0 + \bar{x}_1x_0y_1\bar{y}_0 + x_1\bar{x}_0\bar{y}_1y_0 + x_1x_0y_1y_0$$

■

Given $2^{n-1} < M < 2^n$, a modular adder for any M can be implemented with two n-bit binary adders, using a redundant representation that utilizes the 2^n possible combinations, as illustrated in Fig. 3.2a. When the first adder, S_1, does not generate a carry, the output of the second adder, S_2, is the same as that of the first adder; when a carry is generated in the first adder, the sum $S_1 = S_0 + 2^n$ has to be corrected, so it is transformed into $S_2 = S_1 - M = S_0 + 2^n - M = S_0 + (2^n - M)$, which is what the second adder in Fig. 3.2a carries out for $M = 10$ as example ($16 - 10 = 6 = 4 + 2$).

For the addition with a non-redundant representation, a binary adder can be used for computing $x + y$, followed by a modular reducer, which in this case has to reduce from $n + 1$ to n bits. However, the use of two adders helps to skip the modular reducer, as it is shown below. In fact, for the modular addition $(x + y)$ mod$M$ it holds:

$$(x + y) \textbf{ mod } M = x + y \quad \text{if } x + y < M$$

or

$$(x + y) \textbf{ mod } M = x + y - M \quad \text{if } x + y \geq M$$

Thus, when x and y are n-bit integers, it is enough to compute with two n-bit binary adders, S_1 and S_2, the two possible results, $s_1 = x + y$ or $s_2 = x + y - M$, and select the proper one. This selection would require a comparison, making thus the global circuit more complex. However, it is easy to conclude that, for simplifying this

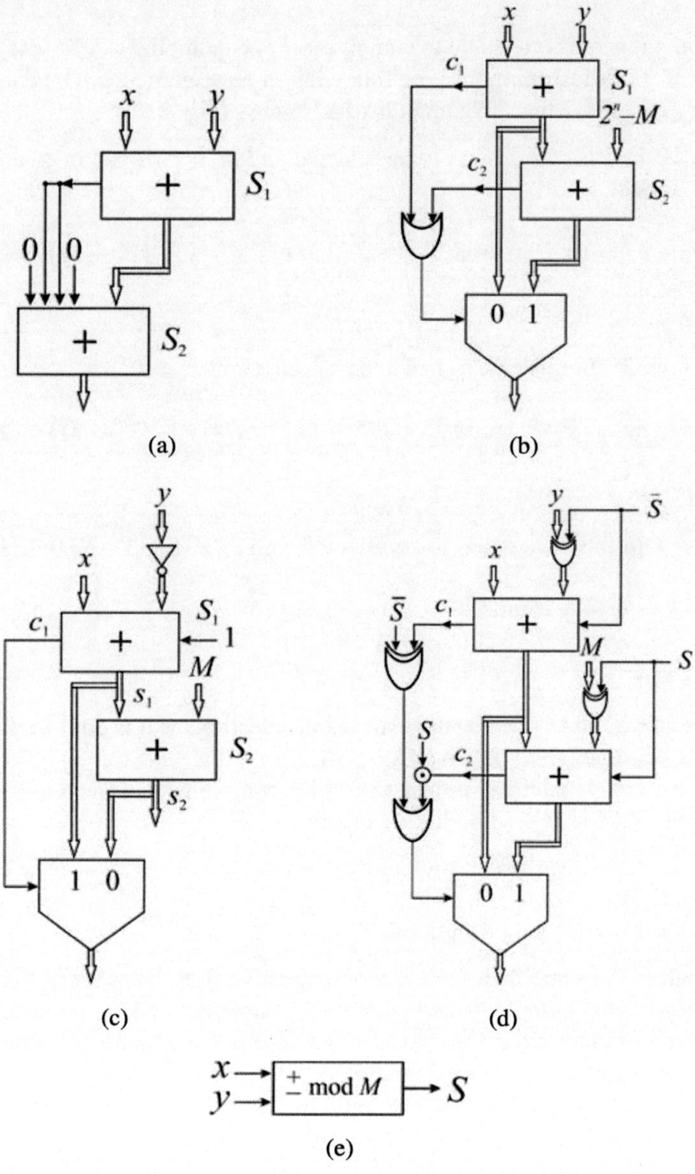

Fig. 3.2 a Redundant modular adder. b Non-redundant modular adder. c Modular subtracter. d Modular adder/subtracter. e Symbol

comparison, it is more convenient to compute $s_2 = (x + y)\text{mod}2^n + (2^n - M)$ instead of $x + y - M$. For illustrating this, the following three scenarios will be considered, assuming c_1 and c_2 are the carry outputs of S_1 and S_2 (Fig. 3.2b):

- $x + y < M$. In this case, s_1 has to be selected and $c_1 = 0$. It is immediate that $c_2 = 0$ too; in fact:

$$s_2 + c_2 = (x + y)\,\mathbf{mod}\,2^n + \left(2^n - M\right) = x + y + \left(2^n - M\right) < 2^n$$

  resulting $s_2 = x + y - M + 2^n$ and $c_2 = 0$.

- $M \leq x + y < 2^n$. In this case s_2 has to be selected and $c_1 = 0$, but $c_2 = 1$; in fact,

$$s_2 + c_2 = (x + y)\,\mathbf{mod}\,2^n + \left(2^n - M\right) = x + y + \left(2^n - M\right) \geq 2^n$$

  resulting $s_2 = x + y - M$; $c_2 = 1$.

- $2^n \leq x + y$. In this case s_2 has to be selected, and $c_1 = 1$ and $c_2 = 0$; in fact,

$$s_2 + c_2 = (x + y)\,\mathbf{mod}\,2^n + \left(2^n - M\right) = \left(x + y - 2^n\right) + \left(2^n - M\right) < 2^n$$

  resulting $s_2 = x + y - M$; $c_2 = 0$.

Thus, carries c_1 and c_2 can be used for result selection, as it is done in the circuit of Fig. 3.2b. It is enough to obtain OR(c_1, c_2).

For $M = 2^n - 1$, addition is just as for 1's complement. Concretely, the two possible scenarios for $(x + y)\text{mod}(2^n - 1)$ are:

a) if $x + y \geq 2^n - 1$, then $(x + y)\text{mod}(2^n - 1) = x + y - (2^n - 1) = (x + y + 1)\text{mod}2^n$
b) if $x + y < 2^n - 1$, then $(x + y)\text{mod}(2^n - 1) = x + y$

This result is the same than for 1's complement addition (see Sects. 1.4.2.4 and 2.3), and thus it can be implemented with a 1's complement adder. In this case, it is clear that two representations for zero are used, as it is the case for 1's complement. If a unique representation for zero is required, it is enough to detect the result $x + y = 11\ldots11$. Obviously, the modular reduction for $M = 2^n - 1$, just an addition (see Sect. 3.7.1), can be implemented with a 1's complement adder.

Regarding subtraction, a combinational solution may be used for $(x - y)\text{mod}M$, as illustrated in the following example. This solution may be recommendable for not too large values of M. The corresponding circuits can be implemented with logic gates, or a ROM. If $M = 2^q$, they will be binary subtracters.

Table 3.11 Modulo 5 subtraction

–	000	001	010	011	100
000	000	100	011	010	001
001	001	000	100	011	010
010	010	001	000	100	011
011	011	010	001	000	100
100	100	011	010	001	000

Example 3.4 Given x and y, as 3-bit binary numbers, design a circuit for computing $(x - y)\text{mod}5$, assuming $x, y < 5$.

Table 3.11 shows the subtraction of the column element from the row element, $d = (x - y)\text{mod}5$. Assuming $d = d_2d_1d_0$, the corresponding combinational functions, which include a lot of don't cares, are

$$d_2 = \sum \text{m}(1, 10, 19, 28, 32) + \text{D}$$

$$d_1 = \sum \text{m}(2, 3, 11, 12, 16, 20, 24, 25, 33, 34) + \text{D}$$

$$d_0 = \sum \text{m}(2, 4, 8, 11, 17, 20, 24, 26, 33, 35) + \text{D}$$

$$\text{D} = \sum \text{m}(5, 6, 7, 13, 14, 15, 21, 22, 23, 29, 30, 31,$$
$$37, 38, 39, 40, 41, 42, 43, 44, 45, 46, 47, 48, 49, 50,$$
$$51, 52, 53, 54, 55, 56, 57, 58, 59, 60, 61, 62, 63)$$

The AND-OR synthesis may be:

$$d_2 = \bar{x}_2\bar{x}_1\bar{x}_0\bar{y}_1y_0 + \bar{x}_1x_0\bar{y}_1y_0 + x_1x_0y_2 + x_2\bar{y}_2\bar{y}_1\bar{y}_0 + x_1\bar{x}_0y_1y_0$$
$$d_1 = \bar{x}_2\bar{x}_1\bar{x}_0y_1 + \bar{x}_2\bar{x}_1y_1y_0 + \bar{x}_1x_0y_2 + x_1\bar{x}_0y_2 + x_1\bar{y}_2\bar{y}_1\bar{y}_0$$
$$+ x_1x_0\bar{y}_2\bar{y}_1 + x_2\bar{y}_1y_0 + x_2y_1\bar{y}_0$$
$$d_0 = \bar{x}_2\bar{x}_1\bar{x}_0y_1\bar{y}_0 + x_0\bar{y}_2\bar{y}_1\bar{y}_0 + x_1\bar{x}_0\bar{y}_1y_0 + \bar{x}_1x_0y_1y_0 + x_1x_0\bar{y}_1y_0$$
$$+ \bar{x}_2\bar{x}_0y_2 + x_2\bar{y}_2y_0$$

■

Adders can be also used for subtraction, as it is shown below. For the modular subtraction $r = (x - y)\text{mod}M$ two cases are possible:

(a) if $x \geq y$, then $r = x - y$
(b) if $x < y$, $x - y$ is negative and then $r = x - y + M$

It is easy to check that the circuit in Fig. 3.2c, with two n-bit binary adders and almost identical to that in Fig. 3.2b, provides the correct result in each case. In fact, adders S_1 and S_2 respectively compute:

$$s_1 + c_1 = x + (2^n - y)$$
$$s_2 + c_2 = x + (2^n - y) + M$$

If $x \geq y$, $s_I = x - y$ and $c_I = 1$, and s_1 is the correct result. If $x < y$, $c_1 = 0$, $s_2 = x - y + M$ and $c_2 = 1$, and s_2 is the correct result. It is evident that c_1 can select the correct result.

Combining both circuits for addition (Fig. 3.2b) and subtraction (Fig. 3.2c), in a similar fashion to the complement adder/subtracter (Fig. 2.6), the modular adder/subtracter in Fig. 3.2d is obtained. ***S*** signals if the operation to be carried out is addition (***S*** = 1) or subtraction (***S*** = 0). The symbol in Fig. 3.2e may be used for representing these operations, independently of how they are synthesized.

For $M = 2^n - 1$, the opposite or additive inverse of any element x is $-x = (2^n - 1) - x$, which is the 1's complement of x. Thus, the 1's complement adder/subtracter (Fig. 2.6f) can be used for adding and subtracting with $M = 2^n - 1$, having in this case two representations for zero. If a single representation for zero is desired, this same circuit can be utilized with a minor modification, as it was noted above for the addition.

3.8.2 *Multiplication and Division*

Given $x, y \in C_M$, assuming that both operands are available in parallel, a combinational solution can be proposed for the modular product $(x \cdot y)\text{mod}M$, as illustrated in the following example. This solution is recommendable for not too large values of M. The corresponding circuits can be implemented using logic gates, or a ROM. In the case of $M = 2^q$, these are binary multipliers.

Example 3.5 Given x and y, as 3-bit binary numbers, design a circuit for computing $(x \cdot y)\text{mod}5$, assuming $x, y < 5$.

Table 3.12 shows the product p = $(x \cdot y)\text{mod}5$. Assuming $p = p_2p_1p_0$, the corresponding combinational functions, which include a lot of don't cares, are

$$d_2 = \sum m(12, 18, 27, 33) + D$$

Table 3.12 Modulo 5 product

·	000	001	010	011	100
000	000	000	000	000	000
001	000	001	010	011	100
010	000	010	100	001	011
011	000	011	001	100	010
100	000	100	011	010	001

$$d_1 = \sum m(10, 11, 17, 20, 25, 28, 34, 35) + D$$
$$d_0 = \sum m(9, 11, 19, 20, 25, 26, 34, 36) + D$$

$$D = \sum m(5, 6, 7, 13, 14, 15, 21, 22, 23, 29, 30, 31, 37, 38, 39, 40, 41, 42, 43, 44, 45, 46, 47, 48, 49, 50, 51, 52, 53, 54, 55, 56, 57, 58, 59, 60, 61, 62, 63)$$

The AND-OR synthesis may be:

$$\begin{aligned} p_2 &= \bar{x}_1 x_0 y_2 + x_2 \bar{y}_1 y_0 + x_1 \bar{x}_0 y_1 \bar{y}_0 + x_1 x_0 y_1 y_0 \\ p_1 &= x_1 y_2 + x_2 y_1 + \bar{x}_1 x_0 y_1 + x_1 \bar{y}_1 y_0 \\ p_0 &= x_2 y_2 + x_2 y_1 \bar{y}_0 + x_1 \bar{x}_0 y_2 + \bar{x}_1 x_0 y_0 \\ &\quad + x_0 \bar{y}_1 y_0 + x_1 x_0 y_1 \bar{y}_0 + x_1 \bar{x}_0 y_1 y_0 \end{aligned}$$

■

For any M, a binary multiplier may be use, followed by an adequate reducer from $2n$ to n bits.

Any of the two solutions above for the product can be adapted to any form or data input, serial or parallel. When one of the data is available in parallel, and the other serially (or a shift register is used for this), the serial-parallel multiplier described in Sect. 2.4.2 may be used, with the corresponding modifications. The product $X \cdot Y$, with $X = x_{n-1}2^{n-1} + \cdots + x_0$, $Y = y_{n-1}2^{n-1} + \cdots + y_0$, can be developed as:

$$\begin{aligned} X \cdot Y &= X \cdot \left(y_{n-1}2^{n-1} + \ldots + y_0\right) \\ &= (\ldots((0 \cdot 2 + X \cdot y_{n-1}) \cdot 2 + X \cdot y_{n-2}) \cdot 2 + \ldots + X \cdot y_1) \cdot 2 + X \cdot y_0 \end{aligned}$$

Once arranged in this way, the computation requires n iterations, each one of them consisting of multiplying by 2 the previous result (initially 0) and adding $X \cdot y_i$. In this case, a modular reduction can be introduced at each iteration, from $n + 1$ to n bits. Thus, using a modular adder, as the one shown in Fig. 3.2b, the corresponding multiplier can be implemented, as illustrated in Fig. 3.3a.

For $M = 2^n - 1$, using all the previous results (Sects. 3.7.1 and 3.8.1) related to modular reduction, an specific multiplier circuit may be built folding a generic multiplier for the same number of bits, as it is done below for $n = 4$ ($M = 15$). The additions that provide the product $X \cdot Y$ ($X = x_3x_2x_1x_0$, $Y = y_3y_2y_1y_0$), shown in Fig. 3.3b, having in mind that $2^4 \text{mod} 15 = 1$, $2^5 \text{mod} 15 = 2$ y $2^6 \text{mod} 15 = 4$, can be folded as shown in Fig. 3.3c, just properly introducing carry propagation. Thus, a multiplier for $M = 15$ can be implemented using 12 full adders, as it is shown in Fig. 3.3d; the carry generated in each row is added to the next one in the less significant position, but the last one, which utilizes the idea of the 1's complement adder; with all this, modular reduction is carried out. A circuit with $n(n - 1)$ full

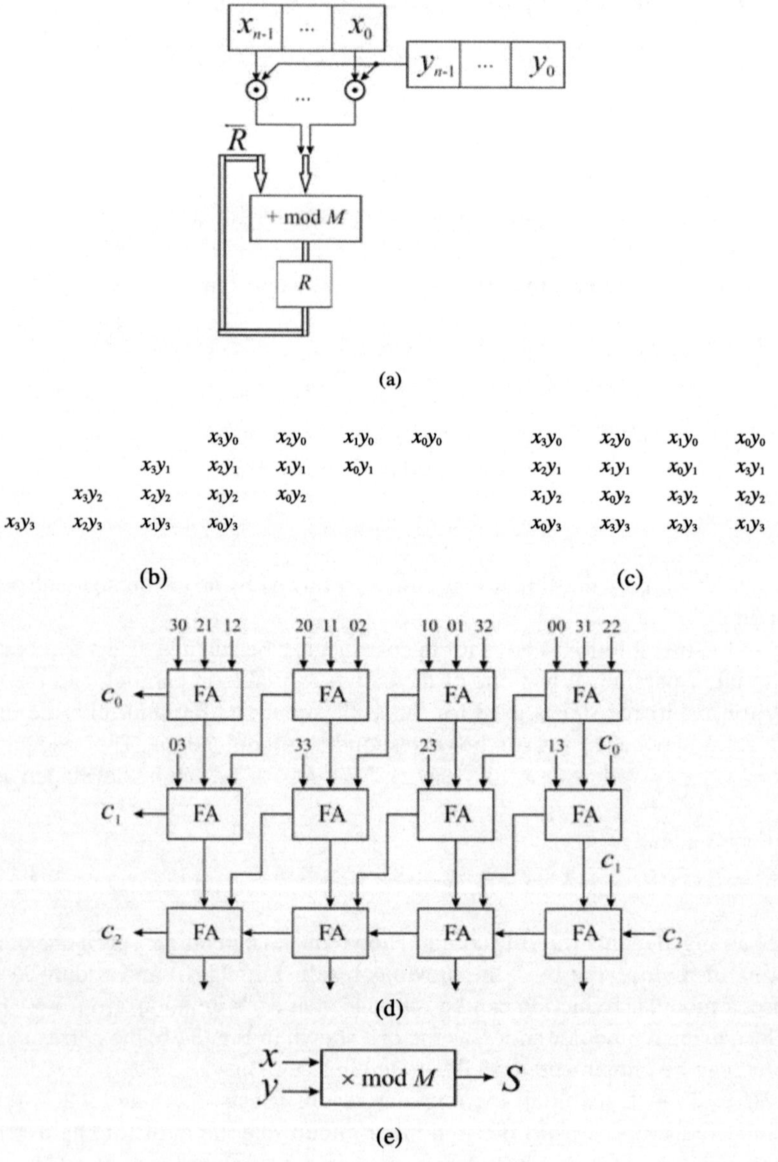

Fig. 3.3 a General modular multiplier. b Generic multiplication. c Folded multiplication. d Multiplier for $M = 15$. e Symbol

adders can be used for any value of n, with the same structure of $(n - 1)$ rows and n columns of the circuit in Fig. 3.3d.

Independently of how they are synthesized, the symbol in Fig. 3.3e will be used for representing the circuits implementing these operations.

Modular multiplication for $M = 2^n - 1$ can be carried out in $n + 1$ iterations using two shift registers, one of them circular, for storing operands, as in the circuit of Fig. 3.4. This has a serial-parallel structure and n AND gates, an n-bit adder, a D flip-flop for storing successive carries, and an n-bit register. The possible final carry is added in the last iteration.

As it has been noted in Sect. 3.2, multiplication for prime M can be transformed into an addition using any primitive element, p. This does not allow multiplying by zero, so the zero value has to be detected in any of the inputs, which is quite simple. An exponent or index, i, is associated to each value, v, so $v = p^i$. This association can be carried out using any logic, such as a ROM (direct ROM). Thus, multiplying two numbers requires to add their indices, and after that the product is obtained using any logic or another ROM (inverse ROM), as illustrated in Fig. 3.5. Using two OR gates and as many AND gates as the number of bits of the result, zero values in the inputs are easily managed, as also illustrated in Fig. 3.5.

The inverse of each non-zero element can be computed when M is prime. A combinational circuit for this purpose can be easily designed, as it is done in the following example.

Example 3.6 Given $x \neq 0$, as a 3-bit binary number, design a circuit for computing x^{-1}mod5, assuming $x < 5$.

Table 3.13 shows the inverse x^{-1}mod5. Assuming $x^{-1} = i_2 i_1 i_0$, the corresponding combinational functions are:

$$i_2 = x_2; \qquad i_1 = x_1; \quad k \quad i_0 = \bar{x}_1 + \bar{x}_0$$

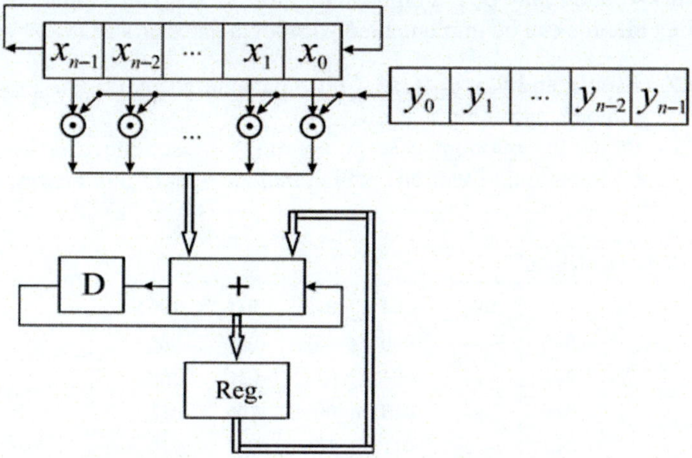

Fig. 3.4 Serial-parallel multiplier

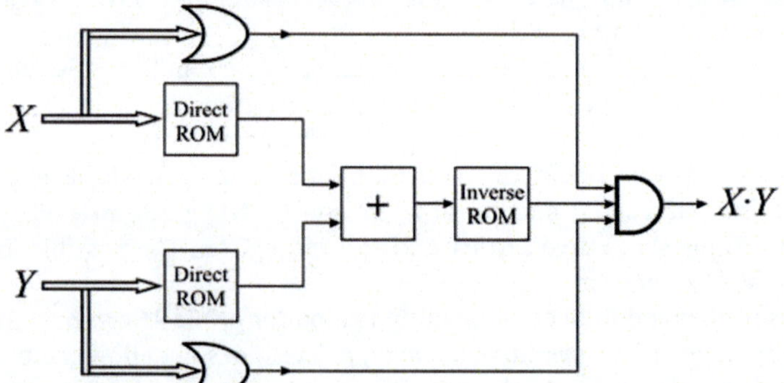

Fig. 3.5 Multiplier using exponents

Table 3.13 Modulo 5 inverses

x	x^{-1}
000	- - -
001	001
010	011
011	010
100	100

Thus, just a two-input NAND gates is required for computing the inverse. ■

Division makes sense for M prime; for implementing it, when both operands are available in parallel, a combinational solution may be used, as in the following example. This solution may be recommendable for not too large values of M. The corresponding circuits can be implemented with logic gates or a ROM.

Example 3.7 Given x and y, $y \neq 0$, as 3-bit binary numbers, design a circuit for computing $(x : y)\text{mod}5$, assuming $x, y < 5$.

Table 3.14 shows the quotient $c = (x : y)\text{mod}5$. Assuming $c = c_2c_1c_0$, the corresponding combinational functions, which include a lot of don't cares, are

Table 3.14 Modulo 5 division

:	000	001	010	011	100
000	- - -	000	000	000	000
001	- - -	001	011	010	100
010	- - -	010	001	100	011
011	- - -	011	100	001	010
100	- - -	100	001	011	001

$$c_2 = \sum m(12, 19, 26, 33) + D$$
$$c_1 = \sum m(10, 11, 17, 20, 25, 28, 35) + D$$
$$c_0 = \sum m(9, 10, 18, 20, 25, 27, 34, 35, 36) + D$$

$$D = \sum m(0, 5, 6, 7, 8, 13, 14, 15, 16, 21, 22, 23, 24, 29, 30, 31, 32, 37, 38, 39, 40, 41, 42, 43, 44, 45, 46, 47, 48, 49, 50, 51, 52, 53, 54, 55, 56, 57, 58, 59, 60, 61, 62, 63)$$

The AND-OR synthesis may be:

$$c_2 = x_2\bar{y}_2\bar{y}_1 + \bar{x}_1 x_0 y_2 + x_1\bar{x}_0 y_1 y_0 + x_1 x_0 y_1 \bar{y}_0$$
$$c_1 = x_1\bar{y}_1 + \bar{x}_1 x_0 y_1 + x_2 y_1 y_0$$
$$c_0 = x_2 y_2 + x_2 y_1 + x_1\bar{x}_0\bar{y}_0 + x_0\bar{y}_2\bar{y}_1 + x_1 x_0 y_0 + \bar{x}_2 x_1 y_2 \bar{y}_1$$

■

A multiplier (followed by an adequate reducer) can be use for division, since dividing by X is the same as multiplying by X^{-1}. Thus, an inverter for generating the inverse of the divisor is enough, and this inverse is then multiplied by the dividend.

3.8.3 Montgomery Multiplier

In modular multiplication (x·y)mod$M$, any multiple of M can be added to the product $x{\cdot}y$ prior to the modular reduction, since this addition does not alter the result. This simple idea may be used to skip division in the reduction process, and it is the basis for the Montgomery multiplier [3], which may be used for any prime M.

Given $M < 2^n$, for A and B ($A, B < M$) it holds:

$$A = a_{n-1}2^{n-1} + a_{n-2}2^{n-2} + \cdots + a_1 2 + a_0$$
$$B = b_{n-1}2^{n-1} + b_{n-2}2^{n-2} + \cdots + b_1 2 + b_0$$

Montgomery multiplication, modulo M, of A and B, represented as $MM(A, B, M)$, is defined as:

$$\mathbf{MM}(A, B, M) = \left(AB2^{-n}\right)\mathbf{mod}M$$

The product $AB2^{-n}$ can be carried out in n iterations, as $Ab_i 2^{-1}$, $i = 0, \ldots, n-1$. If the product Ab_i is even, its multiplication by 2^{-1} is just a shift. If the product Ab_i is odd, M can be added, so $Ab_i + M$ is even and once more multiplication by 2^{-1} is reduced to a shift. Thus, the core of the algorithm is:

(1) $S \leftarrow R + Ab_i$
(2) $R \leftarrow (S + s_0 M)2^{-1}$

With these operations the result in each iteration is translated into the range 0–2^n. The final result will be obtained as:

$$\text{If } R < M, \mathbf{MM}(A, B, M) = R; \mathbf{MM}(A, B, M) = R - M \text{ otherwise}$$

Once $MM(A, B, M)$ is defined and previously computing $Q = 2^{2n} \text{mod} M$, it is easy to show that just two Montgomery multiplications are enough for obtaining $C = AB\text{mod}M$. In fact:

$$\begin{aligned} C = AB \,\mathbf{mod}\, M &= \left((AB2^{-n})2^{2n}2^{-n} \right) \mathbf{mod}\, M \\ &= \mathbf{MM}\left(\mathbf{MM}(A, B, M), 2^{2n}, M \right) \\ &= \mathbf{MM}(\mathbf{MM}(A, B, M), Q, M) \end{aligned}$$

For speeding up the computation, an r-bit block of B can be multiplied in each iteration, with $n = rs$. In this case, there are s blocks and the computation will require s iterations. B can be expressed as:

$$\begin{aligned} B &= \beta_{s-1}2^{r(s-1)} + \beta_{s-2}2^{r(s-2)} + \cdots + \beta_1 2^r + \beta_0 \\ \beta_i &= b_{ir+(r-1)}2^{r-1} + b_{ir+(r-2)}2^{r-2} + \cdots + b_{ir+1}2 + b_{ir} \end{aligned}$$

In this case, the core of the algorithm will be:

(1) $S \leftarrow R + A\beta_i$
(2) $R \leftarrow (S + \Gamma_i M)2^{-r}$

where Γ_i is of r bits. Making $S = \sigma_{s-1}2^{r(s-1)} + \sigma_{s\text{-}2}2^{r(s-2)} + \cdots + \sigma_1 2^r + \sigma_0$, and defining the constant $U = -M^{-1}\text{mod}2^r$, it is easy to conclude that:

$$\Gamma_i = \sigma_0 U \,\mathbf{mod}\, 2^r$$

In fact, the less significant r bits of $S + \Gamma_i M$ will be:

$$\begin{aligned} (\sigma_0 + \sigma_0 UM) \,\mathbf{mod}\, 2^r &= \left(\sigma_0 - \sigma_0 M^{-1} M \right) \mathbf{mod}\, 2^r \\ &= (\sigma_0 - \sigma_0) \,\mathbf{mod}\, 2^r = 0 \end{aligned}$$

The final result will be again obtained as:

$$\text{If } R < M, \mathbf{MM}(A, B, M) = R; \mathbf{MM}(A, B, M) = R - M \text{ otherwise}$$

3.8.4 Exponentiation

As it was detailed in Sect. 2.5, the exponentiation of integer numbers can be iteratively implemented by squaring and multiplying. In the case of modular arithmetic, the same procedure may be used, with modular squaring and multiplication. Thus, once modular multiplication has been revised, squaring will be considered in the following.

Obviously, any of the multiplier described above may be utilized for squaring. However, for a given M, the corresponding multiplier can be simplified when transformed into squarer. For example, the multiplier circuit in Fig. 3.3d is simplified into the circuit of Fig. 3.6 by applying the same ideas used in Sect. 2.4.4, as it can be easily proved.

The most recommendable solution, when feasible (in any case, the easiest to implement), is the use of a look-up table.

Once the multiplier and the squarer are selected, exponentiation will make use of any of the procedures detailed in Sect. 2.5.

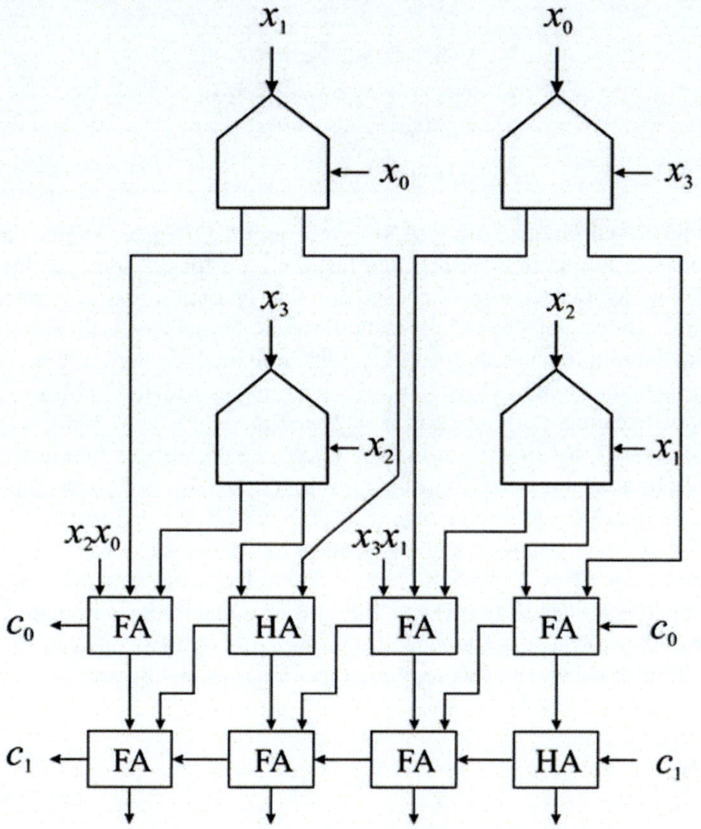

Fig. 3.6 Squarer for $M = 15$

3.8.5 Two Implementation Examples: 3 and 7

Implementation for prime moduli of the form $M = 2^n - 1$ are of great interest: these are the Mersenne prime numbers [6]. It has been noted (Sects. 3.8.1 and 3.8.2) that, in these cases, a 1's complement adder/subtracter can be used for addition and subtraction, while a folded multiplier can be use for multiplication. In this section other circuits will be considered for the two simplest Mersenne primes: 3 and 7. Most of the presented results are easily extended to larger Mersenne prime numbers.

3.8.5.1 $M = 3$

Two bits can be used for $M = 3$ for representing the values 0, 1 and 2 as binary numbers. Thus, using this codification, the addition, S ($s_1 s_0$), and the product, P ($p_1 p_0$), of any two values, X ($x_1 x_0$) and Y ($y_1 y_0$), result in the following Boolean functions:

$$s_1 = x_0 y_0 + x_1 \bar{y}_1 \bar{y}_0 + \bar{x}_1 \bar{x}_0 y_1$$
$$s_0 = x_1 y_1 + x_0 \bar{y}_1 \bar{y}_0 + \bar{x}_1 \bar{x}_0 y_0$$
$$p_1 = x_1 y_0 + x_0 y_1$$
$$p_0 = x_0 y_0 + x_1 y_1$$

A combinational circuit with 8 AND gates and 4 OR gates synthesizes these four functions, resulting in a simpler and faster circuit for the addition than the 1's complement adder that can be also used for this operation and is represented in Fig. 3.7a; this circuit detects and corrects those situations where the addition is 3. For the folded multiplier, the analysis of the different input configurations leads to its simplification to the circuit in Fig. 3.7b, which coincides with the Boolean synthesis.

The opposite element of 1 is 2 for $M = 3$, and the opposite of 1 is 1. Thus, from Table 3.15, the opposite of X ($x_1 x_0$) is $-X$ ($x_0 x_1$), i.e., it suffices to interchange the two bits of a number for obtaining its opposite. Table 3.15 also shows the value of $2X$, and again it suffices to interchange the two bits of X. It is evident that $-2X$ will coincide with X, since negation and duplication imply two successive interchanges.

The inverse for $M = 3$ of each value is the value itself; thus, the quotient, C ($c_1 c_0$), of any two values, X ($x_1 x_0$) and Y ($y_1 y_0$), is equal to their product, P ($p_1 p_0$), and dividing by a constant is equivalent to multiplying by that constant.

Table 3.15 also shows the square, which just requires an OR gate.

3.8.5.2 $M = 7$

For $M = 7$, using three bits for codifying the 7 values, from 0 to 6, Boolean functions can be easily obtained for synthesizing, given any two values, X ($x_2 x_1 x_0$) and Y

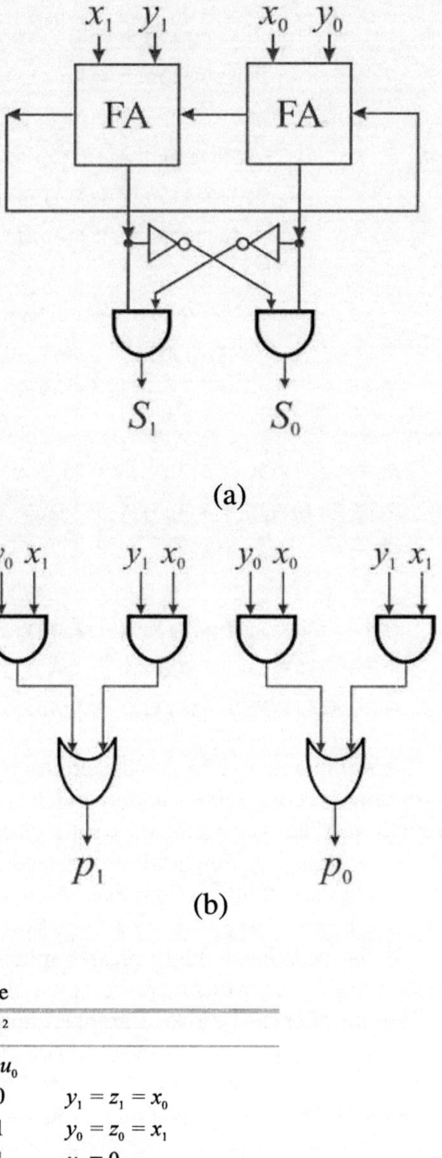

Fig. 3.7 $M = 3$: a 1's complement adder. b Multiplier

Table 3.15 Modulo 3 opposite, double and square

X	$-X$	$2X$	X^2	
x_1x_0	y_1y_0	z_1z_0	u_1u_0	
00	00	00	00	$y_1 = z_1 = x_0$
01	10	10	01	$y_0 = z_0 = x_1$
10	01	01	01	$u_1 = 0$
11	- -	- -	- -	$u_0 = x_1 + x_0$

$(y_2y_1y_0)$, the addition ($S = s_2s_1s_0$), the product ($P = p_2p_1p_0$) and the quotient ($C = X{:}Y = c_2c_1c_0$). A posible implementation of these functions is:

$$s_2 = \bar{x}_2\bar{x}_1y_2\bar{y}_1 + x_2\bar{x}_1\bar{y}_2\bar{y}_1 + \bar{x}_2x_1\bar{y}_2y_1 + \bar{x}_2\bar{x}_1\bar{x}_0y_2$$

$$+ x_2\bar{y}_2\bar{y}_1\bar{y}_0 + x_2x_1y_2y_1 + x_2x_1y_2y_0 + x_2x_0y_2y_1$$
$$+ \bar{x}_2x_0y_1y_0 + \bar{x}_2x_1x_0\bar{y}_2y_0 + \bar{x}_2\bar{x}_0y_2\bar{y}_1\bar{y}_0 + x_2\bar{x}_1\bar{x}_0\bar{y}_2\bar{y}_0$$

$$s_1 = \bar{x}_1\bar{x}_0y_1\bar{y}_0 + \bar{x}_1x_0\bar{y}_1y_0 + x_1\bar{x}_0\bar{y}_1\bar{y}_0 + \bar{x}_2\bar{x}_1\bar{x}_0y_1$$
$$+ x_1\bar{y}_2\bar{y}_1\bar{y}_0 + x_2x_1y_1y_0 + x_2x_0y_1y_0 + x_1x_0y_2y_1$$
$$+ x_2\bar{x}_1y_2y_0 + x_2x_1x_0y_2\bar{y}_1 + \bar{x}_2\bar{x}_1\bar{y}_2y_1\bar{y}_0 + \bar{x}_2x_1\bar{x}_0\bar{y}_2\bar{y}_1$$

$$s_0 = \bar{x}_2\bar{x}_1\bar{x}_0y_0 + x_0\bar{y}_2\bar{y}_1\bar{y}_0 + \bar{x}_2\bar{x}_0\bar{y}_2y_0 + \bar{x}_2x_0\bar{y}_2\bar{y}_0$$
$$+ x_2x_1y_1\bar{y}_0 + x_2x_0y_2y_0 + x_1\bar{x}_0y_2y_1 + x_2\bar{x}_0y_2\bar{y}_0$$
$$+ x_1x_0y_2y_0 + x_2x_0y_1y_0 + \bar{x}_2\bar{x}_1x_0\bar{y}_1\bar{y}_0 + \bar{x}_1\bar{x}_0\bar{y}_2\bar{y}_1y_0$$

$$p_2 = \bar{x}_2x_1y_1\bar{y}_0 + \bar{x}_1x_0y_2\bar{y}_1 + x_1\bar{x}_0\bar{y}_2y_1 + x_2\bar{y}_2\bar{y}_1\bar{y}_0 + x_2\bar{x}_1\bar{x}_0y_0 + \bar{x}_2x_0y_2\bar{y}_0$$
$$p_1 = \bar{x}_1x_0y_1\bar{y}_0 + x_1\bar{x}_0\bar{y}_1y_0 + \bar{x}_2x_0\bar{y}_2y_1 + \bar{x}_2x_1\bar{y}_2y_0 + x_2\bar{x}_1\bar{x}_0y_1 + x_2y_2\bar{y}_1\bar{y}_0$$
$$p_0 = \bar{x}_2x_1y_2\bar{y}_1 + x_2\bar{x}_1\bar{y}_2y_1 + x_1\bar{x}_0y_2\bar{y}_0 + \bar{x}_2\bar{y}_1x_0y_0 + x_0\bar{y}_2\bar{y}_1y_0 + x_2\bar{x}_1\bar{x}_0y_1$$

$$c_2 = x_1\bar{x}_0y_2\bar{y}_1 + x_2\bar{x}_1\bar{y}_2y_1 + \bar{x}_2x_0y_1\bar{y}_0 + \bar{x}_2x_1y_2\bar{y}_0 + \bar{x}_1x_0\bar{y}_2y_1 + x_2\bar{x}_0\bar{y}_1y_0$$
$$c_1 = \bar{x}_2x_0y_2\bar{y}_1 + x_1\bar{x}_0\bar{y}_2y_0 + x_2\bar{x}_1y_1\bar{y}_0 + \bar{x}_2x_1\bar{y}_1y_0 + x_2\bar{x}_0\bar{y}_2y_1 + \bar{x}_1x_0y_2\bar{y}_0$$
$$c_0 = \bar{x}_2x_0\bar{y}_2y_0 + \bar{x}_2x_1\bar{y}_2y_1 + x_1\bar{x}_0y_2\bar{y}_0 + x_2\bar{x}_1y_2\bar{y}_1 + \bar{x}_1x_0\bar{y}_1y_0 + x_2\bar{x}_0y_2\bar{y}_0$$

This implementation for the addition requires more hardware than the corresponding 1's complement adder, which is represented in Fig. 3.8a; the double representation of zero is corrected with a circuit consisting of a NAND gate and three AND gates, whose functioning is immediate to deduce; this corrector circuit for the double representation of zero can be immediately generalized for any adder with $M = 2^n - 1$.

On the other hand, the Boolean implementation of the product is simpler than the folded multiplier, which is represented in Fig. 3.8b.

A multiplier can be also used for computing the quotient; dividing by x is the same as multiplying by its inverse, x^{-1}. Thus, if the inverse of the divisor is previously computed (it will be shown in the following that this just requires to adequately permute the bits of x), division is transformed into a multiplication. In this way, the cost of the divisor is the same as the multiplier's.

Table 3.16 shows the result of multiplying by a constant, as well as the inverse of each element. As it can be easily checked in the Boolean expressions of the different functions, shown in the following, for some cases it is only necessary to interchange the bits of the representation, and for others it suffices to synthesize very simple functions, always the same but for the output order. Concretely, the circuit of Fig. 3.9a can be used for synthesizing $3X$, $5X$ and $6X$; this circuit will be represented as in Fig. 3.9b, adequately ordering the outputs in each case as it is shown below in the expression of these products.

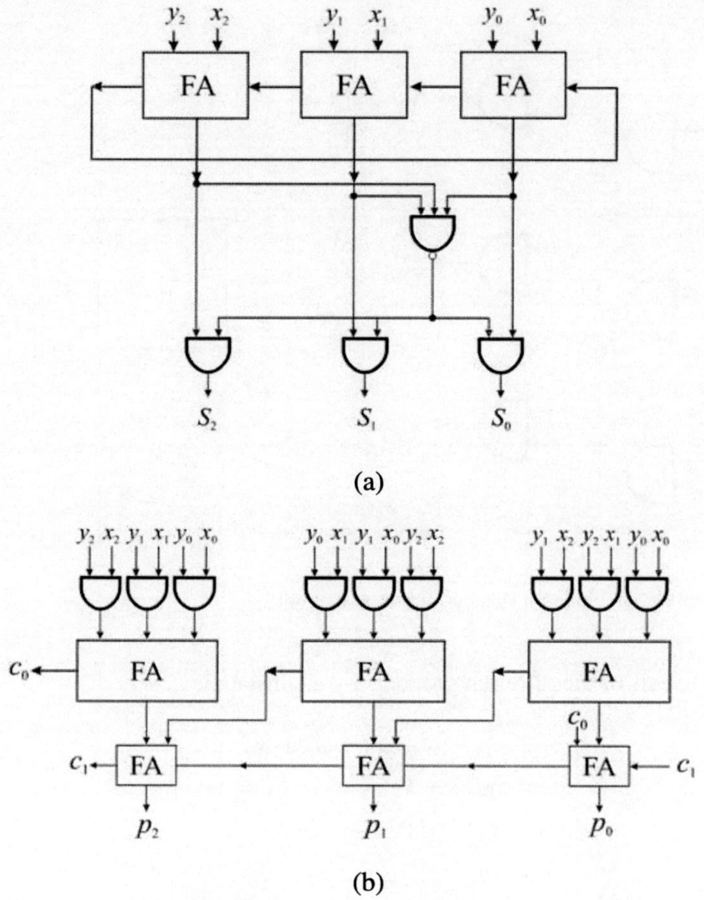

Fig. 3.8 $M = 7$: a 1's complement adder. b Folded multiplier

Table 3.16 Modulo 7 multiplication by a constant and inverse

X	$2X$	$3X$	$4X$	$5X$	$6X$	X^{-1}
$x_2x_1x_0$	$a_2a_1a_0$	$b_2b_1b_0$	$c_2c_1c_0$	$d_2d_1d_0$	$e_2e_1e_0$	$f_2f_1f_0$
000	000	000	000	000	000	- - -
001	010	011	100	101	110	001
010	100	110	001	011	101	100
011	110	010	101	001	100	101
100	001	101	010	110	011	010
101	011	001	110	100	010	011
110	101	100	011	010	001	110
111	- - -	- - -	- - -	- - -	- - -	- - -

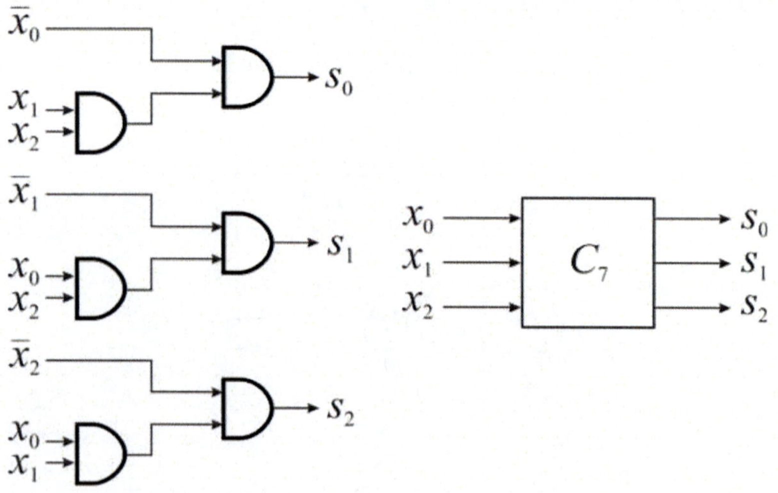

Fig. 3.9 Constant multiplier: a Detailed circuit. b Symbol

The synthesis of the different products by a constant is:

$$
\begin{aligned}
&2X: a_2 = x_1;\quad a_1 = x_0;\quad a_0 = x_2\\
&3X: b_2 = \bar{x}_0(x_2 + x_1) = s_0;\quad b_1 = \bar{x}_2(x_1 + x_0)\\
&\quad = s_2;\quad b_0 = \bar{x}_1(x_2 + x_0) = s_1;\\
&4X: c_2 = x_0;\quad c_1 = x_2;\quad c_0 = x_1;\\
&5X: d_2 = \bar{x}_1(x_2 + x_0) = s_1;\quad d_1 = \bar{x}_0(x_2 + x_1)\\
&\quad = s_0;\quad d_0 = \bar{x}_2(x_1 + x_0) = s_2;\\
&6X: e_2 = \bar{x}_2(x_1 + x_0) = s_2;\quad e_1 = \bar{x}_1(x_2 + x_0)\\
&\quad = s_1;\quad e_0 = \bar{x}_0(x_2 + x_1) = s_0;
\end{aligned}
$$

$$X^{-1}: f_2 = x_1;\quad f_1 = x_2;\quad f_0 = x_0$$

Using that every element has an opposite (see Table 3.2), multiplying by a negative constant just requires applying the following identities:

$$-X = 6X; -2X = 5X; -3X = 4X; -4X = 3X; -5X = 2X; -6X = X$$

In the same way, using that every element (but 0) has an inverse (see Table 3.5), dividing by a constant just requires applying the following identities:

$$X:2 = 4X;\quad X:3 = 5X;\quad X:4 = 2X;\quad X:5 = 3X;\quad X:6 = 6X$$

3.9 Conclusion

This chapter has presented the theoretic fundamentals of the Residue Number Systems, as well as the circuits for implementing the different operations. These circuits correspond to modular arithmetic, which is nothing but the arithmetic in the corresponding Galois fields for prime moduli.

3.10 Exercises

3.1. Obtain the addition table corresponding to CM17.
3.2. Obtain the opposite table corresponding to CM17.
3.3. Obtain the subtraction table corresponding to CM17.
3.4. Obtain the multiplication table corresponding to CM17.
3.5. Obtain the inverse table corresponding to CM17.
3.6. Obtain the division table corresponding to CM17.
3.7. Using (3.9) obtain the inverse of 17 in CM23.
3.8. Find primitives elements in CM17 and CM23.
3.9. Using primitives elements of exercise 8, obtain logarithm tables for CM17 and CM23.
3.10. Using logarithm tables, of exercise 9, obtain Zech logarithm tables for CM17 and CM23.
3.11. Given the set of relatively prime moduli {11, 13, 17, 23}, obtain the representation of 46798 in this RNS.
3.12. Obtain the representation of 46798 in the mixed radix system associated to the RNS {11, 13, 17, 23}.
3.13. Find the decimal value, N, corresponding to the number 28(14)8 represented in the mixed radix system {11, 13, 17, 23}. Find the representation of N in the RNS associated.
3.14. Select some possible set of moduli for the dynamic range $D < 10^4$. Compare the different options with respect to speed and memory requirements.
3.15. Design a modular arithmetic circuit for reducing from 9 to 8 bits, with $M = 249$.
3.16. Design a modular arithmetic circuit for reducing from 16 to 8 bits, with $M = 249$ using adders.
3.17. Design a modular arithmetic circuit for reducing from 16 to 8 bits, with $M = 249$ using multiplicative modular reduction.
3.18. Design a modular arithmetic circuit for addition mod11.
3.19. Design a modular arithmetic circuit for subtraction mod11.
3.20. Design a modular arithmetic circuit for both addition and subtraction mod11.
3.21. Design a modular arithmetic multiplier for $M = 11$ using a multiplier and a reducer.
3.22. Design a modular arithmetic multiplier for $M = 11$ using an adder ROMs.

3.23. Design a modular arithmetic circuit for computing the inverse for $M = 11$.
3.24. Design a modular arithmetic circuit for computing the division for $M = 11$.
3.25. Design a modular arithmetic circuit for squaring for $M = 11$.

References

1. Fraleigh, J.B.: A First Course in Abstract Algebra, Addison-Wesley (2003)
2. Garrett, P.B.: Abstract Algebra, Chapman & Hall (2008)
3. Montgomery, P.L.: Modular multiplication without trial division. Math. Comput. 44(170), 519–521 (1985)
4. Soderstrand, M.A., Jenkins, W.K., Jullien, G.A., Taylor, F.J. (eds.): Residue Number System Arithmetic, IEEE Press (1986)
5. Szabo, N.S., Tanaka, R.I.: Residue Arithmetic and its Applications to Computer Technology, McGraw-Hill (1967)
6. Yan, S.Y.: Number Theory for Computing, Springer (2002)

Chapter 4
Floating Point

This chapter focuses on the floating-point representation of real numbers, which is the number representation format used in most digital systems. It describes the basis of this number representation and analyses the different rounding schemes. The standard IEEE 754 is introduced as well as circuit designs to implement the main floating-point arithmetic operations. To close this chapter, the logarithmic system of real number representation is described with an outline of the circuits implementing it, which may be considered a special case of floating-point representation.

4.1 Introduction

To represent a real number N the fixed-point representation may be used, as shown in Sect. 1.2.5. Also, with a fixed base B, the same N number may be expressed as a product using the following scientific notation:

$$N = R \cdot B^E \tag{4.1}$$

In this case, N may be obtained with the (R, E) pair, where R is a fixed-point real number and E (the exponent) is an integer. R sign is the same of N, which is usually explicitly shown, obtaining:

$$R = (-1)^s \cdot F$$

where s is the sign bit. The N representation is then

$$N = (-1)^s \cdot F \cdot B^E \tag{4.2}$$

Therefore, N is represented by (s, F, E), where F (also known as mantissa or significand) is an unsigned real number.

A. Lloris Ruiz et al., *Arithmetic and Algebraic Circuits*, Intelligent Systems Reference Library 201, https://doi.org/10.1007/978-3-030-67266-9_4

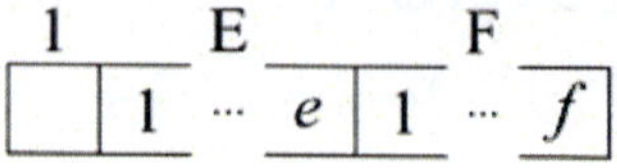

Fig. 4.1 Format for floating-point numbers

The significand F is represented as a fixed-point fractional number (see Sect. 1.2.5); F is also known as fraction or fractional part, which may use any of the number representations of Sect. 1.4: **SM** in complement or biased representation. Usually selecting **SM** or base complement. As it will be noted later on, current standards use **SM** representation for the significand so this is the assumed representation from now on.

The exponent E is an integer hence it may use any of the three previous representations. It will be seen later on that operands' exponents comparison is quite relevant to implement the different arithmetic operations and, for that reason, the biased representation is usually the choice, as explained in Sect. 1.4.4. Thus, from now on this is the assumed representation for the exponent.

Given $\boldsymbol{n}$ bits to represent $\boldsymbol{N}$, one bit (s) will be reserved for the sign, $\boldsymbol{e}$ bits for the exponent E and f bits for the significand F so that $n = 1 + e + f$. From now on, the format of floating-point numbers is the one described in Fig. 4.1.

The number representation systems based on (4.1) are known as floating point systems. As it will be seen in this chapter, this name comes from the decimal point shifts to the left or to the right when operating with real numbers, as if it were floating over. In contrast to the fixed-point representation, the decimal point position here depends on the operations done.

When $E = 0$ in (4.2), it is obvious that a floating point system becomes a fixed point system. When $F = 1$ the resulting representation is the logarithmic one that will be described in Sect. 4.8.

When no additional conditions are imposed to F, an $\boldsymbol{N}$ number may have multiple floating-point representations. For instance, (s, F, E), (s, $F/\boldsymbol{B}$, $E + 1$), (s, $F/\boldsymbol{B}^2$, $E + 2$), (s, $F{\cdot}\boldsymbol{B}$, $E - 1$), (s, $F{\cdot}\boldsymbol{B}^2$, $E - 2$), and so on, are different representations of the same number. To move from one representation to another it is only needed to increment (or decrement) the exponent and, at the same time, divide (or multiply) the significand by the corresponding $\boldsymbol{B}$ power, which is equivalent to shift the decimal point to the left or to the right.

With base $\boldsymbol{B}$, its value is implicitly recognised in any floating-point representation system, so there is no need to explicit it. As base $\boldsymbol{B}$, any real number different from 1 may be used. However, it is clear that if F is a sum of successive power of 2 values, as it is always done in practice, it is a better choice to select $\boldsymbol{B}$ also as a power of 2. This way, to move from one representation to another of the same number, the multiplication or division operations are reduced to shifts. If $\boldsymbol{B} = 2^d$, each increment (decrement) in the E exponent must be followed by d shifts of F to the right (left). Another option is selecting $\boldsymbol{B} = 10$, the same radix we commonly use (see Sect. 1.7).

To avoid redundancies and to simplify comparisons between different values, when the real number representation is taken into account only (without doing arithmetic operations that will be seen later), a limit to F values is usually imposed with either one of the following conditions:

$$1 > F \geq 1/\boldsymbol{B} \tag{4.3}$$

$$\boldsymbol{B} > F \geq 1 \tag{4.4}$$

When F follows one of those conditions the representation is normalised. Normalised representations will be used always in this text unless noted.

Following (4.3) a normalised representation has a significand with a null integer part hence with a fractional part only. If $\boldsymbol{B} = 2^d$, the normalisation condition (4.3) means that among the d more significant bits of F there is at least one 1 digit. For example, with $\boldsymbol{B} = 2$, all representable values have the $0.1x \ldots x$ form, where the most significant bit of the fractional part is 1; with $\boldsymbol{B} = 8$, there is at least one 1 digit among the three more significant bits of the fractional part; with $\boldsymbol{B} = 16$, there is at least one 1 digit among the four more significant bits of the fractional part and so on.

In practice, the normalisation condition (4.4) is used with $\boldsymbol{B} = 2$ and $\boldsymbol{B} = 10$. For $\boldsymbol{B} = 2$ the representable values have the $1.x \ldots x$ form. In other words, the integer part is always 1, so this value may be given implicitly as a *hidden bit* and it does not need to be physically present in the representation, saving a bit in the significand. With condition (4.3) and $\boldsymbol{B} = 2$, the same hidden bit idea is applicable as well.

The most used power of 2 for $\boldsymbol{B}$ are 2 and 16, although other values like 8 or 256 have also been used. Later on, the trade off of different $\boldsymbol{B}$ values are revised. Nowadays the most used values are fundamentally $\boldsymbol{B} = 2$ and $\boldsymbol{B} = 10$.

On the exponent side, with $\boldsymbol{e}$ bits, the maximum value depends on the representation used: $E_{\max} = 2^{e-1}$ or $E_{\max} = 2^{e-1} - 1$ (**SM**, with complement, or biased). Likewise, the minimum possible value is $E_{\min} = -2^{e-1}$ or $E_{\min} = -2^{e-1} - 1$. In any case, for the exponent range **EX**, defined as the difference between the maximum and the minimum exponent values, the following approximation is a very good one:

$$\mathbf{EX} \cong 2^e \tag{4.5}$$

The implementation of floating-point arithmetic operations is the same as with scientific notation. To add or subtract two floating-point numbers, first the exponents must be the same, being this one the exponent of the result value. As expected, equalling the exponents means adjusting the associated significands as well. Next step is to add or subtract the adjusted significands, obtaining the significand of the result value.

To multiply two floating-point numbers, multiply the significands to obtain the product significand, and add the exponents to obtain the product exponent. To divide two floating-point numbers, divide the significands to obtain the quotient significand plus the remainder of the operation, and subtract the exponents to obtain the quotient exponent (the remainder exponent depends on the dividend only). In Sect. 4.5 the arithmetic operations implementations are revised with more detail.

Those are the basic notions of floating-point representations. Next section tackles with the floating-point system representation dilemma, whether it is precision or dynamic range.

4.2 Precision and Dynamic Range

It is clear that with a finite number of bits only a finite number of numeric values may be represented without error. Any other value will be approximated by one of the representable values within an error range. The different methods to round the correct non-representable values to approximated representable values will be seen below.

Being more specific, let's assume a fixed E exponent using normalisation either following (4.3) or (4.4) with hidden bit. In this case, having f bits for the significand (all f bits are the fractional part), the significand precision is 2^{-f} (see Sect. 1.2.5). Let's recall that in fixed-point representation the 2^{-f} value is called unit in the last position, ***ulp***, identifying the increment of representable values or representation precision.

Therefore, the difference between two consecutive representable values with E exponent is $2^{-f} \cdot \boldsymbol{B}^E$, as it may be easily checked. Then, the difference between two consecutive values is bigger when E is bigger as well. Thus, the density of representable values in the normalised range is much bigger at the boundaries of zero (small E values) than at the boundaries of infinite (big E values), as shown in Fig. 4.2. In other words, the precision is greater when small values are represented than when big ones are. The difference between two consecutive representable values with exponent E, $2^{-f} \cdot \boldsymbol{B}^E$, is bigger when $\boldsymbol{B}$ is bigger. Hence, with a bigger base $\boldsymbol{B}$ there is less precision than with a smaller base.

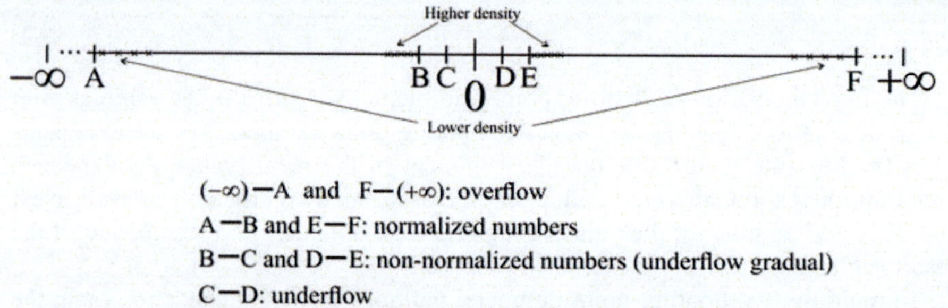

Fig. 4.2 Dynamic range

The floating-point precision directly depends on the significand bits size: if f grows, the difference between consecutive representable values decreases. In fixed-point all bits are dedicated to the significand hence its precision is always higher than in floating-point for a given total bit number of the representation. The ***ulp*** value may be used as an error measure when representing any real value, as seen next.

Let $R_1 = s{\cdot}F{\cdot}\boldsymbol{B}^E$ and $R_2 = s{\cdot}(F + ulp){\cdot}\boldsymbol{B}^E$ be two consecutive representable values and let x be a real number so that $R_2 \geq x \geq R_1$. The x representation in floating-point is $R(x)$ and it will be either R_1 o R_2. Let's now obtain an expression of the error that the floating-point representation has, regardless the sign. The absolute error when representing x is $|R(x) - x|$; and clearly $R(x) - x| \leq |R_1 - R_2| = ulp{\cdot}\boldsymbol{B}^E$ always happens.

The relative error is:

$$RE(x) = \frac{|R(x) - x|}{|x|} \tag{4.6}$$

The maximum value, RE_M, for the relative error $RE(x)$ depends on the rounding procedure used. In any case, RE_M is obtained in (4.6) when the numerator is maximised and the denominator is minimised. Therefore, defining $ULP = ulp{\cdot}\boldsymbol{B}^E$ and $R = F{\cdot}\boldsymbol{B}^E$, the maximum relative error value always follows:

$$RE(x) = RE_\mathrm{M} \leq \frac{ULP}{R} = \frac{ulp}{F} \tag{4.7}$$

The minimum F value is 1/**B** or 1, depending on the used normalisation: (4.3) or (4.4). Thus, *ulp* may be used to define a measure of the corresponding floating-point system precision. Specifically, to increment the measuring precision value when the precision increments, the inverse of *ulp*, *P*, may be used:

$$P = \frac{1}{ulp} = 2^f \tag{4.8}$$

With a normalised representation following (4.3) or (4.4), the maximum (minimum) representable value N_max (N_min) is obtained when both the significand and the exponent have their maximum (minimum) values. In particular, when $1 > F \geq 1/\boldsymbol{B}$, $N_\mathrm{max} \cong \boldsymbol{B}^{E_\mathrm{max}}$ and $N_\mathrm{min} = \boldsymbol{B}^{E_\mathrm{min}}/\boldsymbol{B}$; when $\boldsymbol{B} > F \geq 1$, $N_\mathrm{max} \cong \boldsymbol{B} \times \boldsymbol{B}^{E_\mathrm{max}}$ and $N_\mathrm{min} = \boldsymbol{B}^{E_\mathrm{min}}$. Any value outside those limits would not be representable as a normalised number, leading to an overflow of the exponent. When the value to represent is too big it is called **overflow** and **underflow** when the value to represent is too small, as shown in Fig. 4.2.

For illustration purposes, Table 4.1 shows representable values for a given floating-point format. It has 128 positive representable values (there are other 128 negative values), with 8 bits and base $\boldsymbol{B} = 2$; distributed as 1 sign bit (not shown in the table), 3 exponent bits and 4 significand bits; using normalisation (4.4) with hidden bit (integer part is 1 always). For the exponent it uses a shifted representation and bias

Table 4.1 Representable values with $e = 3$ and bias $= 3, f = 4$, $\boldsymbol{B} = 2$ and a hidden bit

f\e	000	001	010	011	100	101	110	111
0000	16/128	16/64	16/32	1	2	4	8	16
0001	17/128	17/64	17/32	17/16	17/8	17/4	17/2	17
0010	18/128	18/64	18/32	18/16	18/8	18/4	9	18
0011	19/128	19/64	19/32	19/16	19/8	19/4	19/2	19
0100	20/128	20/64	20/32	20/16	20/8	5	10	20
0101	21/128	21/64	21/32	21/16	21/8	21/4	21/2	21
0110	22/128	22/64	22/32	22/16	22/8	22/4	11	22
0111	23/128	23/64	23/32	23/16	23/8	23/4	23/2	23
1000	24/128	24/64	24/32	24/16	3	6	12	24
1001	25/128	25/64	25/32	25/16	25/8	25/4	25/2	25
1010	26/128	26/64	26/32	26/16	26/8	26/4	13	26
1011	27/128	27/64	27/32	27/16	27/8	27/4	27/2	27
1100	28/128	28/64	28/32	28/16	28/8	7	14	28
1101	29/128	29/64	29/32	29/16	29/8	29/4	29/2	29
1110	30/128	30/64	30/32	30/16	30/8	30/4	15	30
1111	31/128	31/64	31/32	31/16	31/8	31/4	31/2	31

= 3. Table 4.2 is similar to Table 4.1 but using $\boldsymbol{B} = 16$ instead. In this case it uses normalisation (4.3), but only 120 values may be represented, as in the 4 significand bits there must be at least one 1.

It is worth noting that it is not possible to represent the zero value as a normalised floating-point value. As the zero value must be used in practice, a non-normalised representation value must be selected for it, as it will be seen below. The same approach is used for huge and tiny values, among other circumstances. Using non-normalised values means using some concrete values for the exponent, which will not be available for the normalised values.

To increment the range of representable values near zero, the **gradual underflow** approach may be used, where it is allowed for the more significant bits of the significand to be 0, which under normalisation conditions should be 1. An example of gradual underflow is the following.

Example 4.1 Let $\boldsymbol{B} = 16$ and $f = 16$. The minimum normalised significand value is $M = 0001000000000000$. Therefore, the minimum representable normalised value is 16^A, with $A = E_{\text{min}} - 1$. Using gradual underflow the minimum non-normalised significand value is M = 0000000000000001 and the minimum representable non-normalised value is 16^B, with $\boldsymbol{B} = E_{\text{min}} - 4$. There are 4095 new representable values, as it is easy to see. ■

Table 4.2 Representable values with $e = 3$ and bias $= 3, f = 4, \boldsymbol{B} = 16$

f\e	000	001	010	011	100	101	110	111
0001	1/65536	1/4096	1/256	1/16	1	16	256	4096
0010	2/65536	2/4096	2/256	2/16	2	32	512	8192
0011	3/65536	3/4096	3/256	3/16	3	48	768	12288
0100	4/65536	4/4096	4/256	4/16	4	64	1024	16384
0101	5/65536	5/4096	5/256	5/16	5	80	1280	20480
0110	6/65536	6/4096	6/256	6/16	6	96	1536	24576
0111	7/65536	7/4096	7/256	7/16	7	112	1792	28672
1000	8/65536	8/4096	8/256	8/16	8	128	2048	32768
1001	9/65536	9/4096	9/256	9/16	9	144	2304	36864
1010	10/65536	10/4096	10/256	10/16	10	160	2560	40960
1011	11/65536	11/4096	11/256	11/16	11	176	2816	45056
1100	12/65536	12/4096	12/256	12/16	12	192	3072	49152
1101	13/65536	13/4096	13/256	13/16	13	208	3328	53248
1110	14/65536	14/4096	14/256	14/16	14	224	3584	57344
1111	15/65536	15/4096	15/256	15/16	15	240	3840	61440

The quotient between the maximum and minimum representable values N_{max}/N_{min} is known as **dynamic range, DR**. It is a measure of the values range that may be represented in a normalised way. It is clear that whatever it is the normalisation criteria:

$$\mathbf{DR} \cong \boldsymbol{B} \cdot \boldsymbol{B}^{E_{max}-E_{min}}$$

But following (4.5), $E_{max} - E_{min} = \mathbf{EX}$ hence:

$$\mathbf{DR} \cong \boldsymbol{B} \cdot \boldsymbol{B}^{EX} = \boldsymbol{B}^{EX+1} \tag{4.9}$$

The dynamic range, following (4.9), is greater when $\boldsymbol{B}$ is greater too. For the 4.1 example, using gradual underflow increments the dynamic range in $16^3 - 1$.

The difference $N_{max} - N_{min}$ is the **representable interval, RI**. As $N_{min} \cong 0$ usually, in this case $\mathbf{RI} \cong N_{max}$.

As an example, in Table 4.1 $\mathbf{EX} = 7$, $\mathbf{DR} = 31 \times 128/16 = 248$, $\mathbf{RI} \cong 31$. For Table 4.2 obtains $\mathbf{EX} = 7$, $\mathbf{DR} = 61440 \times 65536 = 4026531840$, $\mathbf{RI} \cong 61440$.

It is obvious that, for a given number of bits, it is possible to obtain with floating-point representation a much bigger dynamic range than with fixed-point.

Another precision measure that may be used in a floating-point representation is **points' density (PD)**, defined as the quotient between the number of all different possible representable values and the representable interval. For instance, using Table 4.1 the measure is $\mathbf{PD} \cong 128/31 \cong 4'13$, and for Table 4.2 the measure is $\mathbf{PD} \cong 120/61440 \cong 0'0020$. In other words, points' density in Table 4.1 is much greater than in Table 4.2. However, even if one floating-point system has globally a greater points' density than another, locally may happen just the opposite, as it can be seen with Tables 4.1 and 4.2 for values lesser than 1. In Table 4.1 there are 48 points whereas in Table 4.2 there are 60.

With e bits in the exponent and f bits in the fractional part, 2^{e+f} values may be represented. Using $\mathbf{RI} \cong N_{\max} = \boldsymbol{B}^C$, with $C = E_{\max} + 1 = 2^{e-1} + 1$, and having $\boldsymbol{B} = 2^b$, if it is a power of 2 base, we obtain:

$$\mathbf{PD} = \frac{(2^{e+f})}{\boldsymbol{B}^C} = \frac{(2^{e+f})}{(2^b)^{2^{e-1}+1}} = \frac{(2^{e+f})}{2^{b(2^{e-1}+1)}} = 2^{e+f-b(2^{e-1}+1)}$$

From this expression it is clear that points' density decreases when b increases; i.e. when base B is greater, points' density is lower. To sum up, using a big base there is a gain in dynamic range but a loss on precision. If precision is more important to reduce errors in operations, as it is usually the case, it is better to select the minimum radix: base 2.

The precision (P) increases as 2^f (4.8) and the dynamic range (**DR**) increases as $\boldsymbol{B}^{EX}$ (4.9), with $EX = 2^e$. Starting with a fixed number of bits to distribute into the significand and the exponent, $n = 1 + e + f$, it is clear that a precision increment implies a dynamic range reduction and vice versa.

The expressions $\varepsilon = e/n - 1$, $\Delta = f/n - 1$ are defined to get a normalised representation of the precision versus dynamic range relation, obtaining $\varepsilon + \Delta = 1$.

Figure 4.3 shows the precision versus dynamic range for $B = 2$, overlaying the representations of 2^ε and 2^T, with $T = 2^\Delta$.

Focusing again on Fig. 4.3, it is clear the dichotomy between precision and dynamic range: improving one implies worsen the other.

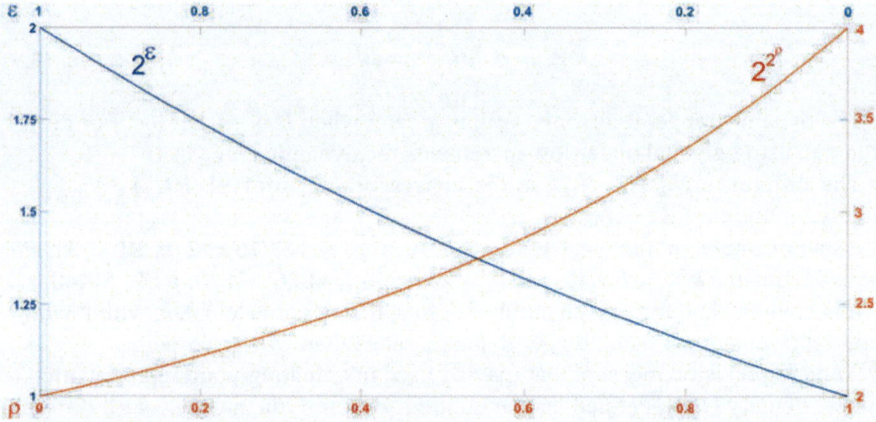

Fig. 4.3 Precision versus dynamic range relation

4.3 Rounding

A real number may have infinite significant digits, but it will be always represented with a finite number of bits in its significand. It is assumed from now on normalised real numbers, hence having all significand bits to represent the fractional part. Moreover, it is also assumed for this fractional part the **SM** representation, because this one is used in standards and the conclusions obtained are extensible to any other representation. All operations related with rounding but rounding overflow , which will be seen later, affects the significand only, hence we will be referring just to the significand in this section.

For a real number *N*, rounding means obtaining *N* representation with a given number of digits from another representation with a greater number of digits. For instance, the result of multiplying two binary numbers with f bits in their significands has $2f$ bits in its significand. If this result has to be represented with only f significand bits, then it must be rounded applying the appropriate rules to discard the f less significant bits and to properly generate the f more significant bits.

In general rounding means obtaining f digits from $f + m$ digits in such a way both represent the same number. For the following significand:

$$N = x_{-1}x_{-2}\ldots x_{-f}\ldots x_{-(f+m)}$$

with rounding the new significand is:

$$\mathbf{RN} = y_{-1}y_{-2}\ldots y_{-f}$$

It is clear that any rounding has an error, as expressed in (4.6). For a more detailed study of the errors, not only for rounding but also for other floating-point operations, see Appendix D.

It is important to know which error is obtained when selecting a rounding method, measured in terms of *ulp*. To compare errors of different rounding methods the following measures may be used:

(a) Maximum absolute error value **MAEV**;
(b) Mean absolute error value, assuming equally probable all the values in the digits to be discarded. This measure is also known as mean bias value, **MBV**.

Let's analyse the different rounding schemes. Given a significand $\mathbf{N} = x_{-1}x_{-2}\ldots x_{-f}\ldots x_{-(f+m)}$, with $f + m$ digits, it will be between two representable significands with f digits, with an *ulp* difference:

$$R_1 = x_{-1}x_{-2}\ldots x_{-f} \text{ and } R_2 = R_1 + ulp = x_{-1}x_{-2}\ldots x_{-f} + ulp = z_{-1}z_{-2}\ldots z_{-f}$$

R_1 and R_2 have the same exponents in all cases but when R_1 has the maximum possible value $R_1 = bb\ldots b$ (11…1 in binary). In the last case $R_2 = 00\ldots0$, and the corresponding exponents will have a difference of one [$\exp(R_2) = \exp(R_1) + 1$]. This is the sole case when the exponents are taken into account, when the selected value for the rounding is R_2 and it is called **rounding overflow**. The right rounded value has to be selected once this situation is detected.

Next are the different rounding schemes, exclusively for the binary case. In Sect. 4.4 it will be described the decimal case, as binary and decimal bases are the ones considered in the standards.

The rounding schemes are rules to select between R_1 and R_2 as N representation. In all cases the function $R(x)$ will be plotted. The selected representable value R_1 or R_2 is obtained with this function depending on the value to round. Specifically, the **RN** bits $y_{-(f-1)}y_{-f}$ are plotted as a function of bits $x_{-(f-1)}x_{-f}\ldots x_{-(f+m)}$ of N. The result is always a stair function.

4.3.1 Rounding Without Halfway Point

The rounding methods without halfway point always approximate the rounding value with only one of the two nearest representable values, without taking into account the difference to any of them, i.e. without considering which representable value, R_1 or R_2, is the nearest to the value to be represented.

As it will be verified later with the two rounding methods without halfway point described below, there is no need to know any of the bits to discard from the rounding number to implement them.

4.3.1.1 Rounding with Truncation or Towards Zero

The simplest rounding method is just to truncate the less significant bits so that from $\mathbf{N} = x_{-1}x_{-2}\ldots x_{-f}\ldots x_{-(f+m)}$ it obtains $\mathbf{RN} = x_{-1}x_{-2}\ldots x_{-f}$. It is called truncation or chopping. The rounding value, **N**, is approximated by the nearest R_1 value, not greater in magnitude than **N**.

Figure 4.4 shows the $R(x)$ function: the height of each step is *ulp* and the transition from one step to the next one is done for every representable value. The rounded value is always less or equal in magnitude than the number to round. Therefore, the errors obtained for rounded signed values have always the same sign.

This rounding procedure is also known as rounding towards zero, as the values to be rounded are closer to zero than the original ones.

The maximum absolute error value **MAEV**, (error = **NR** − **N**) happens when all the bits to discard are 1. When discarding m bits it is immediate that:

$$\mathbf{MAEV} = -\frac{2^m - 1}{2^m} ulp$$

Thus, for a big m, it may approximate to $-ulp$: the more digits to trunk the closer to:

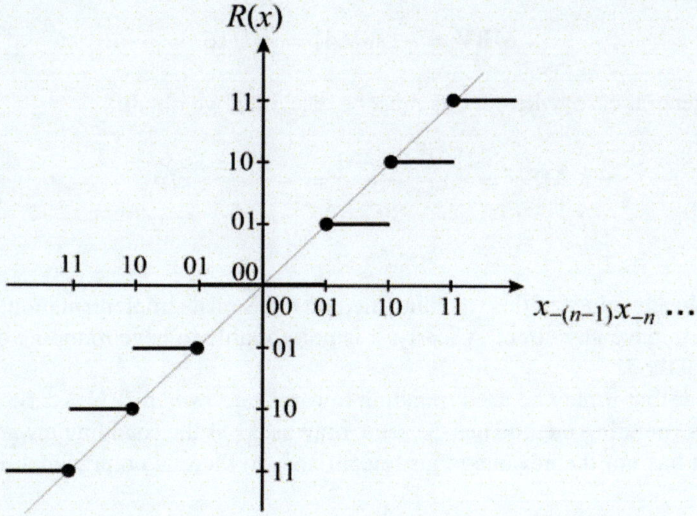

Fig. 4.4 Rounding with truncation

$$\mathbf{MAEV} \cong -ulp$$

For the mean bias value measure, this value is given by the following expression as shown in Example 4.2:

$$\mathbf{MBV} = -\frac{2^m - 1}{2^{m+1}} ulp$$

For big values of m then $\mathbf{MBV} \cong -1/2ulp$.

Example 4.2 Let m = 3. Obtain the medium error found using rounding with truncation.

Table 4.3 gives the corresponding error to any of the 8 possible cases with $m = 3$. Assuming an equal probability to any of those 8 cases, the medium error measured in *ulp* units is:

Table 4.3 Truncation rounding errors (measured in *ulp* units)

$x_{-(f+1)}x_{-(f+2)}x_{-(f+3)}$	Error (*ulp*)
000	0
001	−1/8
010	−2/8
011	−3/8
100	−4/8
101	−5/8
110	−6/8
111	−7/8

$$\mathbf{MBV} = -(28/64) = -7/16$$

In the general case, where *n* bits must be discarded we obtain:

$$\mathbf{MBV} = -\frac{1}{2^m}\frac{2^m - 1}{2} = -\frac{2^m - 1}{2}ulp$$

■

The main advantage of this rounding method relies on its implementation, as there is no need to have any circuit. Clearly, it is not possible to have rounding overflow with this method.

It is clear that it may be used rounding towards one , where $\mathbf{RN} = x_{-1}x_{-2}...x_{-f}$ + *ulp*. This rounding method has the same drawbacks as the rounding towards zero, although it has not the advantage just mentioned, so there is no practical reason to use it.

4.3.1.2 Von Neumann Rounding or Jamming

Given the significand $N = x_{-1}x_{-2}...x_{-f}...x_{-(f+m)}$, between the representable significands with *f* bits $R_1 = x_{-1}x_{-2}...x_{-f}$ and $R_2 = R_1 + ulp = x_{-1}x_{-2}...x_{-f} + ulp$ $= z_{-1}z_{-2}...z_{-f}$, the von Neumann rounding or jamming is the following:

$$\mathbf{RN} = R_1 \text{ if } x_{-f} = 1; \quad \mathbf{RN} = R_2 \text{ if } z_{-f} = 1$$

i.e. it selects the nearest of the two values ending with 1.

Another way to describe this rounding method is as follows: round with truncation and then, if the least significant bit is 0, make it 1.

This rounding method proposed by von Neumann is known also as rounding with forced 1.

Figure 4.5 shows *R*(*x*): the height of each step is 2·*ulp* and the transition from one step to the next one happens with odd representable values. In this case, the errors have both signs. It can be seen graphically how they tend to compensate themselves.

Table 4.4 shows the errors made when 3 bits must be discarded ($x_{-(f+1)}$, $x_{-(f+2)}$, $x_{-(f+3)}$); in each case the error depends on the x_{-f} bit. In all cases $y_{-f} = 1$. It is clear that **MAEV** = *ulp*. The mean bias value is **MBV** = *ulp*/16. In the general case, discarding *m* bits, we obtain:

$$\mathbf{MBV} = \frac{1}{2^{m+1}}ulp$$

The implementation of this rounding method is very simple:

$$y_{-1}y_{-2}\ldots y_{-(f-1)}y_{-f} = x_{-1}x_{-2}\ldots x_{-(f-1)}1$$

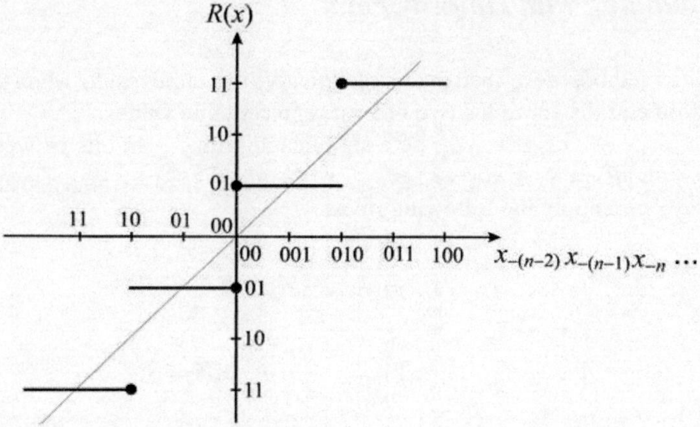

Fig. 4.5 Von Neumann rounding or jamming

Table 4.4 Jamming rounding errors (measured in *ulp* units)

$x_{-f}\ x_{-(f+1)}\ x_{-(f+2)}\ x_{-(f+3)}$	Error (*ulp*)
0000	1
0001	7/8
0010	6/8
0011	5/8
0100	4/8
0101	3/8
0110	2/8
0111	1/8
1000	0
1001	−1/8
1010	−2/8
1011	−3/8
1100	−4/8
1101	−5/8
1110	−6/8
1111	−7/8

It is neither possible a rounding overflow in this case.

Comparing rounding with truncation and jamming rounding, the second one has a better **MBV** value. Both have the same (null) complexity for their implementation. Therefore, jamming rounding is preferable to rounding with truncation.

4.3.2 Rounding with Halfway Point

The rounding methods described in this section give the same results when the value to round is not equidistant to the two nearest representable values.

Let $N = x_{-1}x_{-2}\ldots x_{-f}\ldots x_{-(f+m)}$ be a significand with $f + m$ bits between $R_1 = x_{-1}x_{-2}\ldots x_{-f}$ and $R_2 = R_1 + ulp = x_{-1}x_{-2}\ldots x_{-f} + ulp = z_{-1}z_{-2}\ldots z_{-f}$. The rounding with halfway point apply the following rules:

$$\text{if } x_{-(f+1)}\ldots x_{-(f+m)} < \frac{1}{2}ulp, \quad \mathbf{RN} = R_1$$

$$\text{if } x_{-(f+1)}\ldots x_{-(f+m)} > \frac{1}{2}ulp, \quad \mathbf{RN} = R_2$$

i.e. when $x_{-(f+1)}\ldots x_{-(f+m)}\#1/2\ ulp$, selects the nearest of the two closer representable values for **RN**. The approximation of the halfway point depends on the rounding method.

In rounding with halfway point methods, the rounding value of $x_{-1}x_{-2}\ldots x_{-f}\ldots x_{-(f+m)}$ is obtained truncating the m bits to discard and adding to the remaining f bits in the least significant position 0 if the rounded value is R_1 or 1 if the rounded value is R_2. Therefore, it is enough to use an adder to implement it, obtaining

$$y_{-1}y_{-2}\ldots y_{-(f-1)}y_{-f} = x_{-1}x_{-2}\ldots x_{-(f-1)}x_{-f} + P$$

where $P \in \{0, 1\}$. P is a different expression for each rounding method. Figure 4.6 shows the corresponding circuit. Thus, in contrast with rounding methods without halfway point, here information about the bits to discard is a must.

The rounding overflow is possible with rounding with halfway point methods. It is immediate to detect using Fig. 4.6 circuit, as the adder carry output will be 1 when rounding overflow.

The $R(x)$ graph for these rounding methods has an *ulp* step and a step transition halfway from two representable values.

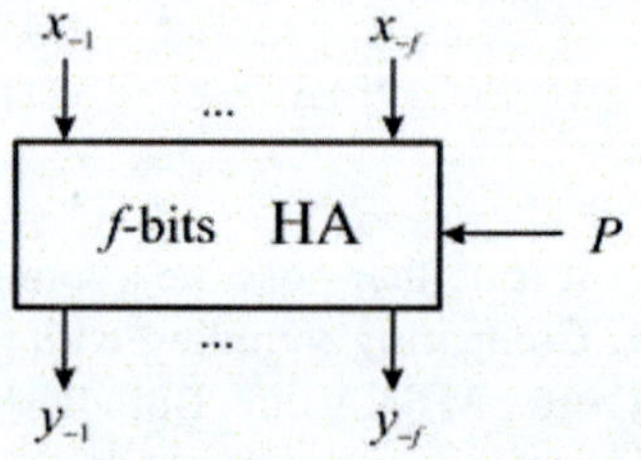

Fig. 4.6 Adder for rounding with halfway point

4.3.2.1 Rounding Half to Even

The rule for rounding to the nearest even in the halfway point

$$\left(x_{-(f+1)} \ldots x_{-(f+m)} = 100\ldots0 = \frac{1}{2}ulp\right)$$

is the following:

$$\mathbf{RN} = R_1 \text{ if } x_{-f} = 0; \quad \mathbf{RN} = R_2 \text{ if } z_{-f} = 0$$

i.e. the nearest value is selected and the halfway point is approximated by its nearest representable value with a 0 as least significant bit. Figure 4.7 represents $R(x)$.

The second column of Table 4.5 gives the corresponding rounding of the value to round. There are 16 possible values depending on bit x_{-f} when there are 3 bits to discard ($m = 3$). This column specifies $+0$ if the rounded number results from $y_{-1}y_{-2}\ldots y_{-(f-1)}y_{-f} = x_{-1}x_{-2}\ldots x_{-(f-1)}x_{-f}$ equality, or $+1$ if it results from $y_{-1}y_{-2}\ldots y_{-(f-1)}y_{-f} = x_{-1}x_{-2}\ldots x_{-(f-1)}\,x_{-f} + 1$.

The third column of Table 4.5 gives the corresponding error measure of each case (error $=$ **RN** $-$ **N**). It is immediate for **MAEV** that

$$\mathbf{MAEV} = \pm\frac{1}{2}ulp$$

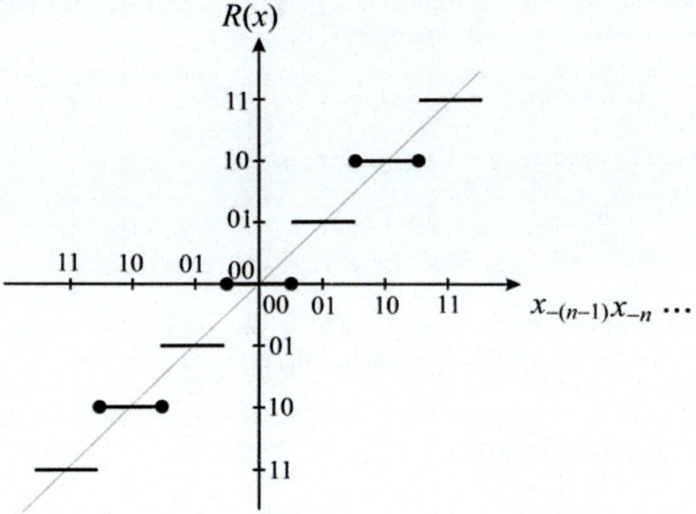

Fig. 4.7 Rounding half to even

Table 4.5 Rounding to even errors (measured in *ulp* units)

$x_{-f}x_{-(f+1)}\ x_{-(f+2)}\ x_{-(f+3)}$	$y_{-1}\ldots y_{-f}$	Error (*ulp*)
0000	+0	0
0001	+0	−1/8
0010	+0	−2/8
0011	+0	−3/8
0100	+0	−4/8
0101	+1	3/8
0110	+1	2/8
0111	+1	1/8
1000	+0	0
1001	+0	−1/8
1010	+0	−2/8
1011	+0	−3/8
1100	+1	4/8
1101	+1	3/8
1110	+1	2/8
1111	+1	1/8

Assuming the same probabilities to all bits to discard combinations, **MBV** = 0 always.

To implement this rounding method using Fig. 4.6 circuit and discarding 3 bits (Table 4.5), it is easy to see that expression *P* is:

$$P = x_{-(f+1)}\left(x_{-f} + x_{-(f+2)} + x_{-(f+3)}\right) = x_{-(f+1)} \quad \text{OR} \quad \left(x_{-f}, x_{-(f+2)}, x_{-(f+3)}\right)$$

The general expression for *m* bits to discard is:

$$\begin{aligned} P &= x_{-(f+1)}\left(x_{-f} + x_{-(f+2)} + \ldots + x_{-(f+m)}\right) \\ &= x_{-(f+1)} \quad \text{OR} \quad \left(x_{-f}, x_{-(\text{f}+2)}, \ldots, x_{-(f+m)}\right) \end{aligned} \tag{4.10}$$

Sometimes this rounding method is known as *rounding to the nearest*. If odd values are preferred to even values, we obtain the next rounding method.

4.3.2.2 Rounding Half to Odd

The rule for rounding to the nearest odd in the halfway point is the following

$$\mathbf{RN} = R_1 \text{ if } x_{-f} = 1; \quad \mathbf{RN} = R_2 \text{ if } z_{-f} = 1$$

i.e. the nearest value is selected and the halfway point is approximated by its nearest representable value with a 1 as least significant bit. Figure 4.8 represents $R(x)$.

The second column of Table 4.6 gives the corresponding rounding of the value to round. There are 16 possible values depending on bit x_{-f} when there are 3 bits to discard ($m = 3$). This column specifies +0 if the rounded number results from $y_{-1}y_{-2}\ldots y_{-(f-1)}y_{-f} = x_{-1}x_{-2}\ldots x_{-(f-1)}x_{-f}$ equality, or +1 if it results from $y_{-1}y_{-2}\ldots y_{-(f-1)}y_{-f} = x_{-1}x_{-2}\ldots x_{-(f-1)}x_{-f} + 1$.

The third column of Table 4.6 gives the corresponding error measure of each case (error = **RN** − **N**). It is immediate for **MAEV** that

$$\mathbf{MAEV} = \pm\frac{1}{2}ulp$$

Assuming the same probabilities to all bits to discard combinations, **MBV** = 0 always.

To implement this rounding method using Fig. 4.6 circuit and discarding 3 bits (Table 4.6), the expression P is:

$$P = x_{-(f+1)}\left(\bar{x}_{-f} + x_{-(f+2)} + x_{-(f+3)}\right) = x_{-(f+1)} \quad \text{OR} \quad \left(\bar{x}_{-f}, x_{-(f+2)}, x_{-(f+3)}\right)$$

The general expression for m bits to discard is:

$$\begin{aligned} P &= x_{-(f+1)}\left(\bar{x}_{-f} + x_{-(f+2)} + \ldots + x_{-(f+m)}\right) \\ &= x_{-(f+1)} \quad \text{OR} \quad \left(\bar{x}_{-f}, x_{-(f+2)}, \ldots + x_{-(f+m)}\right) \end{aligned} \tag{4.11}$$

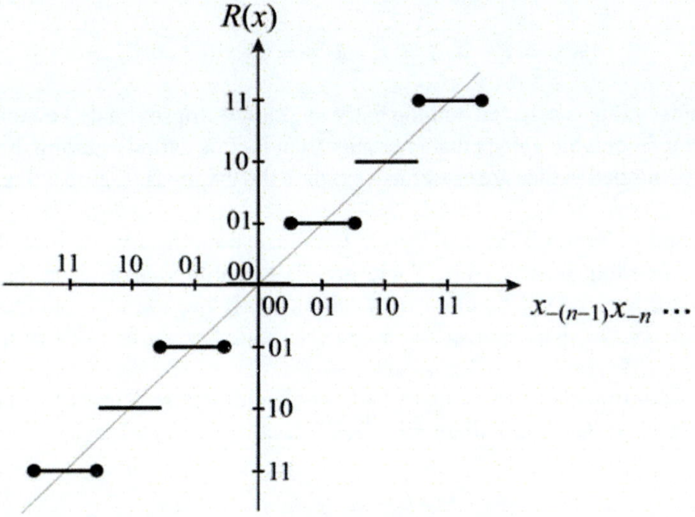

Fig. 4.8 Rounding half to odd

Table 4.6 Rounding to odd errors (measured in *ulp* units)

$x_{-f}\ x_{-(f+1)}\ x_{-(f+2)}\ x_{-(f+3)}$	$y_{-1}\ldots y_{-f}$	Error (*ulp*)
0000	+0	0
0001	+0	−1/8
0010	+0	−2/8
0011	+0	−3/8
0100	+1	4/8
0101	+1	3/8
0110	+1	2/8
0111	+1	1/8
1000	+0	0
1001	+0	−1/8
1010	+0	−2/8
1011	+0	−3/8
1100	+0	−4/8
1101	+1	3/8
1110	+1	2/8
1111	+1	1/8

4.3.2.3 Rounding Half Up or Half Towards $+\infty$

The rule for rounding towards $+\infty$ of the halfway point depends on the sign of the value to round. Having the sign bit s of the real number to round with significand **N**, the rule is:

$$\mathbf{RN} = R_1 \text{ if } s = 1; \quad \mathbf{RN} = R_2 \text{ if } s = 0$$

i.e. the nearest value is selected and the halfway point is approximated by the nearest of the two representable values that is greater than **N**. It is called rounding half up, as the halfway rounded values are closer to $+\infty$ than the originals. Figure 4.9 represents $R(x)$.

The second column of Table 4.7 gives the corresponding rounding of the value to round, depending on the s bit. There are 16 possible values when there are 3 bits to discard ($m = 3$). This column specifies +0 if the rounded number results from $y_{-1}y_{-2}\ldots y_{-(f-1)}y_{-f} = x_{-1}x_{-2}\ldots x_{-(f-1)}x_{-f}$ equality, or +1 if it results from $y_{-1}y_{-2}\ldots y_{-(f-1)}y_{-f} = x_{-1}x_{-2}\ldots x_{-(f-1)}x_{-f} + 1$.

The third column of Table 4.7 gives the corresponding error measure of each case (error = **RN** − **N**). It is immediate for **MAEV** that

$$\mathbf{MAEV} = \pm\frac{1}{2}ulp$$

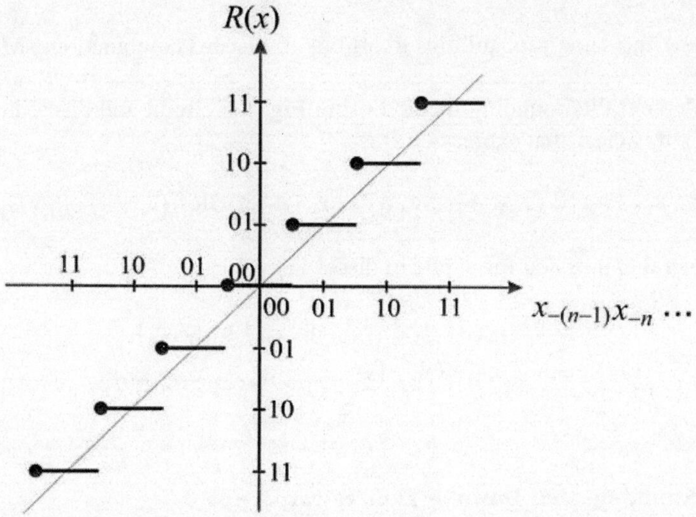

Fig. 4.9 Rounding half up or half towards $+\infty$

Table 4.7 Rounding half up or half towards $+\infty$ errors (measured in *ulp* units)

$sx_{-(f+1)}\,x_{-(f+2)}\,x_{-(f+3)}$	$y_{-1}\ldots y_{-f}$	Error (*ulp*)
0000	+0	0
0001	+0	−1/8
0010	+0	−2/8
0011	+0	−3/8
0100	+1	4/8
0101	+1	3/8
0110	+1	2/8
0111	+1	1/8
1000	+0	0
1001	+0	−1/8
1010	+0	−2/8
1011	+0	−3/8
1100	+0	−4/8
1101	+1	3/8
1110	+1	2/8
1111	+1	1/8

Assuming the same probabilities to all bits to discard combinations, **MBV** = 0 always.

To implement this rounding method using Fig. 4.6 circuit and discarding 3 bits (Table 4.7), it is clear that expression P is:

$$P = x_{-(f+1)}\left(\bar{s} + x_{-(f+2)} + x_{-(f+3)}\right) = x_{-(f+1)} \quad \text{OR} \quad \left(\bar{s}, x_{-(f+2)}, x_{-(f+3)}\right)$$

The general expression for m bits to discard is:

$$\begin{aligned} P &= x_{-(f+1)}\left(\bar{s} + x_{-(f+2)} + \ldots + x_{-(f+m)}\right) \\ &= x_{-(f+1)} \quad \text{OR} \quad \left(\bar{s}, x_{-(f+2)}, \ldots, x_{-(f+m)}\right) \end{aligned} \tag{4.12}$$

4.3.2.4 Rounding Half Down or Half Towards $-\infty$

Again , the rule for rounding towards $-\infty$ of the halfway point depends on the sign of the value to round. Having a sign bit s of the real number to round with significand **N**, the rule is

$$\mathbf{RN} = R_1 \text{ if } s = 0; \quad \mathbf{RN} = R_2 \text{ if } s = 1$$

i.e. the nearest value is selected and the halfway point is approximated by the nearest of the two representable values that is lower than **N**. It is called rounding half down, as the half way rounded values are closer to $-\infty$ than the originals. Figure 4.10 represents $R(x)$.

The second column of Table 4.8 gives the corresponding rounding of the value to round, depending on the s bit. There are 16 possible values when there are 3 bits to discard ($m = 3$). This column specifies +0 if the rounded number results from $y_{-1}y_{-2}\ldots y_{-(f-1)}y_{-f} = x_{-1}x_{-2}\ldots x_{-(f-1)}x_{-f}$ equality, or +1 if it results from $y_{-1}y_{-2}\ldots y_{-(f-1)}y_{-f} = x_{-1}x_{-2}\ldots x_{-(f-1)}x_{-f} + 1$.

The third column of Table 4.8 gives the corresponding error measure of each case (error = **RN** − **N**). It is immediate for **MAEV** that

$$\mathbf{MAEV} = \pm\frac{1}{2}ulp$$

Assuming the same probabilities to all bits to discard combinations, **MBV** = 0 always.

To implement this rounding method using Fig. 4.6 circuit and discarding 3 bits (Table 4.8), it is clear that expression P is:

$$P = x_{-(f+1)}\left(s + x_{-(f+2)} + x_{-(f+3)}\right) = x_{-(f+1)} \quad \text{OR} \quad \left(s, x_{-(f+2)}, x_{-(f+3)}\right)$$

The general expression for m bits to discard is:

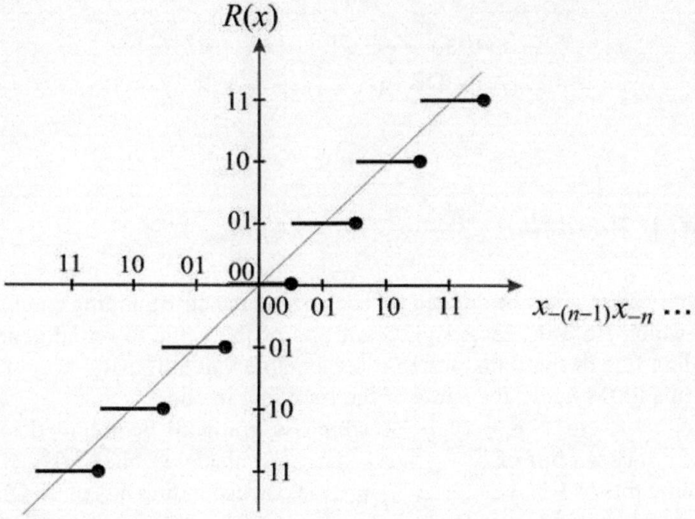

Fig. 4.10 Rounding half down or half towards $-\infty$

Table 4.8 Rounding half down or half towards $-\infty$ errors (measured in *ulp* units)

$sx_{-(f+1)}\, x_{-(f+2)}x_{-(f+3)}$	$y_{-1} \ldots y_{-f}$	Error (*ulp*)
0000	+0	0
0001	+0	−1/8
0010	+0	−2/8
0011	+0	−3/8
0100	+0	−4/8
0101	+1	3/8
0110	+1	2/8
0111	+1	1/8
1000	+0	0
1001	+0	−1/8
1010	+0	−2/8
1011	+0	−3/8
1100	+1	4/8
1101	+1	3/8
1110	+1	2/8
1111	+1	1/8

$$P = x_{-(f+1)}\left(s + x_{-(f+2)} + \ldots + x_{-(f+m)}\right)$$
$$= x_{-(f+1)} \quad \text{OR} \quad \left(s, x_{-(f+2)}, \ldots, x_{-(f+m)}\right) \tag{4.13}$$

4.3.3 ROM Rounding

The rounding criteria may be defined for each particular case building combinational functions, which from the least significant bits of the value to round generate the corresponding bits of the transformed representable value. Usually these functions are stored in a ROM hence the name of the rounding method.

Given $\mathbf{N} = x_{-1}x_{-2}\ldots x_{-f}\ldots x_{-(f+m)}$, which is going to be rounded to $\mathbf{NR} = y_{-1}y_{-2}\ldots y_{-f}$, it uses k bits of **N**, $x_{-j}\ldots x_{-f}\ldots x_{-p}$, to calculate r bits of **RN**, $y_{-k}\ldots y_{-f}$. The remaining bits of **RN**, $y_{-1}\ldots y_{-(k-1)}$, are the corresponding bits of **N**. Obviously, the more bits of **N** used the better to define the possible values for rounding and the rounded result.

As an example, let's assume using 3 bits $x_{-(f-1)}x_{-f}\ x_{-(f+1)}$ of **N** to obtain 2 bits $y_{-(f-1)}y_{-f}$ of **RN**. Table 4.9 has a rounding definition for this case, which could be implemented either with 8 2-bits words ROM or with AND-OR logic for the functions:

$$y_{-(f-1)} = x_{-(f-1)} + x_{-f}x_{-(f+1)}; \quad y_{-f} = x_{-f} \oplus x_{-(f+1)} + x_{-(f-1)}x_{-f}x_{-(f+1)}$$

using a PAL, for instance. Figure 4.11 shows $R(x)$ for this particular case of ROM rounding.

In ROM rounding the possibility of rounding overflow is usually not taken into account. In these cases in the limit, operations are made like rounding without halfway point, as done in Table 4.9.

ROM rounding allows a specific definition for each case. Its implementation is always simple anyways.

Table 4.9 Sample ROM rounding

$x_{-(f-1)}\ x_{-f}\ x_{-(f+1)}$	$y_{-(f-1)}\ y_{-f}$
000	00
001	01
010	01
011	10
100	10
101	11
110	11
111	11

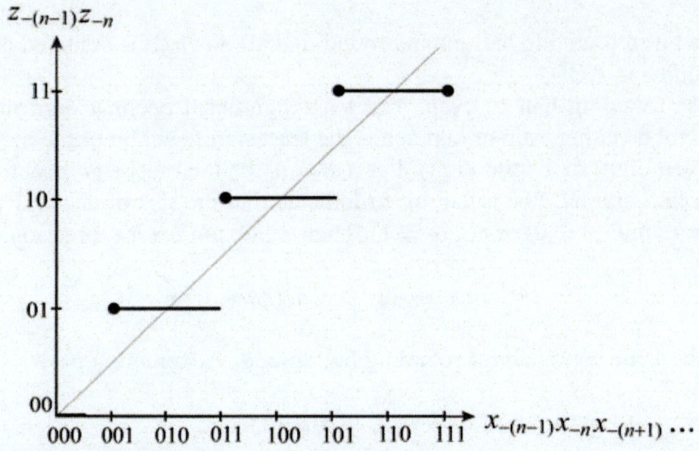

Fig. 4.11 ROM rounding

4.4 Decimal Rounding

All rounding types already seen for the binary case can be immediately applied to a decimal or any other radix. Starting with truncation rounding, it is easy to calculate, using a similar method than for the binary case, the maximum absolute error value (**MAEV**) and the mean bias value (**MBV**) when discarding m digits:

$$\mathbf{MAEV} = -\frac{10^m - 1}{10^m} ulp$$

$$\mathbf{MBV} = -\frac{1}{10^m}\frac{10^m - 1}{2} = -\frac{10^m - 1}{2 \times 10^m} ulp$$

The von Neumann or jamming rounding is applied to a decimal radix as follows: select the nearest of the two values that is odd. When discarding m bits, the maximum absolute error value and the mean bias value for this rounding method are:

$$\mathbf{MAEV} = ulp$$

$$\mathbf{MBV} = \frac{1}{10^{m+1}} ulp$$

Like in the binary case, the implementations of those rounding methods do not need any circuit.

The rounding with halfway point methods may be implemented with a simple adder (Fig. 4.6) like in the binary case. It will be added 0 or 1 depending on the first digit to discard; if the remainder digits to discard are all 0 or not; if the least significant digit is even or odd in rounding half to even or in rounding half to odd; or on the sign bit in rounding half up and rounding half down. It is assumed that each digit is codified in **BCD**.

Let's use rounding half to even. The least significant decimal digit not to be discarded will be either even or odd hence the least significant bit of the digit, b_0, is either 0 (even digit) or 1 (odd digit). Let $a_3a_2a_1a_0$ be the most significant decimal digit to discard and let c be a flag bit to indicate that the rest of decimal digits to discard are all null ($c = 0$) or not ($c = 1$). Then we obtain for P expression:

$$P = a_3 + a_2a_1 + a_2a_0(b_0 + c) \tag{4.14}$$

Using the same approach for rounding half to odd, P expression is:

$$P = a_3 + a_2a_1 + a_2a_0(\bar{b}_0 + c) \tag{4.15}$$

For rounding half up or rounding half down, let s be the sign of the decimal number to round. The P expression for rounding half up is:

$$P = a_3 + a_2a_1 + a_2a_0(\bar{s} + c) \tag{4.16}$$

For rounding half down we obtain:

$$P = a_3 + a_2a_1 + a_2a_0(s + c) \tag{4.17}$$

Therefore, to implement any of the rounding with halfway point methods, it is enough a decimal adder similar to Fig. 4.6 to add either 0 or 1. In this case the carry output of the adder flags if there is rounding overflow or not.

Finally, like with a binary radix, in decimal a **ROM** may be used as well.

4.5 Basic Arithmetic Operations and Rounding Schemes

This section presents the basic concepts to implement addition , subtraction , multiplication and division arithmetic operations and their rounding schemes effects. Before introducing the arithmetic operations, we analyse the comparison operation, as it is needed in some of them. In what follows, it is assumed normalised operands as in (4.3) or in (4.4), and a normalised result.

4.5.1 Comparison

Let $A = s(A)F_1 \cdot 2^D$ and $B = s(B)F_2 \cdot 2^G$ be two floating-point numbers to compare. If A and B have different signs ($s(A)$ # $s(B)$), the number with a positive sign is bigger than the number with negative sign. If both numbers have the same sign, the exponents have to be compared first. If $D \# G$, the inequality relationship between A and B depends only on that of D and G. If $D = G$, the inequality relationship between A and B depends on that of F_1 y F_2, as shown in the Table in Fig. 4.12a. Thus, clearly Fig. 4.12b circuit implements the comparison.

D* # *G			***D* = *G***		
	s(*A*) = s(*B*)			**s(*A*) = s(*B*)**	
$D > G$	+	$A > B$	$F_1 > F_2$	+	$A > B$
$D > G$	–	$A < B$	$F_1 > F_2$	–	$A < B$
$D < G$	+	$A < B$	$F_1 < F_2$	+	$A < B$
$D < G$	–	$A > B$	$F_1 < F_2$	–	$A > B$

(a)

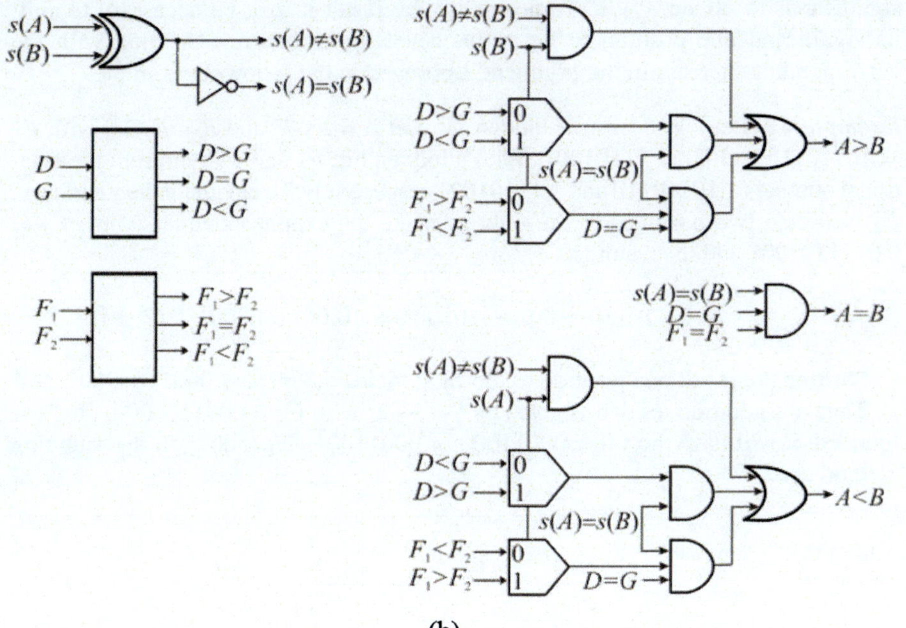

(b)

Fig. 4.12 Comparison: **a** Table; **b** Circuit

4.5.2 Addition and Subtraction

The effective addition or effective subtraction to do for an addition ($A + B$) or a subtraction operation ($A - B$), depends on the signs of A and B, $s(A)$ and $s(B)$. Table 4.10 shows this, where $\bar{s}/r$ represents the required operation ($\bar{s}/r = 0$, addition; $\bar{s}/r = 1$, subtraction), and $\bar{S}/R$ represents the effective operation to do ($\bar{S}/R = 0$, addition; $\bar{S}/R = 1$, subtraction).

Starting with addition, let $A = F_1 \cdot 2^D$ and $B = F_2 \cdot 2^G$, with $D \geq G$, be two addends with the same sign, then the addition result $A + B$ will have that sign. If $D \# G$ then one of the addends representation may be shifted to make the exponents equal. To achieve this the addend significand with a smaller exponent is shifted $D - G$ positions to the right, which leads to an unnormalised addend. The addend significand with a bigger exponent could also be shifted to the left, but this operation is less convenient to implement.

Once the exponents are equal (the result has that exponent), add the corresponding significands to obtain $A + B$. To normalise the result it may be necessary to shift its significand one position to the right, never more than one position with the corresponding increase in the exponent, as proved in the following example.

Example 4.3 Let $f = 8$ using a hidden bit. Let $A = F_1 \cdot 2^6$ and $B = F_2 \cdot 2^4$, with $F_1 = 10110110$ and $F_2 = 11101001$. With a hidden bit, F_1 and F_2 represent the fractional numbers 1.10110110 and 1.11101001 respectively. To calculate their addition, F_2 is shifted two positions to the right to make the exponents equal. Thus, $F_2 = 0.0111101001$ and the result is:

$$1.10110110 + 0.0111101001 = 10.0011000001$$

Shifting the result one position to the right yields the result 1.00011000001.

Thus, the addition result S is given by $S = F_3 \cdot 2^7$ with $F_3 = 00011000001$. If F_3 is rounded, it will have the value 00011000 or 00011001, depending on the rounding method used. ■

Table 4.10 Addition and subtraction

$\bar{s}/r$ $s(A)$ $s(B)$	$\bar{S}/R$
000	0
001	1
010	1
011	0
100	1
101	0
110	0
111	1

It is possible, although less probable, that an overflow occurs during addition, because the exponent may be increased due to the normalisation step after the addition of the significands.

To perform a subtraction the two operands are compared first. The smaller one is subtracted to the bigger one and the result has the sign of the bigger operand. If the operands have different exponents, the exponents are made equal first (the subtraction result will have that exponent) shifting the significand with the smaller exponent to the right. The significands are subtracted next. To normalise the result it may be necessary to shift its significand some positions to the left (and the corresponding decrease in the exponent), as proved in the next example.

Example 4.4 Let $f = 8$ using a hidden bit. Let $A = F_1 \cdot 2^6$ and $B = F_2 \cdot 2^6$, with $F_1 =$ 10110110 and $F_2 =$ 10101001. With a hidden bit, F_1 and F_2 represent the fractional numbers 1.10110110 and 1.10101001 respectively. As the exponents are equal, only the significands have to be subtracted to perform the subtraction and the result is:

$$1.10110110 - 1.10101001 = 0.00001101$$

Shifting the result five positions to the left, yields the result 1.10100000.
Thus, the subtraction result R is given by $R = F_3 \cdot 2^1$ with $F_3 =$ 10100000. ■

When the result of the subtraction needs several left shifts, it is easy to check that the maximum difference between the minuend's exponent and the subtrahend's exponent was initially one. Thus, before performing the subtraction, the subtrahend requires at most one shift right in these cases. That is, if both operands initially present a significand with f digits, when some left shifts are required during subtraction, vacant positions are filled with zeros, except maybe the first one.

After the normalisation step the exponent may be decreased and it is possible, although not very probable, an underflow during subtraction.

4.5.3 Multiplication and Division

The multiplication of two floating-point numbers $A = (-1)^a \cdot F_1 \cdot 2^D$ and $B = (-1)^b \cdot F_2 \cdot 2^G$ is performed by multiplying the significands and adding the exponents:

$$M = A \cdot B = \left[(-1)^a \cdot F_1 \cdot 2^D\right] \cdot \left[(-1)^b \cdot F_2 \cdot 2^G\right] = (-1)^c \cdot [F_1 \cdot F_2] \cdot 2^{D+G}$$

with $c = a \oplus b$.

Obviously, the multiplication of the significands and the addition of the exponents may be done in parallel.

If F_1 and F_2 satisfy (4.3), for product $F = F_1 \cdot F_2$ we have:

$$1 > F \geq 1/\mathbf{B}^2$$

Therefore, after multiplication it may be required one shift (never more than one) to the left, in order to normalise the result, as seen in the following example.

Example 4.5 Let $f = 8$ and normalisation (4.3). Let $A = F_1 \cdot 2^6$ and $B = F_2 \cdot 2^6$, with $F_1 = 10010110$ and $F_2 = 10101001$, i.e. F_1 and F_2 represent the fractional numbers 0.10010110 and 0.10101001, respectively. The result is:

$$0.10010110 \cdot 0.10101001 = 0.0110001100000110$$

Shifting the result one position to the left yields the result 0.110001100000110. Thus, the product M is $M = F_3 \cdot 2^{11}$ with $F_3 = 110001100000110$, and after rounding it will have the value 11000110 or 11000111, depending on the rounding method used. ■

If F_1 and F_2 satisfy (4.4), for product $F = F_1 \cdot F_2$ we have:

$$\mathbf{B}^2 > F \geq 1$$

Therefore, after multiplication it may be required one shift (never more than one) to the right, in order to normalise the result, as shown next.

Example 4.6 Let $f = 8$ using a hidden bit. Let $A = F_1 \cdot 2^6$ and $B = F_2 \cdot 2^6$, with $F_1 = 10110110$ and $F_2 = 10101001$. With a hidden bit, F_1 and F_2 represent the fractional numbers 1.10110110 and 1.10101001, respectively. The result is:

$$1.10110110 \cdot 1.10101001 = 10.1101111100100110$$

Shifting the result one position to the right, yields the result 1.01101111100100110. Thus, the product M is $M = F_3 \cdot 2^{13}$, with $F_3 = 01101111100100110$, and after rounding it will have the value 01101111 or 01111111, depending on the rounding method used. ■

As the product's exponent is the result of the addition of the operands' exponents, besides a left or right shift that may be needed when normalising the result, overflow is also possible during multiplication.

During division a quotient and a remainder are obtained. In this part we are only going to consider the quotient, which in many cases is the interesting result. Section 4.6.6 considers the calculation of the remainder.

A division quotient of two floating-point numbers $A = (-1)^a \cdot F_1 \cdot 2^D$ and $B = (-1)^b \cdot F_2 \cdot 2^G$ is obtained by dividing the significands and subtracting the exponents:

$$C = A/B = \left[(-1)^a \cdot F_1 \cdot 2^D\right] / \left[(-1)^b \cdot F_2 \cdot 2^G\right] = (-1)^c \cdot [F_1/F_2] \cdot 2^{D-G}$$

with $c = a \oplus b$.

Obviously, the division of the significands and the subtraction of the exponents can be done in parallel.

After division it may be required one shift to the left (never more than one), to normalise the result, as seen in the following example.

Example 4.7 Let $f = 8$ using a hidden bit, and let $A = F_1{\cdot}2^6$ and $B = F_2{\cdot}2^3$, with $F_1 = 00110110$ and $F_2 = 11101001$. With a hidden bit, F_1 and F_2 represent the fractional numbers 1.00110110 and 1.11101001, respectively. The quotient result is:

$$1.00110110/1.11101001 = 0.1010001001$$

Shifting the result one position to the left, yields the result 1.01000100. Thus, the quotient in this division C is $C = F_3{\cdot}2^2$, with $F_3 = 10100010$ or $F_3 = 10100011$, depending on the rounding method used. ■

As the quotient's exponent is the result of the subtraction of the operands' exponents, besides a left shift that may be required when normalising the quotient, underflow is also possible during division.

4.5.4 Rounding Bits

As we have seen, the results obtained in arithmetic operations may have significands with many more digits than the number of digits used in a normalised form (f digits in the significand). Once obtained a result, if it needs to be normalised using any of the rounding methods, we may question how many extra digits are needed in the result for a correct normalisation with a minimum error, in addition to the f most significant ones.

As seen in the preceding part, after addition and multiplication there may be one position shift to the left or to the right. After subtraction there may be some positions shifts to the left, and after division there may be one position shift to the left.

Thus, to avoid errors during left shifts, at least one additional digit ($x_{-(f+1)}$) has to be kept, called *guard digit*.

Rounding schemes without halfway point do not need extra digits for a correct rounding. But rounding schemes with halfway point need more information to obtain P using (4.10), (4.11), (4.12) or (4.13) in binary; or (4.14), (4.15), (4.16) or (4.17) in decimal. In binary P expressions are:

$$P = x_{-(f+1)} \quad \text{OR} \quad (\ldots)$$

where $x_{-(f+1)}$ is bit $f + 1$ in the normalised result (before rounding), which may be bit $f + 2$ in the unnormalised result if there is one left shift. In decimal P expressions are:

$$P = a_3 + a_2a_1 + a_2a_0 \quad \text{OR} \quad (\ldots)$$

where $a_3a_2a_1a_0$ are bits of the most significant decimal digit to be discarded before rounding, the $f+1$ decimal digit, which may be digit $f+2$ in the unnormalised result if there is one left shift.

Thus, another extra digit is required in the unnormalised result, digit $f+2$, known as *round digit*.

Finally, the expressions OR, (4.10), (4.11), (4.12) and (4.13), in the different binary rounding schemes with halfway point satisfy:

$$\begin{aligned} &\mathrm{OR}\big(w, x_{-(f+2)} + \ldots + x_{-(f+m)}\big) \\ &\quad = \mathrm{OR}\big(\mathrm{w}, \mathrm{OR}\big(x_{-(f+2)} + x_{-(f+3)} + \ldots + x_{-(f+m)}\big)\big) \end{aligned}$$

with $\mathrm{w} = \{x_{-f}, \bar{x}_{-f}, \bar{s}, s\}$. Here $x_{-(f+2)}, x_{-(f+3)}, \ldots, x_{-(f+m)}$ are bits $f+2, f+3, \ldots, f+m$ in the normalised result. Those bits may be $f+3, f+4, \ldots, f+m$ in the unnormalised result, if there is one left shift during normalisation.

Thus, the $\mathrm{OR}(x_{-(f+3)}, x_{-(f+4)}, \ldots, x_{-(f+m)})$ function may be compacted in a single bit, known as *sticky bit*. In other words, the OR stores in a single bit the remaining bits of the result used for rounding.

In a similar way, using decimal in (4.14), (4.15), (4.16) or (4.17) appears *c*, which is the *sticky bit* for the binary case.

Therefore, we have to calculate $f+2$ digits in the result (the f most significant bits plus the *guard digit* and the *round digit*), to obtain a properly rounded result in any of the four basic arithmetic operations, and with any of the rounding schemes presented before. And finally, identify whether the remaining bits in the result are all zero or not (*sticky bit*).

4.5.5 Leading Zeros Detection

We have seen in previous parts that, to normalise the result obtained in a subtraction and in a quotient, we may need to shift these results some positions to the left. The amount of left shifts is the number of zeros to the left of the most significant 1.

If we want to design a floating-point arithmetic unit to be as fast as possible, we should perform a fast detection of the leading zeros in the result of a subtraction or a quotient. Next, we illustrate this implementation with a simple example.

Example 4.8 Design a leading zeros detector for seven bits ($x_6\ x_5\ x_4\ x_3\ x_2\ x_1\ x_0$), which are supposed to be the most significant bits in the result. With $z_2\ z_1\ z_0$ we indicate the amount of leading zeros following the Table in Fig. 4.13a. Circuit in Fig. 4.13b generates the intermediate signals *abcdefg*. Using these signals as inputs in a codifier with priority, they generate the outputs $z_2\ z_1\ z_0$. These outputs may be used as inputs in a left shifter.■

The method illustrated in Example 4.8 may be used to detect any number of leading zeros.

$x_6x_5x_4x_3x_2x_1x_0$	*abcdefg*	$z_2z_1z_0$
1xxxxxx	0000000	000
01xxxxx	1000000	001
001xxxx	1100000	010
0001xxx	1110000	011
00001xx	1111000	100
000001x	1111100	101
0000001	1111110	110
0000000	1111111	111

(a)

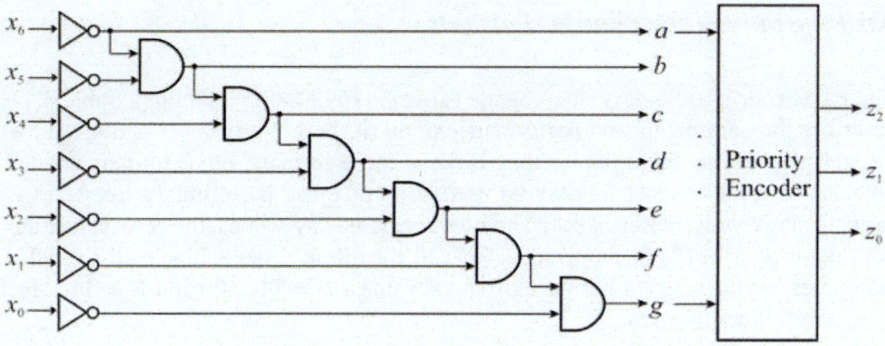

(b)

Fig. 4.13 Leading zeros: **a** Table; **b** Circuit

4.6 The IEEE 754 Standard

The universally used standard for floating-point representation and operations since it was approved in 1985 until it was updated in 2008 was the ***IEEE Standard for Binary Floating-Point Arithmetic*** (**IEEE Std 754-1985**) [1]. Before this standard was introduced, each computer maker used its own representations, not only for real numbers but for any other data type. As the different representations were habitually incompatible, it was very complex to migrate programs or data to other computers.

It was clear the convenience of having a standard for everyone to follow. This standard was IEEE 754-1985. It specified several aspects that must be taken into account, but it did not establish a concrete way to implement them using the following parameters: $\boldsymbol{B} = 2$, normalisation following (4.4) with hidden bit, significand with **SM** and biased exponent representations.

In 1987 another standard was approved, the ***IEEE Standard for Radix-Independent Floating-Point Arithmetic*** (**ANSI/IEEE Std 854-1987**) [2], allowing

$B = 2$ or $B = 10$, and it was less restricted in its specifications. The $B = 10$ radix was of interest for some applications (see Sect. 1.7).

In 2008 the ***IEEE Standard for Floating-Point Arithmetic*** **(IEEE Std 754™-2008)** [3] was approved. It is a revision of 1985 standard. All implementations and systems following the 1985 standard are valid within the 2008 standard.

The following descriptions apply exclusively to the IEEE 754-2008 standard, specifying floating-point formats for binary and decimal numbers. There are two kinds of formats: interchange formats for information sharing among systems, and arithmetic formats to implement arithmetic operations inside a system.

The first ones are more specific hence they will be the ones to start with.

4.6.1 Binary Interchange Formats

There are four basic binary interchange formats (16, 32, 64 *y* 128 bits). Table 4.11 describes the parameters and performance of all of them. Figures 4.1 and 4.2 show the meaning of the Table parameters. Beyond these formats, other formats greater than 128 and multiple of 32 may be used. Any of those basic binary interchange formats may represent normalised and non-normalised N values, the zero value, the not a number value (**NaN**, described later) and the infinity, depending on the E field. All values are normalised but the extreme E values: $E = 00\ldots00$ and $E = 11\ldots11$, using the following cases:

(a) $E \neq 00\ldots00$ and $E \neq 11\ldots11$, N is a normalised number with implicit 1 with F a pure fractional number (without integer part) represented in SM:

$$N = (-1)^s(1 + F)2^{(E-\text{bias})}$$

(b) $E = 00\ldots00$ and $F \neq 00\ldots00$, N is a non-normalised number (subnormal) with implicit 0:

$$N = (-1)^s(0 + F)2^{(-\text{bias})}$$

(c) $E = 00\ldots00$ and $F = 00\ldots00$, N is a signed zero:

$$N = (-1)^s(+0)$$

(d) $E = 11\ldots11$ and $F \neq 00\ldots00$, N is not a number (**NaN**).

(e) $E = 11\ldots11$ and $F = 00\ldots00$, N is a signed infinity:

$$N = (-1)^s(+\infty)$$

Table 4.11 IEEE 754 standard

Parameters				
n	16	32	64	128
e	5	8	11	15
f	10	23	52	112
EX_{max}	+15	+127	+1023	+16383
EX_{min}	−14	−126	−1022	−16382
Bias	+15	+127	+1023	+16383
N_{max}	$\approx 2^{15} = 32768$	$\approx 2^{128} \approx 3.4 \cdot 10^{38}$	$\approx 2^{1024} \approx 1.8 \cdot 10^{308}$	$\approx 2^{16384} \approx 1.2 \cdot 10^{4932}$
N_{min}(N normal)	$2^{-14} \approx 6.1 \cdot 10^{-5}$	$2^{-126} \approx 1.2 \cdot 10^{-38}$	$2^{-1022} \approx 2.2 \cdot 10^{-308}$	$2^{-16382} \approx 3.36 \cdot 10^{-4932}$
N_{min}(N abnormal)	$2^{-25} \approx 3 \cdot 10^{-8}$	$2^{-149} \approx 3.4 \cdot 10^{-38}$	$2^{-1074} \approx 8.29 \cdot 10^{-318}$	$2^{-16495} \approx 3.2 \cdot 10^{-4966}$
DR	$\approx 2^{29} \approx 5.4 \cdot 10^{8}$	$\approx 2^{254} \approx 2.9 \cdot 10^{76}$	$\approx 2^{2046} \approx 8.1 \cdot 10^{615}$	$\approx 2^{32766} \approx 3.6 \cdot 10^{9863}$

Parameters of basic binary interchange formats

4.6.2 Decimal Interchange Formats

There are three basic decimal interchange formats (32, 64 and 128 bits). Table 4.12 describes the parameters and performance of all of them. Figures 4.1 and 4.2 show the meaning of the Table parameters. Beyond these formats, it may be used other formats greater than 128 and multiple of 32. Any of those basic decimal interchange formats may represent *N* finite values, the zero value, the not a number value (**NaN**, described later) and the infinity, depending on the *E* field.

Depending on the implementation used, the significand *F* may be encoded in decimal or in binary (one of those must be selected). In case of decimal encoding, it is a packed encoding **CHC** (see Sect. 1.7.3). In case of binary encoding with *b* bits, the significand has values between 0 and $2^b - 1$. The *N* finite numbers are the result of (s, X, B) following:

$$N = (-1)^s B10^{(X-\text{bias})}$$

where *B* is an integer number made with some *E* field bits concatenated (Fig. 4.1) with field *F*, and *X* is made from some *E* field bits depending on the case, as seen next. The initial *E* field bits specify the type of the represented value.

(a1) If the significand implementation uses decimal encoding and $E_1 = 0$ or $E_1E_2 = 10$, then *N* is a finite number and B most significant digit is $4E_3 + 2E_4 + E_5$ (a value between 0 and 7); the two *X* most significant bits are E_1E_2, being $X = E_1E_2E_6E_7 \ldots E_e$.
This case includes the zero representation: if $E_1E_2 = 00$, 01, 10, and the remainder bits of *E* and *F* are zero, then *N* is a signed zero:

$$N = (-1)^s(+0)$$

Table 4.12 IEEE 754 standard

Parameters			
n	32	64	128
e	11	13	17
f	20	50	110
d (digits)	7	16	34
EX_{max}	+96	+384	+6144
EX_{min}	−95	−383	−6143
Bias	+101	+398	+6176
N_{max}	$\approx 10^{97}$	$\approx 10^{400}$	$\approx 10^{6178}$
N_{min}	10^{-101}	10^{-398}	10^{-6143}
DR	$\approx 10^{198}$	$\approx 10^{798}$	$\approx 10^{12321}$

Parameters of basic decimal interchange formats

(a2) If the significand implementation uses decimal encoding and $E_1E_2E_3 = 110$ or $E_1E_2E_3E_4 = 1110$, then N is a finite number and B most significant digit is 8 + E_5 (i.e. 8 or 9); the two X most significant bits are E_3E_4, with $X = E_3E_4E_6E_7 \ldots E_e$.

(b1) If the significand implementation uses binary encoding and $E_1E_2 = 00$, 01, 10, then N is a finite number and B most significant bits are $E_{e-2}E_{e-1}E_e$; with exponent $X = E_1 \ldots E_{e-3}$.

(b2) If the significand implementation uses binary encoding and $E_1E_2E_3E_4 = 1100$, 1101, 1110, then N is a finite number and B most significant bits are $100E_e$; with exponent $X = E_3 \ldots E_{e-1}$.

(c) If $E_1E_2E_3E_4E_5 = 11111$, N is **NaN**.

(d) If $E_1E_2E_3E_4E_5 = 11110$, N is a signed infinite:

$$N = (-1)^s(+\infty)$$

From the previous description we may see that when the decimal digits to represent are less than the number of available digits, a floating-point decimal number may have different representations, all of them valid. In particular, using d digits for the significand and having n digits to represent, $n < d$, then it is clear that there are $d - n + 1$ different representations, each one with the appropriate exponent value.

In addition, the scientific notation may be used as well. In this case the N finite numbers are described with (s, e, B) following:

$$N = (-1)^s B 10^e$$

But now B is interpreted as a fractional number with the first digit b_0 as the integer part, i.e. $B = b_0.b_1 \ldots b_{p1}$. In order to have the same number with both B interpretations (as an integer or fractional number), it follows that the e exponent has to be:

$$e = X - \text{bias} + (p-1)$$

with p the decimal digits number being used.

When a significand uses binary encoding, the maximum representable value is the same as in decimal encoding (i.e. $10^d - 1$, for d decimal digits). Binary values higher than $10^d - 1$ must be detected and handled as zeros.

If a decimal operation result must be rounded, it is represented with the minimum possible exponent value, to save for the significand the maximum available remaining digits. If that is not the case, it is preferable to select an exponent depending on the operands exponents.

4.6.3 Zero, Infinite and NaNs

To represent the zero and infinite extreme values, needed for arithmetic operations but non-representable with normalised formats, some specific values are reserved for E and F, as described earlier.

When an overflow happens, the result representation is $\pm\infty$, albeit being flagged somehow. A division by 0 also generates $a \pm \infty$, an exceptional case the system implementation may also flag if necessary.

In a similar way, underflow cases generate a 0 result representation, aside of other possible system flags. There are actually two zeros, depending on the sign of the underflow value hence generating $+0$ or -0. The zero sign may be relevant just in some cases, like division by 0. With binary formats there is one zero representation only, with decimal formats there are multiple ones.

The **NaN** *(Not a Number)* value represents special cases, like Table 4.14 invalid operations. In addition, when one of the operands is a **NaN** value, the result is a **NaN** value as well. There are two **NaN** encodings (**sNaN** and **qNaN**), allowing programmers to add more information about the root of those exceptional cases.

4.6.4 Arithmetic Formats

Arithmetic or extended formats may be used to represent either the operands or the results of operations. They are considered transient formats. Before a floating-point operation usually one of the operands needs to be adjusted, i.e. its format has to be modified. Similarly, a floating-point result may have more digits than any of the operands.

In all these cases, to avoid losing precision, it is advised to use formats that increment the precision: the extended formats. In other words, usually before an arithmetic operation it may be useful to change one of the operands from an interchange format to an arithmetic format. After completing the operation, the result in arithmetic format is converted into an interchange format, if that were the case.

As these arithmetic formats are used inside a system only, the designer has a good amount of freedom to define them. They must conform to the minimum requisites of Table 4.13.

Table 4.13 IEEE 754 standard

Parameters	$B = 2$	$B = 2$	$B = 2$	$B = 10$	$B = 10$
Related with	32 bits	64 bits	128 bits	64 bits	128 bits
n digits	≥32	≥64	≥128	≥22	≥40
E_{max}	≥+1023	≥+16383	≥+65535	≥+6144	≥+24576
E_{min}	≤−1022	≤−16382	≤−65534	≤−6143	≤−24575

Extended formats parameters

4.6.5 Formats and Roundings

The standard allows various interchange formats for binary and decimal encodings. Tables 4.11 and 4.12 show those parameters (see Figs. 4.1 and 4.2 for their interpretation). Binary normalised numbers represent $s{\cdot}1.F{\cdot}2^{\mathrm{E-bias}}$ values using a hidden bit hence the significand has $f + 1$ bits.

Underflow is gradual with non-normalised formats ($E = 0$ y F # 0), obtaining $s0.F{\cdot}2^{-\mathrm{bias}+1}$ representable values.

Extended formats have only fixed maximum and minimum values for basic parameters as shown in Table 4.13. Thus, there is a lot of freedom to define a concrete implementation.

Any implementation must include at least the simple normal format. Extended formats may be used for intermediate computations where significands must have more bits.

The default rounding method is rounding half to even, although it is possible to select rounding towards zero, rounding half up or rounding half down.

4.6.6 Operations

All IEEE 754 standard implementations must include a set of operations, including addition ($a + b$), subtraction ($a - b$), multiplication ($a \times b$), division quotient (a/b), division remainder (**remainder** (a, b)), square root $\left(\sqrt{a}\right)$, comparison operators (a.**REL**.b), and conversions among different formats.

With numeric formats configuration of Table 4.1, it is clear that to compare two numeric values using binary interchange format, it may be implemented as a binary comparison of numbers comprising all bits of (s, E, F) fields. This approach cannot be used when using decimal interchange formats, as the same number may have different representations.

The definition of division remainder, $r =$ **remainder** (a, b) with b # 0, is as follows:

$$r = a - b \times n$$

with n the nearest integer to a/b exact value. In the half way point, i.e. when $|n - a/b| = 1/2$, an even n is taken. The remainder is always exact with this definition. When $r = 0$, its sign is that of a.

The square root must be defined for all operands ≥ 0 but -0, which has a special definition: $\sqrt{-0} = -0$.

Table 4.14 shows all cases where operations result is **NaN**.

Table 4.14 IEEE 754 standard

Operation	NaN case
± (addition/subtraction)	$\infty + (-\infty)$
× (multiplication)	$0 \times \infty$
/ (quotient)	0/0, ∞/∞
REM (remainder)	a**REM**0, ∞**REM**a
√ (square root)	$\sqrt{a}$, a negative

Operations generating NaN

4.7 Circuits

In this part we describe circuits for hardware implementations of the different arithmetic operations. Rounding the result has not been taken into account in any case, although it would be very easy to calculate the necessary information to do it (*guard bit, round bit* and *sticky bit*).

In the binary circuits, it is assumed in all cases the availability of the two operands significands incorporating a hidden 1 to the circuits, i.e., with the standard format $1x_{-1}x_{-2}\ldots x_{-f}$. This operation is known as *unpacking* and no circuit is required to implement it. Similarly, *unpacking* is supposed to be implemented in the decimal circuits.

4.7.1 Adder/Subtractor

As both operations (addition and subtraction) share many elements, a unique circuit may be designed. Figure 4.14 shows a possible scheme for this adder/subtractor circuit.

First the signal $\bar{S}/R$ has to be generated to specify which operation is going to be performed. Also, the signals $A < B$ and $D < G$ have to be generated to decide the order in the operands. Block $\boldsymbol{M-m}$ is used to compute the exponents difference and it is an input for the right shifter, where the smaller significand is conveniently shifted. The significands are added or subtracted in block ±, depending on $\bar{S}/R$. The addition can generate a carry, **a**, which will denote that the result has to be right shifted and the resultant exponent has to be increased by one. The significand may be shifted to the right in multiplexor $\boldsymbol{M_d}$, which two inputs are the addition result and the addition result shifted to the right. The exponent is increased in block $\boldsymbol{M+1}$.

In the subtraction the number of leading zeros has to be detected, $\boldsymbol{c_i}$. Its value determines the left shifter operation. On the other hand, the exponent has to be decreased by $\boldsymbol{c_i}$ with subtractor $\boldsymbol{M-c_i}$. Finally, on the significand and the exponent, multiplexor $\boldsymbol{M_e}$ selects the correct resulting exponent. The sign depends on $\bar{S}/R$ and $A < B$, and the bit sign s is properly generated with two multiplexers.

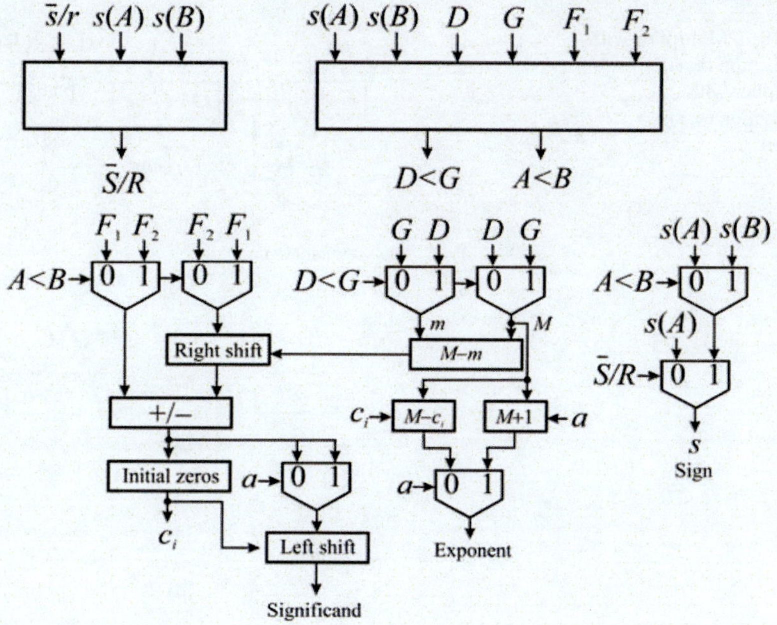

Fig. 4.14 Adder/subtractor

4.7.2 *Multiplier and Divider*

Multiplication implies adding the exponents and multiplying the significands. Thus, when the normalisation is used (4.4), as it is done in binary, this operation can be performed with circuit in Fig. 4.15a.

If the result of $F_1 \times F_2$ has the form 1x.xx…xx, $F_1 \times F_2$, it has to be shifted to the right and the exponent has to be increased by 1. Hence in order to obtain the significand, we choose between $F_1 \times F_2$ or $F_1 \times F_2$ shifted to the right; and the exponent is $D + G$ or $D + G + 1$. It is enough to use an **EXOR** gate to obtain the sign bit.

Multiplication may be performed with circuit of Fig. 4.15b using normalisation (4.3). In this case, if the most significant digit in the significand $F_1 \times F_2$ is 0, $F_1 \times F_2$ has to be shifted to the left and the exponent has to be decreased by 1. Thus, in order to obtain the resulting significand, we choose between $F_1 \times F_2$ or $F_1 \times F_2$ shifted to the left; and the exponent is $D + G$ or $D + G - 1$. It is enough to use an **EXOR** gate to obtain the sign bit, as in normalisation (4.4).

The divider circuit of Fig. 4.15c computes the quotient in a division. It is similar to the multiplier circuit. The significands are divided, F_1/F_2, and the exponents are subtracted. If the most significant bit in F_1/F_2 is zero, F_1/F_2 has to be shifted to the left and $D - G$ has to be decreased. Thus, in order to obtain the resulting significand,

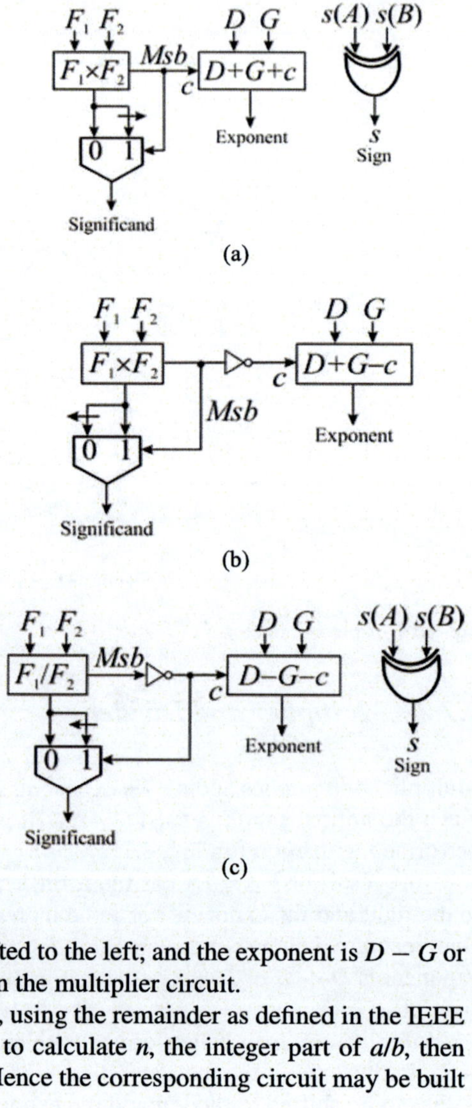

Fig. 4.15 **a** Multiplier with normalisation (4.4); **b** Multiplier with normalisation (4.3); **c** Divider

we choose between F_1/F_2 or F_1/F_2 shifted to the left; and the exponent is $D - G$ or $D - G - 1$. The sign bit is obtained as in the multiplier circuit.

To obtain the remainder in a division, using the remainder as defined in the IEEE 754 standard, $r = a - b \times n$, we have to calculate n, the integer part of a/b, then multiply n by b and subtract it from a. Hence the corresponding circuit may be built joining already described circuits.

4.7.3 *Binary Square-Root*

Let $A = (-1)^a \cdot F \cdot 2^E$ be a non-zero positive real number in floating point binary format, with a = 0 and even E exponent. Its square root may be obtained with:

$$\sqrt{A} = \sqrt{F} \cdot 2^{E/2}$$

To check if E is even (see Sect. 1.4.3) we have to consider the least significant bit: when the bias is odd as it happens for the binary numbers in the IEEE 754 standard, where even numbers end with 1 and odd numbers end with 0. If E is odd, it is decreased (or it is increased, depending on how $\sqrt{F}$ is calculated). In what follows, we will suppose that when E is odd, it is decreased. Then F is properly shifted. Thus, an even E exponent is always processed.

In this case with E even, $E = x_{e-1}x_{e-2}\ldots x_1x_0$, it can be checked that $E/2$ is expressed as $E/2 = x_{e-1}\bar{x}_{e-1}\ldots x_2x_1$. That is, to obtain $E/2$ it is enough to perform one shift, filling with the most significant bit and then to complement the second most significant bit. It is shown in Fig. 4.16a and it can be checked in the Table in Fig. 4.16b, with some values for E even and the corresponding $E/2$ values.

Figure 4.16c shows one possible circuit to calculate the square-root. The inputs in the multiplexor M are F shifted to the left or F. This is selected by the least significant bit of E. The output in the multiplexor is the input in block $\sqrt{}$, which performs the square-root of fractional numbers with a non-zero integer part. It is clear that $\sqrt{F}$ is a fractional number with integer part 1. Thus, the significand of $\sqrt{A}$ is directly obtained. In block -1 the exponent is decreased if it is odd and directly, with the connections depicted in Fig. 4.16a, the exponent of $\sqrt{F}$ is obtained.

4.7.4 Comment

The circuits described and the comparator shown in Sect. 4.5.1, represent the basic blocks for the hardware implementation of a floating-point operation unit. Only conversion circuits are missing, which are easy to implement, and the singular cases processing, such as $\sqrt{-0}$.

4.8 The Logarithmic System

A real number $\boldsymbol{N}$ may be expressed as:

$$\boldsymbol{N} = (-1)^s \cdot \boldsymbol{B}^E$$

with $\boldsymbol{B}$ any predefined radix (usually $\boldsymbol{B} = 2$), s a sign bit, and E a real number. This is the logarithmic representation system. Clearly:

$$E = \log_{\mathbf{B}}|\boldsymbol{N}|$$

In the logarithmic system, a real number $\boldsymbol{N}$ is defined by pair (s, E). If the representable range value has values lesser than 1, E must have negative values.

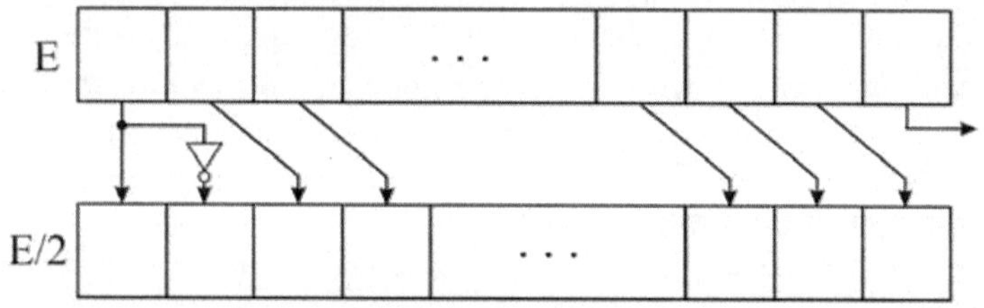

(a)

E	***E/2***	***E***	***E/2***
126→11111101	63→10111110	−126→00000001	−63→01000000
124→11111011	62→10111101	−124→00000011	−62→01000001
122→11111001	61→10111100	−122→00000101	−61→01000010
6→10000101	3→10000010	−6→01111001	−3→01111100
4→10000011	2→10000001	−4→01111011	−2→01111101
2→10000001	1→10000000	−2→01111101	−1→01111110

(b)

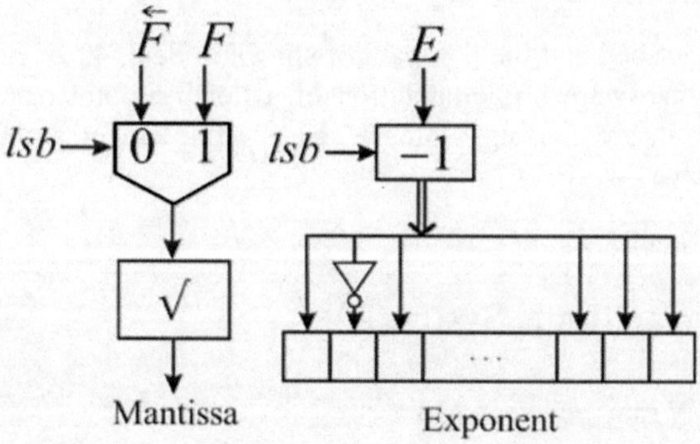

(c)

Fig. 4.16 **a** Shifter; **b** Examples; **c** Square-root

E may use any representation system: **SM**, in complement or biased. Taking into account that the most frequent exponent operations are addition and subtraction, as it is seen next, two's complement representation is the usual one for *E*.

The logarithmic system does not have a standard representation for zero and infinite, like in floating-point. To use those values, a specific definition has to be set for them.

Table 4.15 Positive representable values in the logarithmic system with $e = 3, f = 4$ and $\mathbf{B} = 2$

f\e	000	001	010	011	100	101	110	111
0000	1	2	4	8	0.062500	0.125000	0.250000	0.500000
0001	1.044274	2.088548	4.177095	8.354190	0.065267	0.130534	0.261068	0.522137
0010	1.090508	2.181015	4.362031	8.724062	0.068157	0.136313	0.272627	0.545254
0011	1.138789	2.277577	4.555155	9.110309	0.071174	0.142349	0.284697	0.569394
0100	1.189207	2.378414	4.756828	9.513657	0.074325	0.148651	0.297302	0.594604
0101	1.241858	2.483716	4.967431	9.934862	0.077616	0.155232	0.310464	0.620929
0110	1.296840	2.593679	5.187358	10.374716	0.081052	0.162105	0.324210	0.648420
0111	1.354256	2.708511	5.417022	10.834044	0.084641	0.169282	0.338564	0.677128
1000	1.414214	2.828427	5.656854	11.313708	0.088388	0.176777	0.353553	0.707107
1001	1.476826	2.953652	5.907305	11.814609	0.092302	0.184603	0.369207	0.738413
1010	1.542211	3.084422	6.168843	12.337687	0.096388	0.192776	0.385553	0.771105
1011	1.610490	3.220981	6.441961	12.883923	0.100656	0.201311	0.402623	0.805245
1100	1.681793	3.363586	6.727171	13.454343	0.105112	0.210224	0.420448	0.840896
1101	1.756252	3.512504	7.025009	14.050017	0.109766	0.219532	0.439063	0.878126
1110	1.834008	3.668016	7.336032	14.672065	0.114826	0.229251	0.458502	0.917004
1111	1.915207	3.830413	7.660826	15.321652	0.119700	0.239401	0.478802	0.957603

Example 4.9 Let's have a logarithmic system with $n = 8$, with one sign bit and 7 remaining bits: 3 for the exponent integer part ($e = 3$) and 4 for the exponent fractional part ($f = 4$). The representation of the exponent is binary with two's complement. Tables 4.15 and 4.16 show the positive representable values with $\boldsymbol{B} = 2$ and $\boldsymbol{B} = 16$ respectively (there are corresponding negative values as well).

Comparing both Tables, it is clear it behaves like floating-point: a bigger radix means a bigger dynamic range, albeit with a smaller points' density. ■

4.8.1 Conversions

The usage of the logarithmic system requires the conversion of the more natural fixed-point and floating-point representations into the logarithmic representation and vice versa, i.e. to compute logarithms and antilogarithms. Chapter 8 details the computing procedures for them.

Table 4.16 Representable values in the logarithmic system with $e = 3, f = 4$ and $\mathbf{B} = 16$

f\\e	000	001	010	011	100	101	110	111
0000	1	16	256	4096	0.000015	0.000244	0.003906	0.062500
0001	1.189207	19.02731	304.4370	4870.992	0.000018	0.000290	0.004645	0.074325
0010	1.414214	22.62742	362.0387	5792.619	0.000022	0.000345	0.005524	0.088388
0011	1.681793	26.90869	430.5390	6888.623	0.000026	0.000411	0.006570	0.105112
0100	2.000000	32.00000	512.0000	8192.000	0.000031	0.000488	0.007813	0.125000
0101	2.378414	38.05463	608.8740	9741.985	0.000036	0.000581	0.009291	0.148651
0110	2.828427	45.25483	724.0773	11585.28	0.000043	0.000691	0.011049	0.176777
0111	3.363586	53.81737	861.0779	13777.25	0.000051	0.000821	0.013139	0.210224
1000	4.000000	64.00000	1024.000	16384.00	0.000061	0.000977	0.015625	0.250000
1001	4.756828	76.10926	1217.748	19483.97	0.000073	0.001161	0.018581	0.297302
1010	5.656854	90.50967	1448.155	23170.48	0.000086	0.001381	0.022097	0.353553
1011	6.727171	107.6347	1722.156	27554.49	0.000103	0.001642	0.026278	0.420448
1100	8.000000	128.0000	2048.000	32768.00	0.000122	0.001953	0.031250	0.500000
1101	9.513657	152.2185	2435.496	38967.94	0.000145	0.002323	0.037163	0.594604
1110	11.31371	181.0193	2896.309	46340.95	0.000173	0.002762	0.044194	0.707107
1111	13.45434	215.2695	3444.312	55108.99	0.000205	0.003285	0.052556	0.840896

4.8.2 Arithmetic Operations

In a logarithmic system, multiplication translates into addition, division translates into subtraction, power into multiplication and root into division. However, addition and subtraction operations are more complex than in other systems.

4.8.2.1 Addition and Subtraction

Let $\boldsymbol{B}$ be a radix and E, F so that $M = \boldsymbol{B}^E, N = \boldsymbol{B}^F$, the objective is to compute G so that $R = \boldsymbol{B}^G$, with $R = M \pm N$. From E and F it is not easy to compute G.

A first option to implement those operations is to calculate the equivalent floating-point representation using antilogarithms, perform the operations in floating-point and computing the logarithm of the result. It is obvious that this approach is very expensive.

Another option uses lookup tables as described next. In this case, the first step is to compute the result sign. It is clear that the result sign computation is the same as the **SM** representation one (see Sect. 1.4.1). Thus, the magnitude of the operands must be compared. However, as the logarithmic function is increasing monotone, the comparison may be done with the exponents.

Once computed R sign, it does not affect the rest of the computation, assuming that $M \pm N$ is always positive.

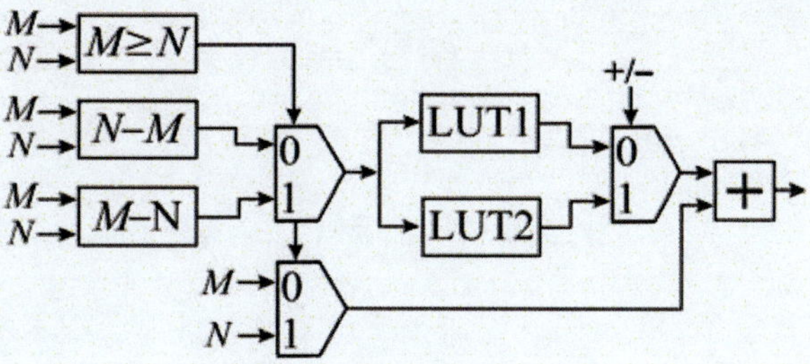

Fig. 4.17 Adder/subtractor in the logarithmic system

$$R = M \pm N = M(1 \pm N/M)$$

it can be obtained

$$G = \lg_B R = \lg_B M(1 \pm N/M) = \lg_B M + \lg_B (1 \pm N/M) = E + \lg_B (1 \pm N/M)$$

On the other hand

$$\lg_B (1 \pm N/M) = \lg_B \left(1 \pm \boldsymbol{B}^{-(E-F)}\right)$$

Functions $\lg_{\mathbf{B}}(1 + \boldsymbol{B}^{-(E-F)})$ and $\lg_{\mathbf{B}}(1 - \boldsymbol{B}^{-(E-F)})$ may be tabulated in lookup tables, so that an adder in the logarithmic system may consist on the blocks shown in Fig. 4.17. The comparator $\boldsymbol{M \geq N}$ selects which difference to use ($\boldsymbol{N - M}$ or $\boldsymbol{M - N}$). If it is an addition it uses the output from **LUT1**, and if it is a subtraction, it uses the output from **LUT2**.

4.8.2.2 Multiplication, Division, Power and Root

The operations of this section are simpler than addition and subtraction, as the logarithmic system ease them a lot. Multiplication translates into addition, division into subtraction, power into multiplication and root into division.

With multiplication, given E and F so that $M = \boldsymbol{B}^E$, $N = \boldsymbol{B}^F$, the objective is to compute G so that $R = \boldsymbol{B}^G$, with $R = M{\cdot}N$. Obviously,

$$M \cdot N = \boldsymbol{B}^E \cdot \boldsymbol{B}^F = \boldsymbol{B}^{E+F} = R = \boldsymbol{B}^G \Rightarrow G = E + F$$

With division we obtain:

$$M/N = \boldsymbol{B}^E / \boldsymbol{B}^F = \boldsymbol{B}^{E-F} = R = \boldsymbol{B}^G \Rightarrow G = E - F$$

A division particular case is the inverse, that is, if $M = \boldsymbol{B}^E$, $M^{-1} = \boldsymbol{B}^{-E}$. Thus, it is immediate that for calculating the inverse of a number it is enough to complement it, assuming that the exponent uses two's complement representation.

With power we have:

$$M^P = \left(B^E\right)^P = B^{E \cdot P} = R = B^G \Rightarrow G = E \cdot P$$

And with root we get:

$$\sqrt[P]{M} = M^{1/P} = \left(B^E\right)^{1/P} = B^{E/P} = R = B^G \rightarrow \mathrm{G} = \mathrm{E/P}$$

4.9 Conclusion

This Chapter reviews all aspects of floating-point representation and its rounding methods, focusing specially on the IEEE 754 standard for binary and decimal encodings. It examines in detail the different arithmetic operations and possible implementation circuits for them, including a section to describe the logarithmic system.

4.10 Exercises

4.1 Build a table with the positive representable values with 8 bits and radix $\boldsymbol{B} = 2$, with one sign bit, four exponent bits and three significand bits, using (4.3) normalisation. Calculate the points' density for the representation.

4.2 How many new representable values will be added when using gradual underflow with $\boldsymbol{B} = 2$ and $f = 23$, using (4.3) normalisation? What is the increment in the dynamic range?

4.3 A simple proof of the stability of a rounding method consists on calculating $S_1 = A + B$ and $S_2 = (A + B) - B$, rounding after each operation with an adequately selected B. For instance, given a floating-point system with (4.3) normalisation, three decimal digits, and $A = 0.728 \times 10^0$ and $B = 0.235 \times 10^{-1}$. When rounding always up after the two previous operations, it is easy to see that $S_2 = 0.729 \times 10^0$, although it should be $S_2 = A = 0.728 \times 10^0$. This anomaly, called drift, happens in some rounding methods. Check with proper examples in what rounding methods this anomaly appears.

4.4 Add the following binary floating-point numbers $A = F_1.2^8$ and $B = F_2.2^{11}$, with $f = 8$ bits, (4.4) normalisation, one hidden bit and von Neumann rounding, with $F_1 = 11010110$ and $F_2 = 00101011$.

4.5 Calculate $A - B$, with A and B two binary floating-point numbers: $A = F_1.2^9$ and $B = F_2.2^7$, with $f = 8$ bits, (4.4) normalisation, one hidden bit and von Neumann rounding, with $F_1 = 01101010$ and $F_2 = 10010101$.

4.6 Multiply the following binary floating-point numbers $A = F_1.2^8$ y $B = F_2.2^{11}$, with $f = 8$ bits, (4.4) normalisation, one hidden bit and von Neumann rounding, with $F_1 = 11010110$ and $F_2 = 00101011$.

4.7 Calculate the quotient of A/B, with A and B binary floating-point numbers: $A = F_1.2^9$ and B $= F_2.2^7$, with $f = 8$ bits, (4.4) normalisation, one hidden bit and von Neumann rounding, with $F_1 = 01101010$ and $F_2 = 10010101$.

4.8 Obtain the corresponding rounding bits to the operations results of exercises 4.4, 4.5, 4.6 and 4.7.

4.9 Calculate the maximum and minimum difference between two consecutive floating-point numbers using the IEEE 754 standard binary interchange format with $n = 16$. Do the same for $n = 32$, 64 and 128.

4.10 Obtain, for exercise 4.9 cases, the maximum and minimum relative spacing, RS, defined as follows, with A and B two consecutive floating-point numbers:

$$RS = \frac{A - B}{A}$$

4.11 Obtain the 18.0375×10^7 number representation in binary interchange format with $n = 16$, using the IEEE 754 standard.

4.12 Obtain the different representations of 18.0375×10^7 and 8.4972×10^{43} in decimal interchange format with $n = 32$ using the IEEE 754 standard, with decimal encoding for the significand.

4.13 Obtain the different representations of 18.0375×10^7 and 8.4972×10^{43} in decimal interchange format with $n = 32$ using the IEEE 754 standard, with binary encoding for the significand.

4.14 In the IEEE 754 standard binary interchange format with $n = 32$, what is the number represented by 10010100011101001001110100101000?

4.15 In the IEEE 754 standard decimal interchange format with $n = 32$, using decimal encoding for the significand, what is the number represented by 10010100011101001001110100101000?

4.16 In the IEEE 754 standard decimal interchange format with $n = 32$, using binary encoding for the significand, what is the number represented by 10010100011101001001110100101000?

4.17 Let be a logarithmic system with $n = 8$, one sign bit, and the remainder 7 bits are distributed as follows: 4 for the integer part ($e = 4$) and 3 for the fractional part ($f = 3$) of the exponent. Obtain Tables similar to 4.15 and 4.16 with $\boldsymbol{B} = 2$ and $\boldsymbol{B} = 16$ respectively.

4.18 Obtain the lookup tables **LUT1** and **LUT2** of Fig. 4.17 using 4 lookup bits.

4.19 Let be a logarithmic system with $\boldsymbol{B} = 10$ and $n = 16$, with one sign bit, 7 bits for the exponent integer part ($e = 7$) and 8 bits for the exponent fractional part ($f = 8$). The exponent representation is in binary, with two's complement. Calculate the maximum and minimum differences between two consecutive representable numbers and the maximum and minimum relative spacing in this system.

References

1. IEEE Standard for Binary Floating-Point Arithmetic (ANSI/IEEE Std 754-1985), IEEE (1985)
2. IEEE Standard for Radix-Independent Floating-Point Arithmetic (ANSI/IEEE Std 754-1987), IEEE (1987)
3. IEEE Standard for Floating-Point Arithmetic (IEEE Std 754™-2008), IEEE (2008)

Chapter 5
Addition and Subtraction

After the presentation of the basics of number systems and the introduction of the first basic arithmetic circuits, this chapter explores in detail addition and the circuits implementing the addition operation in their different variations. Addition is the basic arithmetic operation, so the circuits presented in this chapter are the foundation for the implementation of the remaining arithmetic operations, as it will be discussed in the following chapters. On the other hand, subtraction is just a variation of addition, only replacing the carry concept with that of borrow.

Once the basic concepts of addition are introduced, especially the role of carries, the basic structures for carry propagation will be presented. Carry-look ahead adders, carry-skip adders and prefix adders will be thus discussed. As a variation, conditional adders will be considered next, while the last section will be devoted to multioperand adders. Here, the concept of carry-save adder will be presented along with adder trees, which are key to multiplier implementation and will be thus further detailed in the next chapter. In any case, the reader may find some redundancies with Chap. 2, but the objective has been to make this new Chapter self-contained.

5.1 Introduction

Addition is the basic arithmetic operation, as the rest of operations are computed with algorithms based on addition. It may be possible that many people have not realized this, but the reader only has to remember how arithmetic fundamentals are taught to schoolchildren. Thus, they first learn how to add and, once they have mastered addition, moving to subtraction is straightforward since this operation is based on the same principles and the only difference is replacing the concept of *carry* with that of *borrow*. After that, students learn how to multiply using algorithms based on the generation of partial products and the addition of these. Finally, once addition, subtraction and multiplication are known, it is possible to divide using successive products and subtractions, which generate the different digits of the quotient and the

A. Lloris Ruiz et al., *Arithmetic and Algebraic Circuits*, Intelligent Systems Reference Library 201, https://doi.org/10.1007/978-3-030-67266-9_5

final remainder. In the same way as, didactically, addition and subtraction are the basic operations for performing more complex computations, the same principle is translated to computer arithmetic. As a matter of fact, small microprocessors without specific hardware and instructions for multiplication or division can implement those operations with software algorithms only using addition and subtraction operators. Accordingly, many structures implementing multipliers and dividers are based on adders and subtractors. Thus, any approximation to computer arithmetic, to arithmetic circuits or to how to implement arithmetic operators with digital systems must start from a detailed analysis of addition and the available circuits and algorithms for the efficient realization of this operation.

5.2 Basic Concepts

The way addition is translated to an electronic circuit can be easily understood just having in mind how two figures are added on paper, independently of the radix (given that a weighted, positional representation is used), as illustrated in the following Example.

Example 5.1 The addition in Fig. 5.1 represents the usual way addition is performed on paper, over each digit (units, tens and hundreds in this case). It is evident that an overflow may occur for every digit, *i.e.*, the result of the addition may be a number larger than the radix (for example, $6 + 4 = 10$ or $7 + 9 = 16$). Those readers who have not yet forgotten their first school lessons on arithmetic may remember that this overflow does not prevent the operation, but does indeed introduce one of the fundamental concepts in addition: **carry**. Thus, $6 + 4 = 10$ for the units shows that the addition of both input units results in 10 units, which can be alternatively expressed as one ten and zero units. Consequently, the resulting digit for the units will be 0, while we must add an additional ten to the corresponding result for the next digit. This propagation of an extra figure to the immediate next most significant position as a result of the addition of the two digits of the current position is what we know as carry. Consequently, addition is performed starting from the units (least significant digits) and propagating a possible carry to the tens, which are added in the same way, and so on. In this way, there is no limitation for the length of the numbers to be added (beyond paper availability and our own patience). The human brain makes use of a single adding unit (usually radix-10), which will be called in the following full adder and that is in charge of adding the digits of every position,

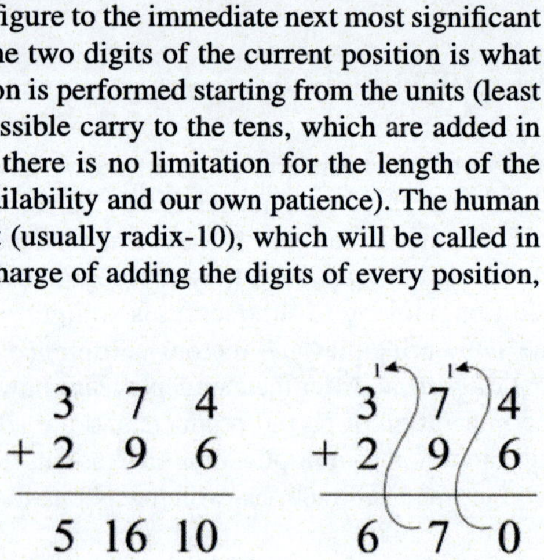

Fig. 5.1 Example of radix-10 addition and the role of carries

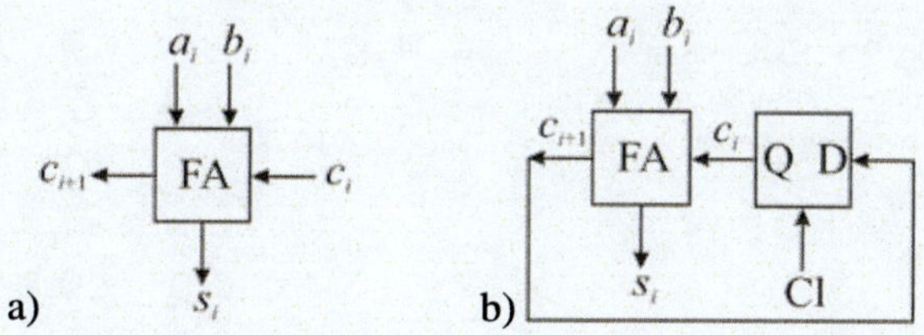

Fig. 5.2 **a** Full adder. **b** Serial adder using a single full adder

along with the storage of the generated carry. This will be used as an extra input when adding the following position.

■

The scheme above, completely independent of the representation radix, can be easily particularized to radix 2 and its direct translation to a digital system. In fact, the simplicity of radix-2 representation is illustrated with a simple addition:

$$
\begin{array}{cccccc}
 & 1 & 0 & 1^1 & 1^1 & 1 \\
+ & 0 & 1 & 0 & 0 & 1 \\
\hline
 & 1 & 0 & 1 & 0 & 0
\end{array}
$$

From the operation above, it can be noted that the addition operation over every position is made of three input bits (the two bits to be added and the carry propagating from the previous digit) and generates two output bits (the result bit itself and the carry to be propagated to the next most significant position). Thus, two three-variable functions define what is known as a full adder (Fig. 5.2a), which was already described in Chap. 2 and corresponds to:

$$
\begin{aligned}
s_i &= a_i \oplus b_i \oplus c_i \\
c_{i+1} &= a_i b_i + a_i c_i + b_i c_i
\end{aligned}
\tag{5.1}
$$

These equations define the basic block in many arithmetic circuits, both for the direct realization of addition and as building block for circuits implementing other operations such as multiplication. This is the reason why, as shown in Fig. 5.2b and discussed in Sect. 2.2.3, a single full adder may implement the addition of n-bit numbers by operating during n clock cycles, only requiring a flip-flop to store the resulting carry from every position, which will be then used as input carry in the next clock cycle when adding the next position. This is the simplest possible adder.

Before continuing further, it is convenient to analyze in detail the similarities existing between addition and subtraction, as illustrated in the following Example.

Example 5.2 Figure 5.3 shows the way a radix-10 subtraction is carried out digit-by-digit, in a similar way to that depicted in Example 5.1 for addition. However, the main difference lies in the fact that subtraction does not propagate a carry, but what is called **borrow**. If carry represents the overflow in the addition over a single digit,

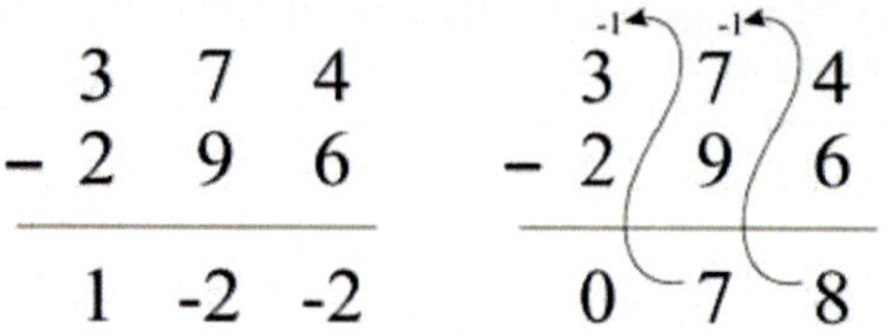

Fig. 5.3 Example of radix-10 subtraction and the role of borrow

borrow corresponds to the case when the magnitude of the subtrahend is larger than that of the minuend ($4 - 6 = -2$), so the operation on that digit is going to influence the next most significant position. In this way, the operation in Fig. 5.3 over the units becomes $14 - 6 = 8$, which requires the extra ten than has been used to avoid a negative result to be discounted when operating over the next position, tens in this case. Once the focus of the operations has shifted to the tens, an additional borrow is generated and propagated to the hundreds.

■

Example 5.2 makes evident that borrow plays the same role for subtraction that carry does for addition, and have to be propagated from one position to the next most significant one. In the case of radix 2, the following operation:

	1^{-1}	0^{-1}	0^{-1}	0	1
−	0	1	1	1	1
	0	0	0	1	0

illustrates again how the subtraction operation over every position requires three input bits (those corresponding to minuend and subtrahend, and the borrow propagated from the preceding position) and generates two output bits (the subtraction result itself, or difference, and the borrow to be propagated to the next position). The following three-variable functions define what is known as full subtractor, already discussed in Sect. 2.3 and whose usual symbol is shown in Fig. 5.4:

$$r_i = a_i \oplus b_i \oplus d_i$$
$$d_{i+1} = \overline{a_i} b_i + \overline{a_i} c_i + b_i d_i \qquad (5.2)$$

Fig. 5.4 Full subtractor

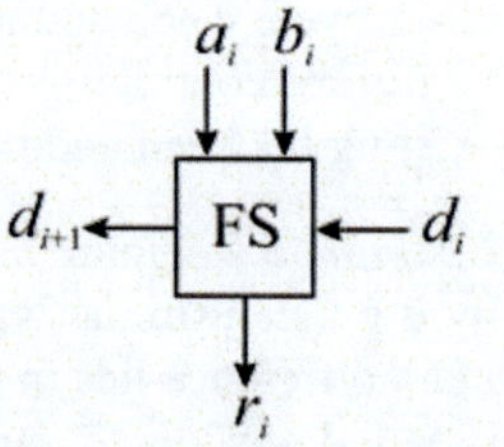

These equations define the basic block for subtracting, and it is evident that the only difference with those of the full adder in Eq. (5.1) lies in the borrow function. In this way, an adder/subtractor can be easily generated by propagating to the next position either the carry or borrow functions, depending on the operation to perform and, thus, the meaning of the input carry/borrow bit.

The main consequence of the similarity of addition and subtraction is that no distinction is usually made between them, which also happens to be the case in this book. As a matter of fact, most of the concepts to be detailed in the following for addition are naturally extended to subtraction with minimal modifications, only replacing carries with borrows. Moreover, the similarity of Eqs. (5.1) and (5.2) shows that the resources required for implementing both operations are almost the same, so specific resources usually found in **FPGAs**, standard-cell libraries, etc., allow implementing both operations. Additionally, once 2's complement representation is considered (as described in Sect. 1.4.2) there is no practical difference between addition and subtraction, as this last operation is nothing else but a sign inversion and an addition.

5.3 Carry Propagation: Basic Structures

From the basic analysis of addition in the previous Section, it is evident that the main characteristic of addition (or subtraction) is the propagation of carries (or borrows) from one digit to the next most significant one. If not for this, the operation could be carried out digit-by-digit without difficulty. In fact, Fig. 5.2b shows the simplest possible structure replicating the addition mechanism in the human brain. However, this option is not the most efficient, since the addition of n-bit operands requires n clock cycles. Consequently, a much smarter structure for implementing addition on a microelectronic circuit is that of Fig. 5.5, where n full adders perform this addition, each over a couple of bits from both operands. While you could think all adders work in parallel, this is not the case as it is necessary to propagate the carry from one digit to the next most-significant. Thus, it is not possible to compute the result bit

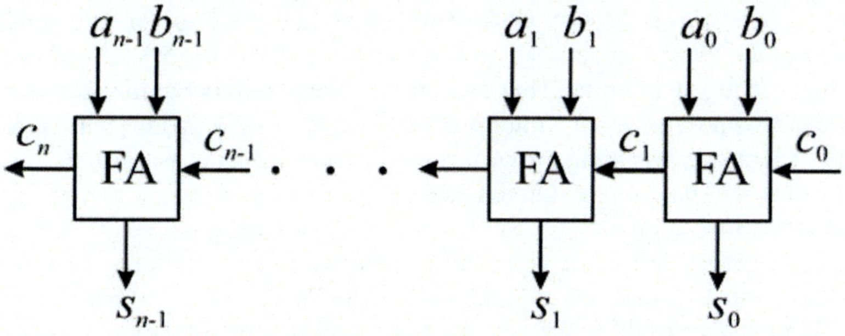

Fig. 5.5 Ripple-Carry Adder (**RCA**), also known as carry-propagation adder (**CPA**)

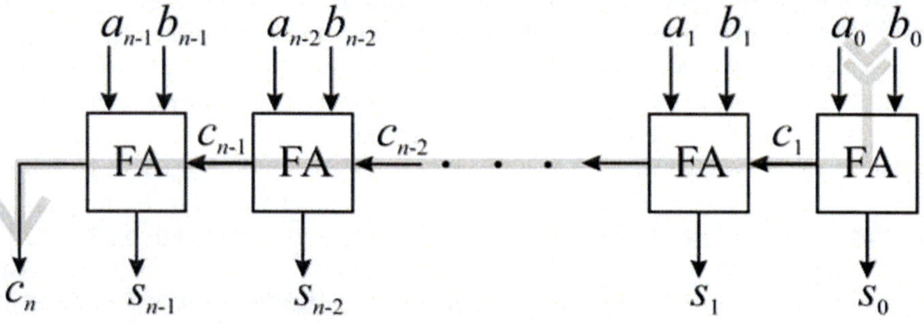

Fig. 5.6 Critical path of a **CPA**

in the 2^1-weight digit, s_1, until the carry from the preceding bit, c_1, has not reached its definitive value. However, once c_1 is established it will be possible to compute s_1 and c_2, which will then make possible the computation of s_2 by the next full adder. Therefore, this structure is based on the propagation of carries from one digit to the next most significant one, so it is known as *Carry Propagation Adder* (**CPA**) or *Ripple-Carry Adder* (**RCA**), and it will be referred to as **CPA** in the following. This is a very popular structure due to its simplicity, and it is present in some form in most of the structures for implementing addition. Even though, it is not adequate for applications requiring great speed or when data size is not small, as the delay associated to the **CPA** is increased linearly with the number of bits, as shown in Fig. 5.6. Thus, the maximum delay associated to a **CPA** of *n* bits is:

$$T_{CPA} = T_{FA}(a_i, b_i \rightarrow c_{i+1}) + (n-2) \times T_{FA}(c_i \rightarrow c_{i+1}) + T_{FA}(c_i \rightarrow s_i) \quad (5.3)$$

where $T_{FA}(a_i, b_i \rightarrow c_{i+1})$ represents the delay associated to a full adder between a change to any of the addition inputs and the corresponding change of the carry-out bit, $T_{FA}(c_i \rightarrow c_{i+1})$ represents the delay of a full adder between the change of the input carry and the subsequent change of the carry output, and $T_{FA}(c_i \rightarrow s_i)$ represents the full-adder delay between a change of the input carry and the settling of the addition bit. The delay in (5.3) obviously corresponds to the worst case scenario, where carry propagation takes place through all the full adders in the **CPA**. This is not usually the case, but the system must take into account the delay in (5.3) to guarantee a correct functioning at any time. However, as this worst case is not the most usual, it is possible to wonder whether or not it is possible to build an adder that takes into account this fact, *i.e.*, it is highly improbable that carry has to be propagated through all digits. This will be discussed in the following subsections. It is worth noting that this **CPA** structure can be used for any radix, particularly for radix 10 (as in Example 5.1). Thus, a decimal full adder is the only block required for building a **BCD CPA** with the very same structure in Fig. 5.5 but for the **BCD** inputs and outputs. In any case, such an adder would behave in the same way described above and Eq. (5.3) would be completely valid taking into account the delays of that decimal full adder.

5.3.1 Considerations on Carry Propagation

It is evident the delay of a **CPA** grows linearly with n, the number of bits of the operands. If the most usual or most probable length of the carry chains (digits where a carry from input c_i is going to be propagated to output c_{i+1}) could be determined, then it would be possible to specify the time required for the completion of the operation. In this way, it is interesting to examine the value of c_{i+1} as a function of c_i from (5.1):

- $c_{i+1} = 1$ even when $c_i = 0$ only if $a_i = 1$ and $b_i = 1$; in this case, the digit under these conditions always generates a carry to the next position.
- if $c_i = 1$, then $c_{i+1} = 1$ if $a_i = 1$ or $b_i = 1$, *i.e.*, carry propagates through this digit.
- even if $c_i = 1$, $c_{i+1} = 0$ when $a_i = 0$ and $b_i = 0$, so even when the digit in question is receiving an input carry, this is not propagated to the next position and the carry chain is thus broken.

A simple statistical analysis of the cases above leads to the conclusion that the probability of generating a carry chain starting from a given position, *i.e.*, generating a carry when $c_i = 0$, is ¼. In the same way, the probability of propagating a carry chain ($c_i = 1$ and $c_{i+1} = 1$) is ½, and the probability of breaking or ending a chain ($c_i = 1$ and $c_{i+1} = 0$) is again ¼. Thus, it can be proved through statistical analysis that the average length of carry chains [3] has a lower bound of $\log_2 n + O(1)$ (keep in mind that the O operator represents the limiting behavior, usually interpreted as complexity, so $O(1)$ implies a constant in this case). In this way, the question arises about whether or not it is possible to build an adder faster than the previously presented **CPA**. In this sense, it must be noted that this result on the average length of carry chains does not imply that the computing time of a **CPA** is similar or even lower than $\log_2 n$, since this a merely statistical result, which assumes that operands are equiprobably distributed among the possible 2^n values. However, any designer may check that assuming $\log_2 n$ as the computation time of a **CPA** will usually lead to incorrect results. Consequently, the statistical analysis of carry propagation exclusively evidences that adders faster than a **CPA** are possible, whose delay boundary is given by $\log_2 n + O(1)$ carry-propagation cycles. In fact, great efforts were dedicated at the beginning of the digital era to the study of the impact of "carry completion" signals on the acceleration of addition, which would later derive into the first alternatives to carry look-ahead.

5.3.2 Basic Carry Look-Ahead

From the analysis above, you may wonder whether it is possible or not to signal the completion of the addition operation of a **CPA**, *i.e.*, all carries have settled to their definitive values and addition bits will thus not present any further change. One of the simplest alternatives is the coding of carry with two bits, rather than one. This allows to explicitly propagate a 0 carry from any point where a carry chain is going to be annihilated, as described in the previous section, and also signal when the operation is complete. Thus, for every digit two bits can be generated, which signal whether the output carry to the next most significant position will depend on the carry input or, on the other hand, only the operand bits define this output carry. With this, these carry-coding bits propagating to the next digit can be defined as:

$$(d_{i+1}, c_{i+1}) = (\overline{a_i + b_i}, a_i b_i) \tag{5.4}$$

so:

- if $d_{i+1} = c_{i+1} = 0$, the output carry will not be defined by the bits of the operands, a_i and b_i, and will depend on the carry arriving to the i-th digit, *i.e.*, the final output carry is not known yet after computing d_{i+1} and c_{i+1}.
- if $d_{i+1} = 0$ and $c_{i+1} = 1$, the output carry will be independent of the carry arriving to the i-th digit and will be 1.
- if $d_{i+1} = 1$ and $c_{i+1} = 0$, the output carry will be independent of the carry arriving to the i-th digit and will be 0.

From the reasoning above, the combination ($d_{i+1} = 0, c_{i+1} = 1$) creates or continues a carry chain, while ($d_{i+1} = 1, c_{i+1} = 0$) signals the termination of carry propagation. In order to define what happens for ($d_{i+1} = 0$, $c_{i+1} = 0$), the carry coding of (5.4) allows to rewrite carry propagation as:

$$\begin{aligned} d_{i+1} &= \overline{a_i + b_i} + d_i \cdot \overline{a_i \cdot b_i} \\ c_{i+1} &= a_i \cdot b_i + c_i \cdot (a_i + b_i) \end{aligned} \tag{5.5}$$

where (d_i, c_i) are the two carry-coding functions arriving to the i-th position. Thus, initializing (d_0, c_0) as $(\overline{c_{in}}, c_{in})$, where c_{in} is the global input carry at the least significant digit, will force the pairs (d_{i+1}, c_{i+1}) in (5.5) to settle to either (0,1) or (1,0) for all positions, once carry propagation is complete through all digits. It is then possible to simply define a carry completion bit as $d_i + c_i$ for each digit, and to combine all of these carry completion bits into a single **AND** gate generating a global signal of the completion of the addition (once all carries have settled to their definitive values). In order to avoid any false completion signal, it is required that no spurious values are allowed for that **AND** gate signal during the propagation of (d_i, c_i) pairs, or to wait for a minimum safeguard delay before it is considered. This allows to signal when the operation is complete, although the computation time of the adder is not yet determinate, since only the operation completion is signaled but the **CPA** mechanism is not modified. Thus, the computation time may be equivalent to the delay of a single full adder, or it may be that of the worst possible case in Eq. (5.3) and the critical path of Fig. 5.6. In any case, this completion signal would make possible to speed up the processing of the next data in an asynchronous system.

If the aim is to speed up addition and not only signal its completion without waiting for the worst case delay, it is required to consider other alternatives that allow to somehow anticipate the input carry to a given digit. In this sense, it is required to observe the effect of the operand digits on the output carry, independently of the resulting addition digit, *i.e.*, to define whether the operand digits generate a carry, propagate to the next most significant position to one arriving from the preceding digit, or, on the opposite, they stop the propagation of any incoming carry to the next position. For this, and having in mind the effect of the bits introduced in Eq. (5.4), carry generation, carry propagation and carry annihilation (or extinction) functions can be defined as:

$$\begin{aligned} g_i &= a_i b_i \\ p_i &= a_i b_i \\ e_i &= \overline{a_i + b_i} = \overline{a_i}\,\overline{b_i} \end{aligned} \tag{5.6}$$

These functions illustrate how the carry propagating to the next most significant position depends on the input carry through the following equation:

$$c_{i+1} = g_i + c_i p_i \tag{5.7}$$

From Eq. (5.7) it is evident that, using these carry generation and propagation functions, fast computation of the carry output from the input carry is possible for every digit. This is the basic idea behind most algorithms for acceleration of addition, since a single **AND-OR** logic level allows to anticipate the output carry for a given digit from the carry generation and propagation functions and the input carry. In fact, these ideas can be extended to any radix using proper expressions for the generation, propagation and annihilation functions, just as Eq. (5.6) defines them for radix 2. Such functions would make the use of Eq. (5.7) straightforward to any radix, so the structures to be presented in the upcoming sections can be easily translated to **BCD** arithmetic.

Finally, one of the basic applications of all the reasoning above is the so called Manchester chain, which relies on an additional function to (5.6) that illustrates the transfer of carry (it is either generated or propagated) through a given position:

$$t_i = g_i + p_i = \overline{e_i} = a_i + b_i \tag{5.8}$$

so (5.7) can be rewritten as:

$$\begin{aligned} c_{i+1} &= g_i + c_i p_i = g_i + c_i g_i + c_i p_i \\ &= g_i + c_i (g_i + p_i) = g_i + c_i t_i \end{aligned} \tag{5.9}$$

Equation (5.9) makes evident that it is possible to work with both the transfer and the propagation functions. The first one is usually generated in a simpler way than the propagation function, as it only requires an **OR** gate rather than an **EXOR**. From (5.7) and (5.9), the Manchester chain and the Manchester adders can be defined. These

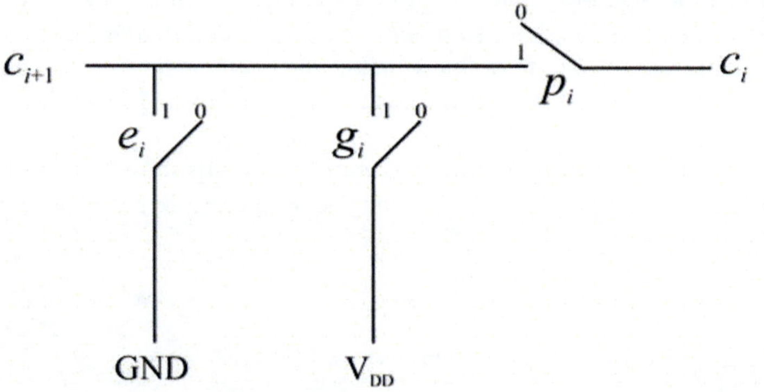

Fig. 5.7 Manchester carry chain main element

are based on the carry generation, propagation, annihilation and transfer functions as switches, which thus allow to connect carry c_{i+1} to c_i as shown in Fig. 5.7, since only one of g_i, p_i and e_i can be 1. Although the delay of a carry chain built concatenating the structure of Fig. 5.7 grows linearly with the number of bits, as it is the case for a **CPA**, the basic delay of this Manchester chain is that of a switch, or a single gate as derived from (5.6) and (5.8), compared to that of several logic gates for the **CPA**.

5.3.3 *Carry Look-Ahead Adders*

From Eq. (5.7) it is easy to compute the functions for the first carry bits of an adder, four of them in this case:

$$
\begin{aligned}
c_4 &= g_3 + g_2p_3 + g_1p_2p_3 + g_0p_1p_2p_3 + c_0p_0p_1p_2p_3 \\
c_3 &= g_2 + g_1p_2 + g_0p_1p_2 + c_0p_0p_1p_2 \\
c_2 &= g_1 + g_0p_1 + c_0p_0p_1 \\
c_1 &= g_0 + c_0p_0
\end{aligned}
\tag{5.10}
$$

where, evidently, c_0 represents the input carry to the adder. Equation (5.10) illustrates, quite graphically, the effect of carry propagation and, on the other hand, the mission of carry generation and carry propagation over each digit. It is evident that the output carry for each digit is defined by the generation function for that digit and the propagation of the input carry arriving to that position, as shown in (5.7), which at the same time depends on the generation and propagation at the preceding digit. Thus, carry at each position only depends on the generation and propagation functions at that digit and those preceding, and the input carry at the least significant position, c_0 in this case. This c_0 will be propagated through every digit and, from the generation and propagation functions, will impact the value of the final output carry.

Equation (5.10) is the basis of carry look-ahead and those adders based on that principle (**CLA**: *Carry Look-ahead Adder*), as it illustrates how the different carry bits can be computed from a single input carry and the generation and propagation functions. However, this cannot be used for long-word additions, as the fan-in resulting from (5.10) increases with the word length and it does not make sense to use it beyond a few bits. Despite that, the recurrence in (5.10) can be aborted at any digit, so c_4 in (5.10) can be simply rewritten in the form of Eq. (5.7):

$$c_4 = g_3 + c_3 p_3 \tag{5.11}$$

Thus, generation and propagation functions may be used once more for the following bits, resulting in:

$$\begin{aligned} c_8 &= g_7 + g_6 p_7 + g_5 p_6 p_7 + g_4 p_5 p_6 p_7 + c_4 p_4 p_5 p_6 p_7 \\ c_7 &= g_6 + g_5 p_5 + g_4 p_5 p_6 + c_4 p_4 p_5 p_6 \\ c_6 &= g_5 + g_4 p_5 + c_4 p_4 p_5 \\ c_5 &= g_4 + c_4 p_4 \end{aligned} \tag{5.12}$$

Equations (5.10) and (5.12) represent the same two-level, **AND-OR** logic block, whose maximum fan-in is 5. Two instances of this block may be combined as shown in Fig. 5.8 for generating all the carries in an addition of 8-bit operands. Thus, addition bits may be computed with a level of **EXOR** gates whose inputs are the operand bits at each position and the corresponding carry. The delay associated to this adder would be the accumulation of:

- the delay of the generation and propagation functions, which is the equivalent delay of a logic level from (5.6);
- the delay associated to the computation of the different carries, which is that of two logic levels (**AND-OR**) as derived from (5.10) and (5.12);
- the delay of the **EXOR** gate required to generate the result bits from the operand bits and the corresponding carries.

From the reasoning above and the structure of Fig. 5.8, the delay associated to the carry look-ahead addition of *n*-bit operands using *k*-bit blocks can be written as:

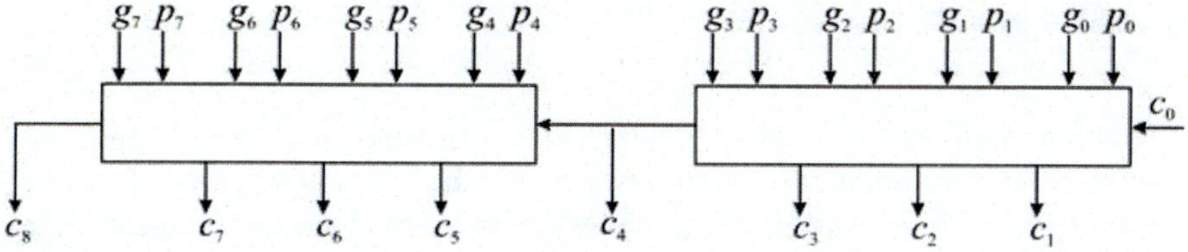

Fig. 5.8 One-level **CLA** 4-bit blocks for the addition of 8-bit words

$$T_{add} = \left\lceil \frac{n}{k} \right\rceil T_{CL} + T_{FA}(c_i \rightarrow s_i) \tag{5.13}$$

where T_{CL} is the delay in the computation, from the input carries, of the output carries for each block in Fig. 5.8. Comparing Eqs. (5.3) and (5.13), the reduction of computing time is evident, especially as n is increased. Despite this reduction, the dependency on the operand size is still linear since, as shown in Fig. 5.8, carry propagation is still present through the output carry of each block acting as carry input for the next one. Thus, it is possible to improve the delay in (5.13) using group generation and propagation signals. Reviewing Eqs. (5.10) and (5.11), the first group carry output, c_4, may be written as:

$$c_4 = (g_3 + g_2p_3 + g_1p_2p_3 + g_0p_1p_2p_3) + c_0(p_0p_1p_2p_3) \tag{5.14}$$

Example 5.3 Consider the addition of two 4-bit words $a_3a_2a_1a_0 = 1101$ and $b_3b_2b_1b_0 = 0010$, with an input carry $c_0 = 1$. From Eq. (5.6) it is clear that the generation and propagation functions are:

- $g_0 = 0$ and $p_0 = 1$,
- $g_1 = 0$ and $p_1 = 1$,
- $g_2 = 0$ and $p_2 = 1$,
- $g_3 = 0$ and $p_3 = 1$,

which define how $c_0 = 1$ will be propagated during the operation. Using the values above, it is clear that the two terms in Eq. (5.14) are:

$$g_3 + g_2p_3 + g_1p_2p_3 + g_0p_1p_2p_3 = 0 + 0 \cdot 1 + 0 \cdot 1 \cdot 1 + 0 \cdot 1 \cdot 1 \cdot 1 = 0$$
$$p_3p_2p_1p_0 = 1 \cdot 1 \cdot 1 \cdot 1 = 1$$

From the first value, it is clear that no carry is generated when adding $a_3a_2a_1a_0$ and $b_3b_2b_1b_0$, as it was expected since each individual carry generation function was 0. At the same time, the second value illustrates how any input carry would be propagated to c_4, as each individual carry generation function was 1. Thus, Eq. (5.14) implies in this case that c_4 will be 1 independently of the final addition result or even the exact input words providing the same values for $g_3 + g_2\,p_3 + g_1\,p_2\,p_3 + g_0\,p_1\,p_2\,p_3$ and $p_3\,p_2\,p_1\,p_0$ terms.

■

Equation (5.14) and Example 5.3 make evident that the group carry output, c_4, depends on the carry input, c_0, through two terms: the first one describes the carry generation at any of the digits in the group, while the second one corresponds to the possibility of having the input carry c_0 directly propagated to the output. It is thus evident that generation and propagation functions can be defined for the group as a whole, and they play the same role as those generation and propagation functions for each digit, as in (5.10). These functions are:

$$g_{[3..0]} = g_3 + g_2 p_3 + g_1 p_2 p_3 + g_0 p_1 p_2 p_3$$
$$p_{[3..0]} = p_0 p_1 p_2 p_3 \tag{5.15}$$

so:

$$c_4 = g_{[3..0]} + c_0 p_{[3..0]} \tag{5.16}$$

This group carry generation and propagation functions may be defined for groups of any size so, in general, Eq. (5.16) may be rewritten as:

$$c_{j+1} = g_{[j..i]} + c_i p_{[j..i]} \quad (j > i) \tag{5.17}$$

where:

$$g_{[j..i]} = g_j + g_{j-1} p_j + g_{j-2} p_{j-1} p_j + \cdots + g_i p_{i+1} p_{i+2} \cdot \cdots \cdot p_{j-1} p_j$$
$$p_{[j..i]} = p_i p_{i+1} \cdot \cdots \cdot p_{j-1} p_j \tag{5.18}$$

From this group functions it is possible to implement multilevel carry look-ahead, which is the most common structure for carry look-ahead adders. For this, these group functions might be combined as in Eqs. (5.10) or (5.12), producing a second level of generation and propagation functions from those in (5.18). In this way, Fig. 5.9 represents the basic structure for multilevel carry look-ahead with 4-bit groups for a 16-bit addition, while Fig. 5.10 adds an additional carry look-ahead level producing a 64-bit adder. Each carry look-ahead block in both Figs. 5.9 and 5.10 implements the logic functions in Eq. (5.10), and each group generation or propagation function corresponds to (5.18).

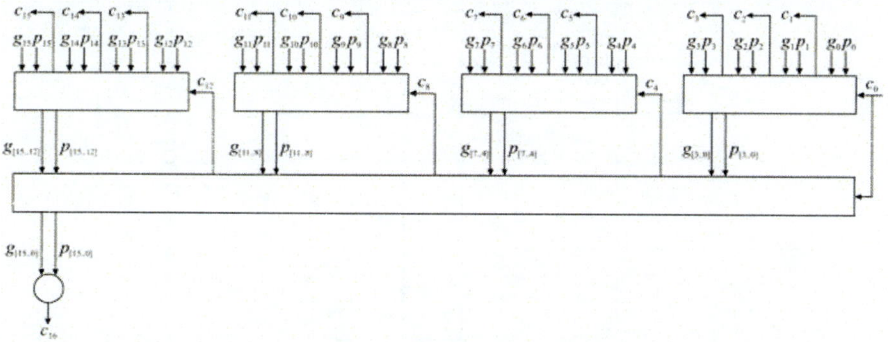

Fig. 5.9 Multilevel **CLA** 4-bit blocks for the addition of 16-bit words

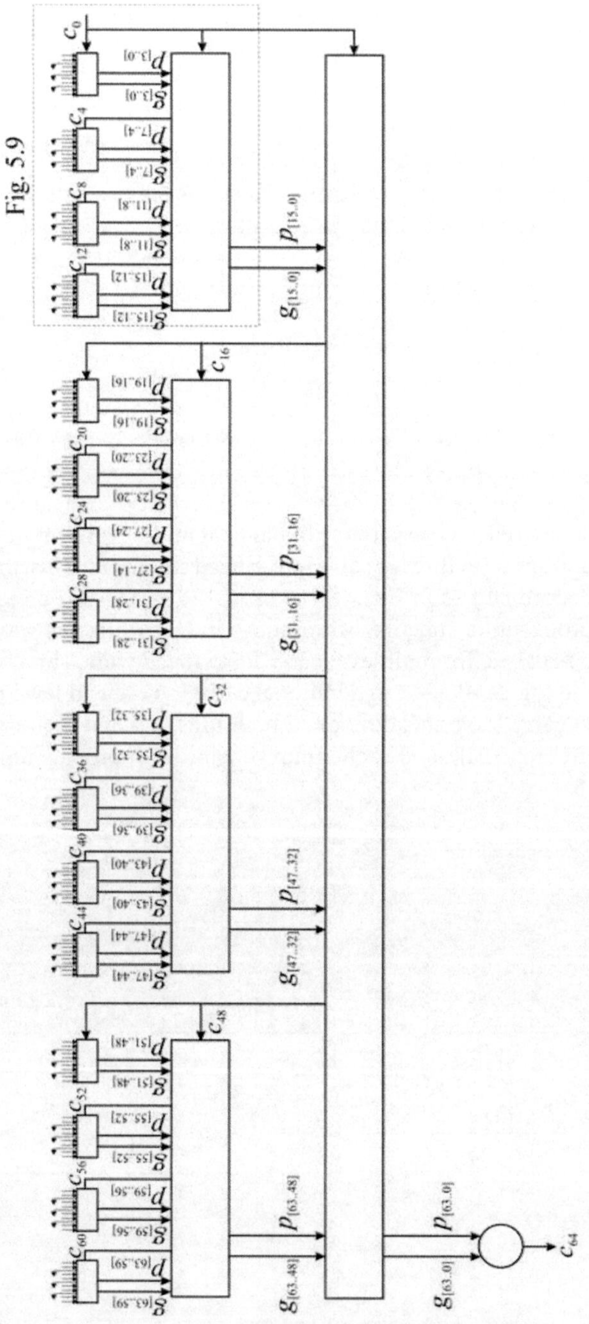

Fig. 5.10 Additional CLA level for 64-bit addition

The partitioning is usually made with 4-bit blocks, even when Eq. (5.18) may be applied to any block size since (5.17) only describes how the block carry output directly depends on the block carry input. However, sizes larger than 4 imply excessive fan-in in the group generation and propagation functions, while sizes not a power of 4 somehow waste inputs of logic gates. Even vintage technology, such as the 74182 carry look-ahead generator [4], makes use of 4-bit carry blocks to manage carry generation and propagation.

Regarding computation time for this multilevel decomposition, the delay associated to the addition of n-bit operands using k-bit blocks for the multilevel carry look-ahead is:

$$T_{add} = \lceil \log_k n \rceil T_{CL-gp} + \lceil \log_k n \rceil T_{CL-carry} + T_{FA}(c_i \rightarrow s_i) \tag{5.19}$$

where $T_{CL\text{-}gp}$ is the delay associated to the generation and propagation functions for each decomposition level from (5.18), $T_{CL\text{-}carry}$ is the delay in computing input carries for each decomposition level using Eq. (5.17) and individual input carries for each digit. This leaves the last term as the delay of the full adder required to generate each output digit from those carries and the corresponding operands' bits.

Example 5.4 For a 16-bit adder with 4-bit carry look-ahead blocks with the structure in Fig. 5.9, the addition time is given by the computation of individual carry generation and propagation functions, the computation of the generation and propagation functions for each 4-bit block, the computation of the input carries for each group (c_4, c_8 and c_{12}) and the overall carry output (c_{16}), the computation of the carries for each individual digit and, finally, the generation of the result in each position, as predicted by (5.19). Extending the adder to 64-bit operands with 4-bit block carry look-ahead would only require to add a new carry look-ahead level, as illustrated in Fig. 5.10, as well as replicating four times the rest of the 16-bit adder.

The inputs to the additional carry look-ahead block would be the generation and propagation functions for each 16-bit group and would include the logic computing the input carries for these 16-bit groups (c_{16}, c_{32} and c_{48}, and the overall output carry c_{64}).

■

It is evident from Eq. (5.19) that it is possible to reduce the linear dependency of the computation time on the number of bits of the operands, as described in Eq. (5.3), while approximating it to a logarithmic relationship. Moreover, this carry look-ahead scheme allows to build any adder from the basic block already shown.

5.3.4 Carry Skip Adders

The ideas in the previous section may be altered to take advantage in different ways of carry generation and propagation functions, thus speeding up addition through carry look-ahead. Specially, the group propagation function:

$$p_{[j..i]} = p_i p_{i+1} \cdot \cdots \cdot p_{j-1} p_j \tag{5.20}$$

requires a single **AND** gate, while at the same time allows to ignore any influence of the group on the output carry when an input carry is present, as illustrated in Eq. (5.17). Thus, when $c_i = 1$ and $p_{[j..i]} = 1$, the generation function is not required and an output carry c_{j+1} will be always produced. Therefore, the computation time of the group output carry can be reduced for this case to just the delay of an **AND** gate when a group input carry is present, while an additional **OR** gate allows to introduce the influence of group carry generation when there is no input carry to that group. This role of the group propagation function can help skip the delay associated to the propagation of a carry through the group in question, thus defining the so-called carry-skip adders, whose basic block is shown in Fig. 5.11. In this way it is possible to combine a normal **CPA** with the advantages offered by the group propagation function for creating a parallel path for rapid propagation of an input carry through all blocks whose group propagation function is 1, as shown in Fig. 5.12a. Regarding the delay of an adder built according to the structure in Fig. 5.12a, the worst case corresponds to the propagation of the output carry of the least significant block to the carry input of the most significant block, so:

$$T_{carry-skip} = T_{CPA}(c_{n-b} \leftarrow c_{n-b-1}) + (n/b - 2)T_{skip} + T_{CPA}(c_b \leftarrow c_0) \tag{5.21}$$

It is assumed in Eq. (5.21) that an n-bit addition is implemented over b-bit blocks, and $T_{CPA}(c_{n-b} \leftarrow c_{n-b-1})$ and $T_{CPA}(c_b \leftarrow c_0)$ correspond to the usual propagation within the least and most significant **CPA** blocks, respectively. T_{skip} represents the delay associated to the block in Fig. 5.11 for the direct propagation through each carry-skip block. A more detailed analysis of Eq. (5.21) makes it possible to equate the delay of each individual **CPA** stage, both T_{CPA} terms in (5.21), to T_{skip}, so (5.21) can be rewritten in terms of T_{skip} as:

$$T_{carry-skip} \approx b + (n/b - 2) + b = n/b + 2b - 2 \tag{5.22}$$

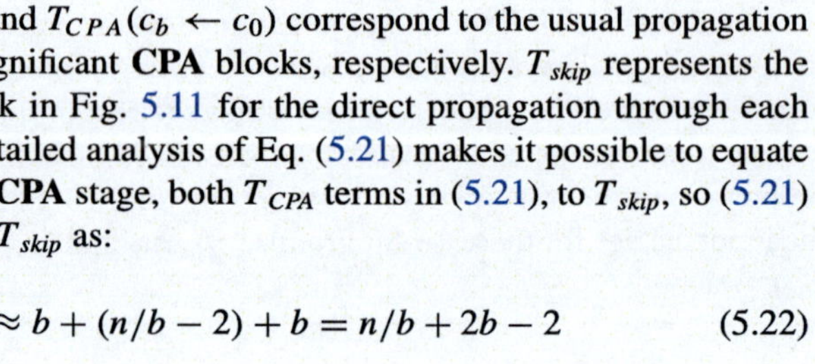

Fig. 5.11 Basic carry-skip adder

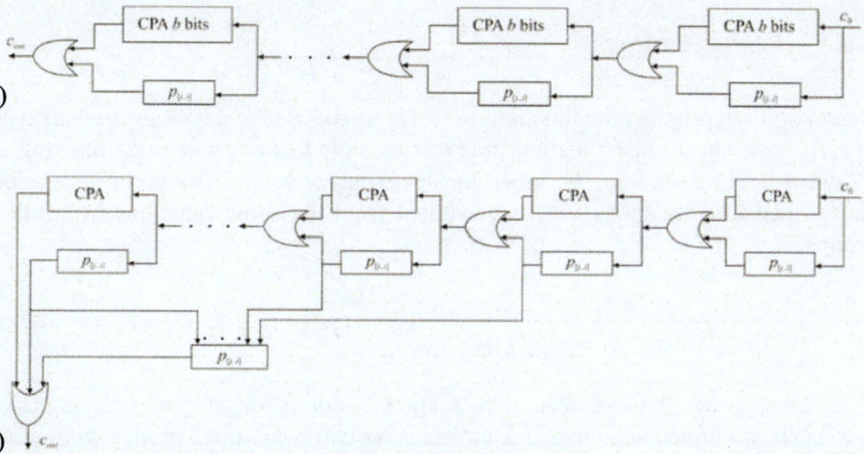

Fig. 5.12 Carry-skip adders: **a** single-level concatenation of carry-skip adders; **b** two-level carry-skip structure

In a similar way to carry look-ahead, this carry-skip technique requires neither a fixed block size nor all blocks to be the same size, since (5.22) is independent of the block size, as well as the adder in Fig. 5.12a. From Eq. (5.22) it can be inferred that the block size b resulting in the minimum delay is $\sqrt{n/2}$. However, different configurations are possible with differing block sizes, depending on target technologies and the particularities they introduce on specific delays.

On the other hand, it is also possible to develop multilevel carry-skip, in a similar way to carry look-ahead. Figure 5.12b shows how a second-level carry skip structure can be added to the original carry-skip adder in Fig. 5.12b. For this, the second-level propagation signal is, as in (5.20), generated with a simple **AND** gate combining the first-level propagation signals. At the same time, the possible carry propagating through this second-level carry-skip structure is fed to the final **OR** gate that defines the input carry to the next block. This same approximation can be used to define as many carry-skip levels as required. It is finally interesting to note that this multi-level carry-skip may be designed with different number of levels at different parts of the adder, *i.e.*, setting different number of **CPA** digits to be "skipped" regarding carry propagation (as a matter of fact, the adder in Fig. 5.12b excludes the first **CPA** block from the second-level carry-skip structure and would be able to propagate its output carry directly to the global c_{out}).

5.3.5 Prefix Adders

Generation and propagation functions in (5.18) can be used to define an alternate type of carry look-ahead. Specifically, while it is possible to work with these function as shown in previous sections, the underlying ideas may be extended so group generation and propagation functions can be generated from the same functions from other groups:

$$\begin{aligned} g_{[j_1..i_0]} &= g_{[j_1..i_1]} + g_{[j_0..i_0]} p_{[j_1..i_1]} \\ p_{[j_1..i_0]} &= p_{[j_0..i_0]} p_{[j_1..i_1]} \end{aligned} \quad (j_1 > j_0 > i_1 > i_0) \tag{5.23}$$

In this way, the generation and propagation functions for the group ranging from j_1 to i_0 are generated as in (5.23) from whatever two overlapped groups within that range.

Example 5.5 In Example 5.3, the addition of two 4-bit words, $a_3a_2a_1a_0 = 1101$ and $b_3b_2b_1b_0 = 0010$, was used to illustrate the concept of group propagation and generation functions. Remember that Eq. (5.6) applied to these operators lead to:

- $g_0 = 0$ and $p_0 = 1$,
- $g_1 = 0$ and $p_1 = 1$,
- $g_2 = 0$ and $p_2 = 1$,
- $g_3 = 0$ and $p_3 = 1$,

while Eqs. (5.15) and (5.18) imply that $g_{[3..0]} = 0$ and $p_{[3..0]} = 1$. Now consider two overlapped groups within the operators $a_3a_2a_1a_0$ and $b_3b_2b_1b_0$, or in terms of Eq. (5.23), assume $j_1 = 3, j_0 = 2, i_1 = 1$ and $i_0 = 0$ $(j_1 > j_0 > i_1 > i_0)$. Thus, Eq. (5.18) results in:

- $g_{[3..1]} = g_3 + g_2\, p_3 + g_1\, p_2\, p_3 = 0$ and $p_{[3..1]} = p_3\, p_2\, p_1 = 1$,
- $g_{[2..0]} = g_2 + g_1\, p_2 + g_0\, p_2\, p_1 = 0$ and $p_{[3..1]} = p_3\, p_2\, p_1 = 1$.

Thus, these values can be introduced in Eq. (5.23), leading to:

$$\begin{aligned} g_{[3..0]} &= g_{[3..1]} + g_{[2..0]} p_{[3..1]} = 0 + 0 \cdot 1 = 0 \\ p_{[3..0]} &= p_{[2..0]} p_{[3..1]} = 1 \cdot 1 = 1 \end{aligned}$$

which, as expected, are the same values that were derived in Example 5.4 and lead Eqs. (5.10) and (5.18).

■

After that, it is possible to define the Brent-Kung operator [1], ·, as:

$$\begin{aligned} \left(g_{[j_1..i_0]}, p_{[j_1..i_0]}\right) &= \left(g_{[j_1..i_1]}, p_{[j_1..i_0]}\right) \cdot \left(g_{[j_0..i_0]}, p_{[j_0..i_0]}\right) \\ &= \left(g_{[j_1..i_1]} + g_{[j_0..i_0]} p_{[j_1..i_1]}, p_{[j_0..i_0]} p_{[j_1..i_1]}\right) \end{aligned} \tag{5.24}$$

which only makes sense for $j_0 \geq i_1$, *i.e.*, when both groups coincide and overlap. This operator has two important properties:

- it is not commutative: this implies that carry can only propagate from the least significant positions to the most significant ones. This is caused by the generation function, since the propagation function is commutative, as it is simply based on the **AND** operator.
- it is associative: this is the most interesting property, as it enables prefix adders. The fact that:

$$\begin{aligned}\left(g_{[j..i]}, p_{[j..i]}\right) &= \left[\left(g_{[j..k]}, p_{[j..k]}\right) \cdot \left(g_{[k-1..l]}, p_{[k-1..l]}\right)\right] \cdot \left(g_{[l-1..i]}, p_{[l-1..i]}\right) \\ &= \left(g_{[j..k]}, p_{[j..k]}\right) \cdot \left[\left(g_{[k-1..l]}, p_{[k-1..l]}\right) \cdot \left(g_{[l-1..i]}, p_{[l-1..i]}\right)\right] \\ &\quad (j > k > l > i)\end{aligned} \tag{5.25}$$

proves that this operator can be evaluated in any order, which allows to define a tree structure with parallelism levels.

The usefulness of the Brent-Kung operator is made evident when the set of carry generation and propagation functions (g_i, p_i) for each digit is transformed into the computation of the group generation and group propagation functions $(g_{[i,0]}, p_{[i,0]})$ for the involved digits. From (5.24) and (5.25), this can be computed in parallel:

$$(g_0, p_0) \cdot (g_1, p_1) \cdot (g_2, p_2) \cdot \cdots \cdot (g_{n-2}, p_{n-2}) \cdot (g_{n-1}, p_{n-1}) \tag{5.26}$$

for n-bit words. This is somehow equivalent to computing the addition of the word $a_{n-1}a_{n-2}\ldots a_1a_0$ from a_0, $a_0 + a_1$, $a_0 + a_1 + a_2$, etc., up to $a_0 + a_1 + a_2 + \ldots + a_{n-2} + a_{n-1}$, thus the prefix adder name.

Optimization for this type of adders is based on the optimization of the prefix computation network. Two of the more popular structures are the Brent-Kung, shown in Fig. 5.13, and the Kogge-Stone, illustrated in Fig. 5.14, both for 16 inputs. The n-input Brent-Kung network consists of $2n - 2 - \log_2 n$ cells distributed in $2\log_2 n - 2$ levels, which define its delay. This structure carries out a reduction by pairs, thus obtaining the addition of odd-indexed inputs that are once again combined with the even-indexed inputs to produce the remaining outputs.

On the other hand, the Kogge-Stone network achieves the fastest possible structure with two-input operators (it is possible to define the Brent-Kung operators with more inputs, whose number is also known as valence in analogy to chemical bond), requiring $\log_2 n$ levels and $n\log_2 n - n + 1$ cells, thus also reducing delay.

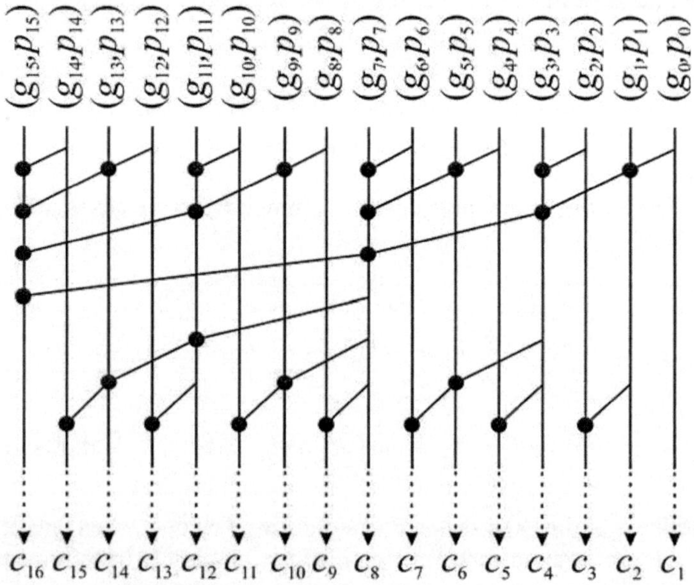

Fig. 5.13 Brent-Kung parallel prefix graph for 16 inputs

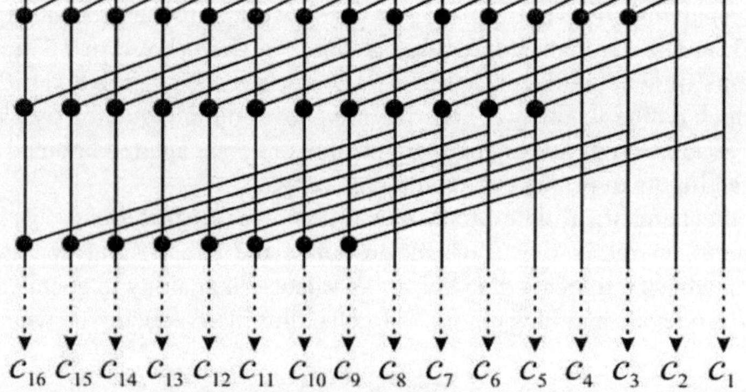

Fig. 5.14 Kogge-Stone parallel prefix graph for 16 inputs

5.4 Carry-Selection Addition: Conditional Adders

It is evident that carry is the only obstacle to simultaneously add all digits involved in an addition. As a matter of fact, all techniques above try to anticipate or suppress the computation of this bit for all or some of the digits. With this, it is possible to reduce the time required to have the final carry value at each position in order to compute the result according to (5.1). A new approximation to addition is to perform two separate additions for a given group of bits, each one assuming one of the two possible input carry values to that group. As a consequence, it is possible to compute the result, which will correspond to one of those two computed, while other parts of the circuit determine the definitive value for this input carry. Figure 5.15 shows the schematic of an adder based on this technique, separating each n-bit operand into two $n/2$-bit blocks, so an $n/2$-bit adder computes the least significant segment of the result. Meanwhile, two additional $n/2$-bit adders compute two different results for the most significant segment, with 0 and 1 input carry respectively. Once the least significant result is set and the input carry to the most significant segment is thus available, the correct result is selected. The delay of such structure is:

$$T_{n\,\text{bits}} = T_{CPA\,n/2\,\text{bits}} + T_{MUX} \tag{5.27}$$

where the T_{MUX} term represents the delay of the multiplexer selecting the correct value of the most significant segment of the result. In this way, T_{MUX} corresponds

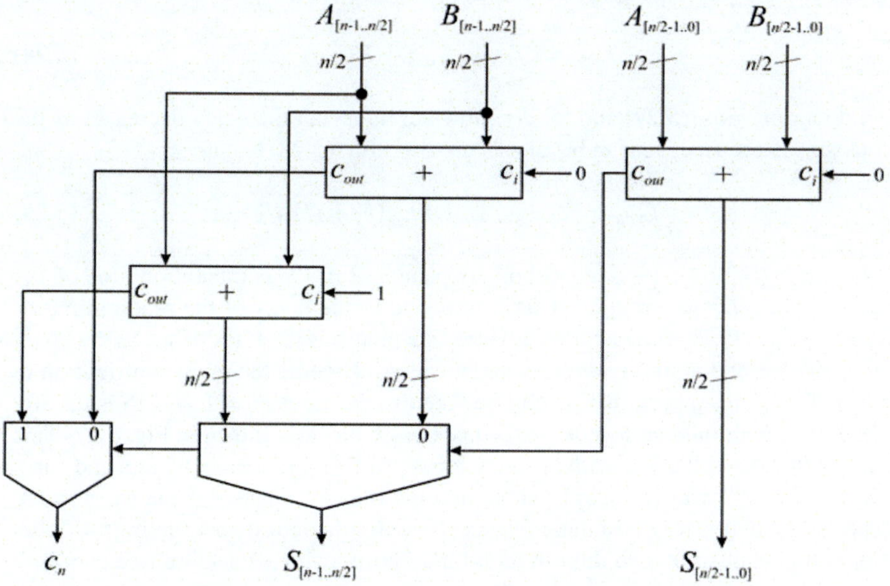

Fig. 5.15 Carry-select adder with $n/2$-bit blocks

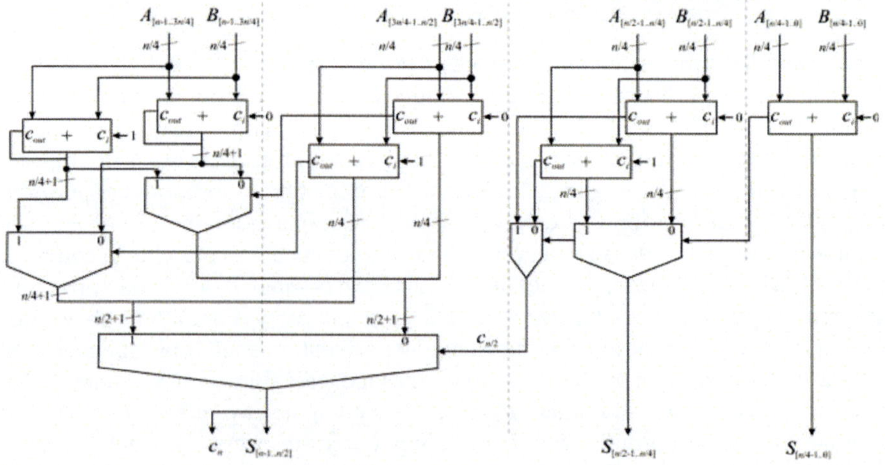

Fig. 5.16 Carry-select adder with $n/4$-bit blocks

to the single **AND-OR** level in a 2-to-1 multiplexer. This approximation is known as carry-select adder. It is evident that this structure can be decomposed into smaller blocks, as shown in Fig. 5.16 with an n-bit adder implemented with $n/4$ blocks. This structure is somehow comparable to carry look-ahead adders with several levels of look-ahead, but with a simpler segmentation made possible by the uniformity in the flow of information. The delay of the adder in Fig. 5.16 is:

$$T_{n\,\mathrm{bits}} = T_{CPA\,n/4\,\mathrm{bits}} + 2T_{MUX} \tag{5.28}$$

Comparing Eqs. (5.27) and (5.28) allows to infer that the delay decreases as the block size is reduced, so if m-bit blocks are used Eq. (5.28) becomes:

$$T_{n\,\mathrm{bits}} = T_{CPA\,m\,\mathrm{bits}} + \log_2(n/m)T_{MUX} \tag{5.29}$$

Equation (5.29) shows a logarithmic dependence of the computation time on the block size, as shown through the term referring to the delay of the required multiplexers. While the dominant term for large size blocks is the **CPA** delay, as the block size is decreased the multiplexer-related term will become the main contribution to (5.29). Thus, this approximation can be taken to the limit $m = 1$, which is usually called the conditional-sum adder and whose basic block is shown in Fig. 5.17. This block includes a full adder with no carry input providing two pairs of sum and carry values, $\left(c_{i+1}^0, s_i^0\right)$ and $\left(c_{i+1}^1, s_i^1\right)$, which assume 0 and 1 values for the carry input, respectively. In this way, n of these blocks allow simultaneously computing two pairs of possible results for each digit in an n-bit addition, with a selection tree choosing the correct pair for each digit.

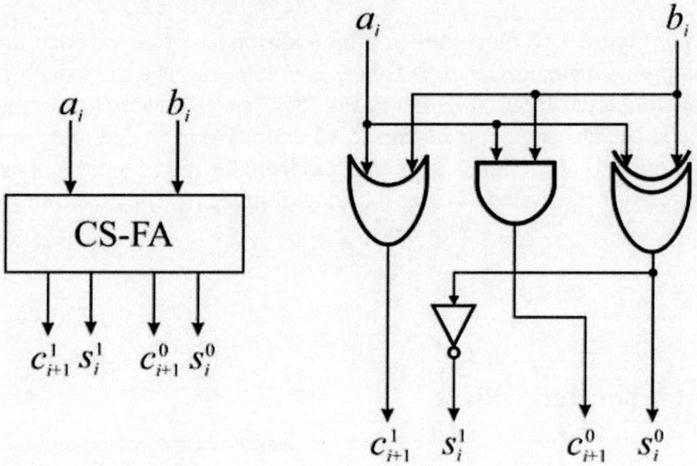

Fig. 5.17 Conditional-sum adder

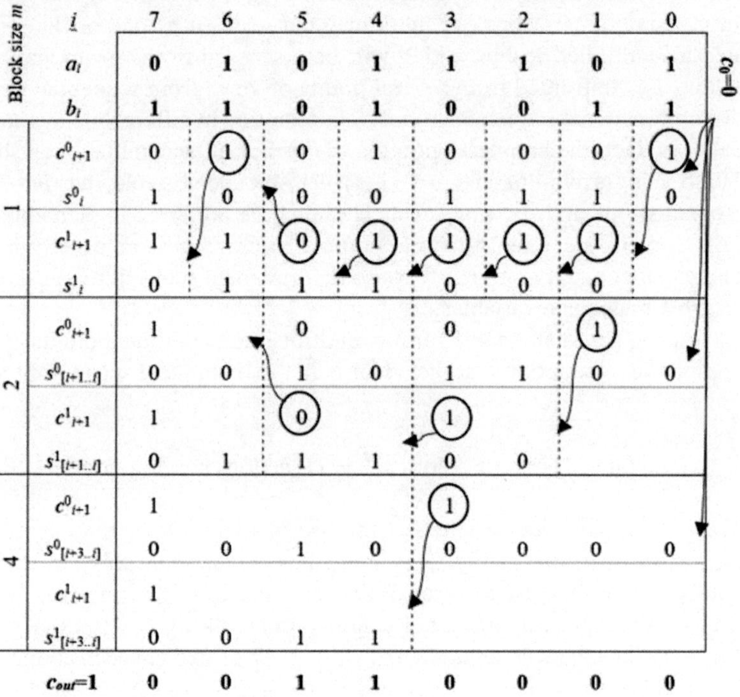

Fig. 5.18 8-bit conditional-sum addition example

Example 5.6 Figure 5.18 illustrates an 8-bit addition based on the conditional-sum adder principle with block sizes $m = 1$, $m = 2$ and $m = 4$. The carry output for each conditional-sum adder block (the one in Fig. 5.17 for $m = 1$, or any type of m-bit adder block in the arrangements of Figs. 5.15 and 5.16) is circled and points to the pair of addition and carry outputs to be selected from the next conditional-sum adder block by means of the corresponding multiplexer-trees (as illustrated in Figs. 5.15 and 5.16).

■

5.5 Multioperand Adders

Up to this point, several structures for adding two operands have been shown, but a usual problem in digital systems is the addition or accumulation of multiple operands. This is the case for a variety of applications, ranging from digital signal processing to scientific computing. Moreover, multioperand addition is one of the structures associated to multiplier design, which will be discussed in the next Chapter. This problem may be confronted from several points of view, from sequential structures to purely combinational implementations, thus resulting in different delays and costs. As a matter of fact, the simplest option is the use of an accumulator, as shown in Fig. 5.19, so a single two-operand, $n + \lceil \log_2(m) \rceil$-bit adder is able, over $m - 1$ clock cycles, to add m operands of n bits. This is in fact the preferred implementation for most programmers, since most general purpose microprocessors usually include a two-operand adder and it is straightforward to implement such addition using a loop instruction over adequate variables.

Regarding the speed attainable for this multioperand addition, both the circuit in Fig. 5.19 and the assumption that the adder is in the theoretical speed limit allow to write:

$$T_{m\text{ operands}} \propto (m - 1)\log_2\big(n + \log_2 m\big) = O\big(m\big(\log_2 n + \log_2 \log_2 m\big)\big) \quad (5.30)$$

On the other hand, it is evident that the use of more adders may speed up the operation of the structure in Fig. 5.19 and, thus, reduce the delay in (5.30). An immediate solution is the use of a tree of two-operand adders, so in a first level each couple of n-bit operands is reduced to a single term of $n + 1$ bits. After that, a second level of $(n + 1)$-bit adders generates a new $(n + 2)$-bit term from each couple of $(n + 1)$-bit terms and so on, until a final result of $n + \lceil \log_2(m) \rceil$ bits is available. This tree structure requires $m - 1$ adders distributed over $\lceil \log_2(m) \rceil$ levels, and the maximum delay that can be achieved, assuming once more that all adders are in the theoretical logarithmic limit, is given by:

$$\begin{aligned} T_{m\text{-operand tree}} &\propto \log_2(n) + \log_2(n + 1) + \cdots + \log_2\big(n + \lceil \log_2(m) \rceil - 1\big) \\ &= O\big(\log_2 m\big(\log_2 n + \log_2 \log_2 m\big)\big) \end{aligned} \quad (5.31)$$

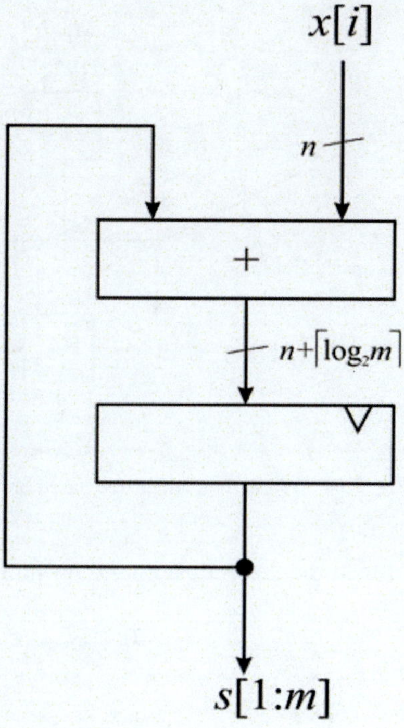

Fig. 5.19 Sequential accumulator

As expected, comparing (5.30) and (5.31) results in a lower computation time for the tree structure. However, few conclusions can be extracted from (5.31) about the maximum possible speed. Specifically, it is even possible to wonder how simple adders, such as carry-propagation adders, would influence the tree structure. Figure 5.20 details the connection between the initial adder level and the following one for the least significant bits of these **CPAs** in the tree. It can be observed that all the inputs to the least significant full adder in the second level, FA_0^1, are fixed as soon as the least significant full adder in the first level, FA_0^0, has set the values of its outputs. In this way, the addition in the second level of the tree can start just a carry-propagation delay after the initial level, since full adders FA_0^1 and FA_1^0 depend, in the same way, on the outputs of FA_0^0. Thus, it is not necessary to wait for the completion of the computation in the first level before the second level is started, as assumed in (5.31). With all this, the delay of a tree of carry-propagation adders for the addition of *m* operands of *n* bits can be written as:

$$T_{m\text{-operand CPA tree}} \propto n + \lceil \log_2(m) \rceil \tag{5.32}$$

Depending on the number of operands, *m*, (5.32) shows how a tree of very simple adders could be even faster than trees with more sophisticated adders. As a matter

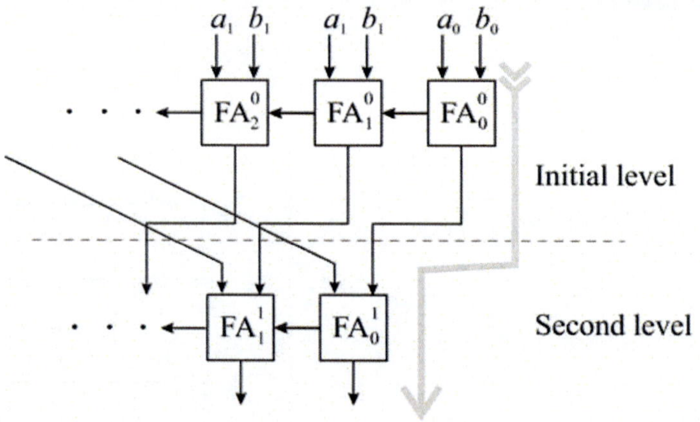

Fig. 5.20 Detail of an adder tree based on **CPAs**

of fact, the theoretical bound for multioperand addition is:

$$T_{m\ \text{operands}} \propto O\big(\log_2(m) + \log_2(n)\big) \tag{5.33}$$

5.5.1 *Carry-Save Adders*

From the analysis above and comparing Eqs. (5.31) and (5.33), it can be deduced that it is possible to improve the delay of a tree of theoretically optimal adders. The key for increasing the speed of this operation is in the own nature of the tree. It is evident that the tree described above, based on two-operand adders, reduces the number of operands until a final couple of terms is added in the final two-operand adder. Thus, the full adder in Fig. 5.2a provides the key, as its basic mission is to add three bits providing a two-bit result. In this way, this full adder reduces a set of three individual input bits to a two-bit output word. It is then possible to modify the **CPA** in Fig. 5.6 in order to have it transformed into a carry-save adder (**CSA**), as shown in Fig. 5.21. Transforming the **CPA** into a **CSA** does only require to suppress carry propagation, so the n-bit **CSA** admits three n-bit input words and generates two n-bit output words. One of these output words corresponds to the addition bits from each full adder, while the other one is formed by the carry bits. Thus, the **CSA** reduces three n-bit input operands to two n-bit output words, but its delay is that of a single full adder as there is no carry propagation. As shown in Fig. 5.21, it is necessary to consider the differing weights of both **CSA** output words, since the word corresponding to the carry bits from the full adders is of more significance than that formed by the addition bits. This fact has to be taken into account to correctly combine the addition and carry words resulting from a **CSA**. Figure 5.22 represents

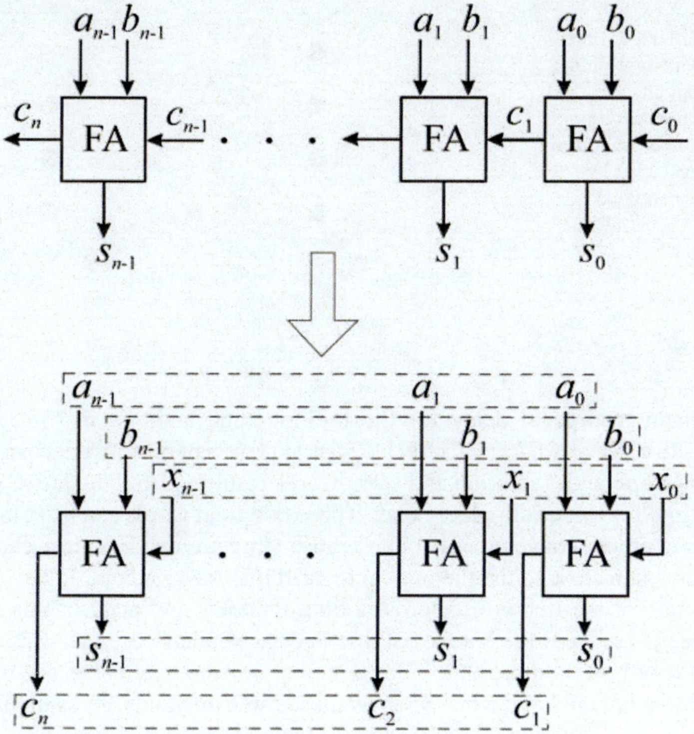

Fig. 5.21 Transforming a **CPA** into a **CSA**

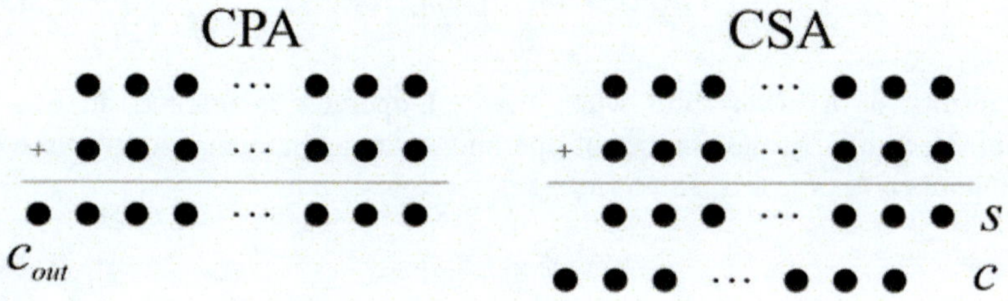

Fig. 5.22 **CPA** (left) *vs* **CSA** (right) in dot notation

this in dot notation, comparing it to a conventional two-operand adder. It must be noted that these dot diagrams are very useful to represent the information flow in mutioperand additions, and they are based on representing each bit with a dot, as shown in Fig. 5.23 for a full adder and a half adder. Bits that are combined in each one of these elements are framed in a box, while output bits (addition and carry) are linked by a solid line and aligned to the corresponding correct weight.

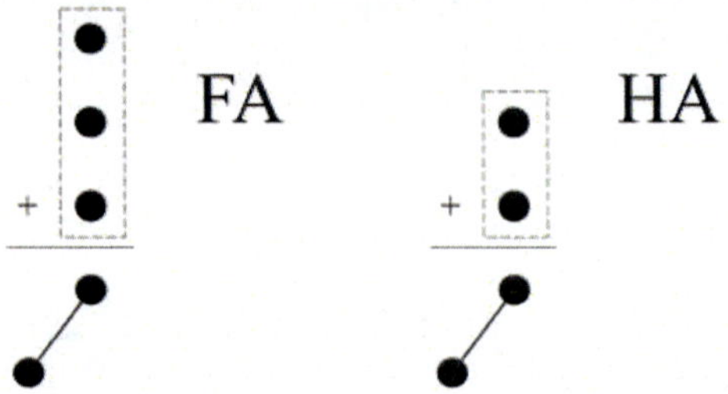

Fig. 5.23 Full adder (left) and half adder (right) in dot notation: input bits are framed by a box, addition and carry outputs are linked by a solid lined and aligned to their proper weights

5.5.2 *Adder Trees*

The minimum theoretical delay for the multioperand addition in (5.33) can be achieved with a tree of **CSAs**, since $O(\lceil \log_2(m) \rceil)$ levels of those adders can produce a final pair of operands (addition and carry words resulting from the last **CSA**), and each level implies just a full-adder delay. These two final words can be combined in a final two-operand, conventional adder, whose structure can be freely chosen and can thus get again close to the theoretical logarithmic lower bound. In any case, the exact structure of the tree will determine both its delay and area. In this way, the Wallace tree [5] reduces the number of operands as soon as possible, thus trying to reduce the length of the final adder. For this, the Wallace tree makes use of a **CSA** to reduce three operands to two as soon as those three operands are available. Since a **CSA** reduces the number of operands in a factor 3/2, the number of **CSA** levels required for an *m*-input tree is given by the following recursion:

$$h(m) = 1 + h(\lceil 2m/3 \rceil) \tag{5.34}$$

which results, as a lower limit when the ceil operator is ignored, in $h(m) \geq \log_{3/2}(m/2)$. Similarly, the number of operands that can be managed by a tree with a given number, *h*, of levels is:

$$m(h) = 3m(h-1)/2 \tag{5.35}$$

so $2(3/2)^{h-1} < m(h) \leq 2(3/2)^h$, which again is in agreement with the lower limit of (5.34). Using the recursion in Eq. (5.35) assumes that $m(0) = 2$, which obviously corresponds to a two-operand addition not requiring any tree structure. Figure 5.24 represents the exact values of *m*(*h*).

Taking advantage of the dot notation introduced in Figs. 5.22 and 5.23, multi-operand addition corresponds to the reduction of a matrix of dots ($m \times n$ for the addition of *m* operands of *n* bits). With this, the strategy of the Wallace tree is to make use of a full adder to reduce three bits of the same weight to two as soon as those are available. Figure 5.25 illustrates this strategy for the addition of 7 operands of 8 bits, which requires two 8-bit **CSAs** in the first level, an 8-bit **CSA** in the second, third

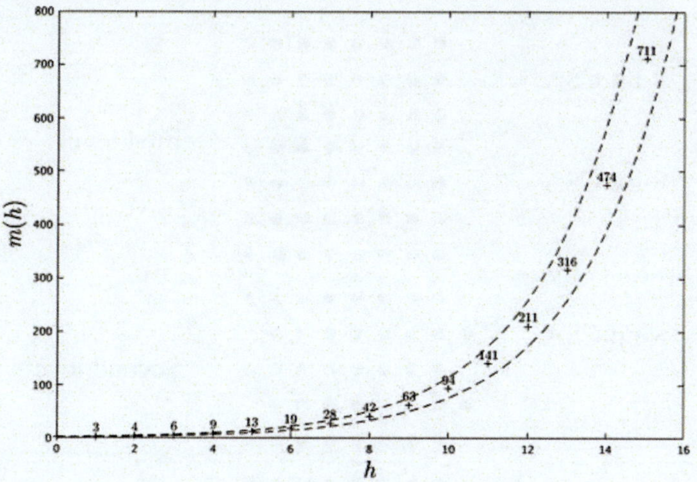

Fig. 5.24 Graphical representation of $m(h)$ values (+ symbols, dotted lines represent $2(3/2)^h$ and $2(3/2)^{h-1}$)

and fourth levels, and a final two-operand, 8-bit adder to obtain the final 11-bit result. The required four levels are in agreement with Eqs. (5.34) and (5.35). However, as it can be deduced from Eq. (5.35) and observed in Fig. 5.24, the same number of levels allows to carry out additions up to 9 operands with this same strategy, while the final adder stage will not linearly increase its speed if length is reduced (as it is usually the case for most of the carry look-ahead structures). Therefore, it is possible to modify the strategy of the Wallace tree in Fig. 5.25, so each level tries to reduce the number of operands to the next level from the value of $m(h)$ provided in (5.35). Figure 5.26 shows the same addition of Fig. 5.25 (7 operands of 8 bits) following this last approach, which is known as the Dadda tree [2]. In this approximation, as opposed to Wallace's, the first level makes use of a single **CSA**, since three levels may process up to 6 operands and this first **CSA** reduces the number of operands in the second level to 6 (this corresponds to $m(3)$ in Eq. (5.35)). The second level makes use of two **CSAs**, although one of them requires only 7 bits, that reduce 6 operands to 4 ($m(2)$ in (5.35)). Those 4 operands will be processed by the following levels as shown in Fig. 5.26. Thus, the third level reduces the number of operands to 3, $m(1)$ in (5.35), with a new **CSA** and, finally, the fourth and final level includes a 7-bit **CSA** that generates the two final operands for the final **CPA**. Even when the cost of the trees in Figs. 5.25 and 5.26 is very similar, due to the simplicity of the example, both illustrate the different approaches to the tree structure. While the Wallace tree tries to reduce as soon as possible the number of operands, the Dadda tree carries out the minimum possible number of reductions in order to reduce the cost of the tree. On the contrary, the Dadda tree can generate a final stage with longer wordwidth than that of Wallace's, but this does not always result in decreased speed unless a final carry propagation adder is used.

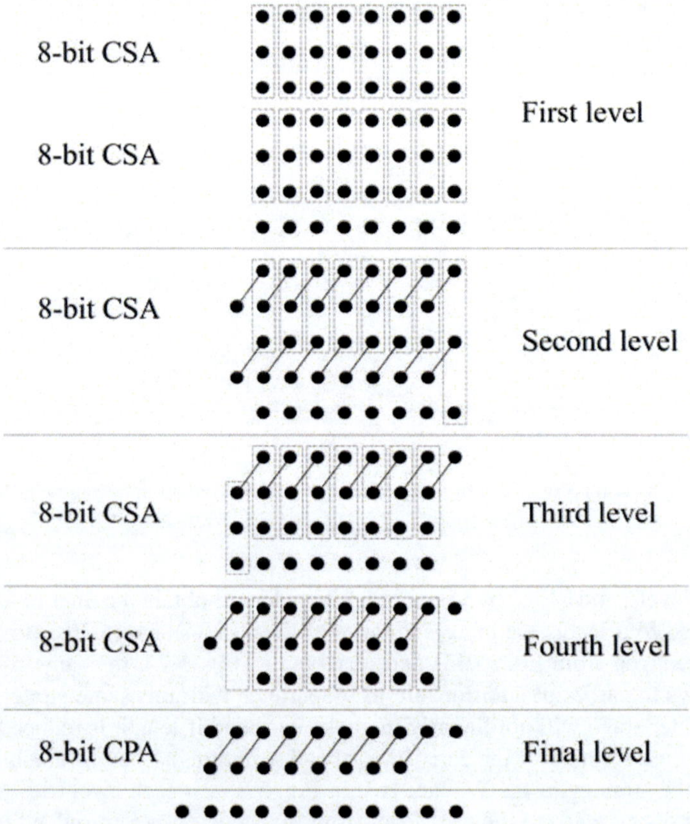

Fig. 5.25 Wallace tree structure for 7-operand addition

From the analysis above, as shown by the 7-operand adders in Figs. 5.25 and 5.26, the set of inputs to multioperand addition may be interpreted as a matrix of bits. With this analogy, different operands do correspond to the matrix rows in the Figures mentioned above, while each column corresponds to a given digit, weight or position. In this way, each full adder in a **CSA** can be viewed as a counter of the number of 1s in three elements in the same column of the matrix, providing a two-bit output in adjacent columns of the resulting matrix. Similarly, each **CSA** make use of three rows of the matrix as inputs, and provides to rows as outputs. Full adders, thus, are also known as 3-to-2 counters, or (3;2) counters, for column reduction, while a **CSA** can be viewed as a 3-to-2 counter for row reduction. However, these should never be mistaken for sequential counters, which may be built using a conventional two-operand adder and a register, and are far different from these counters.

This concept may be further extended, introducing larger blocks performing the same kind of reduction, as the (7;3) counter in Fig. 5.27, which reduces 7 input bits to a 3-bit output word containing the number of 1s in the input bits. In general,

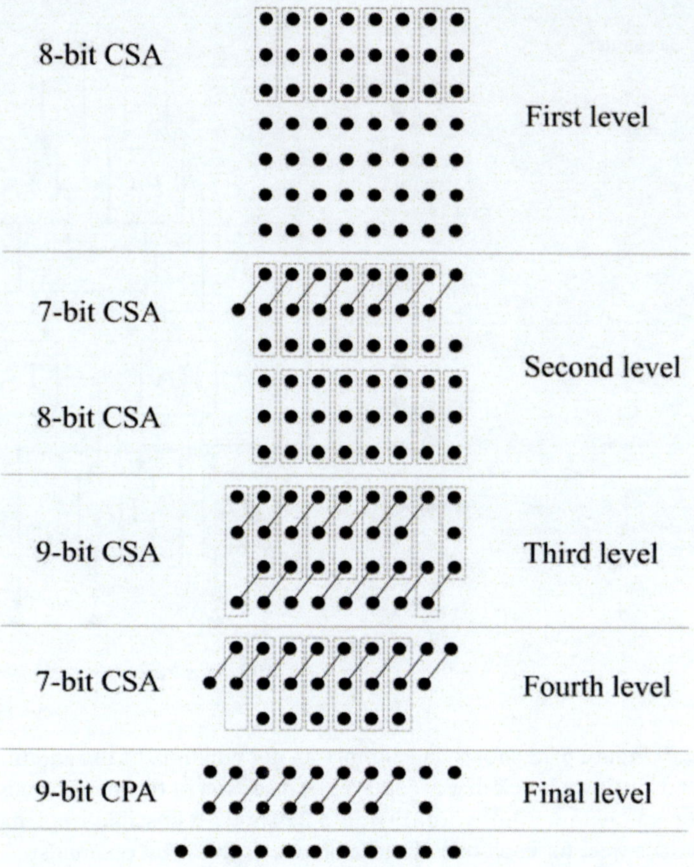

Fig. 5.26 Dadda tree structure for 7-operand addition

$(m; \lceil \log_2(m) \rceil + 1)$ counters, which reduce m input bits to the count of 1s in them, can be defined with a similar structure and can be used as the basic block for the construction of trees in multioperand adders. As an example, two (7;3) counters like the one in Fig. 5.27 may be used as basic block for a 14-operand addition at the first reduction level, thus generating two 3-bit operands for each column in the original matrix of this operation. In this way, these two (7;3) counters would transform two 7-bit sets into two 3-bit sets, *i.e.*, would compress the number of individual bits and, consequently, reduce the dimensions of the bit matrix in the multioperand addition.

If it is possible to use one of those counters to reduce the bit matrix of the multioperand addition, it is also possible in the same way to perform a reduction of rows with an n-bit **CSA**. In this case, the n-bit **CSA** transforms three n-bit words, *i.e.*, three matrix rows, into two words. Thus, different blocks can be defined for the reduction of rows in the bit matrix of multioperand addition. These blocks can comprise all of the matrix columns (several complete operands), or only partial sections.

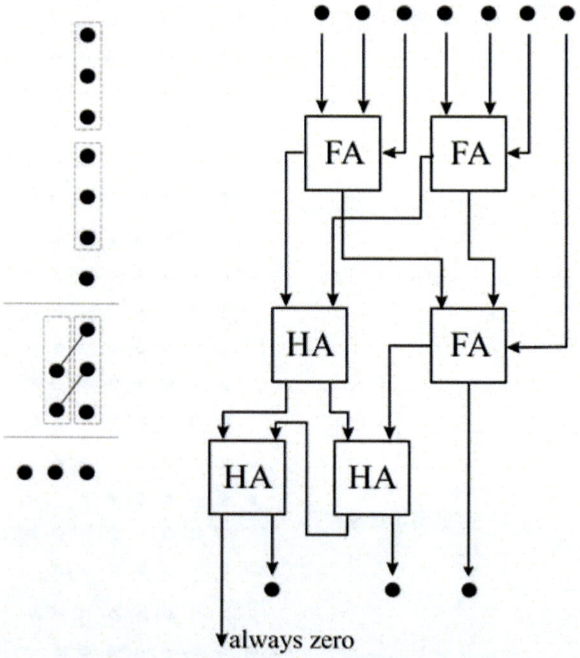

Fig. 5.27 (7;3) counter

Example 5.7 Figure 5.28 shows an example, in dot notation, of the addition of five 8-bit operands, using a block that accepts a 5 × 4 section of the matrix, so it reduces five 4-bit operands to a 7-bit word. Figure 5.29 shows a possible implementation, using a Wallace tree, for each one of these blocks of five 4-bit operands. ■

The final **CPA** stage of these blocks can be skipped, with an output of two words as in a conventional **CSA**, which only accepts three operands. In this way, it is possible to define what is known as a generalized (*p*;2) counter, similar to a **CSA** but admitting

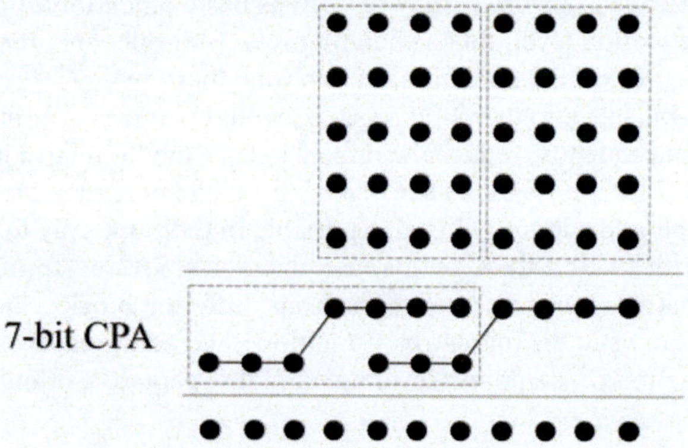

Fig. 5.28 Addition of five 8-bit operands through the reduction of 5 × 4 sections of the dot matrix

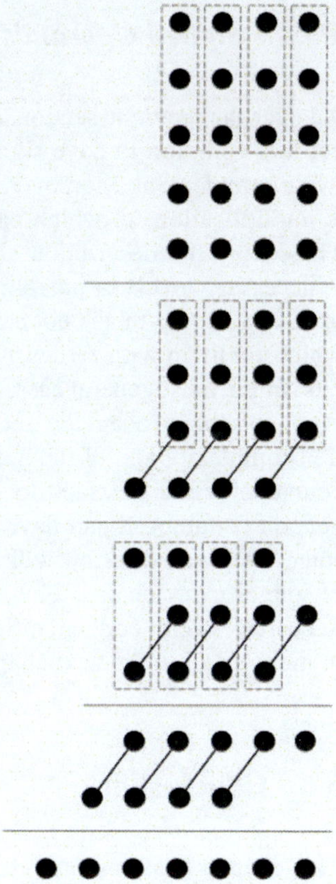

Fig. 5.29 Wallace tree implementation of the reduction of 5 × 4 sections in Fig. 5.28

p operands (rows in the bit matrix) and generating two output words that can help to form a reduced matrix.

Finally, it must be noted that those trees, in any of the described alternatives, can be easily pipelined, both in the direct implementation with **CSAs** and in structures formed by counters or row-reduction blocks. Even if it is possible to register the output of each full adder in the different **CSAs**, it does not make sense in a synchronous datapath and implies an unnecessary area increase. This is caused by the fact that the final **CPA** stage, or any other structure of choice, will not usually match the speed of a **CSA**. Thus, the granularity for the pipelining of the tree will depend on the type of adder chosen to obtain the final addition. With this, the speed of each pipeline stage should be balanced with the speed of this last adder or that of its stages, in case it is also internally pipelined.

5.5.3 Signed Operands

All over the previous section, only unsigned natural operands have been considered, but it is common to have signed integer data in multioperand additions, usually in two's complement. Moreover, one of the basic applications of multioperand addition is multiplication, in which partial products are not aligned in the same positions. Thus, from the description above of multioperand addition as a matrix of bits, it is necessary to fill in positions of that matrix in order to correctly align all input operands. While this is not a major problem in the case of the least significant bits, which are fill in with zeroes, it requires a sign extension to fill in the most significant bits. In the most general case, when the sign of the operands is not known *a priori*, it is not possible to adapt the tree structure and sign extension implies an increase of the tree complexity and of the number of full adders in the **CSAs** or the corresponding counters. While this fact does not affect the speed of each level of full adders or generic counters, it can have an effect for negative operands on the final addition stage, as sign extension will normally introduce more activity and the possibility of longer carry chains. This effect may be mitigated with signed-digit coding (*v.* Sect. 1.8), Booth coding (*v.* Sect. 1.8.3.3) or other techniques that will be described in more detail in the next chapter.

5.6 Conclusion

This chapter has presented the basic ideas on carry propagation, which is the foundation of addition and, consequently, subtraction. After that, the basic structures for fast addition have been presented, including carry-look ahead adders, carry-skip adders, prefix adders and conditional adders. Multioperand adders and the different tree structures, along with the concept of carry-save adders, have been discussed as they are one of the key components of multipliers. In any case, it must be noted that all of these acceleration techniques can be extended to any radix, in particular to radix-10 for decimal adders that were presented in Chaps. 1 and 2.

Adders, in any of these forms, and subtracters will be used in the following Chapters to implement the rest of arithmetic operations.

5.7 Exercises

5.1 Find input operands for a 4-bit **CPA** that require propagating carries from the least significant digit to the **CPA** output carry. Find input operands that do not propagate any carries within the **CPA**. Compare the combinational delay of both cases in light of Eq. (5.3).

5.2 Verify the claim in Sect. 5.3.1 that the probability of generating a carry when $c_i = 0$, is ¼, the probability of propagating a carry chain ($c_i = 1$ and $c_{i+1} = 1$) is ½, and the probability of breaking or ending a chain ($c_i = 1$ and $c_{i+1} = 0$) is again ¼.

5.3 Prove that initializing (d_0, c_0) as (0, 1) or (1, 0) forces all pairs of (d_i, c_i) functions in Eq. (5.5) to settle to either (0, 1) or (1, 0).

5.4 Find, if possible, an alternate layout to the Manchester carry chain in Fig. 5.7 using t_i instead of p_i. Think about the roles of t_i and p_i in carry propagation and Eq. (5.9).

5.5 Rewrite Eq. (5.10) using the carry transfer function defined in Eq. (5.8) and explore any differences that might be introduced in the design of **CLAs**.

5.6 Repeat example 5.3 for input words $a_3a_2a_1a_0 = 1101$ and $b_3b_2b_1b_0 = 1000$, and observe how the carry output is influenced by the input carry.

5.7 Derive the addition delay of a 64-bit adder with the **CLA** structure in Fig. 5.10. Compare this delay to Eq. (5.19).

5.8 Derive the structure of a 256-bit **CLA** based on the 4-bit **CLA** basic block in Fig. 5.9, and repeat the previous exercise for this adder.

5.9 Derive logic expressions for the carry generation, propagation and annihilation functions for radix-10 (decimal adders).

5.10 Using the functions resulting from the previous exercise, check that Eqs. (5.10) and (5.12) hold also for radix 10 and can be used to build a radix-10 **CLA** block as in Fig. 5.8. Compare the values of the generated carries to those obtained with paper and pencil (as in Example 5.1) for the addition of two 8-digit decimal operands of your choice.

5.11 Verify that $\sqrt{n/2}$ is the block size producing the minimum delay in Eq. (5.22).

5.12 Repeat Example 5.5 for the input words in exercise 5.5.

5.13 Derive the computation times for the conditional-sum adders in the different block sizes of Example 5.6 and compare them to Eq. (5.29).

5.14 Define the critical path of an adder tree for three 8-bit operands based on two **CPAs**. Check if its delays holds Eq. (5.32).

5.15 Derive the delay of a 3-operand, 8-bit per operand adder tree based on an 8-bit **CSA**. Compare the result to that of the previous problem.

5.16 Verify the values of $m(h)$ in Fig. 5.24. Derive the first 10 values for $h(m)$ in Eq. (5.34).

5.17 Define Wallace and Dadda trees for the addition of nine 8-bit operands. Compare the resources required for each option.

References

1. Brent, R.T., Kung, H.T.: A regular layout for parallel adders. IEEE Trans. Comput. **31**(3), 260–264 (1982)
2. Dadda, L.: Some schemes for parallel multipliers. Alta Frequenza **34**, 349–356 (1965)

3. Pippenger, N.: Analysis of carry propagation in addition: an elementary approach. J. Algorithms **42**(2), 317–333 (2002)
4. Texas Instruments: SN54S182, SN74N182 Look-Ahead Carry Generators Datasheet, December 1972. Available on-line at http://ti.com/lit/ds/symlink/sn54s182.pdf
5. Wallace, C.S.: A suggestion for a fast multiplier. IEEE Trans. Electron. Comput. **EC-13**(1), 14–17 (1974)

Chapter 6
Multiplication

After the study of the addition as the most basic arithmetic operation, in this chapter the multiplication is approached. First, the basic concepts introduced in Chap. 2 are revised, as a starting point for introducing parallel multipliers. Then, diverse strategies for performing the addition of the partial products generated in these multipliers are introduced, as parallel counters or Wallace and Dadda trees. Next, the circuits for building sequential multipliers are introduced. This multipliers enables the reduction of area resources at expenses of increasing the number of clock cycles. The use of radixes higher than two are also considered, presenting the Booth multiplier, and the chapter ends with some special multipliers that are required in some specific applications.

6.1 Introduction

Multiplication is one of the basic arithmetic operations used in the majority of computing algorithms. Although it can be derived from addition, all modern microprocessors and microcontrollers include a specific multiplication circuit as it enables accelerating the computation of complex algorithms. Multiplication is also a basic mathematical operation for the realization of others such as division, square root, inversion and others. Thus, the study and design of efficient multiplication circuits is a main issue in the field of arithmetic and algebraic circuits.

A. Lloris Ruiz et al., *Arithmetic and Algebraic Circuits*, Intelligent Systems Reference Library 201, https://doi.org/10.1007/978-3-030-67266-9_6

6.2 Basic Concepts

A first approach for implementing a circuit for multiplying two positive binary numbers X and Y is to consider the multiplier as a combinational circuit. If X is an m-bit number, and Y an n-bit one, $P = X \cdot Y$ will take $m + n$ bits, as was shown in Chap. 2:

$$X \leq 2^m - 1$$
$$Y \leq 2^n - 1$$

then:

$$P = X \cdot Y \leq (2^m - 1)(2^n - 1) = 2^{m+n} - 2^m - 2^n + 1.$$

Thus, a combinational multiplier will have $m + n$ inputs, and the same number of outputs (except when $m = 1$ or $n = 1$). When $m = n$=1, the multiplier can be implemented using the **AND** function. In the case of $m = n$=2, it is necessary to implement 4 boolean functions with 4 inputs each, which it is affordable, as was shown in Fig. 2.7 (see Chap. 2). For $m = n$=4, optimization and implementation of 16 functions with 16 inputs each is required. This becomes a hard task, and an alternative approach must be used. If the reader remembers the pencil and paper multiplication algorithm learnt at school, the multiplication of two 2-bit number can be expressed as in Fig. 6.1a, and the corresponding digital circuit in terms of half-adders (**HA**) and full-adders (**FA**), will be the one of Fig. 6.1b.

Note that this implementation presents worst delay figures than direct implementation (see Fig. 2.7, Chap. 2), due to carry propagations. Regarding the area required for both implementations, the pencil-and-paper implementation provides a more compact implementation of the 2 × 2 multiplier. In the case of multipliers with greater number of bits, direct implementation becomes unfeasible, and two approaches can be used in order to implement the pencil-and-paper method:

(1) To generalize implementation presented in Fig. 6.1, thus using **HA**s and **FA**s.
(2) To use multi-bit adders, and perform the addition of partial products column by column.

Figure 6.2b shows the implementation of a 4 × 4 multiplier using option 1), where a systematic procedure for implementing $m \times n$ multipliers can be easily inferred. Moreover, this circuit can be also interpreted as a base-4 multiplier of two 2-digit characters, and synthesized by using 2 × 2 multipliers and a network of adders.

Figure 6.2c details the X by Y multiplication, and Fig. 6.2d presents the circuit with this design strategy, using four 2 × 2-multipliers and a network of adders. Note that this interpretation enables hierarchically organize the design as well as taking advantage of fast adders introduced in Chap. 5.

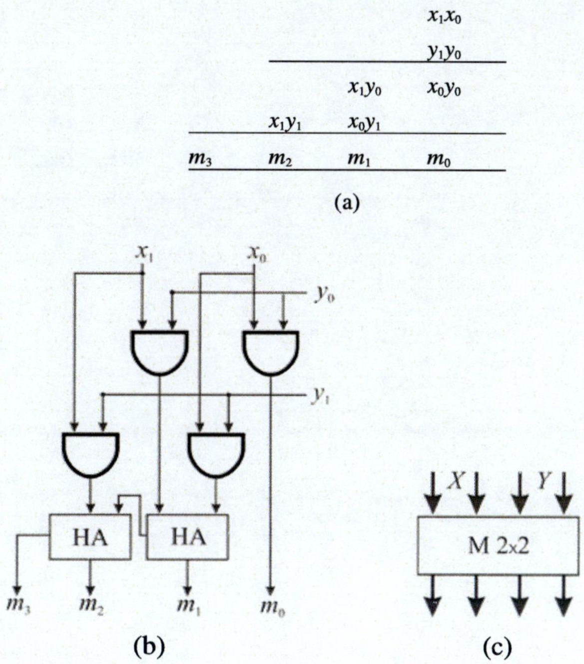

Fig. 6.1 **a** 2-bit *X* by *Y* multiplication. **b** Network for 2-bit character multiplying. **c** 2-bits multiplier

6.3 Combinational Multipliers

Another approach for adding the partial products is to use multi-operand adders (see Sect. 5.5), as presented in Fig. 6.3 for a 3 × 3 multiplier. As can be shown, one (2:2), two (4:3) and one (3:2) counters are needed. To avoid excessive complexity in this type of diagrams, each basic product obtained from the **AND** gates is usually represented by a dot, resulting the dot diagrams introduced in Chap. 5 for multi-operand adders. Figure 6.4 shows a dot diagram for a 4 × 4 multiplier using parallel counters for performing addition.

The use of counters allows diminishing the number of additions, thus accelerating the multiplication. As a drawback, the design of counters with a high number of inputs can become complex, and result in area increasing. Moreover, carry propagation trough the counters do not allow to take full advantage of this structure. Note that in the 4 × 4 multiplier, (5:3) counters are required, which design is more complex than (3:2) or (4:3) ones. In any case, it is possible to limit the number of inputs to the counters, as in Fig. 6.5, where a 5 × 5 multiplier is implemented using (5:3) counters as a maximum.

An alternative for reducing complexity was proposed by Wallace [5], introducing the well-known Wallace-tree (see Chap. 5), where Carry Save Adders (**CSA**), (i.e., (3:2) counters), are used for performing parallel addition of partial products, and a final fast Carry Propagation Adder (**CPA**) completes the addition. This proposal

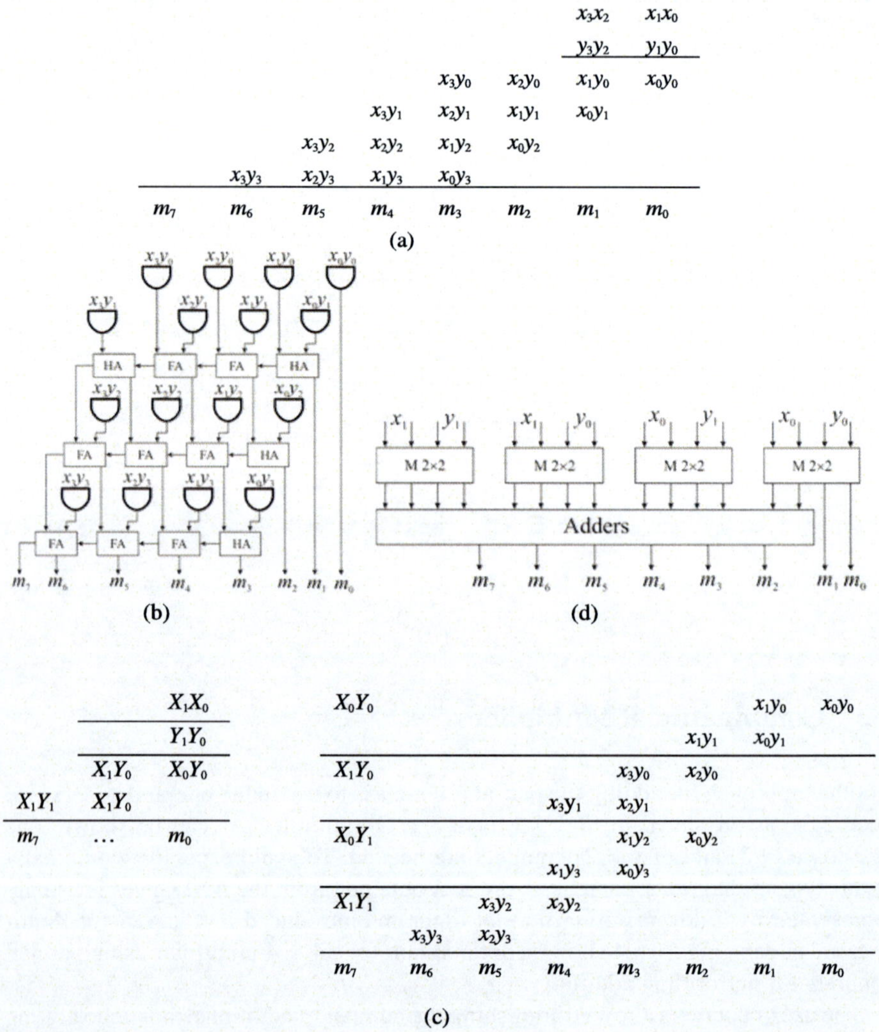

Fig. 6.2 **a** Four-bit characters multiplication. **b** Circuit for multiplication. **c** *X* by *Y* multiplication. **d** Network for multiplying *X* by *Y*

has two advantages: the complexities of counters with a large number of inputs is avoided, and carry propagation affects only to the last stage. Figure 6.6 shows this idea for implementing a 5×5 multiplier. In a first stage, all columns containing the first three rows with three or two summands are reduced by means of (3:2) or (2:2) counters, respectively. This procedure is repeated in successive stages until there are only two rows left. Then, in a final stage, the two rows are added using a fast **CPA**.

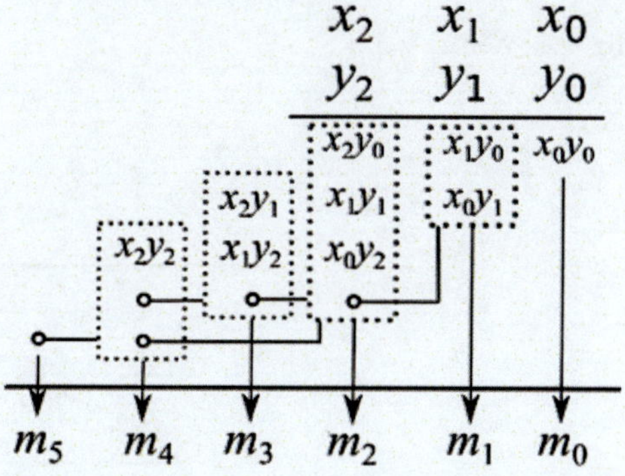

Fig. 6.3 3 × 3 multiplier using counters

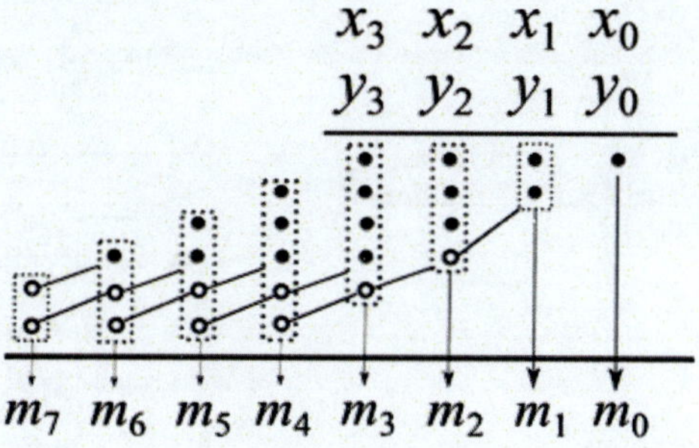

Fig. 6.4 Dot diagram of a 4 × 4 multiplier using counters

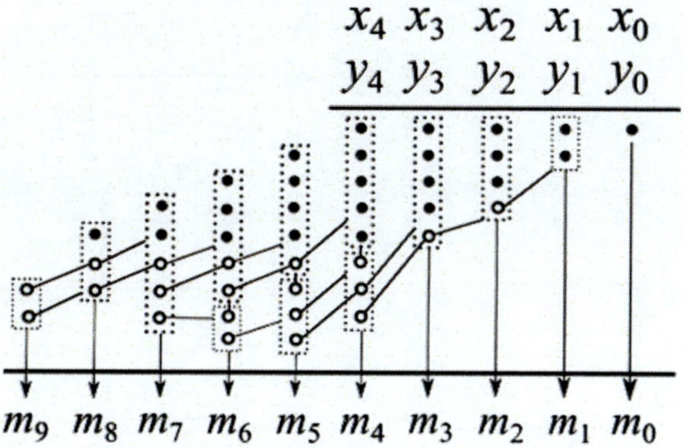

Fig. 6.5 Dot diagram of a 5 × 5 multiplier using (5:3) counters

Example 6.1 **Design a circuit for building a combinational 5 × 5 multiplier using a Wallace's tree.**

Following the stages shown in Fig. 6.6, an array of AND gates will produce the partial products $x_i y_i$. Then, the first level of the Wallace tree is built, following the scheme of the Fig. 6.6, thus requiring 3 (3:2) counters (**FA**s) and 2 (2:2) ones (**HA**s), as shown in Fig. 6.7. The second stage will require 4 (3:2) counters and 2(2:2), and

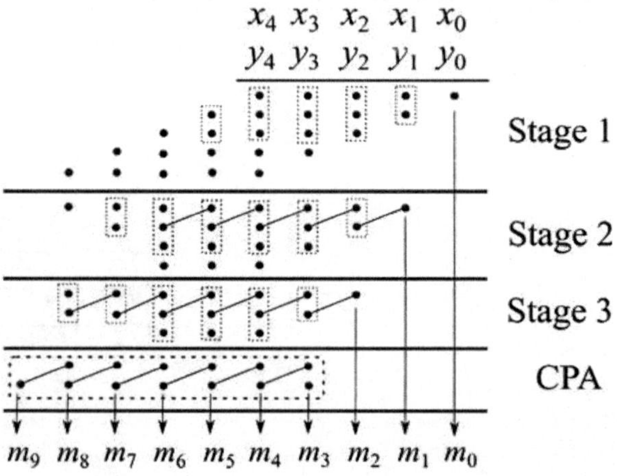

Fig. 6.6 Wallace's 5 × 5 multiplier

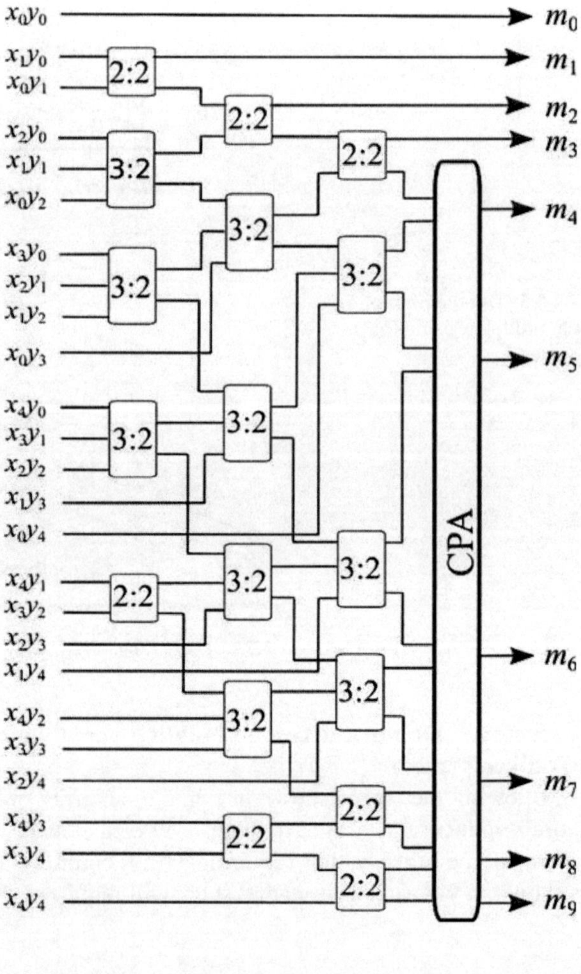

Fig. 6.7 Circuit for a Wallace's 5 × 5 multiplier

the third stage 3 (3:2) and 2 (2:2) counters. The last stage is a 6-bit **CPA**, which will perform the final addition. The circuit corresponding to this implementation is presented in Fig. 6.7, which shows that 10 **FA**s, 7 **HA**s, and 1 6-bit **CPA** are required. ■

A refinement to the Wallace tree is the Dadda multiplier [3], that performs the addition following these steps [4]:

1. Let $d_1 = 2$ and $d_j = \left\lfloor \frac{3}{2} d_{j-1} \right\rfloor$, where d_j is the height of the matrix for the j-th stage. Find the largest j such that at least one column of the bit matrix has more than d_j elements. Note that this expression produces the sequence 2, 3, 4, 6, 9, 13, …
2. Use (3:2) or (2:2) counters as required to achieve a reduced matrix with no column containing more than d_j element. Note that only columns with more than d_j elements are reduced.
3. Repeat step 2 with $j = j - 1$ until a matrix with only two rows is generated ($j = 1$).

Note that the algorithm starts with a value of j such as $d_j < \min\{n, m\}$, where n, m are the number of bits of the multiplicand and the multiplier, respectively. Let's see how this algorithm operates for a 5 × 5 multiplier: In this case, we will start with $j = 3$ ($d_j = 4$), i.e. in stage 3. In this stage, all columns must be reduced to a maximum of 4 bits. In stage 2, the columns will be reduced to 3 bits, and stage 1 to 2 bits. A final stage using a fast **CPA** provides the result of the multiplication. Figure 6.8 shows such a multiplier, where it can be seen the differences with respect to Wallace's tree.

Example 6.2 **Design a circuit for building a combinational 5 x 5 multiplier using a Dadda's tree.**

Following the stages shown in Fig. 6.8, an array of **AND** gates will produce the partial products $x_i y_i$. Then, the first level of the Dadda's tree (corresponding to stage

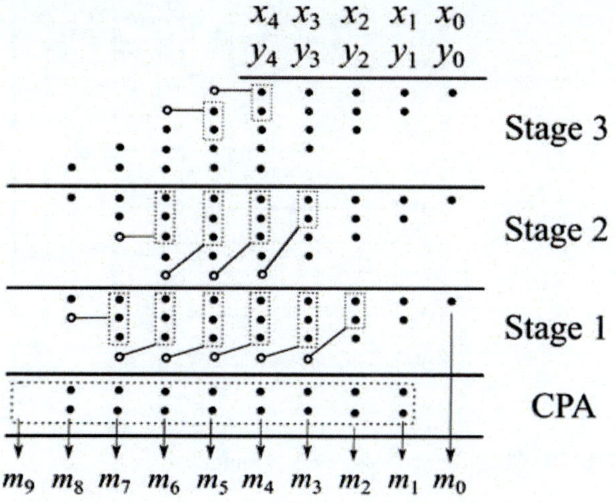

Fig. 6.8 Dadda's 5 × 5 multiplier

number 3) is built, following the scheme of the Fig. 6.8. In this case, only 2 (2:2) counters (**HA**s) are required, as shown in Fig. 6.9. Stage number 2 will require 3 (3:2) and 1 (2:2) counters, and stage number 1 will have 5 (3:2) and 1 (2:2) counters. The last stage is an 8-bit **CPA**, which will perform the final addition. The complete circuit corresponding to this implementation is presented in Fig. 6.9, which shows that 8 **FA**s, 4 **HA**s, and 1 8-bit **CPA** are required.

If we compare the implementations presented in Examples 6.1 and 6.2, in the Daddas's tree, the number of **FA**s and **HA**s is clearly reduced with respect to Wallace's tree realization, at expenses of a more complex **CPA**. Therefore, Dadda's trees provide more compact implementations, but delay results increased.

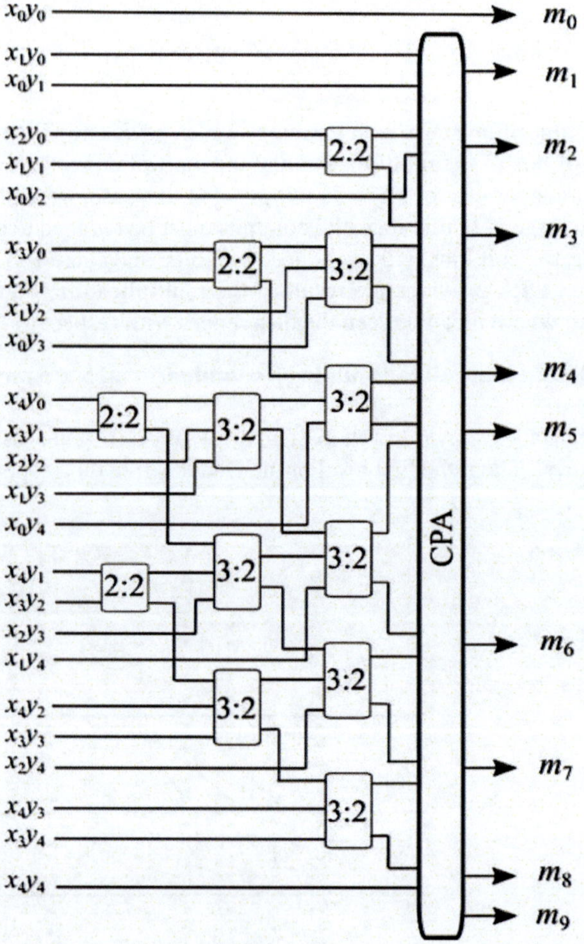

Fig. 6.9 Circuit for a Dadda's 5 × 5 multiplier

6.4 Combinational Multiplication of Signed Numbers

In the case of numbers represented in SM (see Chap. 1), multiplication of two n-bit numbers can be performed multiplying $(n - 1)$ by $(n - 1)$ bits, and generating the sign of the result by means of a **XOR** gate, as introduced in Sect. 1.5.1. Note that the result is a number with 1 sign bit, and $2n - 2$ bits for representing the magnitude.

In the case of 2's complement numbers, the result will have $2n - 1$ bits, where the most significant bit will represent the sign. One option to compute this multiplication is to perform the direct product of the binary representation and correct the result as stablished in Sect. 1.5.2. Another option which avoids the need of corrections, and takes better advantage of the tree adders introduced in previous section, consists on perform direct 2's complements multiplication and addition of the partial products. A first realization of this approach is the Baugh and Wooley method [2], but a more efficient implementation can be obtained as follows.

First, we will establish the result of multiplying a 2's complement number by and individual bit. Consider multiplying an n-bit number A represented in 2's complement as $(a_{n-1}\ a_{n-2},\ a_0)$ where a_{n-1} is the sign-bit. If we multiply A by an individual bit b_i from another 2's complement number B, not being b_i the sign bit, it is clear that:

$$(a_{n-1}, a_{n-2}, \ldots a_0) \times (0, \ldots b_i, \ldots 0) = \left(-a_{n-1}b_i, a_{n-2}b_i, , \ldots a_0b_i, \underbrace{0, \ldots, 0}_{i} \right)$$

Indeed, if the sign bit of A is $a_{n-1} = 0$, the result will be positive, and we have $(0,\ a_{n-2}b_i,\ \ldots,\ a_0b_i,\ 0\ldots0)$. If $a_{n-1} = 1$, the resulting number is negative, and will be: $(-b_i,\ a_{n-2}b_i,\ \ldots,\ a_0b_i,\ 0\ldots0)$. In the case of multiplying by the sign bit b_{n-1} the result is:

$$\begin{aligned}&(a_{n-1}, a_{n-2}, \ldots a_0) \times (b_{n-1}, 0, \ldots, 0)\\ &= \left(a_{n-1}b_{n-1}, -a_{n-2}b_{n-1}, \ldots - a_0b_{n-1}, \underbrace{0, \ldots, 0}_{n-1} \right)\end{aligned}$$

Then, if $b_{n-1} = 0$, the result is $(0, \ldots, 0)$, and if $b_{n-1} = 1$, then we have: $a_{n-1}2^{n-1} - (a_{n-2},\ a_{n-3},\ \ldots,\ a_0)$, which corresponds to the representation in 2's complement of $-A$.

Therefore, the multiplication of two 2's complement 5-bit numbers can be expressed as shown in Fig. 6.10. In order to avoid negative digits, note that:

$$\overline{xy} = 1 - xy; \ -xy = \overline{xy} - 1$$

thus we can obtain the array in Fig. 6.11. Now, the '-1's can be compensated adding two '1's, and introducing a '-1' in the next column, as shown in Fig. 6.12a. The

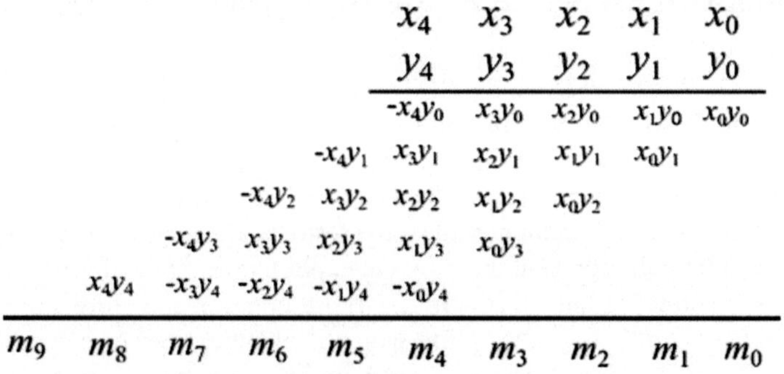

Fig. 6.10 2's complement signed 5 × 5 multiplier

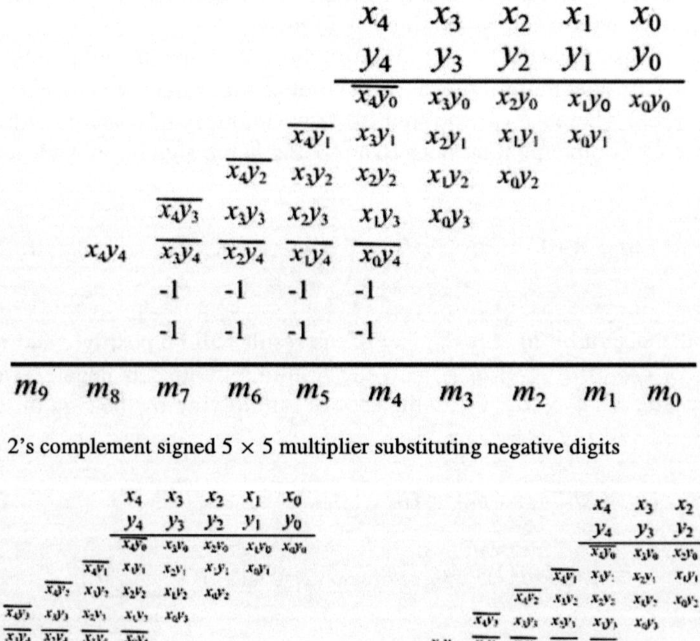

Fig. 6.11 2's complement signed 5 × 5 multiplier substituting negative digits

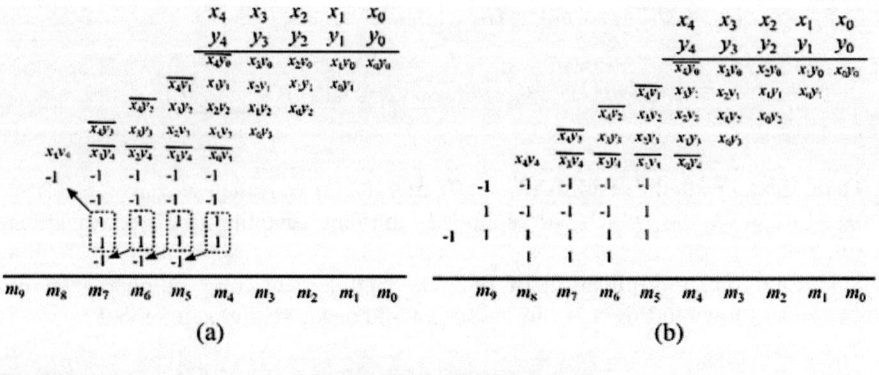

Fig. 6.12 2's complement signed 5 × 5 multiplier introducing '1's

					x_4	x_3	x_2	x_1	x_0
					y_4	y_3	y_2	y_1	y_0
					$\overline{x_4y_0}$	x_3y_0	x_2y_0	x_1y_0	x_0y_0
				$\overline{x_4y_1}$	x_3y_1	x_2y_1	x_1y_1	x_0y_1	
			$\overline{x_4y_2}$	x_3y_2	x_2y_2	x_1y_2	x_0y_2		
		$\overline{x_4y_3}$	x_3y_3	x_2y_3	x_1y_3	x_0y_3			
	x_4y_4	$\overline{x_3y_4}$	$\overline{x_2y_4}$	$\overline{x_1y_4}$	$\overline{x_0y_4}$				
1				1					
m_9	m_8	m_7	m_6	m_5	m_4	m_3	m_2	m_1	m_0

Fig. 6.13 Final 2's complement signed 5 × 5 multiplier structure

procedure can be repeated, obtaining Fig. 6.12b, and finally, Fig. 6.13, where the most significant '−1' has been removed (it corresponds to the -2^{2n} value, defining the 2's complement of the result M, see Chap. 1). Using this technique, the array of partial products in Fig. 6.13 will include only two additional '1's in rows m_n and m_{2n-1} (this is a general result). The corresponding array can be added using any of the trees structures presented in previous section.

Example 6.3 **Multiply −15 by 14 using a 5 × 5 multiplier for 2's complement signed numbers. The same for multiplying −10 by −12.**

Following the structure shown in Fig. 6.13, in the first four rows the most significant partial product has to be complemented. In the fifth row, all partial products have to be complemented except the most significant one, and a sixth row including two '1s' in position n and $2n - 1$ should be added. Figure 6.14a shows the resulting array, and the final result of the multiplication, the 2's complement representation of (−210). For the case or multiplying −10 by −12, the resulting array is the shown in Fig. 6.14b.

					1	0	0	0	1	(-15)
					0	1	1	1	0	(14)
					1	0	0	0	0	
				0	0	0	0	1		
			0	0	0	0	1			
		0	0	0	0	1				
	0	1	1	1	1					
1				1						
1	1	0	0	1	0	1	1	1	0	(-210)

(a)

					1	0	1	1	0	(-10)
					1	0	1	0	0	(-12)
					1	0	0	0	0	
				1	0	0	0	0		
			0	0	1	1	0			
		1	0	0	0	0				
	1	1	0	0	1					
1				1						
0	0	0	1	1	1	1	0	0	0	(+120)

(b)

Fig. 6.14 Examples of 2's complement signed 5 × 5 multiplications

6.5 Basic Sequential Multipliers

When the number of bits of the operands increases, combinational multipliers require excessive area and present high delays. In this case, strategies based on registering partial results may enable an area reduction by means of resources reusing. Regarding the delay, in sequential multipliers the critical path is reduced when compared to combinational ones, thus enabling higher clock frequencies at expenses of increasing the number of clock cycles required for completing the operation. In that follows, some basic structures for implementing sequential multipliers are presented.

6.5.1 Shift and Add Multipliers

A first approach for implementing a sequential $n \times n$ multiplier based on the pencil-and-paper multiplication consists on multiplying the first bit of the multiplier Y by the entire multiplicand X using an **AND** array. Then, the result will be added to the product of the next bit of Y by left-shifted X. This process is repeated for each bit of Y. Note that doing a right shift of Y, the bit to be multiplied is always the less significant bit. This multiplication method can be expressed as in Algorithm 6.1.

Algorithm 6.1

Shift-left multiplication algorithm:
$A \leftarrow X$, $B \leftarrow Y$, $M \leftarrow 0$, $i \leftarrow 0$
while $i<n$ **do**
begin
1) $M \leftarrow M+A \cdot b_0$, $A \leftarrow \overleftarrow{A}$, $B \leftarrow \overrightarrow{B}$, $i \leftarrow i+1$
end
End algorithm

Table 6.1 shows an example of execution of this algorithm with two 5-bit numbers. Note that the resources required are: one $2n$-bit A shift register (A), one n-bit shift register (B), one $2n$-bit register (M), one $2n$-bit adder, one digit-vector multiplier ($2n$ **AND** gates array) and one mod n counter. The algorithm requires $n+1$ clock cycles for completing the operation (one for initialization and n for completing the main loop). Figure 6.15 presents the architecture for the implementation of this algorithm (without including the counter).

Note that the implementation of this algorithm requires a $2n$-bit adder. Nevertheless, if we perform the addition of the partial products as shown in Fig. 6.16, in each step only an $n+1$ bit addition is required. This idea can be expressed as described in Algorithm 6.2.

Table 6.1 Execution example of the left-shift multiplication algorithm

Register	Value
$A[0]$	0000010001
$B[0]$	0111**1**
$M[0]$	0000000000
$A[0] \cdot b_0$	0000010001
$M[1]$	0000010001
$A[1]$	0000100010
$B[1]$	0011**1**
$A[1] \cdot b_0$	0000100010
$M[2]$	0000110011
$A[2]$	0001000100
$B[2]$	0001**1**
$A[2] \cdot b_0$	0001000100
$M[3]$	0001110111
$A[3]$	0010001000
$B[3]$	0000**1**
$A[3] \cdot b_0$	0010001000
$M[4]$	0011111111
$A[4]$	0100010000
$B[4]$	0000**0**
$A[4] \cdot b_0$	0000000000
$M[5]$	0011111111

Algorithm 6.2

Shift-right multiplication algorithm:

$A \leftarrow X$, $B \leftarrow Y$, $M \leftarrow 0$, $i \leftarrow 0$

while $i<n$ **do**

begin

1) $M \leftarrow$ M+$\overrightarrow{(M + 2^n A \cdot b_0)}$, $B \leftarrow \vec{B}$, $i \leftarrow i+1$

end

End algorithm

Table 6.2 shows an execution example for the multiplication of two 5-bit numbers, and Fig. 6.17 the circuit for implementing such a multiplier. The resources required in this case are: one n-bit register (A), one n-bit shift register (B), one $2n$-bit register (M), one ($n + 1$)-bit adder, one digit-vector multiplier (n **AND** gates array) and one mod n counter. Therefore, it requires fewer resources than the shift-left multiplier, while completing the multiplication in $n + 1$ clock cycles. In Table 6.2, $S[i]$ represents the output of the adder at step i.

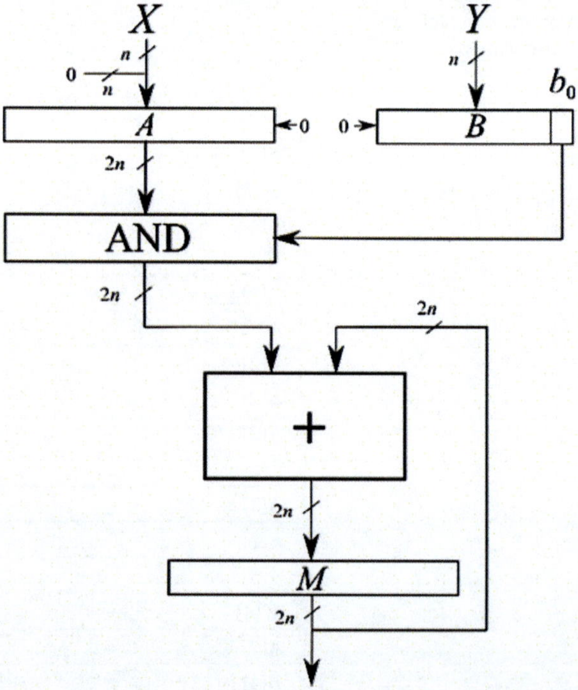

Fig. 6.15 Shift-left sequential multiplier architecture

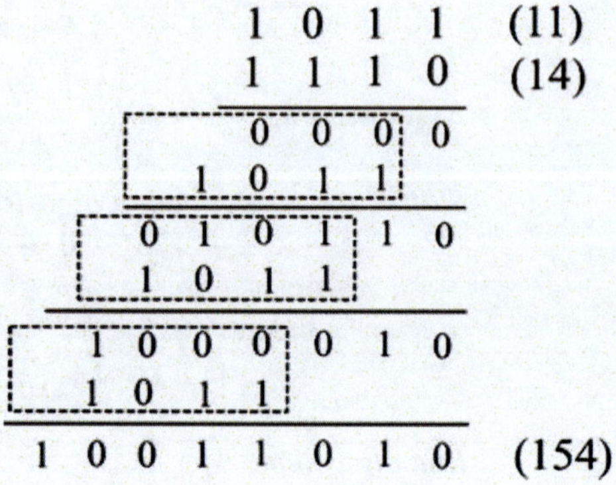

Fig. 6.16 Sequential multiplication using an $(n+1)$-bit adder

A more compact circuit can be obtained including the shift register *B* into the *M* register, as carried out in Fig. 2.10 (Chap. 2, Sect. 2.4.2). Note also that Algorithms 6.1 and 6.2 are equivalent to algorithms 2.1 and 2.2.

Table 6.2 Execution example of the right-shift multiplication algorithm

Register	Value
A	10001
$B[0]$	11111
$M[0]$	0000000000
$M[0]$	0000000000
$2^n A \cdot b_0$	10001
$S[0]$	010001
$M[1]$	0100010000
$B[1]$	01111
$M[1]$	0100010000
$2^n A \cdot b_0$	10001
$S[1]$	011001
$M[2]$	0110011000
$B[2]$	00111
M[2]	0110011000
$2^n A \cdot b_0$	10001
$S[3]$	011101
$M[3]$	0111011100
$B[3]$	00011
$M[3]$	0111011100
$2^n A \cdot b_0$	10001
$S[4]$	011111
$M[4]$	0111111110
$B[4]$	00001
$M[4]$	0111111110
$2^n A \cdot b_0$	10001
$S[5]$	000000
$M[5]$	1000001111

6.5.2 *Shift and Add Multiplication of Signed Numbers*

When the numbers being multiplied are signed numbers expressed in 2's complement, the algorithms presented in previous section are valid, except for the last step. Indeed, if we take into account that the representation in 2's complement of Y is:

$$Y = -y_{n-1}2^{n-1} + \sum_{i=0}^{n-2} y_i 2^i$$

then:

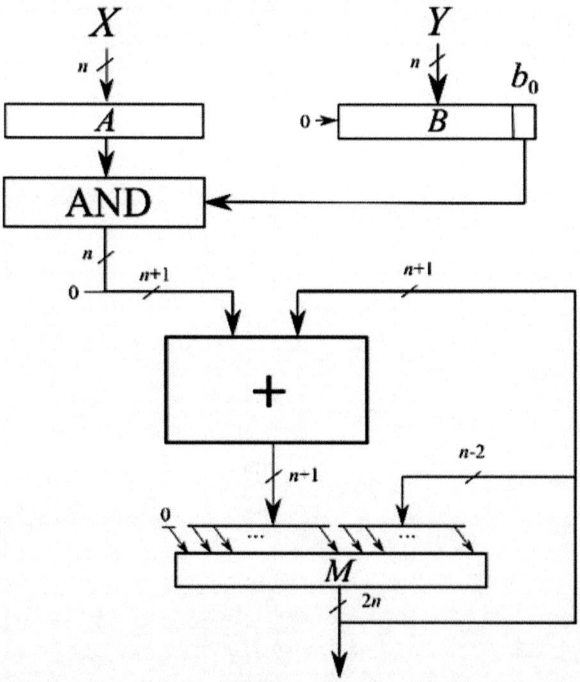

Fig. 6.17 Shift-right sequential multiplier architecture

$$X \cdot Y = X \sum_{i=0}^{n-2} y_i 2^i - X y_{n-1} 2^{n-1}$$

and the computation requires the final step to be a subtraction instead of an addition. Also, the right shift of M has now to be an arithmetic shift, and the additional bit of A has to be a sign extension. The algorithm for shift-right multiplication will be then:

Algorithm 6.3

Shift-right multiplication algorithm for 2's complement numbers:

$A \leftarrow X$, $B \leftarrow Y$, $M \leftarrow 0$, $i \leftarrow 0$
while $i<n$-1 **do**
begin
1) $M \leftarrow M + \overrightarrow{(M + 2^n A \cdot b_0)}$, $B \leftarrow \vec{B}$
End
$M \leftarrow M + \overrightarrow{(M - 2^n A \cdot b_0)}$
End algorithm

Example 6.4 **Multiply −3 by 5 using the shift-right algorithm with a 5-bit multiplier.**

The 2's complement representation of −3 and 5 using 5-bit binary numbers is "11101" and "00101", respectively. The execution of the algorithm, will be then the one presented in Table 6.3.

Example 6.5 **Multiply −3 by −5 using the shift-right algorithm with a 5-bit multiplier.**

The 2's complement representation of −3 and −5 using 5-bit binary numbers is "11101" and "11011", respectively. The execution of the algorithm, will be now the one presented in Table 6.4.

The circuit for performing the signed multiplication requires now and adder/subtractor in 2's complement, having the structure shown in Fig. 6.18. In this

Table 6.3 Execution example of the right-shift multiplication algorithm

Register	Value
A	**1**11101
$B[0]$	0010**1**
$M[0]$	0000000000
$M[0] \rightarrow$	**0**000000000
$2^n A \cdot b_0$	**1**11101
$S[0]$	111101
$M[1]$	1111010000
$B[1]$	0001**0**
$M[1] \rightarrow$	**1**111101000
$2^n\ A \cdot b_0$	**0**00000
$S[1]$	111110
$M[2]$	1111101000
$B[2]$	0000**1**
M[2] $\rightarrow$	**1**111110100
$2^n\ A \cdot b_0$	**1**11101
$S[3]$	111100
$M[3]$	1111000100
$B[3]$	0000**0**
$M[3] \rightarrow$	**1**111100010
$2^n\ A \cdot b_0$	**0**00000
$S[4]$	111110
$M[4]$	1111100010
$B[4]$	0000**0**
$M[4] \rightarrow$	**1**111110001
$2^n\ A \cdot b_0$	**0**00000
$S[5]$	111111
$M[5]$	1111110001

Table 6.4 Execution example of the right-shift multiplication algorithm with two negative numbers

Register	Value
A	111101
$B[0]$	11011
$M[0]$	0000000000
$M[0] \rightarrow$	000000000000
$2^n A \cdot b_0$	111101
$S[0]$	111101
$M[1]$	1111010000
$B[1]$	01101
$M[1] \rightarrow$	1111101000
$2^n A \cdot b_0$	111101
$S[1]$	111011
$M[2]$	1110111000
$B[2]$	00110
$M[2] \rightarrow$	1111011100
$2^n A \cdot b_0$	000000
$S[3]$	111101
$M[3]$	1111011100
$B[3]$	00011
$M[3]$	1111101110
$2^n A \cdot b_0$	111101
$S[4]$	111011
$M[4]$	1110111110
$B[4]$	00001
$M[4] \rightarrow$	1111011111
$2^n A \cdot b_0$	000011
$S[5]$	000000
$M[5]$	0000001111

multiplier, the control unit will have to take into account to perform a subtraction in the last step.

6.6 Sequential Multipliers with Recoding

Basic multipliers introduced in previous Section present low resources requirements, but take $n + 1$ clock cycles for performing a multiplication. One technique for reducing the number of clock cycles may be using radix higher than 2 in order to decreasing the number of digits. A first approach can be using radix-4 digits, but it

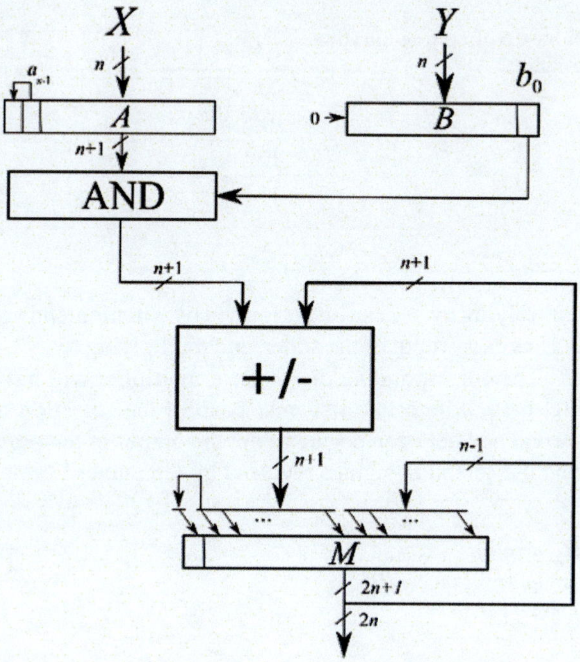

Fig. 6.18 Shift-right sequential 2's complement multiplier architecture

implies designing a more complex digit-vector multiplier. Indeed, when using radix-4, we have four digits, "00", "01", "10" and "11". Multiplying by "00", "01", and "10" does not imply additional arithmetic operations (multiply by "10" is only a shift), but multiplying by "11" requires an addition. In order to avoid multiplication by "11", we can use signed digits coding (Sect. 1.8) thus enabling the use of radix-4 while maintaining an affordable complexity for the digit-vector multiplier, as shown in the next sub-sections.

6.6.1 Multiplication Using Booth Codification

In Chap. 1 (Sect. 1.8.3.3), Booth codifications where introduced. The idea, introduced in [1] is based on detecting '1's chains and substitute them by digit-signed digits. In the case of radix-2, the Booth encoding corresponds to Table 6.5, while in the case of radix-4, the Booth codification is the one shown in Table 6.5. Note that radix-4 Booth codification can be seen as a recodification of digits (0, 1, 2, 3) to (−2, −1, 0, 1, 2), thus presenting the advantage of avoiding multiplication by 3. If we consider the structure of the shift-right multiplier (Fig. 6.18) for implementing multiplier with radix higher than 2, the non-immediate operations required using this recodification are multiplying by −1, 2 and −2. Multiplying by 2 is a left shift of multiplicand X,

Table 6.5 Booth codification in radix 2

b_ib_{i-1}	t_i	Comment
00	0	0's chain
01	1	End of 1's chain
10	$\overline{1}$	Starting of a 1's chain
11	0	1's chain

multiplying by −1 can be performed by complementing each bit of X and introducing a '1' as carry input of the adder, and multiplying by −2, making these operations over 2X. The corresponding digit-vector multiplier will have then the structure shown in Fig. 6.19, where the first mux enables the selection among 0, X and 2X, and the second one between negate or not the output of the previous mux. Table 6.5 presents also the selection signals required by the muxes (Table 6.6).

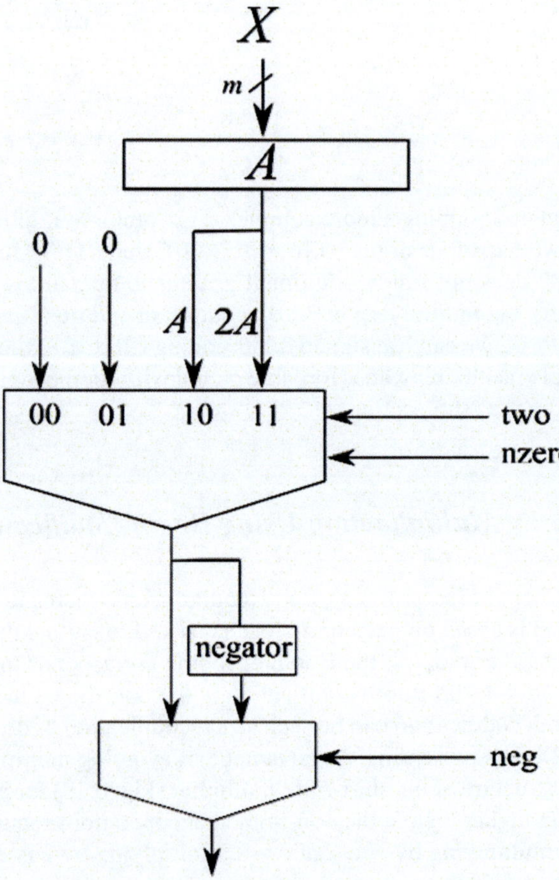

Fig. 6.19 Digit-vector multiplier for radix-4 Booth multiplier

Table 6.6 Booth codification in radix 4

$b_{i+1}b_ib_{i-1}$	$t_{i+1}t_i$	Comment	z_j	nzero two	neg
000	00	0's chain	0	00	0
001	01	End of 1's chain in x_{i-1}	1	10	0
010	01	Isolated 1	1	10	0
011	10	End of 1's chain in x_i	2	11	0
100	$\overline{1}\,0$	Starting of a 1's chain in x_{i+1}	−2	11	1
101	$0\,\overline{1}$	Isolated 0	−1	10	1
110	$0\,\overline{1}$	Starting of 1's chain in x_i	−1	10	1
111	00	1's chain	0	00	0

Using this digital-vector multiplier, a circuit for a radix-4 based on Booth recoding will be the one shown in Fig. 6.20. Note that this multiplier requires $m + 1$ clock cycles for completing a multiplication, where m is the number of radix-4 digits required for representing the operands. In this case, A register is extended with 2 more digits. One of them is for taking into account the increment in the number of bits due to multiplication by 2, and the other for addressing sign extension.

Example 6.3 **Multiply 17 × 30 using radix-4 Booth recoding.**

We are multiplying 2 5-bit numbers, thus in this case, $m = 3$ radix-4 digits are required for representing the operands. Table 6.5 shows the values of the different register in each step of the multiplication (Table 6.7).

6.6.2 *Multiplication Using (−1, 0, 1, 2) Coding*

The $(-1, 0, 1, 2)$ coding represents radix-4 numbers as follows: "00" is coded as "0", "01" as "1", "10" as "2", and "11" as "−1", generating a carry to the next digit. As an example, the number 27, expressed "11011" in binary, and "01 10 11" in radix-4, will be codified as "$2\overline{1}\overline{1}$" in $(-1, 0, 1, 2)$ coding. Note that multiplying a vector X by −1 is equivalent to compute $\overline{X} + 1$. The addition of 1 can be performed introducing a carry input to adder in Fig. 6.21a, thus it is not necessary to introduce additional adders, and complementing the vector X is not a costly operation. Therefore, it is advantageous multiplying by −1 instead of by 3. Nevertheless, this advantage will be lost if the recoding process requires a lot of resources, or increments the number of clock cycles for computing each step. Fortunately, it is easy to implement the coder using only a flip-flop, one AND gate and a full adder, as shown in Fig. 6.21a. In Fig. 6.21b it is detailed the truth table of the encoder, and the result of the multiplication digit-vector in each case. Note that c_{-1} is the carry generated in the previous digit conversion, which is stored in the C flip-flop.

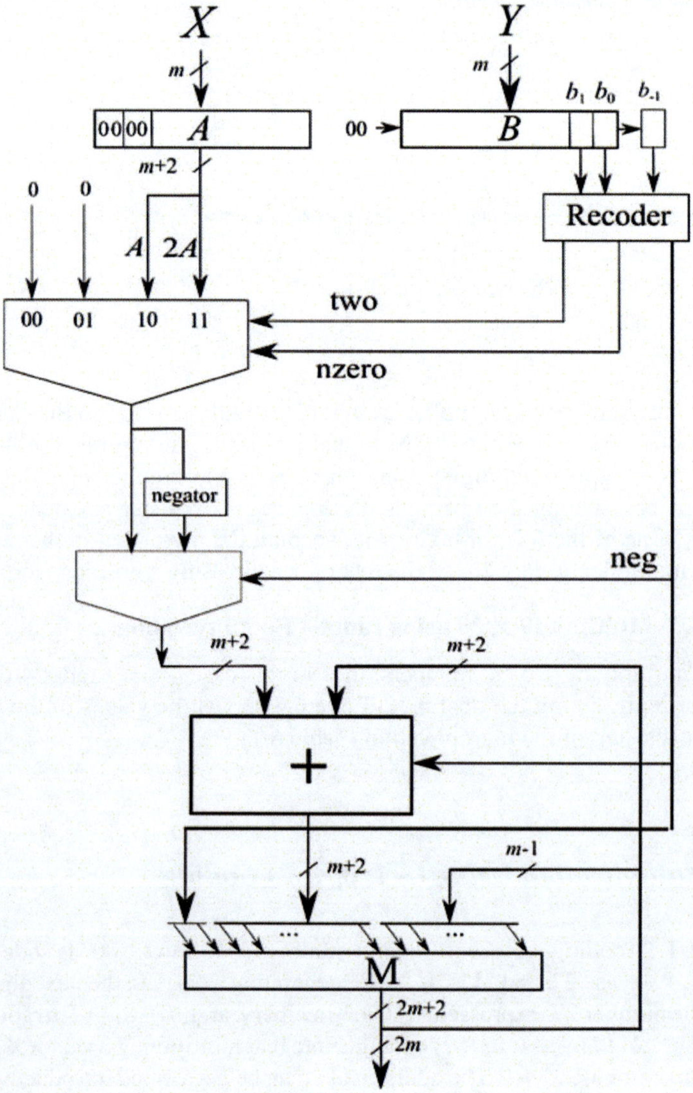

Fig. 6.20 Shift-right radix-4 Booth multiplier

Using this encoder, a inverter and a mux, it is possible to build a digit-vector multiplier in radix-4, and implement the shift-right multiplier shown in Fig. 6.17, which will require $m + 1$ clock cycles for performing the multiplication of two number of m radix-4 digits. Note that A needs to be extended one digit because of computing $2A$, and then, the adder needs to be $m + 2$-digit wide, which implies extending A one more digit.

Table 6.7 Radix-4 Booth multiplication of 17 × 30

Register	Value	z
A	0000010001	
$B[0]$	0111**10** **0**	−2
$M[0]$	0000000000000000	
$M[0] \rightarrow$	**00**00000000000000	
$4^m\ 2\overline{A} \cdot b_1 b_0 + \text{neg}$	1111011110	
$S[0]$	1111011110	
$M[1]$	1111011110000000	
$B[1]$	0001**11** **1**	0
$M[1] \rightarrow$	**11**11110111100000	
$4^m\ A \cdot b_1 b_0 + \text{neg}$	0000000000	
$S[1]$	1111110111	
$M[2]$	1111110111100000	
$B[2]$	000**01** **1**	2
$M[2] \rightarrow$	**00**11111101111000	
$4^m\ 2A \cdot b_1 b_0 + \text{neg}$	0000100010	
$S[3]$	0100011111	
$M[3]$	**0001000111111110**	

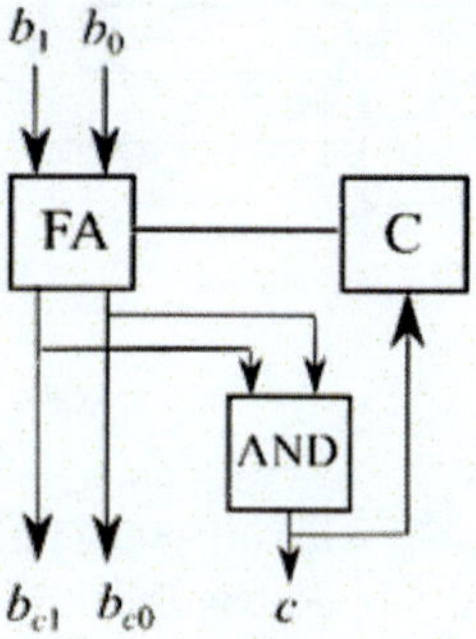

(a)

$b_1 b_0$	c_{-1}	c	$b_{c1} b_{c0}$	Digit-vector result
00	0	0	00	$0 + c$
01	0	0	01	$X + c$
10	0	0	10	$2X + c$
11	0	1	11	$\bar{X} + c$
00	1	0	01	$X + c$
01	1	0	10	$2X + c$
10	1	1	11	$\bar{X} + c$
11	1	0	00	$0 + c$

(b)

Fig. 6.21 (−1, 0, 1, 2) coder for radix-4 digit-vector multiplication **a** and truth table **b**

Example 6.4 **Multiply 17 × 31 using the radix-4 multiplier.**

The number of bits of the two operands is 5, thus 3 radix-4 digits are required for representing them. As mentioned before, A has to be extended 2 more digits, thus resulting the contents of Table 6.8 for the computation of 17 × 31. Note that only four clock cycles (3 + 1 for initialization) are required for completing the operation.

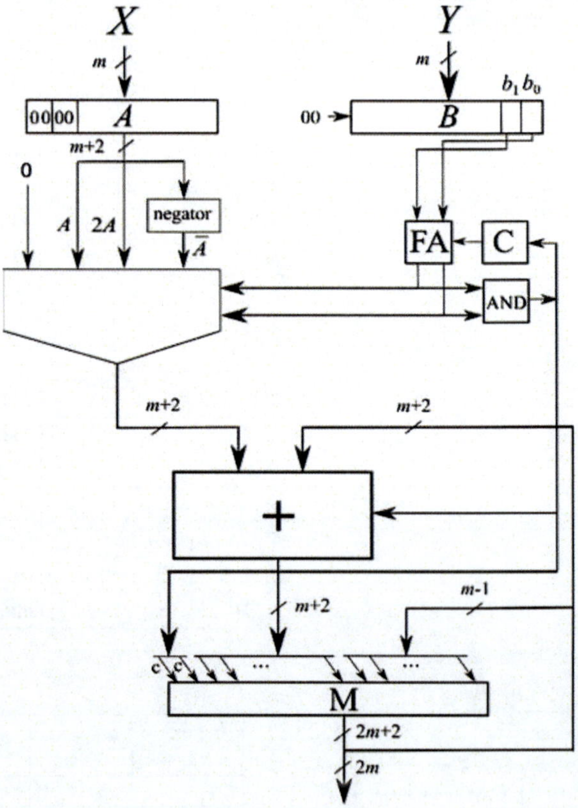

Fig. 6.22 Shift-right sequential radix-4 multiplier architecture

6.7 Special Multipliers

There are situations in which multiplication is concatenated with other operations, such as addition (multiply-and-accumulate), or one of the operands is a constant (see Sect. 2.4.3), or the two operands are the same (see Sect. 2.5), and so on. This particular situations are known as special multipliers, and in the following sections will be approached.

6.7.1 Multipliers with Saturation

Multiplication of two n-bits operands result in a $2n$-bit number. If the system performing these multiplications is designed for processing n-bit numbers, the result of the multiplication is required to fit into n-bit registers. If not, an "overflow" is produced, and it should be signaled. A simple way of generating the overflow signal is to perform the OR operation among the bits m_{2n} to m_n. If any of these bits result

Table 6.8 Radix-4 multiplication of 17 × 31

Register	Value	z
A	0000010001	
$B[0]$	011111	−1
$M[0]$	0000000000000000	
$M[0] \rightarrow$	**00**00000000000000	
$4^m\ \overline{A} \cdot b_1 b_0 + c$	1111101111	
$S[0]$	1111101111	
$M[1]$	1111101111000000	
$B[1]$	000**111**	−1
$BC[1]$	000100	
$M[1] \rightarrow$	**11**111110111100	
$4^m\ A \cdot b_1 b_0 + c$	0000000000	
$S[1]$	1111111011	
$M[2]$	1111111011110000	
$B[2]$	00**001**	1
$BC[2]$	00010	
$M[2] \rightarrow$	**00**1111110111100	
$4^m\ A \cdot b_1 b_0 + c$	0000100010	
$S[3]$	0100100000	
$M[3]$	0001001**000001111**	

set to '1', an "overflow" will be signaled, and the error should be processed by the overall system.

6.7.2 Multiply-and-Accumulate (MAC)

In several algorithms used in Signal Processing (digital filters, as example), are common operations of the type:

$$S = \sum_{i=0}^{n-1} X_i Y_i$$

known as multiply-and-accumulate (MAC) operations. This can be rewritten as:

$$S[i+1] = X_i Y_i + S[i]$$

Thus, in each step a multiplication followed by an addition have to be performed. As in the different multipliers studied the addition of partial products is required, the

					x_4	x_3	x_2	x_1	x_0
					y_4	y_3	y_2	y_1	y_0
					x_4y_0	x_3y_0	x_2y_0	x_1y_0	x_0y_0
				x_4y_1	x_3y_1	x_2y_1	x_1y_1	x_0y_1	
			x_4y_2	x_3y_2	x_2y_2	x_1y_2	x_0y_2		
		x_4y_3	x_3y_3	x_2y_3	x_1y_3	x_0y_3			
	x_4y_4	x_3y_4	x_2y_4	x_1y_4	x_0y_4				
					s_4	s_3	s_2	s_1	s_0
m_9	m_8	m_7	m_6	m_5	m_4	m_3	m_2	m_1	m_0

Fig. 6.23 MAC operation with a 5 × 5 multiplier

addition of $S[i]$ can be incorporated as another partial product. In the case of parallel combinational multipliers, a new row appears then in the array of partial products, that can be reduced using Wallace or Dadda trees. Figure 6.23 shows an example of parallel multiplier for a 5-bit MAC operation. Note that in this type of operations, the value of the result increases in each iteration, being required a careful control of the overflows. In the case of sequential multipliers, it implies introducing a new clock cycle for performing the additional sum, without requiring major changes in the processing unit.

6.7.3 Multipliers with Truncation

As pointed out previously, multiplication of two *n*-bit operands results in a 2*n*-bit number. However, it is usual for the entire system that uses the multiplier to work with a certain number of bits (a 32-bit microprocessor, as an example, truncate all arithmetic operations to 32 bits). If the result of the multiplication has to be truncated to *n* bits, it has no sense to compute all the 2*n* bits, thus allowing to saving resources when implementing the multiplier. Nevertheless, for maintaining an error less than ½ of the less significant bit, the result should not be directly truncated. Indeed, let us consider the multiplication of two 4-bit numbers in fixed point arithmetic with two bits of integer part, and two bits of decimal part: “11.10” × "11.10" (3.5 × 3.5 in decimal). Without truncation, the result is 1010.0100 (12.25). If we multiply only the two most significant bits, i.e. “11” × "11", the result is “1001” (9). Nevertheless, if we truncate only the first three columns of the partial product, as shown in Fig. 6.24, and compute the fourth column for having an additional bit for rounding, the obtained result is “1010” (12), which implies a reasonable error.

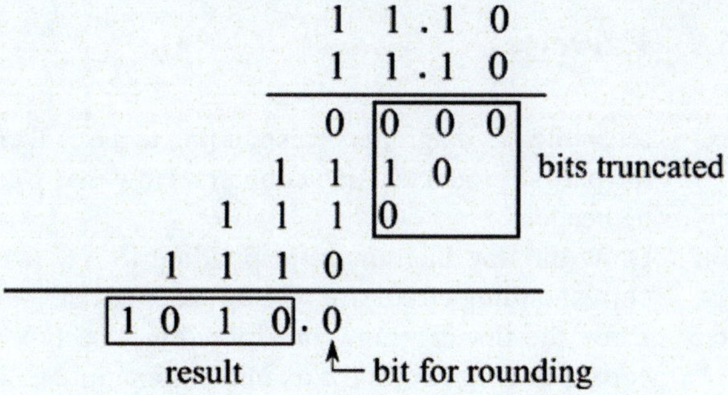

Fig. 6.24 Fixed point 4 × 4 truncated multiplication example

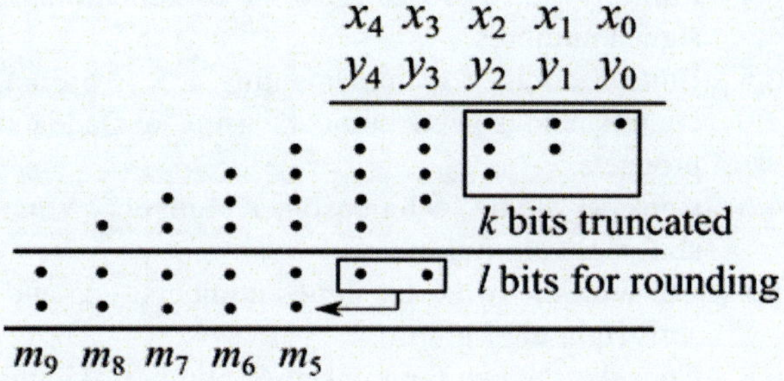

Fig. 6.25 5 × 5 multiplier with truncation

Therefore, in general, the truncation should be realized as shown in Fig. 6.25, first a set of k columns are truncated, and l columns are used for rounding the result, being $k + l = n$. The selection of k and l depends on the maximum error allowed in the system.

6.8 Conclusion

In this chapter, different circuits for performing arithmetic multiplication have been presented. First, parallel multipliers have been introduced, from the most basic structures to implementations based on reductions trees by Wallace and Dadda, considering also the multiplication of signed numbers. Also, sequential multipliers for reducing area requirements have been presented, including radix-4 ones based on recodification to signed digits for reducing the number of clock cycles. Finally, some special multipliers have been considered. The presented circuits will be used in next chapters for performing more complex operations.

6.9 Exercises

6.1 Draw the dot diagram corresponding to a 6 × 6 combinational multiplier. Add the partial products using counters. How many counters, and what size would be needed?

6.2 Draw the dot diagram corresponding to a 6 × 6 Wallace multiplier and the corresponding circuit for its implementation.

6.3 Draw the dot diagram corresponding to a 6 × 6 Dadda multiplier and the corresponding circuit for its implementation. Compare the required resources with the Wallace multiplier developed in exercise 6.2.

6.4 Multiply −25 by 14 using a 6 × 6 combinational multiplier for 2's complement signed numbers.

6.5 Build a circuit for implementing a 5 × 5 combinational multiplier for 2's complement signed numbers using a Dadda's tree for adding the partial products.

6.6 Consider the two 6-bit unsigned numbers 45 and 57. Multiply them using the shift-right algorithm.

6.7 Consider the two 6-bit signed numbers −25 and 14. Multiply them using the shift-right algorithm.

6.8 Consider the two 6-bit unsigned numbers 45 and 57. Multiply them using the Booth radix-4 method.

6.9 Modify the circuit of Fig. 6.20 for building a circuit to multiply 6-bit signed numbers expressed in 2's complement using the radix-4 Booth method.

6.10 Multiply the two 6-bit signed numbers −25 and 14 using the circuit designed in exercise.

6.11 Consider the two 6-bit unsigned numbers 40 and 50. Multiply them using the radix-4 (−1, 0, 1, 2) coding multiplier.

6.12 Modify the circuit of Fig. 6.22 for building a circuit to multiply 6-bit signed numbers expressed in 2's complement using the radix-4 (−1, 0, 1, 2) codification.

6.13 Design a circuit for implementing MAC multiplier in Fig. 6.23 using a Wallace tree.

6.14 Build a 6 × 6 fixed point multiplier (3 bits of integer part) with truncation to obtain a 6-bit result (6 bits of integer part), with a maximum error of 2^{-1}.

References

1. Booth, A.D.: A signed binary multiplication technique. Quart. J. Mech. Appl. Math. **4**(2), 236–240 (1951)
2. Baugh, C.R., Wooley, B.A.: A two's complement parallel array multiplication algorithm. IEEE Trans. Comp. **C-22**(12), 1045–1047 (1973)
3. Dadda, L.: Some schemes for parallel multipliers. Alta Frequenza **34**, 349–356 (1965)

4. Townsend, W.J., Swartzlander Jr, E.E., Abraham, J.A.: A comparison of Dadda and Wallace multiplier delays. In: Advanced Signal Processing Algorithms, Architectures, and Implementations XIII, vol. 5205, pp. 552–560. International Society for Optics and Photonics (2003)
5. Wallace, C.S.: A suggestion for a fast multiplier. IEEE Trans. Electron. Comput. **EC-13**(1), 14–17 (1964)

Chapter 7
Division

Division is the most complex of the basic arithmetic operations, presenting some issues as the management of overflow or avoiding division by 0. These issues are studied at the beginning of the present Chapter, as well as the basic concepts regarding division. Next, circuits for performing restoring and non-restoring division of unsigned and signed numbers will be introduced. The Chapter will finish with the description of SRT division using radix-2 and radix-4 representations.

7.1 Introduction

Division is the last of the four basic arithmetic operations, and the more complex for its computation using circuits. Nevertheless, division is a wide used operation, and it is important to have efficient implementations for computing it. There are two main approaches for performing division, the first one, is the development and implementation of algorithms based in the general expression of the dividend in terms of the product of quotient by the divisor plus the remainder. This will be the focus of the present Chapter, where restoring division will be presented as a basic method for performing division of natural numbers, non-restoring division, as a method for dividing signed operands, and SRT division for performing division of floating point numbers. The other approach consists on multiplying the dividend by the inverse of the divisor. In this case, the difficulty relays on computing the multiplicative inverse of the divisor, which will approached in Chap. 8.

7.2 Basic Concepts

As in the case of multiplication, we will begin considering the division of unsigned number. Thus, as introduced in Sect. 2.6, let consider two unsigned binary numbers:

A. Lloris Ruiz et al., *Arithmetic and Algebraic Circuits*, Intelligent Systems Reference Library 201, https://doi.org/10.1007/978-3-030-67266-9_7

D (m-bit dividend), and d (n-bit divisor). In this case, the division consists on finding two unsigned binary numbers, C (quotient) and r (remainder), $r < d$, such that:

$$D = C \cdot d + r$$

The division is only defined for $d \neq 0$. Therefore, before dividing, the condition $d \neq 0$ has to be checked. With the condition $r < d$, C and r are unique, and the division can be computed. Nevertheless, there are some additional issues that have to take into account when performing a division. As an example, if we assume that $m = 2n$, C and r should be n-bit numbers. But dividing a $2n$-bit number by an n-bit number does not guarantee that the resulting number can be represented using n bits. In this case, an overflow should be reported. Note that the quotient can be represented using n bits if and only if $C < 2^n$. As $r < d$ and D, C, r and d are integers, we have:

$$D < \left(2^n - 1\right) \cdot d + d < 2^n \cdot d$$

Therefore, if this condition is not met, the circuit should report an overflow, and the division will not be performed. As an example, if we want to divide the 8-bit number 200 by the 4-bit number 12, providing 4-bit quotient and remainder, we have to check if $D < 16 * 12$. As $16 * 12 = 192$, the condition is not met, and we have an overflow. Indeed, $200/12 = 16$ (and a remainder of 8), that cannot be represented using 4 bits.

In what follows it is assumed that this condition is met, and we will focus on computing the quotient and the remainder. As in the case of the multiplication, a first approach is to use combinational circuits, using **ROM** memories, or using structures based on school division, as introduced in Sect. 2.6.1. Note that combinational divider shown in Fig. 2.14 have a similar structure than combinational multipliers studied in Chap. 6, but, in this case, instead of **HA**s or **FA**s, we have **CS** and **CR** cells. The complexity of **CR** cells makes, in general, it inefficient to use combinational dividers because of high delay and area requirements.

Regarding sequential dividers, the simplest algorithm, known as restoring division, is based on the pencil-and-paper division, as presented in Algorithm 7.1, where A has $2n + 1$ bits ($2n$ bits from Dividend, and one additional '0' as most significant bit), B has $n + 1$ bits (n from the divisor, and one addition '0' as most significant bit), C has $n + 1$ bits, the bit n stores the overflow flag, and the rest n bits returns the quotient. Finally, the remainder is in the bits $2n - 1$ to n of the A register.

Algorithm 7.1

Restoring division algorithm:
$A \leftarrow (\text{'0'}, D)$, $B \leftarrow (0, d)$, $C \leftarrow 0$, $i \leftarrow 0$
while $i <= n$ **do**
begin
1) $S \leftarrow A - B \cdot 2^n$, $i \leftarrow i+1$
if $S >= 0$ **then**
2) $C_{n-i} \leftarrow 1$
3) $A \leftarrow \overleftarrow{S}$
else
4) $C_{n-i} \leftarrow 0$
5) $A \leftarrow \overleftarrow{A}$
end
end
$r \leftarrow (S_{2n-1} \ldots S_n)$
End algorithm

This algorithm is known as "restoring" because is based on performing a "trial" of setting a '1' bit in the quotient, and in case of carrying out a negative result, "restore" the previous value, set to '0' the corresponding bit in the quotient, and continue with the next bit. The algorithm takes n + 2 clock cycles for completing a division (1 clock cycle is for initialization).

Let us see an execution example of this algorithm.

Example 7.1 Divide 188 by 12 using the restoring algorithm using 8 bits for the dividend, and 4 bits for the divisor, quotient and remainder.

The number 188 can be expressed as an 8-bit number, and 12 as a 4-bit number. The execution of the algorithm, will be then the one presented in Table 7.1, where the quotient obtained is 15 and remain is 8.

Figure 7.1 shows a circuit for implementing this algorithm. In this implementation, A is a $(2n + 1)$-bit register initialized with the dividend, B a $n + 1$-bit register, initialized with the divisor, and C a $n + 1$-bit register storing the quotient and the overflow flag. As at each clock cycle, only a bit of C is computed, it can be implemented as a shift register. Note that A and B have an additional bit set to 0 for performing $(n + 1)$-bit subtractions, and, from Example 7.1, it is clear that only a (n + 1)-bit subtractor is required. The multiplexor enables selecting if A is loaded with the dividend D, with the result of the subtraction (when the "trial" is successful), or with the shifted "restored" value previously in A.

Table 7.1 Execution example for the restoring division algorithm

Register	Value
A	010111100
$2^n \cdot B$	011000000
$S[0]$	111110000
$C[0]$	00000
$A := \overleftarrow{A}$	101111000
$2^n \cdot B$	011000000
S [1]	010111000
C [1]	01000
$A := \overleftarrow{S}$	101110000
$2^n \cdot B$	011000000
S [2]	010110000
C [2]	01100
$A := \overleftarrow{S}$	101100000
$2^n \cdot B$	011000000
S [3]	010100000
C [3]	01110
$A := \overleftarrow{S}$	101000000
$2^n \cdot B$	011000000
S [4]	010000000
C [4]	01111
Overflow	0
C	1111
r	1000

7.3 Non-restoring Division

In order to avoid the "restoring" of the previous value if trial is not successful in Algorithm 7.1, the known as "non-restoring" (NR) algorithm always accept the value of the trial, and correct it in the next clock cycle, adding the corresponding value. In order to compute the quotient, it is then encoded using $\{-1, 1\}$ values instead of $\{0, 1\}$. In this case, a final correction of the quotient will be required, for recoding it to binary values. The algorithm corresponding to non-restoring division is the one shown in Algorithm 7.2:

Algorithm 7.2

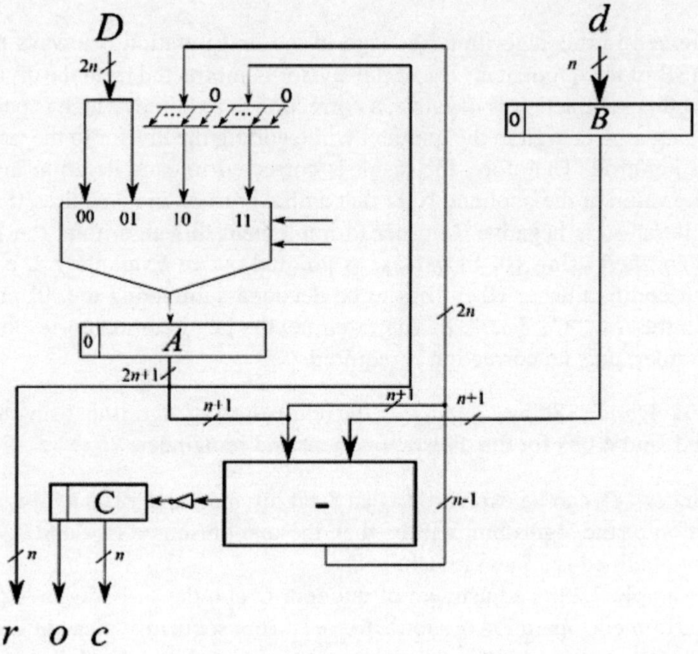

Fig. 7.1 2*n* by *n* restoring divider

Non-Restoring division algorithm:

$A \leftarrow$ (*"00"*, *D*), $B \leftarrow$ (*"00"*, *d*), $C \leftarrow 0$, $i \leftarrow 0$

while $i <= n$ **do**

begin

if sign(*A*)='0' **then**

1) $A \leftarrow \overline{A - B \cdot 2^n}$

2) $C_{n-i} \leftarrow 1$

else

3) $A \leftarrow \overline{A + B \cdot 2^n}$

4) $C_{n-i} \leftarrow -1$

end

5) $i \leftarrow i+1$

end

if sign(*A*)=1 **then**

$C = C - 1$

$A = A + B \cdot 2^n$

end

$r \leftarrow (A_{2n-1} \ldots A_n)$

End algorithm

As expressed in this algorithm, the sign in the first iteration is always positive, thus, the MSB of the quotient is '1' and the divisor is subtracted from the dividend. If the result of this subtraction is negative, a correction is required, which is performed by introducing a '−1' digit in the quotient while adding the divisor to the remainder in the next iteration. Therefore, the result is corrected in each iteration instead of re-trying the value of the quotient. Note that a final correction is required if the sign in the last iteration is negative. In order to implement this algorithm, the $\{-1, 1\}$ set can be codified using $\{0, 1\}$ values, as pointed out in Example 7.2. Note that the quotient codified using $\{0, 1\}$ has to be decoded subtracting the '0' positions, as shown in the row "*C adjust*". In this example, the last iteration corresponds to a positive number, thus no correction is required.

Example 7.2 Divide 188 by 12 applying the non-restoring algorithm using 8 bits for the dividend, and 4 bits for the divisor, quotient and remainder.

The number 188 can be expressed as an 8-bit number, and 12 as a 4-bit number. The execution of the algorithm, will be then the one presented in Table 7.2, where the quotient obtained is 15 and remain is 8.

From Example 7.2 the adjustment of quotient C due the '−1' digits requires an additional arithmetic operation (a subtraction), and consequently, an additional clock cycle. This can be avoided if this adjustment is performed "on-the-fly" at each step. For that, we will update two bits of C in each step, taking into account the previous two bits of the quotient and the sign in the current step. Using this idea, it is possible to build a table relating the quotient codification to the corresponding arithmetic value. As an example, if we are in the second step of the division, previously we obtained c_i = '1' (codifying '1' as '1' and '0' as '0'). If in this step the sign is '−1', the codification for this situation will be "11" (c_i = '1' and sign = '−1') and the corresponding arithmetic value will be "01" ($2^i - 2^{i-1} = 2^{i-1}$). Thus, the codification "11" at input of the recoder corresponds to the output value "01". If in the second step we obtain sign = '0', the codification will be "10", as shown in Table 7.3, where the sign is represented by variable s. Note that c_i always results as '1', and in the next step we will have "10" or "11", not being possible "00" and "01" values.

From Table 7.3, the following equations can be written:

$$c_i = 1$$
$$c_{i+1} = c_{i-1} \cdot \bar{s}$$

and the on-the-fly quotient recoder can be implemented using a shift register and an **AND** gate. The following example shows a non-restoring division using this recodification technique, and requiring a final correction.

Example 7.3 Divide 174 by 12 applying the non-restoring algorithm using 8 bits for the dividend, and 4 bits for the divisor, quotient and remainder. Use on-the-fly computation for the quotient.

Table 7.2 Execution example for the non-restoring division algorithm

Register	Value
A	**0**10111100
$2^n \cdot B$	011000000
$S[0] = A - 2^n \cdot B$	111111100
$C[0]$	**10000**
$A := \overleftarrow{S}$	111111000
$2^n \cdot B$	011000000
$S[1] = A + 2^n \cdot B$	010111000
$C[1]$	**1–1000**
$A := \overleftarrow{S}$	101110000
$2^n \cdot B$	011000000
$S[2] = A - 2^n \cdot B$	010110000
$C[2]$	1–1100
$A := \overleftarrow{S}$	101100000
$2^n \cdot B$	011000000
$S[3] = A - 2^n \cdot B$	010100000
$C[3]$	1–**1110**
$A := \overleftarrow{S}$	101000000
$2^n \cdot B$	011000000
$S[4] = A - 2^n \cdot B$	0**1000**0000
C [4]	**1–1111 (10111)**
Overflow	0
C adjust	10111–01000
C	**01111**
R	**1000**

Table 7.3 Quotient decoding "on-the-fly"

sc_ic_{i-1}	$c_{i+1}c_i$
000	01
001	11
101	01
011	11
111	01

The number 174 can be expressed as an 8-bit number, and 12 as a 4-bit number. The execution of the algorithm, will be then the one presented in Table 7.4, where the quotient obtained is 14 and the remainder is 6.

Figure 7.2 presents a circuit for implementing the non-restoring algorithm. Note that the overflow condition is now detected by comparing the sign of the dividend to the sign (carry out) of the first addition/subtraction. This is performed by an **XOR** gate, and the overflow signal is generated by the control unit, because the condition has to be checked only in the first addition/subtraction. When compared to the restoring algorithm, and the corresponding implementation circuit, non-restoring division seems not presenting any advantage. Nevertheless, NR division establishes the base for performing signed division without performing 2's complement to/from

Table 7.4 Execution example for the non-restoring division algorithm

Register	Value
A	**010101110**
$2^n \cdot B$	011000000
$S[0] = A - 2^n \cdot B$	**111101110**
$C[0]$	0000**1**
$\text{A} := \overleftarrow{S}$	111011100
$2^n \cdot B$	011000000
$S\,[1] = A + 2^n \cdot B$	**010011100**
$C\,[1]$	00**001**
$A := \overleftarrow{S}$	100111000
$2^n \cdot B$	011000000
$S\,[2] = A + 2^n \cdot B$	**001111000**
$C\,[2]$	**00011**
$A := \overleftarrow{S}$	011110000
$2^n \cdot B$	011000000
$S\,[3] = A - 2^n \cdot B$	**000110000**
$C\,[3]$	00111
$A := \overleftarrow{S}$	001100000
$2^n \cdot B$	011000000
$S\,[4] = A - 2^n \cdot B$	110100000
$C\,[4]$	**01111**
Overflow	0
C correction	01111–00001
C	**01110**
R correction	**11010 + 01100**
R	**0110**

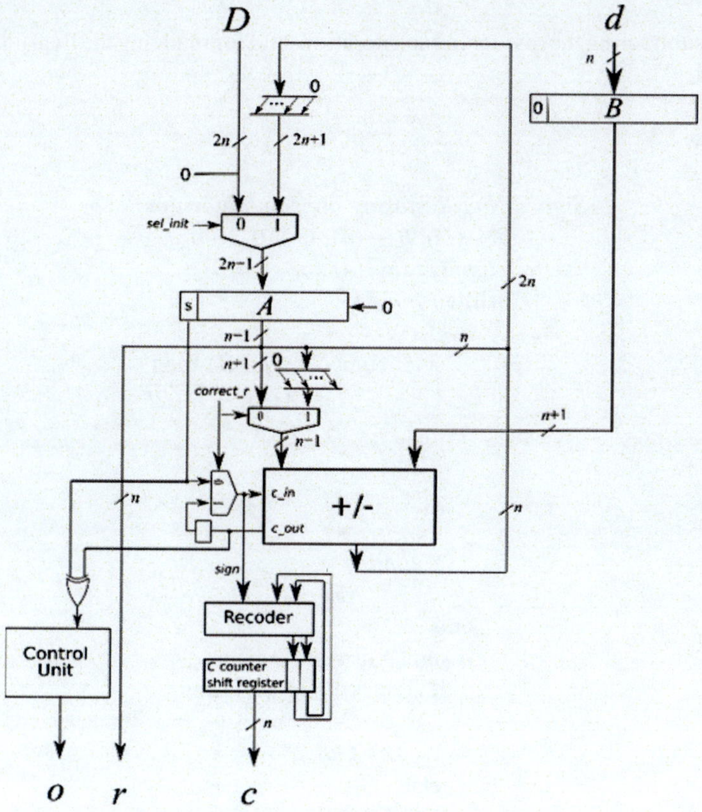

Fig. 7.2 2*n* by *n* non-restoring unsigned divider

sign-magnitude conversions (as required for signed division using the restoring division). Next subsection will present signed division using the NR algorithm.

7.4 Signed Non-restoring Division

Non-restoring division uses 2's complement addition/subtraction as basic operations for performing division. This enables implementing signed division with slightly modifications in the algorithm and the corresponding circuit. Algorithm 7.3 shows this modifications, mainly related to taking into account the sign of the divider for adding or subtracting the divider in each iteration, and for realizing the final correction if required.

Algorithm 7.3

Signed Non-Restoring division algorithm:

$A \leftarrow (D)$, $B \leftarrow (d)$, $C \leftarrow 0$, $i \leftarrow 0$
sign_D=sign(A) xor sign(B)
while $i <= n$ **do**
  begin
    if sign(A)=sign(B) **then**
      1) $A \leftarrow \overleftarrow{A - B \cdot 2^n}$
      2) $C_{n-i} \leftarrow 1$
    else
      3) $A \leftarrow \overleftarrow{A + B \cdot 2^n}$
      4) $C_{n-i} \leftarrow -1$
    end
      5) $i \leftarrow i+1$
  end
  if sign(A)!=sign(D) **then**
    if sign(A)!=sign(B) **then**
      $C=C-1$
      $A=A+B \cdot 2^n$
    else
      $C=C+1$
      $A=A-B \cdot 2^n$
    end
  end
  $r \leftarrow (A_{2n-1} \ldots A_n)$

End algorithm

Note that the NR algorithm looks for maintaining the remainder in each step into the interval $[d, -d)$, as pointed out in the diagram from Fig. 7.3. In this diagram has been represented the new partial remainder $A(i)$ in terms of the previous partial remainder $A(i-1)$ and the selection of the quotient bit C_i. If the remainder $2 \cdot A(i-1) > 0$ then we have to subtract d for maintaining the remainder into $[-d, d]$. Therefore, $c_i = 1$, and we subtract the divisor. If $2 \cdot A(i-1) \leq 0$ it is clear that we have to add the divisor, and then $c_i = -1$. In this case, we have to take the decision of adding or subtracting only in terms of the sign bit.

In the following, some examples for performing signed NR division are presented.

Example 7.4 shows the application of this algorithm for dividing a negative dividend by a positive divisor.

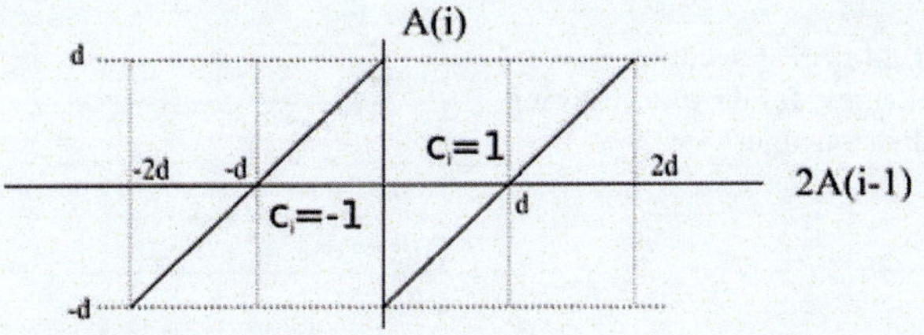

Fig. 7.3 New versus old shifted remainder graph for non-restoring division

Example 7.4 Divide −50 by 12 applying the non-restoring algorithm using 8 bits for the dividend, 4 bits for the divisor, quotient and remainder and 1 sign bit.

The number −50 can be expressed in 2's complement number using 1 + 8 bits as "1 11001110", and 12 as "0 1100". The execution of the algorithm, will be then the one presented in Table 7.5, where the quotient obtained is −4 and remain is −2.

Example 7.5 Divide 50 by −12 applying the non-restoring algorithm using 8 bits for the dividend, 4 bits for the divisor, quotient and remainder and 1 sign bit.

The number 50 can be expressed in 2's complement number using 1 + 8 bits as "**0** 00110010", and −12 as "1 0100". The execution of the algorithm, will be then the one presented in Table 7.6, where the quotient obtained is −4 and remainder is 2.

Example 7.6 Divide −50 by −12 applying the non-restoring algorithm using 8 bits for the dividend, 4 bits for the divisor, quotient and remainder and 1 sign bit.

The number −50 can be expressed in 2's complement number using 1 + 8 bits as "1 11001110", and −12 as "1 0100". The execution of the algorithm, will be then the one presented in Table 7.7, where the quotient obtained is 4 and remainder is −2.

For recoding que quotient on-the-fly, the same table than Table 7.3 can be used adding a new row for "100", which produces '−1' for value '0' with sign = '1'.

The resulting truth table is shown in Table 7.8, and the minimized Boolean functions are:

$$c_i = 1$$
$$c_{i+1} = s \oplus c_{i-1}$$

where s in this case is:

$$s = \text{sign}(A)\,\textbf{XOR}\,\text{sign}(B)$$

In order to design a circuit for performing signed division, the recoder has to include these functions, and the quotient register C will be implemented with an up/down counter for being able to compute all corrections described in Algorithm 7.3. The resulting circuit is the one presented in Fig. 7.4.

Table 7.5 Execution example for the non-restoring division algorithm

Register	Value
A	111001110
$2^n \cdot B$	011000000
$S[0] = A + 2^n \cdot B$	**0**10001110
$C[0]$	**−10000**
$A := \overleftarrow{S}$	100011100
$2^n \cdot B$	011000000
$S[1] = A - 2^n \cdot B$	**0**01011100
$C[1]$	**−11000**
$A := \overleftarrow{S}$	010111000
$2^n \cdot B$	011000000
$S[2] = A - 2^n \cdot B$	**1**11111000
$C[2]$	**−11100**
$A := \overleftarrow{S}$	111110000
$2^n \cdot B$	011000000
$S[3] = A + 2^n \cdot B$	**0**10110000
$C[3]$	**−111–10**
$A := \overleftarrow{S}$	101100000
$2^n \cdot B$	011000000
$S[4] = A - 2^n \cdot B$	**0**10100000
$C[4]$	**−111–11 = −5**
Overflow	0
C	11011
R	**01010**
C correction	**11100 = −4**
R correction	**11110 = −2**

7.5 SRT Division

Restoring and non-restoring division presents difficulties if we try to design dividers with radix higher than two. The advantage in the number of steps required to perform the division is harmed by the need of trials and comparisons. This is obvious in the case of restoring division, and it is also true in non-restoring division because it is not possible to foresee the interval of the remainder without computing it. The SRT (Sweeney-Robertson-Tocher) gives response to this issue by means of normalizing the dividend and the divisor. This algorithm was developed independently by Sweeney, Robertson [2], and Tocher [3], and it is especially suitable for floating point division, because in the case of standard floating point numbers it is possible to avoid

Table 7.6 Execution example for the non-restoring division algorithm

Register	Value
A	**000110010**
$2^n \cdot B$	101000000
$S[0] = \mathrm{A} + 2^n \cdot B$	**101110010**
$C[0]$	**–10000**
$\mathrm{A} := \overleftarrow{S}$	011100100
$2^n \cdot B$	101000000
$S\,[1] = A - 2^n \cdot B$	**110100100**
$C\,[1]$	**–11000**
$A := \overleftarrow{S}$	101001000
$2^n \cdot B$	101000000
$S\,[2] = \mathrm{A} - 2^n \cdot B$	**000001000**
$C\,[2]$	**–11100**
$A := \overleftarrow{S}$	000010000
$2^n \cdot B$	101000000
$S\,[3] = \mathrm{A} + 2^n \cdot B$	**1010100**
$C\,[3]$	**–111–10**
$A := \overleftarrow{S}$	010100000
$2^n\ B$	101000000
$S\,[4] = \mathrm{A} - 2^n \cdot B$	**101100000**
$C\,[4]$	**–111 – 11 = –5**
Overflow	0
C	11011
R	**10110**
Sign(A) = 1 *Sign(B)* = 1 *Sign(D)* = 0	**Correction required:** **C = C + 1** **A = A – B * 2^n**
C correction	**11100 = –4**
R correction	**00010 = 2**

the issues related to normalizing the operands. Indeed, in the widely used IEEE754 standard, normal numbers are represented as "1.xxxx…", thus being the numbers to divide (mantissas) in the interval (–2, 2).

Table 7.7 Execution example for the non-restoring division algorithm

Register	Value
A	**111001110**
$2^n \cdot B$	101000000
$S[0] = A - 2^n \cdot B$	**0**10001110
$C[0]$	**1**0000
$A := \overleftarrow{S}$	100011100
$2^n \cdot B$	101000000
$S[1] = A + 2^n \cdot B$	**0**01011100
$C[1]$	**1–1**000
$A := \overleftarrow{S}$	010111000
$2^n \cdot B$	101000000
$S[2] = A + 2^n \cdot B$	**1**11111000
$C[2]$	1–**1–1**00
$A := \overleftarrow{S}$	111110000
$2^n \cdot B$	101000000
$S[3] = A - 2^n \cdot B$	**0**10110000
$C[3]$	1–1–**11**0
$A := \overleftarrow{S}$	101100000
$2^n \cdot B$	101000000
$S[4] = A - 2^n \bullet B$	**010100000**
$C[4]$	**1–1–11–1 = 5**
Overflow	0
C	00101
R	**01010**
Sign(A) = 0 *Sign(B)* = 1 *Sign(D)* = 1	**Correction required:** **C = C − 1** **A = A + B * 2^n**
C correction	**00100 = 4**
R correction	**11110 = −2**

Table 7.8 Quotient decoding "on the fly"

$sc_i c_{i-1}$	$c_{i+1} c_i$
000	01
001	11
101	01
011	11
111	01
100	11

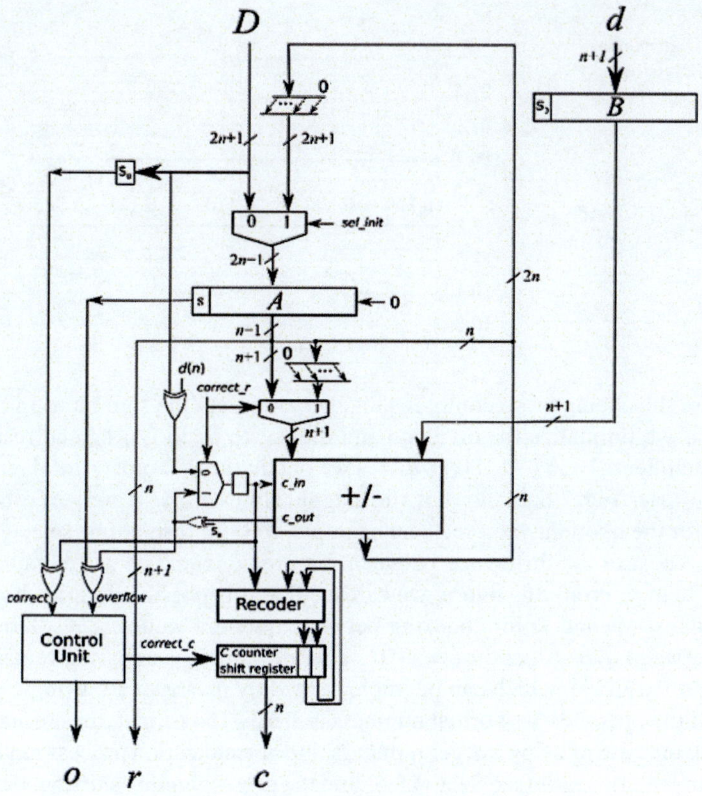

Fig. 7.4 $2n + 1$ by $n + 1$ non-restoring signed divider

7.5.1 Radix-2 SRT

For introducing the ideas of SRT division, we will present first SRT division using radix-2, and the codification $\{-1, 0, 1\}$. Figure 7.5 presents a graph of the new remainder versus the previous shifted partial remainder using this codification which enables performing a NR division, as described in Sect. 7.3. In principle, this not presents any clear advantage because it is required to perform different comparisons

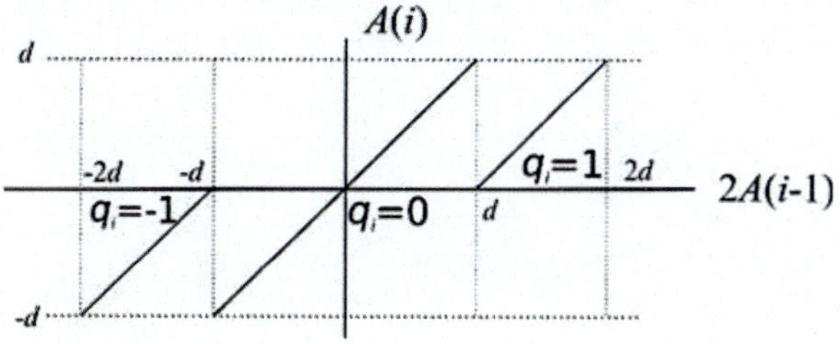

Fig. 7.5 New versus previous shifted remainder graph using $\{-1, 0, 1\}$ codification

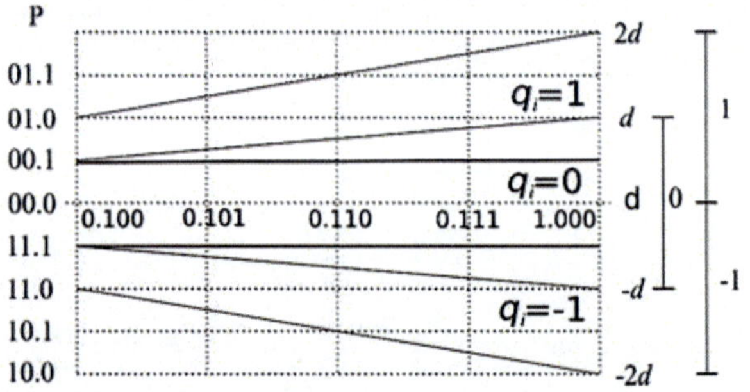

Fig. 7.6 P-D graph for radix-2 SRT division

to –d and d, thus being more complex that checking only the sign bit as in Fig. 7.3. However, if we normalize the dividend and divisor to be in [½, 1), and restrict the partial remainder to be in $[-1, 1]$ ($[-d, d]$), we obtain the P-D plot (Partial remainder vs Divisor, [4]) in Fig. 7.6. In this plot, the diagonal lines d, $2d$, $-d$ and $-2d$ establishes the limits for the quotient values. As an example, $q_i =$ '0' can not be selected upper the d line, because the difference between the previous and actual remainder will be greater than d. From this figure, we can see that it is possible to draw horizontal lines at values $-$½ and ½ for choosing between quotient values $qi =$ '0' and $q_i =$ '1', and between $q_i =$ '0' and $q_i = -$'1'. Therefore, only comparing remainder to fixed values –½ and ½, which can be implemented by means of a few logic gates, is required. If the operands are normal numbers in IEEE-754 format, the normalization is trivial. In the case of using integer numbers, the normalization process requires the detection of leading zeros (see Sect. 4.5.5) and the corresponding shifting, thus introducing additional delay (in the case of using a barrel shifter combinational circuit), or additional clock cycles (when using shift registers). In any case, performance is harmed when not using floating point representations.

Note that there are overlaps for the selection of '1', '0' and '$-$1' values, thus enabling the selection of different limits. As an example, it would be possible to compare to "00.0" for the upper limit of $q_i =$ '0', and to "11.1" as the lower limit instead of to "00.1" and "11.1" as in Fig. 7.6.

Based on Fig. 7.6, an algorithm for performing radix-2 SRT division can be described as in Algorithm 7.4.

Algorithm 7.4

Radix-2 SRT division algorithm:

D floating number into [1/2,1)
d floating number into [1/2,1)

$A \leftarrow (D)$, $B \leftarrow (d)$, $C \leftarrow 0$, $i \leftarrow 0$
while $i<=n$ **do**
  begin
    if $A>1/2$ **then**
      1) $A \leftarrow \overleftarrow{A - B \cdot 2^n}$
      2) $C_{n-i} \leftarrow 1$
    else
      if $A<-1/2$ **then**
        3) $A \leftarrow \overleftarrow{A + B \cdot 2^n}$
        4) $C_{n-i} \leftarrow -1$
      else
        if $A>=-1/2$ **and** $A<=1/2$ **then**
          5) $A \leftarrow \overleftarrow{A}$
          6) $C_{n-i} \leftarrow 0$
          end
        end
      end
        7) $i \leftarrow i+1$
    end
  end
  if sign(A)=1 **then**
    $C=C-1$
    $A=A+B \cdot 2^n$
  end
  $r \leftarrow (A_{2n-1} \dots A_n)$

End algorithm

From Algorithm 7.4, C is codified using the set $\{-1, 0, 1\}$. In order to obtain a binary result, a recodification is required. This recodification can be performed on-the-fly, without introducing additional clock cycles, following the rules presented in Table 7.9. In this table, the operations to perform in the remainder and quotient in each stage in terms of s (sign of the previous remainder), a_0 (integer bit of the previous remainder) and a_{-1} (1/2 bit of the previous remainder) are presented. In the case of the remainder, this operations are **NOP** ($A \leftarrow \overleftarrow{A}$), **SUB** ($A \leftarrow \overleftarrow{A - B \cdot 2^n}$) and **ADD** ($A \leftarrow \overleftarrow{A + B \cdot 2^n}$). Regarding the quotient, in this table, q_i is the quotient digit obtained in stage i, codified using the set $\{-1, 0, 1\}$, L-shift stands for left-shift operation, $+1$ stands for adding '1', and -1 for subtracting '1'. L-shift followed by "+1" can be performed by means a left-shifting introducing a '1' by the left. However, L-shift followed by '-1' requires of a subtractor. In any case, all the sets

Table 7.9 Operations for SRT quotient codification on-the-fly

s	a_0	a_{-1}	Value	Interval	Remainder operation	q_i	Operation for C codification-on-the-fly
0	0	0	00.0xxx	$0 < A < ½$	NOP	0	L-shift
0	0	1	00.1xxx	$½ \leq A$	SUB	1	L-shift, +1
0	1	0	01.0xxx	$½ < A$	SUB	1	L-shift, +1
0	1	1	01.1xxx	$½ < A$	SUB	1	L-shift, +1
1	0	0	10.0xxx	$A < -½$	ADD	−1	L-shift, −1
1	0	1	10.1xxx	$A < -½$	ADD	−1	L-shift, −1
1	1	0	11.0xxx	$A < -½$	ADD	−1	L-shift, −1
1	1	1	11.1xxx	$-½ \leq A < 0$	NOP	0	L-shift

of quotient operations can be computed using only one clock-cycle, thus enabling a codification on-the-fly.

The following example will help the reader to apply all these concepts.

Example 7.7 Divide 1.359375 by 1.5 applying the radix-2 SRT algorithm 8 fractional bits for the dividend, 4 fractional bits for the divisor, quotient and remainder.

The number 1.359375 can be expressed as "1.01011100" using 1 + 8 bits, and 1.5 as "1.1000". First, the dividend and the divisor have to be normalized. In this case, dividing by 2 (right shift), the dividend is in [1/2, 1) interval, and the divisor also by 2 we have the two operands in the desired intervals. After performing the operations, the quotient does not require any correction because dividend and divisor have been divided by the same number, and the remainder has to be divided by 2^8 for positional correction and multiplied by 2 for correcting the initial division of the dividend. As the sign of the remainder is positive, no additions corrections are required. Table 7.10 shows all the required operations.

Figure 7.7 shows a circuit proposal for performing radix-2 SRT division as presented in the previous examples. First, note that the dividend must be a normal IEEE 754 floating point number in the format "$1 \cdot f_{-1}..f_{-2n+1}$", and divisor in the format "$1 \cdot d_{-1}d_{-2}...d_{-n+1}$". When loading into register A and B, both of them are divided by two adding two leading and one leading zeros, respectively. The q_i generator block generates each quotient digit, according to Table 7.9, and it is coded as: '0' → "00", '1' → "01", '−1' → "11". CB is a combinational block which generates the control signals for performing the on-the fly-codification of register C, as described in Table 7.9.

Table 7.10 Execution example for the radix-2 SRT algorithm

Register	Value	Interval	C codification
A	**00.10101110**	$A \geq 1/2$	
$2^n \cdot B$	00.11000000		
$S[0] = A - 2^n \cdot B$	11.11101110		
$C[0]$	**10000**		**00001**
$A := \overleftarrow{S}$	**11.11011100**	$-1/2 \leq A < 0$	
$2^n \cdot B$	00.11000000		
$S\,[1] = A + 0$	11.11011100		
$C\,[1]$	**10000**		**00010**
$A := \overleftarrow{S}$	**11.10111000**	$-1/2 \leq A < 0$	
$2^n \cdot B$	00.11000000		
$S\,[2] = A + 0$	11.10111000		
$C\,[2]$	**10000**		**00100**
$A := \overleftarrow{S}$	**11.01110000**	$A < -1/2$	
$2^n \cdot B$	00.11000000		
$S\,[3] = A + 2^n \cdot B$	00.00110000		
$C\,[3]$	**100–10**		**00111**
$A := \overleftarrow{S}$	**00.01100000**	$0 < A < 1/2$	
$2^n \cdot B$	00.11000000		
$S\,[4] = A$	00.01100000 = 0.375		
$C\,[4]$	**100–10** **(1.00–10 = 0.875)**		**01110**
C	0.1110 = 0.875		
R	**0.0110 ·2^{-7} = 0.046875**		
$Sign(A) = 0$	**No correction required**		

7.5.2 *Radix-4 SRT*

SRT division enables the possibility of using radix higher than two for performing division, thus reducing the number of clock cycles required for completing the operation. In this section radix-4 division will be introduced. As a first approach, the standard set $\{-3, -2, -1, 0, 1, 2, 3\}$ can be selected for codifying the quotient digits, thus obtaining the graph in Fig. 7.8.

As in the case of radix-2 SRT, if the divisor is a floating point mantissa into the interval [1/2, 1), and we restrict the partial remainder to be in the range $[-d, d]$, the P-D plot in Fig. 7.9 is obtained. Note that, in this case, comparisons cannot be limited to fixed values, because it is not possible to draw a straight line between $q_i = 2$ and $q_i = 3$. Additionally, the d_{-2} bit of the divisor needs also to be checked,

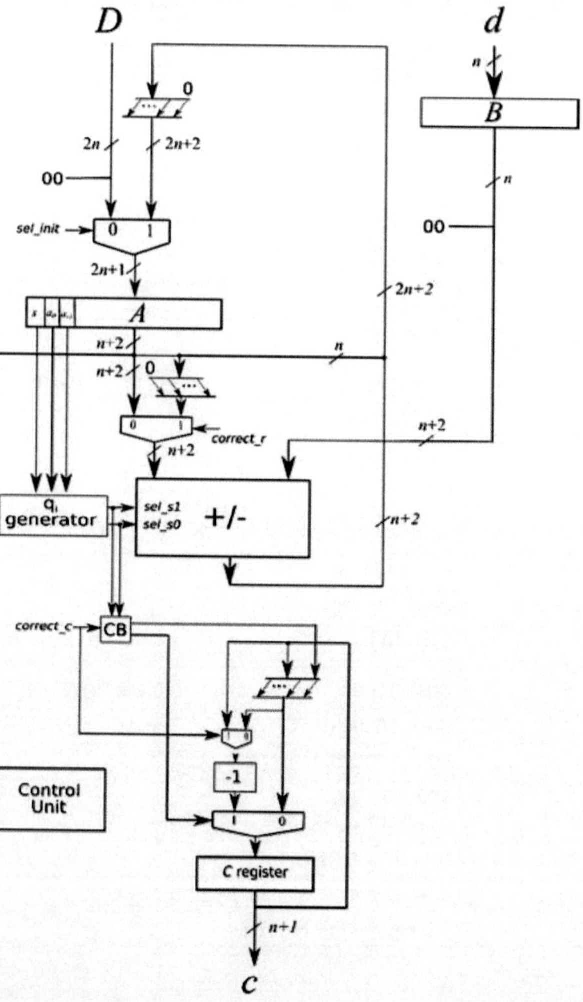

Fig. 7.7 Circuit for implementing radix-2 SRT division

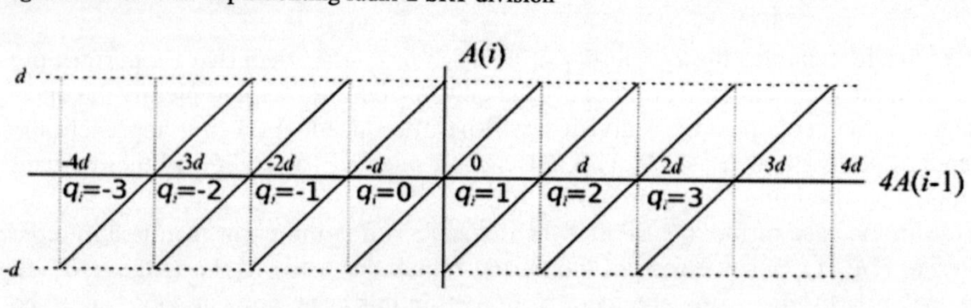

Fig. 7.8 New versus old shifted remainder graph using $\{-3, -2, -1, 0, 1, 2, 3\}$ codification

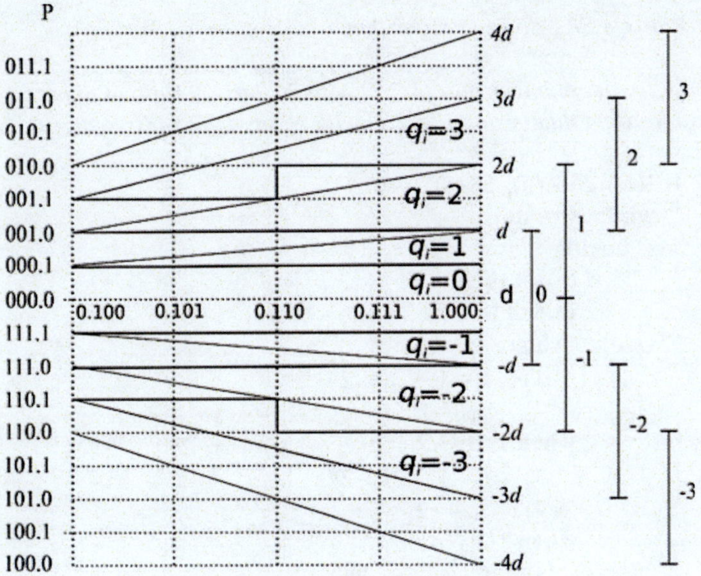

Fig. 7.9 P-D graph for radix-4 SRT division using $\{-3, -2, -1, 0, 1, 2, 3\}$ codification

thus implying to take into account bits a_1, a_0, and a_{-1} of A register, and bit b_{-2} of B register. With this assumptions, Algorithm 7.5 provides a method for performing a radix-4 SRT division using the set $\{-3, -2, -1, 0, 1, 2, 3\}$.

Algorithm 7.5

Radix-4 SRT division algorithm:

D floating number into [1/2,1) with 2N bits (N radix-4 digits)
d floating number into [1/2,1) with N bits (N/2 radix-4 digits)

$A \leftarrow (D)$, $B \leftarrow (d)$, $C \leftarrow 0$, $i \leftarrow 0$
  while $i <= n$ **do**
    begin
      if $B < 3/4$ **then**
        case A **is**
        when $A \geq 3/2$
          1) $A \leftarrow \overleftarrow{A - 3B \cdot 2^n}$
          2) $C_{n-i} \leftarrow 3$
        when $1 \leq A < 3/2$
          3) $A \leftarrow \overleftarrow{A - 2B \cdot 2^n}$
          4) $C_{n-i} \leftarrow 2$
        when $1/2 \leq A < 1$
          5) $A \leftarrow \overleftarrow{A - B \cdot 2^n}$
          6) $C_{n-i} \leftarrow 1$
        when $-1/2 < A < 1/2$
          7) $A \leftarrow \overleftarrow{A}$
          8) $C_{n-i} \leftarrow 0$
        when $-1 < A \leq -1/2$
          9) $A \leftarrow \overleftarrow{A + B \cdot 2^n}$
          10) $C_{n-i} \leftarrow -1$
        when $-3/2 < A \leq -1$
          11) $A \leftarrow \overleftarrow{A + 2B \cdot 2^n}$
          12) $C_{n-i} \leftarrow -2$
        when $A \leq -3/2$
          13) $A \leftarrow \overleftarrow{A + 3B \cdot 2^n}$
          14) $C_{n-i} \leftarrow -3$

end
end
else
case A **is**
when $A \geq 2$
1) $A \leftarrow \overleftarrow{A - 3B \cdot 2^n}$
2) $C_{n-i} \leftarrow 3$
when $1 \leq A < 2$
3) $A \leftarrow \overleftarrow{A - 2B \cdot 2^n}$
4) $C_{n-i} \leftarrow 2$
when $1/2 \leq A < 1$
5) $A \leftarrow \overleftarrow{A - B \cdot 2^n}$
6) $C_{n-i} \leftarrow 1$
when $-1/2 < A < 1/2$
7) $A \leftarrow \overleftarrow{A}$
8) $C_{n-i} \leftarrow 0$
when $-1 < A \leq -1/2$
9) $A \leftarrow \overleftarrow{A + B \cdot 2^n}$
10) $C_{n-i} \leftarrow -1$
when $-2 < A \leq -1$
11) $A \leftarrow \overleftarrow{A + 2B \cdot 2^n}$
12) $C_{n-i} \leftarrow -2$
when $A \leq -2$
13) $A \leftarrow \overleftarrow{A + 3B \cdot 2^n}$
14) $C_{n-i} \leftarrow -3$
end
end
end
15) $i \leftarrow i+1$
end
if sign(A)=1 **then**
$C = C - 1$
$A = A + B \cdot 2^n$
end
$r \leftarrow (A_{2n-1} \ldots A_n)$

End algorithm

Implementation of the above algorithm implies multiplying the divisor by 2 and 3. In order to not increase area, the main adder/subtractor can be used for pre-computing $3 \cdot B$ and store it in an additional register at the initialization clock cycle. Also note that in this algorithm, the shift operation corresponds to a radix-4 digit, thus being equivalent to a 2-bit shift.

Regarding the on-the-fly recodification of quotient digits, Table 7.11 shows the

Table 7.11 Operations for SRT quotient codification on-the-fly

b_{-2}	s	a_1	a_0	a_{-1}	Value	Interval	Remainder operation	q_i	Operation for C codification-on-the-fly
0	0	0	0	0	000.0xxx	$0 < A < ½$	NOP	0	2L-shift
0	0	0	0	1	000.1xxx	$½ \leq A < 1$	SUB	1	2L-shift, +1
0	0	0	1	0	001.0xxx	$1 \leq A < 3/2$	SUB	2	2L-shift, +2
0	0	0	1	1	001.1xxx	$3/2 \leq A < 2$	SUB	3	2L-shift, +3
0	0	1	x	X	01x.xxxx	$A \geq 2$	SUB	3	2L-shift, +3
0	1	0	x	X	10x.xxxx	$A < -2$	ADD	−3	2L-shift, −3
0	1	1	0	0	110.0xxx	$-2 \leq A < -3/2$	ADD	−3	2L-shift, −3
0	1	1	0	1	110.1xxx	$-3/2 \leq A < 1$	ADD	−2	2L-shift, −2
0	1	1	1	0	111.0xxx	$-1 \leq A < -½$	ADD	−1	2L-shift, −1
0	1	1	1	1	111.1xxx	$-½ \leq A < 0$	NOP	0	2L-shift
1	0	0	0	0	000.0xxx	$0 < A < ½$	NOP	0	2L-shift
1	0	0	0	1	000.1xxx	$½ \leq A < 1$	SUB	1	2L-shift, +1
1	0	0	1	0	001.0xxx	$1 \leq A < 3/2$	SUB	2	2L-shift, +2
1	0	0	1	1	001.1xxx	$3/2 \leq A < 2$	SUB	2	2L-shift, +2
1	0	1	x	X	01x.xxxx	$A >=2$	SUB	3	2L-shift, + 3
1	1	0	x	X	10x.xxxx	$A < -2$	ADD	−3	2L-shift, −3
1	1	1	0	0	110.0xxx	$-2 \leq A < -3/2$	ADD	−2	2L-shift, −2
1	1	1	0	1	110.1xxx	$-3/2 \leq A < -1$	ADD	−2	2L-shift, −2
1	1	1	1	0	111.0xxx	$-1 \leq A < -½$	ADD	−1	2L-shift, −1
1	1	1	1	1	111.1xxx	$-½ \leq A < 0$	NOP	0	2L-shift

operations involved for each value of b_{-2}, s, a_1, a_0 and a_{-1} bits, which define the different intervals. In this table, 2L-shift corresponds to a 2-bit left shift.

Summarizing all these considerations, Fig. 7.10 presents a circuit for performing radix-4 division with [−3, 3] codification, and the following example shows the operations to perform accordingly to Algorithm 7.5 and Table 7.11.

Example 7.8 Divide 1.359375 by 1.5 applying the radix-4 SRT algorithm 8 fractional bits for the dividend, 4 fractional bits for the divisor, quotient and remainder.

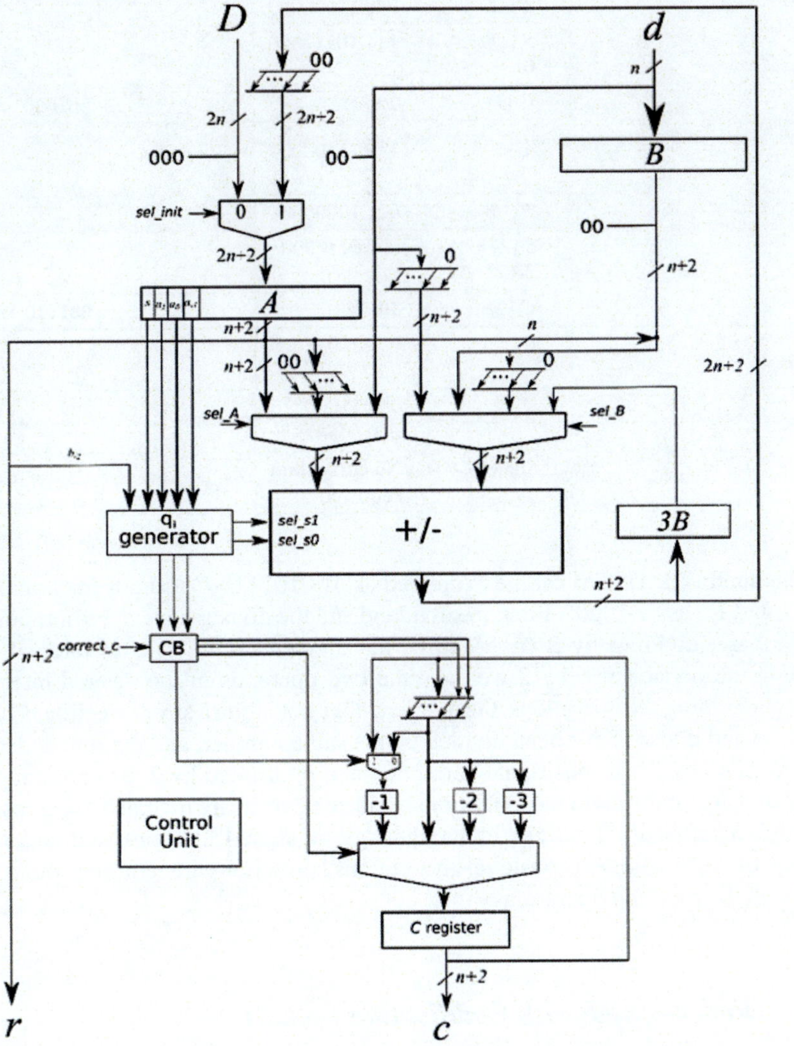

Fig. 7.10 Circuit for implementing radix-4 SRT division using [−3, 3] codification

Table 7.12 Execution example for the radix-4 SRT algorithm

Register	Value	Interval	C codification
A	**000.**10101110	½ ≤ A < 1	
$2^n \cdot B$	000.11000000		
$S[0] =$ A−$2^n \cdot B$	111.11101110		
$C[0]$	**100**		**000001**
A := $\overleftarrow{S}$	**111.**10111000	−1/2 ≤ A < 0	
$2^n \cdot B$	000.11000000		
S [1] = A + 0	111.10111000		
C [1]	**100000**		001**000**
A := $\overleftarrow{S}$	**110.1**110000	−3/2 ≤ A < − 1	
$2^n \cdot B$	001.1000000		
S [2] = A − 2B 2^n	000.0110000		
C [2]	**10–2**		**001110**
C	00.1110 = 0.875		
R	**00.0110 ·2^{-7} = 0.046875**		
$Sign(A) = 0$	**No correction required**		

The number 1.359375 can be expressed as “1.01011100” using 8 fractional bits digits, and 1.5 as “1.1000”. First, the dividend and the divisor have to be normalized. In this case, dividing by 2 (right shift), the dividend is in [1/2, 1) interval, and dividing the divisor also by 2 we have the two operands in the desired intervals. After performing the operations, the quotient does not require any correction because dividend and divisor have been divided by the same number, and the remainder has to be divided by 2^8 for positional correction and multiplied by 2 for correcting the initial division of the dividend. Additionally, leading zeros are included to express the operands in terms of 2-bit digits, and a sign digit for signed additions/subtraction. As the sign of the remainder is positive, no additions corrections are required. Table 7.12 shows all the required operations.

7.5.3 *Radix-4 SRT with Codification [−2, 2]*

In order to avoid operating with 3 and −3 when performing radix-4 SRT division, and taking into account overlaps obtained in Fig. 7.11 it is possible to restrict the

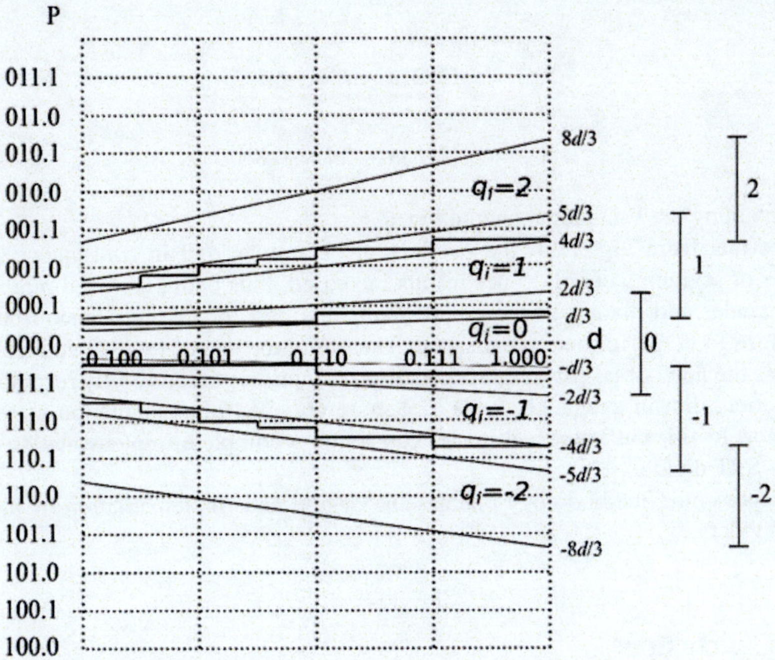

Fig. 7.11 P-D graph using {−2, −1, 0, 1, 2} codification

codification set to [−2, −1, 0, 1, 2]. Note that if we want the partial remainder to be in [$-hd, hd$] from the previous step being in [$4hd, -4hd$] with the selected codification, the maximum value that we can add is $+-2d$, thus:

$$4hd - 2d \leq hd$$

$$h <= 2/3$$

In the case of [−3, 3] codification, $4hd - 3d \leq hd \geq h \leq 1$, and we selected $h = 1$. For [−2, 2], it is usual to use $h = 2/3$ [1, 4]. For computing the limits and corresponding limiting lines, note that the upper and low limit for each quotient value are:

$$\begin{aligned}
2_{up} &= (2/3)d + 2d = 8/3d \\
2_{\text{low}} &= -(2/3)d + 2d = 4/3d \\
1_{\text{up}} &= (2/3)d + d = 5/3d \\
1_{\text{low}} &= -(2/3)d + d = d/3 \\
0_{\text{up}} &= 2d/3 \\
0_{\text{low}} &= -2d/3
\end{aligned}$$

$$-1_{\text{up}} = (2/3)d - d = -d/3$$
$$-1_{\text{low}} = -(2/3)d - d = -5d/3$$
$$-2_{\text{up}} = (2/3)d - 2d = -4d/3$$
$$-2_{\text{low}} = -(2/3)d - 2d = -8d/3$$

thus obtaining the P-D plot shown in Fig. 7.11.

Note that from Fig. 7.11, the overlaps are narrower, and in consequence, the number of segments of solid lines results increased, thus being involved more bits of remainder and divisor in the comparisons required for taking decision on the operations to be performed in each step. Therefore, the use of [−2, 2] codification removes the need of performing 3B operation, simplifies the on-the-fly recoding of the quotient, but in exchange, leads to a more complex implementation of the q_i generator. In [4], the interested reader can found a complete implementation of a radix-4 SRT divider.

The presented methodology enables the design and implementation of higher radix dividers.

7.6 Conclusion

In this chapter, different circuits for performing arithmetic division have been presented. First, basic restoring division has been introduced, enabling the implementation of binary dividers for natural numbers, where a quotient bit per clock cycle is obtained. Later, non-restoring division has been described, allowing to perform signed division of integer numbers. Circuits for signed and unsigned non-restoring division has been presented. Finally, SRT division, intended for floating point number, has been considered. SRT division can be used with different radixes, and circuits for performing SRT division with radix-2 and radix-4 have been described, thus being suitable for reducing the number of clock cycles.

7.7 Exercises

7.1. Divide 500 by 77 using the restoring algorithm

7.2. Divide −150 by 13 using the non-restoring algorithm.

7.3. Divide 150 by −13 using the non-restoring algorithm.

7.4. Divide −150 by −13 using the non-restoring algorithm.

7.5. Divide 9.375 by 3.25 using the radix-2 SRT algorithm. Normalize the dividend and divisor as required.

7.6. Rewrite Table 7.11 for the case of radix-4 SRT using {−2, −1, 0, 1, 2} codification from P-D plot presented in Fig. 7.11.

7.7. Modify circuit of Fig. 7.12 for implementing a radix-4 SRT divider using $\{-2, -1, 0, 1, 2\}$ codification from P-D plot presented in Fig. 7.11.

7.8 Build an alternative P-D plot (different from the one presented in Fig. 7.11) for implementing a radix-4 SRT divider using $\{-2, -1, 0, 1, 2\}$ codification.

References

1. Parhami, D.: Computer Arithmetic, vol. 20, No. 00. Oxford University Press. (2010)
2. J.E. Robertson, *A new class of digital division methods*, IRE Trans. Electronic Computers, vol. EC-7, pp. 218–222, Sept. 1958
3. Tocher, K.D.: Techniques of multiplication and division for automatic binary computers. Quart. J. Mech. Appl. Math. **11**(pt. 3), 364–384 (1958)
4. Wey, C.L., Wang, C.P.: Design of a fast radix-4 SRT divider and its VLSI implementation. IEE Proc. Comput. Digital Tech. **146**(4), 205–210 (1999)

Chapter 8
Special Functions

This chapter is devoted to the computation of the most commonly used mathematical functions such as logarithms, exponentials, and the trigonometric functions. The tools necessary to understand, develop, and implement algorithms for calculating elementary functions will be studied. In this way, the main types of algorithms for the computation of elementary functions will be introduced, focusing on algorithms using table-based methods, algorithms based on polynomial approximations and algorithms that only use shifts and additions also mentioned as hardware-oriented algorithms. In relation to this last type of algorithms, the **CORDIC** algorithm will be also described, which is especially appropriated to hardware implementations since no multiplications are required for its computation. **CORDIC** was originally designed to solution trigonometric problems, and extended later to other applications. Given its age, elegance and originality, **CORDIC** will be used as a general introduction to the calculation of special functions. The main algorithms for some specific functions, such as the square root, the exponential, the logarithm and the sine and cosine functions, will be described. In addition, this chapter also will present the most recent advances in new algorithms for these functions, including hardware structures.

8.1 Introduction

Special functions are mathematical functions, such as exponential, logarithm, square root and trigonometric functions, widely used in the implementation of algorithms. The computation of special functions is required in all fields of industry and science, with the most popular application areas in engineering, computer science, physics, chemistry, and statistics. Concretely, some of these applications are computer aided design (**CAD**), digital signal processing (**DSP**) algorithms, physical simulation, computer graphics and virtual reality.

A. Lloris Ruiz et al., *Arithmetic and Algebraic Circuits*, Intelligent Systems Reference Library 201, https://doi.org/10.1007/978-3-030-67266-9_8

Apart from a few cases, the special functions must be approximated using certain algorithms since these functions cannot be computed exactly. These special functions are often implemented in software routines, which provide results with extremely high accuracy, however they are usually too slow for real-time or numerically intensive applications. In this way, the design and development of hardware structures for the generation of these functions has been an important research field for years, due to it allows to speed up these function approximations.

The implementation of special functions in hardware induces important algorithms concerns that have to be considered. These concerns are easier to understand considering that the function evaluation can be expressed as it is detailed at the following. Let us suppose a function f defined on a finite input interval $I \subset \mathbb{R}$, a positive integer w_i, which indicates the bit length of the input word, and another positive integer, w_o which indicates the bit length of output word. The purpose of function evaluation is to develop a hardware architecture in order to compute an approximation $\check{f}$ of the function f on the interval I. It is possible to particularize the interval such that $I = [0;1)$ and scale f such as $f\ (I) = [0;1)$. Hence, the input word X and the output word Y can be written as follows:

$$X = x_1 x_2 \ldots x_{w_i}$$

$$X = .x_1 x_2 \ldots x_{w_i}$$
$$Y = \check{f}(x) = .y_1 y_2 \ldots y_{w_o}$$

It is also desirable that the evaluation operator guarantees the accuracy of the result. We choose to ensure faithful rounding:

$$\varepsilon = \max_{x \in I} |f(x) - \tilde{f}(x)| < 2^{-w_o}$$

Considering the previous expression, there are several properties that are desirable when building algorithms and architectures for these special functions, as for example speed, accuracy, amount of resources, preservation of mathematical properties (as monotonicity) and the direction of rounding and range limits. Being able to get these properties fulfilled, or some of them, depends on the approximation method selected for the implementation of the special function under consideration. Thus, the computation of special functions can be carried out using a great variety of algorithms oriented to hardware implementations. Among these algorithms it is important to refer to the shift-and-add algorithms, also called convergence algorithms. Additions and shifts are the operations usually required for these algorithms and thus, a digital processor can easily implement them in an efficient way. In addition, these algorithms are very attractive for hardware implementation due to additions and shifts are cost-effective operations for hardware computation and they allow to reduce the computer resources. Another important aspect to mention is that the shift-and-add algorithms give a quite good precision at reduced cost. The most popular shift-and-add algorithm is the **CORDIC** (**CO**ordinate **R**otation **DI**gital **C**omputer), which is of special interest since it was one of the first algorithms implemented into a computer, which allowed the calculation of special functions such as trigonometric and hyperbolic functions [25]. An important feature of **CORDIC** is that the same algorithm makes possible the computation of most mathematical functions requiring operations very basics, which is very convenient for hardware implementation. Thus, it is typically used when there is no hardware available for multiplications, as it for example in microcontrollers and simple **FPGA**s, because it only requires additions, subtractions, bit shifts (bitshift) and look-up tables to prestored constants. The focusing of **CORDIC** can be simplified so that it can be seen as a method that uses powers of two for the calculation of angles.

On the other hand, since the IEEE Standard 754 for floating point arithmetic, many efforts have been devoted for the implementation of special functions with a very high accuracy. Direct table look-up algorithms can be used to reach this accuracy. However, it is important to have into account that a table including function values at equidistant arguments may require considerable storage for hardware implementation. Therefore, the maximum input and output word lengths for direct look-up table algorithms is in the order of 8–10 bit, and thus these tables are limited to low precision computation. Therefore, for larger word lengths, methods including small tables and simple circuitry for arithmetic operations are more effective. These methods are called table-based methods and they achieve high speed operation and minimize hardware resources. Concretely, multipartite table methods is a well-known type of table-based method. As example, the method proposed in [5] is based on first-order (piecewiselinear) approximations that use a few tables and a simple multioperand addition. However, the growth of memory size with required accuracy has limited multipartite table methods to applications with accuracy of up to 16/20 bits. In order to achieve higher order approximations, some methods proposed techniques based on reducing the size of the multipliers or including additional tables [4].

Otherwise, the approximation to special functions based on polynomial functions can be used. In this case, the polynomial represents the function to a desired accuracy over certain range. Compared to algorithms using direct look-up tables, the coefficients of a polynomial may require very little space, and thus the evaluation of the polynomial also may require simple instructions. Nevertheless, in order to provide accurate polynomial approximation in a rather large interval, large degree is required which implies an increment into the computational delay and may also generate some issues related to round-off error propagation. This can be overcome dividing in subintervals the interval in which the special function has to be calculated, and approximating this function with a low-degree polynomial in each subinterval.

After this first view of the different methods for the hardware implementation of special functions we are going to study in detail the most relevant. Firstly, due to the importance of the **CORDIC**, a case study based on the **CORDIC** algorithm is presented.

8.2 A Case Study: The CORDIC

The **CORDIC** algorithm was originally designed to solution trigonometric problems, and extended to other applications. Concretely, **CORDIC** was developed to solve problems for real-time navigation related to the calculation of the trajectory of the planes. Thus, the **CORDIC** solution computed this trajectory approximating the rotation of a two-dimensional vector using basic operations, concretely, shifts and additions. Jack E. Volder was the first describing in 1959 [24] the modern **CORDIC** algorithm, which was developed to replace an analogue resolve in the B-58.1 bomber navigation computer (Convair Aeroelectronics Department), being this solution similar to the methods proposed by Henry Briggs three centuries before [1]. Afterwards, a generalization of the algorithm was carried out by Walther [24], in Hewlett-Packard, which allowed the computation of multiplication, division, square root, hyperbolic, exponential and logarithm functions.

Subsequently, **CORDIC** has been widely used into diverse digital applications as signal processing, telecommunications or robotics, and especially in the design of processors for wireless modems. Although **CORDIC** was firstly designed for binary operations, new applications of different **CORDIC** versions have been proposed recently for decimal computing [16]. It is due the interest in decimal computing has arisen in recent years, as it is really important for some fields of knowledge and applications [2, 6]. For instance, when working with fractional operands, the results are usually rounding if binary operations are used. This is the reason why most applications for financial computations use decimal arithmetic. Other recent application field for decimal computing is optic and magnetic sensors, such that it is possible to directly provide the output in **BCD** format, and thus, the evolution of some magnitudes and the detection of any errors can be monitored easily by the user [9].

The **CORDIC** algorithm will be studied at the following. Firstly, the implementation of the trigonometric functions will be described, expanding it later, by analogy, to the hyperbolic functions. It will be seen that the hyperbolic implementation can be used to calculate both the square and the logarithm of a real number. The so-called linear case of the **CORDIC** algorithm, described in third place, can be used to multiply or divide approximately two real numbers. Finally, a unified description of the three cases (circular, hyperbolic and linear) will be detailed and simple circuits to implement them are proposed.

8.2.1 Circular Case

In 1959 Volder [24] built a computer, **CORDIC**, for the resolution of trigonometric problems. The algorithm that he devised, also known as **CORDIC**, is based on decomposing or canceling (it is the same) a given angle, adding or subtracting prefixed angles, α_i, defined in such a way that:

Table 8.1 Angles α_i

i	2^{-i}	α_i= arctg2^{-i}
0	1	45°
1	0.5	26.5650512°
2	0.25	14.0362435°
3	0.125	7.12501635°
4	0.0625	3.57633437°
5	0.03125	1.78991061°
6	0.015625	0.89517371°
7	0.0078125	0.44761417°
8	0.00390625	0.22381050°
9	0.001953125	0.11190568°
10	0.000976562	0.05595286°
11	0.000488281	0.02797644°
12	0.000244141	0.01398825°

$$\alpha_i = \text{arctg}2^{-i} \tag{8.1}$$

The selection of these angles transforms the calculation associated with the **CORDIC** algorithm into a series of additions/subtractions and shifts, as will be seen later.

The angles defined in (8.1), expressed in sexagesimal grades up to $i = 12$, are given in Table 8.1.

8.2.1.1 Range and Convergence

From Table 8.1 it is easy to calculate the maximum value, $\theta_{\max}$, of the angle that can be developed as an addition of the angles α_i. In fact

$$\theta_{\max} = \sum_{i=0}^{\infty} \text{arctg}2^{-i} \approx \pm 99'88°$$

Therefore, when applying the **CORDIC** algorithm, any other angle outside this range must be transferred into it; without loss of generality it can be assumed that all calculations are transferred to the first or fourth quadrant ($|\theta| \leq 90°$). To develop any angle that is within the range to be used (0–90° or 0–99′88°) as addition/subtraction of angles α_i, all α_i angles must be used, up to a given i or even an i adequate to the desired precision; the objective in each iteration is that the resulting angle be as close as possible to 0°, as can be seen in the following example.

Example 8.1 **Develop the angle θ = 73° 18′ 43″.**

Expressing θ in sexagesimal degrees it results:

$$\theta = 73.31194444°$$

Subtracting or adding α_i, depending on whether the previous result is positive or negative, it results

$$\begin{aligned}
&73.31194444 - 45 = 28.31194444\\
&28.31194444 - 26.5650512 = 1.74689324\\
&1.74689324 - 14.0362435 = -12.28935026\\
&-12.28935026 + 7.12501635 = -5.16433391\\
&-5.16433391 + 3.57633437 = -1.58799954\\
&-1.58799954 + 1.78991061 = 0.20191107\\
&0.20191107 - 0.89517371 = -0.69326264\\
&-0.69326264 + 0.44761417 = -0.24564847\\
&-0.24564847 + 0.22381050 = -0.02183797\\
&-0.02183797 + 0.11190568 = 0.09006771\\
&0.09006771 - 0.05595286 = 0.03411485\\
&0.03411485 - 0.02797644 = 0.00613841\\
&0.00613841 - 0.01398825 = -0.00784984
\end{aligned}$$

It is

$$\theta = \alpha_0 + \alpha_1 + \alpha_2 - \alpha_3 - \alpha_4 - \alpha_5 + \alpha_6 - \alpha_7 - \alpha_8 - \alpha_9 + \alpha_{10} + \alpha_{11} + \alpha_{12} + \cdots$$

In this example it is verified that the magnitude of the resulting angle in a given iteration can be greater than that of the angle resulting in the previous iteration, although globally decreasing. ■

As it has been done in the previous example, any angle $\theta < \theta_{max}$ can be developed according to the α_i angles; this is equivalent to saying that the development is convergent, which means, as indeed it happens, that any α_i is less than or equal to the sum of the remaining α_i of the development (in Table 8.1 it is immediate to verify that the angle in any cell of the last column is greater than half of the angle of the previous cell).

$$\propto_i \leq \sum_{j=i+1}^{\infty} \propto_j$$

8.2.1.2 Trigonometric Functions

Let suppose a point P in the plane (Fig. 8.1), with coordinates (x, y):

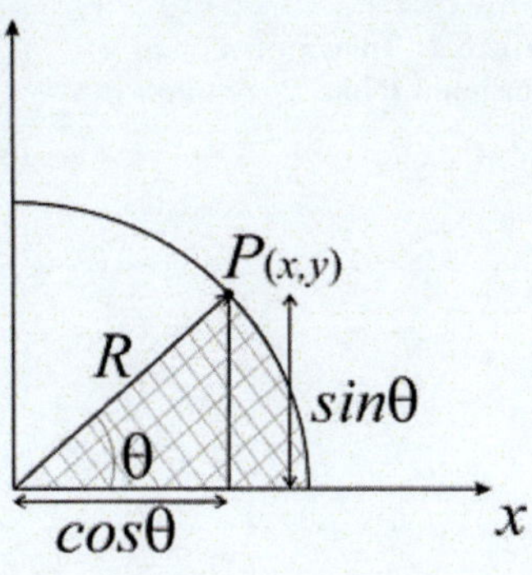

Fig. 8.1 Point P to transform

$$x = R\cos\theta$$

$$y = R\sin\theta$$

These are the parametric equations of a circumference with center at the origin and radius R. The angle θ (which is assumed in the first quadrant) is a measure of the area S of the circular sector of the circumference with center at the origin passing through the point P; specifically

$$\theta = \frac{2S}{R^2}$$

The angle θ can be developed as additions or subtractions of the prefixed α_i (Table 8.1):

$$\theta = \pm\alpha_i \pm \alpha_j \pm \cdots = s_i\alpha_i + s_j\alpha_j + \cdots \tag{8.2}$$

where $s_k = \pm 1$ is the sign corresponding to each α_k in the development of θ.

The **CORDIC** algorithm consists in applying successive transformations of the point P, according to the development of θ, so that in the first transformation (see Fig. 8.2) the point $P_a(x_a, y_a)$ is obtained if $s_i = +1$, or the point $P_b(x_b, y_b)$ if $s_i = -1$. Regardless of whether P is transformed into P_a or P_b, the radius of the transformed point is R_1 in both cases, and for the same it is fulfilled that

$$R_1 = \sqrt{R^2 + R^2 tg^2 \propto_i} = R\sqrt{1 + 2^{-2i}}$$

$$\cos \propto_i = \frac{1}{\sqrt{1 + 2^{-2i}}}$$

$$\sin \propto_i = \frac{2^{-i}}{\sqrt{1 + 2^{-2i}}}$$

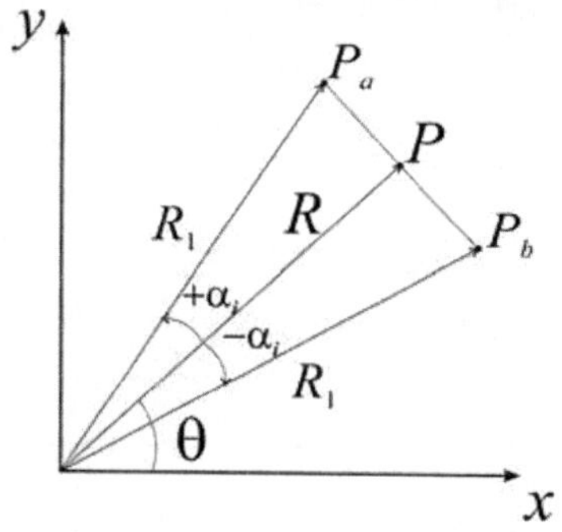

Fig. 8.2 Transformation of the point P into P_a or into P_b

Therefore

$$x_a = R_1 \cos(\theta + \alpha_i) = R_1[\cos\theta \cos\alpha_i - \sin\theta \sin\alpha_i] = R\left(\cos\theta - 2^{-i}\sin\theta\right) = x - 2^{-i}y$$

$$y_a = R_1 \sin(\theta + \alpha_i) = R_1[\sin\theta \cos\alpha_i + \cos\theta \sin\alpha_i] = R\left(\sin\theta + 2^{-i}\cos\theta\right) = y + 2^{-i}x$$

$$x_b = R_1 \cos(\theta - \alpha_i) = R_1[\cos\theta \cos\alpha_i + \sin\theta \sin\alpha_i] = R\left(\cos\theta + 2^{-i}\sin\theta\right) = x + 2^{-i}y$$

$$y_b = R_1 \sin(\theta - \alpha_i) = R_1[\sin\theta \cos\alpha_i - \cos\theta \sin\alpha_i] = R\left(\sin\theta - 2^{-i}\cos\theta\right) = y - 2^{-i}x$$

In general it can be written:

$$x_{i+1} = x_i - 2^{-i}s_i y_i$$

$$y_{i+1} = y_i + 2^{-i}s_i x_i$$

It is immediate to verify that each transformation from (x_i, y_i) consists of a rotation (of $\pm\alpha_i$ degrees) and an elongation. The following section is devoted to this analysis.

8.2.1.3 Elongation

The elongation corresponding to the ith iteration is:

$$\frac{R_i}{R_{i-1}} = \sqrt{1 + 2^{-2i}}$$

After n transformations the elongation is:

$$\frac{R_n}{R} = \prod_{i=0}^{i=n} \sqrt{1 + 2^{-2i}}$$

Table 8.2 Elongation for the first values of i

i	Elongation
0	1.41421356
1	1.58113883
2	1.62980060
3	1.64248407
4	1.64568892
5	1.64649228
6	1.64669325
7	1.64674351
8	1.64675607
9	1.64675921
10	1.64676000
11	1.64676019
12	1.64676024
13	1.64676025
14	1.64676026

The values of the elongation up to $i = 14$ are given in Table 8.2. It is obvious that it quickly takes an almost constant value. For example, for $i = 10$ it results:

$$\frac{R_{10}}{R} = \prod_{i=0}^{i=10} \sqrt{1 + 2^{-2i}} = 1.64676$$

For $n = \infty$ it turns out with good approximation:

$$\frac{R_{\infty}}{R} = \prod_{i=0}^{i=\infty} \sqrt{1 + 2^{-2i}} \approx 1.64676$$

Therefore, if the number of iterations is known, the elongation, E, can be calculated with the desired precision. For example, $E = 1.64676$ can be used as the value of the elongation. If it starts from a vector of magnitude 1, after applying the **CORDIC** transformations a vector of magnitude $E = 1.64676$ will be obtained. Or, if it starts from a vector of magnitude $1/E = 1/1.64676 \approx 0.60725303$, after applying the **CORDIC** transformations a vector of magnitude 1 will be obtained.

Let suppose that, applying the **CORDIC** algorithm, a trigonometric function (sine, cosine, secant, cosecant, tangent or cotangent) of an angle θ, already transferred to the first quadrant and decomposed as in the development (8.2), want to be calculated.

Starting from a vector of magnitude 1/E and angle 0°, after applying the **CORDIC** algorithm (suppose n iterations) a vector of magnitude 1 and angle θ will be obtained, which will allow us to calculate the trigonometric functions.

Let consider a point P_0 (x_0, y_0) in the first quadrant (Fig. 8.1). The **CORDIC** algorithm can be applied to calculate the value of the angle θ, starting from (x_0, y_0) as initial values and trying to cancel the value of y; that is, in each iteration the value of y must be more near to 0. When this is achieved, θ will be given by the development (8.2); the sign s_j of each following addend $s_j\alpha_j$ in this development will be chosen according to the current value of y; specifically, if $y_j \geq 0$, a negative shift will be made ($s_j = -1$), and if $y_j < 0$, a positive shift will be made ($s_j = +1$).

In the **CORDIC** algorithm, therefore, there are two operation modes, which are detailed below, depending on whether the direct trigonometric functions or the angle θ are calculated: the rotation mode and the vectoring mode.

8.2.1.4 Rotation Mode

Given an angle θ, the rotation mode try to calculate a trigonometric function of it. In the application of the algorithm, the z parameter will represent the value of the angle in each iteration. Initially:

$$x_0 = 1/E$$

$$y_0 = 0$$

$$z_0 = \theta$$

Then the algorithm is executed, trying to cancel the value of z:

$$x_{i+1} = x_i - 2^{-i} s_i y_i$$

$$y_{i+1} = y_i + 2^{-i} s_i x_i \tag{8.3}$$

$$z_{i+1} = z_i - s_i \propto_i$$

where

$$s_i = +1 \quad \text{if } z_i \geq 0, \quad \text{and} \quad s_i = -1 \quad \text{if } z_i < 0$$

A vector of length 1 and angle θ will be obtained after n iterations. Therefore:

$$\sin\theta = y_n$$

$$\cos\theta = x_n$$

$$\text{tang}\theta = y_n/x_n$$

$$\text{cotg}\theta = x_n/y_n$$

$$\sec\theta = 1/x_n$$

$$\text{cosec}\,\theta = 1/y_n$$

Example 8.2 **Calculate sin30° and cos30°.**

Initially $x_0 = 1/E$, $y_0 = 0$, $z_0 = 30$. It will reach $i = 14$. Thus, $E = 1.64676026$ or $x_0 = 1/E = 0.607252934316$ are done.

The results for each iteration are shown in Table 8.3. Using only 14 iterations a very good approximation is achieved, resulting the following values:

$$\text{sen}30 = y_{14} = 0.500091900498$$

$$\cos 30 = x_{14} = 0.865972428261$$

■

Table 8.3 Iterations of the Example 8.2

i	x_i	y_i	z_i	s_i
0	0.607252934316	0	30	+
1	0.607252934316	0.607252934316	−15	−
2	0.910879401474	0.303626467158	11.5650512	+
3	0.834972784685	0.531346317527	−2.4711923	−
4	0.901391074376	0.426974719441	4.65382405	+
5	0.874705154411	0.483311661590	1.07748968	+
6	0.859601664986	0.510646197665	−0.71242093	−
7	0.867580511825	0.497214921650	0.18275278	+
8	0.863696020250	0.503992894399	−0.26486139	−
9	0.865664742494	0.500619081820	−0.04105089	−
10	0.866642608004	0.498928330370	0.07085479	+
11	0.866155373556	0.499774660609	0.01490193	+
12	0.865911343085	0.500197587821	−0.01307451	−
13	0.866033461824	0.499986183360	0.00091374	+
14	0.865972428261	0.500091900498		

8.2.1.5 Vectoring Mode

Given a point P (x_p, y_p) in the plane, the vectoring mode try to calculate the angle θ. To apply the **CORDIC** algorithm, the calculation is transferred to the first quadrant, if previously it was not in it (Fig. 8.1). Initially:

$$x_0 = x_\mathrm{p}$$
$$y_0 = y_\mathrm{p}$$
$$z_0 = 0$$

Then the algorithm is executed, trying to cancel the value of y:

$$x_{i+1} = x_i - 2^{-i} s_i y_i$$

$$y_{i+1} = y_i + 2^{-i} s_i x_i$$

$$z_{i+1} = z_i - s_i \propto_i$$

where

$$s_i = +1 \text{ if } y_i < 0, \quad \text{and} \quad s_i = -1 \text{ if } y_i \geq 0$$

A vector of length E and angle θ will be obtained after n iterations, which will correspond to the final value of z.

Example 8.3 **Calculate the angle θ that the point (−5, −3) forms with the X axis.**

This point is in the third quadrant. Moving it to the first quadrant, as it is done in Fig. 8.3, it will be $\theta = 180° + \lambda$, being λ, the angle corresponding to the point (5, 3), which is calculated next. The results of the different iterations up to $i = 12$ are

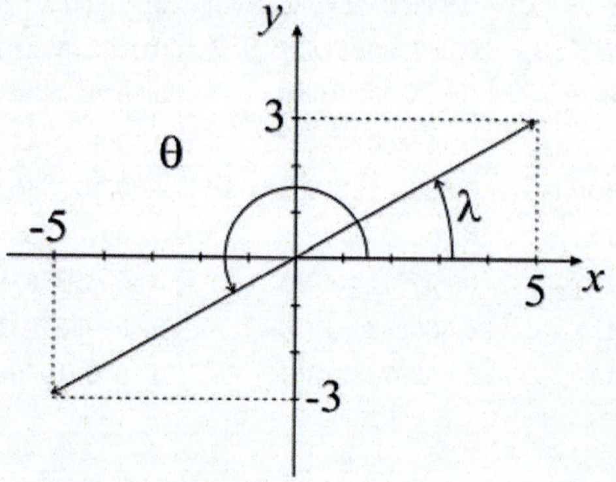

Fig. 8.3 Translation of point P to the first quadrant

Table 8.4 Iterations of the Example 8.3

i	x_i	y_i	λ_i	s_i
0	5	3	0°	–
1	8	−2	45°	+
2	9	2	18.4349488°	–
3	9.5	−0.25	32.4711923°	+
4	9.53125	0.9375	25.34617595°	–
5	9.58984375	0.341796875	28.92251032°	–
6	9.60052490234375	0.0421142578125	30.71242093°	–
7	9.6011829376220703	−0.1078939437866211	31.60759464°	+
8	9.7168918814086914	−0.0328847020864487	31.15998047°	+
9	9.7170203372762166	0.0050722974503040	30.93616997°	–
10	9.7170302441071742	−0.0139055808432026	31.04807565°	+
11	9.7170438237690136	−0.0044162983539568	30.99212279°	+
12	9.7170459801635902	0.00032834952135696	30.96414635°	–

shown in Table 8.4. It results in an angle $\lambda = 30.96414635°$. It really is $\lambda = 30.9638°$. Therefore, with only 12 iterations a very good approximation is achieved. ■

8.2.2 Hyperbolic Case

In 1971 Walther [26] extended the **CORDIC** algorithm to the calculation of hyperbolic functions. As in the circular case, in the hyperbolic case the **CORDIC** algorithm consists of decomposing a given hyperbolic angle by adding or subtracting prefixed hyperbolic angles, β_i, defined as follows:

$$th\beta_i = 2^{-i} \ (i = 1, 2, \ldots)$$

The value $i = 0$ can not be used, since it would be $\beta_0 = \infty$. Table 8.5 shows the first values of β_i. If $th\beta_i = 2^{-i}$, then:

$$sh\beta_i = \frac{2^{-i}}{\sqrt{1 - 2^{-2i}}} \quad \text{and} \quad ch\beta_i = \frac{1}{\sqrt{1 - 2^{-2i}}}$$

Using only the hyperbolic angles of Table 8.5, the resulting development for any hyperbolic angle τ is not convergent [26]; this is corrected by repeating the terms 4, 13, 40, …; that is, the terms to be repeated correspond to the values provided by the following expression:

Table 8.5 Hyperbolic angles β_i

i	2^{-i}	$\beta_i = \text{arctgh}2^{-i}$
1	0.5	0.5493061443
2	0.25	0.2554128119
3	0.125	0.1256572141
4	0.0625	0.0625815715
5	0.03125	0.0312601785
6	0.015625	0.0156262718
7	0.0078125	0.0078126589
8	0.00390625	0.0039062698
9	0.001953125	0.0019531275
10	0.0009765625	0.0009765623
11	0.00048828125	0.0004882813
12	0.000244140625	0.0002441406
13	0.0001220703125	0.0001220703

$$i = \frac{3^{j+1} - 1}{2}, \; j = 1, 2, \ldots$$

Using the sequence 1, 2, 3, 4, 4, 5, 6, …, the development is convergent and the maximum value of the hyperbolic angle that can be developed is $\tau_{\max} \approx 1.13$. However, the range of values commonly used for τ is [26]:

$$|\tau| < \ln 2 = 0.69314718\ldots$$

For hyperbolic angles outside this range, the following identities can be used, which allow to transfer the calculation to the usable range [26]:

$$sh(Aln2 + B) = 2^{A-1}\left[chB + shB - 2^{-2A}(chB - shB)\right], |B| < ln2$$

$$ch(Aln2 + B) = 2^{A-1}\left[chB + shB + 2^{-2A}(chB - shB)\right], |B| < ln2$$

$$th(Aln2 + B) = \frac{sh(Aln2 + B)}{ch(Aln2 + B)}, \quad |B| < \ln 2$$

Example 8.4 **Transfer the hyperbolic angle 1.5 to the usable range.**

It has to be verified that:

$$A \ln 2 + B = 1.5, |B| < \ln 2$$

A suitable and comfortable value for A is chosen and the corresponding value of B is calculated. Given the

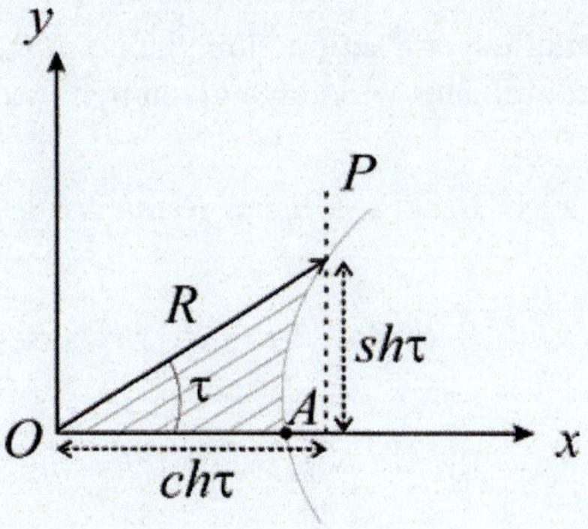

Fig. 8.4 Point P in a hyperbola

$$1.5 : \ln 2 = 1.5 : 0.69314718 = 2.164$$

A can be:

$$A = 2$$

Therefore:

$$2 \ln 2 + B = 1.5 \rightarrow B = 0.113705638$$

■

Let it be a point P (Fig. 8.4), of coordinates (x, y), located in the first quadrant of the plane, with $x > y$. The parametric equations:

$$x = Rch\tau$$

$$y = Rsh\tau$$

represent the positive branch of the equilateral hyperbola of radius R of equation $x^2 - y^2 = R^2$. The parameter τ is a measure of the area S of the AOP sector; specifically:

$$\tau = \frac{2S}{R^2}$$

The hyperbolic angle τ can be developed as addition or subtraction of the prefixed angles β_i (Table 8.5):

$$\tau = \pm\beta_i \pm \beta_j \pm \cdots = s_i\beta_i + s_j\beta_j + \cdots \tag{8.4}$$

where $s_k = \pm 1$ is the sign corresponding to each β_k in the development of τ.

The **CORDIC** algorithm consists in applying successive transformations of the point P, according to the development of τ, so that, depending on s_k, it implies an addition or a subtraction, the point P_a (x_a, y_a) or P_b (x_b, y_b), will be obtained, whose coordinates will come given by:

$$x_a = R_1 ch(\tau + \beta_i) = R_1[ch\tau ch\beta_i + sh\tau sh\beta_i] = R_1\left(ch\tau + 2^{-i} sh\tau\right)\frac{1}{\sqrt{1-2^{-2i}}}$$

$$y_a = R_1 sh(\tau + \beta_i) = R_1[sh\tau ch\beta_i + ch\tau sh\beta_i] = R_1\left(sh\tau + 2^{-i} ch\tau\right)\frac{1}{\sqrt{1-2^{-2i}}}$$

$$x_b = R_1 ch(\tau - \beta_i) = R_1[ch\tau ch\beta_i - sh\tau sh\beta_i] = R_1\left(ch\tau - 2^{-i} sh\tau\right)\frac{1}{\sqrt{1-2^{-2i}}}$$

$$y_b = R_1 sh(\tau - \beta_i) = R_1[sh\tau ch\beta_i - ch\tau sh\beta_i] = R_1\left(sh\tau - 2^{-i} ch\tau\right)\frac{1}{\sqrt{1-2^{-2i}}}$$

It is obvious that, in order to obtain an iterative procedure with only additions/ subtractions and displacements, it must be:

$$R = \frac{R_1}{\sqrt{1-2^{-2i}}}$$

Therefore, the transformations that are used in the hyperbolic case are:

$$\begin{aligned} x_{i+1} &= x_i + 2^{-i} s_i y_i \\ y_{i+1} &= y_i + 2^{-i} s_i x_i \\ z_{i+1} &= z_i - s_i \propto_i \end{aligned} \tag{8.5}$$

In each transformation there is a shortening of the radius; specifically

$$\frac{R_i}{R_{i-1}} = \sqrt{1-2^{-2i}}$$

After n transformations the shortening is:

$$\frac{R_n}{R} = \prod_{i=0}^{i=n} \sqrt{1-2^{-2i}}$$

The values of the shortening up to $i = 14$ are shown in Table 8.6; it is obvious that it quickly takes on an almost constant value. For example, for $i = 10$ it results:

$$\frac{R_{10}}{R} = \prod_{i=0}^{i=10} \sqrt{1-2^{-2i}} = 0.82815989$$

Table 8.6 Shortening for the first values of i

i	Shortening
1	0.866025400
2	0.838525488
3	0.831948716
4	0.830322226
5	0.828698916
6	0.828294179
7	0.828193063
8	0.828167788
9	0.828161470
10	0.828159890
11	0.828159495
12	0.828159396
13	0.828159372
14	0.828159347

For n = ∞ it turns out with good approximation:

$$\frac{R_\infty}{R} = \prod_{i=0}^{i=\infty} \sqrt{1 - 2^{-2i}} \approx 0.828159366$$

Therefore, if the number of iterations is known, the shortening, A, can be calculated with the desired precision. For example, $A = 0.828159366$ can be used as the value of the shortening. If we start from $R_0 = 1$, after applying the **CORDIC** transformations in the hyperbolic case we will have $R_f = 0.828159366$. Or, if we start from $R_0 = 1/A = 1/0.828159366 \approx 1.2074971$, after applying the **CORDIC** transformations in the hyperbolic case it results $R_f = 1$.

In the hyperbolic case there are, again, two modes of operation, which are detailed below: the rotation mode and the vectoring mode.

8.2.2.1 Rotation Mode

In this mode, given an hyperbolic angle τ, an hyperbolic function of it has to be calculated. Initially (z will represent the value of the hyperbolic angle in each iteration):

$$x_1 = 1/A$$

$$y_1 = 0$$

$$z_1 = \tau$$

Then the algorithm is executed, trying to cancel the value of z:

$$x_{i+1} = x_i + 2^{-i} s_i y_i$$

$$y_{i+1} = y + 2^{-i} s_i x_i$$

$$z_{i+1} = z_i - s_i \beta_i$$

where:

$$s_i = +1 \quad \text{if } z_i \geq 0, \quad \text{and} \quad s_i = -1 \quad \text{if } z_i < 0$$

After n iterations a vector of length 1 and hyperbolic angle τ is obtained. So:

$$sh\tau = y_n$$

$$ch\tau = x_n$$

$$th\tau = y_n/x_n$$

$$\coth \tau = x_n/y_n$$

$$\text{sech}\tau = 1/x$$

$$\text{cosech}\tau = 1/y_n$$

Remember that:

$$sh\tau = \frac{e^{\tau} - e^{-\tau}}{2}, \quad ch\tau = \frac{e^{\tau} + e^{-\tau}}{2}$$

Therefore, a procedure to calculate exponentials is obtained; specifically:

$$e^{\tau} = sh\tau + ch\tau$$

$$e^{-\tau} = ch\tau - sh\tau$$

Table 8.7 Iterations of the Example 8.5

i	x_i	y_i	z_i	s_i
1	1.2074971	0	0.14796974	+
2	1.2074971	0.60374855	−0.4013364	−
3	1.0565599625	0.301874275	−0.14592359	−
4	1.01671711875	0.169804279688	−0.02026638	−
5	1.006104351270	0.106259459766	0.04231519	+
6	1.009424959388	0.137700220743	0.01105501	+
7	1.011576525337	0.153472485733	−0.00457117	−
8	1.010377521542	0.145569544129	0.00324149	+
9	1.014343846452	0.149516331323	−0.00066478	−
10	1.014051822367	0.147535190998	0.00128835	+
11	1.014197375423	0.148542197498	0.00031179	+
12	1.014269905614	0.149037409792	−0.00017649	−
13	1.014233519621	0.148789785937	0.00006765	+
14	1.014251682390	0.148913593423	−0.00005442	−

Example 8.5 **Calculate e^5 and e^{-5}.**

The values of sh5 and ch5 must be calculated. Previously, 5 must be transferred to the usable range:

$$5{:}\,0.69314718 = 7.21 \rightarrow A = 7 \rightarrow B = 0.14796974$$

It will reach until $i = 14$. Therefore $A = 0.828159347$, and $1/A = 1.207497088$. The initial values are $x_1 = 1.207497088$, $y_1 = 0$ and $z_1 = 0.14796974$. The results of the iterations up to $i = 14$ are given in Table 8.7.

$$sh(A\ln 2 + B) = 2^{A-1}\left[chB + shB - 2^{-2A}(chB - shB)\right], \quad |B| < \ln 2$$

$$ch(A\ln 2 + B) = 2^{A-1}\left[chB + shB + 2^{-2A}(chB - shB)\right], \quad |B| < \ln 2$$

$$\begin{aligned} sh5 &= 2^6[ch0.14796974 + sh0.14796974 – 2^{-14}(ch0.14796974 – sh0.14796974)] \\ &= 2^6[1.014251682390 + 0.148913593423 - 2^{-14}(1.014251682390 \\ &\quad - 0.148913593423)] = 2^6[1.163165275813 - 2^{-14}(0.865338088967)] \\ &= 2^6[1.16311245990273989155] = 74.43919743377535354592 \end{aligned}$$

$$\begin{aligned} ch5 &= 2^6[ch0.14796974 + sh0.14796974 + 2^{-14}(ch0.14796974 - sh0.14796974)] \\ &= 2^6[1.014251682390 + 0.148913593423 + 2^{-14}(1.014251682390 \\ &\quad - 0.148913593423)] = 2^6[1.163165275813 + 2^{-14}(0.865338088967)] \end{aligned}$$

$$= 2^6[1.1632180917232601008 45] = 74.4459578702886464 5408.$$

$$e^5 = \text{sh}5 + \text{ch}5 = 148.885155304064$$

$$e^{-5} = \text{ch}5 - \text{sh}5 = 0.0067604365132929 0816$$

The values that appear in the corresponding tables are: $e^5 = 148.413159$, $e^{-5} = 0.00673795$. Therefore, with the **CORDIC** algorithm good approximations are achieved. ■

8.2.2.2 Vectoring Mode

In this mode, given a point P (x_p, y_p) in the plane, the hyperbolic angle τ has to be calculated. To apply the **CORDIC** algorithm, the calculation is transferred to the range of usable values, if previously it was not in it. A point is assumed in the first quadrant with $x > y$. Initially:

$$x_1 = x_p$$

$$y_1 = y_p$$

$$z_1 = 0$$

Then the algorithm is executed, trying to cancel the value of y:

$$x_{i+1} = x_i + 2^{-i} s_i y_i$$

$$y_{i+1} = y + 2^{-i} s_i x_i$$

$$z_{i+1} = z_i - s_i \beta_i$$

where:

$$s_i = +1 \quad \text{if } y_i < 0, \quad \text{and} \quad s_i = -1 \quad \text{if } y_i \geq 0$$

After n iterations a vector of length A and angle τ is obtained, which corresponds to the final value of z.

Example 8.6 **Calculate the hyperbolic angle corresponding to the point (6, 2).**

Table 8.8 Iterations of the Example 8.6

i	x_i	y_i	z_i	s_i
1	6	2	0	−
2	5	−1	0.54930614	+
3	4.75	0.25	0.29389333	−
4	4.6875	−0.34375	0.41955054	+
5	4.666015625	−0.05078125	0.35696897	+
6	4.6644287109375	0.09503173828125	0.32570879	−
7	4.662943840026855446875	0.02215003967285156 25	0.34133506	−
8	4.66277079284191131592	−0.01427920907735824585	0.34914772	+
9	4.66271501468145288527	0.00393473933218047023	0.34524102	−
10	4.66270732964369472029	−0.00517212593086924244	0.34719415	+
11	4.66270227874205141876	−0.00061870313561773664	0.34621759	+

The results of the first iterations up to $i = 11$ are shown in Table 8.8. It results $\tau = 0.34621759$. It really is $\tau = 0.34657359$. Therefore, a very good approximation with only 11 iterations is obtained. ■

In the vectoring mode, if:

$$x_1 = a + 1$$

$$y_1 = a - 1$$

$$z_1 = 0$$

it results:

$$R_1 = \sqrt{(a+1)^2 - (a-1)^2} = 2\sqrt{a}$$

After n iterations we will have a vector of length $R_n = x_n = 2A\sqrt{a}$ and angle η, which will correspond to the final value of z. This hyperbolic angle η is easily calculated, since:

$$th\eta = \frac{e^{\eta} - e^{-\eta}}{e^{\eta} + e^{-\eta}} = \frac{a-1}{a+1}$$

From here it results:

$$\eta = 0.5 \ln a$$

Therefore, with these initial values ($x_0 = a + 1, y_0 = a - 1, z_0 = 0$) a procedure to calculate $\sqrt{a}$ and $\ln a$ is obtained.

As an extension of the previous case, if:

$$x_1 = a + b$$

$$y_1 = a - b$$

$$z_1 = 0$$

it results:

$$R_1 = \sqrt{(a+b)^2 - (a-b)^2} = \sqrt{ab}$$

After n iterations we will have a vector of length $R_n = x_n = 2A\sqrt{ab}$ and angle η, which will correspond to the final value of z. This hyperbolic angle η is easily calculated, since:

$$th\eta = \frac{e^{\eta} - e^{-\eta}}{e^{\eta} + e^{-\eta}} = \frac{a-b}{a+b}$$

From here it results:

$$\eta = 0.5 \ln \frac{a}{b}$$

Therefore, with these initial values ($x_0 = a + b, y_0 = a - b, z_0 = 0$) a procedure to calculate $\sqrt{ab}$ and $\ln \frac{a}{b}$ is obtained.

Example 8.7 **Calculate $\sqrt{2}$ and $ln2$.**

It is $a = 2$. It must be done $x_1 = a + 1 = 3, y_1 = a - 1 = 1, z_1 = 0$.

The results of the first iterations up to $i = 12$ are shown in Table 8.9. Using $A = 0.828159396$ in the following equation:

$$2A\sqrt{2} = x_n$$

It results $\sqrt{2} = 1.415$. It is also obtained $\ln 2 = 0.69341174$. Therefore, compared to the real values ($\sqrt{2} = 1.4142$ and $\ln 2 = 0.693147$), good approximations are obtained with only 12 iterations. ■

Table 8.9 Iterations of the Example 8.7

i	x_i	y_i	z_i	s_i
1	3	1	0	−
2	2.5	−0.5	0.54930614	+
3	2.375	0.125	0.29389333	−
4	2.35625	−0.171875	0.41955054	+
5	2.3455078125	−0.024609375	0.35696897	+
6	2.34473876953125	0.048687744140625	0.32570879	−
7	2.343978023529052734375	0.01205120086669921875	0.34133506	−
8	2.34388387352228164673	−0.00626112744212150574	0.34914772	+
9	2.34382713205483742058	0.00289466893882490694	0.34524102	−
10	2.34382147840456627818	−0.001683118428469697397	0.34719415	+
11	2.343819834735067534955	0.0006057685621240226567	0.34621759	−
12	2.343819538949636810335	−0.000538674716555209538	0.34670587	

8.2.3 *Linear Case*

Walther [26] also extended the **CORDIC** algorithm to the linear case, as it is represented in Fig. 8.5. In this case the parameter ν is developed in the form:

$$\nu = \pm\gamma_i \pm \gamma_j \pm \cdots = s_i\gamma_i + s_j\gamma_j + \cdots \tag{8.6}$$

with:

$$\gamma_i = 2^{-i}$$

Depending on the range of values to be used in the data, it is possible to select an appropriate set of values for γ_i. For example, if floating point numbers are to be processed, whose mantissa is smaller than 2, for γ_i, the values corresponding to $i =$ 0, 1, …, can be used.

The point P (x, y) (Fig. 8.5) is transformed into P_a (x_a, y_a) or into P_b (x_b, y_b):

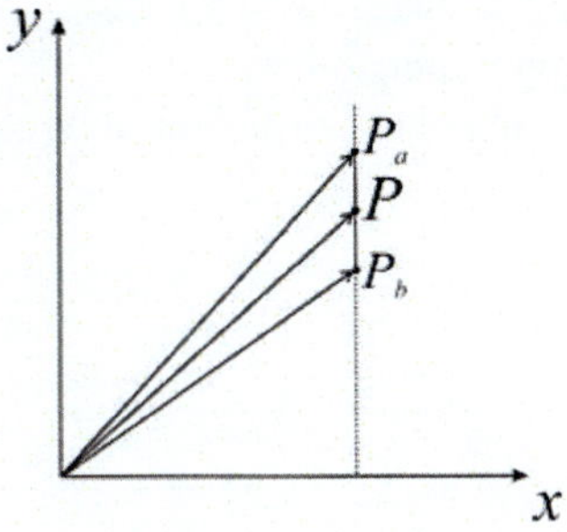

Fig. 8.5 Point P at the linear case

$$x_a = x$$

$$y_a = y + 2^{-i}$$

$$x_b = x$$

$$y_b = x - 2^{-i}$$

It is:

$$x_{i+1} = x_i$$

$$y_{i+1} = y_i + s_i 2^{-i} x_i \tag{8.7}$$

$$z_{i+1} = z_i - s_i 2^{-i}$$

In rotation mode (do $z = 0$), the final values are:

$$x_f = x_0$$

$$y_f = y_0 + x_0 z_0$$

$$z_f = 0$$

In this case, therefore, a procedure to multiply two numbers and add a third has been obtained.

Example 8.8 **Calculate 3.2 × 5.8.**

Since $z = 5.8$, i has been selected such that $2^{-i} < 5.8 < 2^{-(i+1)}$; the value is $i = -2$.

The results of the first iterations up to $i = 12$ are given in Table 8.10. It results $3.2 \times 5.8 = 18.5625$, that compared to the correct result, 18.56, gives a very good approximation. ■

In the vectoring mode (with $y = 0$), the final values are:

$$x_f = x_0$$

$$y_f = 0$$

$$z_f = z_0 + y_0 / x_0$$

Table 8.10 Iterations of the Example 8.8

i	x_i	y_i	z_i	s_i
−2	3.2	0	5.8	+
−1	3.2	12.8	1.8	+
0	3.2	19.2	−0.2	−
1	3.2	16	0.8	+
2	3.2	17.6	0.3	+
3	3.2	18.4	0.05	+
4	3.2	18.8	−0.075	−
5	3.2	18.6	−0.0125	−
6	3.2	18.5	0.01875	+
7	3.2	18.55	0.003125	+
8	3.2	18.575	−0.0046875	−
9	3.2	18.5625	−0.00078125	

In this case, therefore, a procedure to divide two numbers and add a third has been detailed.

Example 8.9 **Calculate** 3.2 : 5.8.

It must be $2^{-i}x \leq y$; that is, $2^{-i}\ 5.8 \leq 3.2$, or $2^{-i} \leq 0.55$. Therefore $i = 1$.

The results of the first iterations up to $i = 12$ are given in Table 8.11. It results 3.2: 5.8 = 0.551269533, which compared to the correct result, 0.551724138, gives a very good approximation. ■

Table 8.11 Iterations of the Example 8.9

i	x_i	y_i	z_i	s_i
1	5.8	3.2	0	−
2	2.9	0.3	0.5	−
3	1.45	−1.15	0.75	+
4	0.725	−0.425	0.625	+
5	0.3625	−0.0625	0.5625	+
6	0.18125	0.11875	0.53125	−
7	0.090625	0.028125	0.546875	−
8	0.0453125	−0.0171875	0.5546875	+
9	0.02265625	0.00546875	0.55078125	−
10	0.011328125	−0.005859375	0.552734375	+
11	0.0056640625	−0.0001953125	0.551757813	+
12	0.00283203125	0.00263671875	0.551269533	−

8.2.4 Unification and Modifications

The three cases of the **CORDIC** algorithm (circular, hyperbolic and linear) can be included in a single description, as it appears in [26]. Using the parameter m to differentiate the three cases ($m = 1$ for the circular case, $m = -1$ for the hyperbolic case, $m = 0$ for the linear case), Eqs. (8.3), (8.5) and (8.7) can be unified as:

$$x_{i+1} = x_i - m2^{-i} s_i y_i$$

$$y_{i+1} = y_i + 2^{-i} s_i x_i$$

$$z_{i+1} = z_i - s_i \varphi_i$$

where

$$\varphi_i = \alpha_i \text{ if } m = 1,\ \varphi_i = \beta_i \text{ if } m = -1,\ \varphi_i = \gamma_i \text{ if } m = 0$$

For elongation or shortening it is obtained:

$$\frac{R_i}{R_{i-1}} = \sqrt{1 + m2^{-2i}}$$

The **CORDIC** algorithm has been the subject of multiple investigations since its inception, and modifications have been proposed to increase performance and to adapt it to certain applications, such as image processing or robotics. For example, to increase performance, in [3] the reduction of the number of iterations to compute the total rotation is presented, and in [7] the rotations are calculated by breaking down the rotation angle into a series of micro rotations. Regarding algorithms, architectures, and applications, in [12] a revision can be consulted.

8.2.5 Implementation

The processing unit to implement the three cases of the **CORDIC** algorithm could be as shown in Fig. 8.6. The multiplication by 2^{-i} is a shift of i positions; therefore, to implement it, a shifter can be used. There are three registers (X, Y, Z) and three adders/subtractors, with control inputs (s_i and m) to select the corresponding operation. The angles α_i, β_i and γ_i will be stored in look-up tables, or will be generated appropriately (for example, γ_i can be generated simply with a shift register). The

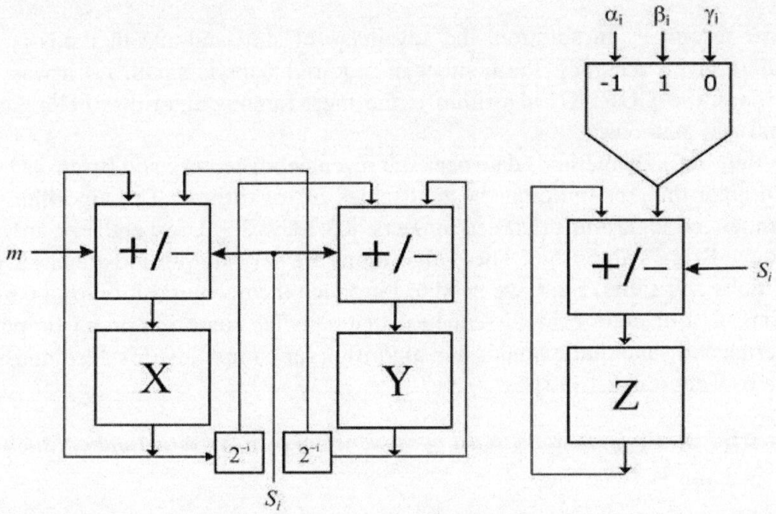

Fig. 8.6 Processing unit for the CORDIC algorithm

control unit must manage the initial values of the *X*, *Y* and *Z* registers, depending on the specific applications that are to be covered by this processing unit.

After this study of the **CORDIC** as a particular shift-and-add algorithm for the implementations of special functions, the different types of algorithms are presented at the following sections. As can be observed from the analysis carried out into the introduction section, for the implementation of special function we can distinguish between shift-and-add algorithms, rational or polynomial approximations and table-based methods. These three types of methods are presented at the following, beginning with shift-and-add and polynomial approximations due to these two approximations can be used as part of the table-based methods.

8.3 Shift-and-Add Algorithms

Methods for computing special functions using only operations with linear complexity, thus without multiplications and divisions, and with a small number of iterations, are very well suited to hardware computations. Among the hardware-efficient algorithms there is a group of iterative solutions for special functions which use shifts and additions only. Shift-and-add algorithms are based on simple elementary steps, decomposing its argument into a number basis such that the decomposition is performed only by means of additions. Furthermore, the function of interest is computed along with the decomposition and requires only additions and multiplications or divisions by 2, which are computed by shifts. Since additions and shifts are very efficiently performed, the elementary iteration is very simple thus saving computer resources. In addition, the advantage of shift-and-add methods is that, to obtain a given accuracy, the number of required steps is small. As it was seen in Sect. 8.1, the **CORDIC** algorithm is the most famous algorithm belonging to shift-and-add methods.

The shift-and-add methods date back the seventeenth century and Briggs [1] used such an algorithm for building the first tables of logarithms. The algorithms for exponential and logarithms detailed in Sects. 8.3.1 and 8.3.2, respectively, are quite close to the Brigg's algorithm. These algorithms are very simple and are based only on additions and shifts. First, we need to introduce the decomposition of a number by means of a discrete base, a sequence that satisfies some important properties considering shift-and-add methods. An algorithm for computing this decomposition is given by Theorem 8.1 [14].

Theorem 8.1 *Let* (ω_n) *be a decreasing sequence of positive real numbers such that the series* $\sum_{i=0}^{\infty} \omega_i$ *converges. If:*

$$\forall n, \omega_n \leq \sum_{k=n+1}^{\infty} \omega_k$$

then, for any $t \in \left[0, \sum_{k=0}^{\infty} \omega_k\right]$*, the sequences* (t_n) *and* (d_n) *defined as:*

$$t_0 = 0$$

$$t_{n+1} = t_n + d_n \omega_n$$

$$d_n = \begin{cases} 1 \text{ if } t_n + \omega_n \leq t \\ 0 \text{ otherwise} \end{cases}$$

Satisfy

$$t = \sum_{n=0}^{\infty} d_n \omega_n = \lim_{n \to \infty} t_n$$

The sequence (ω_n) that meets the previous conditions is called a discrete base. The algorithm given by Theorem 8.1 is called the "restoring algorithm" and its proof is detailed in [14]. The "nonrestoring algorithm" is also presented in [14] and it can be used for the decomposition of a number by means of a discrete base but considering values of d_n equal to -1 or $+1$. This nonrestoring algorithm was used in the **CORDIC** algorithm (see Sect. 8.2).

8.3.1 Algorithm for the Function e^t

We are going to illustrate the computation of the exponential function, e^t, with $t \in [0; b]$ under the "restoring algorithm". Let consider the following sequence:

$$\omega_n = \ln\left(1 + 2^{-n}\right)$$

and

$$t \in \left[0, \sum_{n=0}^{\infty} \omega_n\right] = \left[0, \sum_{n=0}^{\infty} \ln\left(1 + 2^{-n}\right)\right] \approx [0, 1.56202]$$

The sequence ω_n is a discrete base since it satisfies the conditions of Theorem 8.1. Let us assume the exponential function, e^t, for $t \in [0, 1.56202]$. Thus, the algorithm for the computation of this exponential function can be defined as follows:

$$t \in \left[0, \sum_{n=0}^{\infty} \ln\left(1 + 2^{-k}\right)\right]$$

$$t_0 = 0$$

$$E_0 = 1$$

$$t_{n+1} = t_n + d_n \ln\left(1 + 2^{-n}\right)$$

$$E_{n+1} = E_n + d_n E_n 2^{-n}$$

$$d_n = \begin{cases} 1 \text{ if } t_n + \ln\left(1 + 2^{-n}\right) \leq t \\ 0 \text{ otherwise} \end{cases}$$

where E_n is the value of the exponential function at the step n, $E_n = \exp(t_n)$. Due to t satisfies:

$$t = \sum_{n=0}^{\infty} d_n \ln\left(1 + 2^{-n}\right) = \lim_{n \to \infty} t_n$$

t_n goes to t, and thus, E_n goes to $\exp(t)$. This algorithm only requires additions and multiplications by powers of 2 that can be reduced to shifts when binary arithmetic is used. When stopping at step n the algorithm, the relative error on the result is bounded by 2^{-n+1} (see [14] for this estimation of the error), thus, $n - 1$ significant bits are obtained.

Table 8.12 Iterations of the Example 8.13

N	d_{n-1}	t_n	E_n
1	1	0.693147180559945	1.000000000000000
2	0	0.693147180559945	2.000000000000000
3	1	0.916290731874155	2.000000000000000
4	0	0.916290731874155	2.500000000000000
5	0	0.916290731874155	2.500000000000000
6	0	0.916290731874155	2.500000000000000
7	1	0.931794918410120	2.539062500000000
8	0	0.931794918410120	2.539062500000000
9	0	0.931794918410120	2.539062500000000
10	1	0.933746138541382	2.544021606445313
11	0	0.933746138541382	2.544021606445313
12	1	0.934234300620883	2.545263804495335
13	1	0.934478411448411	2.545885206791354
14	0	0.934478411448411	2.545885206791354
15	0	0.934478411448411	2.545885206791354
16	0	0.934478411448411	2.545885206791354
17	1	0.934493670121059	2.545924053916702
18	0	0.934493670121059	2.545924053916702
19	1	0.934497484811049	2.545933765846228
20	1	0.934499392157863	2.545938621829516

Example 8.10 **Calculate the value of $e^{0.9345}$ using the restoring algorithm from $n = 1$ to $n = 20$ and calculate the error.**

Initially $t_0 = 0$ and $E_0 = 1$. The results of the first iteration up to $n = 20$ are shown in Table 8.12. Therefore, compared to the true value, $e^{0.9345} = 2.545940169358760$, a good approximation is obtained with only 20 iterations. ■

A generalization of this algorithm for the computation of the a^t can be obtained replacing the constants $\ln(1+2^{-n})$ by $\log_a(1+2^{-n})$, being $[0, \sum_{n=0}^{\infty} \log_a(1+2^{-n})]$ its convergence domain. Figure 8.7 shows the basic schematic for the implementation of the exponential function using the restoring algorithm.

8.3.2 *Algorithm for the Function* ln(x)

The previous algorithm can be used to deduce another algorithm for logarithms, including a modification. The expression $E_n = \mathrm{e}^{t_n}$, can also be written as $t_n = \ln E_n$,

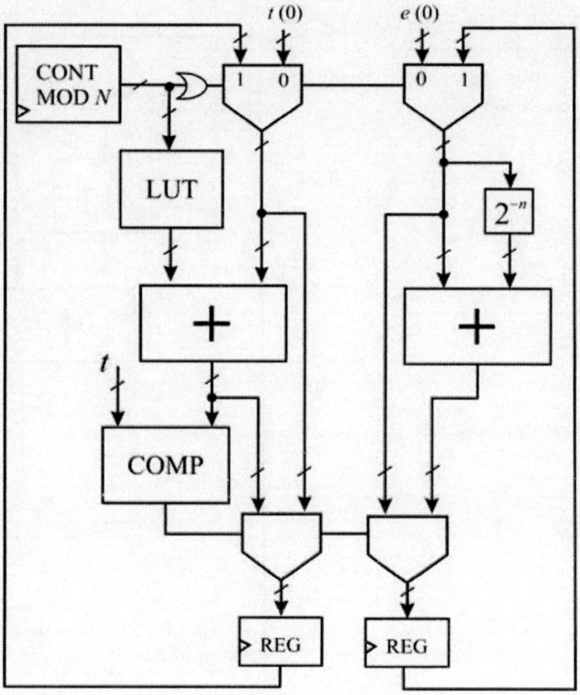

Fig. 8.7 Processing unit for the approximation of the exponential using the restoring algorithm

allowing to compute de logarithm. Thus, let us consider that E^* becomes the input and t^* the result. Since t^* is the result we are looking for, this is unknown, and there is a problem with the comparison $t_n + \ln(1 + 2^{-n})$ that it is necessary on the exponential algorithm. This comparison is equivalent to $E_n(1 + 2^{-n}) \leq E^t = E^*$, since the exponential is an increasing function. Thus, the algorithm for the logarithm can be defined as follows:

$$1 \leq E^* \leq \prod_{n=0}^{\infty}(1 + 2^{-n}) \approx 4.768$$

$$t_0 = 0$$

$$E_0 = 1$$

$$t_{n+1} = t_n + \ln(1 + d_n 2^{-n})$$

$$E_{n+1} = E_n(1 + d_n 2^{-n}) = E_n + d_n E_n 2^{-n}$$

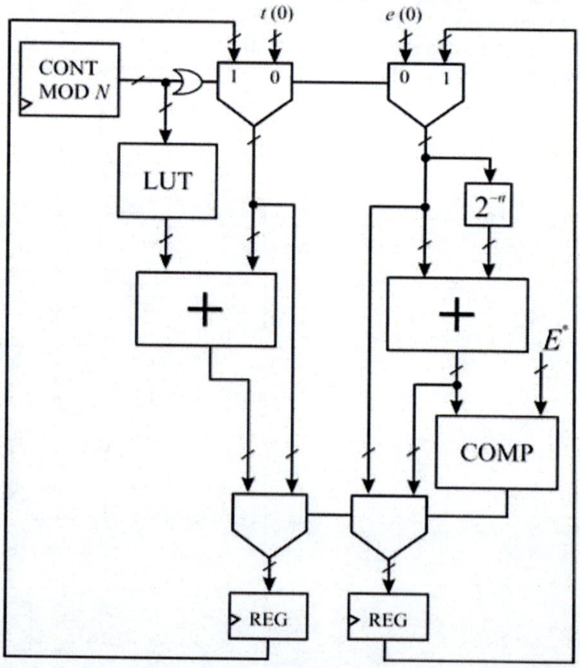

Fig. 8.8 Processing unit for the approximation of the logarithm using the restoring algorithm

$$d_n = \begin{cases} 1 & E_n + E_n 2^{-n} \leq E^* \\ 0 & \text{otherwise} \end{cases}$$

Figure 8.8 shows the basic schematic for the implementation of the logarithm function using the restoring algorithm. It can be observed that it is similar to Fig. 8.7, being different the signals used for the comparison and the final output.

As in the exponential algorithm, we can get a generalization of this algorithm in order to compute the base-a logarithms replacing the constants $\ln(1 + d_n 2^{-n})$ by $\log_a(1 + d_n 2^{-n})$.

8.4 Newton-Raphson Method

The Newton-Raphson method allows to solve equations numerically, and its iterative formula constitutes a powerful tool to approximate special functions as the square-root or the reciprocal, being it implemented on many processors. Newton first introduced a method for finding solutions of a system of non-linear equations, which is known as Newton's technique and did not explicitly use the derivate. Newton applied the method only to polynomials, and did not consider the successive approximations, but calculated a sequence of polynomials to arrive at the approximation of the zeros or roots of a real non-linear function. A few years later, this method was generalized by Raphson establishing an algorithm to find successive approximations of roots of a real function.

Let $f(x)$ be a function, and let r be a root of the equation $f(x) = 0$. The Newton-Raphson iteration is (see Appendix E):

$$x_{i+1} = x_i - \frac{f(x_i)}{f'(x_i)}$$

where $f'(x)$ denotes the first derivate of $f(x)$ with respect to x, x_i is an approximation to the root, and x_{i+1} is a better approximation. Thus, starting with estimate value of r, this is, the initial value x_0, the Newton-Raphson iteration will yield successively better approximations to the root, x_1, x_2, x_3, ... It is important to mention that this convergence is not guaranteed if the initial value is not close enough to the root r.

8.4.1 Square Root

The Newton-Raphson method can be used for evaluating the square root $\sqrt{a}$. By choosing $f(x) = x^2 - a$ the Newton-Raphson method gives the iteration:

$$x_{i+1} = x_i - \frac{x_i^2 - a}{2x_i} = \frac{1}{2}\left(x_i + \frac{a}{x_i}\right)$$

and the obtained sequence of values goes to $\sqrt{a}$.

Example 8.11 **Obtain the Newton-Raphson approximation to $\sqrt{2}$ considering 5 iterations and 1, 1.5 and 2 as initial values.**

For $a = 2$ square root Newton-Raphson iteration is:

$$x_{i+1} = \frac{1}{2}\left(x_i + \frac{2}{x_i}\right)$$

Table 8.13 shows the sequence of approximations for $\sqrt{2}$ using different initial values.

As we can observe from Table 8.13, the number of iterations to reach a precision depends on the initial value. For instance, compared to the real value with 15 decimal numbers, $\sqrt{2} = 1.414213562373095$, for $x_0 = 0$ and $x_0 = 2$, only 5 iterations are required to obtain precision until 15 decimal numbers, while for $x_0 = 1.5$, with only 4 iterations the same precision is reached. These initial values were selected close to the root sought, as there should be selected in order to guarantee the convergence. ■

Newton-Raphson method allows to start with a low precision, and it is able to double the precision with each iteration. However, analyzing the resources required

Table 8.13 Iterations of the Example 8.11

i	$x_0 = 1$	$x_0 = 1.5$	$x_0 = 2$
	x_i	x_i	x_i
1	1.500000000000000	1.416666666666667	1.500000000000000
2	1.416666666666667	1.414215686274510	1.416666666666667
3	1.414215686274510	1.414213562374690	1.414215686274510
4	1.414213562374690	**1.414213562373095**	1.414213562374690
5	**1.414213562373095**	1.414213562373095	**1.414213562373095**

for the hardware implementation of the square root approximation, we can observe from Fig. 8.9, that it requires a division, an addition and a multiplication by $\frac{1}{2}$. For this implementation example, the initial value x_0 is obtained using a **LUT** which input is the value a. The division is usually avoided for hardware implementation due to it is the most complicated and expensive operation. To compute $\sqrt{a}$ and avoid the division, we can take into account that $\sqrt{a} = \frac{a}{\sqrt{a}}$. Thus, the approximation to

Fig. 8.9 Processing unit for the Newton-Raphson approximation to the square root

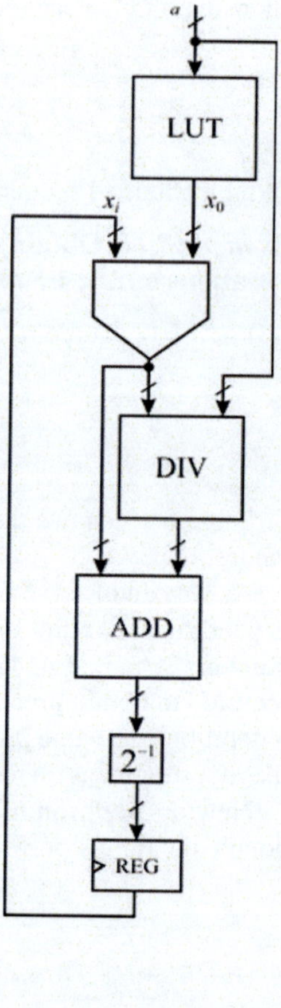

$1/\sqrt{a}$ can be used, so that the final result have to be multiply by a. The Newton-Raphson iteration for the approximation to $1/\sqrt{a}$, is detailed at the following, where it is shown that such iteration avoid the division.

For evaluating $1/\sqrt{a}$ we choose the following function whose root is to be found:

$$f(x) = \frac{1}{x^2} - a$$

Introducing this function in The Newton-Raphson iteration:

$$x_{i+1} = x_i - \frac{\frac{1}{x_i^2} - a}{-\frac{2}{x_i^3}} = \frac{x_i}{2}(3 - ax_i^2)$$

Figure 8.10 shows the diagram block for the implementation of this algorithm, were it can be observed that no division is required.

Example 8.12 **Obtain approximation to $\sqrt{2}$ using the Newton-Raphson iteration for $1/\sqrt{2}$. Consider the initial values 0.1, 0.5 and 1, and the required iterations for reached 15 decimal precision.**

For $a = 2$ inverse square root Newton-Raphson iteration is:

$$x_{i+1} = \frac{x_i}{2}(3 - 2x_i^2)$$

Table 8.14 shows the sequence of approximations for $1/\sqrt{2}$ using different initial values.

As we can observe from Table 8.14, the number of iterations to get a certain precision depends of the initial value. Considering that the real value with 15 decimal numbers is $1/\sqrt{2} = 0.707106781186547$, to obtain this precision, 11 iterations are required for $x_0 = 0.1$, 7 iterations for $x_0 = 0.5$ and 8 iterations for $x_0 = 1$. This value is then multiply by 2, in order to obtain the square root of 2:

$$2 * \left(\frac{1}{\sqrt{2}}\right) = 2 * 0.707106781186547 = 1.414213562373095$$

■

8.4.2 Reciprocal

Another important elementary function is the reciprocal, which can be used for the calculation of the division. Using the Newton-Raphson iterative formula it is possible to find approximate reciprocal $\frac{1}{a}$ of the divisors. In this case, we choose

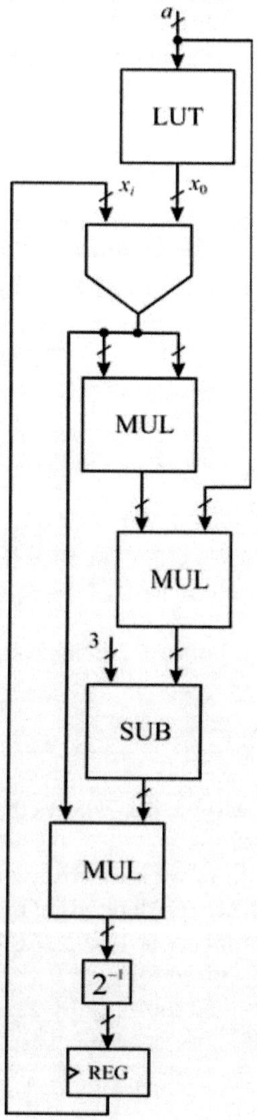

Fig. 8.10 Processing unit for the Newton-Raphson approximation to the inverse of the square root

$$f(x) = a - \frac{1}{x}$$

as the function whose root is to be found. Then, the required Newton-Raphson iteration is:

$$x_{i+1} = x_i - \frac{a - 1/x_i}{1/x_i^2} = x_i(2 - ax_i)$$

This sequence goes to $\frac{1}{a}$, requiring only multiplications and a subtraction.

Table 8.14 Iterations of the Example 8.12

i	$x_0 = 0.1$	$x_0 = 0.5$	$x_0 = 1$
	x_i	x_i	x_i
1	0.149000000000000	0.625000000000000	0.500000000000000
2	0.220192051000000	0.693359375000000	0.625000000000000
3	0.319612166344549	0.706708468496799	0.693359375000000
4	0.446769247675768	0.707106444695907	0.706708468496799
5	0.580977496295658	0.707106781186307	0.707106444695907
6	0.675366091679638	0.707106781186548	0.707106781186307
7	0.705001589508973	**0.707106781186547**	0.707106781186548
8	0.707097389181024	0.707106781186547	**0.707106781186547**
9	0.707106780999427	0.707106781186547	0.707106781186547
10	0.707106781186548	0.707106781186547	0.707106781186547
11	**0.707106781186547**	0.707106781186547	0.707106781186547

Figure 8.11 shows an example of the block diagram of the reciprocal unit based on the Newton-Raphson. For this example, the initial value x_0 is obtained using a **LUT** which input is the value a.

Example 8.13 **Obtain the Newton-Raphson approximation to 1/3 considering 9 iterations and 0.1, 0.3 and 0.6 as initial values.**

For $a = 3$ inverse square root Newton-Raphson iteration is:

$$x_{i+1} = x_i(2 - 3x_i)$$

Table 8.15 shows the sequence of approximations for $1/\sqrt{2}$ using different initial values.

As we can observe from Table 8.15, for $x_0 = 0.1$ $x_0 = 0.3$ and $x_0 = 0.6$, the number of iterations required for obtain precision until 15 decimal numbers is 7, 4 and 8, respectively. ■

8.5 Polynomial Approximation

Polynomials can be calculated using a finite number of operations including additions, subtractions and multiplications, and they are the only functions that fulfill this condition. Thus, the approximation of special functions by polynomials functions emerged naturally and it has been investigated from more than one hundred years. The related algorithms for the approximation of special functions were developed since then, long before the modern electronic computers emerged. Then, the interest

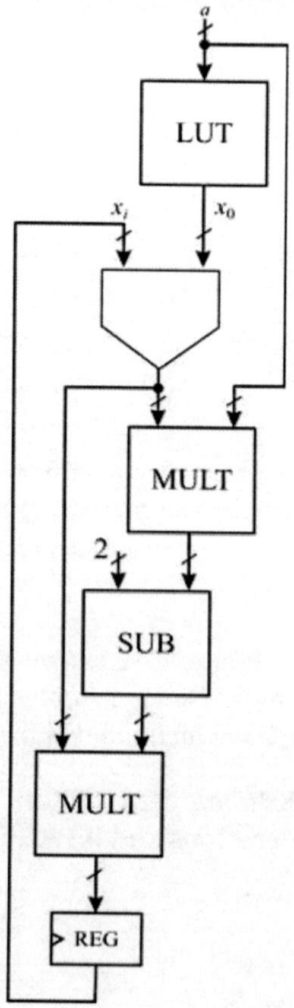

Fig. 8.11 Processing unit for the Newton-Raphson approximation to the reciprocal

Table 8.15 Iterations of the Example 8.13

i	$x_0 = 0.1$	$x_0 = 0.5$	$x_0 = 0.6$
	x_i	x_i	x_i
1	0.170000000000000	0.330000000000000	0.120000000000000
2	0.253300000000000	0.333300000000000	0.196800000000000
3	0.314117330000000	0.333333330000000	0.277409280000000
4	0.332225568981013	**0.333333333333333**	0.323950834109645
5	0.333329651907752	0.333333333333333	0.333069239458286
6	0.333333333292675	0.333333333333333	0.333333124096609
7	**0.333333333333333**	0.333333333333333	0.333333333333202
8	0.333333333333333	0.333333333333333	**0.333333333333333**

in the approximation of functions of a real variable by polynomials revived due to the increasing use of high-speed computing machines.

There are some important aspects to be considered for this approximation as to use the best way to evaluate a polynomial function, in terms of accuracy and/or speed and/or resources, and the way to calculate the tight bounds of the final error. This error is composed by the approximation error and the evaluation error, being the approximation error the distance between the function to approximate and the polynomial. The different types of errors in numerical data, and to the propagations of errors through the different operations are studied in Appendix D.

We are going to do reference to special functions, but the methods studied at this section can be used for any continuous function. Let us suppose that P^n are the set of the polynomials of degree less than or equal to n, including real coefficients. Polynomial approximation for an elementary function, f, tries to approximate f by an element p^* of P^n on an interval a, b. Thus, the objective of this approximation is to find a polynomial p^* of degree $\leq n$

$$p^*(x) = p_n^* x^n + p_{n-1}^* x^{n-1} + p_{n-2}^* x^{n-2} + \cdots + p_1^* x + p_0^*$$

that minimizes a distance $d = \|p^* - f\|$. The expression for this distance will depend on the type of approximation considered. Least squares polynomial approximations and least maximum polynomial approximations can be distinguished, as it is detailed in Appendix E.

8.5.1 Least Squares Methods

For least squares methods the approximation function p^* can be calculated as follows (see Appendix E):

$$p^* = \sum_{i=0}^{n} a_i T_i$$

$$a_i = \frac{f, T_i}{T_i, T_i}$$

being T_m a sequence of orthogonal polynomials of degree $m(m \leq n)$, as for example, Legendre, Chebyshev, Jacobi or Laguerre Polynomials. These polynomials are associated with simple weight w. Although the popular Taylor polynomials have great advantages due their easiness and maturity of analysis, they produce poor maximum error results compared to other approximation polynomials such as minimax, Chebyshev and Legendre approximation polynomials. Chebyshev and Legendre sequences of orthogonal polynomials are detailed in Appendix E.

Example 8.14 **Evaluate the degree-2 Legendre polynomial approximation to the function $\ln(x + 2)$ obtained in the Example E.1 for $x = -0.8$.**

From the Example E.1, we obtained the following degree-2 Legendre polynomial approximation to the function $\ln(x + 2)$:

$$p(x)^* = -0.140611752483767x^2 + 0.528122350496753x + 0.694789017163419$$

Thus,

$$p(-0.8)^* = 0.182299615176406$$

While the value $\ln(-0.8 + 2) = \ln(1.2) = 0.182321556793955$. As we can observe, for this value a 3 decimal numbers accuracy is reached. In order to obtained greater accuracy, high order polynomials have to be considered. ■

8.5.2 Least Maximum Methods

There are several characterizations of the minimax approximation to a function (Appendix E). Concretely the *Chebyshev theorem* states that for any $f(x)$ in closed interval $[a, b]$ a polynomial function p^* is the minimax degree-n approximation to this function if and only if there exist at least $n + 2$ points in $[a, b]$, x_i, at which:

$$p^*(x_i) - f(x_i) = (-1)^i p^*(x_i) - f(x_0) = \pm f - p^*_\infty$$

An example of this approximation method is detailed at the following.

Example 8.15 **Evaluate the minimax approximation to the function e^x obtained in the Example E.2 for $x = 0.5$.**

From the Example E.2, we obtained the following degree-1 minimax approximation to the function e^x, where the *Chebyshev theorem* was applied:

$$p^*(x) = 1.718282x + 0.894067$$

Thus,

$$p(0.5)^* = 1.753208$$

While the true value is $e^{0.5} = 1.648721270700128$. As we can observe low precision for the approximated value is obtained. This is due to a degree-1 polynomial approximation was used. In order to obtained greater accuracy, high order polynomials have to be considered. ■

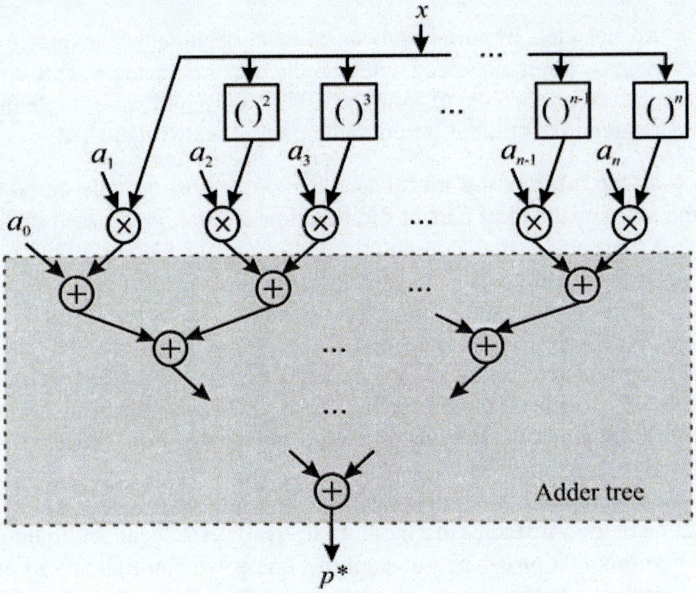

Fig. 8.12 Basic diagram block for degree-n polynomial approximation

Examples 8.14 and 8.15 propose low degree minimax polynomial approximation. However, approximation polynomials may be obliged to use large degrees to satisfy some required approximation error. As Fig. 8.12 shows, the implementation of polynomial approximations requires blocks that raise power (based on multipliers), multipliers and adders. To use large degree implies an important amount of resources for hardware implementation. Rational approximation can possibly produce much lower degree values for numerator and denominator polynomials compared to polynomial approximation assuming the same accuracy. Nevertheless, rational approximation is still unfavorable compared to polynomial approximation due to the known high cost of the division operation.

8.6 Table-Based Methods

Table-based methods have usually been proposed to implement special function. These methods have been used decades ago, but they become especially attractive since memory is less and less expensive. They are the predominant hardware-based techniques for computing high performance floating-point elementary functions. In general, they are considered efficient in terms of speed as well as having the advantage of consuming reasonable amount of memory, especially when the desired accuracy is no more than the faithful rounding case.

Over the last decades, important advances have been achieved related to these methods to increase their accuracy while reducing the memory and computing requirements. Based on the size of required **LUT** in the hardware implementation, three different categories of table-based methods can be distinguished:

- **Mainly look-up table based methods.** There are a type of table-based methods where the main or the total part of the function evaluation is based on a usually rather large table and a small computational (for example, some additions). In the case of algorithms using only look-up tables for the total part of the evaluation, a regularly spaced grid is used to store the values of the function itself.
- **Small look-up tables based methods**. There are table-based methods that use small look-up tables to obtain stored parameters that are used for a polynomial evaluation [22]. In these methods, the main part of the function evaluation is based on arithmetic computations. For high precisions, high-order polynomials are required in these methods.
- **Table-based balanced methods.** Finally, there are also methods that use the combination of look-up tables in a medium size and yet reduced arithmetic computation. This computation can be, for example, one or two multiplications or several small multiplications that use small multipliers. Piecewise polynomial approximation is considered as table-based balanced methods, where mostly degree-1 to 3 piecewise polynomials are used to the approximation with intermediate look-up tables.

8.6.1 Mainly Look-Up Table Based Methods

The first useful type of these methods were introduced in the 80's decade. These methods became implementable thanks to progress in **VLSI** technology, what has managed to reduce the costs of memories. In mainly look-up table based methods, **LUT** size implies a big percentage of the total hardware area, especially for applications requiring high precision. Table-and-addition methods are popular due to only **LUT** and simple adders are required and they do not need multipliers that implies a high time and/or area consume. These methods allow to compute special functions with low accuracy and significantly lower hardware cost compared with an implementation using only **LUT**s. In addition, compared with polynomials approximations or shift and-add algorithms, mainly look-up table based methods are faster. Thus, they are very interesting in digital signal or image processing applications, and to implement addition and subtraction in the logarithm number system. Another interesting application of these methods is for obtaining initial seed values to iterative methods, as for example the Newton-Raphson algorithms (see Sect. 8.4) used for square root and division [8], usually included in the floating-point units of current processors.

A function can be approximated simply storing in a look-up table containing 2^{x_i} values, being x_i the number of bit used for the input representation. However, it was necessary to explore new methods which allowed to approximate functions with much less memory that the required by high input bit. In this sense we can find bipartite and multipartite table methods [17, 19]. Bipartite method was first presented in [17] to compute the reciprocal function, $1/x$, as initial approximation to eventually compute division operation. Evaluation methods based on bipartite function access in parallel to two tables which outputs give an approximation to the function in carry-save form. A carry propagate adder can use these two outputs to generate a two's complement approximation to the function. On the other hand, the multipartite methods are a straightforwardly generalization of bipartite methods. Thus, multipartite methods derive architectures consisting of a number of tables and adders for the evaluation of arbitrary continuous functions, for precisions between 8 and 24 bits, with faithful rounding. For the implementation of the function, these tables are much smaller than a simple table.

Let suppose that f is the function to be evaluated and n-bit, fixed point arguments, between 1/2 and 1. If the output of every fixed-point input argument x with n bits is stored in a look-up table, $n2^n$ bits (2^n words $\times$ n bits) will be needed to store all the outputs.

8.6.1.1 Bipartite Method

Bipartite method consists of split the binary representation of this input x into 3 k-bit numbers, where $k = n/3$, such that x is arranged as follows [13]:

$$x = x_0 + 2^{-k}x_1 + 2^{-2k}x_2$$

where every x_i is a multiple of 2^{-k}. The order-1 Taylor expansion of $f(x)$ centered at the point $x_0 + 2^{-k}x_1$ is calculated as follows:

$$f(x) = f\left(x_0 + 2^{-k}x_1\right) + x_2 2^{-2k} f'\left(x_0 + 2^{-k}x_1\right) + \varepsilon_1$$

where $\varepsilon_1 = x_2^2 2^{-4k} f''(\xi_1)$ and $\xi_1 \in \left[x_0 + x_1 2^{-k}, x\right]$. Using order-0 Taylor's form at x_0, $f'\left(x_0 + 2^{-k}x_1\right)$ is approximated as follows:

$$f'\left(x_0 + 2^{-k}x_1\right) = f'(x_0) + \varepsilon_2$$

where $\varepsilon_2 = x_1 x_2 2^{-3k} f''(\xi_2)$ and $\xi_2 \in \left[x_0, x_0 + x_1 2^{-k}\right]$. From these approximations:

$$f(x) = f\left(x_0 + 2^{-k}x_1\right) + x_2 2^{-2k} f'(x_0) + \varepsilon_1 + \varepsilon_2$$

Thus, this approximation can be expressed as:

$$f(x) = T_1(x_0, x_1) + T_2(x_0, x_2) + \varepsilon$$

where T_1 corresponds to a function whose values can be store in a table with $2n/3$-address bits composed by the bits x_0x_1, that stores $f\left(x_0 + 2^{-k}x_1\right)$. On the other hand, T_2 corresponds to function whose values can be store in a table with $2n/3$-address bits composed by the bits x_0x_2, that stores $x_2 2^{-2k} f'(x_0)$. Hence, bipartite method approximates function values $f(x)$ to $\hat{f}(x) = T_1(x_0, x_1) + T_2(x_0, x_2)$ by adding the contents from these two look-up tables, T_1 and T_2, requiring a total memory of $2 \times 2^{2n/3}$ words $\times$ n bits. Compared to the size of a single table (2^n words $\times$ n bits) it supposes a substantial difference. Figure 8.13 represents a diagram of the implementation of a function using a bipartite method approximation.

Example 8.16 **Compare the size of the tables associated to the functions T_1 and T_2 for bipartite approximation of $1/x$ with a single table approximation. Assume 15-bit input $1.i_1 \ldots i_{15}$ and 15-bit output ($0.z_1 \ldots z_{15}$).**

The reciprocal function can be implemented using single table approximation in $2^{15} \times 15$ bits of memory. Using bipartite approximation, $k = 15/3 = 5$, being, the size for each required table, T_1 and T_2, $2^{10} \times 15$ bits. Thus, the reciprocal function can be implemented using bipartite approximation in $2 \times 2^{10} \times 15 = 2^{11} \times 15$ bits of memory. This is a reduction by a factor of $2^4 = 16$ in area. ■

For piecewise polynomial approximation methods (Sect. 8.6.3.1), low-degree polynomials are used to approximate the function values of the partitioned segments

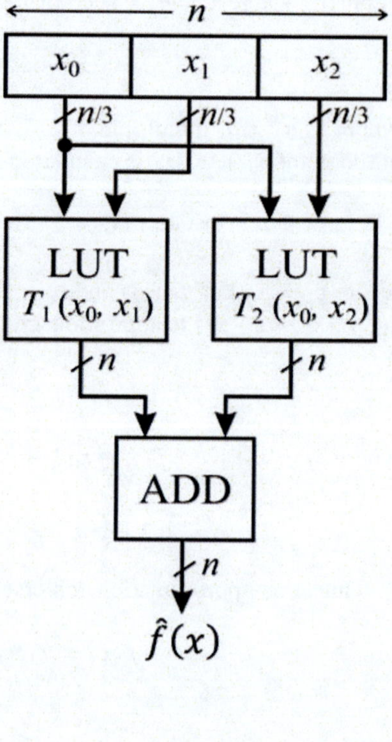

Fig. 8.13 Bipartite method block diagram

[11, 15]. Thus, compared with piecewise polynomial approximation methods, bipartite method does not require multipliers, thus producing a smaller delay. However, high-precision applications require large table sizes. In this way, many variations of the method were developed to increase the accuracy and decrease the size of the look-up tables. One of these variations is multipartite method that is introduced at the following.

8.6.1.2 Multipartite Methods

Let us now assume that the n-bit input number x is divided into $m = 2j + 1$ k-bit numbers, such that x is arranged as follows:

$$x = \sum_{i=0}^{2j} x_i 2^{-ik}$$

where x_i are multiples of 2^{-k}. As in bipartite approximation, Taylor series expansion can be used to obtain the following multipartite approximation:

$$\begin{aligned} f(x) &= T_1(x_0, x_1, \ldots, x_j) + T_2(x_0, x_1, \ldots, x_{j-1}, x_{j+1}) \\ &\quad + T_3(x_0, x_1, \ldots, x_{j-2}, x_{j+2}) + T_4(x_0, x_1, \ldots, x_{j-3}, x_{j+3}) \\ &\quad + \cdots + T_{j+1}(x_0, x_{2j}) + \epsilon \end{aligned}$$

where:

$$T_1 = f\left(\sum_{i=0}^{j} x_i 2^{-ik}\right)$$

$$T_2 = x_{j+1} 2^{-(j+1)k} f'\left(\sum_{i=0}^{j-1} x_i 2^{-ik}\right)$$

$$T_3 = x_{j+2} 2^{-(j+2)k} f'\left(\sum_{i=0}^{j-2} x_i 2^{-ik}\right)$$

$$\vdots$$

$$T_{j+1} = x_{2j} 2^{-2jk} f'(x_0)$$

$$\epsilon = \sum_{i=0}^{j} \varepsilon_i \le \left(2^{-(j+1)k-1} + j2^{-(2j+1)k}\right)\max\left(|f''|\right) \approx j2^{-(2j+1)k}\max\left(|f''|\right)$$

Thus, the function can be approximate by adding the functions T_i, whose values can be stored in separate tables. The block diagram for multipartite method for $j = 2$, this is tripartite method, is shown in Fig. 8.14.

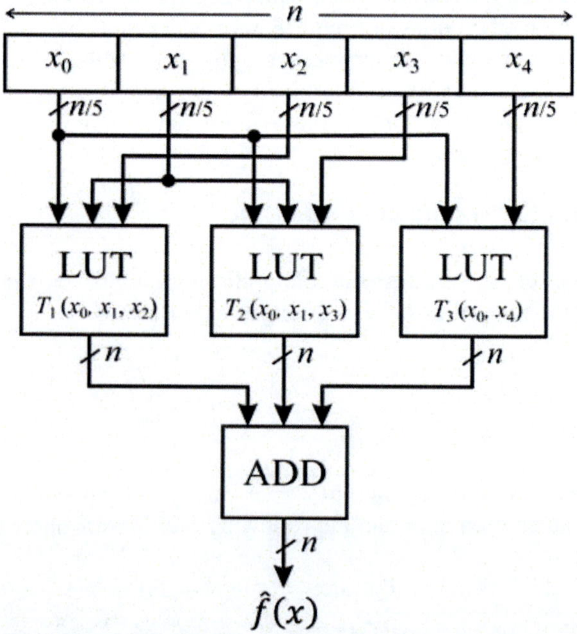

Fig. 8.14 Tripartite method block diagram

Example 8.17 **Calculate the total memory bits required for multipartite approximation of $1/x$ for $k = 3$ assuming 15-bit input, $1.i_1 \ldots i_{15}$, and 15-bit output, $0.z_1 \ldots z_{15}$. Compared it to bipartite approximation and to a single table approximation.**

As it was detailed in Example 8.16, the reciprocal function can be implemented using bipartite approximation in $2^{11} \times 15$ bits of memory and using single table approximation in $2^{15} \times 15$ bits of memory. Using multipartite approximation with $k = 3$, the data is split into five 3-bit numbers as it is shown in Fig. 8.14, requiring three tables T_1, T_2 and T_3 ($j = 2$). T_1 depends of x_0, x_1 and x_2, requiring of $2^{3\times3} \times 15 = 2^9 \times 15$ memory bits. T_2 depends of x_0, x_1 and x_3, thus also requires $2^9 \times 15$ memory bits. T_3 depends of two 3-bit input signal, x_0 and x_4, requiring $2^{2\times3} \times 15 = 2^6 \times 15$ memory bits. Thus, a total of $2 \times 2^9 \times 15 + 2^6 \times 15 = (2^{10} + 2^6) \times 15$ memory bits are required for tripartite method. Compared to bipartite approximation, tripartite method reduces the memory bits by a factor of 1.88, and, compared to single table approximation the factor reduction is 30.12, what suppose and important area reduction. ■

8.6.2 *Small Look-Up Tables Based Methods*

A typical table-based algorithm divides the interval where the function is to be approximated into subintervals, usually of equal size, and an approximation formula, for example, a piece-wise polynomial of certain degree, is applied within the corresponding subinterval. This division of the original wide interval in subintervals have a cost since it obliges using relatively large **LUTs** to store the subinterval polynomial coefficients. Let f be the function to be implemented and $[a, b]$ the domain of interest. Thus, the domain of the function contains a set of breakpoints $c_k, k = 1, 2, \ldots, N$ in $[a, b]$ and a table of N approximations T_k to $f(c_k)$. The size of the table depends on the target precision, the polynomial degree and the required accuracy. In order to maintain certain accuracy, if the size of the subinterval increases, the needed degree of the polynomial also increases. Thus, a higher latency is obtained but less memory is required, and thus shorter **LUTs**, since the number of subinterval decreases. On the other hand, if the size of the subinterval decreases, the needed degree also decreases yet at the expense of using more memory and thus larger **LUTs**. Thus, there exists a trade-off between latency and memory to achieve a result with the designated accuracy.

Different methods based on small **LUTs** have been used on general purpose systems, being the Tang's algorithms [23] very important proposals. These algorithms become easily realizable in hardware due to the table sizes are enough small. Three elementary steps were proposed by Tang as a guideline for the implementation of the special functions using look-up tables methods. Based on this guideline, for a given input $x \in [a, b]$, the algorithm consists on these three steps:

- *Reduction*: The algorithm selects an appropriate breakpoint c_k for this x, Then, a reduction transformation is applied:

$$r = \mathcal{R}(x, c_k)$$

  In general, c_k is the breakpoint closest to x, being the following a typical case:

$$\mathcal{R}(x, c_k) = x - c_k.$$

- *Approximation*: After the reduction step, the algorithm calculate the value of the function for r, $f(r)$, using some approximation formula, that can be for example a polynomial:

$$f(r) \approx p(r)$$

- *Reconstruction*: The algorithm calculates $f(x)$ by a reconstruction formula S. This formula depends on the reduction transformation $\mathcal{R}$, the values and $f(c_k)$ and $f(r)$, and the nature of f:

$$f(x) = S(f(c_k), f(r)) \approx S(f(c_k), p(r)) \approx S(T_k, p(r))$$

Tang's algorithm gives guidelines for several functions as the exponential, logarithm and trigonometric functions. The algorithm for the exponential function is detailed at Appendix E. For this algorithm x is considered to be composed of:

$$x = (m \ln 2) + \frac{j \ln 2}{32} + (r)$$

where m and j are integers, r is a real number in the interval $\left[-\frac{\ln 2}{64}, \frac{\ln 2}{64}\right]$, and $r = R_1 + R_2$. Thus, the exponential function can be approximated as:

$$e^x \approx 2^m \cdot 2^{\frac{j}{32}} \cdot (p(r) + 1)$$

where $p(r)$ is a polynomial such that e^r can be approximated by to $p(r) + 1$. Considering $s_j = 2^{\frac{j}{32}}$, with $j = 0, 1, 2, \ldots, 31$, and $s_j = s_{j1} + s_{j2}$, the computation of the reconstruction for the exponential function can be expressed as (see Appendix E.3):

$$e^x \approx 2^m \cdot \left(s_{j1} + \left(s j_2 + s_j \cdot p(r)\right)\right)$$

This implementation of this algorithm requires:

- The computation of the parameters m, j, R_1 and R_2. These parameters depends on the input argument x:

$$m = N_1/32$$

$$j = N_2$$

$$R_2 = -N \cdot L_2$$

If $N < 2^9$:

$$R_1 = x - N \cdot L_1$$

Else:

$$R_1 = (x - N \cdot L_1) - N_2 \cdot L_1$$

where L_1 and L_2 are constants shown in Table E.2, and:

$$N = \text{round}(x \cdot Inv\_L)$$

$$N_2 = N \bmod 32$$

$$N_1 = N - N_2$$

where $Inv\_L = \frac{32}{\ln 2}$ can be precomputed and preestored.

Figures 8.15 and 8.16 display the blocks that implements the computation of m, j, R_1 and R_2.

- The implementation of the approximation polynomial. The polynomial $p(r)$ is computed as follows:

$$p(r) = r + a_1 r^2 + a_2 r^3 + \cdots$$

The polynomial coefficient a_i can be obtained from an approximation algorithm, as the studied in Sect. 8.5. Table E.2 shows the coefficients that Tang used in its

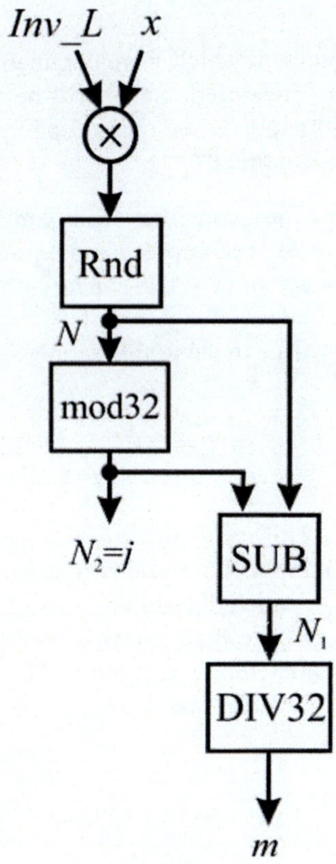

Fig. 8.15 Block diagram for the computation of j and m

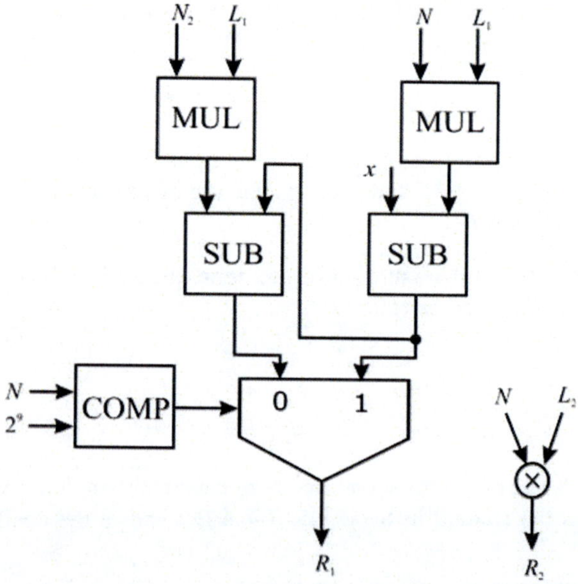

Fig. 8.16 Block diagram for R_1 and R_2

method, which were obtained using Remez algorithm. Thus, these coefficients can be pre-stored, and it will be necessary to include the polynomial implementation structure, based on multipliers and adders, as it is shown in Fig. 8.17 for a 3-degree polynomial.

- The reconstruction implementation. The values $s_j = 2^{\frac{j}{32}}$, with $j = 0, 1, 2, \ldots, 31$ are precomputed and represented by the sum of two working-precision numbers s_{j1} and s_{j2}, that can be stored in look-up tables. Table E.3 shows the constants s_{j1} and s_{j2} for $j = 0, 1, 2, \ldots, 31$. Thus, the computation of the reconstruction for the exponential function can be expressed as (see Appendix E.3):

$$e^x \approx 2^m \cdot \left(s_{j1} + \left(s_{j2} + s_j \cdot p(r)\right)\right)$$

This reconstruction requires the power of two which depends on the value m, two multiplications and two additions, in addition to the two look-up tables for s_{j1} and s_{j2}, which depend on j.

Figure 8.18 shows a block diagram for the implementation of complete exponential Tang's algorithm. This implementation requires the computations shown in Figs. 8.15, 8.16 and 8.17.

Fig. 8.17 Block diagram for polynomial implementation

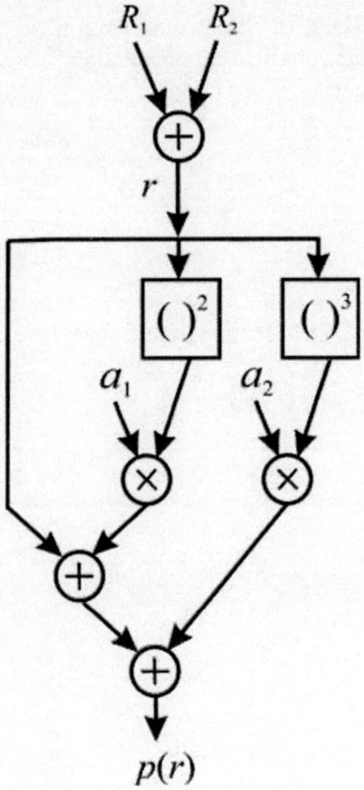

8.6.3 Table-Based Balanced Methods

Among the table-based methods for function approximation, there are methods that achieve a good compromise between table sizes and computation resources. It is possible due to the combination of medium size look-up tables and a reduced computation part. It makes table-based balanced methods appropriate for single-precision computations, reaching fast execution times and admissible hardware requirements. On the other hand, according to the degree of the polynomial approximation, it may be possible to subdivide these methods into linear approximations and second-degree interpolation methods.

Piecewise polynomial approximation is a method in the table-based balanced group. It is based on the division of the interval in which the target function has to be calculated into subintervals such that the special function is approximated in each subinterval with a low-degree polynomial. In addition, a look-up table stores the polynomial coefficients for each subinterval. Non uniform segmentation is very suitable for the approximation of highly nonlinear functions, which it is based on the use of segments on non uniform length. However, non uniform segmentation needs additional resources for the segment index encoder, and thus uniform segmentation may be more adequate for the implementation of special functions. Both methods are analyzed in more detail at the following.

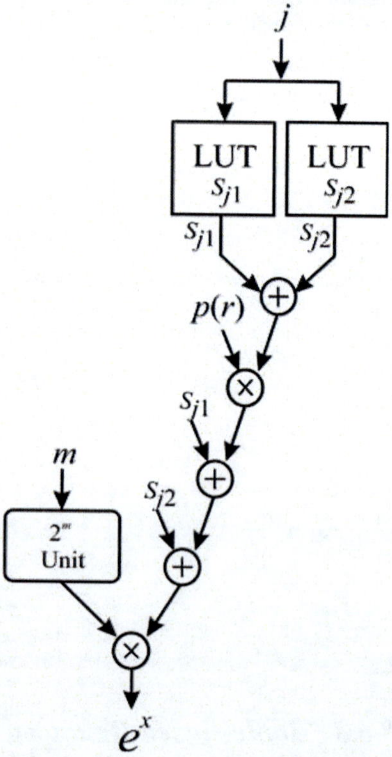

Fig. 8.18 Block diagram for the complete exponentiation unit

8.6.3.1 Piecewise Polynomial Approximation

In order to provide accurate polynomial approximation to an especial function in a rather large interval, a large-degree polynomial is required, as it was analyzed in Sect. 8.3. This implies the performing of many arithmetic operations which requires an increment into the computational delay and may also incurs error problems as round-off error propagation. This is an important drawback that can be overcome by using intermediate tables and low order polynomials, as it is made in piecewise polynomial approximation methods. Thus, given a function $f(x)$ defined on an interval $[a, b]$, piecewise-polynomial function approximation considers n dividing points x_i with $i = 0, \ldots, n-1$ such that:

$$a = x_0 < x_1 < \cdots x_{n-1} < b$$

and thus divides the interval $[a, b]$ into $M = n-1$ segments $\left[x_k, x_{k+1}\right]$, $k = 0, \ldots, n-2$. Uniform and non-uniform segmentation can be distinguished. For uniform segmentation, the number of segments M is a power of 2, $M = 2^m$, and the segment width is $(b-a)/2^{cm}$, simplifying thus the hardware implementation. Then, the function is approximated in each segment with a low-degree polynomial, denominated N-degree (mostly degree-1 to 3 are used):

$$f(x) \approx \hat{f}(x) = \sum_{i=0}^{N} c_{k,i}(x - x_k)^i$$

For:

$$x \in [x_k, x_{k+1})$$

For each segment, a **LUT** stores the polynomial coefficients. An important aspect to be considered is to minimize the **LUT** size but respecting the error constraint. It can be considered by minimizing the bit width of the coefficients. Thereby, the value of M is obtained as the minimum of the all the possible values that get the required precision. To reach this minimum number of segments will allow higher area, speed, and power efficiency. Three different sources of errors can be considered in polynomial approximations [20]:

- e_{∞}: this is an inherent error since the polynomial and the objective function do not match perfectly
- e_Q: this is the quantization error generated by finite-precision effects encountered in the evaluation of the polynomial. Quantization error can be subdivided in two types of errors. On one hand, the error due to polynomial coefficient quantization, e_{Qcoeff}, and on the other hand, the error due to truncation/rounding/approximation carried out in the arithmetic architecture used for polynomial evaluation, e_{Qarith}.
- e_{round}: this is the error that is produced by the final output rounding. It can generate a maximum error of 0.5 ulp.

The total error e_{total} can be expressed as follows:

$$e_{total} \leq e_{\infty} + e_Q + e_{round}$$

where it was considered that the maximum error is obtained if all the error sources contribute additively. Taking into account that $e_{round} \leq$ 0.5 ulp, to ensure faithful rounding, the condition $e_{\infty} + e_Q <$ 0.5 ulp must be met.

The number of segments are related to the approximation accuracy and also to the required table size, so that, this table grows exponentially with the precision. In this way, the maximum approximation error considering all the segments is the parameter that stablish the number of uniform segments. Nevertheless, there exists a high non-linearity in many arithmetic functions, and thus uniform segmentation method is not efficient for these functions. In this case, several non-uniform segmentation methods are recommended [18].

To understand better the problem of the non-linearity and uniform segments [10], let consider the function $\sqrt{-\ln(x)}$. As shown in Fig. 8.19, the regions close to the extremes present the greatest non-linearities of this function. In case of uniform segments are used, and in order to get accurate approximations in the non-linear

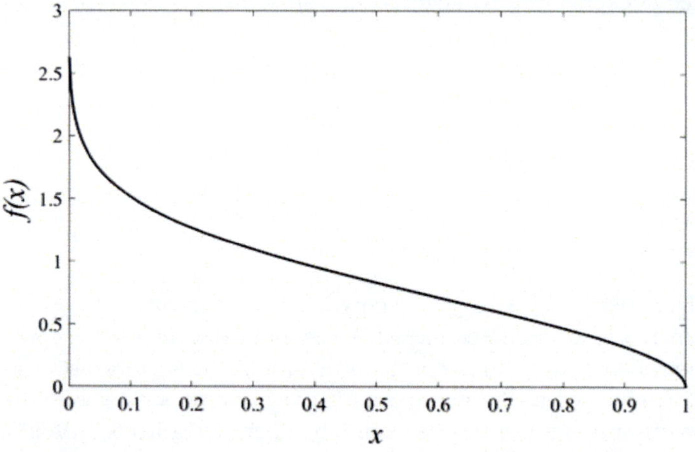

Fig. 8.19 $\sqrt{-\ln(x)}$ over (0, 1]

regions, a large number of small segments are necessary. On the other hand, there is a zone in Fig. 8.19 where the curve is linear, approximately in the middle part of the curve, and it is possible to get accurate approximation using relatively a little number of segments. This analysis shows that using large segments for linear regions and small segments for the non-linear ones it should be possible to have the least error. Nevertheless, due to there is no a fix size address, for a given input, it is necessary to calculate the segment address and the required hardware for this computation can be complex. Thus, it is essential to have into account this fact in order to find an efficient non-uniform method.

Different architectures and scheduling strategies have been proposed for hardware implementation of piecewise-polynomial approximations [15, 21]. Figure 8.20 illustrates the direct architecture for a third-order approximation [21]. The m most significant bits of x are used to address the look-up table, and the remaining bits correspond to the term $(x - a_k)$. On the other hand, the method proposed in [15] is an example of direct architecture of degree-2 piecewise polynomials, which, due its relevance, is studied in more detail at the following.

8.6.3.2 Pineiro's Method

The Pineiro's algorithm is based on the three main blocks or units: look-up tables, an enhanced unit to calculate a minimax quadratic approximation, and a block to obtain an efficient approximation to the second-degree polynomial. The input argument x of width w_i is divided into two sub-words an upper part, X_1, and a lower part, X_2:

$$X_1 = [x_1 x_2 \dots x_m]$$

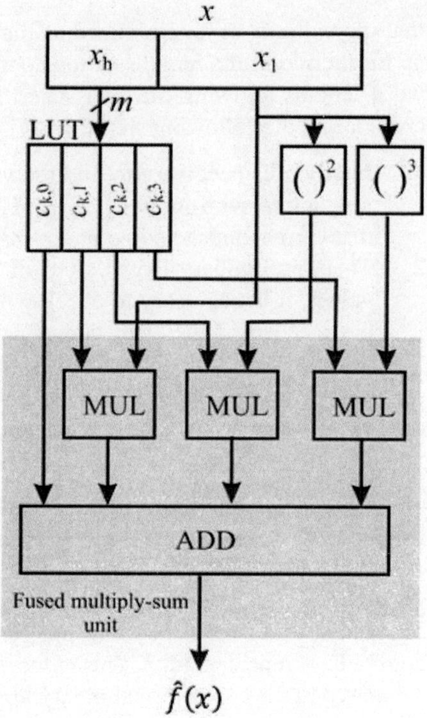

Fig. 8.20 Direct architecture for third-order piecewise-polynomial approximation

$$X_2 = \left[x_{m+1}x_{m+2}\ldots x_{w_i}\right]$$

$$x = X_1 + X_2 \times 2^{-m}$$

The 2-degree approximation piecewise polynomial $P_m(x)$ of $f(x)$ in the range $X_1 \leq x \leq X_1 + 2^{-m}$ can be computed as follows:

$$f(x) \approx P_m(x) = c_1(X_1) + c_2(X_1)X_2 + c_3(X_1)X_2^2$$

with

$$X_2 \in \left[X_1, X_1 + 2^{-m}\right]$$

There are $2m$ sub-intervals and each one has a second degree minimax polynomial coefficients c_1, c_2 and c_3 with a coefficient length w_{c_1}, w_{c_2} and w_{c_3}, respectively. As it can be observed, the values of the coefficients c_1, c_2 and c_3 for each function depend only on X_1, the m most significant bits of x, and they can be stored in look-up tables of 2^m inputs. The exact value of m depends on the target function and on the desirable precision. The method uses the enhanced minimax approximation to ensure that the value of m is minimized as the size of the **LUT** grows exponentially with m. Furthermore, the lengths of the coefficients should be minimized to reasonable word lengths allowing the designated accuracy. The method can be thus applied considering the following steps:

1. Firstly, it is necessary to find the value of m that gives a preliminary minimax polynomial with unrestricted word lengths for the three coefficients c_1, c_2 and c_3 that corresponds to some preliminary designated approximation error.
2. Then, preliminary large values of w_{c_1}, w_{c_2} and w_{c_3} have to be set. For these values, it is necessary to check whether the bound of the approximation error is exceeded; if the error budget is still smaller than the approximation error, decrease w_{c_1}, w_{c_2} and w_{c_3} until the error budget meets the approximation error; otherwise, increment m and go to step 2.
3. Finally, the best rounding constant that can be used as a bias to minimize the overall error have to be found. This can be done by an exhaustive simulation.

As Fig. 8.21 shows, the architecture of this method includes a squaring block to generate X_2^2. A truncation is applied to the squarer as X_2^2 is guaranteed to be less than 2^{-2m} and thus there exists some room to neglect many of the least significant bits without sacrificing the designated accuracy. The terms $c_3X_2^2$, c_2X_2 and c_1 have to be added. The first two components are generated as partial products using the signed-digit radix-4 representation. This reduces the number of partial products to about half compared to the conventional binary approach and subsequently a faster addition is achieved. The third term is added to the function specific rounding constant. A fused

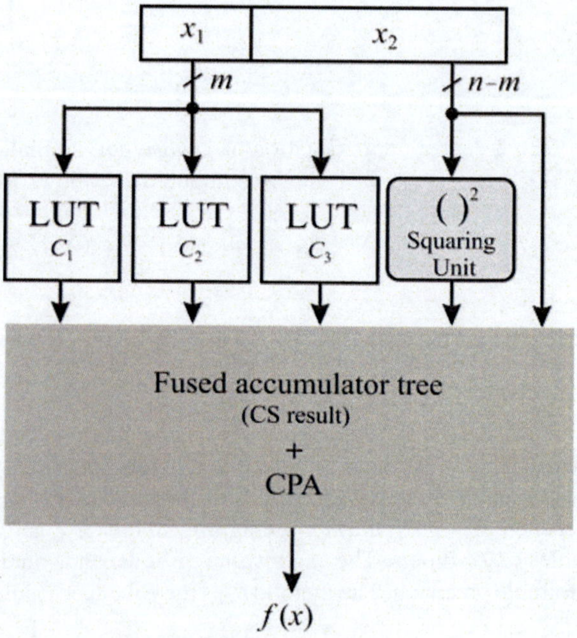

Fig. 8.21 Block diagram of the Piñero's method

accumulation tree is used to compress the partial products using 3:2 and 4:2 carry-save adders (**CSAs**) into a two conventional words that are fed to a fast carry-propagation adder (**CPA**).

Pineiro's method is the most used for single precision elementary functions for GPUs and general-purpose processors. A related work in [21] uses also degree-2 minimax polynomials yet reduces the size of **LUTs** by around 40% by imposing same c_1 for two adjacent subintervals. In [11], degree-1 and 2 interpolations are investigated and compared to their corresponding approximation minimax polynomials for precisions up to single precision.

8.6.3.3 Constrained Piecewise-Polynomial Approximations

The constrained piecewise-polynomial approximation methods group two adjacent segments in a segment pair. Considering the same notation used in piecewise polynomial approximation, the first segment pair is composed by the segments $[x_0, x_1]$, $[x_1, x_2]$; the second segment pair is composed by the segments $[x_2, x_3]$, $[x_3, x_4]$, and so on. The function $f(x)$ is approximated with a polynomial for each segment of the segment pair:

- For the left segment: $\hat{f}(x) = \sum_{i=0}^{N} p_{k,i}\left(x - x_k^*\right)^i$
- For the right segment: $\hat{f}(x) = \sum_{i=0}^{N} q_{k,i}\left(x - x_k^*\right)^i$

Being, x_k^* the midpoint of a segment pair. In order to minimize the **LUT** size required for polynomial coefficients, constraints are then applied to this approximating function. For instance, the following constraints can be considered:

- The continuity of the approximating function for midpoint of each segment pair can be imposed; substituting x by x_k^* in the previous expressions and equaling the two polynomials:

$$p_{k,0} = q_{k,0}$$

This constraint gets an important memory reduction since, for each segment pair, only a single constant coefficient has to be stored.

- The continuity of the jth derivative of the approximating function for de midpoint can be enforced. Substituting x by x_k^* in the derivate of (8.14) and (8.15) and equaling the derivate:

$$p_{k,j} = j$$

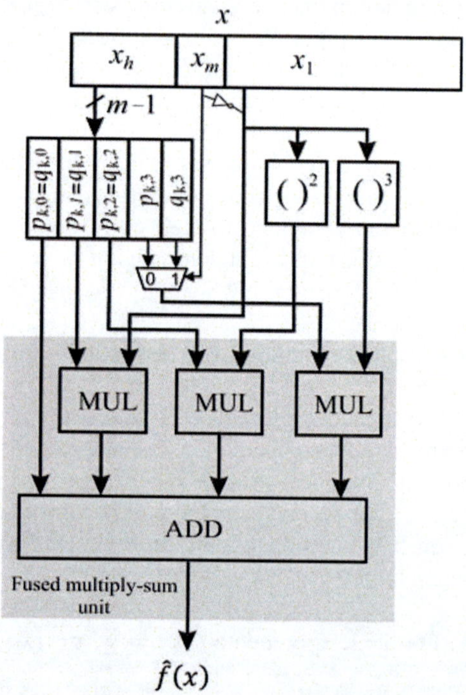

Fig. 8.22 Direct architecture for third-order constrained piecewise-polynomial approximation

Again, for this constraint it is only necessary to store a single constant coefficient instead of two.

In order to further reduce the memory resources, it is possible to simultaneously impose two or more additional constraints to two different j values. As example, Fig. 8.22 shows an implementation structure of this method [21] where additional constraints were considered, concretely, the approximating function and its first and second derivatives are continuous in the midpoint of the segment pair. The coefficients for each segment pair were thus reduced from eight to five.

8.7 Conclusion

This chapter has presented different algorithms for the implementation of special functions. The fundamentals of the **CORDIC** algorithm has been detailed, which is one of the first algorithms developed to calculate special functions and that has been widely used since its introduction. The main types of algorithms for the computation of special functions have been studied, distinguishing among algorithms that only use additions and shifts, algorithms based on polynomial approximations and algorithms using table-based methods. Hardware structures for these algorithms are also presented in order to compare these methods in terms of resources and throughput.

8.8 Exercises

8.1. Develop the angle $\theta = 12° \, 23' \, 47''$ as addition/subtraction of the angles α_i, to $i = 12$.

8.2. Calculate sin45° and cos45° using the **CORDIC** algorithm and compare both results, as well as with the correct values. Evaluate the relative error according to the number of iterations.

8.3. Determine which trigonometric angle corresponds to the point (6, −2), using the **CORDIC** algorithm.

8.4. Develop the hyperbolic angle 0.43 as addition/subtraction of the angles β_i, until $i = 12$.

8.5. Transfer the hyperbolic angle 6.28 to the usable range.

8.6. Calculate *sh*2.23 and *ch*2.23 using the **CORDIC** algorithm and compare the results with the values consulted in tables. Evaluate the relative error according to the number of iterations.

8.7. Determine, using the **CORDIC** algorithm, which hyperbolic angle corresponds to the point (5.34, 2.18).

8.8. Calculate, using the **CORDIC** algorithm, $\sqrt{24}$ and ln(8/3).

8.9. Calculate, using the **CORDIC** algorithm, 0.37×87.65.

8.10. Calculate, using the **CORDIC** algorithm, 0.065:48.03.

8.11. Use the Newton-Raphson method, with 2 as initial value, to find an approximation of $\sqrt{5}$ correct to 8 decimal places.

8.12. Obtain an approximation to $\sqrt{5}$ using the Newton-Raphson iteration for $1/\sqrt{5}$. Consider the initial value 0.5, and indicates the reached precision for the same number of iterations required in Exercise 8.11.

8.13. Obtain the Newton-Raphson approximation to 1/7 considering 0.2 as initial values and a precision of 5 decimal places.

8.14. Using minimax approximation method, approximate the function ln x by degree-2.

8.15. Using least square method, approximate the function ln(x + 2) by degree-2 Chebyshev polynomials and compare this approximation with the degree-2 Legendre approximation.

8.16. Calculate the value of ln(3.2457) using the shift and add algorithm of Sect. 8.6.1 and calculate the error using $n = 1$ to $n = 20$.

8.17. Obtain the expression of the functions T_1 and T_2 for bipartite approximation of $1/x$.

8.18. Obtain the expression of the functions T_1, T_2 and T_3 for tripartite approximation of $1/x$.

8.19. Using Tang's algorithm, approximate the function ln(x) for $x \in [1, 2]$.

References

1. Briggs, H.: Arithmetica Logarithmica. London (1624) (trans.: [1] Archived 4 Mar 2016 at the Wayback Machine)
2. Castillo, E., Lloris, A., Morales, D.P., Parrilla, L., García, A., Botella, G.: A new area-efficient BCD-digit multiplier. Digit. Signal Proc. **62**, 1–10 (2017)
3. Changela, A., Zaveri, M., Lakhlani, A.: ASIC implementation of high performance Radix-8 CORDIC algorithm. In: 2018 International Conference on Advances in Computing, Communications and Informatics (ICACCI), Bangalore, 2018, pp. 699–705
4. Detrey, J., De Dinechin, F.: Table-based polynomials for fast hardware function evaluation. In: Proceedings of 16th International Conference on Application Specific Systems, Architecture and Processors, 2005
5. de Dinechin, F., Tisserand, A.: Multipartite table methods. IEEE Trans. Comput. **54**(3), 319–330 (2005)
6. Garai, S.K.: An all-optical frequency encoded BCD data addition. Opt. Quantum Electron. 48–49 (2016)
7. Garrido, M., Källström, P., Kumm, M., Gustafsson, O.: CORDIC II: a new improved CORDIC algorithm. IEEE Trans. Circuits Syst. II Express Briefs **63**(2), 186–190 (2016)
8. Iordache, C., Matula, D.W.: Analysis of reciprocal and square root reciprocal instructions in the AMD K6-2 implementation of 3DNow! Electron. Notes Theor. Comput. Sci. **24** (1999)
9. Kim, S., Kwon, J., Kim, S., Lee, B.: Multiplexed strain sensor using fiber gratin-tuned fiber laser with a semiconductor optical amplifier. IEEE Photonics Technol. Lett. **13**(4), 350–351 (2001)
10. Lee, D.U., Luk, W., Villasenor, J., Cheung, P.Y.K.: Non-uniform segmentation for hardware function evaluation. In: Cheung, P.Y.K., Constantinides, G.A. (eds.) Field Programmable Logic and Application. FPL 2003. Lecture Notes in Computer Science, vol. 2778. Springer, Berlin, Heidelberg (2003)
11. Lee, D., Villasenor, J.D.: Optimized custom precision function evaluation for embedded processors. IEEE Trans. Comput. **58**(1), 46–59 (2009)
12. Meher, P.K., Valls, J., Juang, T., Sridharan, K., Maharatna, K.: 50 years of CORDIC: algorithms, architectures, and applications. IEEE Trans. Circuits Syst. I Regul. Pap. **56**(9), 1893–1907 (2009)
13. Muller, J.M.: A few results on table-based methods. Reliab. Comput. **5**(3), 279–288 (1999)
14. Muller, J.M.: Elementary Functions: Algorithms and Implementations, 2nd edn. Birkhauser (2006)
15. Pineiro, J.A., Muller, J.M., Bruguera, J.D.: High-speed function approximation using a minimax quadratic interpolator. IEEE Trans. Comput. **54**(3), 304–318 (2005)
16. Sanchez-Romero, J.L., Jimeno-Morenilla, A., Molina-Carmona, R., Perez-Martinez, J.: An approach to the application of shift-and-add algorithms on engineering and industrial processes. Math. Comput. Model. **57**(7–8), 1800–1806 (2013)
17. Sarma, D.D., Matula, D.W.: Faithful bipartite ROM reciprocal tables. In: Proceedings of the 12th Symposium on Computer Arithmetic, pp. 17–29 (1995)
18. Sasao, T., Nagayama, S., Butler, J.T.: Numerical function generators using LUT cascades. IEEE Trans. Comput. **56**(6), 826–838 (2007)
19. Schulte, M.J., Stine, J.E.: Approximating elementary functions with symmetric bipartite tables. IEEE Trans. Comput. **48**(8), 842–847 (1999)
20. Strollo, A.G.M., De Caro, D., Petra, N., Napoli, E., Garofalo, V.: Constrained piecewise polynomial approximation for hardware implementation of elementary functions. In: 15th IEEE International Conference on Electronics, Circuits and Systems, St. Julien's, pp. 698–701 (2008)
21. Strollo, A.G.M., De Caro, D., Petra, N.: Elementary functions hardware implementation using constrained piecewise-polynomial approximations. IEEE Trans. Comput. **60**(3), 418–432 (2011)
22. Tang, P.T.P.: Table-lookup algorithms for elementary functions and their error analysis. In: Proceedings of 10th IEEE Symposium on Computer Arithmetic, pp. 232–236 (1991)

23. Tang, P.T.P.: Table driven implementation of the Expml function in IEEE floating point arithmetic. ACM Trans. Math. Softw. **18**(2), 211–222 (1992)
24. Volder, J.E.: The CORDIC trigonometric computing technique. In: IRE Trans. Electron. Comput. **EC-8**(3), 330–334. Reproducido en Swartzlander, E.E.: Computer Arithmetic, vol. 1. IEEE Comp. Soc., pp. 226–230 (1990)
25. Walther, J.S.: The story of unified CORDIC. J. VLSI Signal Process. **25**(2), 107–112 (2000)
26. Walther, J.S.: A unified algorithm for elementary functions. In: Spring Joint Computer Conference, 1971. Proceedings, pp. 379–385. Reproducido en Swartzlander, E.E.: Computer Arithmetic, vol. 1. IEEE Comp. Soc., pp. 272–278 (1990)

Chapter 9
Basic Algebraic Circuits

The basic algebraic circuits, linear feedback shift registers and cellular automata, are studied in this Chapter. These circuits present a temporal evolution that can be described through some algebraic structures: concretely, a polynomial can be associated to each circuit, so the properties of the polynomial allow anticipating the temporal evolution of the circuit.

The core of the Chapter consists of four sections. Binary **LFSR**s, or **LFSRmod2**, are studied in the first section and are the basic blocks in the construction of many modulo 2 algebraic circuits; **LFSRmod2^n** are also described as an extension of **LFSRmod2**. In the second section, the **LFSRmod2** are generalized by defining the **LFSRmodp**, with a similar behavior of the previous, for any prime integer ***p***. Several circuits, which profusely make use of **LFSRs**, for adding, subtracting, multiplying and dividing polynomials are presented in the third section; also, the consideration of multiplier and divisor as filters is introduced. Finally, **CA**s are studied in the fourth section; those are blocks similar to **LFSR**s, with many applications.

Appendices A and B are related to the contents of this Chapter; they include the basic concepts on finite, or Galois, fields and on the polynomial algebra, respectively. These are previous concepts required for the development of the following. Thus, it is convenient to review these Appendices and get back to them whenever it may be necessary.

9.1 LFSR

Binary **LFSR**s are introduced in this Section, which are the main building block for designing algebraic circuits. When there is no room for confusion, these will be simply referred to as **LFSR**, but in a more general context they will be called **LFSRmod2**, as it will be justified below. A polynomial will be associated to each

A. Lloris Ruiz et al., *Arithmetic and Algebraic Circuits*, Intelligent Systems Reference Library 201, https://doi.org/10.1007/978-3-030-67266-9_9

LFSR, and the properties of the **LFSR** will be related to those of the associated polynomial. There are two types of **LFSRs**, which justifies the structure of this section.

9.1.1 Type 1 LFSR

An **LFSR** is a shift register with linear feedback, as its acronym reads **L**inear **F**eedback **S**hift **R**egister. In this way, the feedback is a linear combination of the contents of the register and, thus, the structure of a length-*n* **LFSR (nLFSR)** is that in Fig. 9.1. This register stores *n* bits, b_0, ..., b_{n-1}, and has *n* feedback coefficients, a_0, ..., a_{n-1}, which can be 0 (this term is not included in the feedback) or 1 (the term is included in the feedback). The usual is to assume $a_0 = 1$, as it will be done in the following. Thus, an **nLFSR** consists of *n* **D** flip-flops, for building the shift register, and **EXOR** gates, for generating the linear feedback. In fact, the **LFSR** in Fig. 9.1 is known as type 1 **LFSR**, since in subsection 9.1.4 an alternate structure with **D** flip-flops and **EXOR** gates will be described, with an almost identical behavior to this type 1 **LFSR (LFSR1** for brevity) and that will be called type 2 **LFSR (LFSR2)**. As an extension, the shift register used for building the **LFSR1** is also known as type 1 shift register.

The **LFSR1** in Fig. 9.1 works autonomously, so it is often referred to as **ALFSR** (Autonomous **LFSR**). Its behavior, i.e., its temporal evolution, when working autonomously is described by the following equations:

$$b_j(t+1) = b_{j+1}(t), \quad j = 0, \ \ldots, \ n-2$$
$$b_{n-1}(t+1) = \sum_{i=0}^{n-1} a_i b_i(t) \tag{9.1}$$

that can be expressed in matrix forms as:

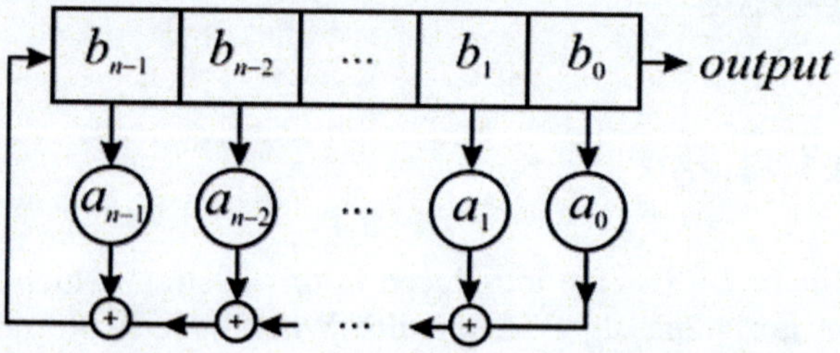

Fig. 9.1 **LFSR1** structure

$$\begin{bmatrix} b_0(t+1) \\ b_1(t+1) \\ \cdots \\ b_{n-2}(t+1) \\ b_{n-1}(t+1) \end{bmatrix} = \begin{bmatrix} 0 & 1 & 0 & \dots & 0 & 0 & 0 \\ 0 & 0 & 1 & \dots & 0 & 0 & 0 \\ \dots & \dots & \dots & \dots & \dots & \dots & \dots \\ 0 & 0 & 0 & \dots & 0 & 0 & 1 \\ a_0 & a_1 & a_2 & \dots & a_{n-3} & a_{n-2} & a_{n-1} \end{bmatrix} \begin{bmatrix} b_0(t) \\ b_1(t) \\ \cdots \\ b_{n-2}(t) \\ b_{n-1}(t) \end{bmatrix}$$

or:

$$B(t+1) = T1B(t)$$

where $T1$ is an $n \times n$ square matrix whose elements are 0 or 1:

$$T1 = \begin{bmatrix} 0 & 1 & 0 & \dots & 0 & 0 & 0 \\ 0 & 0 & 1 & \dots & 0 & 0 & 0 \\ \dots & \dots & \dots & \dots & \dots & \dots & \dots \\ 0 & 0 & 0 & \dots & 0 & 0 & 1 \\ a_0 & a_1 & a_2 & \dots & a_{n-3} & a_{n-2} & a_{n-1} \end{bmatrix} \tag{9.2}$$

In the **LFSR1** of Fig. 9.1 its output will present a sequence of zeros and ones, b_0, b_1, ..., b_{n-1}, b_n, ..., b_{r-1}, b_r, ..., that is, obviously, determined by the feedback coefficients a_j and by the initial content of the register (which is sometimes known as the **seed**), since it is an autonomous system whose only external input is the clock signal that synchronizes its functioning and induces the changes of state. It is immediate to check that the output sequence is given by the following recurrence, obtained from Eq. (9.1):

$$b_{j+n} = \sum_{i=0}^{n-1} a_i b_{j+i}$$

Thus, known n consecutive bits of the sequence, the $n + 1$ bit is determined. Or equivalently, the content of the register in a given instant determines all of the following bits of the sequence. In any case, since an **LFSR** is a synchronous sequential system with n memory elements and can have a maximum of 2^n states, and since it is also autonomous, the output sequence will always be periodical. Strictly, for any autonomous synchronous sequential system, given any initial state, the output sequence will always be cyclical after a certain number of initial transitions, which can be referred as *transist states*, when the system gets to an state that is part of a cycle. In the case of **LFSRs**, due to the type of feedback, all the states are part of a cycle, so there are no transit states. The period of the resulting sequence is known as the **sequence length**. Each state of this synchronous sequential system can be characterized by the contents of the **LFSR**, and the transition between states will be cyclical.

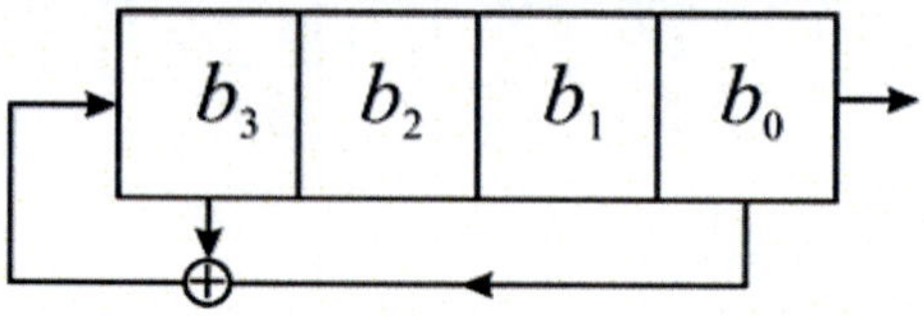

Fig. 9.2 **LFSR1** of Example 9.1

Every **LFSR** has a particular state whose behavior is perfectly predictable: if all the flip-flops contain the 0 value (0…0 state, or zero state), the system will keep indefinitely that state, since the term $\sum_{i=0}^{n-1} a_i b_j(t)$ in (9.1) will always be zero. Thus, this seed generates a sequence of length 1. On the other hand, if the number of non-zero coefficients a_i in the term $\sum_{i=0}^{n-1} a_i b_j(t)$ is odd, the state 1…1 (all flip-flops containing the value 1) also generates a sequence of length 1, since obviously the term $\sum_{i=0}^{n-1} a_i b_j(t)$ will always be 1 and the system will not get out of that state. Any other initial state has a periodic output sequence, with period $p > 1$. From all the above, it is evident that for any **nLFSR** it is $p \leq 2^n - 1$, since at least the zero state has to be discounted from the maximum of 2^n states. In the following there are several illustrative examples.

Example 9.1 The **LFSR1** of Fig. 9.2 has the following transition matrix:

$$T_A = \begin{bmatrix} 0100 \\ 0010 \\ 0001 \\ 1001 \end{bmatrix} \tag{9.3}$$

The initial state 1000 generates the output sequence 000111101011001, as it is illustrated in the b_0 column of Table 9.1, which shows the ordered states of the **LFSR** during its evolution. It is a length 15 sequence.

Since the **LFSR** has $n = 4$ flip-flops, the length of the generated sequence is the maximum. If the initial state is 0111, it is immediate to check that the generated output sequence is 111010110010001. It is obvious that, but for the four clock cycle shift in time, these two sequences coincide and will be thus considered as the same sequence. In this way, this **LFSR** can generate two output sequences, one of them all zeros (length 1) with seed 0000, and another one of maximum length ($2^n - 1$) with any other seed. The sum of the lengths of both sequences is 2^n. The 16 states of this **LFSR1** are structured in two cycles, one for the zero state and another for the remaining 15 states.

■

Example 9.2 The **LFSR1** of Fig. 9.3a has the following transition matrix:

Table 9.1 States of the **LFSR1** of Example 9.1

State	b_3	b_2	b_1	b_0	$p_i(3)$	$b_3 \oplus b_0$
8	1	0	0	0	−1	1
12	1	1	0	0	−1	1
14	1	1	1	0	−1	1
15	1	1	1	1	1	0
7	0	1	1	1	−1	1
11	1	0	1	1	1	0
5	0	1	0	1	−1	1
10	1	0	1	0	−1	1
13	1	1	0	1	1	0
6	0	1	1	0	1	0
3	0	0	1	1	1	0
9	1	0	0	1	1	0
4	0	1	0	0	1	0
2	0	0	1	0	1	0
1	0	0	0	1	−1	1
8	1	0	0	0	−1	1

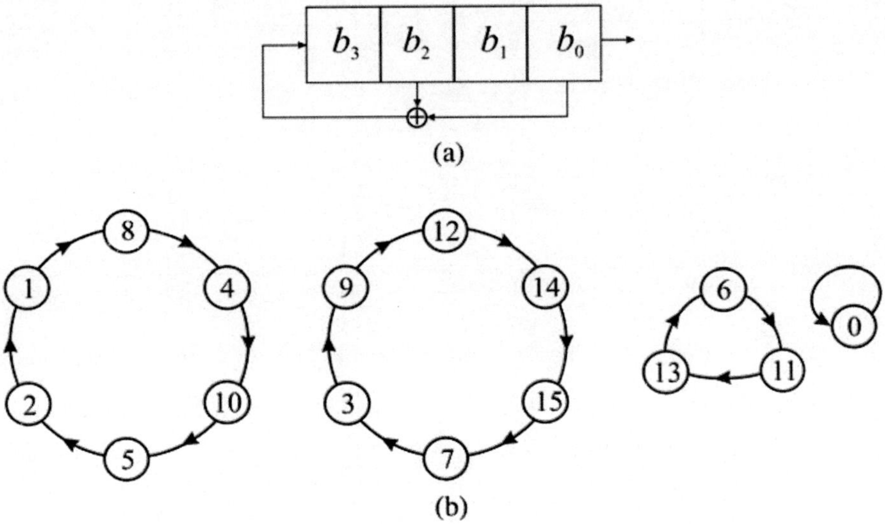

Fig. 9.3 **a LFSR1** of Example 9.2; **b** state diagram of Example 9.2

$$T_B = \begin{bmatrix} 0 & 1 & 0 & 0 \\ 0 & 0 & 1 & 0 \\ 0 & 0 & 0 & 1 \\ 1 & 0 & 1 & 0 \end{bmatrix} \tag{9.4}$$

The initial state 1000 generates the output sequence 000101, as it is illustrated in the b_0 column of Table 9.2. This is a sequence of length 6, so it is not of maximum length. If the initial state is 1100, the generated sequence is 001111, once more of length 6 (*v.* b_0 column of Table 9.2). Finally, the sequence 011, of length 3, is generated when the initial state is 0110. Moreover, the length 1 all-zero sequence is also generated. The sum of the lengths of all the generated sequences is again 2^n. The state diagram of this **LFSR1** is shown in Fig. 9.3b, with four cycles.

■

Example 9.3 The **LFSR1** of Fig. 9.4 has the following transition matrix:

Table 9.2 States of the **LFSR1** of Example 9.2

State	b_3	b_2	b_1	b_0
8	1	0	0	0
4	0	1	0	0
10	1	0	1	0
5	0	1	0	1
2	0	0	1	0
1	0	0	0	1
12	1	1	0	0
14	1	1	1	0
15	1	1	1	1
7	0	1	1	1
3	0	0	1	1
9	1	0	0	1
6	0	1	1	0
11	1	0	1	1
13	1	1	0	1

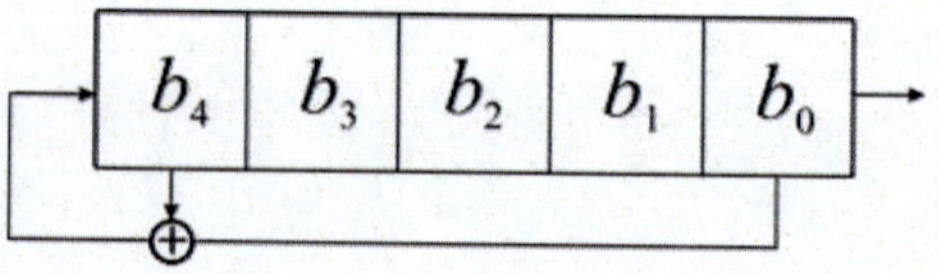

Fig. 9.4 **LFSR1** of Example 9.3

$$T_C = \begin{bmatrix} 0\,1\,0\,0\,0 \\ 0\,0\,1\,0\,0 \\ 0\,0\,0\,1\,0 \\ 0\,0\,0\,0\,1 \\ 1\,0\,0\,0\,1 \end{bmatrix} \tag{9.5}$$

The output sequence 000011111010100110001 is generated by the initial state 10000, as it can be easily checked. The seed 00111 generates the sequence 1110010, while the seed 01101 generates the sequence 101. Thus, the mentioned seeds generate sequences of different lengths, one of length 21, a second of length 7 and a last one of length 3. Apart from those, the length 1 all-zero sequence is also generated. The sum of the lengths of all the generated sequences is again 2^n.

■

From the examples above, it results that, depending on the feedback coefficients, the **LFSR1** will generate or not sequences of maximum length. Usually, for most applications of these systems, **LFSR1** generating sequences of maximum length will be of interest. In the following it will be shown how to select feedback coefficient for this purpose.

9.1.2 M Sequences

Regarding sequences of maximum length (also known as **M sequences**), these have a series of properties [1] that are very interesting in many applications.

Given a sequence **M** of length $2^n - 1$, this has $2^{n-1} - 1$ zeros and 2^{n-1} ones. Thus, the probabilities of a bit of the sequence being zero, P_0, or one, P_1, are respectively:

$$P_0 = \frac{2^{n-1} - 1}{2^n - 1} = \frac{1}{2} - \frac{1}{2^{n+1} - 2}$$

$$P_1 = \frac{2^{n-1}}{2^n - 1} = \frac{1}{2} + \frac{1}{2^{n+1} - 2}$$

These probabilities can be made as close to ½ as desired, chosing large enough value of n.

An **nLFSR1** generating a sequence of maximum length goes through all $2^n - 1$ states. Thus, the n flip-flops will have all the 2^n configurations of zeros and ones but the all-zero. This implies that the bits at the output $\boldsymbol{b_0}$ appear grouped in 2^{n-1} bursts of zeros and ones, half of those of length 1, a quarter of length 2, and successively up to one of length $n - 1$ (all zeros) and another of length n (all ones). Half of the 2^{n-1} bursts are zeros and the other half are ones, as it can be checked in Table 9.1. The fact that there are 2^{n-1} bursts in a sequence implies that there are in the sequence 2^{n-1} transitions from 0 to 1 or from 1 to 0.

Given a sequence of zeros and ones of length L, in order to determine its degree of randomness (*i.e.*, given a bit of the sequence, is it predetermined any other bit of the sequences, shifted from the original d positions?), one of the used parameters is the value of the *autocorrelation* function of the sequence for a shift d, $A(d)$, defined as:

$$A(d) = \frac{1}{L}\sum_{i}^{L} p_i(d); \quad 1 \leq d \leq L-1$$

where $p_i(d)$ is a function that shows if a given bit and the one d positions further (shift d) are equal or not, and is computed for every bit of the sequence; concretely, $p_i(d) = 1$ if the compared bits are equal, and $p_i(d) = -1$ if they are different. The autocorrelation measures the degree of randomness of a sequence, so for long enough and absolutely random sequences, it tends to zero.

Example 9.4 For the **LFSR1** of Fig. 9.2, $L = 15$ and the sequences in $\boldsymbol{b_0}$ and $\boldsymbol{b_3}$ are shifted 3 positions, $d = 3$. Table 9.1 shows the value of $\boldsymbol{p_i(3)}$ for the different bits of the output sequence in $\boldsymbol{b_0}$. Applying the definition, it results that $A(3) = -1/15$. Computing $A(d)$ for all the other possible values of d, it always results $A(d) = -1/15$:

■

In general, for **M** sequences it results $A(d)$ = - $1/L$, $\forall d$. Thus, making L large enough (i.e., since $L = 2^n - 1$, making L large enough consists of choosing an **LFSR** long enough) can make $A(d)$ as close to zero as desired.

Since **M** sequences satisfy properties that are quite close to those of random sequences, even though they are not strictly random, they are known as *pseudorandom sequences*.

Apart from the properties above, due to the linearity of the system, any linear combination of **M** sequences generated by a given **LFSR** (in all cases, it will be the same sequence with different shifts) is another **M** sequence generated by the same **LFSR**. This property is known as *shift and add*: given an **M** sequence, and the same shifted sequence, their bit-to-bit addition (**EXOR** addition) results in the same **M** sequence with another shift. This property can be generalized to **M** sequences generated by two **LFSRs** with different lengths. Concretely, the **EXOR** addition of two **M** sequences of lengths $L_p = 2^p - 1$ and $L_q = 2^q - 1$ generates another pseudorandom sequence whose length is L_r= l cm(L_p, L_q). Thus, if L_p and L_q are relatively prime, a pseudorandom sequence of length $(2^p - 1) \cdot (2^q - 1)$ is generated. All this is checked in the following examples.

Example 9.5 For the **LFSR1** of Fig. 9.2, the addition of sequences $\boldsymbol{b_0}$ and $\boldsymbol{b_3}$ provides the sequence $\boldsymbol{b_3 \oplus b_0}$ in Table 9.1, which is the same $\boldsymbol{b_0}$ sequence with a shift $d = 4$.

■

Example 9.6 The **LFSR1** of Fig. 9.5a generates the **M** sequence of length 7 0010111 with seed 100; the **LFSR1** of Fig. 9.5b generates the **M** sequence of length 3 011

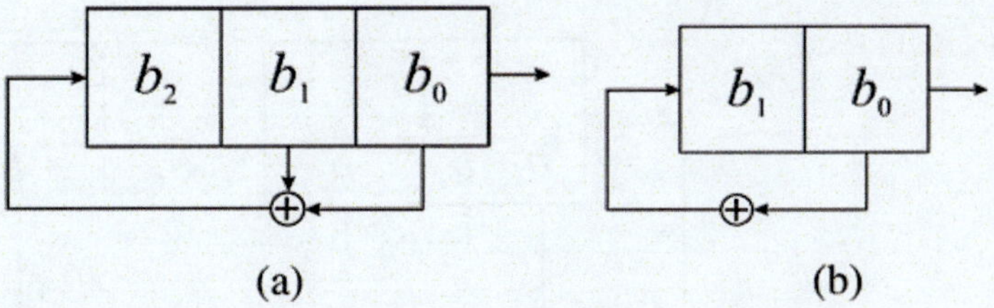

Fig. 9.5 **a** First **LFSR1** of Example 9.6; **b** Second **LFSR1** of Example 9.6

Table 9.3 Addition of the sequences generated by the **LFSR1s** of Example 9.6

Sequence 1	0	0	1	0	1	1	1	0	0	1	0	1	1	1	0	0	1	0	1	1	1
Sequence 2	0	1	1	0	1	1	0	1	1	0	1	1	0	1	1	0	1	1	0	1	1
⊕	0	1	0	0	0	0	1	1	1	1	1	0	1	0	1	0	0	1	1	0	0

with seed 10. The **EXOR** addition of these two sequences, as shown in Table 9.3, generates a sequence of length 21.

■

The **M** sequences generated by an **nLFSR1** lack the burst of *n* zeros (0…0 state). A minor modification of the corresponding **nLFSR1** can add this state, so the resulting sequences are of length $L = 2^n$. The resulting system is not linear. This modification, proposed by de Bruijn [3], consists of adding, as a new term, the product $\prod_{i=1}^{n-1} \overline{bi}$ to the addition in (9.1), resulting in

$$b_{n-1}(t+1) = \sum_{i=0}^{n-1} a_i b_i(t) \oplus \prod_{i=1}^{n-1} \overline{b_i} \tag{9.6}$$

Example 9.7 The **LFSR1** of Fig. 9.2, corresponding to Example 9.1, generates an **M** sequence. The modification proposed by de Bruijn leads to:

$$b_3(t+1) = b_3(t) \oplus b_0(t) \oplus \overline{b}_3(t)\overline{b}_2(t)\overline{b}_1(t)$$

which results in the circuit of Fig. 9.6. Assuming any initial state, 1000 for example, the output sequence in Table 9.4 is generated. This is the same **M** sequence generated by the non-modified **LFSR1**, but extended with the 0000 state. The properties of this extended sequence are very similar to those of the non-extended **M** sequence.

■

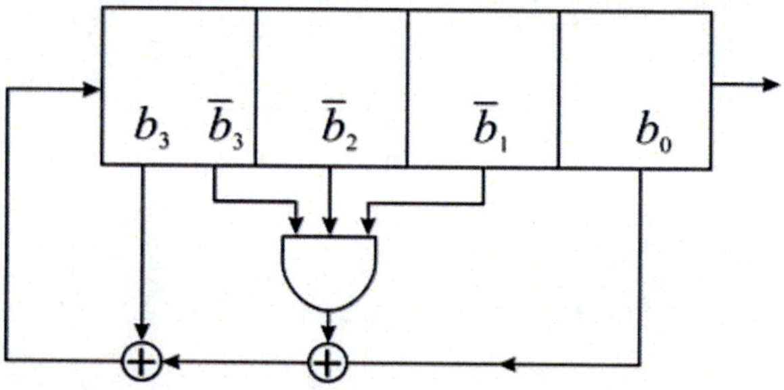

Fig. 9.6 Modified **LFSR1**

Table 9.4 States of the modified **LFSR1** of Example 9.7

State	b_3	b_2	b_1	b_0
8	1	0	0	0
12	1	1	0	0
14	1	1	1	0
15	1	1	1	1
7	0	1	1	1
11	1	0	1	1
5	0	1	0	1
10	1	0	1	0
13	1	1	0	1
6	0	1	1	0
3	0	0	1	1
9	1	0	0	1
4	0	1	0	0
2	0	0	1	0
1	0	0	0	1
0	0	0	0	0

9.1.3 Polynomials Associated to LFSR1s

A square matrix $T1$ (9.2) is associated to each **LFSR1**, completely describing its behavior. It is well-known in linear algebra [2] that a square matrix corresponds to a linear application whose eigenvalues (and thus most of the properties of the matrix) can be obtained from the characteristic equation or characteristic polynomial associated to the matrix. Taking into account that operations are carried out over **GF(2)**, it is immediate that the characteristic polynomial for matrix $T1$ results from the following equation:

$$\det(T1 + xI) = 0$$

Thus,

$$\det(T1 + xI) = \begin{vmatrix} x & 1 & 0 & \dots 0 & 0 \\ 0 & x & 1 & \dots 0 & 0 \\ \dots & \dots & \dots & \dots & \dots \\ 0 & 0 & 0 & \dots x & 1 \\ a_0 & a_1 & a_2 & \dots a_{n-2} & a_{n-1} + x \end{vmatrix}$$

Expanding this determinant by the first column, it leads to:

$$\det(T1 + xI) = x \cdot \begin{vmatrix} x & 1 & 0 & \dots 0 & 0 \\ 0 & x & 1 & \dots 0 & 0 \\ \dots & \dots & \dots & \dots & \dots \\ 0 & 0 & 0 & \dots x & 1 \\ a_1 & a_2 & a_3 & \dots a_{n-2} & a_{n-1} + x \end{vmatrix} + a_0 \begin{vmatrix} 1 & 0 & 0 & \dots 0 & 0 \\ x & 1 & 0 & \dots 0 & 0 \\ \dots & \dots & \dots & \dots & \dots \\ 0 & 0 & 0 & \dots 1 & 0 \\ 0 & 0 & 0 & \dots x & 1 \end{vmatrix}$$

$$= x \cdot \begin{vmatrix} x & 1 & 0 & \dots 0 & 0 \\ 0 & x & 1 & \dots 0 & 0 \\ \dots & \dots & \dots & \dots & \dots \\ 0 & 0 & 0 & \dots x & 1 \\ a_1 & a_2 & a_3 & \dots a_{n-2} & a_{n-1} + x \end{vmatrix} + a_0$$

Repeating this for the successive determinants, each time of lesser dimensions, the final expression is:

$$\begin{aligned} \det(T1 + xI) &= x(x(\dots (x(x + a_{n-1}) + a_{n-2}) + \dots + a_2) + a_1) + a_0 \\ &= x^n + a_{n-1}x^{n-1} + a_{n-2}x^{n-2} + \dots + a_2x^2 + a_1x + a_0 \end{aligned}$$

Thus, the characteristic polynomial associated to the **nLFSR1** of Fig. 9.1 is

$$P(x) = x^n + a_{n-1}x^{n-1} + a_{n-2}x^{n-2} + \dots + a_2x^2 + a_1x + a_0$$

In this way, a polynomial is associated to each **nLFSR1** and, vice versa, given any polynomial with $a_0 = 1$, an **nLFSR1** is associated to it.

The Cayley-Hamilton theorem [2] establishes that every square matrix satisfies its own characteristic equation. Thus, for the matrix $T1$ it holds that:

$$T1^n + a_{n-1}T1^{n-1} + a_{n-2}T1^{n-2} + \dots + a_2T1^2 + a_1T1 + a_0I = 0$$

Example 9.8 The characteristic polynomial of the **LFSR1** of Fig. 9.2, corresponding to Example 9.1, is given by:

$$\det(T_A + xI) = \begin{vmatrix} x & 1 & 0 & 0 \\ 0 & x & 1 & 0 \\ 0 & 0 & x & 1 \\ 1 & 0 & 0 & x + 1 \end{vmatrix} = x^4 + x^3 + 1$$

This is a primitive polynomial, as it can be checked in the table of primitive polynomials of Appendix B (Table B.10). The matrix T_A satisfies:

$$T_A^4 + T_A^3 + I = 0$$

or also:

$$T_A^4 = T_A^3 + I$$

or:

$$T_A^4 + T_A^3 = I$$

It is easy to check that $T_A^{15} = I$, so whatever the initial state is, after 15 iterations the state sequence will be repeated. In fact,

$$\begin{aligned}
T_A^5 &= T_A^4 T_A = (T_A^3 + I)\,T_A = T_A^4 + T_A \\
T_A^8 &= T_A^4 T_A^4 = (T_A^3 + I)(T_A^3 + I) = T_A^6 + I \\
T_A^9 &= T_A^8 T_A = (T_A^6 + I)\,T_A = T_A^7 + T_A \\
T_A^{11} &= T_A^8 T_A^3 = (T_A^6 + I)\,T_A^3 = T_A^9 + T_A^3 \\
T_A^{13} &= T_A^8 T_A^5 = (T_A^6 + I)\,T_A^5 = T_A^{11} + T_A^5 \\
T_A^{15} &= T_A^8 T_A^7 = (T_A^6 + I)\,T_A^7 = T_A^{11} + T_A^7 + T_A^5 = T_A^9 + T_A^7 + T_A^5 + T_A^3 \\
&= T_A^7 + T_A^7 + T_A^5 + T_A^3 + T_A = T_A^4 + T_A^3 + T_A + T_A = I
\end{aligned}$$

Moreover, it can be checked that for any other value of $p < 15$, $T_A^p \neq I$. Thus, using its characteristic polynomial, it has been proved again that this **LFSR1** generates sequences of maximum length.

■

Example 9.9 The characteristic polynomial of the **LFSR1** of Fig. 9.3a, corresponding to Example 9.2, is given by:

$$\det(T_B + xI) = \begin{vmatrix} x & 1 & 0 & 0 \\ 0 & x & 1 & 0 \\ 0 & 0 & x & 1 \\ 1 & 0 & 1 & x \end{vmatrix} = x^4 + x^2 + 1$$

This is an irreducible polynomial, but it is not primitive (*v.* Appendix B). Moreover, it can be easily checked that it is a divisor of $x^6 + 1$; in fact:

$$x^6 + 1 = \left(x^4 + x^2 + 1\right) \cdot \left(x^2 + 1\right)$$

The matrix T_B (9.4) satisfies:

$$T_B^4 + T_B^2 + I = 0$$

or also

$$T_B^4 = T_B^2 + I$$

or

$$T_B^4 + T_B^2 = I$$

It is easy to check that $T_B^6 = I$, so whatever the initial state is, after 6 iterations the state sequence will be repeated. In fact,

$$T_B^6 = T_B^4 T_B^2 = (T_B^2 + I)T_B^2 = T_B^4 + T_B^2 = I$$

Thus, this **LFSR1** generates sequences whose maximum length is six, as it had been shown in Example 9.2.

■

Example 9.10 The characteristic polynomial of the **LFSR1** of Fig. 9.4, corresponding to Example 9.3, is given by:

$$\det(T_C + xI) = \begin{vmatrix} x & 1 & 0 & 0 & 0 \\ 0 & x & 1 & 0 & 0 \\ 0 & 0 & x & 1 & 0 \\ 0 & 0 & 0 & x & 1 \\ 1 & 0 & 0 & 0 & x+1 \end{vmatrix} = x^5 + x^4 + 1$$

This is a reducible polynomial; concretely:

$$x^5 + x^4 + 1 = \left(x^3 + x + 1\right) \cdot \left(x^2 + x + 1\right)$$

The polynomials $x^3 + x + 1$ and $x^2 + x + 1$ are primitive (*v.* Table B.10 in Appendix B) and are associated to the **LFSR1s** of Fig. 9.5a, b. It is immediate that the length 21 sequence generated with the polynomial $x^5 + x^4 + 1$ is the **EXOR** addition of the **M** sequences generated with $x^3 + x + 1$ and $x^2 + x + 1$. The length 7 sequence generated with $x^5 + x^4 + 1$ is the same **M** sequence generated with $x^3 + x + 1$ (**LFSR1** of Fig. 9.5a), and the length 3 sequence is the one corresponding to $x^2 + x + 1$ (**LFSR1** of Fig. 9.5b). ■

What happens to the **LFSR1s** of Examples 9.8 and 9.9 can be generalized: if the associated polynomial $P(x)$ is a primitive polynomial, the **LFSR1** generates **M** sequences; if the polynomial $P(x)$ is irreducible but not primitive, the sequences generated by the **LFSR1** are not **M** sequences, and their maximum length is the minimum value k satisfying that the binomial $x^k + 1$ is a multiple of $P(x)$; if $P(x)$

is the product of several primitive polynomials, then the output sequences generated with $P(x)$ can be easily related to the **M** sequences generated with the primitive polynomials that are factors of $P(x)$, as in Example 9.10.

If an **M** sequence of a given length has to be generated with an **LFSR1**, it suffices to find, in the primitive polynomial table, a polynomial of adequate degree and then make use of the associated **LFSR1**. Of all the primitive polynomials of a given degree, those with the smallest number of terms will be the most interesting for implementing the sequence generator circuit, since the feedback will require less **EXOR** gates. This is reflected in the corresponding table, which shows the primitive polynomials with the smallest number of terms.

9.1.4 Type 2 LFSR

The structure of a type 2 **LFSR** (**LFSR2**) is shown in Fig. 9.7. It has again n **D** flip-flops and n feedback coefficients, assuming $a_0 = 1$. This circuit is not strictly a shift register, since the **EXOR** gates are within the flip-flop chain (this **LFSR** is also known as **LFSR** with internal **EXOR**). Nonetheless, this flip-flop chain with internal **EXOR** gates is also known as type 2 shift register.

The autonomous behavior of the type 2 **nLFSR** (Fig. 9.7) is described by the following equations:

$$\begin{aligned} b_i(t+1) &= a_i b_{n-1}(t) + b_{i-1}(t), \quad i = 1, \ldots, n-1 \\ b_0(t+1) &= a_0 b_{n-1}(t) \end{aligned} \tag{9.7}$$

that can be expressed in matrix form as:

$$\begin{bmatrix} b_{n-1}(t+1) \\ b_{n-2}(t+1) \\ \cdots \\ b_1(t+1) \\ b_0(t+1) \end{bmatrix} = \begin{bmatrix} a_{n-1} & 1 & 0 & \ldots & 0 & 0 & 0 \\ a_{n-2} & 0 & 1 & \ldots & 0 & 0 & 0 \\ \cdots & \cdots & \cdots & \cdots & \cdots & \cdots & \cdots \\ a_1 & 0 & 0 & \ldots & 0 & 0 & 1 \\ a_0 & 0 & 0 & \ldots & 0 & 0 & 0 \end{bmatrix} \begin{bmatrix} b_{n-1}(t) \\ b_{n-2}(t) \\ \cdots \\ b_1(t) \\ b_0(t) \end{bmatrix}$$

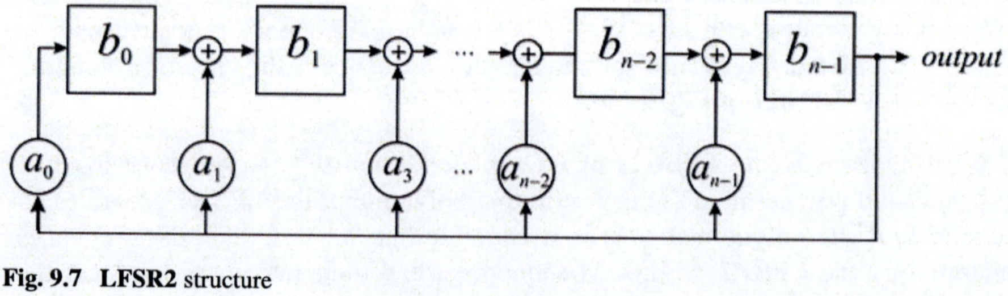

Fig. 9.7 **LFSR2** structure

or

$$B(t + 1) = T2B(t)$$

where:

$$T2 = \begin{bmatrix} a_{n-1} & 1 & 0 & \dots & 0 & 0 & 0 \\ a_{n-2} & 0 & 1 & \dots & 0 & 0 & 0 \\ \dots & \dots & \dots & \dots & \dots & \dots & \dots \\ a_1 & 0 & 0 & \dots & 0 & 0 & 1 \\ a_0 & 0 & 0 & \dots & 0 & 0 & 0 \end{bmatrix} \tag{9.8}$$

is an $n \times n$ square matrix, whose elements are 0 or 1, with an structure practically identical to that of $T1$ (9.2). Thus, the behavior of the **LFSR2** will be practically identical to that of the **LFSR1**, as it is detailed in the following.

The characteristic polynomial associated to the **LFSR2** will result from the following equation:

$$\det(T2 + xI) = 0$$

Thus,

$$\det(T2 + xI) = \begin{vmatrix} a_{n-1} + x & 1 & 0 & \dots & 0 & 0 \\ a_{n-2} & x & 1 & \dots & 0 & 0 \\ \dots & \dots & \dots & \dots & \dots & \dots \\ a_1 & 0 & 0 & \dots & x & 1 \\ a_0 & 0 & 0 & \dots & 0 & x \end{vmatrix}$$

Expanding this determinant by the first row, it leads to:

$$\det(T2 + xI) = (a_{n-1} + x) \cdot \begin{vmatrix} x & 1 & 0 & \dots & 0 & 0 \\ 0 & x & 1 & \dots & 0 & 0 \\ \dots & \dots & \dots & \dots & \dots & \dots \\ 0 & 0 & 0 & \dots & x & 1 \\ 0 & 0 & 0 & \dots & 0 & x \end{vmatrix} + \begin{vmatrix} a_{n-2} & 1 & 0 & \dots & 0 & 0 \\ a_{n-3} & x & 1 & \dots & 0 & 0 \\ \dots & \dots & \dots & \dots & \dots & \dots \\ a_1 & 0 & 0 & \dots & x & 1 \\ a_0 & 0 & 0 & \dots & 0 & x \end{vmatrix}$$

Successive expanding leads to the following expression for the characteristic polynomial:

$$P(x) = x^n + a_{n-1}x^{n-1} + a_{n-2}x^{n-2} + \cdots + a_2x^2 + a_1x + a_0$$

As for an **LFSR1**, the output of an **LFSR2** will generate a sequence of zeros and ones determined by the coefficients a_i and the initial state. Again, and for the same reasons, the output sequence will be periodical, with period $p \leq 2^n - 1$. The **LFSR2** cannot get out of the state 0…0.

Example 9.11 The **LFSR2** of Fig. 9.8, with initial state 1000, generates the output sequence 000111101011001, as illustrated in Table 9.5, which shows the ordered states during the evolution of the **LFSR**. It is a sequence of length 15. Since the **LFSR** has $n = 4$ flip-flops, the length of the generated sequence is the maximum: it is an **M** sequence. If the initial state is 1001, it is immediate to check that the generated output sequence is 111010110010001, identical to the previous one, but for a shift. Thus, this **LFSR2** can generate two output sequences, one all zeros (length 1) and another of maximum length ($2^n - 1$), being 2^n the addition of the lengths of both sequences.

It is easy to obtain the polynomial associated to this **LFSR2**, which is $x^4 + x^3 + 1$ and is the same as for the **LFSR1** of Examples 9.1 and 9.8. Thus, it is not a surprise that the **LFSR1** of Fig. 9.2 and the **LFSR2** of Fig. 9.8 generate the same output sequences, although their state sequences are not the same.

■

Example 9.12 The **LFSR2** of Fig. 9.9, with initial state 1000, generates the output

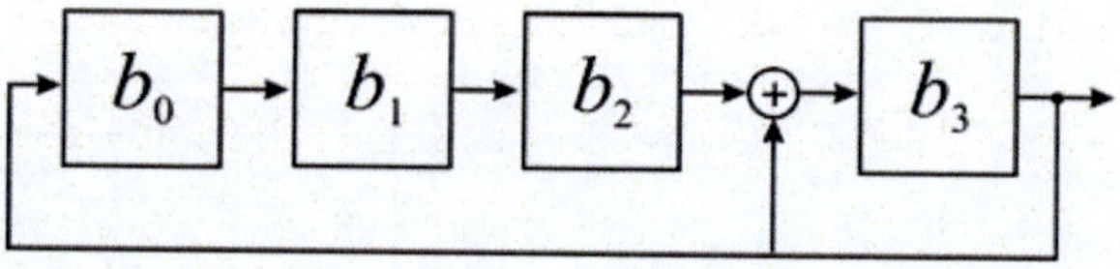

Fig. 9.8 LFSR2 of Example 9.11

Table 9.5 States of the **LFSR2** of Example 9.11

State	b_0	b_1	b_2	b_3
8	1	0	0	0
4	0	1	0	0
2	0	0	1	0
1	0	0	0	1
9	1	0	0	1
13	1	1	0	1
15	1	1	1	1
14	1	1	1	0
7	0	1	1	1
10	1	0	1	0
5	0	1	0	1
11	1	0	1	1
12	1	1	0	0
6	0	1	1	0
3	0	0	1	1

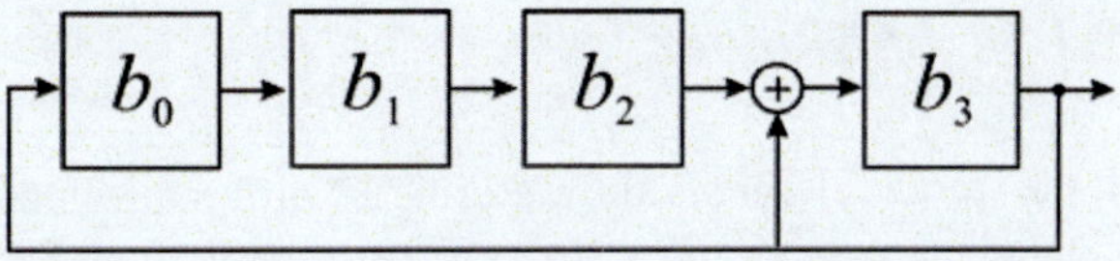

Fig. 9.9 **LFSR2** of Example 9.12

sequence 000101, as illustrated in Table 9.6. It is a sequence of length 6, thus it is not of maximum length. If the initial state is 1100, the generated output sequence is 001111, of length 6 again. Finally, the initial state 1110 generates the output sequence, of length 3, 011. Moreover, the length 1 all-zero sequence is also generated. The sum of the lengths of the generated sequences is again 2^n.

The polynomial associated to this **LFSR2** is $x^4 + x^2 + 1$, which is the same as for the **LFSR1** of Examples 9.2 and 9.9. Once more, the same output sequences are generated by the **LFSR1** of Fig. 9.3 and the **LFSR2** of Fig. 9.9, as they are associated to the same polynomial.

■

For each **LFSR1** it can be found an **LFSR2**, and viceversa, with identical behavior regarding the generated output sequences, since both **LFSRs** are described by the same equations and are associated to the same polynomials.

Table 9.6 States of the **LFSR2** of Example 9.12

State	b_0	b_1	b_2	b_3
8	1	0	0	0
4	0	1	0	0
2	0	0	1	0
1	0	0	0	1
10	1	0	1	0
5	0	1	0	1
12	1	1	0	0
6	0	1	1	0
3	0	0	1	1
11	1	0	1	1
15	1	1	1	1
13	1	1	0	1
14	1	1	1	0
7	0	1	1	1
9	1	0	0	1

9.1.5 LFSRmod2^m

If the storage elements allow storing 2^m different values, and the related operation are carried out over GF(2^m), structures similar to the **LFSRmod2** studied above can be defined; **LFSRmod2** is generalized in this subsection, introducing the **LFSRmod2$^{\mathbf{m}}$**.

The structure of a type 1 **LFSRmod2$^{\mathbf{m}}$** of length n (**nLFSR1mod2$^{\mathbf{m}}$**) is the same as that of the **LFSR1mod2** (Fig. 9.1), but for each cell stores now an m-bit digit. Thus, this register stores n digits, b_0, ..., $b_{n\text{-}1}$, and has n feedback coefficients, a_0, ..., $a_{n\text{-}1}$, each one of them with values ranging from 0 (this term does not exist in the feedback) to (2^m − 1) (the term is part of the feedback with the corresponding weight). It is logical to assume $a_0 \neq 0$, so this will be done in the following. An **nLFSR1mod2 $^{\mathbf{m}}$** consists of n storage cells, each one of size m, that can be built with m **D** flip-flops (a total of $m \times n$ **D** flip-flops for building the shift register), and constant multipliers and adders for generating the linear feedback [both operations, multiplication and addition, are carried out over GF(2^m)]. When an **nLFSR1mod2$^{\mathbf{m}}$** works autonomously, its behavior is described by Eq. (9.1), resulting in the transition matrix (9.2), but with the elements of this matrix now belonging to GF(2^{m}). A characteristic polynomial can be associated to each **nLFSR1mod2$^{\mathbf{m}}$**, with coefficients in GF(2^m), so the behavior of the **nLFSR1mod2$^{\mathbf{m}}$** can be derived from the properties of the characteristic polynomial, as illustrated in the following example.

Example 9.13 The **4LFSR1mod2$^{\mathbf{2}}$** of Fig. 9.10a consists of four cells, B3B2B1B0, each one of them with two flip-flops; it is built over GF(2^2){$x^2 + x + 1$}, and its associated polynomial is $Y^4 + Y^3 + Y^2 + CY + C$. Each cell can contain the values 00, 01, 10 and 11, represented as A, B, C and D, respectively, as it is done in Appendix B, section B.6. The multiplication and addition tables are shown in Example B.27 and are also shown, for commodity, in Fig. 9.10b. The circuit comprises adders and constant multipliers (concretely, it is required to multiply by C). It is immediate that the adders—with four inputs, A (a_1 and a_0) and B (a_1 and a_0) and two outputs, S (s_1 and s_0)– are synthesized using just two **EXOR** gates, as it is illustrated in Fig. 9.10c. Multiplication by C requires the circuit described by the table of Fig. 9.10d, also showing the resulting functions; a single **EXOR** gate is required.

Assuming that its initial state is AAAB, the **4LFSR1mod2$^{\mathbf{2}}$** runs autonomously, generating the 255 states given in Table 9.7. Thus, this circuit generates a sequence of maximum length, which indicates that its associated polynomial is primitive.

■

The structure of a type-2 **LFSRmod2$^{\mathbf{m}}$** of length n (**nLFSR2mod2$^{\mathbf{m}}$**) is that of the **LFSR2mod2** (Fig. 9.7), but for each cell now storing an m-bit digit; thus, this register stores n digits, b_0, ..., $b_{n\text{-}1}$, and has n feedback coefficients, a_0, ... ,$a_{n\text{-}1}$, each one of them with values ranging from 0 (this term is not present in the feedback) to (2^m − 1) (the term takes part in the feedback with the corresponding weight). It is logical to assume $a_0 \neq 0$, so it will be done in the following. An **nLFSR2mod2$^{\mathbf{m}}$** consists of n storage cells, each one of size m and that can be built with m type **D** flip-flops (a total of $m \times n$ type **D** flip-flops for the shift register), constant multipliers and adders

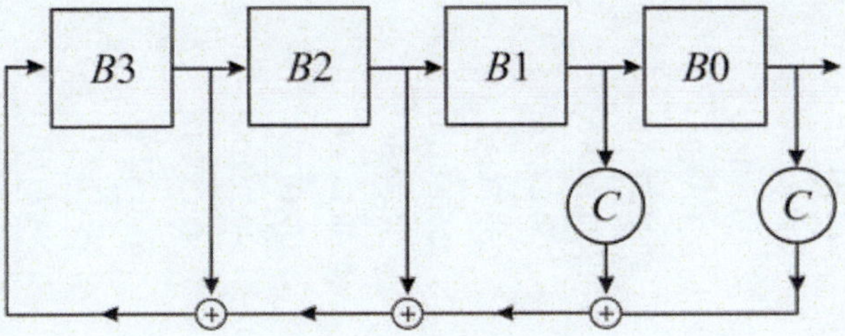

(a)

⊗	A	B	C	D
A	A	B	C	D
B	B	A	D	C
C	C	D	A	B
D	D	C	B	A

⊗	A	B	C	D
A	A	A	A	A
B	A	B	C	D
C	A	C	D	B
D	A	D	B	C

(b)

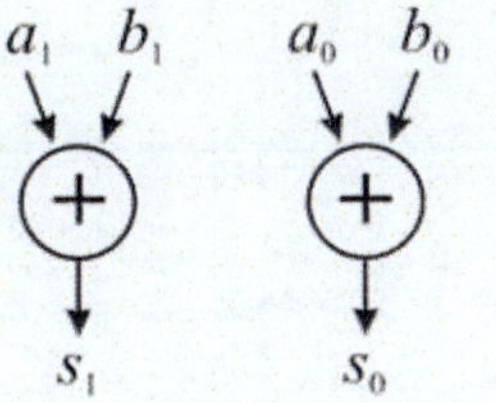

(c)

e_1	e_0	s_1	s_0	
0	0	0	0	
0	1	1	0	$s_0 = e_1$
1	0	1	1	$s_1 = e_1 \otimes e_0$
1	1	0	1	

(d)

Fig. 9.10 Example 9.13: **a 4LFSR1mod22**; **b** addition and multiplication tables; **c** adders; **d** multiplier by C

for generating the linear feedback [both operations, multiplication and addition, are carried out over GF(2^m)]. When an **nLFSR2mod2^m** works autonomously, its behavior is described by Eq. (9.7), resulting the transition matrix (9.8), its elements in GF(2^m). A characteristic polynomial, with coefficients in GF(2^m), can be associated to each **nLFSR2mod2^m**, so the behavior of the **nLFSR2mod2^m** can be obtained from the properties of this characteristic polynomial, such as it is proofed in the following example.

Table 9.7 States of the **4LFSR1mod2²** in Example 9.13

1:AAAB	33:BBCD	65:CDAD	97:BDAB	129:ADBD	161:DBAC	193:BCBA	225:BDCD
2:CAAA	34:CBBC	66:ACDA	98:ABDA	130:AADB	162:BDBA	194:BBCB	226:ABDC
3:CCAA	35:CCBB	67:DACD	99:AABD	131:DAAD	163:ABDB	195:BBBC	227:DABD
4:ACCA	36:ACCB	68:BDAC	100:DAAB	132:CDAA	164:CABD	196:BBBB	228:ADAB
5:BACC	37:DACC	69:BBDA	101:BDAA	133:BCDA	165:BCAB	197:ABBB	229:BADA
6:BBAC	38:DDAC	70:BBBD	102:CBDA	134:CBCD	166:BBCA	198:BABB	230:ABAD
7:DBBA	39:DDDA	71:DBBB	103:CCBD	135:BCBC	167:DBBC	199:BBAB	231:AABA
8:ADBB	40:BDDD	72:CDBB	104:DCCB	136:CBCB	168:DDBB	200:CBBA	232:CAAB
9:DADB	41:CBDD	73:BCDB	105:ADCC	137:CCBC	169:ADDB	201:BCBB	233:ACAA
10:ADAD	42:DCBD	74:ABCD	106:DADC	138:BCCB	170:AADD	202:DBCB	234:CACA
11:CADA	43:CDCB	75:DABC	107:BDAD	139:CBCC	171:AAAD	203:DDBC	235:BCAC
12:DCAD	44:ACDC	76:CDAB	108:DBDA	140:DCBC	172:BAAA	204:BDDB	236:ABCA
13:ADCA	45:AACD	77:DCDA	109:DDBD	141:ADCB	173:BBAA	205:BBDD	237:CABC
14:AADC	46:CAAC	78:ADCD	110:DDDB	142:CADC	174:ABBA	206:ABBD	238:DCAB
15:CAAD	47:BCAA	79:BADC	111:DDDD	143:ACAD	175:DABB	207:CABB	239:DDCA
16:DCAA	48:DBCA	80:DBAD	112:ADDD	144:DACA	176:DDAB	208:CCAB	240:DDDC
17:BDCA	49:BDBC	81:DDBA	113:DADD	145:ADAC	177:CDDA	209:CCCA	241:CDDD
18:BBDC	50:DBDB	82:CDDB	114:DDAD	146:AADA	178:ACDD	210:DCCC	242:BCDD
19:CBBD	51:BDBD	83:CCDD	115:BDDA	147:BAAD	179:CACD	211:BDCC	243:DBCD
20:ACBB	52:BBDB	84:ACCD	116:DBDD	148:ABAA	180:ACAC	212:CBDC	244:ADBC
21:CACB	53:DBBD	85:AACC	117:CDBD	149:BABA	181:BACA	213:BCBD	245:CADB
22:DCAC	54:BDBB	86:AAAC	118:CCDB	150:DBAB	182:CBAC	214:ABCB	246:BCAD

(continued)

Table 9.7 (continued)

23:CDCA	55:CBDB	87:DAAA	119:DCCD	151:ADBA	183:ACBA	215:AABC	247:CBCA
24:CCDC	56:ACBD	88:DDAA	120:DDCC	152:BADB	184:AACB	216:BAAB	248:ACBC
25:CCCD	57:BACB	89:ADDA	121:ADDC	153:CBAD	185:BAAC	217:DBAA	249:DACB
26:CCCC	58:ABAC	90:CADD	122:BADD	154:CCBA	186:CBAA	218:CDBA	250:CDAC
27:ACCC	59:CABA	91:CCAD	123:BBAD	155:CCCB	187:DCBA	219:DCDB	251:CCDA
28:CACC	60:ACAB	92:BCCA	124:BBBA	156:BCCC	188:DDCB	220:CDCD	252:BCCD
29:CCAC	61:AACA	93:ABCC	125:CBBB	157:DBCC	189:BDDC	221:DCDC	253:BBCC
30:DCCA	62:DAAC	94:BABC	126:DCBB	158:CDBC	190:ABDD	222:DDCD	254:ABBC
31:CDCC	63:ADAA	95:ABAB	127:BDCB	159:ACDB	191:BABD	223:CDDC	255:AABB
32:BCDC	64:DADA	96:DABA	128:DBDC	160:BACD	192:CBAB	224:DCDD	1:AAAB

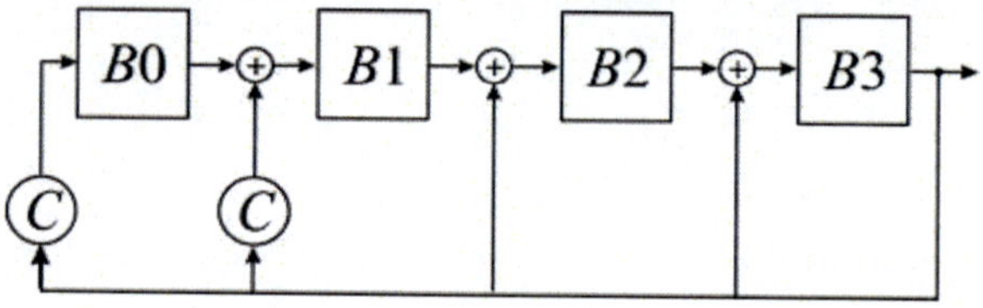

Fig. 9.11 **4LFSR1mod2²** of Example 9.14

Example 9.14 The **4LFSR2mod2²** of Fig. 9.11 consists of four cells, each one of them with two flip-flops; it is built over $GF(2^2)\{x^2 + x + 1\}$, and its associated polynomial is $Y^4 + Y^3 + Y^2 + CY + C$. The relevant operations are carried out as in Example 9.13.

Assuming any initial state other than AAAA, the **4LFSR2mod2²** runs autonomously, generating the 255 states in Table 9.7, but in a different order. Thus, this circuit generates a sequence of maximum length, since its associated polynomial is primitive.

■

9.2 LFSRmodp

If the storage cells allow storing p different values (p being a prime) and the involved operations are carried out in GF(p), structures similar to the **LFSRmod2** studied in Sect. 9.1 can be defined; **LFSRmod2** are generalized in this section, with the introduction of the **LFSRmodp**.

9.2.1 *Type 1 LFSRmodp*

The structure of an **LFSRmodp** of length n (**nLFSRmodp**) is that of Fig. 9.12; this register stores n digits, $b_0, \ldots, b_{n\text{-}1}$, and has n feedback coefficients, $-a_0, \ldots, -a_{n\text{-}1}$, each one of them in the range from 0 (this term is not present in the feedback) to –(p

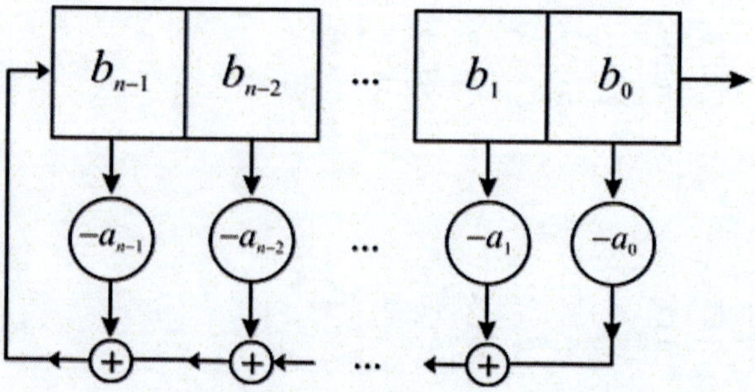

Fig. 9.12 Structure of **LFSR1modp**

– 1) (the term is included in the feedback with the corresponding weight). It is usual to assume $a_0 \neq 0$, so it will be done in the following. Thus, an **nLFSRmodp** consists of n storage cells, of size p and that can be built with r ($r =$ élog$_2$pù, éxù being the lower integer greater or equal to x) **D** flip-flops (a total of $r \times n$ **D** flip-flops for the shift register), constant multipliers (negative, in this case) and adders for generating the linear feedback [both operations, multiplication and addition, are carried out over **GF(p)**]. In fact, the **LFSRmodp** of Fig. 9.12 is known as of type 1 (**LFSR1modp**), since a similar structure known as type 2 **LFSRmodp** (or **LFSR2modp**) will be shown later. As an extension, the shift register used for building the **LFSR1modp** is also known as type 1 shift register.

The temporal evolution of the **nLFSR1modp** of Fig. 9.12 is described by the following equations:

$$b_j(t+1) = b_{j-1}(t), \quad j = 0, \ldots, \; n-2$$

$$b_{n-1}(t+1) = -\sum_{i=0}^{n-1} a_i b_i(t) \tag{9.9}$$

that can be expressed in matrix form as:

$$\begin{bmatrix} b_0(t+1) \\ b_1(t+1) \\ \cdots \\ b_{n-2}(t+1) \\ b_{n-1}(t+1) \end{bmatrix} = \begin{bmatrix} 0 & 1 & 0 & \ldots & 0 & 0 & 0 \\ 0 & 0 & 1 & \ldots & 0 & 0 & 0 \\ \cdots & \cdots & \cdots & \cdots & \cdots & \cdots & \cdots \\ 0 & 0 & 0 & \ldots & 0 & 0 & 1 \\ -a_0 & -a_1 & -a_2 & \ldots\ldots & -a_{n-3} & -a_{n-2} & -a_{n-1} \end{bmatrix} \begin{bmatrix} b_0(t) \\ b_1(t) \\ \cdots \\ b_{n-2}(t) \\ b_{n-1}(t) \end{bmatrix}$$

or

$$B(t+1) = T3B(t)$$

where $T3$ is an $n \times n$ square matrix:

$$T3 = \begin{bmatrix} 0 & 1 & 0 & \ldots & 0 & 0 & 0 \\ 0 & 0 & 1 & \ldots & 0 & 0 & 0 \\ \cdots & \cdots & \cdots & \cdots & \cdots & \cdots & \cdots \\ 0 & 0 & 0 & \ldots & 0 & 0 & 1 \\ -a_0 & -a_1 & -a_2 & \ldots & -a_{n-3} & -a_{n-2} & -a_{n-1} \end{bmatrix} \tag{9.10}$$

It is immediate that the characteristic polynomial for matrix $T3$ derives from the following equation:

$$\det(xI - T3) = 0$$

i.e.,

$$\det(xI - T3) = \begin{vmatrix} x & -1 & 0 & \dots & 0 & 0 \\ 0 & x & -1 & \dots & 0 & 0 \\ \dots & \dots & \dots & \dots & \dots & \dots \\ 0 & 0 & 0 & \dots & x & -1 \\ a_0 & a_1 & a_2 & \dots & a_{n-2} & a_{n-1} + x \end{vmatrix}$$

Developing this determinant leads to the characteristic polynomial associated to the **nLFSR1modp** of Fig. 9.12, which is:

$$P(x) = x^n + a_{n-1}x^{n-1} + a_{n-2}x^{n-2} + \cdots + a_2x^2 + a_1x + a_0$$

In the following, two examples of **LFSR1modp** are used to show the similarity of its behavior to that of the **LFSR1mod2**.

Example 9.15 The characteristic polynomial of the **4LFSR1mod3** of Fig. 9.13(a) is $x^4 + x + 2$, which is primitive (*v.* Table B.14). Table 9.8 shows the states the system goes through, starting from 1000. The column $(c_1 + 2c_0)$ shows the opposite to the feedback value, for computing the next state.

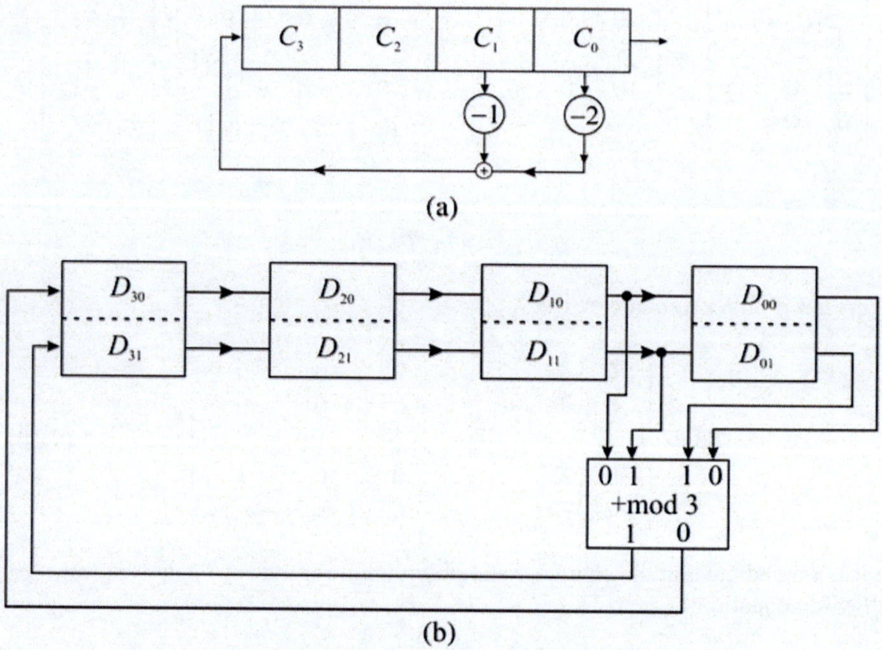

Fig. 9.13 **a 4LFSR1mod3** of Example 9.15; **b** detailed circuit of the **4LFSR1mod3**

Table 9.8 States of the **4LFSR1mod3** in Example 9.15

$c_3c_2c_1c_0$	(c_1+2c_0)	$c_3c_2c_1c_0$	(c_1+2c_0)	$c_3c_2c_1c_0$	(c_1+2c_0)	$c_3c_2c_1c_0$	(c_1+2c_0)
1000	0	1122	0	2000	0	2211	0
0100	0	0112	2	0200	0	0221	1
0010	1	1011	0	0020	2	2022	0
2001	2	0101	2	1002	1	0202	1
1200	0	1010	1	2100	0	2020	2
0120	2	2101	2	0210	1	1202	1
1012	2	1210	1	2021	1	2120	2
1101	2	2121	1	2202	1	1212	2
1110	1	2212	2	2220	2	1121	1
2111	0	1221	1	1222	0	2112	2
0211	0	2122	0	0122	0	1211	0
0021	1	0212	2	0012	2	0121	1
2002	1	1021	1	1001	2	2012	2
2200	0	2102	1	1100	0	1201	2
0220	2	2210	1	0110	1	1120	2
1022	0	2221	1	2011	0	1112	2
0102	1	2222	0	0201	2	1111	0
2010	1	0222	0	1020	2	0111	0
2201	2	0022	0	1102	1	0011	0
1220	2	0002	1	2110	1	0001	2

The system goes through 80 of the 81 possible states, all but 0000, which the system could not get out of. Thus, any initial state different of 0000 generates an output sequence of maximum length.

As it is shown in Sect. 3.8.5.1, multiplying by −1 in GF(3) requires interchanging the two bits of the corresponding digit, while multiplying by −2 does not change the digit. The construction of the **4LFSR1mod3** requires, apart from the four storage cells, each one of them consisting of two **D** flip-flops, a modulo 3 adder (v. Sect. 3.8.5.1). Figure 9.13b shows in detail the circuit corresponding to the **4LFSR1mod3**.

■

Example 9.16 The characteristic polynomial of the **4LFSR1mod3** of Fig. 9.14 is $x^4 + x^2 + 2$, which is not a primitive polynomial. Table 9.9 shows the states the system goes through, starting from 1000, 1100, 1110, 1200 and 0110. The column $(c_2 + 2c_0)$ shows the opposite to the feedback value, for computing the next state.

Depending on the initial content, the system enters and goes through one of five 16-state cycles, apart from the cycle containing just the zero state.

■

Fig. 9.14 4LFSR1mod3 of Example 9.16

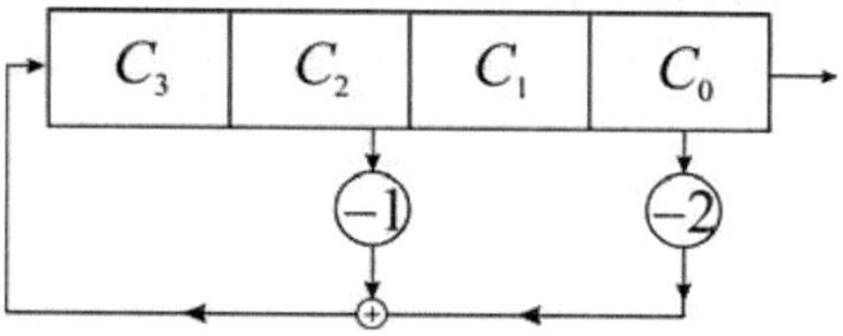

Table 9.9 States of the **4LFSR1mod3** in Example 9.16

$c_3c_2c_1c_0$	$(c_2 + 2c_0)$	$c_3c_2c_1c_0$	$(c_2 + 2c_0)$	$c_3c_2c_1c_0$	$(c_2 + 2c_0)$	$c_3c_2c_1c_0$	$(c_2 + 2c_0)$
1000	0	2222	0	2220	2	1212	0
0100	1	0222	0	1222	0	0121	0
2010	0	0022	1	0122	2	0012	1
0201	1	2002	1	1012	1	2001	2
2020	0	2200	2	2101	0	0110	1
0202	0	1220	2	0210	2	2011	2
0020	0	1122	2	1021	2	1201	1
0002	1	1112	2	1102	2	2120	1
2000	0	1111	0	1200	2	2212	0
0200	2	0111	0	1120	1	0221	1
1020	0	0011	2	2112	2	2022	1
0102	2	1001	2	1211	1	2202	0
1010	0	1110	1	2121	0	0220	2
0101	0	2111	0	0212	0	1022	1
0010	0	0211	1	0021	2	2102	2
0001	2	2021	2	1002	1	1210	2
1100	1	1202	0	2100	1	1121	0
2110	1	0120	1	2210	2	0112	2
2211	1	2012	1	1221	1	1011	2
2221	1	2201	1	2122	2	1101	0

As in Example 9.15, any **LFSR1modp** whose characteristic polynomial is primitive generates two sequences: one of a single state, the zero state, and another one of maximum length, with all the remaining states. If the polynomial is not primitive several independent sequences are generated, depending on the initial state, as it is shown in Example 9.16.

9.2.2 Type 2 LFSRmodp

The structure of a type-2 **LFSRmodp (LFSR2modp)** is that of Fig. 9.15. Once more, it has n storage cells able to store p different values, and n feedback coefficients, and it is assumed $a_0 \neq 0$. This circuit is not strictly a shift register, since the arithmetic operators are within the flip-flop chain. However, this storage cell chain with internal operations is also known as type-2 shift register.

The behavior of the **nLFSR2modp** (Fig. 9.15) working autonomously is described by the following equations:

$$\begin{aligned} b_i(t+1) &= b_{i-1}(t) - a_i b_{n-1}(t)\,, \quad i = 1, \ldots, n-1 \\ b_0(t+1) &= -a_0 b_{n-1}(t) \end{aligned} \tag{9.11}$$

that can be expressed in matrix form as:

$$\begin{bmatrix} b_{n-1}(t+1) \\ b_{n-2}(t+1) \\ \ldots \\ b_1(t+1) \\ b_0(t+1) \end{bmatrix} = \begin{bmatrix} -a_{n-1} & 1 & 0 & \ldots & 0 & 0 & 0 \\ -a_{n-2} & 0 & 1 & \ldots & 0 & 0 & 0 \\ \ldots & \ldots & \ldots & \ldots & \ldots & \ldots & \ldots \\ -a_1 & 0 & 0 & \ldots & 0 & 0 & 1 \\ -a_0 & 0 & 0 & \ldots & 0 & 0 & 0 \end{bmatrix} \begin{bmatrix} b_{n-1}(t) \\ b_{n-2}(t) \\ \ldots \\ b_1(t) \\ b_0(t) \end{bmatrix}$$

or:

$$B(t+1) = T4B(t)$$

where

$$T4 = \begin{bmatrix} -a_{n-1} & 1 & 0 & \ldots & 0 & 0 & 0 \\ -a_{n-2} & 0 & 1 & \ldots & 0 & 0 & 0 \\ \ldots & \ldots & \ldots & \ldots & \ldots & \ldots & \ldots \\ -a_1 & 0 & 0 & \ldots & 0 & 0 & 1 \\ -a_0 & 0 & 0 & \ldots & 0 & 0 & 0 \end{bmatrix} \tag{9.12}$$

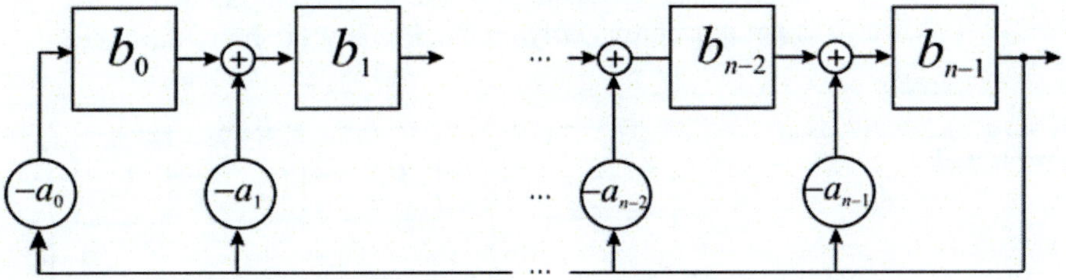

Fig. 9.15 Structure of the **LFSR2modp**

is an $n \times n$ square matrix with structure practically identical to that of $T3$. Thus, the behavior of the **LFSR2modp** will be practically identical to that of the **LFSR1modp**, as it is detailed in the following.

The characteristic polynomial associated to the **LFSR2modp** will result from the following equation:

$$\det(xI - T4) = 0$$

i.e.,

$$\det(xI - T4) = \begin{vmatrix} a_{n-1} + x & -1 & 0 & \ldots & 0 & 0 \\ a_{n-2} & x & -1 & \ldots & 0 & 0 \\ \ldots & \ldots & \ldots & \ldots & \ldots & \ldots \\ a_1 & 0 & 0 & \ldots & x & -1 \\ a_0 & 0 & 0 & \ldots & 0 & x \end{vmatrix} = 0$$

Developing this determinant leads to the following expression of the characteristic polynomial:

$$P(x) = x^n + a_{n-1}x^{n-1} + a_{n-2}x^{n-2} + \cdots + a_2x^2 + a_1x + a_0$$

Next, two examples of **LFSR2modp** are presented for showing the similarities of their behavior to that of the **LFSR2mod2** and the **LFSR1modp**.

Example 9.17 The characteristic polynomial of the **4LFSR2mod3** of Fig. 9.16 is $x^4 + x + 2$, which is primitive (*v.* Table B.14). Table 9.10 shows the states the system goes through, starting from 1000. The column C_0C_1 shows the next values ($C_0 = -2c_3$, $C_1 = c_0 - c_3$) of cells c_0c_1. The system goes through 80 of the 81 possible states, once more all but 0000, which the system could not get out of. Thus, any initial state different of 0000 generates an output sequence of maximum length. Comparing with the results from Example 9.15, the behavior of both systems is practically identical, but with different output sequences.

■

Example 9.18 The characteristic polynomial of the **4LFSR2mod3** of Fig. 9.17 is $x^4 + x^2 + 2$, which is not a primitive polynomial. Table 9.11 shows the states the

Fig. 9.16 4LFSR2mod3 of Example 9.17

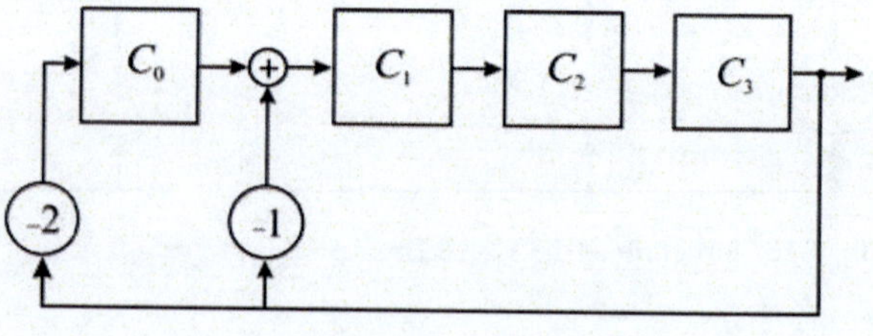

Table 9.10 States of the **4LFSR2mod3** in Example 9.17

$c_0c_1c_2c_3$	$C_0\ C_1$	$c_0c_1c_2c_3$	$C_0\ C_1$	$c_0c_1c_2c_3$	$C_0\ C_1$	$c_0c_1c_2c_3$	$C_0\ C_1$
1000	0 1	0122	2 1	2000	0 2	0211	1 2
0100	0 0	2112	2 0	0200	0 0	1221	1 0
0010	0 0	2011	1 1	0020	0 0	1022	2 2
0001	1 2	1101	1 0	0002	2 1	2202	2 0
1200	0 1	1010	0 1	2100	0 2	2020	0 2
0120	0 0	0101	1 2	0210	0 0	0202	2 1
0012	2 1	1210	0 1	0021	1 2	2120	0 2
2101	1 1	0121	1 2	1202	2 2	0212	2 1
1110	0 1	1212	2 2	2220	0 2	2121	1 1
0111	1 2	2221	1 1	0222	2 1	1112	2 2
1211	1 0	1122	2 2	2122	2 0	2211	1 1
1021	1 0	2212	2 0	2012	2 0	1121	1 0
1002	2 2	2021	1 1	2001	1 1	1012	2 2
2200	0 2	1102	2 2	1100	0 1	2201	1 1
0220	0 0	2210	0 2	0110	0 0	1120	0 1
0022	2 1	0221	1 2	0011	1 2	0112	2 1
2102	2 0	1222	2 2	1201	1 0	2111	1 1
2010	0 2	2222	2 0	1020	0 1	1111	1 0
0201	1 2	2022	2 0	0102	2 1	1011	1 0
1220	0 1	2002	2 0	2110	0 2	1001	1 0

Fig. 9.17 4LFSR2mod3 of Example 9.18

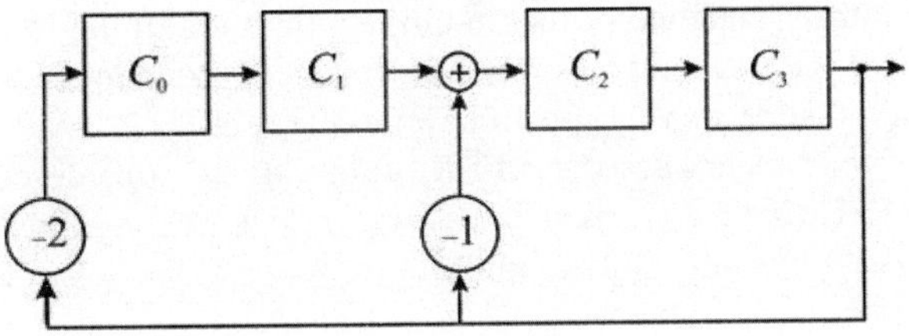

system goes through, starting from 1000, 1100, 1110, 1200 and 2001. The column C_0C_2 shows the next values ($C_0 = -2c_3$, $C_2 = c_1 - c_3$) of cells c_0c_2.

Depending on the initial content, the system enters and goes through one of five 16-state cycles, apart from the cycle containing just the zero state. Comparing to the results of Example 9.15, the behavior of both systems is practically the same, but with different output sequences. ■

As in Example 9.17, any **LFSR2modp** whose characteristic polynomial is primitive generates two sequences: a single-state one, containing just the zero state, and

Table 9.11 States of the **4LFSR2mod3** in Example 9.18

$c_0c_1c_2c_3$	C_0 C_2	$c_0c_1c_2c_3$	C_0 C_2	$c_0c_1c_2c_3$	C_0 C_2	$c_0c_1c_2c_3$	C_0 C_2
1000	0 0	1122	2 2	2220	0 2	2112	2 2
0100	0 1	2122	2 2	0222	2 0	2221	1 1
0010	0 0	2222	2 0	2002	2 1	1212	2 0
0001	1 2	2202	2 0	2210	0 2	2101	1 0
1020	0 0	2200	0 2	0221	1 1	2001	1 2
0102	2 2	0220	0 2	1012	2 1	1220	0 2
2020	0 0	0022	2 1	2111	1 0	0122	2 2
0202	2 0	2012	2 1	1201	1 1	2022	2 1
2000	0 0	2211	1 1	1200	0 2	2212	2 0
0200	0 2	1211	1 1	0120	0 1	2201	1 1
0020	0 0	1111	1 0	0012	2 1	1210	0 2
0002	2 1	1101	1 0	2011	1 2	0121	1 0
2010	0 0	1110	0 1	1221	1 1	1002	2 1
0201	1 1	0111	1 0	1112	2 2	2110	0 1
1010	0 0	1001	1 2	2121	1 0	0211	1 1
0101	1 0	1120	0 1	1202	2 0	1011	1 2
1100	0 1	0112	2 2	2100	0 1	1121	1 0
0110	0 1	2021	1 2	0210	0 2	1102	2 2
0011	1 2	1222	2 0	0021	1 2	2120	0 1
1021	1 2	2102	2 2	1022	2 1	0212	2 0

another sequence of maximum length, with all the remaining states. If the polynomial is not primitive, several independent sequences are generated, depending on the initial state, as it is shown in Example 9.18.

It is clear that the **LFSRmodp** can be considered as a generalization of the **LFSRmod2** and, as it is shown in Sect. 9.3, can be applied in the same way to operations with polynomials.

9.2.3 *LFSRmodp^m^*

In the same way that the **LFSRmod2^m^** are defined from the **LFSRmod2**, the **LFSR-modp^m^** can be defined from the **LFSRmodp**. The storage cells of the **LFSRmodp^m^** allow storing p^m different values and involved operations are carried out over **GF(p^m)**.

The structure of a type-1 **LFSRmodp^m^** of length n (**nLFSR1modp^m^**) is the same as that of the **LFSR1modp** (Fig. 9.12), but for each cell now storing an m-figure digit. Thus, this register stores n digits, b_0, …, b_{n-1}, and has n feedback coefficients, $-a_0$, …, $-a_{n-1}$, each one of them ranging from 0 (the term is not present in the feedback) to $-(p^m - 1)$ (the term is part of the feedback, with the corresponding weight). It is logical and usual to assume $a_0 \neq 0$, and it will be done in the following. An **nLFSR1modpm** consists of n storage cells, each one able to store p^m different values and that can be built with r ($r =$ élog$_2$ p^mù) **D** flip-flops (a total of $r \times n$ **D** flip-flops for building the shift register), of constant multipliers, and of adders for generating the linear feedback [both operations, multiplication and addition, are carried out over **GF(pm)**]. When an **nLFSR1modpm** works autonomously, its behavior is described by Eq. (9.9), resulting the transition matrix (9.10), whose elements belong to **GF(pm)**. A characteristic polynomial with coefficients in **GF(pm)** can be associated to each **nLFSR1modpm**, so the behavior of the **nLFSR1modpm** can be obtained from the properties of this characteristic polynomial.

The structure of a type-2 **LFSRmodpm** of length n (**nLFSR2modpm**) is that of the **LFSR2modp** (Fig. 9.15), but for each cell now storing an m-figure digit. Thus, this register stores n digits, b_0, …, b_{n-1}, and has n feedback coefficients, $-a_0$, …, $-a_{n-1}$, each one of them ranging from 0 (the term is not present in the feedback) to -($p^m - 1$) (the term is part of the feedback, with the corresponding weight). It is logical and usual to assume $a_0 \neq 0$, and it will be done in the following. An **nLFSR2modpm** consists of n storage cells, each one able to store p^m different values and that can be built with r ($r =$ élog_2 p^mù) **D** flip-flops (a total of $r \times n$ **D** flip-flops for building the shift register), of constant multipliers, and of adders for generating the linear feedback. When an **nLFSR2modpm** works autonomously, its behavior is described by Eq. (9.11), resulting the transition matrix (9.12), whose elements belong to **GF(pm)**. A characteristic polynomial with coefficients in **GF(pm)** can be associated to each **nLFSR2modpm**, so the behavior of the **nLFSR2modpm** can be obtained from the properties of this characteristic polynomial.

9.3 Circuits for Operating with Polynomials

Different systems for adding, subtracting, multiplying and dividing polynomials defined over **GF(p)** or over **GF(pm)** are discussed in this Section. These operations are interesting for different applications of finite or Galois fields, such as cryptography, information coding (including error correction) and circuit test. Also the analysis of multipliers and divisors as filters is introduced.

As it is noted in Appendix B, a polynomial in x, $P(x)$, defined over **GF(p)** or over **GF(pm)**

$$P(x) = a_n x^n + a_{n-1} x^{n-1} + a_{n-2} x^{n-2} + \cdots + a_2 x^2 + a_1 x + a_0$$

will be defined for a series of $n + 1$ values $a_i \in$**GF(p)** or **GF(pm)**, which are interpreted as the polynomials coefficients, a_n, …, a_0, in this same order unless otherwise stated. It will be noted as $P = \{a_i\}$. These $n + 1$ coefficients can be given serially or in parallel and, consequently, the corresponding operations will be carried out with this in mind. For many polynomial operations, different circuits built with shift registers will be used, with structures similar to the **LFSR** previously discussed.

9.3.1 Circuits for Polynomial Addition and Subtraction

Lets remind that the addition $S = P + Q$ (or subtraction $R = P - Q$) of the polynomials $P = \{a_i\}$ and $Q = \{b_i\}$ is defined as $S = \{a_i + b_i\}$ ($R = \{a_i - b_i\}$), where $a_i + b_i$ ($a_i - b_i$) is computed over **GF(p)** or over **GF(p^m^)**. Independently of the serial or parallel nature of information, the addition or subtraction of two polynomials is carried out directly over the *n*-tuples of integers that represent them. As an example, with parallel information, let it be the polynomials over **GF(2)** $P = (1, 0, 0, 1)$ and $Q = (1, 0, 1, 0, 1)$. Equalling the lengths of P and Q, it is immediate that:

$$P + Q = (0, 1, 0, 0, 1) + (1, 0, 1, 0, 1) = (1, 1, 1, 0, 0)$$

This addition can be carried out over **GF(2)** with binary half-adders neglecting carries (*i.e.*, with two-input **EXOR** gates, as it is represented in Fig. 9.18a), as it follows:

$$\begin{array}{r} 01001 \\ +10101 \\ \hline 11100 \end{array}$$

Subtraction $P - Q$ over **GF(2)** will lead to the same result as addition, as addition and subtraction are the same operation over **GF(2)**.

With serial information, the operation can be carried out over **GF(2)** with a single two-input **EXOR** gate, applying to its input the same degree coefficients, as it is shown in Fig. 9.18b.

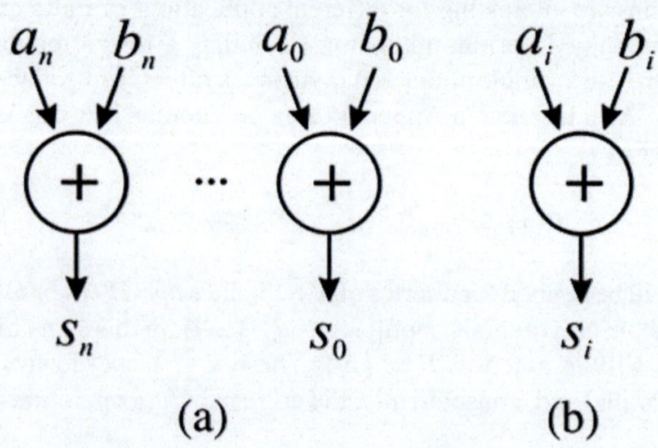

Fig. 9.18 Circuits for polynomial addition: **a** parallel; **b** serial

In the general case, for **GF(p)** or **GF(p^m)**, the additions and subtractions for each one of the coefficients are carried out as it is detailed in Chap 3.

9.3.2 *Circuits for Polynomial Multiplication*

The product of the polynomials $P = \{a_i\}$ and $Q = \{b_i\}$ is defined as $M = \left\{c_i = \sum_{j+k=i} a_j b_k\right\}$, where $a_j b_k$ are computed over **GF(p)** or over **GF(p^m)**. It is usually assumed that both polynomials P and Q are of the same degree, $n - 1$, so M will be of degree $2n - 2$. It is immediate that the following matrix expression can be used for the product:

$$\begin{bmatrix} c_0 \\ \cdots \\ c_{2n-2} \end{bmatrix} = \begin{bmatrix} a_0 & 0 & \cdots & 0 & 0 \\ a_1 & a_0 & \cdots & 0 & 0 \\ \cdots & \cdots & \cdots & \cdots & \cdots \\ a_{n-2} & a_{n-3} & \cdots & a_0 & 0 \\ a_{n-1} & a_{n-2} & \cdots & a_1 & a_0 \\ 0 & a_{n-1} & \cdots & a_2 & a_1 \\ \cdots & \cdots & \cdots & \cdots & \cdots \\ 0 & 0 & \cdots & a_{n-1} & a_{n-2} \\ 0 & 0 & \cdots & 0 & a_{n-1} \end{bmatrix} \begin{bmatrix} b_0 \\ \cdots \\ b_{n-1} \end{bmatrix}$$

Applying this definition to the product of the following polynomials defined over **GF(2)**, $P = (1, 0, 0, 1)$ and $Q = (1, 0, 1, 0, 1)$, it leads to:

$$\begin{aligned} c_0 &= a_0 b_0 \\ c_1 &= a_0 b_1 + a_1 b_0 \\ c_2 &= a_0 b_2 + a_1 b_1 + a_2 b_0 \\ c_3 &= a_0 b_3 + a_1 b_2 + a_2 b_1 + a_3 b_0 \\ c_4 &= a_0 b_4 + a_1 b_3 + a_2 b_2 + a_3 b_1 \\ c_5 &= a_1 b_4 + a_2 b_3 + a_3 b_2 \\ c_6 &= a_2 b_4 + a_3 b_3 \\ c_7 &= a_3 b_4 \end{aligned}$$

i.e.:

$$PQ = (1, 0, 0, 1) \cdot (1, 0, 1, 0, 1) = (1, 0, 1, 1, 1, 1, 0, 1)$$

This result is also obtained multiplying [over **GF(2)**] the numbers 1001 ad 10101, as it follows:

$$1001 \times 10101 = 10111101$$

With parallel information, this result can be also obtained using simple binary circuits, only **AND** and **EXOR** gates over **GF(2)**. Once more, carries are neglected. Figure 9.19 shows, as an example, the combinational circuit for multiplying, over **GF(2)**, two polynomials of third degree, $A = (a_3, a_2, a_1, a_0)$ y $B = (b_3, b_2, b_1, b_0)$, which requires 16 **AND** gates and 9 **EXOR** gates. In general, it is necessary to use adequate multipliers and adders, as those presented in Chap. 3, for **GF(p)** or for **GF(p^m^)**.

If the resulting multiplication circuit is of an excessive size, the polynomials to be multiplied can be decomposed into fragments that allow utilizing simpler multipliers, as it is done in Sect. 2.4.1 for integers. As an example, dividing into two fragments the polynomials $A(x)$ and $B(x)$ to be multiplied it results in:

$$\begin{array}{ccccccc}
 & & & a_3 & a_2 & a_1 & a_0 \\
 & & \times & b_3 & b_2 & b_1 & b_0 \\
\hline
 & & & a_3b_0 & a_2b_0 & a_1b_0 & a_0b_0 \\
 & & a_3b_1 & a_2b_1 & a_1b_1 & a_0b_1 & \\
 & a_3b_2 & a_2b_2 & a_1b_2 & a_0b_2 & & \\
a_3b_3 & a_2b_3 & a_1b_3 & a_0b_3 & & & \\
\hline
c_6 & c_5 & c_4 & c_3 & c_2 & c_1 & c_0
\end{array}$$

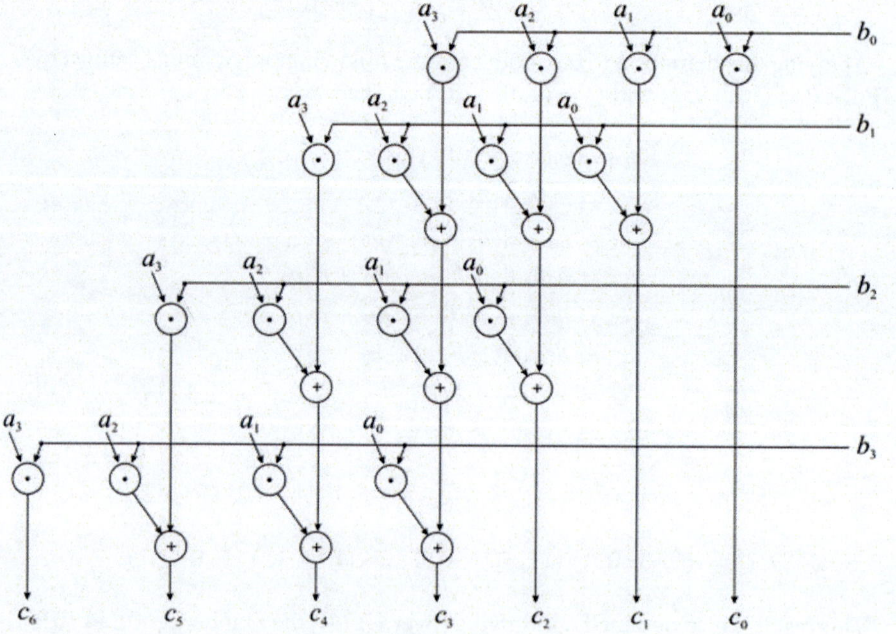

Fig. 9.19 Combinational circuit for polynomial multiplication

$$A(x) = a_m x^m + a_{m-1}x^{m-1} + \cdots + a_1 x + a_0 = x^p A_H + A_L$$
$$B(x) = b_n x^n + b_{n-1}x^{n-1} + \cdots + b_1 x + b_0 = x^p B_H + B_L$$
$$A(x) \cdot B(x) = \left(x^p A_H + A_L\right) \cdot \left(x^p B_H + B_L\right)$$
$$= x^{2p} A_H B_H + x^p (A_H B_L + A_L B_H) + A_L B_L$$

With the development above, apart from the corresponding adders, four multipliers would be required for obtaining $A_H B_H$, $A_H B_L$, $A_L B_H$ and $A_L B_L$. If the number of multipliers has to be reduced, at the cost of incrementing the number of adders, the product $A(x) \cdot B(x)$ can be also expressed as:

$$A(x) \cdot B(x) = x^{2p} A_H B_H + x^p ((A_H + A_L) \cdot (B_H + B_L) + A_H B_H + A_L B_L) + A_L B_L$$

Thus, only three multipliers would be required for obtaining $A_H B_H$, $(A_H + A_L) \cdot (B_H + B_L)$ and $A_L B_L$. The most favorable situation is when two or more fragments of the polynomials are of the same length, so the three or four required multipliers are identical.

With serial information, lets assume that a circuit has to be built for multiplying any multiplicand polynomial, P, of degree n, $(a_n, a_{n-1}, \ldots, a_1, a_0)$, by a fixed multiplier polynomial, M, of degree r, $(m_r, m_{r-1}, \ldots, m_1, m_0)$. For this multiplication a type-1 shift register (Fig. 9.20), similar to an **LFSR1**, can be used, so the actual output, o_t, is a linear combination (determined by the multiplier coefficients) of the current input, i_t, and r previous inputs, i_{t-j}, $j = 1, \ldots, r$:

$$o_t = \sum_{j=0}^{j=r} i_{(t-j)} m_{(r-j)} \tag{9.12}$$

If an initial state with all storage cells set to zero is assumed, and after that multiplier coefficients $(a_n, a_{n-1}, \ldots, a_1, a_0)$ are fed to the input $\boldsymbol{i}$ (a_n first, a_0 last) followed by r zeros, it is easy to check that the output $\boldsymbol{o}$ will produce the product coefficients $PM = (p_{n+r}, p_{n+r-1}, \ldots, p_1, p_0)$. In fact, the linear combination (9.12) corresponds to the definition of the product of two polynomials and it is immediate

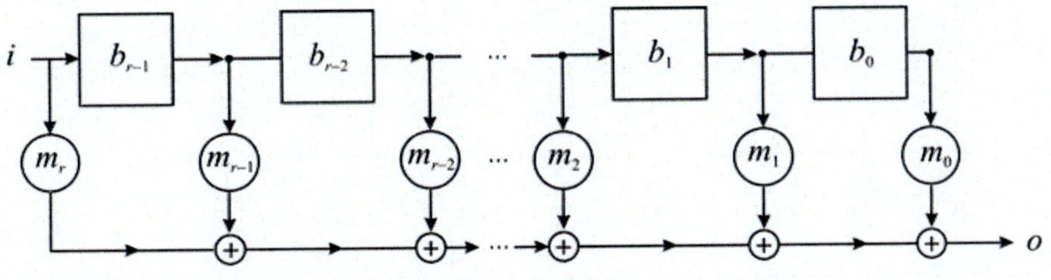

Fig. 9.20 Polynomial multiplier with a type-1 shift register

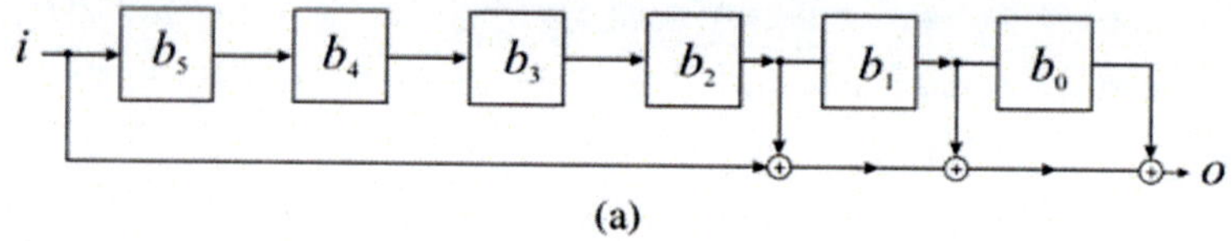

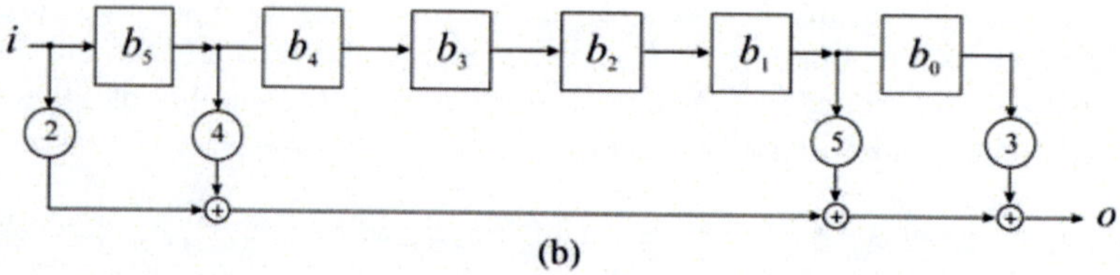

Fig. 9.21 **a** Type-1 LFSR for multipliying by $x^6 + x^2 + x + 1$ over **GF(2)**; **b** Type-1 LFSR for multiplying by $2x^6 + 4x^5 + 5x + 3$ over **GF(7)**

that the first output is $a_n m_r$, the second is $a_n m_{r\text{-}1} + a_{n\text{-}1} m_r$, and so on successively, as it is proofed in the following example. Thus, $n + r$ clock cycles have to be applied to the register for obtaining the product at the output.

Example 9.19 The type-1 multiplier circuit of Fig. 9.21a can be used for multiplying any multiplicand polynomial by the fixed polynomial $M(x) = x^6 + x^2 + x + 1$ over **GF(2).**

The type-1 multiplier circuit of Fig. 9.21b can be used for multiplying any multiplicand polynomial by the fixed polynomial $M(x) = 2x^6 + 4x^5 + 5x + 3$ over **GF(7)**. Table 9.12 shows the evolution of this circuit when the multiplicand is $x^3 + 2x + 4$; the result of this multiplication is $2x^9 + 4x^8 + 4x^7 + 2x^6 + 2x^5 + 5x^4 + 3x^3 + 3x^2 + 5x + 5$, as it is obtained as output o.

Having in mind the circuits designed in Sect. 3.8.5.2 for operating **mod7**, it is easy to design the adder and constant multiplier circuits of Fig. 9.21b.

Table 9.12 Results of multiplying $x^3 + 2x + 4$ by $2x^6 + 4x^5 + 5x + 3$

i	b_5	b_4	b_3	b_2	b_1	b_0	$o = 2i + 4b_5 + 5b_1 + 3b_0$
1	0	0	0	0	0	0	2×1 = 2
0	1	0	0	0	0	0	4×1 = 4
2	0	1	0	0	0	0	2×2 = 4
4	2	0	1	0	0	0	2×4 + 4×2 = 2
0	4	2	0	1	0	0	4×4 = 2
0	0	4	2	0	1	0	5×1 = 5
0	0	0	4	2	0	1	3×1 = 3
0	0	0	0	4	2	0	5×2 = 3
0	0	0	0	0	4	2	5×4 + 3×2 = 5
0	0	0	0	0	0	4	3×4 = 5

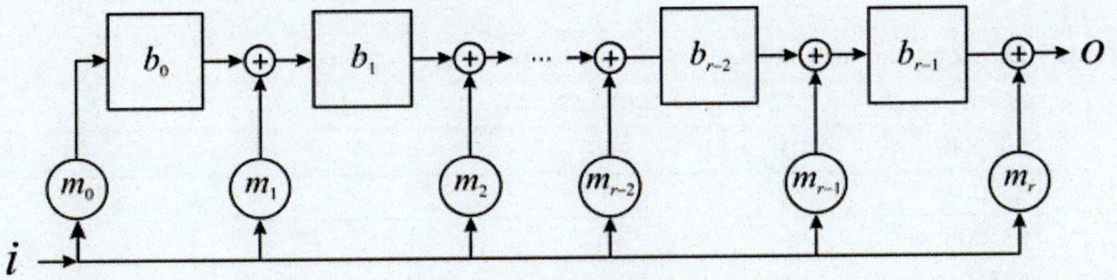

Fig. 9.22 Polynomial multiplier with a type-2 shift register

■

A type-2 shift register, as the one of Fig. 9.22 (similar to the **LFSR2**), can be used for multiplying any multiplicand P by a fixed multiplier M. It is assumed again an initial state with all the storage cells set to zero, and the multiplicand coefficients are fed to $\boldsymbol{i}$ followed by r zeros. It is immediate that the first output is $a_n m_r$, the second is $a_n m_{r\text{-}1} + a_{n\text{-}1} m_r$, and successively all the product coefficients are generated, as it is illustrated in the following example. Once more, $n + r$ clock cycles have to be applied for obtaining the product.

Example 9.20 The type-2 multiplier circuit of Fig. 9.23a can be used for multiplying any multiplicand polynomial by the fixed polynomial $M(x) = x^6 + x^2 + x + 1$ over **GF(2)**.

The type-2 multiplier circuit of Fig. 9.23b can be used for multiplying any multiplicand polynomial by the fixed polynomial $M(x) = 2x^6 + 4x^5 + 5x + 3$ over **GF(7)**. Table 9.13 shows the evolution of this circuit when the multiplicand is $x^3 + 2x + 4$; with the same result of Example 9.19.

■

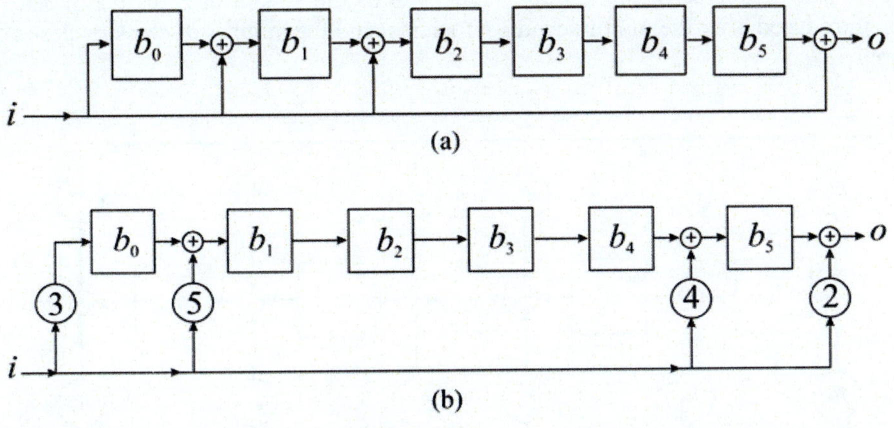

Fig. 9.23 a Type-1 **LFSR** for multiplying by $x^6 + x^2 + x + 1$ over **GF(2)**; **b** type-1 **LFSR** for multiplying by $2x^6 + 4x^5 + 5x + 3$ over **GF(7)**

Table 9.13 Results of multiplying $x^3 + 2x + 4$ by $2x^6 + 4x^5 + 5x + 3$

i	$b_0b_1b_2b_3b_4b_5$	$b_0^+ = 3i$	$b_1^+ = 5i + b_0$	$b_5^+ = 4i + b_4$	$o = 2i + b_5$
1	000000	3	5	4	2
0	350004	0	3	0	4
2	035000	6	3	1	4
4	633501	5	5	2	2
0	553352	0	5	5	2
0	055335	0	0	3	5
0	005533	0	0	3	3
0	000553	0	0	5	3
0	000055	0	0	5	5
0	000005	0	0	0	5

The structure with a type-2 shift register for computing the product by a fixed multiplier has the advantage of also allowing generating any polynomial expressed as a sum of products, in the form:

$$M(x) = \sum P_i(x) \cdot Q_i(x) \qquad (9.13)$$

As an example, for the case of the sum of two products, $M(x) = P_1(x)Q_1(x) + P_2(x)Q_2(x)$, where P_1 and P_2 are any multiplicands and Q_1 and Q_2 are fixed multipliers, the circuit of Fig. 9.24 can be used and is immediately extended to the sum of any number of products. Both multiplicands and multipliers in expression (9.13) can be polynomials of any degree. The register structure, as well as its behavior, will adapt to the degree of the multipliers; concretely, the register length will correspond to the higher degree of the multipliers, as well as the stream of zeros that has to be introduced after the multiplicand coefficients, and the number of clock cycles to apply.

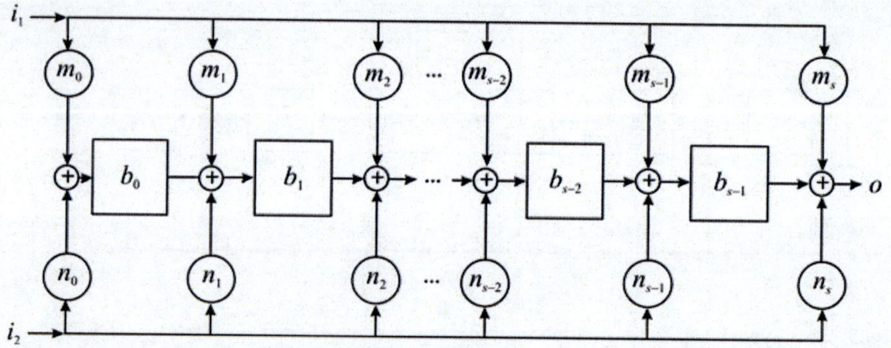

Fig. 9.24 Circuit for adding two products

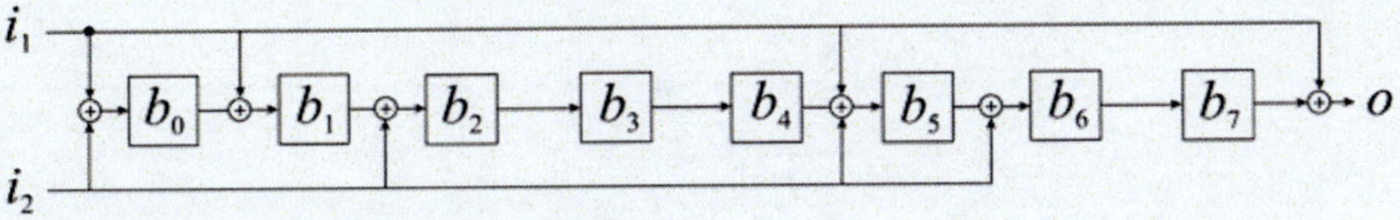

Fig. 9.25 Example circuit for sum of products

Example 9.21 The circuit of Fig. 9.25 can be used for obtaining $M(x) = P_1(x)Q_1(x) + P_2(x)Q_2(x)$ over **GF(2)**, where $P_1(x)$, of degree r_1, and $P_2(x)$, of degree r_2, are any multiplicands and $Q_1(x) = x^8 + x^5 + x + 1$ and $Q_2(x) = x^6 + x^5 + x^2 + 1$. Coefficients of $P_1(x)$ are introduced through i_1. Coefficients of $P_2(x)$ are introduced through i_2. After the corresponding coefficients, zeros are introduced and $p + u$ clock cycles are applied, with $p = \max(r_1, r_2)$ and $u = \max[g(Q_1), g(Q_2)] = \max(8, 6) = 8$.

■

9.3.3 *Circuits for Polynomial Division*

The division of the polynomial P (dividend) by the polynomial Q (divider) is defined by the quotient polynomial, C, and the remainder polynomial, R, such as:

$$P = QC + R$$

Imposing the restriction $g(R) < g(Q)$, C and R are unique.

In the same way as for the preceding operations, the coefficients of the quotient and the remainder of the division of two polynomials can be obtained from the coefficients of the dividend and divider. With parallel information, any of the circuits shown for the division of unsigned binary numbers can be easily adapted to the polynomial division. For example, from the divider circuit of Fig. 2.14d, Fig. 9.26a shows a combinational circuit for dividing, over **GF(2)** or over **GF(2^m)**, any polynomial of degree less than or equal to seven (D_7x^7+···+ D_0) by any polynomial of third degree (x^3+ d_2x^2+···+ d_0); from this división will result a quotient of fourth degree (c_4x^4+···+ c_0) and a degree-2 remainder (r_2x^2+···+ r_0).

With serial information, the **LFSR2** of Fig. 9.26b, with an external input ***i*** (thus, not working autonomously), can be used for dividing, over GF(2) or over GF(2^m), any polynomial P by a fixed divider polynomial Q of degree s,

$$Q = m_s x^s + m_{s-1} x^{s-1} + \cdots + m_1 x + m_0$$

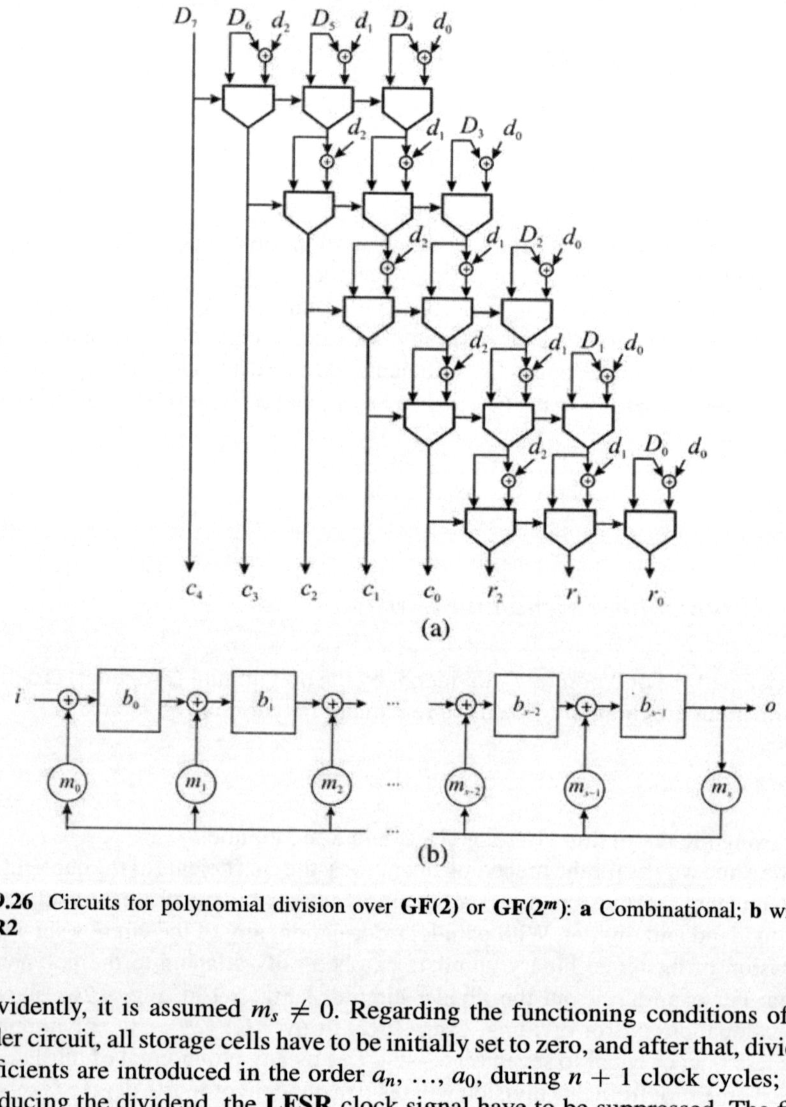

Fig. 9.26 Circuits for polynomial division over **GF(2)** or **GF(2^m)**: **a** Combinational; **b** with an **LFSR2**

Evidently, it is assumed $m_s \neq 0$. Regarding the functioning conditions of this divider circuit, all storage cells have to be initially set to zero, and after that, dividend coefficients are introduced in the order a_n, …, a_0, during $n + 1$ clock cycles; after introducing the dividend, the **LFSR** clock signal have to be suppressed. The first s output bits are zero, and in the instant $s + 1$, the coefficient a_n will be in $\boldsymbol{b}_{s\text{-}1}$, a_{n-1} will be in $\boldsymbol{b}_{s-2}$, and so on; if $a_n \neq 0$, the corresponding quotient bit has to be a_n, and the divider has to be subtracted (that over GF(2) or over GF(2^{m}) is equivalent to be added) from the most significant bits of the dividend; in the following iterations, the most significant bit of the current remainder will be in $\boldsymbol{b}_{s-1}$, and once more, every time the content of $\boldsymbol{b}_{s-1}$ is different from zero, the quotient bit has to be such content of $\boldsymbol{b}_{s-1}$ and the divider has to be subtracted from the current remainder. Thus, after the first stream of s zeros in the output $\boldsymbol{o}$, the quotient will appear, with $n - s$ bits, and once the clock has been suppressed, the s bits of the remainder will be in the **LFSR**.

For GF(p) or GF(p^m), divider circuits can be built that are similar, but more complex, to those that have just been described for GF(2) or GF(2^m). From the divider circuit in Fig. 2.14d, Fig. 9.27a shows the combinational circuit for dividing, over GF(p)or GF(p^m), any polynomial of degree equal or less than four (D_4x^4+⋯+ D_0) by any second degree polynomial ($d_2x^2 + d_1x + d_0$); from this division will result a second degree quotient (c_2x^2+⋯+ c_0) and a first degree remainder ($r_1x + r_0$). This circuit requires the subtracters, dividers and multipliers described in Sect. 3.8.

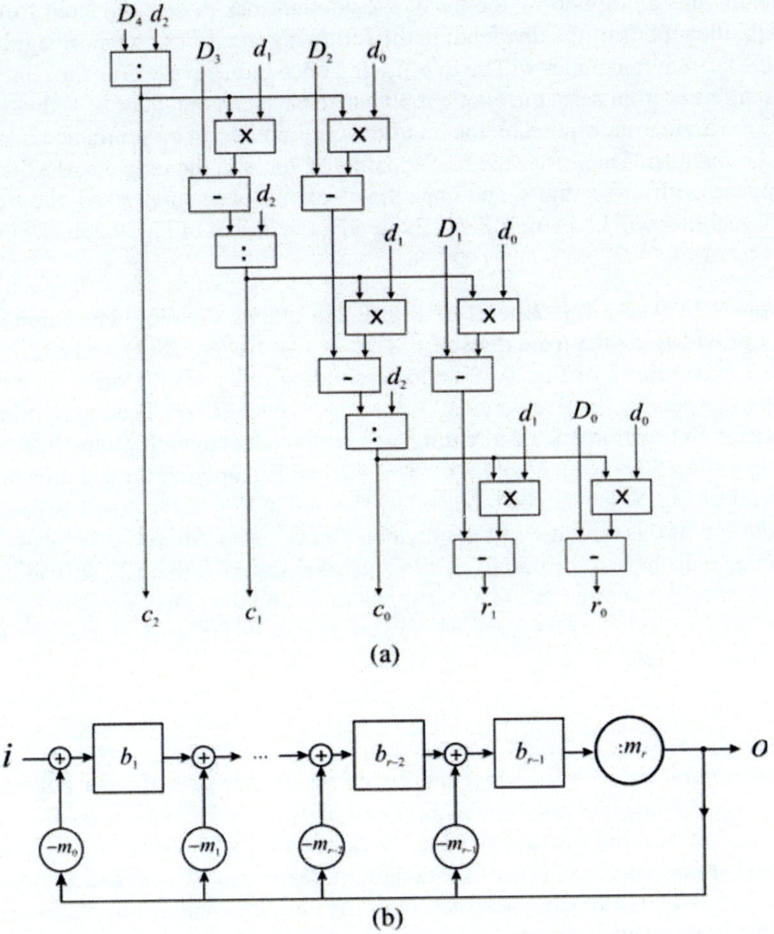

Fig. 9.27 Circuits for polynomial division over **GF(p)** or **GF(p^m)**: **a** combinational; **b** with an **LFSR2modp**

With serial information, the **LFSR2modp** of Fig. 9.27b, with an external input $\boldsymbol{i}$ (thus, not working autonomously), can be used for dividing, over GF(p) or over GF(p^m), any polynomial P by a fixed divider polynomial Q of degree r,

$$Q = m_r x^r + m_{r-1} x^{r-1} + \cdots + m_1 x + m_0$$

Evidently, it is assumed $m_r \neq 0$. Regarding the functioning conditions of this divider circuit, all storage cells have to be initially set to zero, and after that, dividend coefficients are introduced in the order $a_n, \ldots, a_0$, during $n + 1$ clock cycles; after introducing the dividend, the **LFSR** clock signal have to be suppressed. The first r output bits are zero, and in the instant $r + 1$, the coefficient a_n will be in $\boldsymbol{b}_{r-1}$, a_{n-1} will be in $\boldsymbol{b}_{r-2}$, and so on; if $a_n \neq 0$, the corresponding quotient bit has to be $a_n{:}m_r$, and the divider multiplied by the bit of the quotient has to be subtracted from the most significant bits of the dividend; in the following iterations, the most significant bit of the current remainder will be in $\boldsymbol{b}_{r-1}$, and once more, every time the content of $\boldsymbol{b}_{r-1}$ is different from zero, the quotient bit has to be such content of $\boldsymbol{b}_{r-1}$ divided by m_r and the divider multiplied by the bit of the quotient has to be subtracted from the current remainder. Thus, after the first stream of r zeros in the output $\boldsymbol{o}$, the quotient will appear, with $n - r$ digits, and once the clock has been suppressed, the r digits of the remainder will be in the **LFSR**. Evidently, the circuit of Fig. 9.26b is included in the one of Fig. 9.27b.

Example 9.22 The **LFSR2mod2** of Fig. 9.28a allows dividing any polynomial, serially provided, by the fixed divisor $x^7 + x^4 + x + 1$ over **GF(2)**.

The **LFSR2mod7** of Fig. 9.28b allows dividing, over **GF(7)**, any polynomial, serially provided, by the fixed divisor $3x^3 + x + 4$, over **GF(7)**. Having in mind the results from Subsection 3.8.5.2, dividing by 3 is equivalent to multiplying by 5; at the same time $-1{\times}5 = -5 = 2$, and $-4{\times}5 = -20 = 1$. Applying these equivalences, the circuit of Fig. 9.28b can be replaced by the one in Fig. 9.28c. Table 9.14 shows the evolution of this circuit when the dividend is $2x^7 + 4x^5 + 3x^4 + 5x^2 + 4x + 3$; the storage cells have to be initially set to zero, and input $\boldsymbol{i}$ will be 20,430,543; after 3 null outputs, the quotient 30,543 will appear in $\boldsymbol{o}$, and the remainder 265 will be in $b_2 b_1 b_0$; in fact, $(3x^4 + 5x^2 + 4x + 3) \cdot (3x^3 + x + 4) + 2x^2 + 6x + 5 = 2x^7 + 4x^5 + 3x^4 + 5x^2 + 4x + 3$.

■

It is possible, using an **LFSR2modp**, to simultaneously multiply by a fixed multiplier polynomial $(m_r x^r + \cdots + m_0)$ and divide by another fixed divider polynomial $(d_t x^t + \cdots + d_0)$ an input polynomial $E(x) = a_p x^p + a_{p-1} x^{p-1} + \cdots + a_0$. For this, it suffices to combine the structures of Figs. 9.22 and 9.27b. The length of the register to be used is $L = \max(r, t)$. If $r < t$, a quotient of degree $p - (t - r)$ and a remainder of degree $(t - 1)$ will result; the structure of the circuit is that of Fig. 9.29a, as it is illustrated in Example 9.23. If $r \geq t$, a quotient of degree $p + (r - t)$ and a remainder of degree $(t - 1)$ will result; the structure of the circuit is that of Fig. 9.29b, which requires cells with negative sub indexes, and a tail with as many zeros as these cells

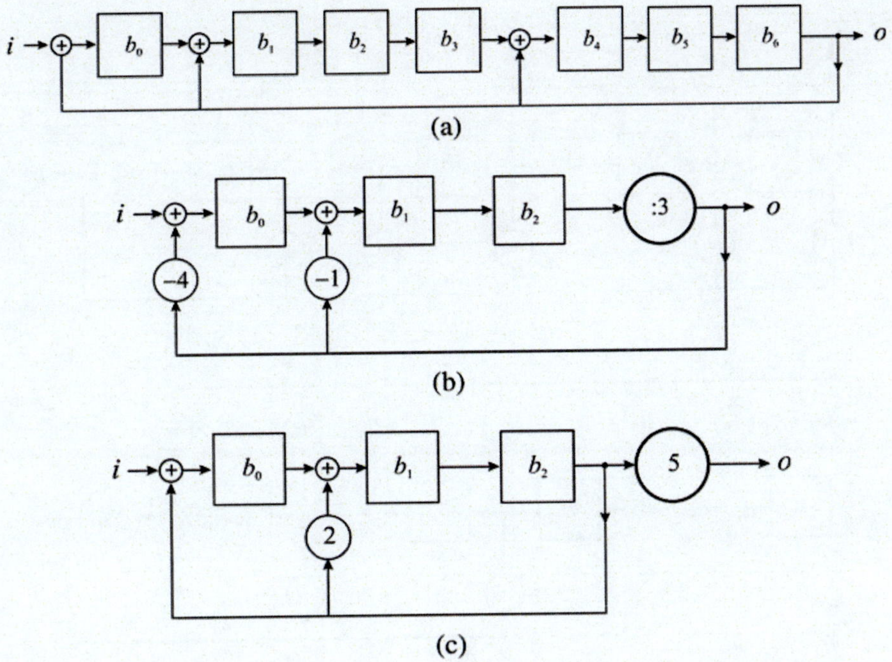

Fig. 9.28 **a** Circuit for dividing by $x^7 + x^4 + x + 1$ over **GF(2)**; **b** Circuit for dividing by $3x^3 + x + 4$ over **GF(7)**; **c** Modification of the previous circuit

Table 9.14 Results of dividing $2x^7 + 4x^5 + 3x^4 + 5x^2 + 4x + 3$ by $3x^3 + x + 4$

i	$b_0b_1b_2$	$b_0^+ = i + b_2$	$b_1^+ = b_0 + 2b_2$	$o = 5b_2$
2	000	2	0	0
0	200	0	2	0
4	020	4	0	0
3	402	5	1	3
0	510	0	5	0
5	051	6	2	5
4	625	2	2	4
3	222	5	6	3
	562			

has to be added to the input polynomial $E(x)$, while the corresponding extra clock cycles have also to be introduced, as it is illustrated in Example 9.24.

Example 9.23 The **LFSR2** of Fig. 9.30(a) allows simultaneously dividing by $x^8 + x^7 + x^3 + 1$ and multiplying by $x^5 + x^2 + x + 1$, over GF(2), any polynomial $E(x)$ provided serially.

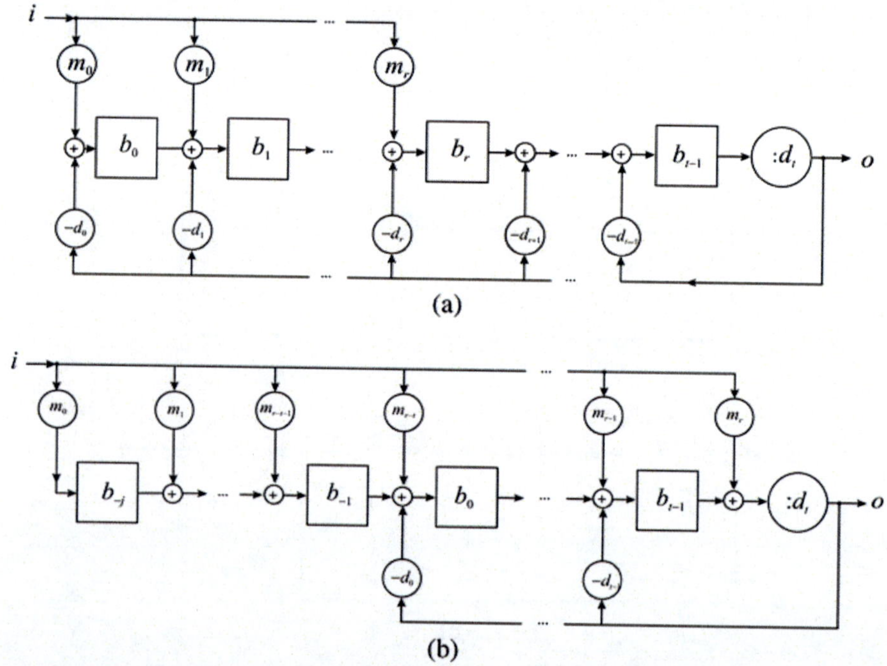

Fig. 9.29 Circuits for simultaneous multiplication and division: **a** Divider of higher degree than multiplier; **b** Divider of lesser degree than the multiplier

The **LFSR2**mod3 of Fig. 9.30b allows simultaneously dividing by $2x^5 + x^3 + x^2 + 2$ and multiplying by $2x^3 + x^2 + 1$, over GF(3), any polynomial $E(x)$ provided serially. The register cells are initially set to zero, and the bits of the dividend/multiplicand polynomial are introduced, namely $a_p x^p + a_{p-1}x^{p-1} + \cdots + a_0$. A $p - 2$ degree quotient will result, along with a 4th degree remainder. For its correct functioning, the dividend/multiplicand polynomial coefficients are introduced during $p + 1$ clock cycles, and the clock is stopped after that. Output o will initially show two zeroes, followed by the $p - 2$ quotient coefficients, while the remainder coefficients will be in $b_4 \ldots b_0$. Having in mind that $X{:}2 = 2X$, that—$X = 2X$, and that—$2X = X$, the circuit of Fig. 9.30b can be transformed in the one in Fig. 9.30c. If $E(x) = x^6 + 2x^4 + x^3 + 2x + 1$, as an example, is applied to this circuit, the following will result. Multiplying $E(x)$ by $2x^3 + x^2 + 1$, the result is $2x^9 + x^8 + x^7 + 2x^6 + x^5 + 2x^3 + x^2 + 2x + 1$, and dividing this polynomial by $2x^5 + x^3 + x^2 + 2$ results in the quotient polynomial $x^4 + 2x^3 + x + 1$ and the remainder $2x^3 + 2$. Table 9.15 shows the evolution of the circuit of Fig. 9.30c under the conditions above, generating at output o, from the third time instant, the expected quotient, and providing in the register the correct remainder.

■

Example 9.24 The **LFSR2** of Fig. 9.31a allows simultaneously dividing by $x^6 +$

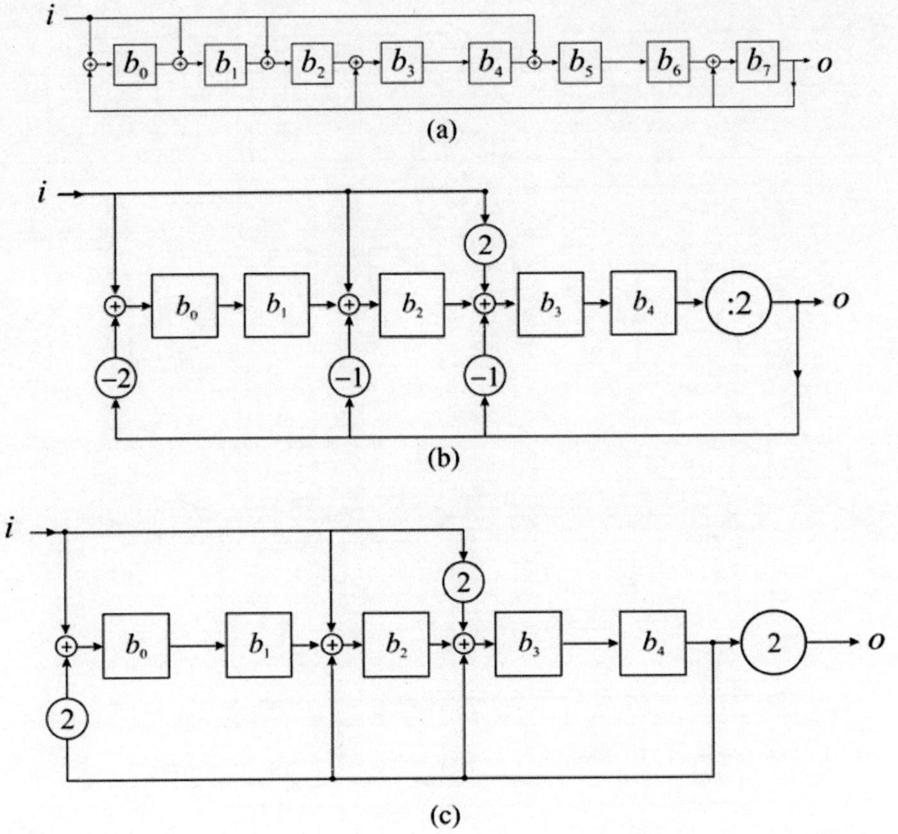

Fig. 9.30 **a** Circuit multiplying by $x^5 + x^2 + x+1$ and dividing by $x^8 + x^7 + x^3 + 1$ over GF(2); **b** Circuit multiplying by $2x^3 + x^2 + 1$ and dividing by $2x^5 + x^3 + x^2 + 2$ over GF(3); **c** Modification of the previous circuit

Table 9.15 Results for $E(x) = x^6 + 2x^4 + x^3 + 2x + 1$

i	$b_0b_1b_2b_3b_4$	$b_0^+ = i + 2b_4$	$b_2^+ = i + b_1 + b_4$	$b_3^+ = 2i + b_2 + b_4$	$o = 2b_4$
1	00000	1	1	2	0
0	10120	0	0	1	0
2	01012	0	2	0	1
1	00201	0	2	2	2
0	00220	0	0	2	0
2	00022	0	1	0	1
1	00102	2	0	2	1
	20020				

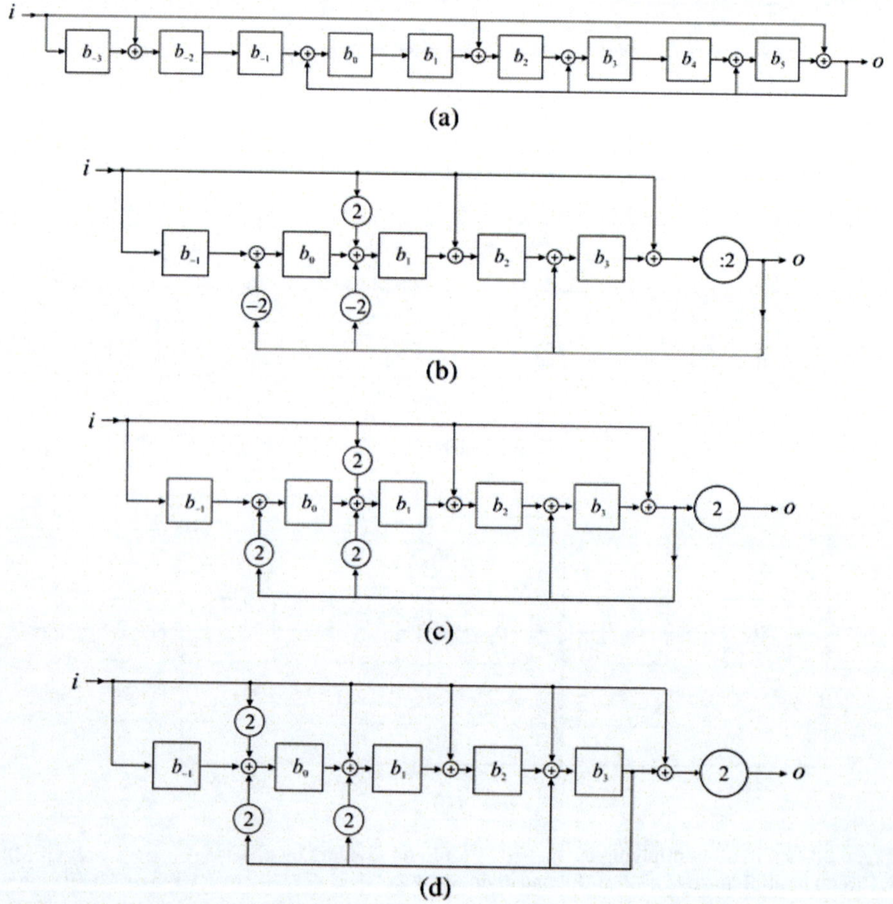

Fig. 9.31 **a** Circuit multiplying by x^9+x^5+x+1 y and dividing by $x^6+x^5+x^3+1$ *over* GF(2); **b** Circuit multiplying by $x^5 + x^3 + kx^2 + 1$ and dividing by $2x^4 + x^3 + 2x + 2$ over GF(3); **c** Modification of the previous circuit

x^5+x^3+1 and multiplying by $x^9 + x^5+x+1$ over GF(2) any polynomial provided serially.

The **LFSR2**mod3 of Fig. 9.31b allows simultaneously dividing by $2x^4+x^3+2x+2$ and multiplying by $x^5+x^3+2x^2+1$ over GF(3) any polynomial $E(x)$ provided serially. The register cells are initially set to zero, and after that the bits of the dividend/multiplicand polynomial, namely $a_px^p + a_{p-1}x^{p-1}+ \cdots + a_0$. It will result in a quotient of degree $p+1$, and a degree 3 remainder. For its correct functioning, the coefficients of the dividend/multiplicand polynomial are introduced during $p+2$ clock cycles, and the clock is stopped after that. Output o will show the $p+1$ quotient coefficients, and the remainder coefficients will be in $b_3 \ldots b_0$. Having in

Table 9.16 Results for $E(x) = x^6 + 2x^4 + x^3 + 2x + 1$ in Example 9.24

i	$b_{-1}b_0b_1b_2b_3$	$b_{-1}^+ = i$	$b_0^+ = 2i + b_{-1} + 2b_3$	$b_1^+ = i + b_0 + 2b_3$	$b_2^+ = i + b_1$	$b_3^+ = i + b_2 + b_3$	$o = 2(e + b_3)$
1	00000	1	2	1	1	1	2
0	12111	0	0	1	1	2	2
2	00112	2	2	0	0	2	2
1	22002	1	2	1	1	0	0
0	12110	0	1	2	1	1	0
2	01211	2	0	2	1	1	0
1	20211	1	0	0	0	0	1
0	10000	0	1	0	0	0	0
	01000						

mind that $X:2 = 2X$, that—$X = 2X$, and that—$2X = X$, the circuit o Fig. 9.31b can be transformed into the one of Fig. 9.31c, which can be also represented as in Fig. 9.31d.

Let it be, as an example, $E(x) = x^6 + 2x^4 + x^3 + 2x + 1$ in the circuit of Fig. 9.31c or d. Multiplying $E(x)$ by $x^5 + x^3 + 2x^2 + 1$ results in $x^{11} + 2x^7 + 2x^6 + x^4 + 2x^2 + 2x + 1$, and dividing this polynomial by $2x^4 + x^3 + 2x + 2$ results in $2x^7 + 2x^6 + 2x^5 + x$ as quotient and 1 as remainder. Table 9.16 shows the evolution of the circuit of Fig. 9.31c or d under the conditions above, generating the quotient at output o and having the correct remainder in the register.

■

9.3.4 *Multipliers and Divisors as Filters*

LFSR multiplier and divisor circuits are sequential circuits with an associated transfer function. The action of these circuits may be seen as transformers of input sequences into output sequences, independently of the interpretation of these sequences. So, the output sequence can be obtained multiplying the input sequence by the transfer function. Circuits transforming input sequences into output sequences are called **filters**. The purpose now is to obtain the transfer function for multipliers and divisors build using LFSRs.

Input (I) and output (O) sequences (see Sect. 2.1.1) are polynomials of D:

$$I(D) = i_0 + i_1 D + i_2 D^2 + \cdots + i_n D^n + \cdots$$
$$O(D) = o_0 + o_1 D + o_2 D^2 + \cdots + o_n D^n + \cdots$$

Let us consider the multiplier circuit of Fig. 9.20. Taking into account that each memory element introduce a delay D, it is clear that, if the initial content of all memory elements is zero, the output sequence $O(D)$ generate by the input sequence

$I(D)$ is:

$$\begin{aligned} O(D) &= m_r I(D) + m_{r-1} I(D) D + m_{r-2} I(D) D^2 + \cdots + m_0 I(D) D^r \\ &= (m_r + m_{r-1} D + m_{r-2} D^2 + \cdots + m_0 D^r) I(D) \end{aligned}$$

The transfer function in this case is the following polynomial of D:

$$T(D) = m_r + m_{r-1} D + m_{r-2} D^2 + \cdots + m_0 D^r$$

The circuit of Fig. 9.20, interpreted as a multiplier, multiply any input polynomial by $M(x)$:

$$M(x) = m_r x^r + m_{r-1} x^{r-1} + m_{r-2} x^{r-2} + \cdots + m_0$$

Polynomials $T(D)$ and $M(x)$ are inverses ($M(x) = T(x) \cdot x^{-r}$). This is logical, given our different interpretation of the action of a memory element. When storing the coefficient of a polynomial of x, going from a memory element to the next in the LFSR imply to multiply by x^{-1}, as the first bit of the sequence is the coefficient of the grader power of x. When storing a sequence of bits, going from a memory element to the next in the **LFSR** imply to multiply by D, as the first bit of the sequence is the coefficient of the lower power of D (see Appendix B for more details).

Similarly, for the multiplier circuit of Fig. 9.22, it is easy to see that, if the initial content of all memory element is zero, the output sequence $O(D)$ generate by the input sequence $I(D)$ is:

$$\begin{aligned} O(D) &= m_r I(D) + m_{r-1} I(D) D + m_{r-2} I(D) D^2 + \cdots + m_0 I(D) D^r \\ &= (m_r + m_{r-1} D + m_{r-2} D^2 + \cdots + m_0 D^r) I(D) \end{aligned}$$

given the same transfer function than the multiplier circuit of Fig. 9.20.

Obviously the transfer function of the circuit of Fig. 9.24, if $i_1 = i_2 = i$, is the addition of two polynomials in D. If $i_1 \neq i_2$, the output sequence will be obtained as the sum of two products, as it is tested in the following example.

Example 9.25 The output sequence of the circuit of Fig. 9.25, when $i_1 \neq i_2$, is as follows:

$$\begin{aligned} O(D) &= I_1(D)\left(1 + D^3 + D^7 + D^8\right) + I_2(D)\left(D^2 + D^3 + D^6 + D^8\right) \\ &= I_1(D) T_1(D) + I_2(D) T_2(D) \end{aligned}$$

If $i_1 = i_2$, then,

$$\begin{aligned} O(D) &= I(D)\left(1 + D^3 + D^7 + D^8 + D^2 + D^3 + D^6 + D^8\right) \\ &= I(D)\left(1 + D^2 + D^6 + D^7\right) \end{aligned}$$

The transfer function in this case is:

$$T(D) = (1 + D^2 + D^6 + D^7)$$

■

For the divisor circuit of Fig. 9.26b, it is clear that, if the initial content of the memory elements is zero, the output sequence is:

$$O(D) = m_s O(D)[m_0 D^s + m_1 D^{s-1} + \cdots + m_{s-1} D] + I(D) D^s$$

That is,

$$O(D) = I(D)\frac{D^S}{1 - m_s\left(m_0 D^S + \cdots + m_{s-1} D\right)}$$

The transfer function in this case is:

$$T(D) = \frac{D^S}{1 - m_s\left(m_0 D^S + \cdots + m_{s-1} D\right)}$$

If $m_s = 1$, and operating in GF(2^n), it results:

$$T(D) = \frac{D^S}{m_0 D^S + \cdots + m_{s-1} D + 1}$$

That is, this circuit, after a delay of s clock pulses, divides the input sequence by the polynomial $m_0 D^S + m_1 D^{S-1} + \cdots + m_{s-1} D + 1$. The circuit of Fig. 9.26b, interpreted as a divider, divides any input polynomial by $x^S + m_{s-1} x^{S-1} + m_{s-2} x^{S-2} + \cdots + m_0$

Example 9.26 The output sequence of the circuit of Fig. 9.28a is the following one:

$$O(D) = I(D) D^7 + O(D)(D^7 + D^6 + D^3)$$

That is,

$$O(D)(D^7 + D^6 + D^3 + 1) = I(D) D^7$$

The transfer function in this case is:

$$T(D) = \frac{D^7}{D^7 + D^6 + D^3 + 1}$$

■

For the divisor circuit of Fig. 9.27b, it is clear that, if the initial content of the memory elements is zero, the output sequence is:

$$m_r O(D) = O(D)[-m_0 D^r - m_1 D^{r-1} - \cdots - m_{r-1} D] + I(D) D^r$$

That is,

$$O(D)(m_0 D^r + m_1 D^{r-1} + \cdots + m_{r-1} D + m_r) = I(D) D^r$$

Or,

$$O(D) = I(D) \frac{D^r}{m_0 D^r + \cdots + m_{r-1} D + m_r}$$

The transfer function in this case is:

$$T(D) = \frac{D^r}{m_0 D^r + \cdots + m_{r-1} D + m_r}$$

That is, this circuit, after a delay of r clock pulses, divides the input sequence by the polynomial $m_0 D^r + m_1 D^{r-1} + \cdots + m_{r-1} D + 1$. The circuit of Fig. 9.27b, interpreted as a divider, divides any input polynomial by $x^r + m_{r-1} x^{r-1} + m_{r-2} x^{r-2} + \cdots + m_0$. Newly, these two polynomials are inverses.

Example 9.27 The output sequence of the circuit of Fig. 9.28c is the following one:

$$O(D) = 5\{I(D) D^3 + \frac{O(D)}{5}(D^3 + 2D^2)\}$$

Or,

$$O(D)(-D^3 - 2D^2 + 1) = 5I(D) D^3$$

The transfer function in this case is:

$$T(D) = \frac{5D^3}{-D^3 - 2D^2 + 1} = \frac{D^3}{-3D^3 - 6D^2 + 3} = \frac{D^3}{4D^3 + D^2 + 3}$$

■

Example 9.28 The output sequence of the circuit of Fig. 9.30a is the following one:

$$O(D) = I(D)(D^8 + D^7 + D^6 + D^3) + O(D)(D^8 + D^5 + D)$$

That is,

$$O(D)(D^8 + D^5 + D + 1) = I(D)(D^8 + D^7 + D^6 + D^3)$$

The transfer function in this case is:

$$T(D) = \frac{5D^3(D^5 + D^4 + D^3 + 1)}{D^8 + D^5 + D + 1}$$

■

Example 9.29 The output sequence of the circuit of Fig. 9.31a is the following one:

$$O(D) = I(D)(D^9 + D^8 + D^4 + 1) + O(D)(D^6 + D^3 + D)$$

That is,

$$O(D)(D^6 + D^3 + D + 1) = I(D)(D^9 + D^8 + D^4 + 1)$$

The transfer function in this case is:

$$T(D) = \frac{D^9 + D^8 + D^4 + 1}{D^6 + D^3 + D + 1}$$

■

Each **LFSR** multiplier/divisor circuit has its transfer function. Also, given a transfer function in the form of the ratio of two polynomials, it is easy to draw the corresponding LFSR multiplier/divisor circuit.

Filters can be arranged in series (or cascade), as in Fig. 9.32a, or in parallel, as in Fig. 9.32b. It is obvious that the transfer function corresponding to the arrangement in series is the product of the transfer function of the arranged filters, and the transfer

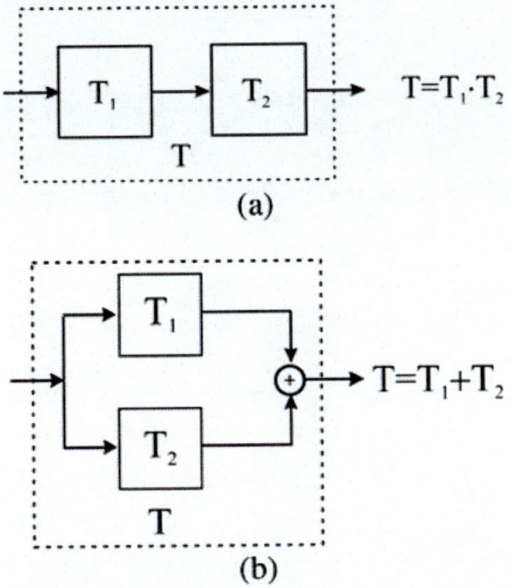

Fig. 9.32 **a** Series connection of filters; **b** parallel connection of filters

function corresponding to the arrangement in parallel is the sum of the transfer function of the arranged filters.

An important characteristic of a filter is its response to the impulse sequence when the content of all memory element of the LFSR is zero: That is, its response to the sequence 1000…, or $I(D) = 1$. If the filter has no feedback, its transfer function is a polynomial, and its impulse response is a finite sequence followed by zeros. This finite sequence corresponds to the transfer function, as:

$$O(D) = T(D)I(D) = T(D)$$

If the filter includes some feedback, its transfer function is the ratio of two polynomials, and its impulse response will include an initial transitory sequence followed by a periodic sequence. In fact, the filter, as sequential circuit, will has a finite number of states; as the input sequence is zero, except the initial 1, the filter will go to some previous state, and after that the sequence will be periodic. In the following, some examples are developed.

Example 9.30 Let us consider the divisor circuit in GF(2) of Fig. 9.33a, which transfer function is:

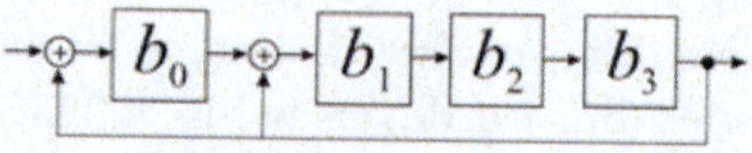

(a)

D^4 | $1 + D^3 + D^4$

$D^4 + D^7 + D^8$ | $D^4 + D^7 + D^8 + D^{10} + D^{12} + D^{13} + D^{14} + D^{15} + D^{19} + \ldots$

$D^7 + D^8$

$D^7 + D^{10} + D^{11}$

$D^8 + D^{10} + D^{11}$

$D^8 + D^{11} + D^{12}$

$D^{10} + D^{12}$

$D^{10} + D^{13} + D^{14}$

$D^{12} + D^{13} + D^{14}$

$D^{12} + D^{15} + D^{16}$

$D^{13} + D^{14} + D^{15} + D^{16}$

$D^{13} + D^{16} + D^{17}$

$D^{14} + D^{15} + D^{17}$

$D^{14} + D^{17} + D^{18}$

$D^{15} + D^{18}$

$D^{15} + D^{18} + D^{19}$

D^{19}

•

(b)

Fig. 9.33 Divisor-divisor (Example 9.30): **a** Circuit; **b** division

Table 9.17 Evolution of the circuit of Fig. 9.33a

T	b_0	b_1	b_2	b_3	O
0	1	0	0	0	0
1	0	1	0	0	0
2	0	0	1	0	0
3	0	0	0	1	0
4	1	1	0	0	1
5	0	1	1	0	0
6	0	0	1	1	0
7	1	1	0	1	1
8	1	0	1	0	1
9	0	1	0	1	0
10	1	1	1	0	1
11	0	1	1	1	0
12	1	1	1	1	1
13	1	0	1	1	1
14	1	0	0	1	1
15	1	0	0	0	1
16	0	1	0	0	0
17	0	0	1	0	0
18	0	0	0	1	0

$$T(D) = \frac{D^4}{1 + D^3 + D^4}$$

Table 9.17 gives the evolution of the divisor, with nulls initial contents, when the input is an impulse. The same output sequence may be obtained making the division corresponding to the transfer function, as it is done in Fig. 9.33b. Initially there are four nulls outputs, concluding from this that the input is not acting directly on the output; on the other hand, the output sequence has a periodicity of 15, indicating that there are feedbacks, with as least 4 memory elements. As the first four outputs are zeroes, it is obvious that the circuit is a divisor of length four. In conclusion, from the response to the impulse we can determine the topology of the circuit. ■

Example 9.31 Let us consider the multiplier-divisor circuit in GF(2) of Fig. 9.34a, which transfer function is:

$$T(D) = \frac{D\left(1 + D + D^3\right)}{1 + D^3 + D^4} = \frac{D + D^2 + D^4}{1 + D^3 + D^4}$$

Table 9.18 gives the evolution of the multiplier-divisor, with nulls initial contents,

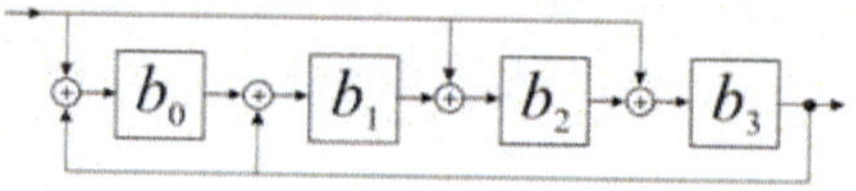

(a)

$$D + D^2 + D^4 \quad | \quad 1 + D^3 + D^4$$

$$\underline{D + D^4 + D^5} \quad\quad D + D^2 + D^6 + D^9 + D^{10} + D^{12} + D^{14} + D^{15} + D^{16} + D^{17} + D^{21} + \ldots$$

$$D^2 + D^5$$
$$\underline{D^2 + D^5 + D^6}$$
$$D^6$$
$$\underline{D^6 + D^9 + D^{10}}$$
$$D^9 + D^{10}$$
$$\underline{D^9 + D^{12} + D^{13}}$$
$$D^{10} + D^{12} + D^{13}$$
$$\underline{D^{10} + D^{13} + D^{14}}$$
$$D^{12} + D^{14}$$
$$\underline{D^{12} + D^{15} + D^{16}}$$
$$D^{14} + D^{15} + D^{16}$$
$$\underline{D^{14} + D^{17} + D^{18}}$$
$$D^{15} + D^{16} + D^{17} + D^{18}$$
$$\underline{D^{15} + D^{18} + D^{19}}$$
$$D^{16} + D^{17} + D^{19}$$
$$\underline{D^{16} + D^{19} + D^{20}}$$
$$D^{17} + D^{20}$$
$$\underline{D^{17} + D^{20} + D^{21}}$$
$$D^{21}$$
$$\cdot$$

(b)

Fig. 9.34 Multiplier-divisor (Example 9.31): **a** Circuit; **b** division

when the input is an impulse. The same output sequence may be obtained making the division corresponding to the transfer function, as it is done in Fig. 9.34b. Initially there is one null output, concluding from this that the input is not acting directly on the output; on the other hand, the output sequence has a periodicity of 15, indicating that there are feedbacks, with as least 4 memory elements in the feedback loop. As only the first output is zero, it is obvious that the circuit is a multiplier-divisor, with the multiplier of lower degree than the divisor. In conclusion, from the response to the impulse we can determine the topology of the circuit.

■

Example 9.32 Let us consider the multiplier-divisor circuit in GF(2) of Fig. 9.35a, which transfer function is:

$$T(D) = \frac{1 + D + D^4}{1 + D^2 + D^3}$$

Table 9.19 gives the evolution of the multiplier-divisor, with nulls initial contents,

Table 9.18 Evolution of the circuit of Fig. 9.34a

T	$b0$	$b1$	$b2$	$b3$	O
0	1	0	1	1	0
1	1	0	0	1	1
2	1	0	0	0	1
3	0	1	0	0	0
4	0	0	1	0	0
5	0	0	0	1	0
6	1	1	0	0	1
7	0	1	1	0	0
8	0	0	1	1	0
9	1	1	0	1	1
10	1	0	1	0	1
11	0	1	0	1	0
12	1	1	1	0	1
13	0	1	1	1	0
14	1	1	1	1	1
15	1	0	1	1	1
16	1	0	0	1	1
17	1	0	0	0	1
18	0	1	0	0	0

when the input is an impulse. The same output sequence may be obtained making the division corresponding to the transfer function, as it is done in Fig. 9.35b. The initial output is 1, concluding from this that the input is acting directly on the output; on the other hand, the output sequence has a periodicity of 7, indicating that there are feedbacks, with as least 3 memory elements in the feedback loop. In conclusion, from the response to the impulse we can determine the topology of the circuit.

■

9.4 Cellular Automata

An other type of system that are often used in a wide range of applications and whose behavior can also be algebraically described are cellular automata (**CA** in the following). Their structure is very similar to the **LFSR**s above. Most simple **CA**s, one-dimensional and linear, are going to be describe first, and they will be generalized by defining the multidimensional and the non-linear **CA**s.

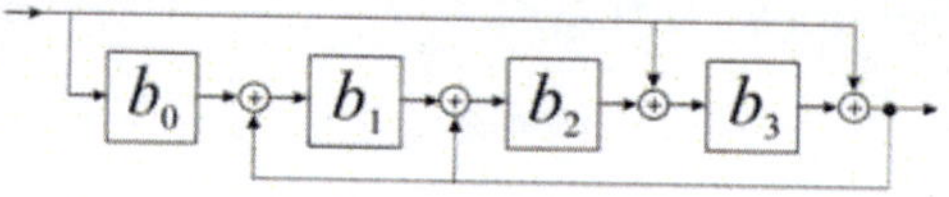

(a)

$1 + D + D^4 \;\; | \;\; 1 + D^2 + D^3$

$1 + D^2 + D^3 \qquad 1 + D + D^2 + D^4 + D^5 + D^6 + D^9 + D^{11} + D^{12} + D^{13} + \ldots$

$D + D^2 + D^3 + D^4$

$D + D^3 + D^4$

D^2

$D^2 + D^4 + D^5$

$D^4 + D^5$

$D^4 + D^6 + D^7$

$D^5 + D^6 + D^7$

$D^5 + D^7 + D^8$

$D^6 + D^8$

$D^6 + D^8 + D^9$

D^9

$D^9 + D^{11} + D^{12}$

$D^{11} + D^{12}$

$D^{11} + D^{13} + D^{14}$

$D^{12} + D^{13} + D^{14}$

$D^{12} + D^{14} + D^{15}$

$D^{13} + D^{15}$

•

(b)

Fig. 9.35 Multiplier-divisor (Example 9.32): **a** Circuit; **b** division

Table 9.19 Evolution of the circuit of Fig. 9.35a

T	*b*0	*b*1	*b*2	*b*3	*O*
0	1	1	1	1	1
1	0	0	0	1	1
2	0	1	1	0	1
3	0	0	1	1	0
4	0	1	1	1	1
5	0	1	0	1	1
6	0	1	0	0	1
7	0	0	1	0	0
8	0	0	0	1	0
9	0	1	1	0	1
10	0	0	1	1	0
11	0	1	1	1	1
12	0	1	0	1	1
13	0	1	0	0	1

9.4.1 *One-Dimensional Linear Cellular Automata*

A one-dimensional linear cellular automata (usually known just as cellular automata) of size n is formed by a linear concatenation of n **D** flip-flops, or **CA** cells. These stores n bits, b_0, ..., b_{n-1}, so the state of the **CA** in any instant is given by b_0, ..., b_{n-1}. The cells of the **CA** are connected in such a way that the next state of each cell in the chain is a linear function of only the present state of this cell and its two neighbors, *i.e.*,

$$b_i(n+1) = \mathbf{EXOR}[a_{i-1}b_{i-1}(n), a_i b_i(n), a_{i+1}b_{i+1}(n)] \tag{9.14}$$

where, as in previous cases, the coefficients a_j can have values 0 or 1. Thus, from an initial state or *seed*, the **CA** evolutes autonomously through different states. This concept of neighborhood (each cell has only two neighbors) can be generalized: the state of each cell can be made to depend on itself and four neighbors, or six, or, in general, on r neighbors, so each neighborhood is formed by $r + 1$ cells (the central cell plus r neighbors). This generalization leads to a more varied behavior. Nevertheless, given the introductory nature of this section, the following will be limited to the simplest case of three-cell neighborhoods.

In order to simplify the graphic representation of the **CA**s, the concatenation of an **EXOR** gate to the input of a **D** flip-flop will be represented as in Fig. 9.36a. Using this symbol, Fig. 9.36b shows a simple example of **CA** with four cells.

Expression (9.14), which determines the evolution of the automata, can be expressed as a truth table, as it is done in the marked column of Table 9.20 for the case:

$$b_i(n+1) = \mathbf{EXOR}[b_{i-1}(n), b_i(n), b_{i+1}(n)] \tag{9.15}$$

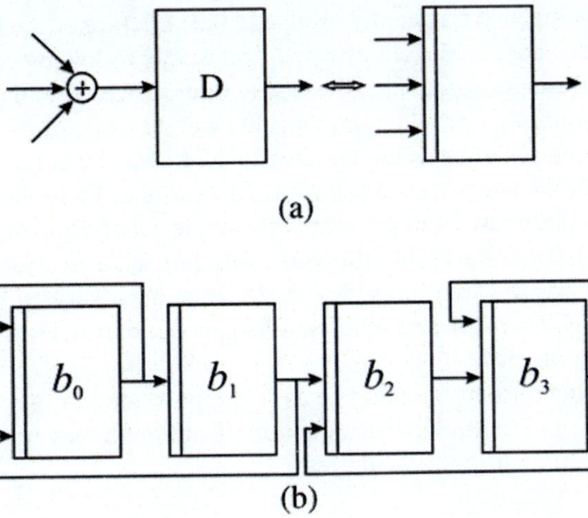

Fig. 9.36 **a** EXOR-D concatenation; **b** CA example

Table 9.20 **EXOR** rules

	$b^n_{i-1}b^n_i b^n_{i+1}$	b_i^{n+1}	b_i^{n+1}	b_i^{n+1}	b_i^{n+1}	b_i^{n+1}	b_i^{n+1}	b_i^{n+1}
1	000	0	0	0	0	0	0	0
2	001	1	0	1	1	1	0	0
4	010	1	1	0	1	0	1	0
8	011	0	1	1	0	1	1	0
16	100	1	1	1	0	0	0	1
32	101	0	1	0	1	1	0	1
64	110	0	0	1	1	0	1	1
128	111	1	0	0	0	1	1	1
	Rule no.	150	60	90	102	170	204	240

Concretely, for this case of (9.15), it is the switching function {0, 1, 1, 0, 1, 0, 0, 1}, which interpreted as a binary number with the last bit as most significant would be $10010110_2 = 150_{10}$. The usual is to designate expression (9.14) using its decimal value, so (9.15) is known as **rule 150**. Strictly sticking to relation (9.14), apart from rule 0 that has no interest, there are seven rules (60, 90, 102, 150, 170, 204 and 240), corresponding to the different combinations of the values of coefficients a_j, such as they are shown in Table 9.20. Among these, the two rules producing interesting results are 90 and 150, which are the only ones that are normally used in practice.

If a measurement of the cost of each rule has to be established, this would be given by the number of inputs of the **EXOR** gate. With this measurement, rule 150 would be the costliest, since it requires te use of a three-input **EXOR** gate; rules 60, 90 and 102 require a two-input **EXOR** gate, while rules 170, 204 and 240 do not require any gate.

The simplest **CA**s are those with all cells following the same rule; in this case, the **CA** structure is known as **uniform**. Cellular automata with each cell following its own rule are also frequently used, and the structure of the **CA** is known as **hybrid structure** ; concretely, circuits with some cells following rule 90 and other following rule 150 can be used, since these structures provide good behaviors. There are other rule couples that provide good results, and the couples of rules that provide good results acting together as known as **autopletic rules**. In the extreme case, each cell in the cellular automata can follow a different rule. The following will start considering the simplest case of application of a single rule (uniform **CA**).

Expression (9.14) allows determining the next state of all the cells of the automata, but for the two ends. For these, it is necessary to define some boundary conditions. There are two types of relevant boundaries, **constant boundaries** and **periodic boundaries**.

In constant boundaries, as it is represented in Fig. 9.37a, the terminal values are constant (c_1 and c_2), usually both 0, although they might be 1. Thus, there can be up to four constant boundaries.

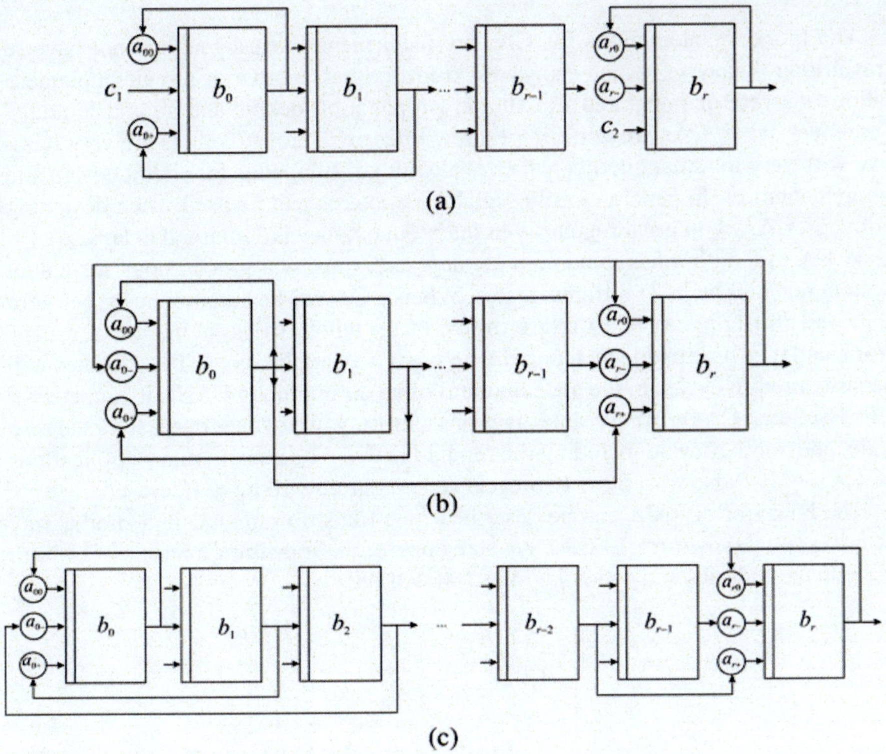

Fig. 9.37 **a** Constant boundaries; **b** periodic boundaries; **c** local boundaries

In periodic boundaries, as it is represented in Fig. 9.37b, the cellular automata is considered to have a circular structure, so the first and last elements are neighbors. Having in mind the implementation of the corresponding circuits, periodic boundaries in one-dimensional cellular automata would naturally lead to arranging the cells in two rows, so the first and last cells are close and the feedback signals do not require long paths.

Apart from these two most relevant boundaries, **local boundaries** are occasionally used, considering that the two neighbors of each terminal cell are the two cells closer ones, as it is represented in Fig. 9.37c.

Observing Fig. 9.36b, the similarity of **CA**s and **LFSR**s is quite clear; in both cases, a set of memory elements evolutes autonomously from an initial state or **seed**. A significant difference between both structures is the range of the interactions between memory elements; for **LFSR**s, the next state of a cell can depend on the current states of other cells that are very distant from it in the flip-flop chain, while for **CA**s interactions are always closer (even in the case of periodic boundaries, the flip-flop chain can be easily arranged with the two terminals as neighbors).

The fact that interactions for **CA**s are local means that signals do not have to travel long distances, which translates, from the perspective of circuit implementation (discrete or integrated), in shorter propagation delays and simpler circuits. Moreover, when **CA**s are uniform or a few rules are used, structures are very iterative with very modular circuits, which is always advantageous for circuit design and implementation. In general, easily scalable structures can be used when designing with con **CA**s, with great regularity in interconnections and minimal delays.

A **CA** of length n has n memory element and, thus, is a synchronous sequential system that can be in 2^n different states. When a **CA** works autonomously, the zero state (all flip-flops set to 0) is a terminal state, where the system cannot escape from, and it will skipped when possible in many **CA** applications. The 2^n states will be structured in cycles in the state diagram of an autonomous **CA**, as it happens for **LFSR**s. Some **CA**s will have state diagrams with only two cycles: one just for the zero state, and another cycle including the remaining $2^n - 1$ states; in many applications of **CA**s, only those with these two-cycle state diagram will be of interest.

The behavior of a **CA** can be described by a transition matrix, in a similar way as it has been shown for **LFSR**s. As an example, the transition matrix of **CA** with periodic boundaries and following rule 150 will be:

$$T = \begin{bmatrix} 1 & 1 & 0 & \dots & 0 & 0 & 1 \\ 1 & 1 & 1 & \dots & 0 & 0 & 0 \\ \dots & \dots & \dots & \dots & \dots & \dots & \dots \\ 0 & 0 & 0 & \dots & 1 & 1 & 1 \\ 1 & 0 & 0 & \dots & 0 & 1 & 1 \end{bmatrix}$$

In general, the transition matrix for an uniform **CA**, whose rule is given by (9.14), and with periodic boundaries, will be:

$$T = \begin{bmatrix} a_0 & a_+ & 0 & \dots & 0 & 0 & a_- \\ a_- & a_0 & a_+ & \dots & 0 & 0 & 0 \\ \dots & \dots & \dots & \dots & \dots & \dots & \dots \\ 0 & 0 & 0 & \dots & a_- & a_0 & a_+ \\ a_+ & 0 & 0 & \dots & 0 & a_- & a_0 \end{bmatrix}$$

Some properties of the **CA** can be extracted from matrix T, whose characteristic equation or polynomial, as it has been previously discussed for **LFSR**s, is given by:

$$\det(T + xI) = 0$$

Example 9.33 The transition matrix for the CA of size 4 in Fig. 9.38a, with null boundaries and all cells following rule 90, is:

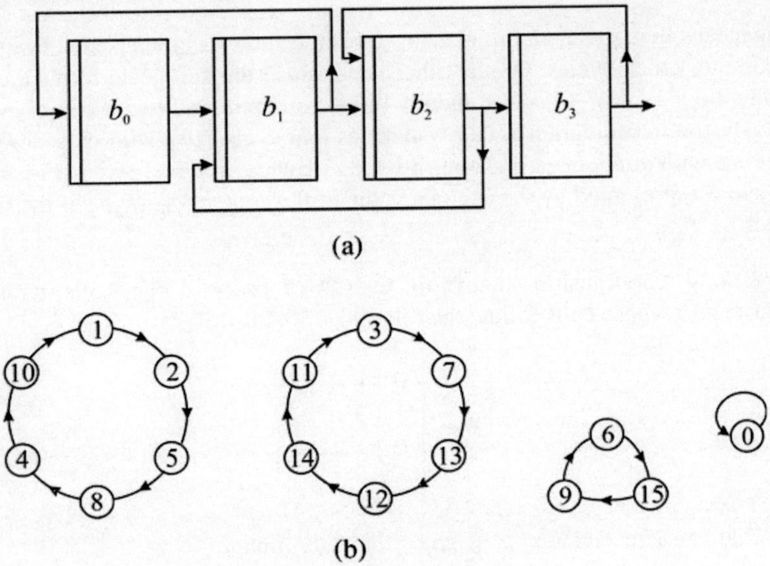

Fig. 9.38 **a** CA of Example 9.33; **b** State diagram for Example 9.33

$$T = \begin{bmatrix} 0 & 1 & 0 & 0 \\ 1 & 0 & 1 & 0 \\ 0 & 1 & 0 & 1 \\ 0 & 0 & 1 & 0 \end{bmatrix}$$

The characteristic equation is given by the determinant

$$T = \begin{vmatrix} x & 1 & 0 & 0 \\ 1 & x & 1 & 0 \\ 0 & 1 & x & 1 \\ 0 & 0 & 1 & x \end{vmatrix} = x^4 + x^2 + 1$$

This is not a primitive polynomial, since it divides $x^6 + 1$. As it is known, T satisfies its own characteristic equation, i.e.:

$$T^4 + T^2 + I = 0$$

From this, $T^4 = T^2 + I$, or also $T^4 + T^2 = I$. With all this, it is easy to see that $T^6 = I$ (this can be also directly checked, computing T^6); in fact:

$$T^6 = T^4 \cdot T^2 = \left(T^2 + I\right) \cdot T^2 = T^4 + T^2 = I$$

This means that, given an initial state, the same state will be repeated every six, as maximum, clock cycles. On the other hand, since the zero state forms a cycle, and there are a total of 16 states, there will be, as maximum, two six-state cycles. Effectively, the state diagram of this system has four cycles, two with six states each, another one with three states, and one with the zero state, as it is shown in Fig. 9.38b. Each state is represented by the decimal value of the content $\boldsymbol{b_0b_1b_2b_3}$ of the **CA**.

■

Example 9.34 The transition matrix for the CA of size 4 in Fig. 9.39a, with null boundaries and whose cells follow the rules 90, 150, 90, 150, is:

$$T = \begin{bmatrix} 0\,1\,0\,0 \\ 1\,1\,1\,0 \\ 0\,1\,0\,1 \\ 0\,0\,1\,1 \end{bmatrix}$$

The characteristic equation is given by the determinant

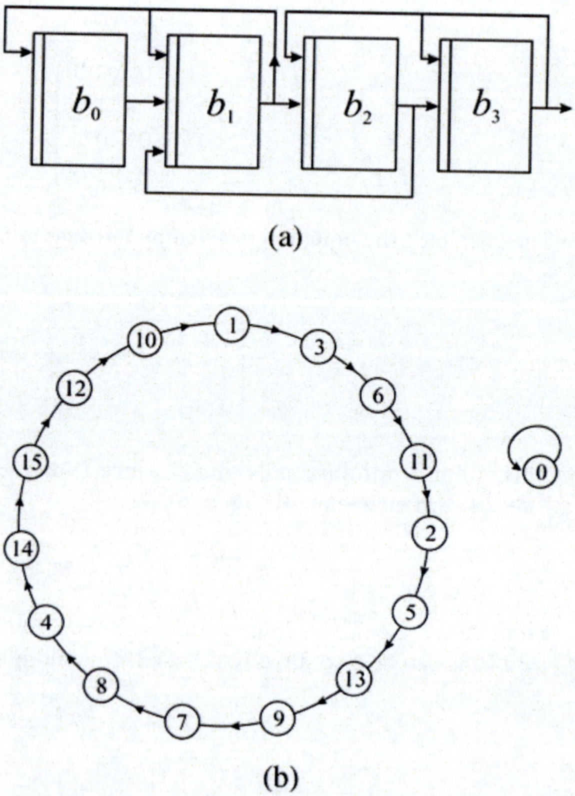

Fig. 9.39 **a** CA of Example 9.34; **b** State diagram for Example 9.34

$$\begin{vmatrix} x & 1 & 0 & 0 \\ 1 & x+1 & 1 & 0 \\ 0 & 1 & x & 1 \\ 0 & 0 & 1 & x+1 \end{vmatrix} = x^4 + x + 1$$

This is a primitive polynomial, and the state diagram of this system has two cycles, one with fifteen states and another single-state cycle (zero state), as it is represented in Fig. 9.39b. Each state is represented by the decimal value of the content $b_0b_1b_2b_3$ of the **CA**.

■

Example 9.35 The transition matrix for the CA of size 4 in Fig. 9.40a, with null boundaries and whose cells follow rules 150, 60, 90 150, is:

$$T = \begin{bmatrix} 1\,1\,0\,0 \\ 1\,1\,0\,0 \\ 0\,1\,0\,1 \\ 0\,0\,1\,1 \end{bmatrix}$$

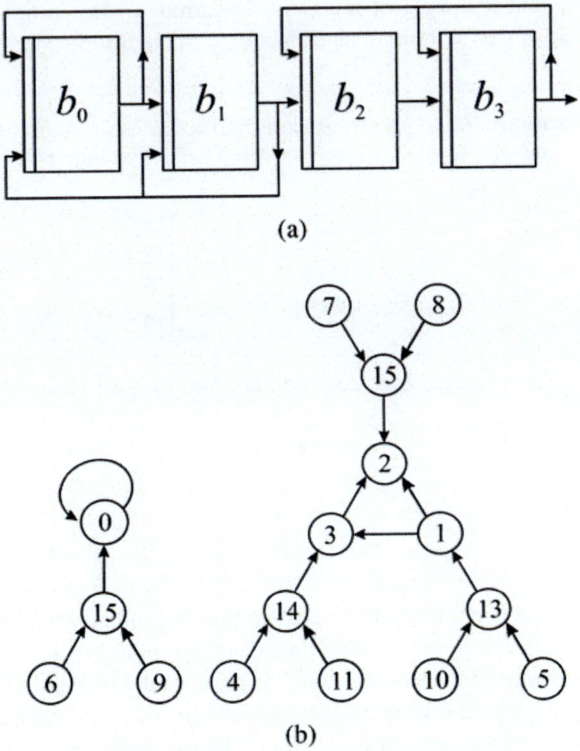

Fig. 9.40 **a** CA of Example 9.35; **b** state diagram for Example 9.35

The characteristic equation is given by the determinant

$$\begin{vmatrix} x+1 & 1 & 0 & 0 \\ 1 & x+1 & 0 & 0 \\ 0 & 1 & x & 1 \\ 0 & 0 & 1 & x+1 \end{vmatrix} = x^4 + x^3 + x^2 = x^2\left(x^2 + x + 1\right)$$

This is not a primitive polynomial. T satisfies its own characteristic equation, i.e.:

$$T^4 + T^3 + T^2 = 0$$

From this, $T^4 = T^2 + T^2$, or also $T^4 + T^3 = T^2$. Wit all this it is easy to see that $T^5 = T^2$ (this can be also directly checked, computing T^2 and T^5); in fact:

$$T^5 = T^4 \cdot T = \left(T^3 + T^2\right) \cdot T = T^4 + T^3 = T^2$$

Correspondingly, the state diagram of this system has a tree structure, with two branching levels, and includes a cycle of three states. Figure 9.40b represents the corresponding state diagram. Each state is represented by the decimal value of the content $\boldsymbol{b_0 b_1 b_2 b_3}$ of the **CA**. Depending on the initial state, the system ends, after one or two intermediate transitions, in the three-state cycle, or in the zero state.

■

Example 9.36 The transition matrix for the CA of size 4 in Fig. 9.41a, with null boundaries and whose cells follow rules 150, 60, 150 150, is:

$$T = \begin{bmatrix} 1\,1\,0\,0 \\ 1\,1\,0\,0 \\ 0\,1\,1\,1 \\ 0\,0\,1\,1 \end{bmatrix}$$

The characteristic equation is given by the determinant

$$\begin{vmatrix} x+1 & 1 & 0 & 0 \\ 1 & x+1 & 0 & 0 \\ 0 & 1 & x+1 & 1 \\ 0 & 0 & 1 & x+1 \end{vmatrix} = x^4$$

This is not a primitive polynomial. It holds that $T^4 = 0$ (as it can be easily check with the proper computations). Correspondingly, the state diagram of this system has an 4-level tree structures, as it is represented in Fig. 9.41b. Each state is represented by the decimal value of the content $\boldsymbol{b_0 b_1 b_2 b_3}$ of the **CA**. Independently of the initial state, the system always ends in the zero state after a maximum of four clock cycles.

■

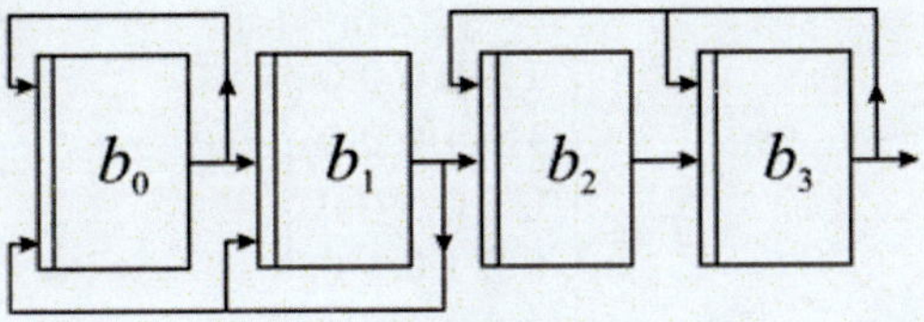

(a)

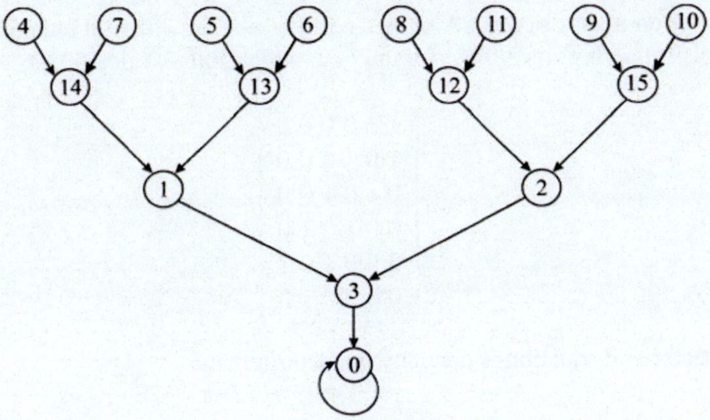

(b)

Fig. 9.41 **a CA** of Example 9.36; **b** state diagram for Example 9.36

Example 9.37 The transition matrix of the **CA** of size 6 in Fig. 9.42a, with null boundaries and with all cells following rule 90 but for the first, which follow rule 150, is:

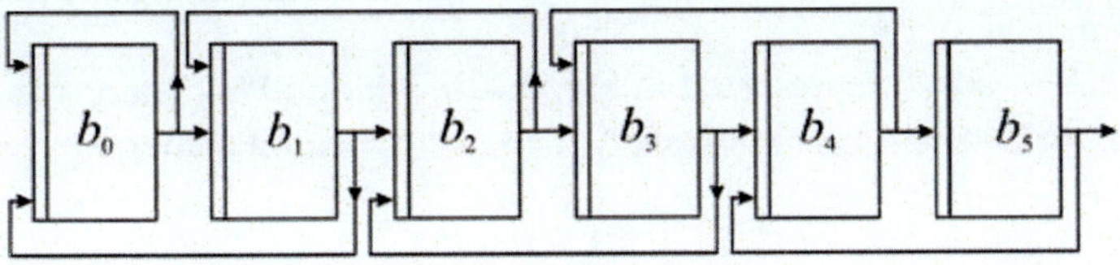

(a)

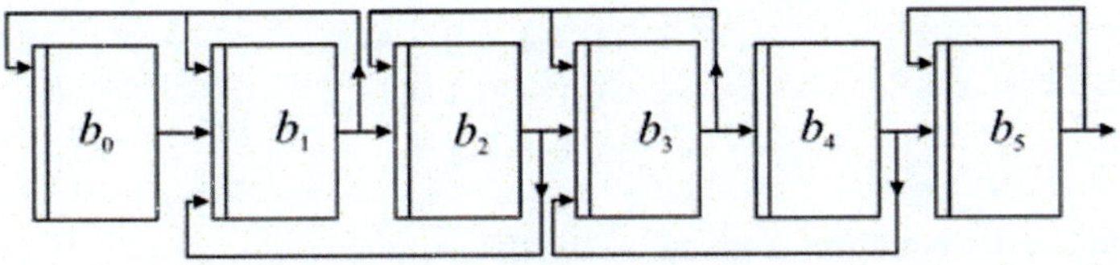

(b)

Fig. 9.42 **a** First **CA** of Example 9.37; **b** second **CA** of Example 9.37

$$T = \begin{bmatrix} 1\,1\,0\,0\,0\,0 \\ 1\,0\,1\,0\,0\,0 \\ 0\,1\,0\,1\,0\,0 \\ 0\,0\,1\,0\,1\,0 \\ 0\,0\,0\,1\,0\,1 \\ 0\,0\,0\,0\,1\,0 \end{bmatrix}$$

The transition matrix of the **CA** of size 6 in Fig. 9.42a, with null boundaries and with all cells following rule 90 but for the first, which follow rule 150, is:

$$T = \begin{bmatrix} 1\,1\,0\,0\,0\,0 \\ 1\,0\,1\,0\,0\,0 \\ 0\,1\,0\,1\,0\,0 \\ 0\,0\,1\,0\,1\,0 \\ 0\,0\,0\,1\,0\,1 \\ 0\,0\,0\,0\,1\,0 \end{bmatrix}$$

so the characteristic equation is given by the determinant:

$$\begin{vmatrix} 1+x & 1\,0\,0\,0\,0 \\ 1 & x\,1\,0\,0\,0 \\ 0 & 1\,x\,1\,0\,0 \\ 0 & 0\,1\,x\,1\,0 \\ 0 & 0\,0\,1\,x\,1 \\ 0 & 0\,0\,0\,1\,x \end{vmatrix} = x^6 + x^5 + x^4 + x + 1$$

This is a primitive polynomial and, thus, this **CA** will generate sequences of maximum length, $2^6 - 1 = 63$, as it can be easily checked assuming any initial state different to state zero.

For the **CA** of size 6 represented in Fig. 9.42b, with null boundaries and whose cells follow the rules 90, 150, 90, 150, 90, 150, the transition matrix is:

$$T = \begin{bmatrix} 0\,1\,0\,0\,0\,0 \\ 1\,1\,1\,0\,0\,0 \\ 0\,1\,0\,1\,0\,0 \\ 0\,0\,1\,1\,1\,0 \\ 0\,0\,0\,1\,0\,1 \\ 0\,0\,0\,0\,1\,1 \end{bmatrix}$$

so the characteristic equation is given by the determinant:

$$\begin{vmatrix} x & 1 & 0 & 0 & 0 & 0 \\ 1 & x+1 & 1 & 0 & 0 & 0 \\ 0 & 1 & x & 1 & 0 & 0 \\ 0 & 0 & 1 & x+1 & 1 & 0 \\ 0 & 0 & 0 & 1 & x & 1 \\ 0 & 0 & 0 & 0 & 1 & x+1 \end{vmatrix} = x^6 + x^5 + x^3 + x^2 + 1$$

Once more this is a primitive polynomial and, thus, this **CA** will generate sequences of maximum length, $2^6 - 1 = 63$, as it can be easily checked assuming any initial state different to state zero.

■

From the **CA** examples above, it is concluded that the behavior of each one of these systems can be mainly deduced from the associated polynomial. The behavior of the **CA**s can be more varied than that of the **LFSR**s. There are **CA**s that are associated to a primitive polynomial and whose behavior is practically identical to that of the **LFSR1** or **LFSR2** associated to the same polynomial; thus, they generate **M** sequences with properties that are practically identical to those generated by the **LFSR**s, and the **CA** goes through all the possible states (but the zero state) and, in this way, the **CA** will hold all the *n*-bit combinations but the all-zero. Other **CA**s evolute in such a way that they always end in the zero state (in general, in a **drain** state); this behavior is of interest in the design of some applications of the **CA**s.

Comparing the two **CA**s of size 6 presented in Example 9.37, the first one would be preferable since it employs less the rule 150, which is the most costly. This more economic solution will be usually preferred. Using this idea, Table 9.21 shows a minimal solution, making use of only rules 90 and 150, for building **CA*s*** of size up to 50 that generate sequences of maximum length. It is supposed that all cells follow rule 90 but those indicated, which follow rule 150. This table is organized in five columns and provides one of the possible solutions. For example, for a **CA** of length 32 generating sequences of maximum length, rule 90 can be used for all cells but for

Table 9.21 CAs generating sequences of maximum length

1:1	11:01	21:1, 10	31:11	41:01:00
2:1	12:3, 7	22:05	32:1, 15	42:19:00
3:1	13:05	23:01	33:1	43:03:00
4:1, 3	14:01	24:8, 12	34:1, 19	44:4, 26
5:1	15:03	25:09:00	35:1	45:09:00
6:1	16:1, 15	26:01:00	36:6	46:2, 10
7:3	17:05	27:1, 20	37:09:00	47:13:00
8:2, 3	18:1, 17	28:03:00	38:07:00	48:15:00
9:1	19:03	29:01:00	39:01:00	49:1, 10
10:2, 7	20:2, 3	30:01:00	40:08:00	50:11:00

Table 9.22 **EXNOR** rules

	$b^n_{i-1}b^n_i b^n_{i+1}$	b_i^{n+1}	b_i^{n+1}	b_i^{n+1}	b_i^{n+1}	b_i^{n+1}	b_i^{n+1}	b_i^{n+1}
1	000	1	1	1	1	1	1	1
2	001	1	1	0	0	0	0	1
4	010	1	0	1	0	0	1	0
8	011	1	0	0	1	1	0	0
16	100	0	1	1	0	1	0	0
32	101	0	1	0	1	0	1	0
64	110	0	0	1	1	0	0	1
128	111	0	0	0	0	1	1	1
	Rule no.	15	51	85	105	153	165	195

cell 1 and cell 15, which have to follow rule 150. In [4] a more complete table can be found.

9.4.2 *One-Dimensional Non-linear Cellular Automata*

One-dimensional **CAs** shown above make use of only the **EXOR** connector for generating the interaction signals between neighboring cells: these are **linear one-dimensional CA (LOCA)**. Obviously, the **EXNOR** connector, which does not suppose any cost increase compared to the **EXOR** connector, could be also used. The **EXNOR**, apart from rule 255, which is of no interest, leads to rules 15, 51, 85, 105, 153, 165 y 195, as it is shown in Table 9.22, and that are known as **complementary rules**. **CAs** only using linear or complementary rules are known as **additive CAs**.

Moreover, rules derived from the connectors **AND-OR** can be used; with these it is possible to have any relationship between neighbor cells. In this case, it is said that **non-additive rules** are used.

9.4.3 *Bidimensional Cellular Automata*

One-dimensional **CA** cells are linearly structured, and each internal cell has two neighbors. The generalization of this idea is immediate, so the cells are arranged in a plain. The most simple structure is that corresponding to the nodes of an square network, as it is represented in Fig. 9.43a. In this case, each internal cell **C** is considered to have four neighbors: north, east, south, west, thus resulting in a five-cell neighborhood, as it is represented in Fig. 9.43b. It is also possible to consider that

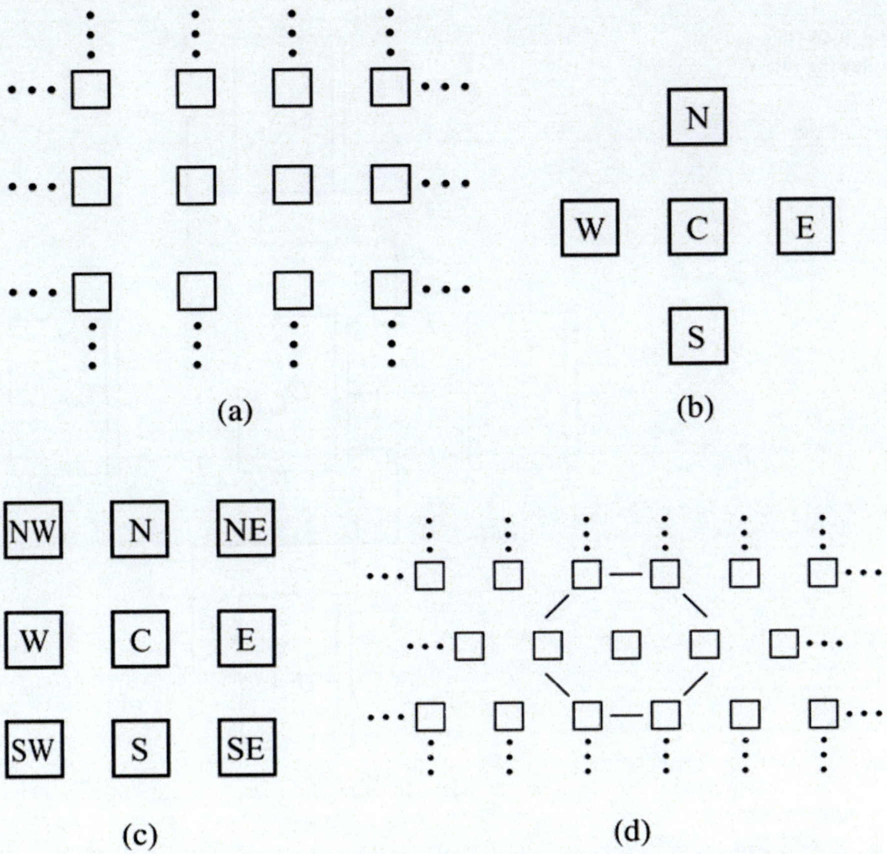

Fig. 9.43 Bidimensional networks: **a** Square; **b** five-cell neighborhood; **c** nine-cell neighborhood; **d** hexagonal network

each internal cell **C** has eight neighbors: north, northeast, east, southeast, south, southwest, west, northwest, resulting in a nine-cell neighborhood, as it is represented in Fig. 9.43c. Another option is to arrange the cells in an hexagonal network, such as the one represented in Fig. 9.43d: each internal cell **C** has six neighbors (seven-cell neighborhood). From all the possible bidimensional spatial structures, rectangular networks will be the only considered in the following, with each cell having four neighbors (Fig. 9.43b).

For the linear bidimensional **CA** (**LBCA**) with five-cell neighborhood, which are the only ones to be considered in this section, functioning rules for each cell are defined in a similar way to how it has been done for the **LOCA**s: in a five-cell neighborhood, the functioning rules will be five-variable functions. Nevertheless, when limited to the **EXOR** connector (*i.e.*, only linear automata are considered), it is enough to indicate which neighbors are in the sum for defining the rule. If $c_{i,j}(t)$ is the content, in instant t, of the cell in row i, column j, it holds:

Fig. 9.44 Internal cell following rule 11

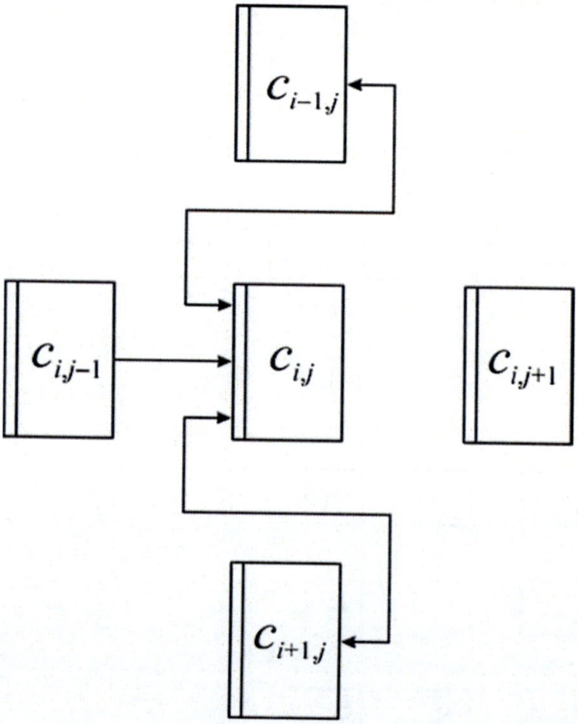

$$c_{i,j}(t+1) = \mathbf{EXOR}(a_{i,j}c_{i,j}(t), a_{i-1,j}c_{i-1,j}(t), a_{i,j+1}c_{i,j+1}(t), a_{i+1,j}c_{i+1,j}(t), a_{i,j-1}c_{i,j-1}(t))$$

The coefficients $a_{i,j}$ $a_{i-1,j}$ $a_{i,j+1}$ $a_{i+1,j}$ $a_{i,j-1}$, which can have values 0 or 1, in the given order, are used to numerate the rules. For example, the cell in Fig. 9.44, whose next value is given by **EXOR**($c_{i-1,j}(t)$, $c_{i+1,j}(t)$, $c_{i,j-1}(t)$), is said to follow rule 11 ($a_{i,j}a_{i-1,j}a_{i,j+1}a_{i+1,j}a_{i,j-1} = 01011_2 = 11_{10}$).

The boundary conditions for the **LBCA** can be as for one-dimensional **CA**: constant, periodic or local. It is usual to consider null constant boundary conditions, since these lead to the simplest analysis. Periodic boundaries are not practical for bidimensional automata when its physical implementation is considered, due to the necessity to carry signals from end to end of the structure, apart from the long paths and consequent delays. Local boundaries are easily implemented, and the definition of the rules corresponding to the boundary cells is done as noted above, with adequate substitutions: Table 9.23 shows the substitute cells of those neighbors not present for boundary cells.

In order to describe the behavior of the **LBCAs**, the same tools that in the case of **LOCAs** can be used, as it can be checked in the following example.

Example 9.38 Figure 9.45 shows a square 3×3 **LBCA**, with local boundaries, whose cells satisfy the rules in Table 9.24. Table 9.25 details the connections between the

Table 9.23 Cell substitution in local boundaries

Non- existing cell	Substitute cell
$c_{i\text{-}1,j}$	$c_{i+2,j}$
$c_{i+1,j}$	$c_{i\text{-}2,j}$
$c_{i,j\text{-}1}$	$c_{i,j+2}$
$c_{i,j+1}$	$c_{i,j\text{-}2}$

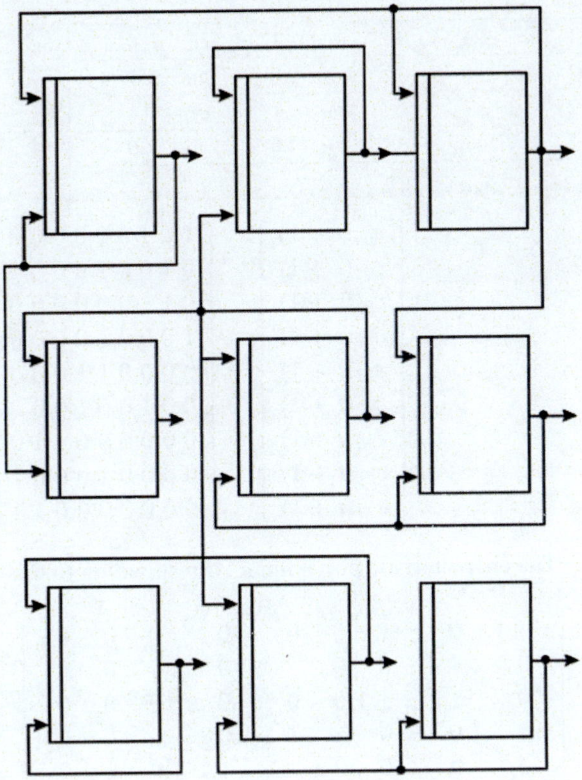

Fig. 9.45 LBCA example circuit

Table 9.24 Rules followed by the LBCA of Fig. 9.45

17	18	17
12	20	24
20	12	17

different cells of the **LBCA**. The functioning of this autonomous system can be described in matrix form as:

Table 9.25 LBCA connection details

i	j	Rule	a_{ij}	c_{ij}	$a_{i\text{-}1j}$	$c_{i\text{-}1j}$	a_{ij+1}	c_{ij+1}	a_{i+1j}	c_{i+1j}	$a_{ij\text{-}1}$	$c_{ij\text{-}1}$
0	0	17	1	c_{00}	0	c_{20}	0	c_{01}	0	c_{10}	1	c_{02}
0	1	18	1	c_{01}	0	c_{21}	0	c_{02}	1	c_{11}	0	c_{00}
0	2	17	1	c_{02}	0	c_{22}	0	c_{00}	0	c_{12}	1	c_{01}
1	0	12	0	c_{10}	1	c_{00}	1	c_{11}	0	c_{20}	0	c_{12}
1	1	20	1	c_{11}	0	c_{01}	1	c_{12}	0	c_{21}	0	c_{10}
1	2	24	1	c_{12}	1	c_{02}	0	c_{10}	0	c_{22}	0	c_{11}
2	0	20	1	c_{20}	0	c_{10}	1	c_{21}	0	c_{00}	0	c_{22}
2	1	12	0	c_{21}	1	c_{11}	1	c_{22}	0	c_{01}	0	c_{20}
2	2	17	1	c_{22}	0	c_{12}	0	c_{20}	0	c_{02}	1	c_{21}

$$\begin{bmatrix} c_{00}(t+1) \\ c_{01}(t+1) \\ c_{02}(t+1) \\ c_{10}(t+1) \\ c_{11}(t+1) \\ c_{12}(t+1) \\ c_{20}(t+1) \\ c_{21}(t+1) \\ c_{22}(t+1) \end{bmatrix} = \begin{bmatrix} 1\,0\,1\,0\,0\,0\,0\,0\,0 \\ 0\,1\,0\,0\,1\,0\,0\,0\,0 \\ 0\,1\,1\,0\,0\,0\,0\,0\,0 \\ 1\,0\,0\,0\,1\,0\,0\,0\,0 \\ 0\,0\,0\,0\,1\,1\,0\,0\,0 \\ 0\,0\,1\,0\,0\,1\,0\,0\,0 \\ 0\,0\,0\,0\,0\,0\,1\,1\,0 \\ 0\,0\,0\,0\,1\,0\,0\,0\,1 \\ 0\,0\,0\,0\,0\,0\,0\,1\,1 \end{bmatrix} \begin{bmatrix} c_{00}(t) \\ c_{01}(t) \\ c_{02}(t) \\ c_{10}(t) \\ c_{11}(t) \\ c_{12}(t) \\ c_{20}(t) \\ c_{21}(t) \\ c_{22}(t) \end{bmatrix}$$

The characteristic polynomial corresponding to this transition matrix is:

$$\begin{vmatrix} x+1 & 0 & 1 & 0 & 0 & 0 & 0 & 0 & 0 \\ 0 & x+1 & 0 & 0 & 1 & 0 & 0 & 0 & 0 \\ 0 & 1 & x+1 & 0 & 0 & 0 & 0 & 0 & 0 \\ 1 & 0 & 0 & x & 1 & 0 & 0 & 0 & 0 \\ 0 & 0 & 0 & 0 & x+1 & 1 & 0 & 0 & 0 \\ 0 & 0 & 1 & 0 & 0 & x+1 & 0 & 0 & 0 \\ 0 & 0 & 0 & 0 & 0 & 0 & x+1 & 1 & 0 \\ 0 & 0 & 0 & 0 & 1 & 0 & 0 & x & 1 \\ 0 & 0 & 0 & 0 & 0 & 0 & 0 & 1 & x+1 \end{vmatrix} = x^5(x+1)^2(x^2+x+1)$$

The corresponding state diagram, as it happened with the **CA** of Example 9.35 (Fig. 9.40), has eight independent sub diagrams: four as the one in Fig. 9.46, and four as the one in Fig. 9.47. Independently of the initial state, the **LBCA** of Fig. 9.45 ends in a three-state cycle or in a drain state.

■

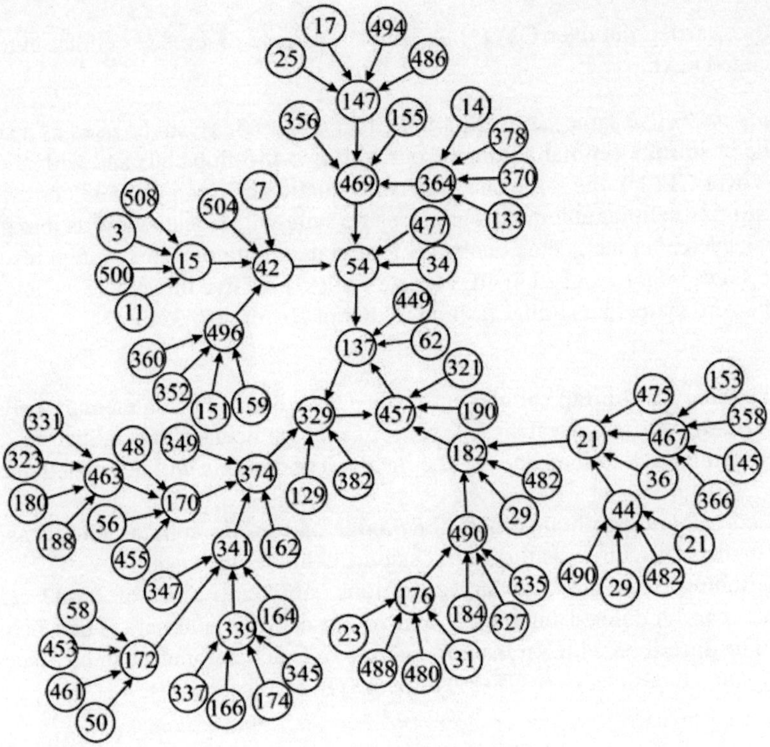

Fig. 9.46 First partial state diagram corresponding to Example 9.38

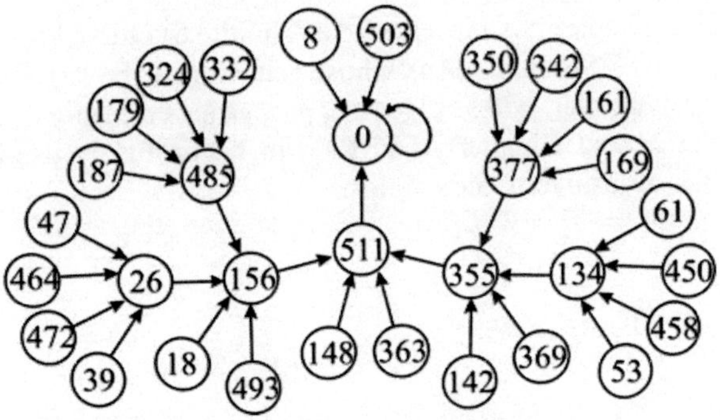

Fig. 9.47 Second partial state diagram corresponding to Example 9.38

9.4.4 Mod2^n and Modp Cellular Automata

Using memory cells able to store 2^n different values, **mod2ⁿ** cellular automata can be defined as an immediate generalization of the cellular automata above, which might be called **mod2** cellular automata. Operations in **mod2ⁿ** cellular automata have to be, of course, carried out over **GF(2ⁿ)**. A simple example of **mod2²** cellular automata is presented next.

Example 9.39 The same CA of Example 9.33 (Fig. 9.38a) can be used as a simple example of **mod2ⁿ** cellular automata, but with two-flip-flop cells and with addition defined over GF(2^2); the same characteristic equation of Example 9.33 corresponds to this **mod2²** cellular automata, and the same state will repeat every, as maximum, six clock cycles; in fact, detail analysis of the state diagram of this system results in the 256 states being grouped in 40 six-state cycles, plus five three-state cycles, apart from the zero state, in a similar fashion to that of Example 9.33.

■

In the same way, **modp** cellular automata can be defined using memory cells that allow storing *p* different values (*p* prime), as an immediate generalization of the cellular automata described above. The involved operations will be carried out over **GF(p)**.

The corresponding circuits, both for **mod2*ⁿ*** and **modp** cellular autómatas, will be rectangular structures, as the bidimensional cellular automata.

Functioning rules and boundary conditions, similar to those of **mod2** cellular automata, can be defined for **mod2ⁿ** and **modp** cellular automata. Their behavior can also be described with a transition matrix, and each automata can be associated to a polynomial defined over **GF(2ⁿ)** or over **GF(p)**, respectively.

9.5 Conclusion

This Chapter has presented two circuit structures based on **D** flip-flops and **EXOR** gates (**LFSR**s and **CA**s), whose behavior can be described associating to each circuit a polynomial, whose algebraic properties correspond to the evolution of the circuit. In the next Chapters, **LFSR**s will be profusely used for implementing different operations over Galois fields.

9.6 Exercises

9.1 Obtain the transfer function for the **LFSR1** circuit of Fig. 9.48. Generate the

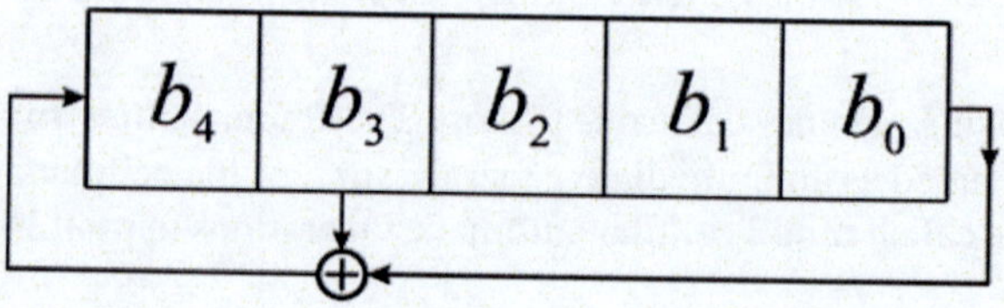

Fig. 9.48 Circuit for exercise 9.1

output sequence for the initial state 10000.

9.2 Obtain the transfer function for the **LFSR1** circuit of Fig. 9.49. Generate all the possible output sequences, depending of the initial state.

9.3 Obtain the transfer function for the **LFSR1** circuit of Fig. 9.50. Generate all the possible output sequences, depending of the initial state.

9.4 Obtain the polynomial corresponding to **LFSR1** of Fig. 9.48 and verify that it is a primitive polynomial.

9.5 Obtain the polynomial corresponding to **LFSR1** of Fig. 9.49 and verify that it is an irreducible polynomial, but not primitive.

9.6 Obtain the polynomial corresponding to **LFSR1** of Fig. 9.50 and verify that it is the product of two primitive polynomials.

9.7 Draw the **LFSR1** corresponding to the primitive polynomial $x^8 + x^4 + x^3 + x^2 + 1$.

9.8 Obtain the transition matrix corresponding to **LFSR2** of Fig. 9.51. Generate the output sequence corresponding to the initial state 10000.

9.9 Obtain the transition matrix corresponding to **LFSR2** of Fig. 9.52. Generate all possible output sequences, depending on the initial state.

9.10 Obtain the transition matrix corresponding to **LFSR2** of Fig. 9.53. Generate

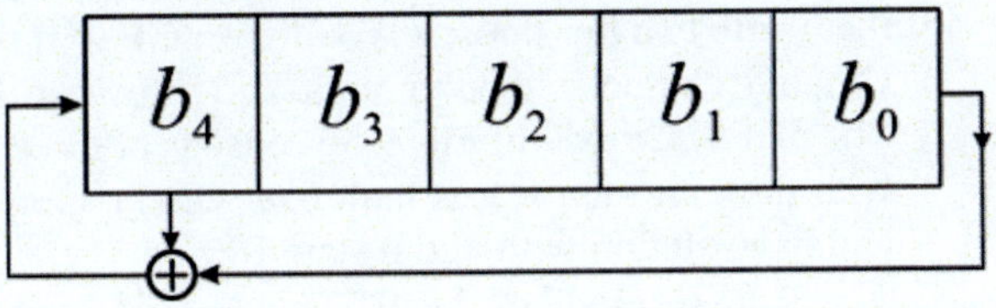

Fig. 9.49 Circuit for exercise 9.2

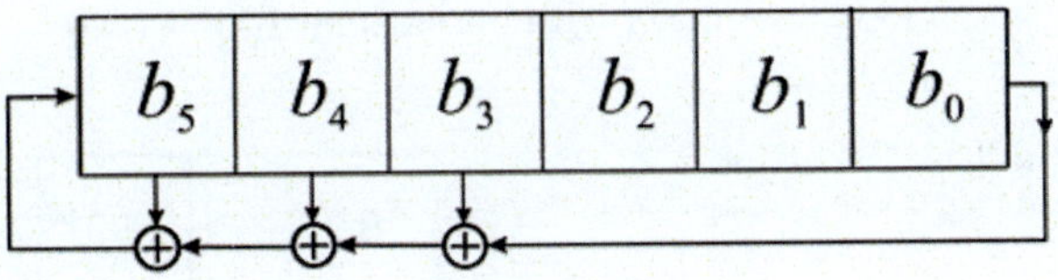

Fig. 9.50 Circuit for exercise 9.3

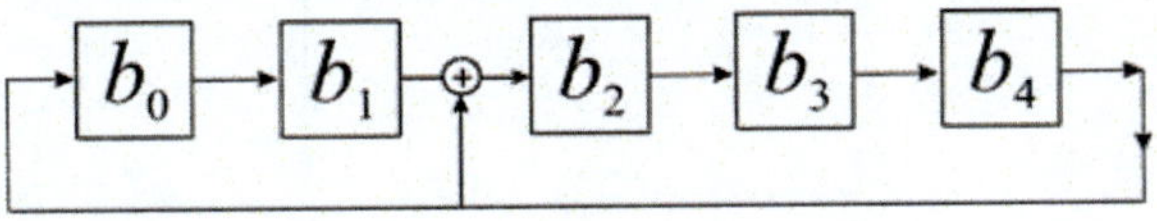

Fig. 9.51 Circuit for exercise 9.8

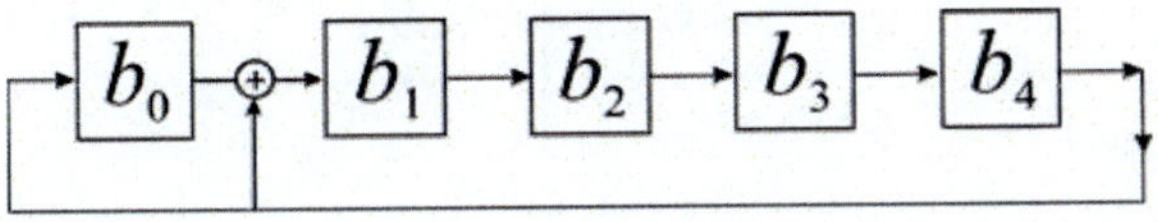

Fig. 9.52 Circuit for exercise 9.9

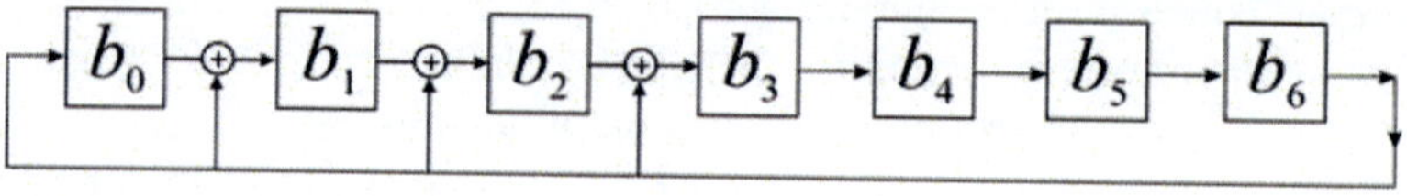

Fig. 9.53 Circuit for exercise 9.10

all possible output sequences, depending on the initial state.

9.11 Obtain the polynomial corresponding to **LFSR2** of Fig. 9.51 and verify that it is a primitive polynomial.

9.12 Obtain the polynomial corresponding to **LFSR2** of Fig. 9.52 and verify that it is an irreducible polynomial, but not primitive.

9.13 Obtain the polynomial corresponding to **LFSR2** of Fig. 9.53 and verify that it is the product of two primitive polynomials.

9.14 Draw the **LFSR2** corresponding to the primitive polynomial $x^8 + x^4 + x^3 + x^2 + 1$.

9.15 The **4LFSR1mod2²** of Fig. 9.54 consists of four cells, B3B2B1B0, each of them with two flip-flops; it is built over GF(2^2){x^2+ x +1} (see Example 9.13). Obtain its associated polynomial and its evolution with initial state AAAB.

9.16 The **4LFSR2mod2²** of Fig. 9.55 consists of four cell, B3B2B1B0, each of them with two flip-flops; it is built over GF(2^2){x^2+ x +1} (see Example 9.14). Obtain its associated polynomial and its evolution with initial state AAAB.

9.17 The **3LFSR1mod7** of Fig. 9.56 consists of three cells, B2B1B0, each of them with three flip-flops; it is built over GF(7). Obtain its associated polynomial and its evolution with initial state 100.

9.18 The **3LFSR1mod7** of Fig. 9.57 consists of three cells, B2B1B0, each of them with three flip-flops; it is built over GF(7). Obtain its associated polynomial and its evolution with initial state 100.

Fig. 9.54 Circuit for exercise 9.15

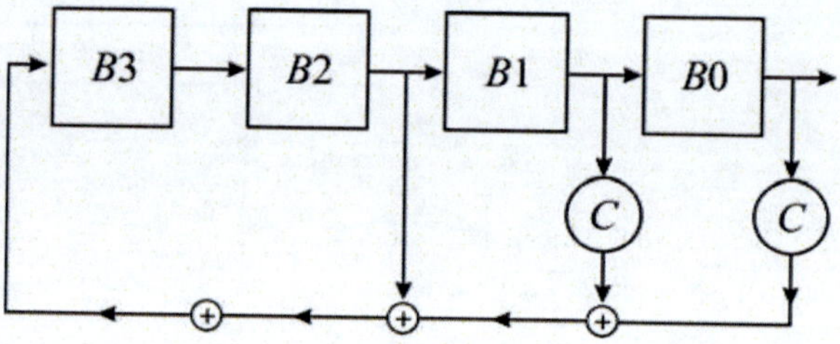

Fig. 9.55 Circuit for exercise 9.16

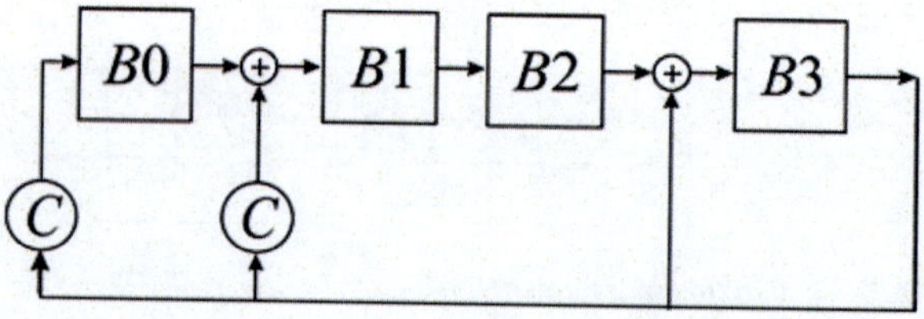

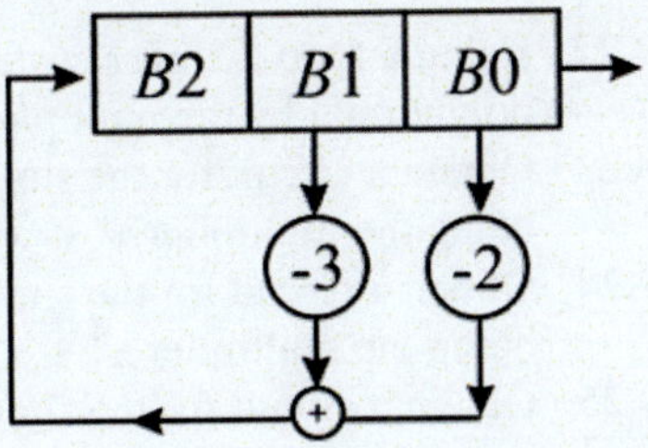

Fig. 9.56 Circuit for exercise 9.17

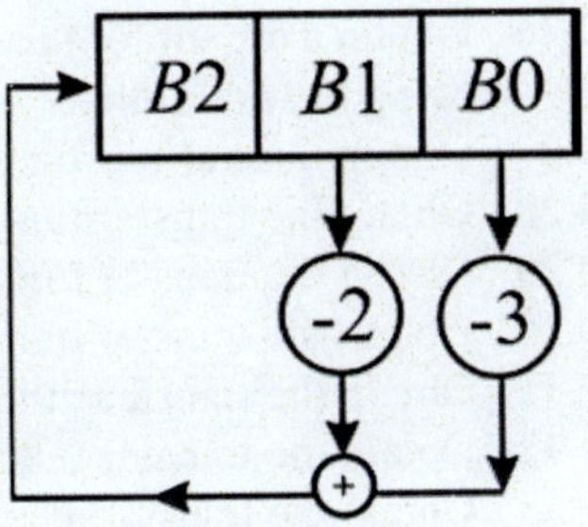

Fig. 9.57 Circuit for exercise 9.18

9.19 The **3LFSR1mod7** of Fig. 9.58 consists of three cells, B2B1B0, each of them with three flip-flops; it is built over GF(7). Obtain its associated polynomial and its evolution with initial state 100.

9.20 The **3LFSR1mod7** of Fig. 9.59 consists of three cells, B2B1B0, each of them with three flip-flops; it is built over GF(7). Obtain its associated polynomial and its evolution with initial state 100.

9.21 Using a type-1 shift register, obtain a circuit for multiplying over GF(2) any polynomial by the multiplier $x^8 + x^5 + x^2 + x + 1$.

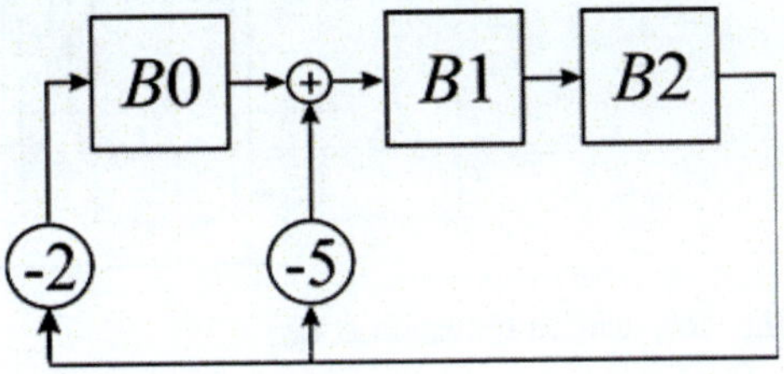

Fig. 9.58 Circuit for exercise 9.19

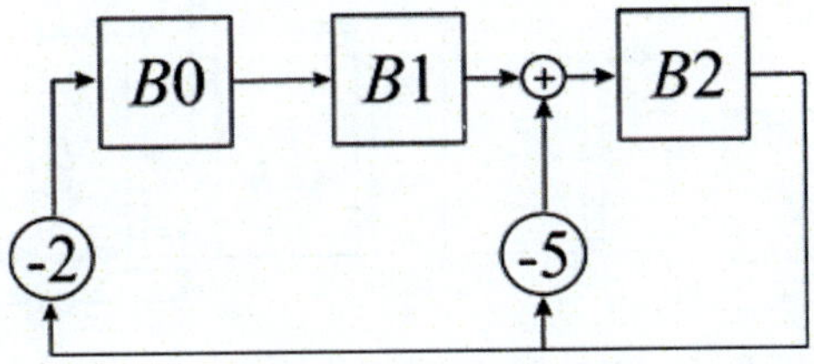

Fig. 9.59 Circuit for exercise 9.20

9.22 Using a type-2 shift register, obtain a circuit for multiplying over GF(2) any polynomial by the multiplier $x^9 + x^6 + x^3 + x + 1$.
9.23 Obtain a circuit for the simultaneous multiplication and division, over GF(2), being the multiplier $x^9 + x^5 + x^3 + x + 1$, and the divisor $x^7 + x^6 + x^2 + 1$.
9.24 Obtain a circuit for the simultaneous multiplication and division, over GF(2), being the multiplier $x^6 + x^4 + x^3 + 1$, and the divisor $x^8 + x^6 + x^3 + x + 1$.
9.25 Obtain a circuit for the simultaneous multiplication and division, over GF(7), being the multiplier $2x^9 + 3x^5 + x^3 + 4x + 1$, and the divisor $3x^7 + 2x^6 + x^2 + 5$.
9.26 Obtain a circuit for the simultaneous multiplication and division, over GF(7), being the multiplier $x^6 + 5x^4 + 2x^3 + 4$, and the divisor $4x^8 + 2x^6 + x^3 + 1$.
9.27 Obtain the transfer function for the circuit of exercise 9.21.
9.28 Obtain the transfer function for the circuit of exercise 9.22.
9.29 Obtain the transfer function for the circuit of exercise 9.23.
9.30 Obtain the transfer function for the circuit of exercise 9.24.
9.31 Obtain the transfer function for the circuit of exercise 9.25.
9.32 Obtain the transfer function for the circuit of exercise 9.26.
9.33 Obtain the transition matrix, the characteristic equation and the state diagram for the **CA** of Fig. 9.60.
9.34 Obtain the transition matrix, the characteristic equation and the state diagram for the **CA** of Fig. 9.61.
9.35 Obtain the transition matrix, the characteristic equation and the state diagram for the **CA** of Fig. 9.62.
9.36 Obtain the transition matrix, the characteristic equation and the state diagram for the **CA** of Fig. 9.63.

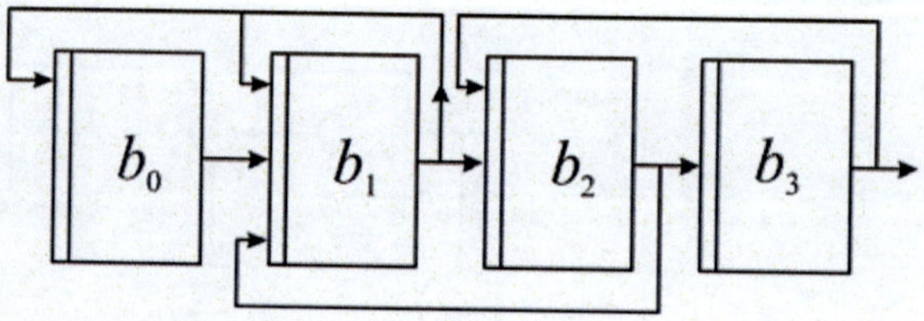

Fig. 9.60 Circuit for exercise 9.33

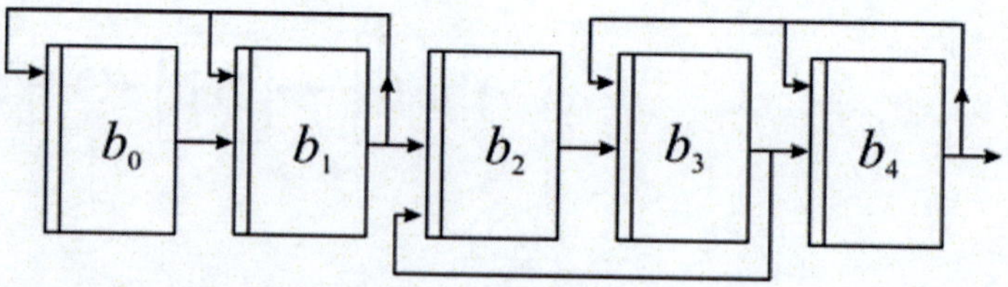

Fig. 9.61 Circuit for exercise 9.34

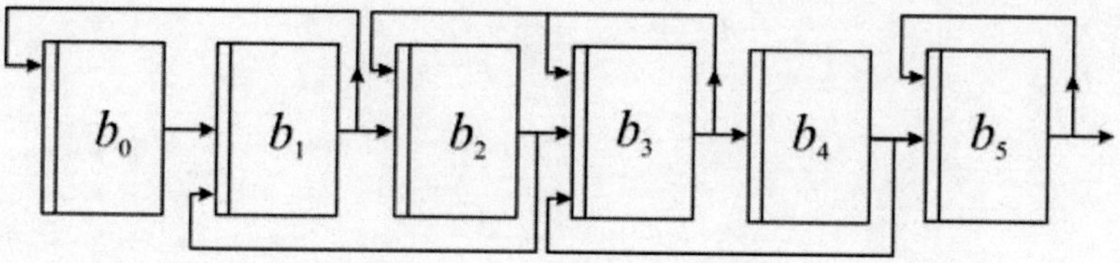

Fig. 9.62 Circuit for exercise 9.35

Fig. 9.63 Circuit for exercise 9.36

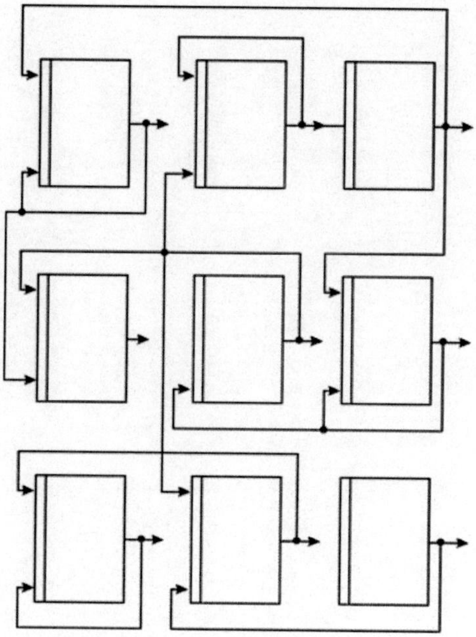

References

1. Golom SW (1980) On the classification of balanced binary sequences of period $2^n - 1$. IEEE Trans Inf Theor IT-26(6):730–732
2. Larson R, Edwards BH, Falvo DC (2004) Elementary linear algebra. Houghton Mifflin Company
3. Lempel A (1970) On the a homomorphism of the de Bruijn graph and its applications to the design of feedback shift registers. IEEE Trans Compt C-19(12):1204–1209
4. Zhang, S., Miller, D.M., Muzio, J.C.: Determination of minimal cost one-dimensional linear hybrid cellular automata. Electron. Lett. **27**(18), 1625–1627 (1991)

Chapter 10
Galois Fields GF(2^m)

After the presentation of the basic algebraic circuits, this chapter is specifically dedicated to the circuits based on the finite fields GF(2^m). Finite fields or Galois fields have a variety of applications in several areas such as cryptography, coding and digital signal processing. The theoretical foundations of the Galois fields, algebra of polynomials, and GF(2^m) are summarized in Appendices A and B; it is advisable to review them before reading this chapter, and also when necessary. This chapter is devoted mainly to show some circuits for the more relevant operations over GF(2^m). Concretely, the addition, the product, the square, the exponentiation, the inversion and the division over GF(2^m) are studied.

Given GF(2^m) and the polynomials $A(x)$ y $B(x)$, with degrees lower than m ($g(A) < m$, $g(B) < m$), thus being elements of GF(2^m), the objective is to add/subtract, to multiply or to divide two polynomials, or raise to a power, or to obtain its inverse, all over GF(2^m). It will be seen that the complexity of these operations depends strongly on the base used to represent the operand polynomials; therefore it is advisable to select the most suitable base. In the following in relation with GF(2^m), it will be considered only the implementation of the various operations with the power representation and using standard, normal and dual bases. These bases are introduced in Appendix B, and are the most frequently used.

Finally, operations in the composite fields GF($(2^n)^m$){$P_1(x)${$P_2(Y)$}} are also considered as a generalization of operations over GF(2^m){$P(x)$}.

10.1 Addition Over GF(2^m)

Any operation over GF(2^m){$P(x)$} can be calculated as a normal operation between polynomials, such as described in Sect. 9.3, followed by modular reduction with respect to $P(x)$. In the case of the addition $A(x) + B(x)$, such modular reduction is not needed: as both polynomials have lower degree than m, the addition has also lower degree, independently of the representation used.

A. Lloris Ruiz et al., *Arithmetic and Algebraic Circuits*, Intelligent Systems Reference Library 201, https://doi.org/10.1007/978-3-030-67266-9_10

The addition using polynomial representation, whatever the base (standard or normal in our case), is straightforward. It consists only of adding the coefficients of each power over GF(2). Then, the modular reduction is not necessary, because there is no generation of carries in this addition. Therefore, any of the studied circuits for the addition of polynomials in Sect. 9.3.1 can be used for addition over GF(2^m).

The power representation is clearly an inadequate representation for the addition. However, if for some reason it has to be used, a table for the addition can be constructed, if the number of elements of GF(2^m) is not too high, as done bellow in Example 10.1. Although it is perhaps advisable converting the power representation of each polynomial into a polynomial representation, adding the polynomials in polynomial representation, and if desired, finally converting the polynomial representation results in power representation. Example 10.2 shows how to go from power representation to a standard base and vice versa.

Example 10.1 **Obtain the addition table for power representation over GF(2^4)$\{x^4 + x^3 + 1\}$.**

The addition table is given in Table 10.1 and it is easily obtained using Table B.20. It is easy to build a combinational circuit to implement this addition. This circuit has 8 inputs and 4 outputs.

■

Table 10.1 Addition table for Example 10.1

+	0	1	x	x^2	x^3	x^4	x^5	x^6	x^7	x^8	x^9	x^{10}	x^{11}	x^{12}	x^{13}	x^{14}
0	0	1	x	x^2	x^3	x^4	x^5	x^6	x^7	x^8	x^9	x^{10}	x^{11}	x^{12}	x^{13}	x^{14}
1	1	0	x^{12}	x^9	x^4	x^3	x^{10}	x^8	x^{13}	x^6	x^2	x^5	x^{14}	x	x^7	x^{11}
x	x	x^{12}	0	x^{13}	x^{10}	x^5	x^4	x^{11}	x^9	x^{14}	x^7	x^3	x^6	1	x^2	x^8
x^2	x^2	x^9	x^{13}	0	x^{14}	x^{11}	x^6	x^5	x^{12}	x^{10}	1	x^8	x^4	x^7	x	x^3
x^3	x^3	x^4	x^{10}	x^{14}	0	1	x^{12}	x^7	x^6	x^{13}	x^{11}	x	x^9	x^5	x^8	x^2
x^4	x^4	x^3	x^5	x^{11}	1	0	x	x^{13}	x^8	x^7	x^{14}	x^{12}	x^2	x^{10}	x^6	x^9
x^5	x^5	x^{10}	x^4	x^6	x^{12}	x	0	x^2	x^{14}	x^9	x^8	1	x^{13}	x^3	x^{11}	x^7
x^6	x^6	x^8	x^{11}	x^5	x^7	x^{13}	x^2	0	x^3	1	x^{10}	x^9	x	x^{14}	x^4	x^{12}
x^7	x^7	x^{13}	x^9	x^{12}	x^6	x^8	x^{14}	x^3	0	x^4	X	x^{11}	x^{10}	x^2	1	x^5
x^8	x^8	x^6	x^{14}	x^{10}	x^{13}	x^7	x^9	1	x^4	0	x^5	x^2	x^{12}	x^{11}	x^3	x
x^9	x^9	x^2	x^7	1	x^{11}	x^{14}	x^8	x^{10}	x	x^5	0	x^6	x^3	x^{13}	x^{12}	x^4
x^{10}	x^{10}	x^5	x^3	x^8	x	x^{12}	1	x^9	x^{11}	x^2	x^6	0	x^7	x^4	x^{14}	x^{13}
x^{11}	x^{11}	x^{14}	x^6	x^4	x^9	x^2	x^{13}	x	x^{10}	x^{12}	x^3	x^7	0	x^8	x^5	1
x^{12}	x^{12}	x	1	x^7	x^5	x^{10}	x^3	x^{14}	x^2	x^{11}	x^{13}	x^4	x^8	0	x^9	x^6
x^{13}	x^{13}	x^7	x^2	x	x^8	x^6	x^{11}	x^4	1	x^3	x^{12}	x^{14}	x^5	x^9	0	x^{10}
x^{14}	x^{14}	x^{11}	x^8	x^3	x^2	x^9	x^7	x^{12}	x^5	x	x^4	x^{13}	1	x^6	x^{10}	0

10.2 Multiplication Over GF(2^m) with Power Representation

Many research works have been dedicated to the multiplication of polynomials, since it is a frequently used operation and usually costly in time and resources. Moreover, the multiplication of polynomials is widely used in the implementations of the remaining operations that will be discussed later on. Some of the main circuit for the multiplication over GF(2^m) are introduced bellow. These circuits are structured according to the representation used, starting in this section with power representation.

The power representation is the most appropriate for the multiplication, because adding the exponents and reducing the result to the proper range is basically required. It is important to remember that each polynomial $P(x)$ will be represented by the binary value of the exponent e of x^e such that $P(x) = x^e$ (see Appendix B).

Given two nonzero polynomials $A(x)$ and $B(x)$ respectively represented by x^a and x^b ($a, b < 2^m$), their product $A(x)\cdot B(x)$ is represented by $x^a \cdot x^b = x^{a+b} = x^e$. To reduce the result to the proper range, the exponent e in $A(x)\cdot B(x) = x^e$ must meet $e < 2^m - 1$; if $e \geq 2^m - 1$, it suffices to subtract $2^m - 1$ from $a + b$, because $x^{2^m-1} = 1$, as seen in Appendix B. This correction in the exponent of the result must be carried out in two cases: when a carry is generated in the addition $(a + b)$ and when $a + b = 2^m - 1$. Moreover, if the correction does not use the carry that may be generated, subtracting $2^m - 1$ from $a + b$ is equivalent to adding 1 to the m least significant bits of $a + b$, this is:

$$(a + b) - (2^m - 1) = (a + b + 1) - 2^m$$

where -2^m is canceled by the carry that is generated by adding $a + b$.

That is, using power representation, the multiplication of two non-zero polynomials is converted into the addition of the exponents. If both exponents a and b are parallel data, it is easy to see that the circuit of Fig. 10.1a can be used to multiply over GF(2^m). This circuit includes two m-bit binary adders and an additional control logic. It is possible to detect if one or both of the multiplier operands is zero. A zero is represented as all the bits in the signals to one. Thus, a null operand can be detected using an **AND** gate; in this case the exponent of the result is zero and no the addition of the operand exponents. Thus, adding the circuit of Fig. 10.1b to the circuit of Fig. 10.1a, it is possible to detect if one of the operands is zero.

It is easy to design circuits for adding $a + b$ when one or both exponents are serial data. It is proposed as an exercise.

If no power representation is used, an option to multiply consists on converting each operand to the power representation, then multiplying them as indicated above, and finally, if necessary, converting the result to the original representation. The transition from one representation to another one can be done as in the following example which involves polynomial representation.

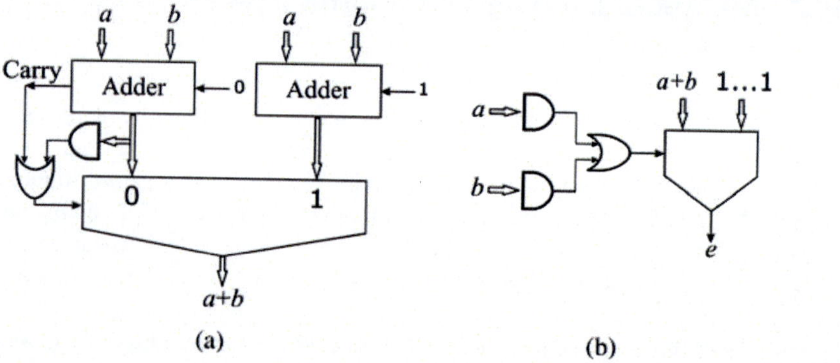

Fig. 10.1 Multiplier circuit using power representation. **a** With no null operands. **b** With the possibility of null operands

Example 10.2 **Design the circuits to go from the polynomial representation to the power representation and vice versa over GF(2^4)$\{x^4 + x^3 + 1\}$ for any non-zero element.**

Using Table B.20, the Table 10.2 is the truth table to go from the polynomial representation (*xyzu*) towards the power representation (*abcd*):

$$a = \bar{x}y(u + z) + \bar{x}zu + xy\bar{z}\bar{u} + x\bar{y}z$$
$$b = xy + x\bar{z}u + \bar{x}\bar{y}zu + yz\bar{u}$$

Table 10.2 Going from the polynomial representation towards the power representation

	xyzu	*abcd*
0	0000	- - - -
1	0001	0011
2	0010	0010
3	0011	1110
4	0100	0001
5	0101	1010
6	0110	1101
7	0111	1000
8	1000	0000
9	1001	0100
10	1010	1001
11	1011	1011
12	1100	1100
13	1101	0101
14	1110	0111
15	1111	0110

Table 10.3 Going from the power representation towards the polynomial representation

	abcd	*xyzu*
0	0000	1000
1	0001	0100
2	0010	0010
3	0011	0001
4	0100	1001
5	0101	1101
6	0110	1111
7	0111	1110
8	1000	0111
9	1001	1010
10	1010	0101
11	1011	1011
12	1100	1100
13	1101	0110
14	1110	0011
15	1111	----

$$c = \overline{xy} + \overline{xz}u + \overline{y}zu + xyz$$

$$d = \overline{xyz} + \overline{xzu} + xy\overline{z}u + yz\overline{u} + x\overline{y}z$$

On the other hand, Table 10.3 is the truth table corresponding to the transformation of the power representation (*abcd*) into the polynomial representation (*xyzu*):

$$x = \overline{a}b + (\overline{a} + b)\overline{c}\,\overline{d} + a\overline{b}d$$

$$y = a\overline{bd} + ab\overline{c} + \overline{a}bc + \overline{ac}d$$

$$z = ad + bc + a\overline{b}\,\overline{c} + \overline{a}c\overline{d}$$

$$u = \overline{b}cd + a\overline{bd} + \overline{a}b\overline{c} + bc\overline{d}$$

■

10.3 Multiplication Over GF(2^m) Using Standard Base

This section presents the most common multiplier circuits, which corresponds to the standard base representation. The structure of the circuit depends on the availability of the data as serial or parallel data. The processing can be divided into two parts: the multiplication of the operands and the modular reduction of the result. It consists on applying the multiplier circuits described in Sect. 9.3.2, some of which have an

LFSR as core, for the multiplication over GF(2^m). The possible provisions of the operands (serial or parallel data) are considered. A deep study of the implementation of the modular reduction is carried out at the following.

10.3.1 Modular Reduction

The modular reduction can be implemented using an **LFSR**, since the modular reduction is just a division, from which we are interested in the remainder. Therefore an **LFSR2**, such as in Fig. 9.26b, can be used as it will be made later on. But for the modular reduction a combinational solution for each case can be raised. A simple example is considered at the following.

Example 10.3 **Design a combinational circuit over GF(2^4){x^4+ x^3+ 1} implementing the modular reduction after the multiplication of two elements of GF(2^4)**

The product of two elements of GF(2^4) is given by $P(x) = p_6x^6 + p_5x^5 + p_4x^4 + p_3x^3 + p_2x^2 + p_1x + p_0$. Dividing $P(x)$ by $x^4 + x^3 + 1$, it is easy to check the remainder:

$$R(x) = (p_3 + p_4 + p_5 + p_6)x^3 + (p_2 + p_6)x^2 + (p_1 + p_5 + p_6)x + (p_0 + p_4 + p_5 + p_6)$$

Therefore, after the multiplication and once calculated p_0, …, p_6, the modular reduction can be performed in this case using four **EXOR** gates: one of 2-input (p_2 and p_6), one of 3-inputs (p_1, p_5 and p_6), and two of 4-inputs (p_3, p_4, p_5 and p_6 one hand, and p_0, p_4, p_5 and p_6 by another).

Let suppose the most general case GF(2^4)$\{x^4 + f_3x^3 + f_2x^2 + f_1x + f_0\}$ and being $x^4 + f_3x^3 + f_2x^2 + f_1x + f_0$ any primitive polynomial of degree 4 (obviously, $f_0 \neq 0$). The coefficients of the remainder $R(x) = r_3x^3 + r_2x^2 + r_1x + r_0$ can be expressed as follows:

$$\begin{aligned}
r_3 &= p_3 + p_4f_3 + p_5(f_2 + f_3f_3) + p_6(f_1 + f_3f_2 + f_2f_3 + f_3f_3f_3)\\
r_2 &= p_2 + p_4f_2 + p_5(f_1 + f_2f_3) + p_6(f_0 + f_3f_1 + f_2f_2 + f_3f_3f_2)\\
r_1 &= p_1 + p_4f_1 + p_5(f_0 + f_1f_3) + p_6(f_3f_0 + f_2f_1 + f_3f_3f_1)\\
r_0 &= p_0 + p_4f_0 + p_5f_0f_3 + p_6(f_2f_0 + f_3f_3f_0)
\end{aligned}$$

In matrix notation:

$$\begin{bmatrix} r_0 \\ r_1 \\ r_2 \\ r_3 \end{bmatrix} = \begin{bmatrix} 1\,0\,0\,0 & c_{00} & c_{01} & c_{02} \\ 0\,1\,0\,0 & c_{10} & c_{11} & c_{12} \\ 0\,0\,1\,0 & c_{20} & c_{21} & c_{22} \\ 0\,0\,0\,1 & c_{30} & c_{31} & c_{32} \end{bmatrix} \begin{bmatrix} p_0 \\ p_1 \\ p_2 \\ p_3 \\ p_4 \\ p_5 \\ p_6 \end{bmatrix}$$

The coefficients c_{ij} are given by:

$$c_{00} = f_0;\ c_{01} = f_3 f_0;\ c_{02} = f_0(f_2 + f_3 f_3)$$
$$c_{10} = f_1;\ c_{11} = f_0 + f_3 f_1;\ c_{12} = f_3 f_0 + f_1(f_2 + f_3 f_3)$$
$$c_{20} = f_2;\ c_{21} = f_1 + f_3 f_2;\ c_{22} = f_0 + f_3 f_1 + f_2(f_2 + f_3 f_3)$$
$$c_{30} = f_3;\ c_{31} = f_2 + f_3 f_3;\ c_{32} = f_1 + f_3 f_2 + f_3(f_2 + f_3 f_3)$$

Let suppose $c_{-1,j}$=0, $c_{i,-1}$=0, it is easy to verify that the coefficients c_{ij} can be expressed as follows:

$$c_{i0} = f_i,\ i = 0, \ldots, 3$$
$$c_{ij} = c_{i-1,j-1} + c_{i0}c_{3,i-1},\ i = 0, \ldots, 3,\ j = 1, 2$$

For GF(2^4)$\{x^4 + x^3 + 1\}$:

$$c_{00} = 1;\ c_{01} = 1;\ c_{02} = 1;$$
$$c_{10} = 0;\ c_{11} = 1;\ c_{12} = 1;$$
$$c_{20} = 0;\ c_{21} = 0;\ c_{22} = 1;$$
$$c_{30} = 1;\ c_{31} = 1;\ c_{32} = 1.$$

■

The conclusion reached in Example 10.3 can also be obtained considering that modular reduction consists on transforming the summands p_6x^6, p_5x^5 and p_4x^4 of the polynomial product in other expressions including only the powers x^3, x^2, x^1 and x^0. Now, if the reduction is made respect to $x^4 + f_3x^3 + f_2x^2 + f_1x + f_0$, it is clear that:

$$x^4\textbf{mod}\left(x^4 + f_3x^3 + f_2x^2 + f_1x + f_0\right) = f_3x^3 + f_2x^2 + f_1x + f_0$$
$$= c_{00} + c_{10}x + c_{20}x^2 + c_{30}x^3$$

$$x^5\textbf{mod}\left(x^4 + f_3x^3 + f_2x^2 + f_1x + f_0\right)$$
$$= \left(f_3x^4 + f_2x^3 + f_1x^2 + f_0x\right)\textbf{mod}\left(x^4 + f_3x^3 + f_2x^2 + f_1x + f_0\right)$$

$$= (f_2 + f_3 f_3)x^3 + (f_1 + f_3 f_2)x^2 + (f_0 + f_3 f_1)x + f_3 f_0$$
$$= c_{01} + c_{11}x + c_{21}x^2 + c_{31}x^3$$

$$x^6\mathbf{mod}\left(x^4 + f_3x^3 + f_2x^2 + f_1x + f_0\right)$$
$$= \left(f_3x^5 + f_2x^4 + f_1x^3 + f_0x^2\right)\mathbf{mod}\left(x^4 + f_3x^3 + f_2x^2 + f_1x + f_0\right)$$
$$= (f_1 + f_3 f_2 + f_3(f_2 + f_3 f_3))x^3 + (f_0 + f_3 f_1 + f_2(f_2 + f_3 f_3))x^2$$
$$+ (f_3 f_0 + f_1(f_2 + f_3 f_3))x + f_0(f_2 + f_3 f_3) = c_{02} + c_{12}x + c_{22}x^2 + c_{32}x^3$$

That is:

$$c_{00} = f_0;\ c_{10} = f_1;\ c_{20} = f_2;\ c_{30} = f_3$$
$$c_{01} = f_3 f_0;\ c_{11} = f_0 + f_3 f_1;\ c_{21} = f_1 + f_3 f_2;\ c_{31} = f_2 + f_3 f_3$$
$$c_{02} = f_0(f_2 + f_3 f_3);\ c_{12} = f_3 f_0 + f_1(f_2 + f_3 f_3);$$
$$c_{22} = f_0 + f_3 f_1 + f_2(f_2 + f_3 f_3);\ c_{32} = f_1 + f_3 f_2 + f_3(f_2 + f_3 f_3)$$

They are the same expressions obtained in Example 10.3, as expected. In the general case GF(2^m)$\{x^m + f_{m-1}x^{m-1} + \cdots + f_1x + f_0\}$, we must express the powers $x^j (j = m + 1, \ldots, 2m{-}2)$ in terms of the base x^i ($i = m - 1, \ldots, 0$). It is obvious that for the modular reduction the matrix expression is as follows:

$$\begin{bmatrix} r_0 \\ r_1 \\ \cdots \\ r_{m-2} \\ r_{m-1} \end{bmatrix} = \begin{bmatrix} 1 & 0 & \ldots & 0 & 0 & c_{00} & \ldots & c_{0,m-2} \\ 0 & 1 & \ldots & 0 & 0 & c_{10} & \ldots & c_{1,m-2} \\ \ldots & \ldots & \ldots & \ldots & \ldots & \ldots & \ldots & \ldots \\ 0 & 0 & \ldots & 1 & 0 & c_{m-2,0} & \ldots & c_{m-2,m-2} \\ 0 & 0 & \ldots & 0 & 1 & c_{m-1,0} & \ldots & c_{m-1,m-2} \end{bmatrix} \begin{bmatrix} p_0 \\ p_1 \\ \cdots \\ p_{2m-3} \\ p_{2m-2} \end{bmatrix}$$

It is easy to check that the coefficients c_{ij}, with $c_{-1,j}$=0, $c_{i,-1}$=0 are given by:

$$c_{i0} = f_i,\ i = 0, \ldots, m-1$$
$$c_{ii} = c_{i-1,j-1} + c_{i0}c_{m-1,j-1}, = 0, \ldots, m-1,\ j = 1, \ldots, m-2$$

10.3.2 Parallel Multiplication

Let suposse the polynomials $A(x) = a_{m-1}x^{m-1} + a_{m-2}x^{m-1} + \cdots + a_1x + a_0$ and $B(x) = b_{m-1}x^{m-1} + b_{m-2}x^{m-2} + \cdots + b_1x + b_0$ are given as parallel data. A solution strictly combinational for the product $A(x) \cdot B(x)$ can be implemented, joining matrix multiplication (seen in Sect. 9.3.2) and the modular reduction (just studied). It is obtained:

$$\begin{bmatrix} r_0 \\ r_1 \\ \cdots \\ r_{m-2} \\ r_{m-1} \end{bmatrix} = \begin{bmatrix} 1 & 0 & \ldots & 0 & 0 & c_{00} & \ldots & c_{0,m-2} \\ 0 & 1 & \ldots & 0 & 0 & c_{10} & \ldots & c_{1,m-2} \\ \ldots & & & & & \ldots & \ldots & \ldots \\ 0 & 0 & \ldots & 1 & 0 & c_{m-2,0} & \ldots & c_{m-2,m-2} \\ 0 & 0 & \ldots & 0 & 1 & c_{m-1,0} & \ldots & c_{m-1,m-2} \end{bmatrix}$$

$$\begin{bmatrix} a_0 & 0 & \ldots & 0 & 0 \\ a_1 & a_0 & \ldots & 0 & 0 \\ \ldots & \ldots & \ldots & \ldots & \ldots \\ a_{m-2} & a_{m-3} & \ldots & a_0 & 0 \\ a_{m-1} & a_{m-2} & \ldots & a_1 & a_0 \\ 0 & a_{m-1} & \ldots & a_2 & a_1 \\ \ldots & \ldots & \ldots & \ldots & \ldots \\ 0 & 0 & \ldots & a_{m-1} & a_{m-2} \\ 0 & 0 & \ldots & 0 & a_{m-2} \end{bmatrix} \begin{bmatrix} b_0 \\ \ldots \\ b_{m-1} \end{bmatrix}$$

A combinational circuit will result in each case, such as is obtained in the following examples.

Example 10.4 **Design a combinational circuit over GF(2^4){x^4+ x^3+ 1} implementing the multiplication and the modular reduction of two elements of GF(2^4).**

For this case:

$$\begin{bmatrix} r_0 \\ r_1 \\ r_2 \\ r_3 \end{bmatrix} = \begin{bmatrix} 1&0&0&0&1&1&1 \\ 0&1&0&0&0&1&1 \\ 0&0&1&0&0&0&1 \\ 0&0&0&1&1&1&1 \end{bmatrix} \begin{bmatrix} a_0 & 0 & 0 & 0 \\ a_1 & a_0 & 0 & 0 \\ a_2 & a_1 & a_0 & 0 \\ a_3 & a_2 & a_1 & a_0 \\ 0 & a_3 & a_2 & a_1 \\ 0 & 0 & a_3 & a_2 \\ 0 & 0 & 0 & a_3 \end{bmatrix} \begin{bmatrix} b_0 \\ b_1 \\ b_2 \\ b_3 \end{bmatrix}$$

$$\begin{bmatrix} 1&0&0&0&1&1&1 \\ 0&1&0&0&0&1&1 \\ 0&0&1&0&0&0&1 \\ 0&0&0&1&1&1&1 \end{bmatrix} \begin{bmatrix} a_0 & 0 & 0 & 0 \\ a_1 & a_0 & 0 & 0 \\ a_2 & a_1 & a_0 & 0 \\ a_3 & a_2 & a_1 & a_0 \\ 0 & a_3 & a_2 & a_1 \\ 0 & 0 & a_3 & a_2 \\ 0 & 0 & 0 & a_3 \end{bmatrix} = \begin{bmatrix} p_{00} & p_{01} & p_{02} & p_{03} \\ p_{10} & p_{11} & p_{12} & p_{13} \\ p_{20} & p_{21} & p_{22} & p_{23} \\ p_{30} & p_{31} & p_{32} & p_{33} \end{bmatrix}$$

The following coefficients are obtained:

$p_{00} = a_0$; $p_{01} = a_3$; $p_{02} = a_2 + a_3$; $p_{03} = a_1 + a_2 + a_3$;

$p_{10} = a_1$; $p_{11} = a_0$; $p_{12} = a_3$; $p_{13} = a_2 + a_3$;

$$p_{20} = a_2;\ p_{21} = a_1;\ p_{22} = a_0;\ p_{23} = a_3;$$
$$p_{30} = a_3;\ p_{31} = a_2 + a_3;\ p_{32} = a_1 + a_2 + a_3;\ p_{03} = a_0 + a_1 + a_2 + a_3;$$

This is:

$$\begin{bmatrix} r_0 \\ r_1 \\ r_2 \\ r_3 \end{bmatrix} = \begin{bmatrix} p_{00} & p_{01} & p_{02} & p_{03} \\ p_{10} & p_{11} & p_{12} & p_{13} \\ p_{20} & p_{21} & p_{22} & p_{23} \\ p_{30} & p_{31} & p_{32} & p_{33} \end{bmatrix} \begin{bmatrix} b_0 \\ b_1 \\ b_2 \\ b_3 \end{bmatrix}$$
$$= \begin{bmatrix} a_0 & a_3 & a_2 + a_3 & a_1 + a_2 + a_3 \\ a_1 & a_0 & a_3 & a_2 + a_3 \\ a_2 & a_1 & a_0 & a_3 \\ a_3 & a_2 + a_3 & a_1 + a_2 + a_3 & a_0 + a_1 + a_2 + a_3 \end{bmatrix} \begin{bmatrix} b_0 \\ b_1 \\ b_2 \\ b_3 \end{bmatrix}$$

Therefore, with a circuit as the one of Fig. 10.2a, with 16 **AND** gates and 4 **EXOR** gates (4-inputs, 5-inputs, 7-inputs and 10-inputs) this multiplier can be build.

Another modular circuit can be obtained. Figure 10.2b, c show a module with 9 inputs and 8 outputs, related as follows:

$$S_1 = e_7;\ S_2 = e_0e_1 + e_2;\ S_3 = e_1;\ S_4 = e_0e_3 + e_4;$$
$$S_5 = e_3;\ S_6 = e_0e_5 + e_6;\ S_7 = e_5 + e_7;\ S_8 = e_0e_7 + e_8$$

Even outputs and odd outputs are given in Fig. 10.2b, c respectively. It is easy to check [10] that, using 4 of these modules and concatenating them properly, a multiplier circuit for GF(2^4) can be constructed, such as depicted in Fig. 10.2d.

■

Example 10.5 Design a combinational circuit which implement the multiplication and modular reduction of two elements of GF(2^4) in the more general case GF(2^4){($x^4 + f_3x^3 + f_2x^2 + f_1x + f_0$}, being $x^4 + f_3x^3 + f_2x^2 + f_1x + f_0$ any primitive polynomial of degree 4 (obviously $f_0 \neq 0$).

As seen in Example 10.3, the modular reduction matrix in this case is the following one:

$$R = \begin{bmatrix} 1\,0\,0\,0 & f_0 & f_3f_0 & f_0(f_2 + f_3f_3) \\ 0\,1\,0\,0 & f_1 & f_0 + f_3f_1 & f_3f_0 + f_1(f_2 + f_3f_3) \\ 0\,0\,1\,0 & f_2 & f_1 + f_3f_2 & f_0 + f_3f_1 + f_2(f_2 + f_3f_3) \\ 0\,0\,0\,1 & f_3 & f_2 + f_3f_3 & f_1 + f_3f_2 + f_3(f_2 + f_3f_3) \end{bmatrix}$$

Multiplying this matrix by the matrix multiplication:

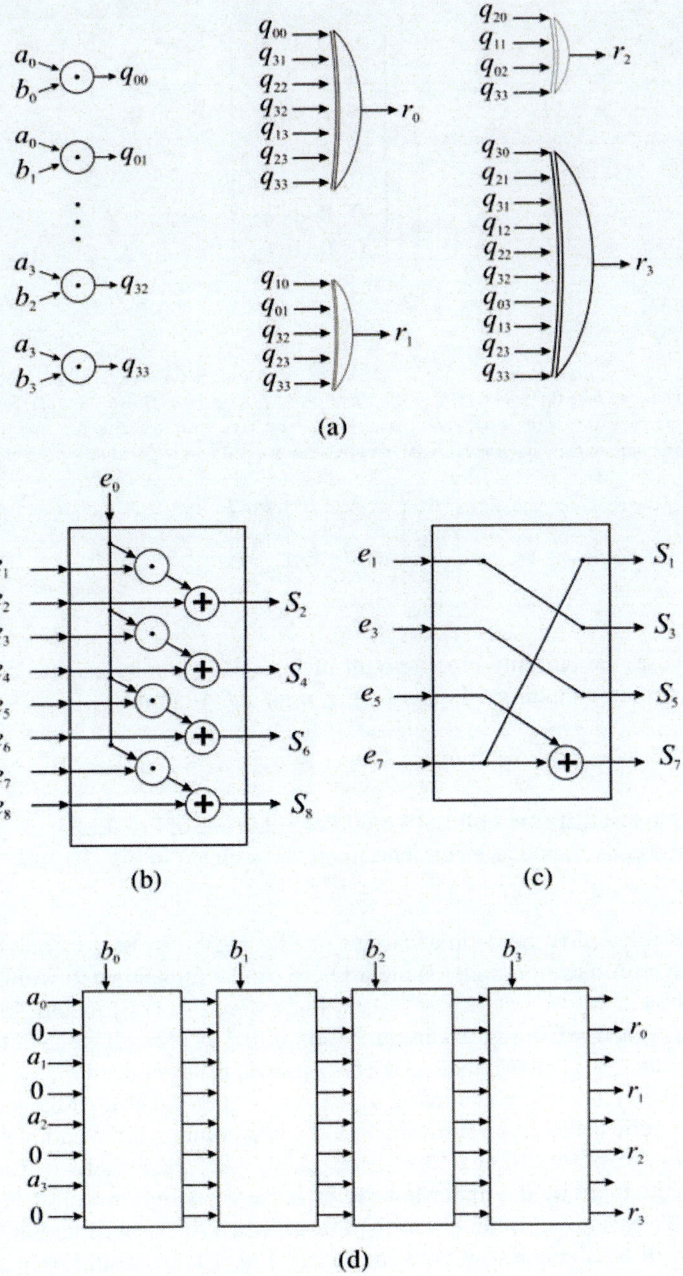

Fig. 10.2 Example 10.4. **a** Immediate circuit. **b** Constructive module (even outputs). **c** Constructive module (odd outputs). **d** Modular circuit

$$M = \begin{bmatrix} a_0 & 0 & 0 & 0 \\ a_1 & a_0 & 0 & 0 \\ a_2 & a_1 & a_0 & 0 \\ a_3 & a_2 & a_1 & a_0 \\ 0 & a_3 & a_2 & a_1 \\ 0 & 0 & a_3 & a_2 \\ 0 & 0 & 0 & a_3 \end{bmatrix}$$

For $R \cdot M$:

$$\begin{bmatrix} a_0 & a_3 f_0 & (a_2 + a_3 f_3) f_0 & (a_1 + a_3 f_2 + (a_2 + a_3 f_3) f_3) f_0 \\ a_1 & a_0 + a_3 f_1 & a_3 f_0 + (a_2 + a_3 f_3) f_1 & (a_2 + a_3 f_3) f_0 + (a_1 + a_3 f_2 + (a_2 + a_3 f_3) f_3) f_1 \\ a_2 & a_1 + a_3 f_2 & a_0 + a_3 f_1 + (a_2 + a_3 f_3) f_2 & a_3 f_0 + (a_2 + a_3 f_3) f_1 + (a_1 + a_3 f_2 + (a_2 + a_3 f_3) f_3) f_2 \\ a_3 & a_2 + a_3 f_3 & a_1 + a_3 f_2 + (a_2 + a_3 f_3) f_3 & a_0 + a_3 f_1 + (a_2 + a_3 f_3) f_2 + (a_1 + a_3 f_2 + (a_2 + a_3 f_3) f_3) f_3 \end{bmatrix}$$

$$\begin{bmatrix} r_0 \\ r_1 \\ r_2 \\ r_3 \end{bmatrix} = R \cdot M \begin{bmatrix} b_0 \\ b_1 \\ b_2 \\ b_3 \end{bmatrix}$$

In this case, the building block circuit of Fig. 10.3a can be used to construct a multiplier circuit, including 5-inputs (x, y, z, u, v) and 4-outputs:

$$S_1 = z \oplus x(y \oplus uv); \quad S_2 = y \oplus uv; \quad S_3 = x; \quad S_4 = v$$

It is easy to test that assembling 16 modules as the one of Fig. 10.3a, the multiplier circuit can be constructed as a cellular circuit, as depicted in Fig. 10.3b.

■

It is immediate that the cellular circuit of Fig. 10.3b can be generalized to any GF(2^m): the multiplication under standard base can be implemented with a cellular circuit with $m \times m$ cells as in Fig. 10.3a. For a given $P(x)$ (it means, known the coefficients f_i), the cell in Fig. 10.3a can be simplified by removing one of the inputs (corresponding to f_i) and originating the two generic cells depicted in Fig. 10.4: one when $f_i = 0$ (Fig. 10.4a), and another when $f_i = 1$ (Fig. 10.4b). At the same time, these two generic cells can be simplified for the first column of the complete circuit, providing the cell of Fig. 10.4c with only one **AND** gate, and for the first row, so that $f_0 = 1$ and the input by the upper left vertex is zero, giving the cell of Fig. 10.4d. Obviously the easier multiplier circuit will result when $P(x)$ is a trinomial, in which case there will be a single row with the cell of Fig. 10.4b), resulting in a total of m^2 **AND** gates and $m^2 - 1$ **EXOR** gates, all of two inputs. The cellular circuit for GF(2^m) is shown in Fig. 10.4e.

The multiplication can also be made using a circuit as in Fig. 9.19, storing the result in a register, then the reduction can be done with an **LFSR2** modular divider (Fig. 9.26b), being the result in the **LFSR**.

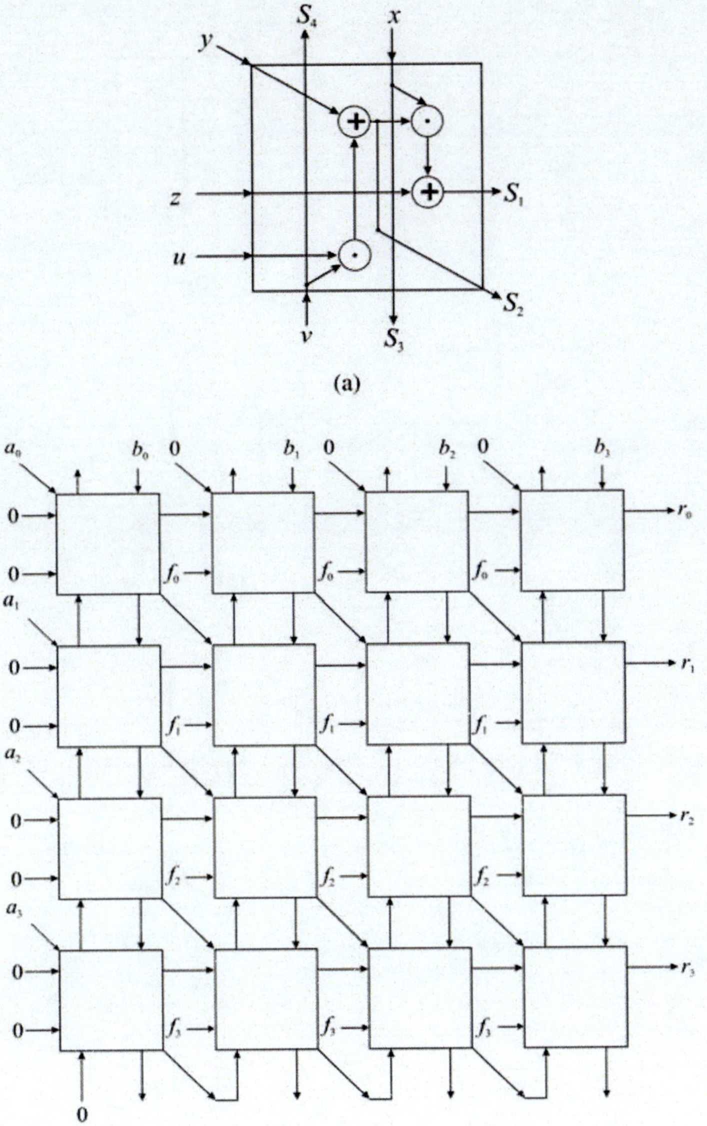

Fig. 10.3 Example 10.5. **a** Constructive module. **b** Cellular circuit

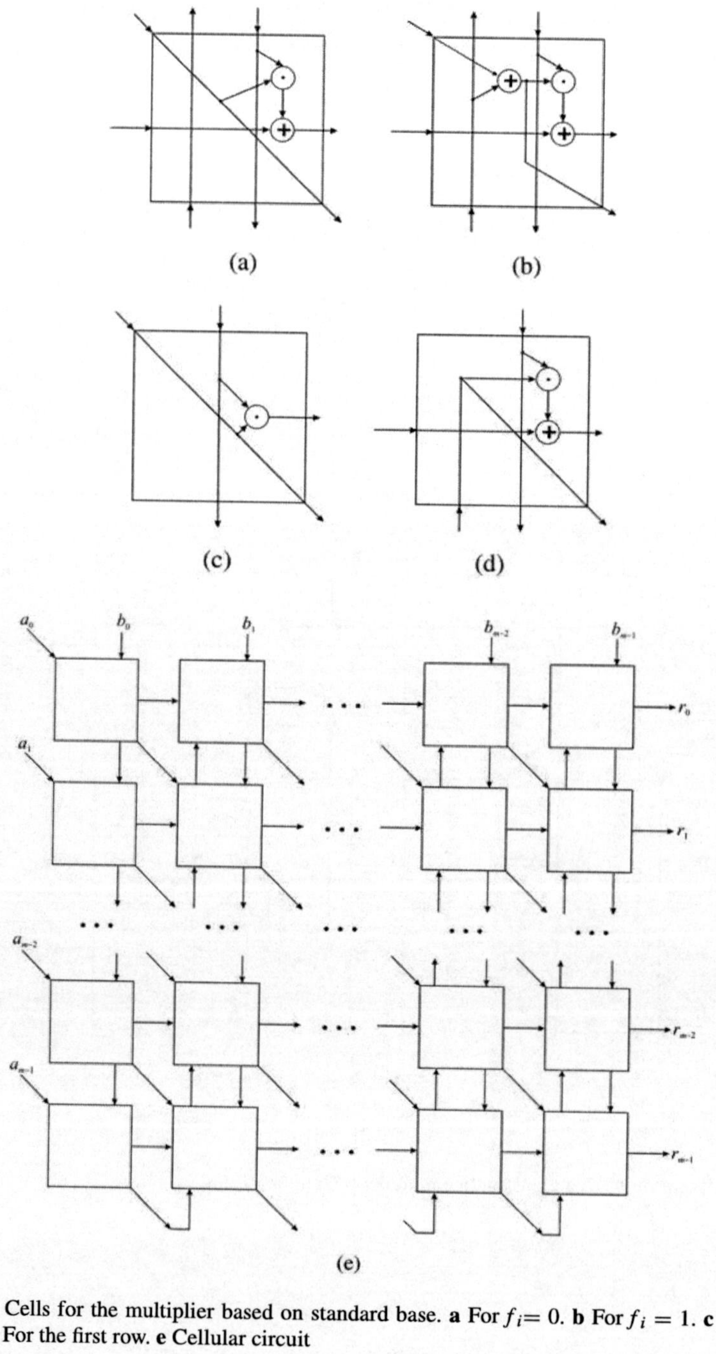

Fig. 10.4 Cells for the multiplier based on standard base. **a** For $f_i = 0$. **b** For $f_i = 1$. **c** For first column. **d** For the first row. **e** Cellular circuit

10.3.3 Serial-Parallel Multiplication

Let suppose now the case where the polynomial $A(x) = a_{m-1}x^{m-1} + a_{m-2}\,x^{m-2} + \cdots + a_1x + a_0$ is provided as serial data and $B(x) = b_{m-1}\,x^{m-1} + b_{m-2}\,x^{m-2} + \cdots + b_1x + b_0$ is provided as parallel data. Below are various circuits for implementing the product of both inputs, some of them using **LFSR**s.

It is easy to see that the cellular circuit of Fig. 10.4e may be adapted to consider this situation. It is only necessary to include a shift register of length m as a buffer for the inputs a_i, so that $A(x)$ is introduced in serial and becomes available in parallel, as depicted in Fig. 10.5a. After application of $B(x)$ in parallel form and the introduction of $A(x)$ in serial form, the result can be read in the output r_i, taking into account the delay of the cellular circuit. A register with parallel load can also be used for the result. The polynomial $A(x)$ can be introduced in both upstream and downstream (first coefficient a_0, or first the coefficient a_{m-1}): it depends on the serial input.

It is easy to conclude that this solution based on the adaptation of the combinational cellular circuit has no advantage, because more circuits are needed and it is not faster than **LFSR**s-based circuits showed in the following sections.

10.3.3.1 Multiplier/Divider

To calculate $M(x) = A(x) \cdot B(x)\,\mathbf{mod}P(x)$ (i.e., to multiply $A(x)$ and $B(x)$ over GF(2^m)) is equivalent to multiply $A(x) \cdot B(x)$ as seen in Sect. 9.3, and then get the remainder dividing $A(x) \cdot B(x)$ by $P(x)$. Using a standard base and in appropriate conditions, these operations can be performed with an **LFSR** multiplier-divider as shown in Fig. 9.29. In this case it is assumed that the polynomials $B(x)$ and $P(x)$ are given in parallel and $A(x)$ is serie-input of the **LFSR**, first the higher coefficient. The product $M(x)$ is the remainder of the division and is stored in the flip-flops of the **LFSR**, as we have seen in Sect. 9.3.3. Recall that the biestables of the **LFSR** must be initially to zero.

Let $A(x) = a_{m-1}x^{m-1} + \cdots + a_0$, $B(x) = b_{m-1}x^{m-1} + \cdots + b_0$, $P(x) = x^m + p_{m-1}x^{m-1} + \cdots + p_0$. For a given m, $A(x) \cdot B(x)\,\mathbf{mod}P(x)$ can be calculated with the **LFSR** of Fig. 10.5b. $B(x)$ and $P(x)$ coefficients have to be applied in parallel at the labeled points, before the coefficients of $A(x)$ are introduced as serial data to the **LFSR** (highest first).

It is immediate that in reality, the circuit of Fig. 10.5b can be used for the multiplication over GF(2^i){$Q(x)$}, for any $i \leq m$. For this it is enough to apply the coefficients b_j and p_j, starting by the highest, from the highest positions (entries b_{m-1} and p_{m-1}, respectively), resetting the lower positions that become vacant.

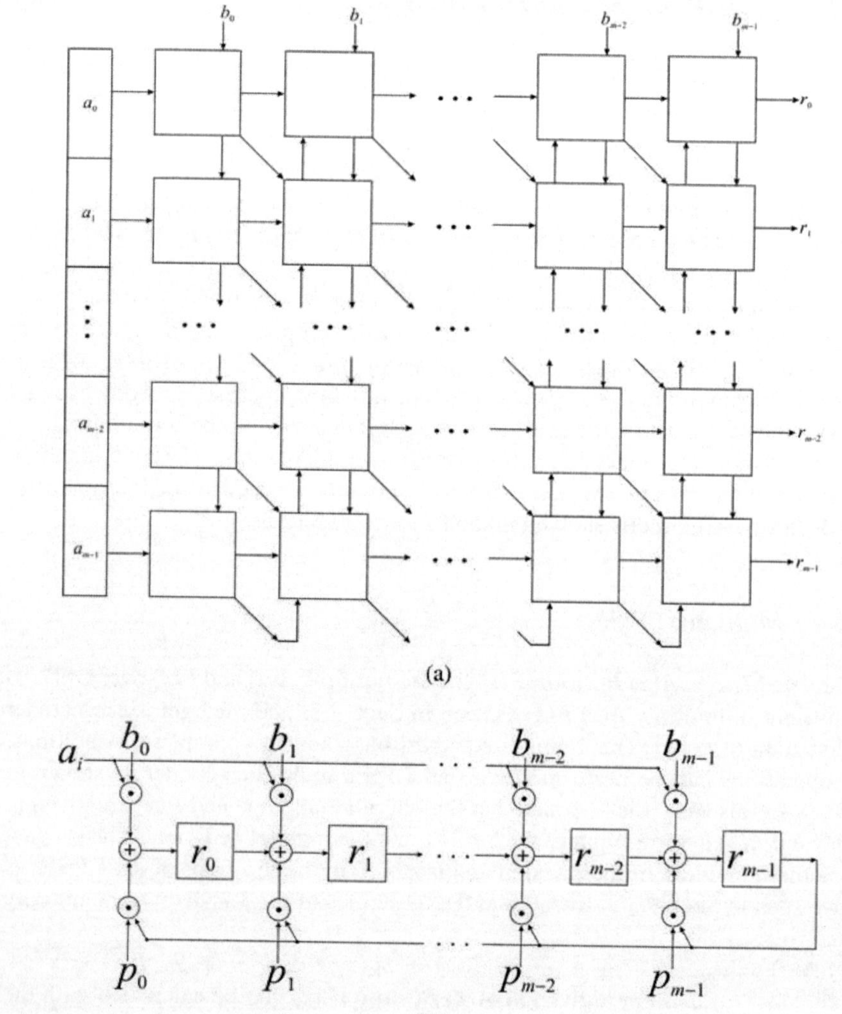

Fig. 10.5 Serial-parallel multiplication. **a** Cellular circuit, **b** multiplier/divider

10.3.3.2 Divider/Accumulator

Let suppose again the case $B(x)$ and $P(x)$ are provided in parallel form and $A(x)$ in serial form, but now it is first provided the least significant coefficient, where $A(x) = a_{m-1}x^{m-1} + \cdots + a_0$, $B(x) = b_{m-1}x^{m-1} + \cdots + b_0$, $P(x) = x^m + p_{m-1}\, x^{m-1} + \cdots + p_0$. The product $M(x) = A(x){\cdot}B(x)$ can be expanded as follows:

$$M(x) = A(x) \cdot B(x) \bmod P(x) = a_0 B(x) + a_1\{B(x)x \bmod P(x)\}$$

$$+ a_2\{B(x)x^2 \bmod P(x)\} + \cdots + a_{m-1}\{B(x)x^{m-1} \bmod P(x)\}$$

The calculation can be performed in m iterations. Indeed, the factors $B(x)x^i$**mod**$P(x)$ can be obtained with an **LFSR** divider by $P(x)$, with a development as in Example B.5 (Figure B.1). Figure 10.6a shows the full circuit for calculating $A(x){\cdot}B(x)$**mod**$P(x)$, which includes an **LFSR** and an **M** register, which has to be

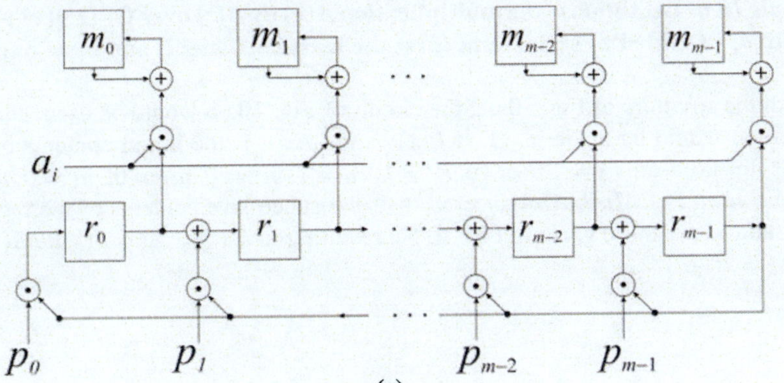

(a)

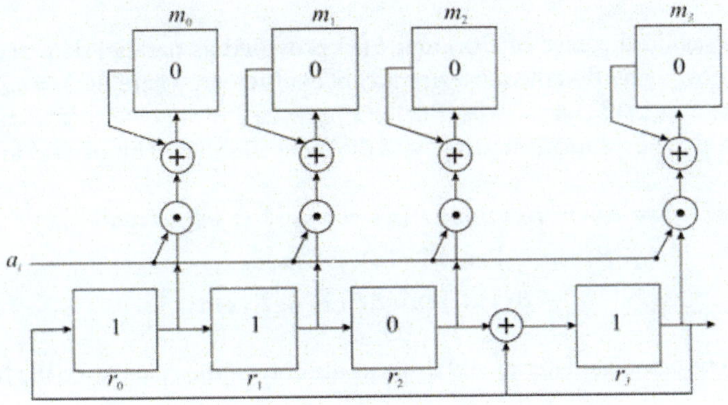

(b)

t	a_i	m_3	m_2	m_1	m_0	r_3	r_2	r_1	r_0
0		0	0	0	0	1	0	1	1
1	$a_0 = 1$	1	0	1	1	1	1	1	1
2	$a_1 = 0$	1	0	1	1	0	1	1	1
3	$a_2 = 0$	1	0	1	1	1	1	1	0
4	$a_3 = 1$	0	1	0	1				

(c)

Fig. 10.6 **a** Divider/accumulator. **b** Circuit of Example 10.6. **c** Results for Example 10.6

initialized to 0. Initially, the **LFSR** is loaded with $B(x)$ and at each iteration the **LFSR** content is multiplied by x and is reduced **mod**$P(x)$. In order to calculate $A(x)\cdot B(x)$**mod**$P(x)$, at each iteration the **LFSR** content is multiplied by a_i, is added to the previous result stored in **M**, and is stored in **M**, where the final result is saved.

If the initial content of **M** is the polynomial $C(x)$, then $M(x) = \{A(x)\cdot B(x) + C(x)\}$**mod**$P(x)$ would be calculated.

Example 10.6 **Let suppose the multiplication $A(x)$ by $B(x)$ over GF(2^4)$\{x^4+x^3+1\}$, with a_i provided as serial data (first the least coefficient) and b_i as parallel data.**

With the structure of Fig. 10.6a, the circuit of Fig. 10.6b would be used, and the calculation would be made in 4T. If $B(x) = x^3 + x + 1$, the initial contents of the various flip-flops are $r_3 = 1, r_2 = 0, r_1 = 1, r_0 = 1, m_3 = 0, m_2 = 0, m_1 = 0, m_0 = 0$, as shown in Fig. 10.6b. If $A(x) = x^3 + 1$, the successive contents of the register **LFSR** and M are those given in Fig. 10.6c, resulting in $M(x) = A(x)\cdot B(x)$**mod**$(x^4 + x^3 + 1) = x^2 + 1$.

■

10.3.3.3 LFSR Accumulator

Let suppose again the case of $B(x)$ and $P(x)$ provided as parallel data and $A(x)$ as serial data, now again first the most significant coefficient, where $A(x) = a_{n-1}x^{n-1} + \dots + a_0$, $B(x) = b_{n-1}x^{n-1} + \dots + b_0$, $P(x) = x^n + p_{n-1}x^{n-1} + \dots + p_0$. The calculation of $M(x)$ can also be arranged starting with the most significant bit of $A(x)$ as follows:

$$\begin{aligned} M(x) = A(x)\cdot B(x)\mathbf{mod}P(x) = \{\dots\{B(x)a_{n-1}x\mathbf{mod}P(x) \\ +B(x)a_{n-2}\}x\mathbf{mod}P(x) + B(x)a_{n-3}\} \\ +\dots+ B(x)a_1\}x\mathbf{mod}P(x) + B(x)a_0 \end{aligned}$$

$M(x)$ can be calculated iteratively using as the core of the calculation the following expression:

$$M^k(x) = M^{k-1}(x)x\mathbf{mod}P(x) + B(x)a_{n-k}, \quad k = 1,\dots,n; \quad M^0 = 0$$

The computation of $M(x)$ could be made using the circuit of Fig. 10.7. Each iteration in this computation would be made in two clock cycles: one for the shift in the **LFSR**, i.e., for the multiplication of the previous contents of the **LFSR** and x, and the modular correction of the result, $M^{k-1}(x)x$mod$P(x)$; and another clock cycle for the addition of $B(x)a_{n-k}$ and for loading the result in parallel in the **LFSR**.

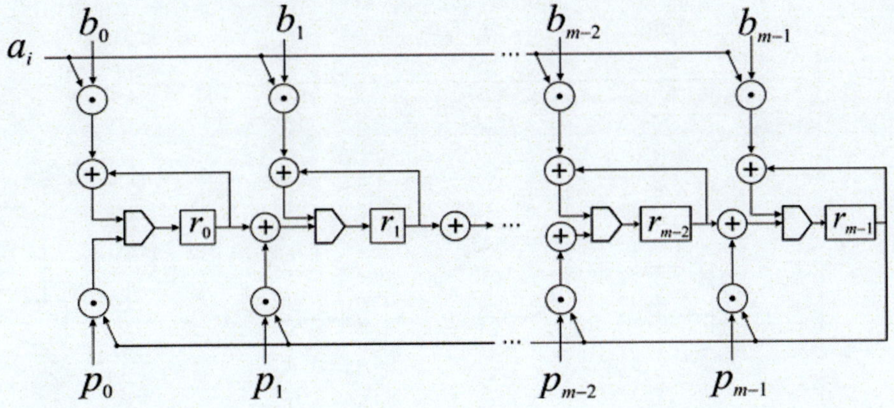

Fig. 10.7 LFSR accumulator

10.3.3.4 Montgomery Multiplier

The Montgomery multiplication defined over GF(p) [5, 7, 9] may be extended to GF(2^m). Given a primitive polynomial $P(x)$ of degree m, and given the polynomials $A(x)$ and $B(x)$ such that $g(A) < m$ and $g(B) < m$, the Montgomery multiplication of $A(x)$ and $B(x)$ modulo $P(x)$, noted as $MM\{A(x), B(x), P(x)\}$, is defined as follows:

$$MM\{A(x),\ B(x),\ P(x)\} = \left\{A(x) \cdot B(x)x^{-m}\right\} \bmod P(x)$$

The product $A(x)\cdot B(x)x^{-m}$ can be developed as follows:

$$A(x) \cdot B(x)x^{-m} = \{\{\ldots\{\{A(x)b_0x^{-1} + A(x)b_1\}x^{-1} + A(x)b_2\} + \cdots + A(x)b_{m-2}\}x^{-1} + A(x)b_{m-1}\}x^{-1}$$

Therefore this product may be performed using m iterations, as $\{R(x) + A(x)b_i\}x^{-1}$, for $i = 0, \ldots, m-1$, where initially $R = 0$. If in the addition $S(x) = \{R(x) + A(x)b_i\}$, the independent term is zero ($s_0 = 0$), the multiplication by x^{-1} is reduced to a shift such as the independent term disappears; if the independent term is $s_0 = 1$, $\{R(x) + A(x)b_i\}$ can be added to $P(x)$, such that $\{R(x) + A(x)b_i\} + P(x)$ have no independent term and, again, the multiplication by x^{-1} is reduced to a shift. Thus the algorithm core will be the following one:

$$1)\ S \leftarrow R + Ab_i$$
$$2)\ R \leftarrow (S + s_0P)x^{-1}$$

With the circuit of Fig. 10.8 the Montgomery multiplication is performed using 2^m iterations. Initially, the register $R = r_{m-1}, \ldots, r_0$ must be zero, and the result is loaded in R.

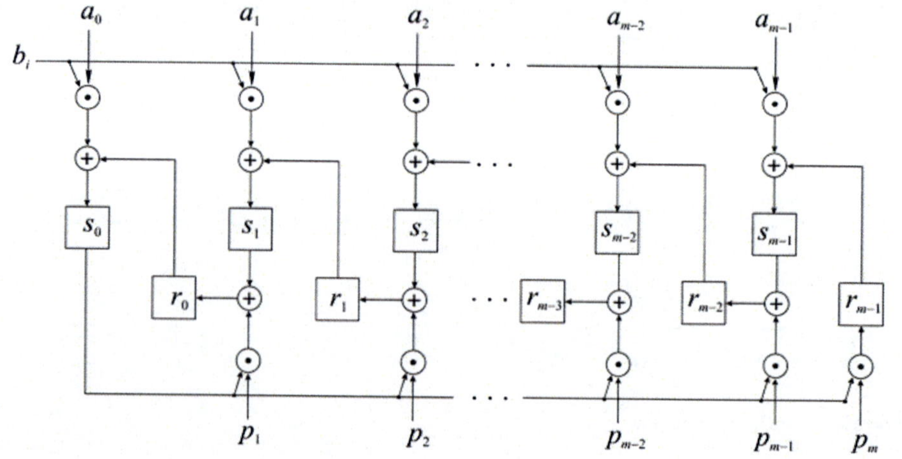

Fig. 10.8 Montgomery multiplier circuit

Once defined $MM\{A(x), B(x), P(x)\}$, it is easy to see that for obtaining $Z(x) = X(x)\cdot Y(x)\mathbf{mod}P(x)$ is enough to perform two Montgomery multiplications. In fact, with the previous computation of:

$$Q(x) = x^{2m} \bmod P(x)$$

it results:

$$\begin{aligned} Z(x) &= X(x)\cdot Y(x)\mathbf{mod}P(x) \\ &= \left\{\{X(x)\cdot Y(x)x^{-m}\}x^{2m}x^{-m}\right\}\mathbf{mod}P(x) \\ &= MM\left\{MM\{X(x), Y(x), P(x)\}, x^{2m}, P(x)\right\} \\ &= MM\{MM\{X(x), Y(x), P(x)\}, Q(x), P(x)\} \end{aligned}$$

Therefore, the use of the Montgomery multiplier avoids the modular reduction: it is only necessary to make additions and shifts.

10.3.4 Serial Multiplication

The cellular circuit of Fig. 10.4e may be adapted again to consider the situation in which both operands are provided as serial data. Simply attach two shift registers of m length as the buffers for a_i, and b_i, input data, so that $A(x)$ and $B(x)$ are inserted as serial data, and thus are available in parallel, similar to that depicted in Fig. 10.5a.

Addressing this situation from another perspective, when the polynomials $A(x)$ and $B(x)$ are provided as serial data, a coefficient of the product $A(x)\cdot B(x)$ can be computed at each clock cycle, using an appropriate circuit for each m. The modular reduction can be done with an **LFSR** divider. The details of these solutions (one for each m) are illustrated with the following example.

Example 10.7 **This example consists on multiplying over GF(2^4)\{$P(x)$\}.**

With the two operands $A(x)$ and $B(x)$ being serial data, first the most significant coefficients, with $A(x) = a_3x^3 + a_2x^2 + a_1x + a_0$, $B(x) = b_3x^3 + b_2x^2 + +b_1x + b_0$, $P(x) = x^4 + p_3x^3 + p_2x^2 + p_1x + p_0$. The product is given by:

$$\begin{aligned} A(x) \cdot B(x) = a_3b_3x^6 + (a_3b_2 + a_2b_3)x^5 + (a_3b_1 + a_1b_3 + a_2b_2)x^4 \\ + (a_3b_0 + a_0b_3 + a_2b_1 + a_1b_2)x^3 + (a_2b_0 + a_0b_2 + a_1b_1)x^2 \\ + (a_0b_1 + a_1b_0)x + a_0b_0 \end{aligned}$$

That is,

$$\begin{aligned} c_6 &= a_3b_3; \\ c_5 &= a_3b_2 + a_2b_3; \\ c_4 &= a_3b_1 + a_1b_3 + a_2b_2; \\ c_3 &= a_3b_0 + a_0b_3 + a_2b_1 + a_1b_2; \\ c_2 &= a_2b_0 + a_0b_2 + a_1b_1; \\ c_1 &= a_0b_1 + a_1b_0; \\ c_0 &= a_0b_0 \end{aligned}$$

It is necessary to compute $A(x) \cdot B(x)$ and then to make the modular reduction using an **LFSR**. The coefficients a_i and b_i are applied in four clock cycles (first the most significant), so that using the circuit of Fig. 10.9a the values x, y, z, u of the table in Fig. 10.9b can be obtained, some of which are stored in $\boldsymbol{D_x}$, $\boldsymbol{D_y}$ and $\boldsymbol{D_z}$ (Fig. 10.9c). When the coefficients are generated they are applied to the **LFSR** (Fig. 10.9d). The calculation is done using seven clock cycles, according to the calculation scheme of Fig. 10.9e, being the coefficients r_i of the result in the **LFSR**.

■

To obtain directly the final result can be more interesting (multiplication and modular reduction), such as in the following example.

Example 10.8 This example consists on multiplying over GF(2^4)\{$x^4 + x^3 + 1$\}, with the two operands $A(x)$ and $B(x)$ provided as serial data, first the most significant coefficients, with $A(x) = a_3x^3 + a_2x^2 + a_1x + a_0$, $B(x) = b_3x^3 + b_2x^2 + b_1x + b_0$.

From Example 10.4:

$$\begin{bmatrix} r_0 \\ r_1 \\ r_2 \\ r_3 \end{bmatrix} = \begin{bmatrix} a_0 & a_3 & a_2 + a_3 & a_1 + a_2 + a_3 \\ a_1 & a_0 & a_3 & a_2 + a_3 \\ a_2 & a_1 & a_0 & a_3 \\ a_3 & a_2 + a_3 & a_1 + a_2 + a_3 & a_0 + a_1 + a_2 + a_3 \end{bmatrix} \begin{bmatrix} b_0 \\ b_1 \\ b_2 \\ b_3 \end{bmatrix}$$

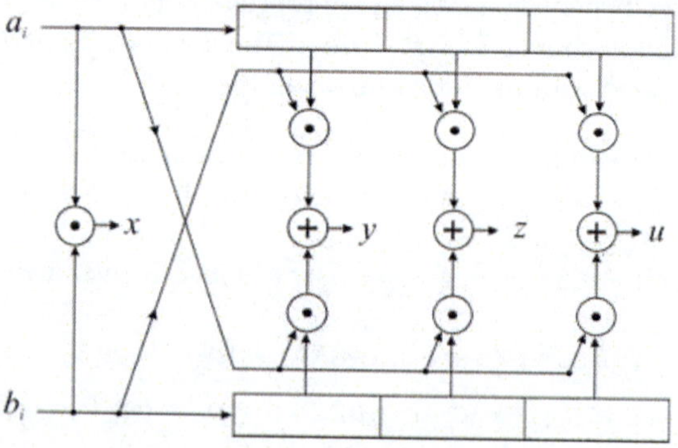

(a)

	x	y	z	u
1	a_3b_3			
2	a_2b_2	$a_3b_2 + a_2b_3$		
3	a_1b_1	$a_2b_1 + a_1b_2$	$a_3b_1 + a_1b_3$	
4	a_0b_0	$a_1b_0 + a_0b_1$	$a_2b_0 + a_0b_2$	$a_3b_0 + a_0b_3$

(b)

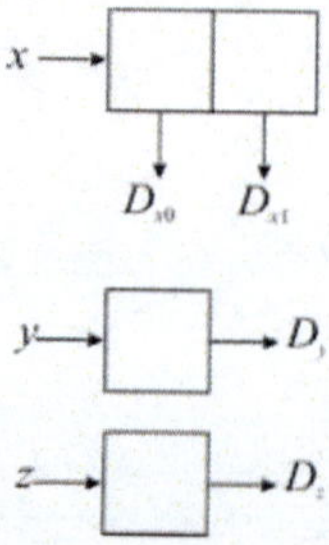

(c)

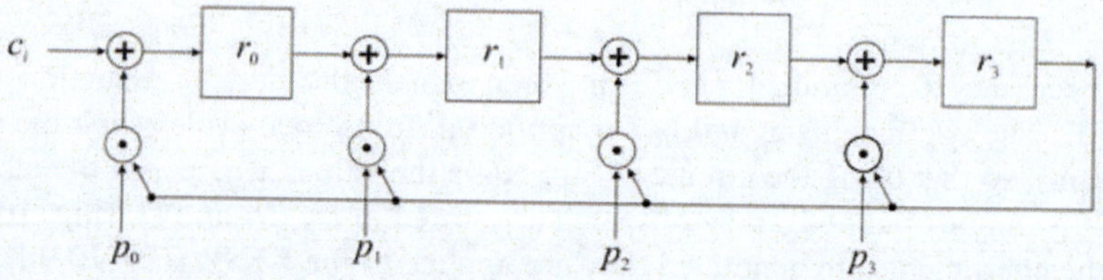

(d)

1) $c_6 \bullet x$
2) $c_5 \bullet y, D_x \bullet x$
3) $c_4 \bullet z + D_{x0}, D_x \bullet x, D_y \bullet y$
4) $c_3 \bullet u + D_y, D_z \bullet z, D_x \bullet x, D_y \bullet y$
5) $c_2 \bullet D_{x1} + D_z$
6) $c_1 \bullet D_y$
7) $c_0 \bullet D_{x0}$

(e)

Fig. 10.9 Serial multiplication (Example 10.7). **a** Circuit, **b** Results. **c** Memory elements. **d** LFSR. **e** Algorithm for the computation of the coefficient

Each r_i can be expressed as follows:

$$r_0 = [a_3b_3] + [a_3b_2 + a_2b_3 + a_2b_2] + [a_3b_1 + a_1b_3] + [a_0b_0]$$
$$r_1 = [a_3b_3] + [a_3b_2 + a_2b_3] + [a_1b_0 + a_0b_1]$$
$$r_2 = [a_3b_3] + [a_1b_1] + [a_2b_0 + a_0b_2]$$
$$r_3 = [a_3b_3] + [a_3b_2 + a_2b_3 + a_2b_2] + [a_2b_1 + a_3b_1 + a_1b_2 + a_1b_3] + [a_3b_0 + a_0b_3]$$

Introducing the coefficients of both polynomials in the respective shift registers, as shown in Fig. 10.10a, the sum of three or four products for each r_i can be generated using four clock cycles. Seven products have to be generated: $x_1y_1, x_1y_2, x_2y_1, x_3y_1, x_1y_3, x_1y_4, x_4y_1$, i.e., it takes seven 2-input **AND** gates. Adding these products appropriately the following expression are obtained:

$$E = x_1y_1;\ F = x_1y_2 + x_2y_1;\ G = E + F;$$
$$H = x_1y_3 + x_3y_1;\ I = F + H;\ J = x_1y_4 + x_4y_1$$

These signals can be generated using five *EXOR* gates, as shown in Fig. 10.10b. The coefficients r_i are calculated by adding the appropriate signals $E \ldots J$ at each instant, as given in the table of Fig. 10.10c. The coefficients r_i are obtained in four cycles using four D flip-flops, initialized to zero and whose inputs are from 4-to-1 multiplexers, as depicted in Fig. 10.10d.

■

10.4 Multiplication Over GF(2^m) Using the Normal Base

The use of a normal base in multiplication has no advantage over a standard base, but not a special difficulty. For each *m*, multiplier circuits can be easily designed, depending on the data are provided as serial or parallel form. For parallel data, a strictly combinational multiplier can be implemented, as has been done with standard base, or to implement intermediate solutions. To see how to proceed, a simple example is detailed at the following.

Example 10.9 **Design a circuit to multiply over GF(2^4){x^4+ x^3 +1}, with the normal base {x, x^2, x^4, x^8}.**

In this case it holds that $x^{16} = x$. For other power contained in the product it is obtained:

$$x^{12} = x^8 + x^4 + x^2;\quad x^{10} = x^8 + x^2;\quad x^9 = x^8 + x^4 + x;$$
$$x^6 = x^4 + x^2 + x;\quad x^5 = x^4 + x;\quad x^3 = x^8 + x^2 + x$$

The following expression results for the multiplication of $A(x) = a_3x^8 + a_2x^4 + a_1x^2 + a_0x$ and $B(x) = b_3x^8 + b_2x^4 + b^2 + b_0x$:

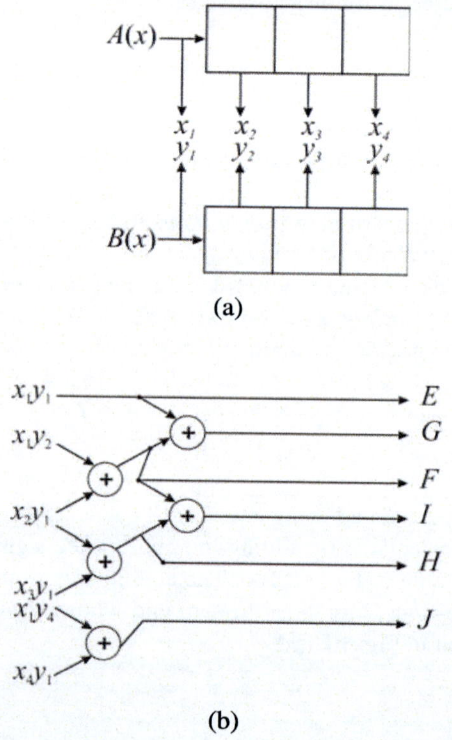

	1	2	3	4
r_0	E	G	H	E
r_1	E	F	0	F
r_2	E	0	E	H
r_3	E	G	I	J

(c)

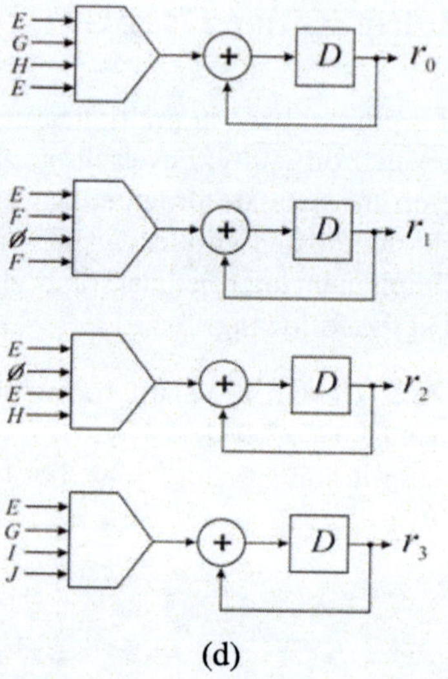

(d)

Fig. 10.10 Serial multiplication (Example 10.8). **a** Shift registers. **b** Intermediate additions. **c** Table for the final computation. **d** Circuits for the final additions

$$
\begin{aligned}
C(x) = A(x) \cdot B(x) &= a_3b_3x^{16} + (a_3b_2 + a_2b_3)x^{12} + (a_3b_1 + a_1b_3)x^{10} \\
&+ (a_3b_0 + a_0b_3)x^9 + a_2b_2x^8 + (a_2b_1 + a_1b_2)x^6 + (a_0b_2 + a_2b_0)x^5 + a_1b_1x^4 \\
&+ (a_1b_0 + a_0b_1)x^3 + a_0b_0x^2 = c_3x^8 + c_2x^4 + c_1x^2 + c_0x
\end{aligned}
$$

After appropriate substitutions, the following coefficients of C(x) results:

$$
\begin{aligned}
c_3 &= (a_3b_2 + a_2b_3) + (a_3b_1 + a_1b_3) + (a_3b_0 + a_0b_3) + (a_1b_0 + a_0b_1) + a_2b_2 \\
&= F + G + H + I + J \\
c_2 &= (a_3b_2 + a_2b_3) + (a_3b_0 + a_0b_3) + (a_2b_1 + a_1b_2) + (a_0b_2 + a_2b_0) + a_1b_1 \\
&= F + H + K + L + M \\
c_1 &= (a_3b_2 + a_2b_3) + (a_3b_1 + a_1b_3) + (a_1b_0 + a_0b_1) + (a_2b_1 + a_1b_2) + a_0b_0 \\
&= F + G + I + K + N \\
c_0 &= (a_3b_0 + a_0b_3) + (a_1b_0 + a_0b_1) + (a_2b_1 + a_1b_2) + (a_0b_2 + a_2b_0) + a_3b_3 \\
&= H + I + K + L + O
\end{aligned}
$$

where:

$$
\begin{aligned}
&F = a_3b_2 + a_2b_3; \; G = a_3b_1 + a_1b_3; \; H = a_3b_0 + a_0b_3; \; I = a_1b_0 + a_0b_1; \\
&J = a_2b_2; \quad K = a_2b_1 + a_1b_2; \; L = a_0b_2 + a_2b_0; \; M = a_1b_1; \\
&N = a_0b_0; \quad O = a_3b_3
\end{aligned}
$$

The calculation of these coefficients can be done using the circuit of Fig. 10.11a. The coefficients a_3, ..., a_0, b_3, ..., b_0, of the two operands are inserted in respective ring shift registers; after this, the calculation is completed using four iterations. At each iteration, the circuit outputs x, y, z of Fig. 10.11a generate the values given in the table of the Fig. 10.11b. The contribution of each coefficient at each iteration, is given by the table of Fig. 10.11c. The computation is performed using the calculation scheme shown in Fig. 10.11d. The loading of a_3, ..., a_0, b_3, ..., b_0, in the ring shift registers can be done in parallel, if all these coefficients are available at the same time, or in serial, in this last case the calculation is completed in eight cycles.

Of course, if the coefficients of the two operands are available in parallel, a combinational circuit may be designed to obtain the coefficients of $C(x)$; and if the coefficients of the two operands are available in series, the coefficients of $C(x)$ may be calculated in four cycles using a circuit different of that of Fig. 10.1a (see Exercise 13).

The cross products for the normal base $\{\alpha, \alpha^2, \alpha^4, \alpha^8\}$ are given in Table 10.4. In each column of this table there are 9 ones, and therefore for the calculation of each coefficient of $C(x)$ it is necessary the addition of 9 products.

It is also possible the use of a multiplication matrix and a reduction to the base. In this case:

Fig. 10.11 Multiplier using normal base. **a** Circuit. **b** Generated data. **c** Calculation of the coefficients. **d** Calculation scheme

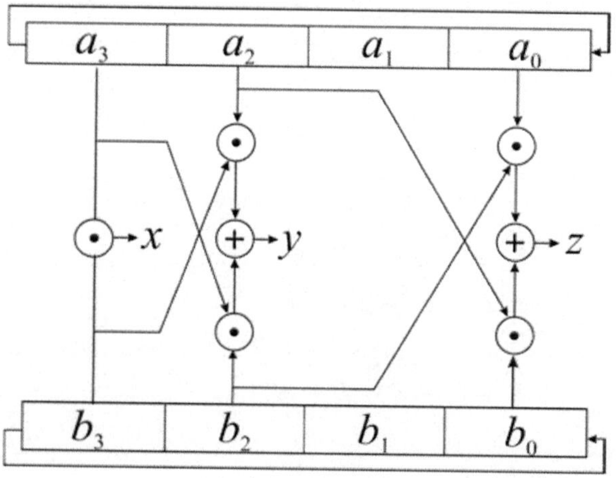

(a)

	x	y	z
1	O	F	L
2	J	K	G
3	M	I	L
4	N	H	G

(b)

	1	2	3	4
c_3	y	$x+z$	y	y
c_2	y	y	$x+z$	y
c_1	y	y	y	$x+z$
c_0	$x+z$	y	y	y

(c)

1) $c_3 \bullet y$, $c_2 \bullet y$, $c_1 \bullet y$, $c_0 \bullet x+z$;

2) $c_3 \bullet c_3+(x+z)$, $c_2 \bullet c_2+y$, $c_1 \bullet c_1+y$, $c_0 \bullet c_0+y$;

3) $c_3 \bullet c_3+y$, $c_2 \bullet c_2+(x+z)$, $c_1 \bullet c_1+y$, $c_0 \bullet c_0+y$;

4) $c_3 \bullet c_3+y$, $c_2 \bullet c_2+y$, $c_1 \bullet c_1+(x+z)$, $c_0 \bullet c_0+y$;

(d)

Table 10.4 Table of cross products for the normal base {α, $α^2$, $α^4$, $α^8$}

		A·B			
A	B	$α^8$	$α^4$	$α^2$	α
α	α	0	0	1	0
α	$α^2$	1	0	1	1
α	$α^4$	0	1	0	1
α	$α^8$	1	1	0	1
$α^2$	α	1	0	1	1
$α^2$	$α^2$	0	1	0	0
$α^2$	$α^4$	0	1	1	1
$α^2$	$α^8$	1	0	1	0
$α^4$	α	0	1	0	1
$α^4$	$α^2$	0	1	1	1
$α^4$	$α^4$	1	0	0	0
$α^4$	$α^8$	1	1	1	0
$α^8$	α	1	1	0	1
$α^8$	$α^2$	1	0	1	0
$α^8$	$α^4$	1	1	1	0
$α^8$	$α^8$	0	0	0	1

$$\begin{aligned} A(x)\cdot B(x) &= \left(a_3x^8 + a_2x^4 + a_1x^2 + a_0x\right)\cdot\left(b_3x^8 + b_2x^4 + b_1x^2 + b_0x\right) \\ &= M(x) = m_9x^{16} + m_8x^{12} + m_7x^{10} + m_6x^9 + m_5x^8 + m_4x^6 \\ &\quad + m_3x^5 + m_2x^4 + m_1x^3 + m_0x^2 \end{aligned}$$

where:

$$\begin{bmatrix} m_0 \\ m_1 \\ \ldots \\ m_8 \\ m_9 \end{bmatrix} = \begin{bmatrix} a_0 & 0 & 0 & 0 \\ a_1 & a_0 & 0 & 0 \\ 0 & a_1 & 0 & 0 \\ a_2 & 0 & a_0 & 0 \\ 0 & a_2 & a_1 & 0 \\ 0 & 0 & a_2 & 0 \\ a_3 & 0 & 0 & a_0 \\ 0 & a_3 & 0 & a_1 \\ 0 & 0 & a_3 & a_2 \\ 0 & 0 & 0 & a_3 \end{bmatrix} \begin{bmatrix} b_0 \\ b_1 \\ b_2 \\ b_3 \end{bmatrix}$$

On the other hand:

$$C(x) = c_3x^8 + c_2x^4 + c_0x$$

$$= M(x) = m_9x^{16} + m_8x^{12} + m_7x^{10} + m_6x^9 + m_5x^8$$
$$+ m_4x^6 + m_3x^5 m_2x^4 + m_1x^3 + m_0x^2$$

resulting:

$$\begin{bmatrix} c_0 \\ c_1 \\ c_2 \\ c_3 \end{bmatrix} = \begin{bmatrix} 0\,1\,0\,1\,1\,0\,1\,0\,0\,1 \\ 1\,1\,0\,0\,1\,0\,0\,1\,1\,0 \\ 0\,0\,1\,1\,1\,0\,1\,0\,1\,0 \\ 0\,1\,0\,0\,0\,1\,1\,1\,1\,0 \end{bmatrix} \begin{bmatrix} m_0 \\ m_1 \\ \ldots \\ m_8 \\ m_9 \end{bmatrix}$$

Putting together the expressions:

$$\begin{bmatrix} c_0 \\ c_1 \\ c_2 \\ c_3 \end{bmatrix} = \begin{bmatrix} 0\,1\,0\,1\,1\,0\,1\,0\,0\,1 \\ 1\,1\,0\,0\,1\,0\,0\,1\,1\,0 \\ 0\,0\,1\,1\,1\,0\,1\,0\,1\,0 \\ 0\,1\,0\,0\,0\,1\,1\,1\,1\,0 \end{bmatrix} \begin{bmatrix} a_0 & 0 & 0 & 0 \\ a_1 & a_0 & 0 & 0 \\ 0 & a_1 & 0 & 0 \\ a_2 & 0 & a_0 & 0 \\ 0 & a_2 & a_1 & 0 \\ 0 & 0 & a_2 & 0 \\ a_3 & 0 & 0 & a_0 \\ 0 & a_3 & 0 & a_1 \\ 0 & 0 & a_3 & a_2 \\ 0 & 0 & 0 & a_3 \end{bmatrix} \begin{bmatrix} b_0 \\ b_1 \\ b_2 \\ b_3 \end{bmatrix}$$

And operating:

$$\begin{bmatrix} c_0 \\ c_1 \\ c_2 \\ c_3 \end{bmatrix} = \begin{bmatrix} a_1 + a_2 + a_3 & a_0 + a_2 & a_0 + a_1 & a_0 + a_3 \\ a_0 + a_1 & a_0 + a_2 + a_3 & a_1 + a_3 & a_1 + a_2 \\ a_2 + a_3 & a_1 + a_2 & a_0 + a_1 + a_3 & a_0 + a_2 \\ a_1 + a_3 & a_0 + a_3 & a_2 + a_3 & a_0 + a_1 + a_2 \end{bmatrix} \begin{bmatrix} b_0 \\ b_1 \\ b_2 \\ b_3 \end{bmatrix}$$

For combinational multiplication, it is easy to assess the necessary hardware. Specifically each c_i can be obtained by adding 9 of the 16 possible products a_ib_j; therefore, to make the calculation it is easy to build a circuit with 32 **EXOR** gates and 16 **AND** gates, being all these two-input gates. It is also possible to do first the internal additions of the matrix multiplication, for which it takes 10 two-input **EXOR** gates; once these elements of the matrix have been added, the rows of the matrix and the terms b_i are multiplied by means of 16 two-input **AND** gates. Finally, the four products for each coefficient are added, requiring $3 \times 4 = 12$ two-input **EXOR** gates. In total, with this new organization, 22 **EXOR** gates and 16 **AND** gates are needed, all two-input gates.

Considering the coefficients with a circular structure (i.e., $a_4 = a_0$, $a_5 = a_1$, etc.), it is immediate the following expression:

$$c_i = (a_{i+1} + a_{i+2} + a_{i+3})b_i + (a_i + a_{i+2})b_{i+1} + (a_i + a_{i+1})b_{i+2} + (a_i + a_{i+3})b_{i+3}$$

■

From Example 10.9 it is concluded that, for the multiplication over GF(2^4)$\{x^4 + x^3 + 1\}$ with normal base operands $\{x, x^2, x^4, x^8\}$, the coefficients of the product are obtained from the coefficients of the operands, using the same function for all of them. It is easy to see that this is a general conclusion: *for the multiplication over GF(2^m) on any normal base, the coefficients of the product can be obtained from those of the operands, using all the same function.* The property that given the vector representation of an element E of GF(2^m) with a normal base, $E = (e_{m-1}, e_{m-2}, \ldots, e_1, e_0)$, E^2 is obtained with a rotation to the left of E, $E^2 = (e_{m-2}, e_{m-3}, \ldots, e_1, e_0, e_{m-1})$, is used (see Section B.4.2).

Let $A = (a_{m-1}, a_{m-2}, \ldots, a_1, a_0)$ and $B = (b_{m-1}, b_{m-2}, \ldots, b_1, b_0)$. The product will be:

$$\begin{aligned} C &= (c_{m-1}, c_{m-2}, c_1, c_0) = A \cdot B \\ &= (a_{m-1}, a_{m-2}, \ldots, a_1, a_0) \cdot (b_{m-1}, b_{m-2}, \ldots, b_1, b_0) \end{aligned}$$

Each coefficient of C, for example the most significant coefficient, c_{m-1}, was obtained as a combinational function of $a_{m-1}, a_{m-2}, \ldots, a_1, a_0, b_{m-1}, b_{m-2}, \ldots, b_1, b_0$:

$$c_{m-1} = f(a_{m-1}, a_{m-2}, \ldots, a_1 a_0, \ldots, b_{m-1}, \ldots, b_{m-2}, \ldots, b_1, b_0)$$

Squaring $C = A{\cdot}B$, $C^2 = A^2{\cdot}B^2$. That is:

$$\begin{aligned} C^2 &= (c_{m-2}, c_{m-3}, \ldots, c_1, c_0, c_{m-1}) = A^2 \cdot B^2 \\ &= (a_{m-2}, a_{m-3}, \ldots, a_1, a_0, a_{m-1}) \cdot (b_{m-2}, b_{m-3}, \ldots, b_1, b_0, b_{m-1}) \end{aligned}$$

For the most significant coefficient of C^2, c_{m-2}, due to it is obtained as a product, it has the same dependence on the coefficients of A^2 and B^2 as c_{m-1}. That is,

$$c_{m-2} = f(a_{m-2}, a_{m-3}, \ldots, a_1, a_0, a_{m-1}, b_{m-2}, b_{m-3}, \ldots, b_1, b_0, b_{m-1})$$

Squaring $C^2 = A^2{\cdot}B^2$ (i.e., computing $C^4 = A^4{\cdot}B^4$) the same conclusion for c_{m-3} is obtained, and so on: always the same function is used, rotating the arguments.

From Example 10.9 it is also concluded that the complexity of the function f () depends on the base that has been used. Specifically, using an optimal base, the function f () will have the minimum number of products, such as seen in the following example.

Example 10.10 **Design a circuit to multiply over GF(2^4)$\{x^4 + x^3 + 1\}$, with the normal base $\{\alpha^3, \alpha^6, \alpha^{12}, \alpha^{24}\}$, which from Example B.9 is concluded as optimal.**

Multiplying $A(x) = a_3\alpha^{24} + a_2\alpha^{12} + a_1\alpha^6 + a_0\alpha^3$ and $B(x) = b_3\alpha^{24} + b_2\alpha^{12} + b_1\alpha^6 + b_0\alpha^3$ it results:

$$
\begin{aligned}
C(x) = A(x) \cdot B(x) = &(a_1b_0 + a_2b_0 + a_0b_1 + a_3b_1 + a_0b_2 + a_2b_2 + a_1b_3)\alpha^{24} \\
&+ (a_2b_0 + a_3b_0 + a_1b_1 + a_3b_1 + a_0b_2 + a_0b_3 + a_1b_3)\alpha^{12} \\
&+ (a_0b_0 + a_2b_0 + a_3b_1 + a_0b_2 + a_3b_2 + a_1b_3 + a_2b_3)\alpha^{6} \\
&+ (a_2b_0 + a_2b_1 + a_3b_1 + a_0b_2 + a_1b_2 + a_1b_3 + a_3b_3)\alpha^{3}
\end{aligned}
$$

The cross products for the normal base $\{\alpha^3, \alpha^6, \alpha^{12}, \alpha^{24}\}$ are given in Table 10.5. In each column of this table there are 7 ones, and therefore for the calculation of each coefficient of $C(x)$, 7 products must be added.

Using matrix notation it results:

$$
\begin{bmatrix} c_0 \\ c_1 \\ c_2 \\ c_3 \end{bmatrix} = \begin{bmatrix} a_2 & a_2 + a_3 & a_0 + a_1 & a_1 + a_3 \\ a_0 + a_2 & a_3 & a_0 + a_3 & a_1 + a_2 \\ a_2 + a_3 & a_1 + a_3 & a_0 & a_0 + a_1 \\ a_1 + a_2 & a_0 + a_3 & a_0 + a_2 & a_1 \end{bmatrix} \begin{bmatrix} b_0 \\ b_1 \\ b_2 \\ b_3 \end{bmatrix}
$$

A combinational circuit to implement this multiplier, as it is expressed by the above matrix relationship, needs 6 **EXOR** gates for the additions ($a_i + a_j$) internal to the matrix, 16 **AND** gates for the products of the four rows of the matrix by the b_i, and 12 **EXOR** gates for the additions of the four products, a total of 18 **EXOR**

Table 10.5 Table of cross products for the normal base $\{\alpha^3, \alpha^6, \alpha^{12}, \alpha^{24}\}$

		$A \cdot B$			
A	B	α^{24}	α^{12}	α^6	α^3
α^3	α^3	0	0	1	0
α^3	α^6	1	0	0	0
α^3	α^{12}	1	1	1	1
α^3	α^{24}	0	1	0	0
α^6	α^3	1	0	0	0
α^6	α^6	0	1	0	0
α^6	α^{12}	0	0	0	1
α^6	α^{24}	1	1	1	1
α^{12}	α^3	1	1	1	1
α^{12}	α^6	0	0	0	1
α^{12}	α^{12}	1	0	0	0
α^{12}	α^{24}	0	0	1	0
α^{24}	α^3	0	1	0	0
α^{24}	α^6	1	1	1	1
α^{24}	α^{12}	0	0	1	0
α^{24}	α^{24}	0	0	0	1

gates and 16 **AND** gates, all two-input gates, with a circuit in three levels **EXOR-AND-EXOR**. It is also possible to calculate the 16 products a_ib_j, and then get each c_k with the addition of the corresponding seven products; structured the circuit as **AND-EXOR** (**AND** gates feeding **EXOR** gates) 24 **EXOR** gates and 16 **AND** gates are required, all two-input gates.

■

Comparing the cost of the Examples 10.9 and 10.10 implementations, it is clear the advantage of using an optimal base. In general, a combinational circuit for the multiplication in GF(2^n) in the **AND-EXOR** implementation, using an optimal base, needs $2n(n-1)$ **EXOR** gates and n^2 **AND** gates, all 2-inputs.

10.5 Multiplication Over GF(2^m) Using the Dual Base

Let suppose a standard base $\{b_0, b_1, \ldots, b_{m-1}\} = \{1, \alpha, \alpha^2, \ldots, \alpha^{m-1}\}$ in GF(2^m) and its dual base $\{d_0, d_1, \ldots, d_{m-1}\}$. Any element A of GF(2^m), using the dual base $\{d_0, d_1, \ldots, d_{m-1}\}$, can be expressed as follows:

$$A = \sum_{i=0}^{m-1} a_i d_i$$

The coordinate a_i can be obtained from the expression $a_i = Tr(Ab_i)$. In fact:

$$Tr(Ab_i) = Tr\left(\left(\sum_{j=0}^{m-1} a_j d_j\right) b_i\right) = \sum_{j=0}^{m-1} a_j Tr\left(d_j b_i\right) = a_i$$

For a given L and N ε GF(2^m) and using the standard base for L and the dual base for N, the following development can be made:

$$L = \sum_{i=0}^{m-1} l_i b_i; \quad N = \sum_{i=0}^{m-1} n_i d_i;$$

The product $V = LN$ can be expressed in dual base as follows [3]:

$$V = \sum_{i=0}^{m-1} Tr(Vb_i)d_i = \sum_{i=0}^{m-1} Tr(LNb_i)d_i$$

Thus:

$$v_0 = Tr(LNb_0) = Tr(LN) = Tr(l_0N) + Tr(l_1\alpha N) + \cdots + Tr\left(l_{m-1}\alpha^{m-1}N\right)$$

$$= l_0Tr(N) + l_1Tr(\alpha N) + \cdots + l_{m-1}Tr\left(\alpha^{m-1}N\right) = l_0n_0 + l_1n_1 + \cdots + l_{m-1}n_{m-1}$$

$$v_1 = Tr(LNb_1) = Tr(\alpha LN) = Tr(l_0\alpha N) + Tr\left(l_1\alpha^2 N\right) + \cdots + Tr\left(l_{m-1}\alpha^m N\right)$$

$$= l_0Tr(\alpha N) + l_1Tr\left(\alpha^2 N\right) + \cdots +_{m-1} Tr\left(\alpha^m N\right) = l_0n_1 + l_1n_2 + \cdots + l_{m-1}n_m$$

where:

$$n_m = n_0p_0 + n_1p_1 + \cdots + n_{m-1}p_{m-1}$$

That is, v_1 can be obtained from v_0 using the circuit for multiplying by α, given in Figure B.3, which is repeated for convenience in Fig. 10.12a. N, αN, $\alpha^2 N$, …, $\alpha^{m-1}N$ can be obtained successively using this circuit. These terms have always to be multiplied by L to calculate the successive coefficients of the product LN, which are given in the dual base.

In general, for the coefficients v_i, $i = 0, \ldots, m - 1$, it results:

$$\begin{aligned} v_i &= Tr(LNbi) = Tr\left(\alpha^i LN\right) \\ &= Tr\left(l_0\alpha^i N\right) + Tr\left(l_1\alpha^{i+1}N\right) + \cdots + Tr\left(l_{m-1}\alpha^{i+m-1}N\right) \\ &= l_0Tr\left(\alpha^i N\right) + l_1Tr\left(\alpha^{i+1}N\right) + \cdots + l_{m-1}Tr\left(\alpha^{i+m-1}N\right) \\ &= l_0n_i + l_1n_{i+1} + \cdots + l_{m-1}n_{i+m-1} \end{aligned} \tag{10.1}$$

where, for $k > m - 1$

$$n_k = n_{k-m}p_0 + n_{k-m+1}p_1 + \cdots + n_{k-1}p_{m-1}$$

In conclusion, using as core the circuit for multiplying by α, the circuit for the multiplication in dual base (i.e. an operand in standard base and, the other operand and the result in dual base) is the one in Fig. 10.12b due to Berlekamp [1]. Once introduced the N coefficients into the **LFSR** (as serial or parallel data), and the coefficients of L applied as parallel data, the coefficients of the product are obtained as serial data through v in m cycles, the first v_0.

Dual base multiplication, as described, has the disadvantage that standard and dual base are used simultaneously. Therefore, transformations between both bases should be implemented, which generally increases the hardware and prolong the computation. Last drawbacks can be overcoming using an almost self-dual base, in which case the transformation between bases can be a single permutation of the coefficients, as shown in Section B.4.3. Moreover, it is known [3] that, if not taken into account or not be used converters between the bases, the less hardware multiplier is the Berlekamp multiplier. Therefore, this multiplier is recommended with a suitable polynomial that supports almost self-dual base.

If a multiplier with parallel output wants to be used, it is easy to put in matrix notation the dependence (10.1) of the output coefficients with respect to the input coefficients. Concretely, it results:

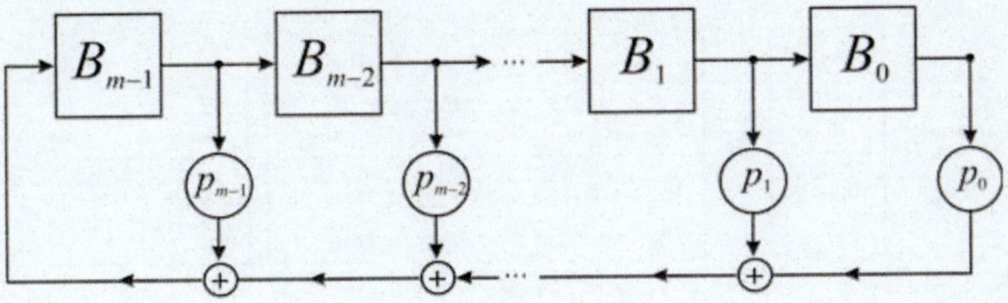

(a)

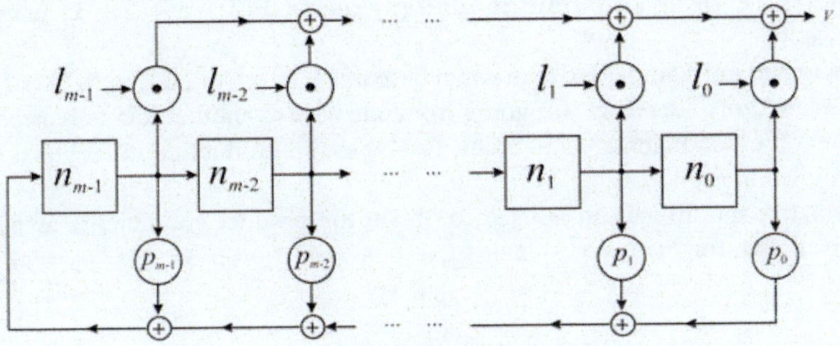

(b)

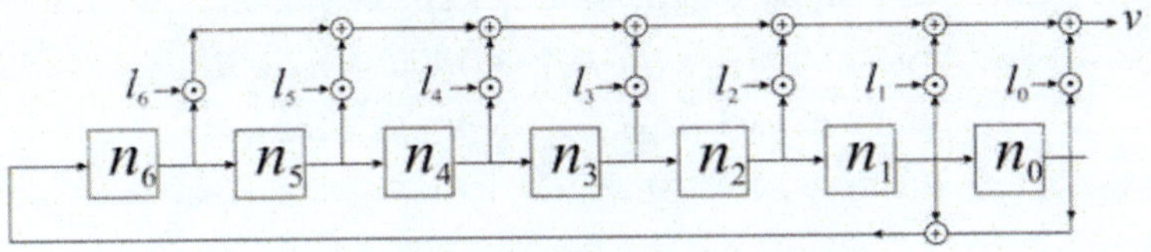

(c)

t	$n_6 \ldots n_0$	$l_6 \ldots l_0$	$\Sigma n_i l_i$
1	0101000	0101011	$c_0 = 0$
2	0010100	0101011	$c_1 = 0$
3	0001010	0101011	$c_2 = 0$
4	1000101	0101011	$c_3 = 1$
5	1100010	0101011	$c_4 = 0$
6	1110001	0101011	$c_5 = 0$
7	1111000	0101011	$c_6 = 0$

(d)

Fig. 10.12 Multiplication using dual base. **a** Circuit for multiplying by α. **b** Full circuit with serial output. **c** Serial circuit of Example 10.11. **d** Results of Example 10.11

$$\begin{bmatrix} v_0 \\ v_1 \\ \dots \\ v_{m-2} \\ v_{m-1} \end{bmatrix} = \begin{bmatrix} n_0 & n_1 & \dots & n_{m-2} & n_{m-1} \\ n_1 & n_2 & \dots & n_{m-1} & n_m \\ \dots & \dots & \dots & \dots & \dots \\ n_{m-2} & n_{m-1} & \dots & n_{2m-4} & n_{2m-3} \\ n_{m-1} & n_m & \dots & n_{2m-3} & n_{2m-2} \end{bmatrix} \begin{bmatrix} l_0 \\ l_1 \\ \dots \\ l_{m-2} \\ l_{m-1} \end{bmatrix}$$

The above equation could be detailed for each GF(2^m) and the corresponding circuit could be obtained, such as in the following example.

Example 10.11 **Design a circuit to multiply over GF(2^7){x^7+ x+ 1} using the dual base.**

As obtained in Example B.16, the dual base of $\{1, \alpha, \alpha^2, \alpha^3, \alpha^4, \alpha^5, \alpha^6\}$ is $\{1, \alpha^6, \alpha^5, \alpha^4, \alpha^3, \alpha^2, \alpha\}$. Therefore, for going from one base to another it is only necessary to permute the coefficients. The circuit to obtain the product coefficients as serial data is shown in Fig. 10.12c.

The matrix multiplication is used to obtain the product coefficients as parallel data. Given that, for $P(x) = x^7 + x + 1$ ($p_0 = p_1 = 1; p_2 = p_3 = p_4 = p_5 = p_6 = 0$), it results:

$$n_7 = n_0 p_0 + n_1 p_1 + \dots + n_6 p_6 = n_0 + n_1$$
$$n_8 = n_1 p_0 + n_2 p_1 + \dots + n_7 p_6 = n_1 + n_2$$
$$n_9 = n_2 p_0 + n_3 p_1 + \dots + n_8 p_6 = n_2 + n_3$$
$$n_{10} = n_3 p_0 + n_4 p_1 + \dots + n_9 p_6 = n_3 + n_4$$
$$n_{11} = n_4 p_0 + n_5 p_1 + \dots + n_{10} p_6 = n_4 + n_5$$
$$n_{12} = n_5 p_0 + n_6 p_1 + \dots + n_{11} p_6 = n_5 + n_6$$

The matrix multiplication is:

$$\begin{bmatrix} v_0 \\ v_1 \\ v_2 \\ v_3 \\ v_4 \\ v_5 \\ v_6 \end{bmatrix} = \begin{bmatrix} n_0 & n_1 & n_2 & n_3 & n_4 & n_5 & n_6 \\ n_1 & n_2 & n_3 & n_4 & n_5 & n_6 & n_0+n_1 \\ n_2 & n_3 & n_4 & n_5 & n_6 & n_0+n_1 & n_1+n_2 \\ n_3 & n_4 & n_5 & n_6 & n_0+n_1 & n_1+n_2 & n_2+n_3 \\ n_4 & n_5 & n_6 & n_0+n_1 & n_1+n_2 & n_2+n_3 & n_3+n_4 \\ n_5 & n_6 & n_0+n_1 & n_1+n_2 & n_2+n_3 & n_3+n_4 & n_4+n_5 \\ n_6 & n_0+n_1 & n_1+n_2 & n_2+n_3 & n_3+n_4 & n_4+n_5 & n_5+n_6 \end{bmatrix} \begin{bmatrix} l_0 \\ l_1 \\ l_2 \\ l_3 \\ l_4 \\ l_5 \\ l_6 \end{bmatrix}$$

It is easy to implement the multiplier circuit corresponding to this matrix relation using an **EXOR-AND-EXOR** structure and parallel output through 49 **AND** gates (in general m^2) and 48 two-input **EXOR** gates (usually the number of **EXOR** gates is $\geq m^2 - 1$). The circuit is minimum in this case. It is easy to see that for an **AND-EXOR** structure, the corresponding circuit requires 49 two-input **AND** gates and 63 two-input **EXOR** gates.

To multiply $A(x) = x^5 + x^3 + x + 1$ by $B(x) = x^4 + x^2$ using the circuit of Fig. 10.12c, for example, the different values involved are given in the table of Fig. 10.12d. It is assumed that $A(x)$ is given in standard base ($l_6 = 0, l_5 = 1, l_4 = 0, l_3 = 1, l_2 = 0, l_1 = 1, l_0 = 1$) and $B(x)$ in dual base ($n_6 = 0, n_5 = 1, n_4 = 0, n_3 = 1, n_2 = 0, n_1 = 0, n_0 = 0$). The result $C(x) = A(x) \cdot B(x)$ is given in dual base. Using that $x^9 = x^3 + x^2$, the product is $C(x) = (x^5 + x^3 + x + 1)(x^4 + x^2) = x^9 + x^4 + x^3 + x^2 = x^4$. And indeed, this is the result provided by the circuit of Fig. 10.12c.

■

The self-dual normal bases are optimal bases and therefore require a minimum hardware, as it is demonstrated in the following example.

Example 10.12 **Design a circuit to multiply over GF(2^5)$\{x^5 + x^4 + x^2 + x + 1\}$ using the self-dual normal base $\{\alpha, \alpha^2, \alpha^4, \alpha^8, \alpha^{16}\}$, obtained in Example B.18.**

The product of $A(x) = a_4\alpha^{16} + a_3\alpha^8 + a_2\alpha^4 + a_1\alpha^2 + a_0\alpha$ by $B(x) = b_4\alpha^{16} + b_3\alpha^8 + b_2\alpha^4 + b_1\alpha^2 + b_0\alpha$ can be expressed in matrix notation as follows:

$$\begin{bmatrix} c_0 \\ c_1 \\ c_2 \\ c_3 \\ c_4 \end{bmatrix} = \begin{bmatrix} a_1 & a_0 + a_3 & a_3 + a_4 & a_1 + a_2 & a_2 + a_4 \\ a_0 + a_3 & a_2 & a_1 + a_4 & a_0 + a_4 & a_2 + a_3 \\ a_3 + a_4 & a_1 + a_4 & a_3 & a_0 + a_2 & a_0 + a_1 \\ a_1 + a_2 & a_0 + a_4 & a_0 + a_2 & a_4 & a_1 + a_3 \\ a_2 + a_4 & a_2 + a_3 & a_0 + a_1 & a_1 + a_3 & a_0 \end{bmatrix} \begin{bmatrix} b_0 \\ b_1 \\ b_2 \\ b_3 \\ b_4 \end{bmatrix}$$

A combinational circuit to implement this multiplier using an **AND-EXOR** structure, as it is expressed by the above matrix relation, requires 40 **AND** gates and 25 **EXOR** gates, all two-input gates. On the other hand, using an **EXOR-AND-EXOR** structure the multiplier with parallel output corresponding to this matrix relation uses 25 two-input **AND** gates and 30 two-input **EXOR** gates.

■

10.6 Square and Square Root Over GF(2^m)

Given the similarity between these two operations, square and square root, both operations are studied together in this section.

10.6.1 Square

Obviously any multiplier circuit can be used to obtain the square: it is just only necessary the two operands are the polynomial whose square wants to be obtained. But it is obvious that a circuit designed specifically for obtaining the square will be simpler than the corresponding multiplier.

Using power representation, if a polynomial is represented by x^a, the square is:

$$\left(x^a\right)^2 = x^{2a}$$

Therefore, a is shifted one position to the left for squaring, filling with 0 the free position, and modularly reduced. Remind that if $x^a \neq 0$, a can not be all ones. The modular reduction must be done when a carry is generated in the shift, and the reduction, as seen in Sect. 10.2, consists on adding 1. That is, if the most significant bit of a is 0, it is required to shift and not to do modular reduction; if 1, it is require the shift and to make the modular reduction. In short, to shift and to reduce a, it is only necessary a rotation to the left (i.e. to shift with feedback, so the most significant bit of a becomes the least significant).

With polynomial representation using standard base, for squaring over GF(2^n) the following expression can be used:

$$(a+b)^2 = a^2 + b^2$$

Thus, given $B(x) = b_{n\text{-}1}x^{n-1} + b_{n\text{-}2}x^{n-2} + \cdots + b_1x + b_0$, it results:

$$B^2(x) = b_{n-1}x^{2(n-1)} + b_{n-2}x^{2(n-2)} + \cdots + b_1x^2 + b_0$$

$B^2(x)$ is a polynomial with only even powers. Therefore, it is sufficient to make modular reduction to the polynomial with coefficients $b_{n-1}\, 0\, b_{n-2}\, 0 \ldots 0\, b_1\, 0\, b_0$. This modular reduction can be done with an **LFSR** divider, as in the following example.

Example 10.13 **Obtain $A^2(x)$ with $P(x) = x^4 + x^3 + 1$, where $A(x) = x^3 + x + 1$ (i.e. 1011).**

Obviously:

$$A^2(x) = x^6 + x^2 + 1 = x^3 + x$$

This is the result that remains in the **LFSR** of Fig. 10.13 after introducing as dividend the sequence 1000101.

■

Also the modular reduction can be implemented from the corresponding matrix, such as obtained in Sect. 10.3.1; it is applied in the following examples.

Example 10.14 **Design a combinational circuit for calculating the square of elements of GF(2^4)$\{x^4 + f_3x^3 + f_2x^2 + f_1x + f_0\}$, being $P(x) = x^4 + f_3x^3 + f_2x^2 + f_1x + f_0$ any primitive polynomial of degree 4.**

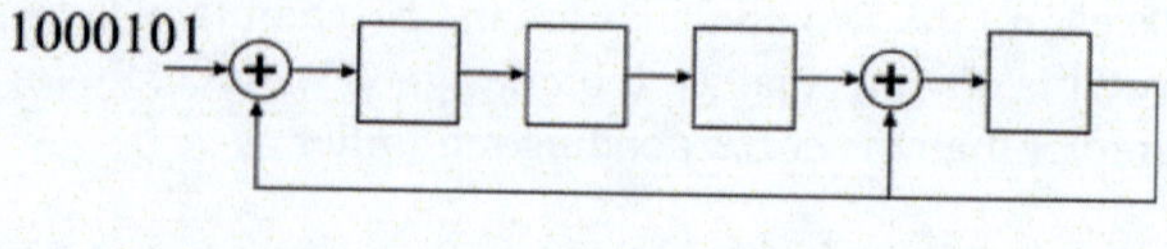

Fig. 10.13 Squaring using standard base

As seen in Example 10.3, the matrix of modular reduction in this case is:

$$R = \begin{bmatrix} 1\,0\,0\,0 & f_0 & f_3 f_0 & f_0(f_2 + f_3 f_3) \\ 0\,1\,0\,0 & f_1 & f_0 + f_3 f_1 & f_3 f_0 + f_1(f_2 + f_3 f_3) \\ 0\,0\,1\,0 & f_2 & f_1 + f_3 f_2 & f_0 + f_3 f_1 + f_2(f_2 + f_3 f_3) \\ 0\,0\,0\,1 & f_3 & f_2 + f_3 f_3 & f_1 + f_3 f_2 + f_3(f_2 + f_3 f_3) \end{bmatrix}$$

Therefore, $B(x) = b_3x^3 + b_2x^2 + b_1x + b_0$, is obtained from $A(x) = a_3x^3 + a_2x^2 + a_1x + a_0$, where $B(x) = A^2(x)$, using the following matrix relation:

$$\begin{bmatrix} b_0 \\ b_1 \\ b_2 \\ b_3 \end{bmatrix} = \begin{bmatrix} 1\,0\,0\,0 & f_0 & f_3 f_0 & f_0(f_2 + f_3 f_3) \\ 0\,1\,0\,0 & f_1 & f_0 + f_3 f_1 & f_3 f_0 + f_1(f_2 + f_3 f_3) \\ 0\,0\,1\,0 & f_2 & f_1 + f_3 f_2 & f_0 + f_3 f_1 + f_2(f_2 + f_3 f_3) \\ 0\,0\,0\,1 & f_3 & f_2 + f_3 f_3 & f_1 + f_3 f_2 + f_3(f_2 + f_3 f_3) \end{bmatrix} \begin{bmatrix} a_0 \\ 0 \\ a_1 \\ 0 \\ a_2 \\ 0 \\ a_3 \end{bmatrix}$$

That is:

$$\begin{aligned} b_0 &= a_0 + a_2 f_0 + a_3 f_0(f_2 + f_3) \\ b_1 &= a_2 f_1 + a_3(f_3 f_0 + f_1(f_2 + f_3)) \\ b_2 &= a_1 + a_2 f_2 + a_3(f_0 + f_3 f_1 + f_2(f_2 + f_3)) \\ b_3 &= a_2 f_3 + a_3(f_1 + f_3 f_2 + f_3(f_2 + f_3)) \end{aligned}$$

In matrix notation is results:

$$\begin{bmatrix} b_0 \\ b_1 \\ b_2 \\ b_3 \end{bmatrix} = \begin{bmatrix} 1\,0 & f_0 & f_0(f_2 + f_3) \\ 0\,0 & f_1 & f_3 f_0 + f_1(f_2 + f_3) \\ 0\,1 & f_2 & f_0 + f_3 f_1 + f_2(f_2 + f_3) \\ 0\,0 & f_3 & f_1 + f_3 f_2 + f_3(f_2 + f_3) \end{bmatrix} \begin{bmatrix} a_0 \\ a_1 \\ a_2 \\ a_3 \end{bmatrix}$$

For instance, for $P(x) = x^4 + x^3 + 1$:

$$\begin{bmatrix} b_0 \\ b_1 \\ b_2 \\ b_3 \end{bmatrix} = \begin{bmatrix} 1\,0\,1\,1 \\ 0\,0\,0\,1 \\ 0\,1\,0\,1 \\ 0\,0\,1\,1 \end{bmatrix} \begin{bmatrix} a_0 \\ a_1 \\ a_2 \\ a_3 \end{bmatrix}$$

To implement this circuit 4 two-input **EXOR** gates are only needed. It is a much simpler circuit than the multiplier obtained in Example 10.4.

■

Example 10.15 **One of the primitive polynomials that the standard FIPS 186 [6] proposes is $P(x) = x^{233} + x^{74} + 1$. Design a combinational circuit for calculating the square of elements of GF(2^{233}).**

Let $A = a_{232}x^{232} + a_{231}x^{231} + \cdots + a_1x + a_0$ be an element of GF(2^{233})$\{x^{233} + x^{74} + 1\}$, and $B(x) = b_{232}x^{232} + b_{231}x^{231} + \cdots + b_1x + b_0$ its square ($B(x) = A^2(x)$). For $B(x)$ it is obtained:

$$b_i = a_{i/2} + a_j, \quad j = 196 + i/2, i = 0, 2, 4, \ldots, 72$$
$$b_i = a_j, \quad j = 117 + (i-1)/2, i = 1, 3, 5, \ldots, 73$$
$$b_i = a_{i/2} + a_j, \quad j = 159 + i/2, i = 74, 76, \ldots, 146$$
$$b_i = a_j + a_k, \quad j = 80 + (i-1)/2, k = 117 + (i-1)/2, i = 75, 77, \ldots, 231$$
$$b_i = a_{i/2}, \quad i = 148, 150, \ldots, 232$$

Only 153 two-input **EXOR** gates are required to implement this circuit. ■

In general, using standard base, the following matrix relation is obtained to calculate $B(x)$ from $A(x)$, where $B(x) = A^2(x)$,

$$\begin{bmatrix} b_0 \\ \ldots \\ b_{n-1} \end{bmatrix} = [C] \begin{bmatrix} a_0 \\ \ldots \\ a_{n-1} \end{bmatrix}$$

where $[C]$ is called the matrix of the square.

Using the normal base, as detailed in Section B.4.2, squaring is reduced to a rotation to the left, as it is obtained with power representation. For GF(2^n), it can be implemented with a ring shift register of length n, with a cost of n flip flops. Comparing with the cost using standard base in a particular case, for $n = 233$, as obtained in Example 10.15, 153 **EXOR** gates are needed only, against the 233 flip flops using normal base.

The result of the multipliers using dual base can be used to square, such as in the following example.

Example 10.16 **Design a circuit for squaring over GF(2^7)$\{x^7 + x + 1\}$.**

Let use the results of Example 10.11 and assume that both operands are the same polynomial, given using standard base. The square ($c_6x^6 + c_5x^5 + c_4x^4 + c_3x^3 + c_2x^2 + c_1x + c_0$) is obtained from the given polynomial ($l_6x^6 + l_5x^5 + l_4x^4 + l_3x^3 + l_2x^2 + l_1x + l_0$) using the following matrix relation, as can be easily checked:

$$\begin{bmatrix} c_0 \\ c_6 \\ c_5 \\ c_4 \\ c_3 \\ c_2 \\ c_1 \end{bmatrix} = \begin{bmatrix} l_0 & l_6 & l_5 & l_4 & l_3 & l_2 & l_1 \\ l_6 & l_5 & l_4 & l_3 & l_2 & l_1 & l_0 + l_6 \\ l_5 & l_4 & l_3 & l_2 & l_1 & l_0 + l_6 & l_5 + l_6 \\ l_4 & l_3 & l_2 & l_1 & l_0 + l_6 & l_5 + l_6 & l_4 + l_5 \\ l_3 & l_2 & l_1 & l_0 + l_6 & l_5 + l_6 & l_4 + l_5 & l_3 + l_4 \\ l_2 & l_1 & l_0 + l_6 & l_5 + l_6 & l_4 + l_5 & l_3 + l_4 & l_2 + l_3 \\ l_1 & l_0 + l_6 & l_5 + l_6 & l_4 + l_5 & l_3 + l_4 & l_2 + l_3 & l_1 + l_2 \end{bmatrix} \begin{bmatrix} l_0 \\ l_1 \\ l_2 \\ l_3 \\ l_4 \\ l_5 \\ l_6 \end{bmatrix}$$

Operating and arranging becomes:

$$\begin{bmatrix} c_0 \\ c_1 \\ c_2 \\ c_3 \\ c_4 \\ c_5 \\ c_6 \end{bmatrix} = \begin{bmatrix} 1\,0\,0\,0\,0\,0\,0 \\ 0\,0\,0\,0\,1\,0\,0 \\ 0\,1\,0\,0\,1\,0\,0 \\ 0\,0\,0\,0\,0\,1\,0 \\ 0\,0\,1\,0\,0\,1\,0 \\ 0\,0\,0\,0\,0\,0\,1 \\ 0\,0\,0\,1\,0\,0\,1 \end{bmatrix} \begin{bmatrix} l_0 \\ l_1 \\ l_2 \\ l_3 \\ l_4 \\ l_5 \\ l_6 \end{bmatrix}$$

Therefore, only 3 two-input **EXOR** gates are required to obtain the square.

■

10.6.2 Square Root

Using power representation, if a polynomial is represented by x^a, its square root is:

$$\left(x^a\right)^{1/2} = x^{a/2}$$

Therefore, if a is even, to obtain the square root a is simply shifted to the right. If a is odd, a shift to the right eliminates the one least significant, thereby producing an error. However, since $x^{2^m-1} = 1$, the exponent may be added to $(2^m - 1)$, so that $a + (2^m - 1)$ is even, and this exponent can be shifted to the right, as it is checked in the following example.

Example 10.17 **Obtain the square root of x^3 over GF(2^4).**

Since 3 is odd, 15 is added, resulting: $0011 + 1111 = 10010$. With a right shift is 1001, so that $\sqrt{x^3} = x^9$.

■

To obtain the square root using standard base the previous result related to the square can be used, as is done in the following example. Obviously, using the inverse matrix $[C]^{-1}$, is:

$$\begin{bmatrix} a_0 \\ \dots \\ a_{n-1} \end{bmatrix} = [C]^{-1} \begin{bmatrix} b_0 \\ \dots \\ b_{n-1} \end{bmatrix}$$

Example 10.18 **Obtain the square root of any element over GF(2^4){x^4+x^3+1}.**

Calling $R(x) = r_3x^3 + r_2x^2 + r_1x + r_0 = \sqrt{a_3x^3 + a_2x^2 + a_1x + a_0}$, and considering that in Example 10.14 it was obtained:

$$[C] = \begin{bmatrix} 1\,0\,1\,1 \\ 0\,0\,0\,1 \\ 0\,1\,0\,1 \\ 0\,0\,1\,1 \end{bmatrix}$$

For $[C]^{-1}$:

$$\begin{bmatrix} r_0 \\ r_1 \\ r_2 \\ r_3 \end{bmatrix} = \begin{bmatrix} 1\,0\,0\,1 \\ 0\,1\,1\,0 \\ 0\,1\,0\,1 \\ 0\,1\,0\,0 \end{bmatrix} \begin{bmatrix} a_0 \\ a_1 \\ a_2 \\ a_3 \end{bmatrix}$$

■

With representation using normal base, the square root is reduced to a rotation to the right, as detailed in section B.4.2.

10.7 Exponentiation Over GF(2^m)

Given a polynomial B(x) belonging to GF(2^m), the objective is to calculate $B^k(x)$, which also belongs to GF(2^m), with k integer.

First, it is immediate that k can be reduced to the range $2^m - 1$. Indeed, since $B^{2^m-1}(x) = 1$, it is clear that $B^k(x) = B^q(x)$, where $q = k\,\mathbf{mod}(2^m - 1)$.

The computation of B^q can be reduced to multiplications and squares. For this, q is developed as a binary number:

$$\begin{aligned} q &= q_{m-1}2^{m-1} + q_{m-2}2^{m-2} + \dots + q_1 2 + q_0 \\ &= ((\dots(q_{m-1}2 + q_{m-2})2 + \dots +) + q_1)2 + q_0 \end{aligned}$$

That is:

$$B^q = \left(\dots\left(\left(B^{q_{m-1}}\right)^2 \cdot B^{q_{m-2}}\right)^2 \dots \cdot B^{q_1}\right)^2 \cdot B^{q_0}$$

With this development for B^q, the calculation involves squaring and multiplication. The core of the calculation would be:

$$1)\ R \leftarrow R^2$$
$$2)\ \text{if } q_i = 1,\ R \leftarrow RB$$

The result is obtained in register *R*, which initially has to be $R \leftarrow 1$. The algorithm could be as follows:

Algorithm 10.1

```
First algorithm to calculate B^q
    R ← 1, i ← m−1;
    while i > 0 do
        begin
            1) if q_i = 1 then R ← RB;
            2) R ← R^2, i ← i−1;
        end
    if q_0 = 1 then R ← RB;
End algorithm
```

A possible circuit for the exponentiation using the above algorithm is represented in Fig. 10.14a. This circuit includes a register, *R*, a multiplier and a squarer and can be used with any representations for *B*: just implementing the multiplication and the squaring in the representation being used.

Another possible development of B^q is:

$$B^q = \left(B^{2^0}\right)^{q_0} \cdot \left(B^{2^1}\right)^{q_1} \cdot \cdots \cdot \left(B^{2^{m-2}}\right)^{q_{m-2}} \cdot \left(B^{2^{m-1}}\right)^{q_{m-1}}$$

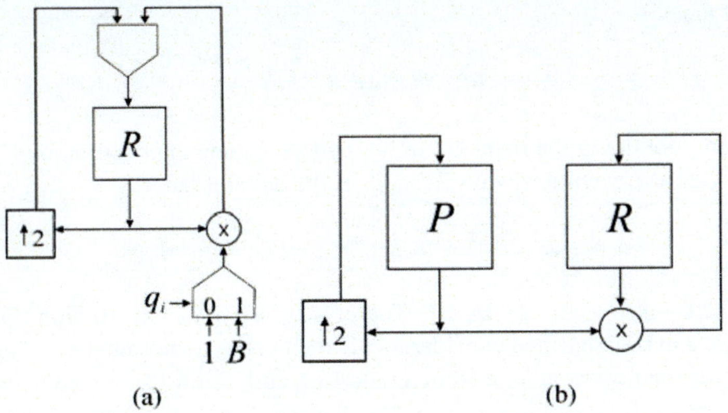

Fig. 10.14 Exponentiation: **a** first circuit; **b** second circuit

Again, the calculation involves squaring and multiplication, according to the following core for the calculations:

$$1)\ P \leftarrow P^2$$
$$2)\ \text{if } q_{m-1-i} = 1,\ R \leftarrow RP$$

Initially must be $P \leftarrow B$, $R \leftarrow 1$. The result remains in R. The algorithm in this case could be as follows:

Algorithm 10.2

Second algorithm to calculate B^q

```
P←B, R←1, i←m−1;
while i>0 do
    begin
        1) if q_{m−1−i} =1 then R←RP;
        2) P←P^2, i←i−1;
    end
```

End algorithm

A possible circuit for the exponentiation using last algorithm is shown in Fig. 10.14b. This circuit includes two registers, a multiplier and a squarer, and can be used with any representation of B: just implementing multiplication and squaring in the representation being used.

10.8 Inversion and Division Over GF(2^m)

Given a polynomial $B(x)$ belonging to GF(2^m), to calculate the inverse $B^{-1}(x)$ it can be applied that:

$$B^{2^m-2}(x) = B^{-1}(x)$$

Therefore, obtaining the inverse can be reduced to one exponentiation [2, 8]. As in this case the exponent is always $2^m - 2$, in the development:

$$q = q_{m-1}2^{m-1} + q_{m-2}2^{m-2} + \cdots + q_1 2 + q_0$$

must be $q_i = 1, i = m - 1, \ldots, 1; q_0 = 0$. Therefore, the two exponentiation algorithms of Sect. 10.7 can be simplified considerably if $B^{-1}(x)$ has to be calculated. This is the foundation of the inversion algorithm due to Itoh and Tsujii [4], which is described below.

Obviously $B^{2^m-2} = B^{(2^{m-1}-1)2} = \left(B^{(2^{m-1}-1)}\right)^2$. That is, $B^{(2^{m-1}-1)}$ is calculated, and is squared. Then, the calculation will depend on the development of $m - 1$. Let start with the case $m - 1 = 2^p$. The exponent of B can be expressed as follows:

$$2^{m-1} - 1 = 2^{2^p} - 1 = \left(2^{2^{p-1}} - 1\right)2^{2^{p-1}} + \left(2^{2^{p-1}} - 1\right)$$

Using $e = 2^{2^{p-1}}$, is $B^{(2^{m-1}-1)} = \left(B^{e-1}\right)^e B^{e-1}$. Remind that, using a normal base, squaring is reduced to a rotation. Thus, elevation to exponent $e = 2^{2^{p-1}}$ is reduced to $p - 1$ rotations. If B^{e-1} is available, to calculate $B^{(2^{m-1}-1)}$, $p - 1$ rotations and one multiplication must be performed. In turn, B^{e-1} can be calculated iteratively, with the same procedure as it is demonstrated in the following example.

Example 10.19 **Calculate the inverse $B^{-1}(x)$ for any element $B(x)$ over GF(2^9).**

$$B^{-1}(x) = B^{2^9-2} = \left(B^{2^8-1}\right)^2$$

$$B^{2^{2^3}-1} = \left(B^{2^{2^2}-1}\right)^{2^{2^2}}\left(B^{2^{2^2}-1}\right) \quad \leftrightarrow \quad B^{255} = \left(B^{15}\right)^{2^{2^2}} B^{15}$$

$$B^{2^{2^2}-1} = \left(B^{2^{2^1}-1}\right)^{2^{2^1}}\left(B^{2^{2^1}-1}\right) \quad \leftrightarrow \quad B^{15} = \left(B^3\right)^{2^{2^1}} B^3$$

$$B^3 = B^2 B$$

Using a normal base, one rotation and one multiplication is needed to calculate B^3; B^{15} is calculated with two rotations and one multiplication; B^{255} is obtained with four rotations and one multiplication; and finally $B^{-1}(x)$ requires only one rotation. ■

Now continue with the case $m - 1 = 2^p + 2^r$, with $p > r$. In this case it is easy to verify that:

$$B^{(2^{m-1}-1)} = B^{\left(2^{2^p} \cdot 2^{2^r}-1\right)} = B^{\left(2^{2^r}-1\right)}\left(B^{\left(2^{2^p}-1\right)}\right)^{2^{2^r}}$$

With a procedure similar to the previous case, again, with rotations and multiplications, $B^{-1}(x)$ is obtained, as is demonstrated in the following example.

Example 10.20 **Calculate the inverse $B^{-1}(x)$ for any element $B(x)$ over GF(2^{11}).**

$$B^{-1}(x) = B^{2^{11}-2} = \left(B^{2^{10}-1}\right)^2$$

$$B^{2^{10}-1} = B^{2^{2^3}2^{2^1}-1} = \left(B^{2^{2^1}-1}\right)\left(B^{2^{2^3}-1}\right)^{2^{2^1}} \quad \leftrightarrow \quad B^{1023} = B^3\left(B^{255}\right)^{2^{2^1}}$$

Moreover, in the Example 10.19 it holds:

$$B^{255} = \left(B^{15}\right)^{2^{2^2}} B^{15}; \quad B^{15} = \left(B^3\right)^{2^{2^1}} B^3; \quad B^3 = B^2 B$$

Using a normal base, one rotation and one multiplication is needed to calculate B^3; B^{15} is calculated with two rotations and one multiplication; B^{255} is obtained with four rotations and one multiplication; B^{1023} requires two rotations and one multiplication; and finally $B^{-1}(x)$ requires only one rotation.

■

In general it will be $m-1=2^p+2^r+\ldots+2^t+2^u$, con $p>r>\cdots>t>u$. In this case it can also be checked that:

$$B^{(2^{m-1}-1)}=B^{\left(2^{2^p}\cdot 2^{2^r}\ldots 2^{2^t}\cdot 2^{2^u}-1\right)}$$

$$=B^{\left(2^{2^u}-1\right)}\left[\left(B^{\left(2^{2^t}-1\right)}\right)\left[\cdots\left[\left(B^{\left(2^{2^r}-1\right)}\right)\left(B^{\left(2^{2^p}-1\right)}\right)^{2^{2^r}}\right]\cdots\right]^{2^{2^t}}\right]^{2^{2^u}}$$

This development is applied to the following example.

Example 10.21 **Calculate the inverse $B^{-1}(x)$ for any element $B(x)$ over GF(2^{15}).**

$$B^{-1}(x)=B^{2^{15}-2}=\left(B^{(2^{14}-1)}\right)^2$$

$$B^{\left(2^{2^3}2^{2^2}2^{2^1}-1\right)}=B^{\left(2^{2^1}-1\right)}\left(B^{\left(2^{2^2}-1\right)}\left(B^{\left(2^{2^3}-1\right)}\right)^{2^{2^2}}\right)^{2^{2^1}}$$

$$\leftrightarrow B^{16383}=B^3\left(B^{15}\left(B^{255}\right)^{2^{2^2}}\right)^{2^{2^1}}$$

On the other hand, from example 10.20:

$$B^{255}=\left(B^{15}\right)^{2^{2^2}}B^{15};\qquad B^{15}=\left(B^3\right)^{2^{2^1}}B^3;\qquad B^3=B^2B$$

Using a normal base, once calculated B^3, B^{15} and B^{255}, B^{16383} is obtained with 6 rotations and 2 multiplications; finally $B^{-1}(x)$ only requires another rotation. In total, to obtaining $B^{-1}(x)$ it is necessary to square 14 times and multiply 5 times.

■

The above development can be generalized. Calling $F_n(A)=A^{2^n-1}$, it is immediate that:

$$F_{n+m}(A)=\left(F_m(A)\right)^{2^n}F_n(A)=\left(F_n(A)\right)^{2^m}F_m(A)$$

Actually:

$$F_{n+m}(A)=A^{2^{n+m}-1}=A^{2^n2^m+2^n-2^n-1}=A^{(2^m-1)2^n}A^{2^n-1}$$

This facilitates the use of any development, particularly the additive chains, such as it is checked in the following example.

Example 10.22 **Calculate the inverse $B^{-1}(x)$ of any element $B(x)$ over GF(2^{15}) using additive chains.**

$$B^{-1}(x) = B^{2^{15}-2} = \left(B^{(2^{14}-1)}\right)^2$$

It consists on calculating $F_{14}(B)$ and then squaring. For 14, the Bauer chain {1, 2, 3, 4, 7, 14} can be used, thus:

$$\begin{aligned}
F_1(B) &= B \\
F_2(B) &= F_{1+1}(B) = (F_1(B))^2 F_1(B) \\
F_3(B) &= F_{2+1}(B) = (F_2(B))^2 F_1(B) \\
F_4(B) &= F_{3+1}(B) = (F_3(B))^2 F_1(B) \\
F_7(B) &= F_{4+3}(B) = (F_4(B))^{2^3} F_3(B) \\
F_{14}(B) &= F_{7+7}(B) = (F_7(B))^{2^7} F_7(B)
\end{aligned}$$

Therefore, to calculate $B^{-1}(x)$, it is necessary to square 14 times and to multiply 5 times, the same as in Example 10.21.

■

Another method for calculating the inverse is based on that (as seen in Appendix B), if $M(x)$ and $N(x)$ are two polynomials prime to each other, then two other polynomial $R(x)$ and $S(x)$ can be found, such that:

$$1 = R(x) \cdot M(x) + S(x) \cdot N(x)$$

With $M(x) = P(x)$ and $N(x) = B(x)$, and operating over GF(2^m), it remains:

$$1 = \{R(x) \cdot P(x) + S(x) \cdot B(x)\} \bmod P(x) = S(x) \cdot B(x) \bmod P(x)$$

since $R(x) \cdot P(x) \, \mathbf{mod} \, P(x) = 0$. Therefore,

$$S(x) = B^{-1}(x)$$

$S(x)$ is calculated using the procedure seen in Appendix B, such as in the following example.

Example 10.23 **In GF(2^4){$x^4 + x^3 + 1$}, obtain the inverse of $B(x) = x^3 + x^2 + 1$.**
Through successive divisions becomes:

$$x^4 + x^3 + 1 = \left(x^3 + x^2 + 1\right)x + (x + 1) \quad \Rightarrow \quad C_1 = x$$

$$x^3 + x^2 + 1 = (x + 1)x^2 + 1 \qquad \Rightarrow \quad C_2 = x^2$$

From the calculation of $S(x)$:

$$S_{-1} = 0, \quad S_0 = 1, \quad S_1 = S_{-1} - S_0C_1 = x, \quad S_2 = S_0 - S_1C_2 = 1 + xx^2 = x^3 + 1$$

Thus, $B^{-1}(x) = x^3 + 1$.

■

The calculation can be done by successive subtractions, as in Algorithm (B.6) of Appendix B. This algorithm to calculate the inverse of a polynomial, $Q(x)$, is reproduced bellow:

Algorithm 10.3

Algorithm to calculate the inverse of a polynomial

1) $R_1 \leftarrow P(x)$, $R_2 \leftarrow Q(x)$, $B_1 \leftarrow 0$, $B_2 \leftarrow 1$;
while $R_2 \neq 0$ **do**
begin
  2) $e \leftarrow g(R_1) - g(R_2)$
    while $e \geq 0$ **do**
    begin
      3) $R_1 \leftarrow R_1 - x^e R_2$, $B_1 \leftarrow B_1 - x^e B_2$;
      4) $e \leftarrow g(R_1) - g(R_2)$;
    end
  5) $R_1 \leftarrow R_2$, $B_1 \leftarrow B_2$;
end

End algorithm

The inverse $Q^{-1}(x)$ remains in the register B_1.

To divide $A(x)$ by $B(x)$ is the same as multiplying $A(x)$ by $B^{-1}(x)$. Therefore, it is sufficient to divide using any of the above algorithms to calculate $B^{-1}(x)$, and then use any of the multiplication algorithms.

10.9 Operations Over GF($(2^n)^m$)

All operations for GF(2^m) seen in previous sections of this chapter are transferred without any formal modification to GF($(2^n)^m$), unless the involved operations are performed in GF(2^m) instead of in GF(2). Therefore, all circuits seen for GF(2^m), with logical modifications due to changing circumstances in which operations are performed, are valid for GF($(2^n)^m$), as it is demonstrated in the following examples.

Example 10.24 **Draw a multiplier circuit for one operand provided as serial data and other as parallel data with the structure multiplier/divider (Fig. 10.5b) over GF($(2^2)^4$)\{$P_1(x)$\{$P_2(Y)$\}\}, $P_1(x) = x^2 + x + 1$, $P_2(Y) = Y^4 + Y^3 + Y^2 + CY + C$.**

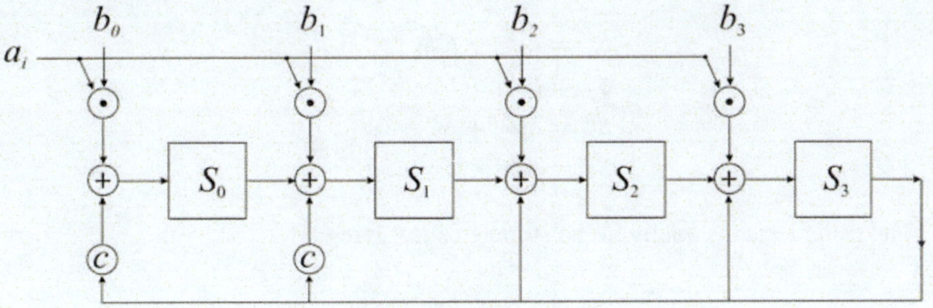

(a)

i	S_0	S_1	S_2	S_3	a_i	a_ib_0	a_ib_1	a_ib_2	a_ib_3	CS_3	S_0^+	S_1^+	S_2^+	S_3^+
3	A	A	A	A	C	D	B	C	A	A	D	B	C	A
2	D	B	C	A	A	A	A	A	A	A	A	D	B	C
1	A	D	B	C	B	C	D	B	A	D	B	A	A	D
0	B	A	A	D	D	B	C	D	A	B	A	C	A	D
	A	C	A	D										

(b)

Fig. 10.15 Example 10.24. **a** Circuit. **b** Table

The circuit of Fig. 10.15a, with the structure multiplier/divider of Fig. 10.5b, allows to multiply two polynomials belonging to $\mathrm{GF}((2^2)^4)\{P_1(x)\{P_2(Y)\}\}$. Each of the four cells $S_0S_1S_2S_3$ consists of 2 flip-flop. Remind (Example 9.3) that each adder is constructed with 2 **EXOR** gates, and just 1 **EXOR** gate is needed to multiply by C. The multiplier over $\mathrm{GF}(2^2)\{x^2 + x + 1\}$ can be implemented using a combinational design. Specifically, if $(m_1x + m_0)$ is multiplied by $(n_1x + n_0)$, it results:

$$\begin{aligned}(m_1x + m_0)(n_1x + n_0) &= m_1n_1x^2 + (m_1n_0 + m_0n_1)x + m_0n_0 \\ &= m_1n_1(x + 1) + (m_1n_0 + m_0n_1)x + m_0n_0 \\ &= (m_1n_1 + m_1n_0 + m_0n_1)x + (m_1n_1 + m_0n_0)\end{aligned}$$

Thus, the multiplication over $\mathrm{GF}(2^2)\{x^2 + x + 1\}$ is implemented using a circuit with four 2-input **AND** gates and three 2-input **EXOR** gates.

Counting the use of elements in this circuit with four storage cells, four multipliers, seven adders and two multipliers by C, there are 8 flip-flops, sixteen **AND** gates and twenty eight **EXOR** gates.

The table of Fig. 10.15b includes the contents of the storage cells when, as a concrete example, $M(Y) = CY^3 + BY + D$ is multiplied by $N(Y) = BY^2 + DY + C$; in this case, $a_0 = D$, $a_1 = B$, $a_2 = A$, $a_3 = C$, $b_0 = C$, $b_1 = D$, $b_2 = B$, $b_3 = A$, and for the next values it results:

$$S_0^+ = a_i b_0 + CS_3$$
$$S_1^+ = a_i b_1 + CS_3 + S_0$$
$$S_2^+ = a_i b_2 + S_1 + S_3$$
$$S_3^+ = a_i b_3 + S_2 + S_3$$

The result remains finally in the storage cells, giving:

$$P(Y) = M(Y) \times N(Y) = DY^3 + CY$$

Calculating it manually:

$$\begin{aligned} P(Y) &= M(Y) \times N(Y) = \left(CY^3 + BY + D\right) \times \left(BY^2 + DY + C\right) \\ &= CY^5 + CY^4 + CY^3 + B \end{aligned}$$

Given $Y^4 = Y^3 + Y^2 + CY + C$, and $Y^5 = DY^2 + C$, the same result is obtained ($DY^3 + CY$).

■

Example 10.25 **Obtain a multiplier circuit for the case that both operands are given in parallel over GF((2^2)^4){$P_1(x)${$P_2(Y)$}}, $P_1(x) = x^2 + x + 1$, $P_2(Y) = Y^4 + Y^3 + Y^2 + CY + C$.**

Let suposse $A(Y) = a_3Y^3 + a_2Y^2 + a_1Y + a_0$, $B(Y) = b_3Y^3 + b_2Y^2 + b_1Y + b_0$. Given that $Y^4 = Y^3 + Y^2 + CY + C$, $Y^5 = DY^2 + C$, $Y^6 = DY^3 + CY$, the modular reduction matrix is:

$$R = \begin{bmatrix} 1\,0\,0\,0\,C & C & 0 \\ 0\,1\,0\,0\,C & 0 & C \\ 0\,0\,1\,0\,1 & D & 0 \\ 0\,0\,0\,1\,1 & 0 & D \end{bmatrix}$$

In this matrix the constants 0 and 1 are used, with the understanding that, in this case $0 = A$ and $1 = B$. Operating as it has been done previously, the following matrix relation is obtained as the result:

$$\begin{bmatrix} r_0 \\ r_1 \\ r_2 \\ r_3 \end{bmatrix} = \begin{bmatrix} a_0 & Ca_3 & C(a_2 + a_3) & C(a_1 + a_2) \\ a_1 & a_0 + Ca_3 & Ca_2 & C(a_1 + a_3) \\ a_2 & a_1 + a_3 & a_0 + a_2 + Da_3 & a_1 + Da_2 \\ a_3 & a_2 + a_3 & a_1 + a_2 & a_0 + a_1 + Da_3 \end{bmatrix} \begin{bmatrix} b_0 \\ b_1 \\ b_2 \\ b_3 \end{bmatrix}$$

To multiply by C (see Example 9.13) just 1 **EXOR** gate is needed, and likewise, the multiplication by D is made with only 1 **EXOR** gate, as it is detailed in Table 10.6. Defining $S_{ij} = a_i + a_j$, the following expression is obtained for the matrix multiplication:

Table 10.6 Multiplication by D

e_1	e_0	s_1	s_0	
0	0	0	0	
0	1	1	1	$s_1 = e_0$
1	0	0	1	$s_0 = e_1 \oplus e_0$
1	1	1	0	

$$\begin{bmatrix} r_0 \\ r_1 \\ r_2 \\ r_3 \end{bmatrix} = \begin{bmatrix} a_0 & Ca_3 & CS_{23} & CS_{12} \\ a_1 & a_0 + Ca_3 & Ca_2 & CS_{13} \\ a_2 & S_{13} & S_{02} + Da_3 & a_1 + Da_2 \\ a_3 & S_{23} & S_{12} & S_{01} + Da_3 \end{bmatrix} \begin{bmatrix} b_0 \\ b_1 \\ b_2 \\ b_3 \end{bmatrix}$$

Implementing the multiplication using three operation levels, nine additions and seven multiplications by a constant (C or D) are made at a first level. At a second level sixteen multiplications have to be made. At the third level there are twelve additions. 45 **EXOR** gates are required for the 19 additions and 7 multiplications by a constant. 80 **AND** gates and 48 **EXOR** gates are required for the 16 multiplications (see Example 10.23); thus, 80 **AND** gates and 93 **EXOR** gates are required for implementing the multiplication in GF($(2^2)^4$)$\{P_1(x)\{P_2(Y)\}\}$.

■

Given that GF($(2^2)^4$)$\{P_1(x)\{P_2(Y)\}\}$ is isomorphic to GF($(2^4)^2$)$\{P_3(x)\{P_4(Y)\}\}$ and to GF(2^8)$\{P_5(z)\}$, a comparison is relevant. Specifically, the comparison is made with respect to multiplication, which can be considered to be the most important operation. To this effect the following examples are developed.

Example 10.26 **Draw a multiplier circuit for one operand as serial data and other as parallel data with the structure multiplier/divider (Fig. 10.5b) over GF($(2^4)^2$) $\{P_3(x)\{P_4(Y)\}\}$, $P_3(x) = x^4 + x + 1$, $P_4(Y) = Y^2 + Y + J$.**

Addition and product operations over GF(2^4)$\{x^4 + x + 1\}$ are detailed in Table 10.7 and Table 10.8. First column of Table 10.7 gives the correspondence between the binary coordinates and the alphabetical designation that is used for the 16 elements of GF(2^4).

The circuit of Fig. 10.16a, with the multiplier/divider structure of Fig. 10.5b, allows to multiply two any polynomials belonging to GF($(2^4)^2$)$\{P_3(x)\{P_4(Y)\}\}$. Each of the two cells S_0S_1 contains four flip-flops.

Each adder is constructed with four **EXOR** gates over GF(2^4)$\{x^4 + x + 1\}$. To multiply by J, as obtained from Table 10.8, the functions given by table of Fig. 10.16b are synthesized: just one **EXOR** gate is needed.

The multiplier over GF(2^4)$\{x^4 + x + 1\}$ can be implemented using combinational logic. For this multiplier the following matrix multiplication is obtained:

Table 10.7 Addition table over GF(2^4){$x^4 + x + 1$}

	+	A	B	C	D	E	F	G	H	I	J	K	L	M	N	O	P
0000	**A**	A	B	C	D	E	F	G	H	I	J	K	L	M	N	O	P
0001	**B**	B	A	D	C	F	E	H	G	J	I	L	K	N	M	P	O
0010	**C**	C	D	A	B	G	H	E	F	K	L	I	J	O	P	M	N
0011	**D**	D	C	B	A	H	G	F	E	L	K	J	I	P	O	N	M
0100	**E**	E	F	G	H	A	B	C	D	M	N	O	P	I	J	K	L
0101	**F**	F	E	H	G	B	A	D	C	N	M	P	O	J	I	L	K
0110	**G**	G	H	E	F	C	D	A	B	O	P	M	N	K	L	I	J
0111	**H**	H	G	F	E	D	C	B	A	P	O	N	M	L	K	J	I
1000	**I**	I	J	K	L	M	N	O	P	A	B	C	D	E	F	G	H
1001	**J**	J	I	L	K	N	M	P	O	B	A	D	C	F	E	H	G
1010	**K**	K	L	I	J	O	P	M	N	C	D	A	B	G	H	E	F
1011	**L**	L	K	J	I	P	O	N	M	D	C	B	A	H	G	F	E
1100	**M**	M	N	O	P	I	J	K	L	E	F	G	H	A	B	D	C
1101	**N**	N	M	P	O	J	I	L	K	F	E	H	G	B	A	D	C
1110	**O**	O	P	M	N	K	L	I	J	G	H	E	F	D	D	A	B
1111	**P**	P	O	N	M	L	K	J	I	H	G	F	E	C	C	B	A

Table 10.8 Multiplier table over GF(2^4){$x^4 + x + 1$}

•	B	C	D	E	F	G	H	I	J	K	L	M	N	O	P
B	B	C	D	E	F	G	H	I	J	K	L	M	N	O	P
C	C	E	G	I	K	M	O	D	B	H	F	L	J	P	N
D	D	G	F	M	P	K	J	L	I	N	O	H	E	B	C
E	E	I	M	D	H	L	P	G	C	O	K	F	B	N	J
F	F	K	P	H	C	N	I	O	L	E	B	J	M	D	G
G	G	M	K	L	N	H	B	F	D	J	P	O	I	C	E
H	H	O	J	P	I	B	G	N	K	D	E	C	F	M	L
I	I	D	L	G	O	F	N	M	E	P	H	K	C	J	B
J	J	B	I	C	L	D	K	E	N	F	M	G	P	H	O
K	K	H	N	O	E	J	D	P	F	I	C	B	L	G	M
L	L	F	O	K	B	P	E	H	M	C	J	N	G	I	D
M	M	L	H	F	J	O	C	K	G	B	N	P	D	E	I
N	N	J	E	B	M	I	F	C	P	L	G	D	O	K	H
O	O	P	B	N	D	C	M	J	H	G	I	E	K	L	F
P	P	N	C	J	G	E	L	B	O	M	D	I	H	F	K

Fig. 10.16 Example 10.26. **a** Circuit. **b** Multiplication by J

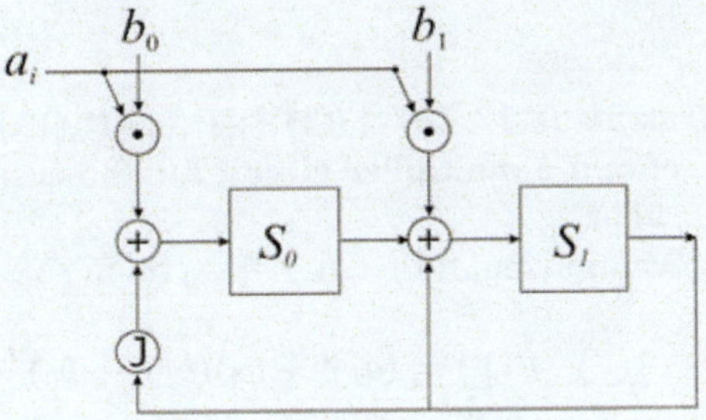

(a)

$e_3e_2e_1e_0$	$t_3t_2t_1t_0$	
0000	0000	
0001	1001	
0010	0001	
0011	1000	
0100	0010	
0101	1011	
0110	0011	$t_3 = e_0$
0111	1010	$t_2 = e_3$
1000	0100	$t_1 = e_2$
1001	1101	$t_0 = e_1 \oplus e_0$
1010	0101	
1011	1100	
1100	0110	
1101	1111	
1110	0111	
1111	1110	

(b)

$$\begin{bmatrix} r_0 \\ r_1 \\ r_2 \\ r_3 \end{bmatrix} = \begin{bmatrix} a_0 & a_3 & a_2 & a_1 \\ a_1 & a_0 + a_3 & a_2 + a_3 & a_1 + a_2 \\ a_2 & a_1 & a_0 + a_3 & a_2 + a_3 \\ a_3 & a_2 & a_1 & a_0 + a_3 \end{bmatrix} \begin{bmatrix} b_0 \\ b_1 \\ b_2 \\ b_3 \end{bmatrix}$$

Implementing the circuit using three levels of gates, in the first level there are three adders, in the second level 16 multipliers, and in the third level 12 adders. Therefore, in total 16 **AND** gates and 15 **EXOR** gates are required to implement the multiplication over GF(2^4)$\{x^4 + x + 1\}$.

Counting the use of elements in this circuit of Fig. 10.16a, with 2 storage cells, two multipliers, 3 adders and 1 multiplier by J, there are 8 flip-flops, 32 **AND** gates and 43 **EXOR** gates.

■

Example 10.27 **In GF($(2^4)^2$)$\{P_3(x)\{P_4(Y)\}\}$, $P_3(x) = x^4 + x + 1$, $P_4(Y) = Y^2 + Y + J$, obtain a multiplier circuit for the case that both operands are available in parallel.**

Multiplying $A(Y) = a_1 Y + a_0$ by $B(Y) = b_1 Y + b_0$, it results:

$$\begin{aligned} A(Y) \cdot B(Y) &= (a_1 Y + a_0)(b_1 Y + b_0) = a_1 b_1 Y^2 + (a_1 b_0 + a_0 b_1) Y + a_0 b_0 \\ &= a_1 b_1 (Y + J) + (a_1 b_0 + a_0 b_1) Y + a_0 b_0 \\ &= (a_1 b_1 + a_1 b_0 + a_0 b_1) Y + (J a_1 b_1 + a_0 b_0) \end{aligned}$$

With 4 adders, 3 multipliers, and 1 multiplier by a constant, the circuit of Fig. 10.17 implements the multiplication $A(Y) \cdot B(Y)$. All the above operations are performed on GF(2^4)$\{x^4 + x + 1\}$, as detailed in Example 10.26.

Counting the use of elements in this circuit of Fig. 10.17, there are 48 **AND** gates and 78 **EXOR** gates.

■

Example 10.28 **Obtain multipliers over GF(2^8)$\{z^8 + z^4 + z^3 + z^2 + 1\}$.**

The circuit of Fig. 10.18 allows the multiplication using an operand as serial data and the other as parallel data. In this circuit, 8 flip-flops, 11 **EXOR** gates and 8 **AND** gates are used.

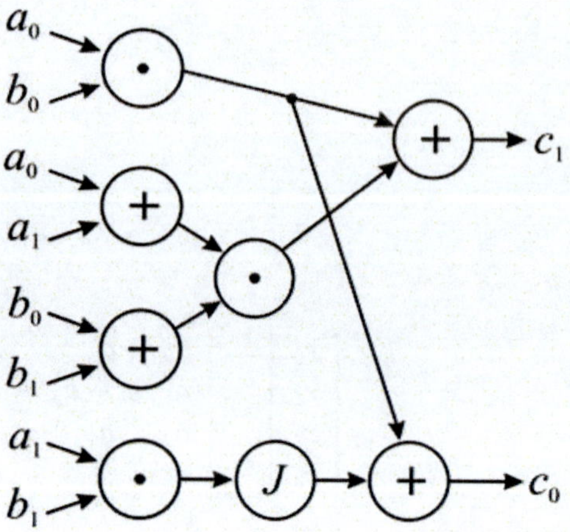

Fig. 10.17 Multiplier for Example 10.27

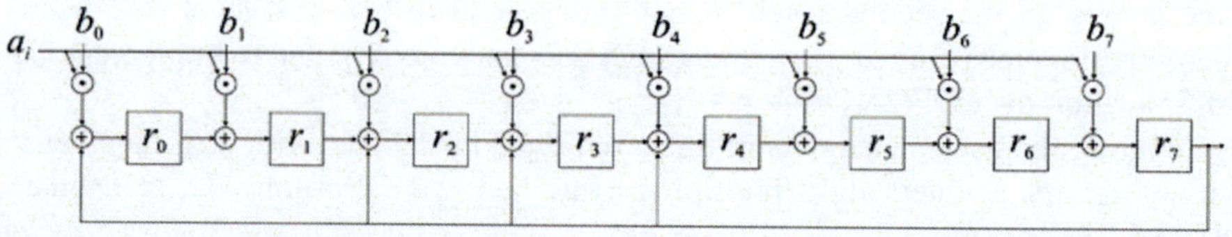

Fig. 10.18 Circuit of Example 10.28

Let consider now the case in which both operands are provided as parallel data. Given that:

$$
\begin{aligned}
z^8 &= z^4 + z^3 + z^2 + 1 \\
z^9 &= z^5 + z^4 + z^3 + z \\
z^{10} &= z^6 + z^5 + z^4 + z^2 \\
z^{11} &= z^7 + z^6 + z^5 + z^3 \\
z^{12} &= z^7 + z^6 + z^3 + z^2 + 1 \\
z^{13} &= z^7 + z^2 + z + 1 \\
z^{14} &= z^4 + z + 1
\end{aligned}
$$

the modular reduction matrix is:

$$
R = \begin{bmatrix}
1&0&0&0&0&0&0&0&1&0&0&0&1&1&1 \\
0&1&0&0&0&0&0&0&0&1&0&0&0&1&1 \\
0&0&1&0&0&0&0&0&1&0&1&0&1&1&0 \\
0&0&0&1&0&0&0&0&1&1&0&1&1&0&0 \\
0&0&0&0&1&0&0&0&1&1&1&0&0&0&1 \\
0&0&0&0&0&1&0&0&0&1&1&1&0&0&0 \\
0&0&0&0&0&0&1&0&0&0&1&1&1&0&0 \\
0&0&0&0&0&0&0&1&0&0&0&1&1&1&0
\end{bmatrix}
$$

For the multiplication matrix it results:

$$
\begin{bmatrix} r_0 \\ r_1 \\ r_2 \\ r_3 \\ r_4 \\ r_5 \\ r_6 \\ r_7 \end{bmatrix} =
\begin{bmatrix}
a_0 & a_7 & a_6 & a_5 & a_4 & a_3 + a_7 & a_2 + a_6 + a_7 & a_1 + a_5 + a_6 + a_7 \\
a_1 & a_0 & a_7 & a_6 & a_5 & a_4 & a_3 + a_7 & a_2 + a_6 + a_7 \\
a_2 & a_1 + a_7 & a_0 + a_6 & a_5 + a_7 & a_4 + a_6 & a_3 + a_5 + a_7 & a_2 + a_4 + a_6 + a_7 & a_1 + a_3 + a_5 + a_6 \\
a_3 & a_2 + a_7 & a_1 + a_6 + a_7 & a_0 + a_5 + a_6 & a_4 + a_5 + a_7 & a_3 + a_4 + a_6 + a_7 & a_2 + a_3 + a_5 + a_6 & a_1 + a_2 + a_4 + a_5 \\
a_4 & a_3 + a_7 & a_2 + a_6 + a_7 & a_1 + a_5 + a_6 + a_7 & a_0 + a_4 + a_5 + a_6 & a_3 + a_4 + a_5 & a_2 + a_3 + a_4 & a_1 + a_2 + a_3 + a_7 \\
a_5 & a_4 & a_3 + a_7 & a_2 + a_6 + a_7 & a_1 + a_5 + a_6 + a_7 & a_0 + a_4 + a_5 + a_6 & a_3 + a_4 + a_5 & a_2 + a_3 + a_4 \\
a_6 & a_5 & a_4 & a_3 + a_7 & a_2 + a_6 + a_7 & a_1 + a_5 + a_6 + a_7 & a_0 + a_4 + a_5 + a_6 & a_3 + a_4 + a_5 \\
a_7 & a_6 & a_5 & a_4 & a_3 + a_7 & a_2 + a_6 + a_7 & a_1 + a_5 + a_6 + a_7 & a_0 + a_4 + a_5 + a_6
\end{bmatrix}
\begin{bmatrix} b_0 \\ b_1 \\ b_2 \\ b_3 \\ b_4 \\ b_5 \\ b_6 \\ b_7 \end{bmatrix}
$$

Implementing multiplication operations on three levels, on a first level can be calculated:

$$
\begin{aligned}
S_{17} &= a_1 + a_7, \\
S_{27} &= a_2 + a_7, \\
S_{37} &= a_3 + a_7, \\
S_{06} &= a_0 + a_6, \\
S_{57} &= a_5 + a_7, \\
S_{46} &= a_4 + a_6, \\
S_{167} &= S_{17} + a_6, \\
S_{267} &= S_{27} + a_6, \\
S_{056} &= S_{06} + a_5, \\
S_{457} &= S_{57} + a_4, \\
S_{357} &= S_{37} + a_5, \\
S_{345} &= S_{457} + S_{37}, \\
S_{3467} &= S_{37} + S_{46}, \\
S_{234} &= S_{3467} + S_{267}, \\
S_{1567} &= S_{167} + a_5, \\
S_{0456} &= S_{056} + a_4, \\
S_{2467} &= S_{27} + S_{46}, \\
S_{2356} &= S_{267} + S_{357}, \\
S_{1356} &= S_{1567} + S_{37}, \\
S_{1245} &= S_{1567} + S_{2467}, \\
S_{1237} &= S_{1567} + S_{2356}
\end{aligned}
$$

A total of 21 additions. On a second level 64 products would be made, and in a third level 56 additions. That is, $21 + 56 = 77$ **EXOR** gates and 64 **AND** gates are required for the implementation of this multiplication over GF(2^8)$\{P_3(z)\}$.

■

Comparing the results of Examples 10.24 to 10.28, the three circuits with multiplier/divider structure obviously require the same 8 flip-flops, but the corresponding to the GF(2^8) also uses 8 **AND** gates and 11 **EXOR** gates, to the GF($(2^2)^4$) uses 16 **AND** gates and 28 **EXOR** gates, and to the GF($(2^4)^2$) 32 **AND** gates and 43 and **EXOR** gates. Obviously, given the number of gates, the circuit of Fig. 10.18 is the most advantageous, and the circuit of Fig. 10.15a is advantageous compared to the circuit of Fig. 10.16. Now the circuit of Fig. 10.18 performs the calculation on 8 clock cycles, compared to the 4 clock cycles that the circuit of Fig. 10.15a and the 2 clock cycles for the circuit of Fig. 10.16.

Regarding the three combinational circuits, the one corresponding to GF(2^8) uses 64 **AND** gates and 77 **EXOR** gates, the corresponding to GF($(2^2)^4$) uses 80 **AND** gates and 48 **EXOR** gates, and the corresponding to GF($(2^4)^2$), uses 48 **AND** gates and 78 **EXOR** gates. For a clearer comparison, let suppose a **CMOS** implementation with 1 two-input **AND** gate constructed with 6 transistors and 1 two-input **EXOR** gate with 8 transistors. In this case, the circuit corresponding to GF(2^8), GF($(2^2)^4$) and GF($(2^4)^2$) requires, 1000, 864 and 912 transistors, respectively. The second is the most economical, although there is no a big difference with the third.

10.10 Conclusion

In this chapter the main circuits to implement the various operations on the Galois fields GF(2^n) and GF($(2^n)^m$) have been presented, which are the most used in the different applications of algebraic circuits. Moreover, the circuits for Galois Fields of the form GF(p^n) and GF($(p^n)^m$), for which the next chapter is dedicated, will be introduced as a generalization of those presented here.

10.11 Exercises

10.1 Obtain the addition table for power representation over GF(2^5)$\{x^5 + x^2 + 1\}$.

10.2. Design a circuit for adding $a + b$ when both addents are serial data.

10.3. Design a circuit for adding $a + b$ when a is serial data and b is parallel data.

10.4. Design the circuits to go from the polynomial representation to the power representation and vice versa over GF(2^5)$\{x^5 + x^2 + 1\}$.

10.5. Design a combinational circuit over GF(2^5)$\{x^5 + x^2 + 1\}$ implementing the modular reduction after the multiplication of two elements of GF(2^5).

10.6. Design a combinational circuit over GF(2^5)$\{x^5 + x^2 + 1\}$ implementing the multiplication and the modular reduction of two elements of GF(2^5).

10.7. Design a cellular circuit over GF(2^5)$\{x^5 + f_4x^4 + f_3x^3 + f_2x^2 + f_1x^1 + 1\}$ implementing the multiplication and the modular reduction of two elements of GF(2^5).

10.8. Using the **LFSR** accumulator (Fig. 10.7), obtain $A(x)\cdot B(x)$ over GF(2^4)$\{x^4 + x^3 + 1\}$.

10.9. Using the Montgomery multiplier circuit (Fig. 10.8), obtain $A(x)\cdot B(x)$ over GF(2^4)$\{x^4 + x^3 + 1\}$.

10.10. Design a circuit for multiplying over GF(2^4)$\{P(x)\}$ similar to that in Fig. 10.9 (Example 10.7), but first the least significant coefficients.

10.11. Design a circuit for multiplying over GF(2^4)$\{x^4 + x^3 + 1\}$ similar to that in Fig. 10.10 (Example 10.8), but first the least significant coefficients.

10.12. Design a circuit for multiplying over GF(2^5)$\{x^5 + x^2 + 1\}$ with the normal base $\{\alpha^5, \alpha^{10}, \alpha^{20}, \alpha^{40}\}$.

10.13. Design a circuit for multiplying over GF(2^4)$\{x^4 + x^3 + 1\}$ in four cycles with the normal base $\{x, x^2, x^4, x^8\}$, being the coefficients provided as serial data, first a_3 and b_3.

10.14. Design a circuit for multiplying over GF(2^4)$\{x^4 + x^3 + 1\}$ using the dual base (from Example B.12, the dual base of $\{1, \alpha, \alpha^2, \alpha^3\}$ is $\{\alpha^{12}, \alpha^{11}, \alpha^{10}, \alpha^{13}\}$).

10.15. Design a combinational circuit to obtain the square of any element of GF(2^5)$\{x^5 + x^4 + x^2 + x + 1\}$.

10.16. Design a combinational circuit to obtain the square root of any element of GF(2^5)$\{x^5 + x^4 + x^2 + x + 1\}$.

10.17. Calculate the inverse $B^{-1}(x)$ of any element $B(x)$ over GF(2^{17}) using additive chains.

References

1. Berlekamp, E.R.: Bit-Serial reed-solomon encoders. IEEE Trans. Inf. Theory, **IT-28**(6), 869–874 (1982)
2. Brunner, H., Curiger, A., Hofstetter, M.: On Computing Multiplicative Inverses in GF(2^{m}). IEEE Trans. Compt. **42**(8), 1010–1015 (1993)
3. Hsu, I.S., Troung, T.K., Deutsch, L.J., Reed, I.S.: A comparison of VLSI architecture of finite field multipliers using dual, normal, or standard bases. IEEE Trans. Compt. **37**(6), 735–739 (1988)
4. Itoh, T., Tsujii, S.: A fast algorithm for computing multiplicative inverses in GF(2^{m}) using normal bases. Inf. Comput. **78**, 171–177 (1988)
5. Montgomery, P.L.: Modular multiplication without trial division. Math. Comput. **44**(170), 519–521 (1985)
6. National Institute of Standards and Technology: FIPS 186-4: Digital Signature Standard (DSS), Gaithersburg, July 2013
7. Savas, E., Koç, Ç.K.: The montgomery modular inverse-revisited. IEEE Trans. Compt. **49**(7), 763–766 (2000)
8. Wang, C.H.; Lin, J.L.: A systolic architecture for computing inverses and divisi6n in finite fields GF(2^{m}), IEEE Trans. Compt. **42**(9), 1141–1146 (1993)
9. Wu, H.: Montgomery multiplier and squarer for a class of finite fields. IEEE Trans. Compt. **51**(5), 521–529 (2002)
10. Yeh, C.S., Reed, I.S., Troung, T.K., Systolic multiplier for finite fields GF(2^{m}), IEEE Trans. on Compt. **33**(4), 357–360 (1984)

Chapter 11
Galois Fields GF(p^n)

After the presentation of the basic algebraic circuits in Chap. 9 and the finite field or Galois field GF(2^m) in Chap. 10, this chapter is devoted specifically to the circuits related to the finite fields GF(p) and GF(p^m), being p prime, following the same structure of Chap. 10. The theoretical foundations related to the Galois fields, algebra of polynomials and, particularly so now interested, related to GF(p) and to GF(p^m) are summarized in Appendices A and B. Therefore, it is advisable to review these Appendices when necessary.

The first section is devoted to GF(p), nevertheless the more relevant about GF(p) has already seen in Chap. 3, when study modular arithmetic. Some special cases for certain values of p are detailed in the present chapter. The rest of the chapter is devoted to GF(p^m), following a parallel development of Chap. 10, since the greater part of the considered circuits for GF(2^m) are moved to GF(p^m), with appropriate modifications.

Given GF(p^m), and the polynomials $A(x)$ y $B(x)$, with degrees lower than m ($g(A) < m$, $g(B) < m$), thus being elements of de GF(p^m), the objective is to add/subtract, to multiply or to divide two polynomials, or raise to a power, or to obtain its inverse, all over de GF(p^m). It will be seen that the complexity of these operations depends strongly on the base that is used to represent the operand polynomials; therefore it is advisable to select the most suitable base. In the following in relation with GF(p^m), it will be considered only the implementation of the various operations with the power representation and using standard, normal and dual bases. These bases are introduced in Appendix B, and are the most frequently used.

Finally, operations in the composite fields GF($(p^n)^m$)$\{P_1(x)\{P_2(Y)\}\}$ are also considered.

A. Lloris Ruiz et al., *Arithmetic and Algebraic Circuits*, Intelligent Systems Reference Library 201, https://doi.org/10.1007/978-3-030-67266-9_11

11.1 GF(p)

As stated in Sect. 3.2, the set **CMp** (p prime) with the operations of product modulo p and addition modulo p, is a Galois field. Consequently, operations over GF(p) are modular arithmetic operations, already seen in Chap. 3.

Values of p with special features are used in some applications. For example, in the standard FIPS 186 [1] the use of generalized Mersenne numbers [2] which can be expressed as a sum or difference of different powers of 2 is proposed. Specifically, it proposes the use of the five prime numbers given in Table 11.1 which are particularly suitable, as detailed below.

11.1.1 Modular Reduction

The great advantage of generalized Mersenne numbers is that, after a multiplication, the modular reduction can be performed more easily than for others primes. It consists on reducing to the proper range any integer $N < p^2$, that is, any N that results from multiplying two integers less than p. Then, for each p_i of Table 11.1 expressions to obtain $R < N$, $R = N$ **mod** p are given, so that to complete the modular reduction it is sufficient to subtract p a few times.

For p_1 ($2^{192} - 2^{64} - 1$), of 192-bit, p^2 has a maximum of 384 bits, and any integer $N < p^2$ can be expressed as concatenation of six 64-bit words (n_5, …, n_0):

$$N = n_5 2^{320} + n_{54} 2^{256} + n_3 2^{192} + n_2 2^{128} + n_1 2^{64} + n_0$$

In this case, applying the multiplicative reduction (see Sect. 1.2.4) with:

$$b^k - a = 2^{192} - (2^{64} + 1)$$

It results:

$$N = (n_5 2^{128} + n_4 2^{64} + n_3) 2^{192} + (n_2 2^{128} + n_1 2^{64} + n_0)$$

Table 11.1 Generalized mersenne numbers

	Bits	Value
p_1	192	$2^{192} - 2^{64} - 1$
p_2	224	$2^{224} - 2^{96} + 1$
p_3	256	$2^{256} - 2^{224} + 2^{192} + 2^{96} - 1$
p_4	384	$2^{384} - 2^{128} - 2^{96} + 2^{32} - 1$
p_5	521	$2^{521} - 1$

This is:

$$c_0 = n_5 2^{128} + n_4 2^{264} + n_3$$

$$r_0 = n_2 2^{128} + n_1 2^{64} + n_0$$

$$ac_0 = n_5 2^{192} + \left[(n_4 + n_5)2^{128} + (n_3 + n_4)2^{64} + n_3\right]$$

$$c_1 = n_5$$

$$r_1 = (n_4 + n_5)2^{128} + (n_3 + n_4)2^{64} + n_3$$

$$ac_1 = n_5 2^{64} + n_5$$

$$c_2 = 0$$

$$r_2 = n_5 2^{64} + n_5$$

For R ($R = N$ **mod** p) it results:

$$R = (n_2 + n_4 + n_5)2^{128} + (n_1 + n_3 + n_4 + n_5)2^{64} + n_0 + n_3 + n_5$$

That can be expressed as:

$$R = (S_1 + S_2 + S_3 + S_4)\mathbf{mod}\ p_1$$

Being:

$$\begin{aligned} S_1 &= n_2 2^{128} + n_1 2^{64} + n_0 \\ S_2 &= n_3 2^{64} + n_3 \\ S_3 &= n_4 2^{128} + n_4 2^{64} \\ S_4 &= n_5 2^{128} + n_5 2^{64} + n_5 \end{aligned}$$

Therefore, to reduce N to the proper range it is just necessary to calculate $\Sigma = S_1 + S_2 + S_3 + S_4$, and if $\Sigma \geq p_1$, p_1 is subtracted while $\Sigma \geq p_1$.

For p_2 $(2^{224} - 2^{96} + 1)$, of 224 bits, p^2 has a maximum of 448 bits, and any integer $N < p^2$ can be expressed as concatenation of fourteen 32-bit words (n_{13}, …, n_0):

$$\begin{aligned} N = {} & n_{13}2^{416} + n_{12}2^{384} + n_{11}2^{352} + n_{10}2^{320} + n_9 2^{288} \\ & + n_8 2^{256} + n_7 2^{224} + n_6 2^{192} + n_5 2^{160} + n_4 2^{128} \\ & + n_3 2^{96} + n_2 2^{64} + n_1 2^{32} + n_0; \end{aligned}$$

In this case, applying again the multiplicative reduction, for R ($R = N$ **mod** p) it results:

$$R = (S_1 + S_2 + S_3 - S_4 - S_5)\textbf{mod } p_2$$

Being:

$$\begin{aligned} S_1 &= n_6 2^{192} + n_5 2^{160} + n_4 2^{128} + n_3 2^{96} + n_2 2^{64} + n_1 2^{32} + n_0 \\ S_2 &= n_{10}2^{192} + n_9 2^{160} + n_8 2^{128} + n_7 2^{96} \\ S_3 &= n_{13}2^{160} + n_{12}2^{128} + n_{11}2^{96} \\ S_4 &= n_{13}2^{192} + n_{12}2^{160} + n_{11}2^{128} + n_{10}2^{96} + n_9 2^{64} + n_8 2^{32} + n_7 \\ S_5 &= n_{13}2^{64} + n_{12}2^{32} + n_{11} \end{aligned}$$

Therefore, to reduce N to the proper range it is just necessary to calculate $\Sigma = S_1 + S_2 + S_3 - S_4 - S_5$, and if $\Sigma \geq p_2$, p_2 is subtracted while $\Sigma \geq p_2$.

For p_3 ($2^{256} - 2^{224} + 2^{192} + 2^{96} - 1$), p^2 has a maximum of 512 bits, and any integer $N < p^2$ can be expressed as concatenation of sixteen 32-bit words ($n_{15}, \ldots, n_0$):

$$\begin{aligned} N = {} & n_{15}2^{480} + n_{14}2^{448} + n_{13}2^{416} + n_{12}2^{384} \\ & + n_{11}2^{352} + n_{10}2^{320} + n_9 2^{288} + n_8 2^{256} \\ & + n_7 2^{224} + n_6 2^{192} + n_5 2^{160} + n_4 2^{128} \\ & + n_3 2^{96} + n_2 2^{64} + n_1 2^{32} + n_0 \end{aligned}$$

In this case, for R ($R = N$ **mod** p) it results:

$$R = (S_1 + 2S_2 + 2S_3 + S_4 + S_5 - S_6 - S_7 - S_8 - S_9) \bmod p_3$$

Being:

$$\begin{aligned} S_1 &= n_7 2^{224} + n_6 2^{192} + n_5 2^{160} + n_4 2^{128} + n_3 2^{96} + n_2 2^{64} + n_1 2^{32} + n_0 \\ S_2 &= n_{15}2^{224} + n_{14}2^{192} + n_{13}2^{160} + n_{12}2^{128} + n_{11}2^{96} \\ S_3 &= n_{15}2^{192} + n_{14}2^{160} + n_{13}2^{128} + n_{12}2^{96} \\ S_4 &= n_{15}2^{224} + n_{14}2^{192} + n_{10}2^{64} + n_9 2^{32} + n_8 \end{aligned}$$

$$S_5 = n_8 2^{224} + n_{13} 2^{192} + n_{15} 2^{160} + n_{14} 2^{128} + n_{13} 2^{96} + n_{11} 2^{64} + n_{10} 2^{32} + n_9$$
$$S_6 = n_{10} 2^{224} + n_8 2^{192} + n_{13} 2^{64} + n_{12} 2^{32} + n_{11}$$
$$S_7 = n_{11} 2^{224} + n_9 2^{192} + n_{15} 2^{96} + n_{14} 2^{64} + n_{13} 2^{32} + n_{12}$$
$$S_8 = n_{12} 2^{224} + n_{10} 2^{160} + n_9 2^{128} + n_8 2^{96} + n_{15} 2^{64} + n_{14} 2^{32} + n_{13}$$
$$S_9 = n_{13} 2^{224} + n_{11} 2^{160} + n_{10} 2^{128} + n_9 2^{96} + n_{15} 2^{32} + n_{14}$$

Thus, for reducing N to the proper range it is enough to calculate $\Sigma = S_1 + 2\,S_2 + 2\,S_3 + S_4 + S_5 - S_6 - S_7 - S_8 - S_9$, and if $\Sigma \geq p_3$, p_3 have to be subtracted while $\Sigma \geq p_3$.

For p_4 ($2^{384} - 2^{128} - 2^{96} + 2^{32} - 1$), p^2 have a maximum of 768 bits, and any integer $N < p^2$ can be expressed as the concatenation of twenty four 32-bits words ($n_{23}, \ldots, n_0$):

$$\begin{aligned} N = {} & n_{23} 2^{736} + n_{22} 2^{704} + n_{21} 2^{672} + n_{20} 2^{640} + n_{19} 2^{608} \\ & + n_{18} 2^{576} + n_{17} 2^{544} + n_{16} 2^{512} + n_{15} 2^{480} + n_{14} 2^{448} \\ & + n_{13} 2^{416} + n_{12} 2^{384} + n_{11} 2^{352} + n_{10} 2^{320} + n_9 2^{288} + n_8 2^{256} \\ & + n_7 2^{224} + n_6 2^{192} + n_5 2^{160} + n_4 2^{128} + n_3 2^{96} + n_2 2^{64} + n_1 2^{32} + n_0 \end{aligned}$$

For this case, for R ($R = N\,\mathbf{mod}\,p$) it results:

$$R = (S_1 + 2S_2 + S_3 + S_4 + S_5 + S_6 + S_7 - S_8 - S_9 - S_{10})\,\mathbf{mod}\ p_4$$

Being:

$$\begin{aligned} S_1 = {} & n_{11} 2^{352} + n_{10} 2^{320} + n_9 2^{288} + n_8 2^{256} + n_7 2^{224} + n_6 2^{192} \\ & + n_5 2^{160} + n_4 2^{128} + n_3 2^{96} + n_2 2^{64} + n_1 2^{32} + n_0 \\ S_2 = {} & n_{23} 2^{192} + n_{22} 2^{160} + n_{21} 2^{128} \\ S_3 = {} & n_{23} 2^{352} + n_{22} 2^{320} + n_{21} 2^{288} + n_{20} 2^{256} \\ & + n_{19} 2^{224} + n_{18} 2^{192} + n_{17} 2^{160} + n_{16} 2^{128} \\ & + n_{15} 2^{96} + n_{14} 2^{64} + n_{13} 2^{32} + n_{12} \\ S_4 = {} & n_{20} 2^{352} + n_{19} 2^{320} + n_{18} 2^{288} + n_{17} 2^{256} \\ & + n_{16} 2^{224} + n_{15} 2^{192} + n_{14} 2^{160} + n_{13} 2^{128} \\ & + n_{12} 2^{96} + n_{23} 2^{64} + n_{22} 2^{32} + n_{21} \\ S_5 = {} & n_{19} 2^{352} + n_{18} 2^{320} + n_{17} 2^{288} + n_{16} 2^{256} + n_{15} 2^{224} \\ & + n_{14} 2^{192} + n_{13} 2^{160} + n_{12} 2^{128} + n_{20} 2^{96} + n_{23} 2^{32} \\ S_6 = {} & n_{23} 2^{224} + n_{22} 2^{192} + n_{21} 2^{160} + n_{20} 2^{128} \\ S_7 = {} & n_{23} 2^{160} + n_{22} 2^{128} + n_{21} 2^{96} + n_{20} \end{aligned}$$

$$S_8 = n_{22}2^{352} + n_{21}2^{320} + n_{20}2^{288} + n_{19}2^{256} + n_{18}2^{224} + n_{17}2^{192} + n_{16}2^{160} + n_{15}2^{128} + n_{14}2^{96} + n_{13}2^{64} + n_{12}2^{32} + n_{23}$$
$$S_9 = n_{23}2^{128} + n_{22}2^{96} + n_{21}2^{64} + n_{20}2^{32}$$
$$S_{10} = n_{23}2^{128} + n_{23}2^{96}$$

Thus, for reducing N to the proper range it is enough to calculate $\Sigma = S_1 + 2S_2 + S_3 + S_4 + S_5 + S_6 + S_7 - S_8 - S_9 - S_{10}$, and if $\Sigma \geq p_4$, p_4 have to be subtracted while $\Sigma \geq p_4$.

For p_5 ($2^{521} - 1$), which is strictly a Mersenne number, p^2 have a maximum of 1042 bits, and any integer $N < p^2$ can be expressed as the concatenation of two 521-bit words (n_1, n_0):

$$N = n_1 2^{521} + n_0$$

For this case, for R ($R = N$ **mod** p) it results:

$$R = (n_1 + n_0)\textbf{mod } p_5$$

Thus, for reducing N to the proper range it is enough to calculate $\Sigma = n_1 + n_0$, and if $\Sigma \geq p_5$, p_5 have to be subtracted.

The operations of addition, subtraction and multiplication can be implemented as it is done in Sect. 3.8, specifying the circuits for each case. The calculation of the inverse (and therefore the division) can be done as follows.

11.1.2 Inversion and Division

For $a \in$GF(p), $a^{-1} = a^{p-2}$, as established in (3.9). Thus, the exponentiation can be used to compute the inverse (see Sect. 3.8.4).

Another possibility is to apply the extended Euclidean algorithm (Algorithm B.1, see Section B.1.1), repeated below:

Algorithm 11.1

Algorithm to calculate the inverse of a ∈GF(p)

```
1) R1 ← p, R2 ← a, B1 ← 0, B2 ← 1;
      while R2 ≠ 0 do
         begin
            2) Q ← COC(R1, R2), TEMP ← R2;
            3) R2 ← R1 - QR2, R1 ← TEMP, TEMP ← B2;
            4) B2 ← B1 - QB2, B1 ← TEMP;
         end
```

End algorithm

After applying this algorithm, a^{-1} remains in $B1$, as it is demonstrated in the following example.

Example 11.1 **Calculate 54^{-1} over GF (233) by applying the Algorithm** 11.1.

Table 11.2 gives the content of the different registers when applying to this case the Algorithm 11.1. It results $54^{-1} = 82$. ■

If successive subtractions want to be used instead of the division, the following modified algorithm (Algorithm 11.2) can be used.

Table 11.2 Calculation of 54^{-1} over GF(233)

Step	R1	R2	B1	B2	Q	TEMP
1	233	54	0	1		
2					4	54
3	54	17				1
4			1	−4		
2					3	17
3	17	3				−4
4			−4	13		
2					5	3
3	3	2				13
4			13	−69		
2					1	2
3	2	1				−69
4			−69	82		
2					2	1
3	1	0				82
4			82	−233		

Algorithm 11.2

Modified algorithm to calculate the inverse of a ∈GF(p)

```
1)R1 ← p, R2 ← a, B1 ← 0, B2 ← 1;
    while R2 ≠ 0 do
begin
   2) e ← g(R1)-g(R2)
        while e≥0 do
           begin
             3) if R1-2^eR2 ≥ 0, R1←R1-2^eR2, B1←B1-2^eB2,else
                if e>0,R1 ← R1-2^(e-1)R2, B1 ← B1-2^(e-1)B2,else go to 5;
             4) e ← g(R1)-g(R2);
        end
   5) R1 ↔ R2, B1 ↔ B2;
end
```

End algorithm

After applying this algorithm, a^{-1} remains in $B1$, as is demonstrated in the following examples.

Example 11.2 **Calculate 54^{-1} over GF(233) by applying the Algorithm** 11.2.

Table 11.3 gives the content of the different registers when applying to this case the Algorithm 11.2. Two's complement is used for the operations. Again, it results $54^{-1} = 82$. ■

11.2 Addition and Subtraction Over GF(p^n)

The addition and subtraction of polynomials over GF(p^n) are implemented simply using a modular adder/subtractor for each coefficient, without requiring subsequent modular reduction, if polynomial representation is used (i.e., using standard bases, or normal or dual). Using power representation, the implementation of a table of add/subtract would be required (or to change to polynomial representation), as it is done over GF(2^n). The procedures to be used are those seen in the previous chapter, with appropriate adjustments for each value of p.

11.3 Product Over GF(p^n) Using Power Representation

The product of two polynomials with power representation over GF(p^n) consists of adding the exponents and then reducing the result to the appropriate range, using that $x^{p^n-1} = 1$. It is a modular addition with non-redundant representation with $M = p^n - 1$; therefore, the adder circuit in Fig. 3.2b can be used. To contemplate the possibility that either operand is null, just use the circuit of Fig. 10.1b.

Table 11.3 Calculation of 54^{-1} over GF(233)

Step	R1	R2	B1	B2	e
1	11101001	00110110	00000000	00000001	
2					2
3	00010001		11111100		
4					−1
5	00110110	00010001	00000001	11111100	
2					1
3	00010100		00001001		
4					0
3	00000011		00001101		
4					−3
5	00010001	00000011	11111100	00001101	
2					3
3	00000101		11001000		
4					1
3	00000010		10111011		
4					0
5	00000011	00000010	00001101	10111011	
2					0
3	00000001		01010010		
4					−1
5	00000010	00000001	10111011	01010010	
2					1
3	00000000		01101001		
4					−1
5	00000001	00000000	01010010	01101001	

11.4 Product Over GF(p^n) Using the Standard Base

The product using an standard base can be decomposed into two parts: the multiplication of the operands, as described in Sect. 9.3.2, and the modular reduction of the result. Regarding the modular reduction, a combinational circuit that implements it can easily design in each case, as it is done in the following example.

Example 11.3 **Design a combinational circuit over GF(p^4){x^4+ f_3x^3+ f_2x^2+ f_1x+ f_0} that implements the modular reduction after multiplication of two elements of GF(p^4), where x^4+ f_3x^3+ f_2x^2+ f_1x+ f_0 is any 4-grade primitive polynomial modulo p (obviously $f_0 \neq 0$).**

The product of two 3-grade polynomials shall be as $p_6x^6 + p_5x^5 + p_4x^4 + p_3x^3 + p_2x^2 + p_1x + p_0$. We must reduce x^6, x^5 and x^4, using that

$$x^4 = -\left(f_3x^3 + f_2x^2 + f_1x + f_0\right) = F_3x^3 + F_2x^2 + F_1x + F_0$$

being $F_i = -f_i$.

Recalling the example 10.3, it is evident that the coefficients of the reduced polynomial r_i are obtained using the following matrix relationship:

$$\begin{bmatrix} r_0 \\ r_1 \\ r_2 \\ r_3 \end{bmatrix} = \begin{bmatrix} 1\,0\,0\,0 & c_{00} & c_{01} & c_{02} \\ 0\,1\,0\,0 & c_{10} & c_{11} & c_{12} \\ 0\,0\,1\,0 & c_{20} & c_{21} & c_{22} \\ 0\,0\,0\,1 & c_{30} & c_{31} & c_{32} \end{bmatrix} \begin{bmatrix} p_0 \\ p_1 \\ p_2 \\ p_3 \\ p_4 \\ p_5 \\ p_6 \end{bmatrix}$$

where the coefficients c_{ij} are the following ones:

$$\begin{aligned} &c_{00} = F_0;\, c_{01} = F_3F_0;\, c_{02} = F_0(F_2 + F_3F_3) \\ &c_{10} = F_1;\, c_{11} = F_0 + F_3F_1;\, c_{12} = F_3F_0 + F_1(F_2 + F_3F_3) \\ &c_{20} = F_2;\, c_{21} = F_1 + F_3F_2;\, c_{22} = F_0 + F_3F_1 + F_2(F_2 + F_3F_3) \\ &c_{30} = F_3;\, c_{31} = F_2 + F_3F_3;\, c_{32} = F_1 + F_3F_2 + F_3(F_2 + F_3F_3) \end{aligned}$$

■

In the general case GF(p^m)$\{x^m + f_{m-1}x^{m-1} + \ldots + f_1x + f_0\}$, to obtain the reduced matrix after multiplication (i.e., moving from a polynomial of grade $2m - 2$ to another of grade m) powers x^j ($j = m + 1, \ldots, 2m - 2$) must be expressed in terms of the base x^i ($i = m - 1, \ldots, 0$). It is clear that for the modular reduction is the same matrix expression obtained for GF(2^m):

$$\begin{bmatrix} r_0 \\ r_1 \\ \ldots \\ r_{m-2} \\ r_{m-1} \end{bmatrix} = \begin{bmatrix} 1 & 0 & \ldots & 0 & 0 & c_{00} & \ldots & c_{0,m-2} \\ 0 & 1 & \ldots & 0 & 0 & c_{10} & \ldots & c_{1,m-2} \\ \ldots & \ldots & \ldots & \ldots & \ldots & \ldots & \ldots & \ldots \\ 0 & 0 & \ldots & 1 & 0 & c_{m-2,0} & \ldots & c_{m-2,m-2} \\ 0 & 0 & \ldots & 0 & 1 & c_{m-1,0} & \ldots & c_{m-1,m-2} \end{bmatrix} \begin{bmatrix} p_0 \\ p_1 \\ \ldots \\ p_{2m-3} \\ p_{2m-2} \end{bmatrix}$$

It is easy to check that for the coefficients c_{ij} with $c_{-1,j} = 0$, $c_{i,-1} = 0$, it results:

$$c_{i0} = F_i,\, i = 0, \ldots, m-1$$

$$c_{ij} = c_{i-1,j-1} + c_{i0}c_{m-1,j-1},\, i = 0, \ldots, m-1,\ j = 1, \ldots, m-2$$

11.4.1 Parallel Multiplication

Multiplying two polynomials given in parallel can be performed with a combinational circuit joining matrix multiplication and modular reduction, as was done over GF(2^n), and tested in the following example.

Example 11.4 **Design a combinational circuit over GF(3^4){$x^4 + x + 2$} implementing the multiplication and the modular reduction of two elements of GF(3^4), $A(x) = a_3x^3 + a_2x^2 + a_1x + a_0$ and B(x) $= b_3x^3 + b_2x^2 + b_1x + b_0$.**

In this case, considering that $x^4 = -(x+2)$, $x^5 = -(x^2+2x)$, $x^6 = -(x^3+2x^2)$ the modular reduction matrix is:

$$\begin{bmatrix} 1 & 0 & 0 & 0 & -2 & 0 & 0 \\ 0 & 1 & 0 & 0 & -1 & -2 & 0 \\ 0 & 0 & 1 & 0 & 0 & -1 & -2 \\ 0 & 0 & 0 & 1 & 0 & 0 & -1 \end{bmatrix}$$

With this:

$$\begin{bmatrix} r_0 \\ r_1 \\ r_2 \\ r_3 \end{bmatrix} = \begin{bmatrix} 1 & 0 & 0 & 0 & -2 & 0 & 0 \\ 0 & 1 & 0 & 0 & -1 & -2 & 0 \\ 0 & 0 & 1 & 0 & 0 & -1 & -2 \\ 0 & 0 & 0 & 1 & 0 & 0 & -1 \end{bmatrix} \begin{bmatrix} a_0 & 0 & 0 & 0 \\ a_1 & a_0 & 0 & 0 \\ a_2 & a_1 & a_0 & 0 \\ a_3 & a_2 & a_1 & a_0 \\ 0 & a_3 & a_2 & a_1 \\ 0 & 0 & a_3 & a_2 \\ 0 & 0 & 0 & a_3 \end{bmatrix} \begin{bmatrix} b_0 \\ b_1 \\ b_2 \\ b_3 \end{bmatrix}$$

Operating it results:

$$\begin{bmatrix} r_0 \\ r_1 \\ r_2 \\ r_3 \end{bmatrix} = \begin{bmatrix} a_0 & -2a_3 & -2a_2 & -2a_1 \\ a_1 & a_0 - a_3 & -a_2 - 2a_3 & -a_1 - 2a_2 \\ a_2 & a_1 & a_0 - a_3 & -a_2 - 2a_3 \\ a_3 & a_2 & a_1 & a_0 - a_3 \end{bmatrix} \begin{bmatrix} b_0 \\ b_1 \\ b_2 \\ b_3 \end{bmatrix}$$

Directly, using adders, subtractors and multipliers modulo 3, it is easy to construct a multiplier circuit. As in Example 10.4, it can also be used a modular structure. Figures 11.1a and b represent a module with 9 inputs and 8 outputs (Fig. 11.1a includes the even outputs, and Fig. 11.1b includes the odd outputs), related as follows:

$$S_1 = -2e_7;\ S_2 = e_0e_1 + e_2;\ S_3 = e_1 - e_7;\ S_4 = e_0e_3 + e_4;$$
$$S_5 = e_3;\ S_6 = e_0e_5 + e_6;\ S_7 = e_5;\ S_8 = e_0e_7 + e_8.$$

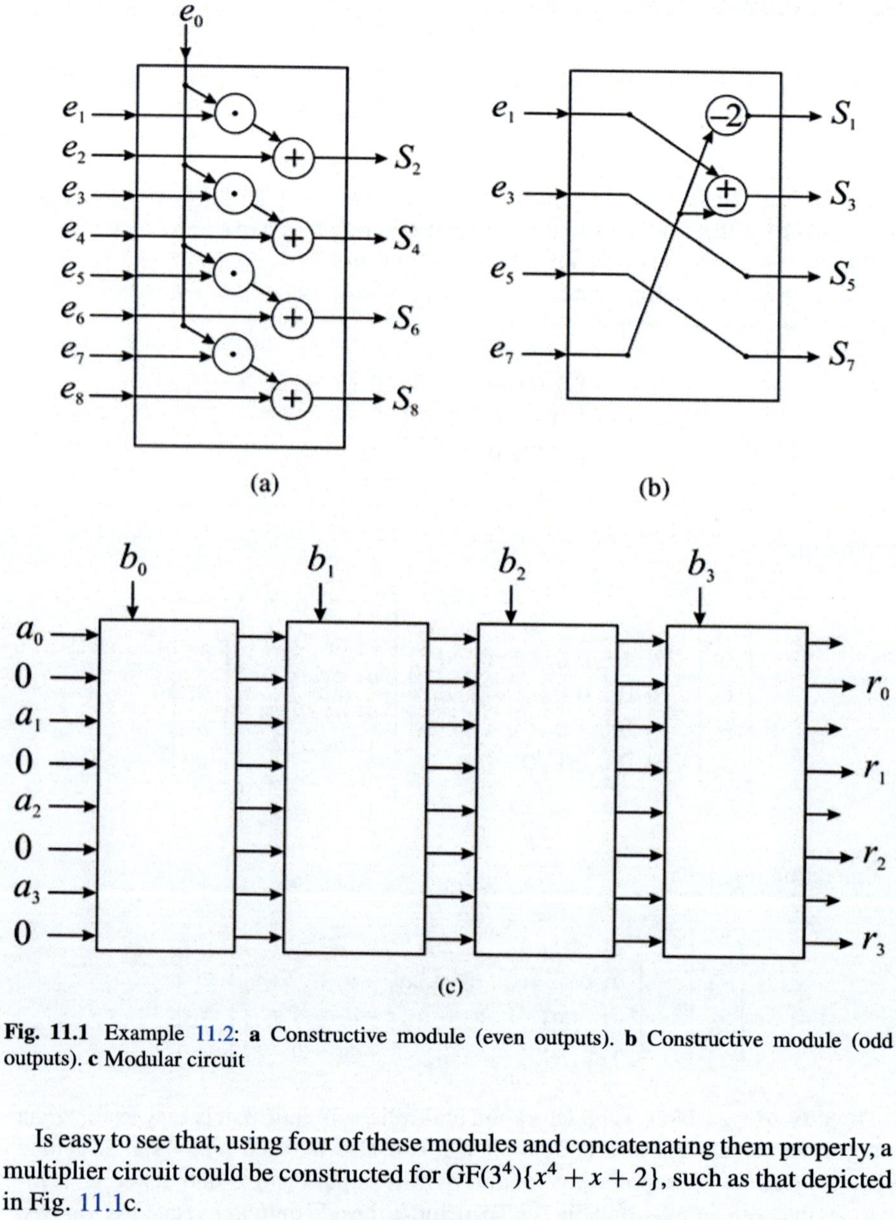

Fig. 11.1 Example 11.2: **a** Constructive module (even outputs). **b** Constructive module (odd outputs). **c** Modular circuit

Is easy to see that, using four of these modules and concatenating them properly, a multiplier circuit could be constructed for GF(3^4)$\{x^4+x+2\}$, such as that depicted in Fig. 11.1c.

Obviously, for any polynomial $P(x)$ the corresponding modular multiplier circuit can be designed in the same way as that of Fig. 11.1. ■

Example 11.5 **Design a combinational circuit implementing the multiplication and the modular reduction of two elements of GF(3^4){P(x)}, for the general case of $P(x) = x^4 + f_3x^3 + f_2x^2 + f_1x + f_0$, being $P(x)$ any primitive 4-grade polynomial modulo p (obviously $f_0 \neq 0$).**

As seen above (Example 11.3), the array of modular reduction in this case is obtained in Example 10.5, replacing f_i by $F_i = -f_i$. Therefore, the corresponding circuit will be the one in Fig. 10.3, replacing f_i by $F_i = -f_i$, and operating modulo p. ■

It is immediate that the cellular circuit of Fig. 10.3b can be generalized to any GF(p^m)$\{P(x)\}$: the standard based multiplication can be implemented with a cellular circuit, with $m \times m$ cells as those in Fig. 10.3a. For a given $P(x)$ (i.e. known the coefficients f_i), the cell of Fig. 10.3a may be simplified by eliminating one of the inputs (the corresponding to f_i) and generating p generic cells, one for each possible value of f_i. In turn, these generic cells can be simplified to the first column of the complete circuit, giving the cell in Fig. 10.4c, with only a multiplier, and for the first row, for which the top left vertex input is zero. Evidently the simpler multiplier circuit will result when $P(x)$ is a trinomial. The cellular circuit for GF(p^m) can be the one in Fig. 10.4e, with the corresponding adjustments.

Two polynomials over GF(p^m) can also be multiply using a circuit as in Fig. 9.19, storing the result in a register, then the reduction can be made using a **LFSR2 mod**p modular divider (Fig. 9.27b), being the result in the **LFSR**.

11.4.2 Serial-Parallel Multiplication

Let suppose now the case where the polynomial $A(x) = a_{m-1}x^{m-1} + a_{m-2}x^{m-2} + \ldots + a_1x + a_0$ is provided as serial data and $B(x) = b_{m-1}x^{m-1} + b_{m-2}x^{m-2} + \ldots + b_1x + b_0$ is provided as parallel data. Below are various circuits for implementing the product of both inputs, some of them using **LFSR**s.

It is easy to see that the cellular circuit of Fig. 10.4e may be adapted to consider this situation, as it was made for GF(2^m) and represented in Fig. 10.5a. This solution has no advantage against circuits based on **LFSR**s presented in the following sections.

11.4.2.1 Multiplier/Divider

To calculate $M(x) = A(x) \cdot B(x)$**mod**$P(x)$ is equivalent to multiply $A(x) \cdot B(x)$ as seen in Sect. 9.3.2, and then get the remainder dividing $A(x) \cdot B(x)$ by $P(x)$. Using a standard base and in appropriate conditions, these operations can be performed with an **LFSR** multiplier-divider as shown in Fig. 9.29. In this case it is assumed that the polynomials $B(x)$ and $P(x)$ are given in parallel and $A(x)$ is serie-input of the **LFSR**, first the higher coefficient. Moreover, it has to be applied that $P(x)$ is monic, due to $P(x)$ being primitive. The product $M(x)$ is the remainder of the division and is stored in the flip-flops of the **LFSR**.

Let $A(x) = a_{m-1}x^{m-1} + \ldots + a_0$, $B(x) = b_{m-1}x^{m-1} + \ldots + b_0$, $P(x) = x^m + p_{m-1}x^{m-1} + \ldots + p_0$. For a given m, $A(x) \cdot B(x)$**mod**$P(x)$ can be calculated with the **LFSR** of Fig. 11.2. $B(x)$ and $P(x)$ coefficients have to be applied in parallel at the

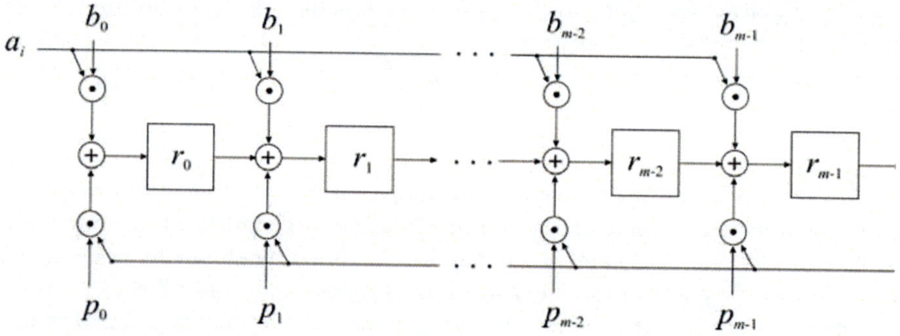

Fig. 11.2 Multiplier/divider using a serial operand and a parallel operand

labeled points, before the coefficients of $A(x)$ are introduced as serial data to the **LFSR** (highest first).

It is immediate that in reality, the circuit of Fig. 11.2 can be used for the multiplication over GF(p^i){$Q(x)$}, for any $i \le m$. For this it is enough to apply the coefficients b_j and p_j, starting by the highest, from the highest positions (entries b_{m-1} and p_{m-1}, respectively), resetting the lower positions that become vacant.

11.4.2.2 Divider/Accumulator

Let suppose again the case $B(x)$ and $P(x)$ are provided in parallel form and $A(x)$ in serial form, but now it is first provided the least significant coefficient, where $A(x) = a_{m-1}x^{m-1} + \ldots + a_0$, $B(x) = b_{m-1}x^{m-1} + \ldots + b_0$, $P(x) = x^m + p_{m-1}x^{m-1} + \ldots + p_0$. The product $M(x) = A(x)\cdot B(x)$ can be expanded as follows:

$$M(x) = A(x)\cdot B(x)\mathbf{mod}\ P(x) = a_0B(x) + a_1\{B(x)x\mathbf{mod}\ P(x)\}$$
$$+ a_2\left\{B(x)x^2\mathbf{mod}\ P(x)\right\} + \ldots + a_{m-1}\left\{B(x)x^{m-1}\mathbf{mod}\ P(x)\right\}$$

The calculation can be performed in m iterations. Indeed, the factors $B(x)x^i$**mod**$P(x)$ can be obtained with an **LFSR** divider by $P(x)$, with a development as in Example B.7 (Figure B.2). Figure 11.3a shows the full circuit for calculating $A(x)\cdot B(x)$**mod**$P(x)$, which includes an **LFSR** and an **M** register, which has to be initialized to 0. Initially, the **LFSR** is loaded with $B(x)$ and at each iteration the **LFSR** content is multiplied by x and is reduced **mod**$P(x)$. In order to calculate $A(x)\cdot B(x)$**mod**$P(x)$, at each iteration the **LFSR** content is multiplied by a_i, is added to the previous result stored in **M**, and is stored in **M**, where the final result is saved.

If the initial content of M is the polynomial $C(x)$, then $M(x) = \{A(x)\cdot B(x) + C(x)\}$**mod**$P(x)$ would be calculated.

Example 11.6 **Let suppose the multiplication $A(x)$ by $B(x)$ over GF(3^4){x^4+x^3+1}, with a_i provided as serial data (first the lower coefficient) and b_i as parallel data.**

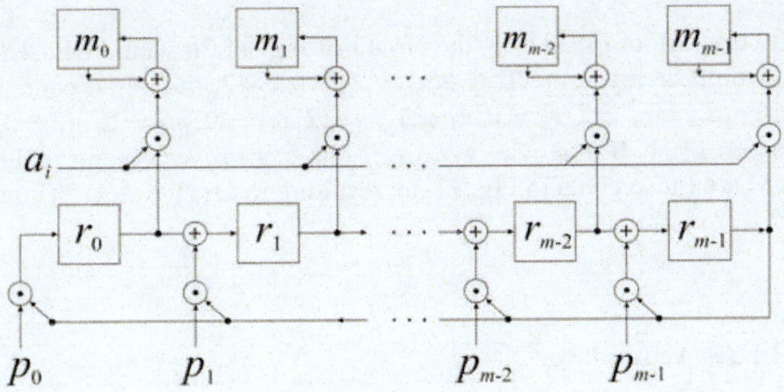

(a)

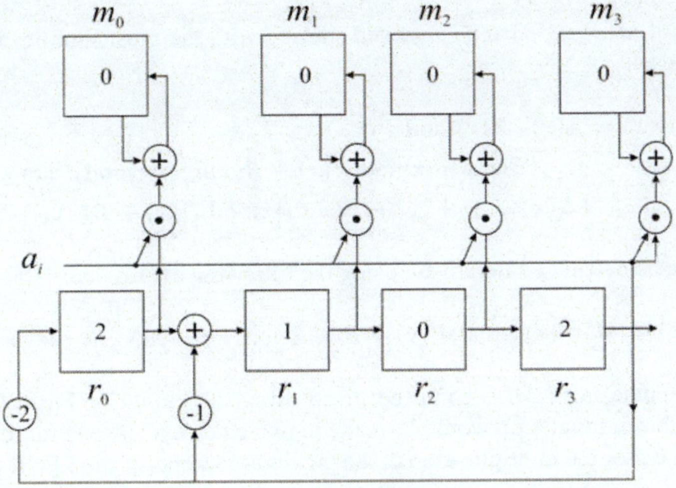

(b)

t	a_i	m_3	m_2	m_1	m_0	r_3	r_2	r_1	r_0
0		0	0	0	0	2	0	1	2
1	$a_0 = 1$	2	0	1	2	0	1	0	2
2	$a_1 = 0$	2	0	1	2	1	0	2	0
3	$a_2 = 2$	1	0	2	2	0	2	2	1
4	$a_3 = 1$	1	2	1	0				

(c)

Fig. 11.3 **a** Divider/accumulator. **b** Circuit of example 11.6. **c** Results in example 11.6

With the structure of Fig. 11.3a, the circuit of Fig. 11.3b would be used, and the calculation would be made in 4T. If $B(x) = 2x^3 + x + 2$, the initial contents of the various flip-flops are $r_3 = 2, r_2 = 0, r_1 = 1, r_0 = 2, m_3 = 0, m_2 = 0, m_1 = 0, m_0 = 0$, as shown in Fig. 11.3b. If $A(x) = x^3 + 2x^2 + 1$, the successive contents of the register **LFSR** and M are those given in Fig. 11.3c, resulting in $M(x) = A(x) \cdot B(x)\mathbf{mod}(x^4 + x + 2) = x^3 + 2x^2 + x$. ■

11.4.2.3 LFSR Acumulator

Let suppose again the case of $B(x)$ and $P(x)$ provided as parallel data and $A(x)$ as serial data, now again first the most significant coefficient, where $A(x) = a_{n-1}x^{n-1} + \ldots + a_0$, $B(x) = b_{n-1}x^{n-1} + \ldots + b_0$, $P(x) = x^n + p_{n-1}x^{n-1} + \ldots + p_0$. The calculation of $M(x)$ can also be arranged starting with the most significant digit of $A(x)$ as follows:

$$\begin{aligned} M(x) &= A(x) \cdot B(x)\mathbf{mod}\ P(x) \\ &= \{\ldots\{B(x)a_{n-1}x\mathbf{mod}\ P(x) + B(x)a_{n-2}\}x\mathbf{mod}\ P(x) \\ &\quad + B(x)a_{n-3}\} + \ldots + B(x)a_1\}x\mathbf{mod}\ P(x) + B(x)a_0 \end{aligned}$$

$M(x)$ can be calculated iteratively using the following expression:

$$M^k(x) = M^{k-1}(x)x\mathbf{mod}\ P(x) + B(x)a_{n-k}, \quad k = 1, \ldots, n; \quad M^0 = 0$$

The computation of $M(x)$ could be made using the circuit of Fig. 11.4. Each iteration in this computation would be made in two clock cycles: one for the shift in the **LFSR**, i.e., for the multiplication of the previous contents of the **LFSR** by x, and

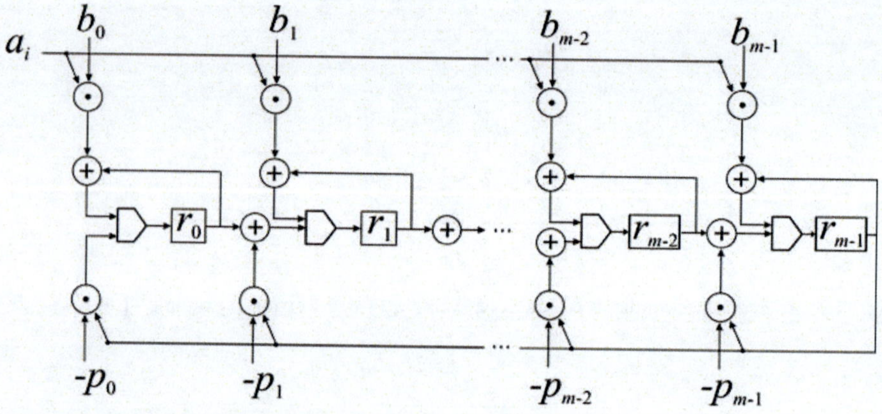

Fig. 11.4 LFSR accumulator

the modular correction of the result, $M^{k-1}(x)x\mathbf{mod}P(x)$; and another clock cycle for the addition of $B(x)a_{n-k}$ and for loading the result in parallel in the **LFSR**.

11.4.2.4 Montgomery Multiplier

The Montgomery multiplication defined over GF(p) [3] may be extended to GF(p^m). Given a primitive polynomial $P(x)$ of degree m, and given the polynomials $A(x)$ and $B(x)$ such that $g(A) < m$ and $g(B) < m$, the Montgomery multiplication of $A(x)$ and $B(x)$ modulo $P(x)$, noted as $MM\{A(x), B(x), P(x)\}$, is defined as follows:

$$MM\{A(x),\ B(x),\ P(x)\} = \left\{A(x)\cdot B(x)x^{-m}\right\}\mathbf{mod}\ P(x)$$

The product $A(x)\cdot B(x)x^{-m}$ can be developed as follows:

$$A(x)\cdot B(x)x^{-m} = \{\{\ldots\{\{A(x)b_0x^{-1} + A(x)b_1\}x^{-1} + A(x)b_2\}\ x^{-1} + \ldots + A(x)b_{m-2}\}x^{-1} + A(x)b_{m-1}\}x^{-1}$$

Therefore this product may be performed using m iterations, as $\{R(x) + A(x)b_i\}x^{-1}$, for $i = 0, \ldots, m-1$, where initially $R = 0$. If in the addition $S(x) = \{R(x) + A(x)b_i\}$, the independent term is zero ($s_0 = 0$), the multiplication by x^{-1} is reduced to a shift such as the independent term disappears; if the independent term is $s_0 \neq 0$, $(s_0{:}p_0)P(x)$ can be subtracted from $\{R(x) + A(x)b_i\}$, such that $\{R(x) + A(x)b_i\} - (s_0{:}p_0)P(x)$ have no independent term and, again, the multiplication by x^{-1} is reduced to a shift. Thus the algorithm core will be the following one:

$$\begin{aligned}&(1)\quad S \leftarrow R + Ab_i\\&(2)\quad R \leftarrow (S - (s_0 : p_0)P)x^{-1}\end{aligned}$$

With the circuit of Fig. 11.5 the Montgomery multiplication is performed using $2m$ iterations. Initially, the register R must be zero, and the result is loaded in R.

Once defined $MM\{A(x), B(x), P(x)\}$, it is easy to see that for obtaining $Z(x) = X(x)\cdot Y(x)\mathbf{mod}P(x)$ it is enough to perform two Montgomery multiplications. In fact, with the previous computation of:

$$Q(x) = x^{2m}\mathbf{mod}\ P(x)$$

it results:

$$\begin{aligned}Z(x) = X(x)\cdot Y(x)\mathbf{mod}\ P(x) &= \left\{\left\{X(x)\cdot Y(x)x^{-m}\right\}x^{2m}x^{-m}\right\}\mathbf{mod}\ P(x)\\&= MM\left\{MM\{X(x), Y(x), P(x)\}, x^{2m}, P(x)\right\}\\&= MM\{MM\{X(x), Y(x), P(x)\}, Q(x), P(x)\}\end{aligned}$$

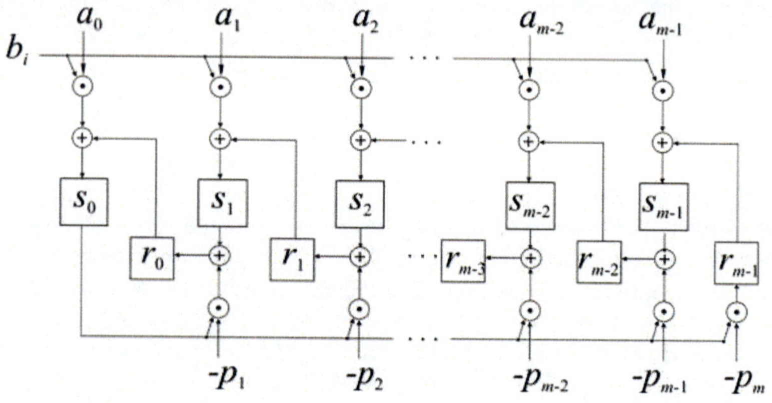

Fig. 11.5 Montgomery multiplier circuit

Therefore, the use of the Montgomery multiplier avoids the modular reduction: it is only necessary to make additions and shifts.

11.4.3 Serial Multiplication

The cellular circuit of Fig. 10.4e may be adapted again to consider the situation in which both operands are provided as serial data. Simply attach two shift registers of m length as the buffers for a_i, and b_i, input data, so that $A(x)$ and $B(x)$ are inserted as serial data, and thus are available in parallel, similar to that depicted in Fig. 10.5a.

When the polynomials $A(x)$ and $B(x)$ are provided as serial data, a coefficient of the product $A(x) \cdot B(x)$ can be computed at each clock cycle, using an appropriate circuit for each m. The modular reduction can be done with an **LFSR** divider. The details of these solutions (one for each m) are illustrated with the Example 10.7, since there is no formal difference between GF(2^n) y GF(p^n).

It could result more interesting to obtain the final result (multiplication and modular reduction) as it is made in the following example.

Example 11.7 **This example consists on multiplying over GF(3^4)$\{x^4 + x + 2\}$, with the two operands $A(x)$ and $B(x)$ provided as serial data, first the most significant coefficients, with $A(x) = a_3x^3 + a_2x^2 + a_1x + a_0$, $B(x) = b_3x^3 + b_2x^2 + b_1x + b_0$.**

From example 11.4:

$$\begin{bmatrix} r_0 \\ r_1 \\ r_2 \\ r_3 \end{bmatrix} = \begin{bmatrix} a_0 & -2a_3 & -2a_2 & -2a_1 \\ a_1 & a_0 - a_3 & -a_2 - 2a_3 & -a_1 - 2a_2 \\ a_2 & a_1 & a_0 - a_3 & -a_2 - 2a_3 \\ a_3 & a_2 & a_1 & a_0 - a_3 \end{bmatrix} \begin{bmatrix} b_0 \\ b_1 \\ b_2 \\ b_3 \end{bmatrix}$$

Each r_i can be expressed as follows:

$$r_0 = [-2a_2b_2] + [-2a_3b_1 - 2a_1b_3] + [a_0b_0]$$
$$r_1 = [-a_2b_2 - 2a_3b_2 - 2a_2b_3] + [-a_3b_1 - a_1b_3] + [a_1b_0 + a_0b_1]$$
$$r_2 = [-2a_3b_3] + [-a_3b_2 - a_2b_3] + [a_1b_1] + [a_2b_0 + a_0b_2]$$
$$r_3 = [-a_3b_3] + [a_2b_1 + a_1b_2] + [a_3b_0 + a_0b_3]$$

Introducing the coefficients of both polynomials in the respective shift registers, the sum of the three or four products for each r_i, can be generated using four clock cycles, similarly to as in Example 10.8. ■

11.5 Multiplication Over GF(p^m) Using the Normal Base

The use of a normal base in multiplication has no advantage over a standard base, but not a special difficulty. For each m, multiplier circuits can be easily designed, depending on the data are provided as serial or parallel form. For parallel data, a strictly combinational multiplier can be implemented, as has been done with standard base, or to implement intermediate solutions. To see how to proceed, a simple example is detailed at the following.

Example 11.8 **It is easy to probe that {A, A^3, A^9, A^{27}} is a normal base over GF(3^4){$x^4 + x + 2$}, being $A = x + 1$. Design a circuit to multiply two polynomials using this base.**

The following expression results for the multiplication of $A(x) = a_3A^{27} + a_2A^9 + a_1A^3 + a_0A$ and $B(x) = b_3A^{27} + b_2A^9 + b_1A^3 + b_0A$:

$$\begin{aligned} C(x) = A(x) \cdot B(x) &= a_3b_3A^{54} + (a_3b_2 + a_2b_3)A^{36} \\ &+ (a_3b_1 + a_1b_3)A^{30} + (a_3b_0 + a_0b_3)A^{28} + a_2b_2A^{18} \\ &+ (a_2b_1 + a_1b_2)A^{12} + (a_0b_2 + a_2b_0)A^{10} + a_1b_1A^6 + \\ &+ (a_1b_0 + a_0b_1)A^4 + a_0b_0A^2 = m_9A^{54} + m_8A^{36} \\ &+ m_7A^{30} + m_6A^{28} + m_5A^{18} + m_4A^{12} + m_3A^{10} \\ &+ m_2A^6 + m_1A^4 + m_0A^2 = c_3A^{27} + c_2A^9 + c_1A^3 + c_0A \end{aligned}$$

Using Table 11.4 the following developments for the different powers involved in the product are obtained:

$$A^2 = A^9 + 2A^3 + A$$
$$A^4 = A^{27} + A^9 + 2A^3 + A$$
$$A^6 = A^{27} + 2A^9 + A^3$$

Table 11.4 Elements of GF(3^4){$x^4 + x + 2$}

0	1	16	$2x^3 + x + 2$	32	$x^3 + 2x^2 + 2$	48	$2x^2 + 2x + 2$	64	$2x^2 + 2$
1	x	17	$x^2 + 2$	33	$2x^3 + x + 1$	49	$2x^3 + 2x^2 + 2x$	65	$2x^3 + 2x$
2	x^2	18	$x^3 + 2x$	34	$x^2 + 2x + 2$	50	$2x^3 + 2x^2 + x + 2$	66	$2x^2 + x + 2$
3	x^3	19	$2x^2 + 2x + 1$	35	$x^3 + 2x^2 + 2x$	51	$2x^3 + x^2 + 2$	67	$2x^3 + x^2 + 2x$
4	$2x + 1$	20	$2x^3 + 2x^2 + x$	36	$2x^3 + 2x^2 + 2x + 1$	52	$x^3 + 2$	68	$x^3 + 2x^2 + x + 2$
5	$2x^2 + x$	21	$2x^3 + x^2 + x + 2$	37	$2x^3 + 2x^2 + 2x + 2$	53	$x + 1$	69	$2x^3 + x^2 + x + 1$
6	$2x^3 + x^2$	22	$x^3 + x^2 + 2$	38	$2x^3 + 2x^2 + 2$	54	$x^2 + x$	70	$x^3 + x^2 + 2x + 2$
7	$x^3 + x + 2$	23	$x^3 + x + 1$	39	$2x^3 + 2$	55	$x^3 + x^2$	71	$x^3 + 2x^2 + x + 1$
8	$x^2 + x + 1$	24	$x^2 + 1$	40	2	56	$x^3 + 2x + 1$	72	$2x^3 + x^2 + 1$
9	$x^3 + x^2 + x$	25	$x^3 + x$	41	$2x$	57	$2x^2 + 1$	73	$x^3 + 2x + 2$
10	$x^3 + x^2 + 2x + 1$	26	$x^2 + 2x + 1$	42	$2x^2$	58	$2x^3 + x$	74	$2x^2 + x + 1$
11	$x^3 + 2x^2 + 1$	27	$x^3 + 2x^2 + x$	43	$2x^3$	59	$x^2 + x + 2$	75	$2x^3 + x^2 + x$
12	$2x^3 + 1$	28	$2x^3 + x^2 + 2x + 1$	44	$x + 2$	60	$x^3 + x^2 + 2x$	76	$x^3 + x^2 + x + 2$
13	$2x + 2$	29	$x^3 + 2x^2 + 2x + 2$	45	$x^2 + 2x$	61	$x^3 + 2x^2 + 2x + 1$	77	$x^3 + x^2 + x + 1$
14	$2x^2 + 2x$	30	$2x^3 + 2x^2 + x + 1$	46	$x^3 + 2x^2$	62	$2x^3 + 2x^2 + 1$	78	$x^3 + x^2 + 1$
15	$2x^3 + 2x^2$	31	$2x^3 + x^2 + 2x + 2$	47	$2x^3 + 2x + 1$	63	$2x^3 + 2x + 2$	79	$x^3 + 1$

$$A^{10} = A^{27} + A^3$$
$$A^{12} = A^{27} + 2A^9 + A^3 + A$$
$$A^{18} = 2A^{27} + A^9 + A$$
$$A^{28} = A^{27} + A^9 + A^3 + 2A$$
$$A^{30} = A^9 + A$$
$$A^{36} = 2A^{27} + A^9 + A^3 + A$$
$$A^{54} = A^{27} + A^3 + 2A$$

Thus, the reduction of the product C(x) to the base $\{A, A^3, A^9, A^{27}\}$ can be made applying the following matrix relation:

$$\begin{bmatrix} c_0 \\ c_1 \\ c_2 \\ c_3 \end{bmatrix} = \begin{bmatrix} 1\,1\,0\,0\,1\,1\,2\,1\,1\,2 \\ 2\,2\,1\,1\,1\,0\,1\,0\,1\,1 \\ 1\,1\,2\,0\,2\,1\,1\,1\,1\,0 \\ 0\,1\,1\,1\,1\,2\,1\,0\,2\,1 \end{bmatrix} \begin{bmatrix} m_0 \\ m_1 \\ \cdots \\ m_8 \\ m_9 \end{bmatrix}$$

Using the multiplication matrix it results:

$$\begin{bmatrix} c_0 \\ c_1 \\ c_2 \\ c_3 \end{bmatrix} = \begin{bmatrix} 1\,1\,0\,0\,1\,1\,2\,1\,1\,2 \\ 2\,2\,1\,1\,1\,0\,1\,0\,1\,1 \\ 1\,1\,2\,0\,2\,1\,1\,1\,1\,0 \\ 0\,1\,1\,1\,1\,2\,1\,0\,2\,1 \end{bmatrix} \begin{bmatrix} a_0 & 0 & 0 & 0 \\ a_1 & a_0 & 0 & 0 \\ 0 & a_1 & 0 & 0 \\ a_2 & 0 & a_0 & 0 \\ 0 & a_2 & a_1 & 0 \\ 0 & 0 & a_2 & 0 \\ a_3 & 0 & 0 & a_0 \\ 0 & a_3 & 0 & a_1 \\ 0 & 0 & a_3 & a_2 \\ 0 & 0 & 0 & a_3 \end{bmatrix} \begin{bmatrix} b_0 \\ b_1 \\ b_2 \\ b_3 \end{bmatrix}$$

And operating:

$$\begin{bmatrix} c_0 \\ c_1 \\ c_2 \\ c_3 \end{bmatrix} = M \begin{bmatrix} b_0 \\ b_1 \\ b_2 \\ b_3 \end{bmatrix}$$

$$M = \begin{bmatrix} a_0 + a_1 + 2a_3 & a_0 + a_2 + a_3 & a_1 + a_2 + a_3 & 2a_0 + a_1 + a_2 + 2a_3 \\ 2a_0 + 2a_1 + a_2 + a_3 & 2a_0 + a_1 + a_2 & a_0 + a_1 + a_3 & a_0 + a_2 + a_3 \\ a_0 + a_1 + a_3 & a_0 + 2a_1 + 2a_2 + a_3 & 2a_1 + a_2 + a_3 & a_0 + a_1 + a_2 \\ a_1 + a_2 + a_3 & a_0 + a_1 + a_2 & a_0 + a_1 + 2a_2 + 2a_3 & a_0 + 2a_2 + a_3 \end{bmatrix}$$

Table 11.5 Table of cross-products for the Example 11.8

		A^{27}	A^9	A^3	A
A	A	0	1	2	1
A	A^3	1	1	2	1
A	A^9	1	0	1	0
A	A^{27}	1	1	1	2
A^3	A	1	1	2	1
A^3	A^3	1	2	1	0
A^3	A^9	1	2	1	1
A^3	A^{27}	0	1	0	1
A^9	A	1	0	1	0
A^9	A^3	1	2	1	1
A^9	A^9	2	1	0	1
A^9	A^{27}	2	1	1	1
A^{27}	A	1	1	1	2
A^{27}	A^3	0	1	0	1
A^{27}	A^9	2	1	1	1
A^{27}	A^{27}	1	0	1	2

The direct calculation of each coefficient of the result involves adding thirteen products, three of them multiplied by 2. The same conclusion can be obtained with cross-product table, which is in Table 11.5. In each column of this table are ten ones and three twos: in total thirteen non-null values.

Considering the coefficients with a circular structure (i.e., $a_4 = a_0$, $a_5 = a_1$, etc.), it is immediate that:

$$c_i = (a_i + a_{i+1} + 2a_{i+3})b_i + (a_i + a_{i+2} + a_{i+3})b_{i+1} \\ + (a_{i+1} + a_{i+2} + a_{i+3})b_{i+2} + (2a_i + a_{i+1} + a_{i+2} + 2a_{i+3})b_{i+3}$$

For calculating the internal additions of the matrix multiplication, it is defined:

$$S_{01} = a_0 + a_1$$
$$S_{03} = a_0 + a_3$$
$$S_{12} = a_1 + a_2$$
$$S_{23} = a_2 + a_3$$

The matrix M can be expressed as follows:

$$\begin{bmatrix} S_{01} + 2a_3 & S_{23} + a_0 & S_{23} + a_1 & S_{12} + 2S_{03} \\ S_{23} + 2S_{01} & S_{12} + 2a_0 & S_{01} + a_3 & S_{23} + a_0 \\ S_{01} + a_3 & S_{03} + 2S_{12} & S_{23} + 2a_1 & S_{01} + a_2 \\ S_{23} + a_1 & S_{01} + a_2 & S_{01} + 2S_{23} & S_{03} + 2a_2 \end{bmatrix}$$

Having in mind that there are repeated elements, the internal additions can be implemented using 16 adders. Thus, with 16 multipliers and another 12 adders (28 in total) the full calculation can be made. ■

From Example 11.8 it is concluded that, for the multiplication over GF(3^4)$\{x^4 + x + 2\}$ with normal base operands $\{A, A^3, A^9, A^{27}\}$, the coefficients of the product are obtained from the coefficients of the operands, using the same function for all of them. It is easy to see that this is a general conclusion: *for the multiplication over GF(p^m) on any normal base, the coefficients of the product can be obtained from those of the operands, using all the same function.* The property that given the vector representation of an element E of GF(p^m) with a normal base, $E = (e_{m-1}, e_{m-2}, \ldots, e_1, e_0)$, E^p is obtained with a rotation to the left of E, $E^p = (e_{m-2}, e_{m-3}, \ldots, e_1, e_0, e_{m-1})$, is used (see Section B.4.2).

Let $A = (a_{m-1}, a_{m-2}, \ldots, a_1, a_0)$ and $B = (b_{m-1}, b_{m-2}, \ldots, b_1, b_0)$. The product will be:

$$\begin{aligned} C &= (c_{m-1}, c_{m-2}, c_1, c_0) = A \cdot B \\ &= (a_{m-1}, a_{m-2}, \ldots, a_1, a_0) \cdot (b_{m-1}, b_{m-2}, \ldots, b_1, b_0) \end{aligned}$$

Each coefficient of C, for example the most significant coefficient, c_{m-1}, was obtained as a combinational function of $a_{m-1}, a_{m-2}, \ldots, a_1, a_0, b_{m-1}, b_{m-2}, \ldots, b_1, b_0$:

$$c_{m-1} = f(a_{m-1}, a_{m-2}, \ldots, a_1, a_0, b_{m-1}, b_{m-2}, \ldots, b_1, b_0)$$

Raising $C = A \cdot B$ to the power of p, $C^p = A^p \cdot B^p$. That is:

$$\begin{aligned} C^p &= (c_{m-2}, c_{m-3}, \ldots, c_1, c_0, c_{m-1}) = A^p \cdot B^p \\ &= (a_{m-2}, a_{m-3}, \ldots, a_1, a_0, a_{m-1}) \cdot (b_{m-2}, b_{m-3}, \ldots, b_1, b_0, b_{m-1}) \end{aligned}$$

For the most significant coefficient of C^p, due to it is obtained as a product, it has the same dependence on the coefficients of A^p and B^p. That is,

$$c_{m-2} = f(a_{m-2}, a_{m-3}, \ldots, a_1, a_0, a_{m-1}, b_{m-2}, b_{m-3}, \ldots, b_1, b_0, b_{m-1})$$

Raising $C^p = A^p \cdot B^p$ to the power p, the same conclusion for c_{m-3} is obtained, and so on: always the same function is used, rotating the arguments.

The complexity of the function $f()$ depends on the base that has been used. Specifically, using an optimal base, the function $f()$ will have the minimum number of products, such as seen in the following example.

Example 11.9 **From Example B.20 it is concluded that $\{x^{16}, x^{48}, x^{144}, x^{432}\}$ is a optimal normal base over GF(3^4)$\{x^4+ x + 2\}$. Design a circuit to multiply two polynomials of GF(3^4) using this base.**

Multiplying $A(x) = a_3x^{432} + a_2x^{144} + a_1x^{48} + a_0x^{16}$ and $B(x) = b_3x^{432} + b_2x^{144} + b_1x^{48} + b_0x^{16}$ it results:

$$\begin{aligned} C(x) = A(x)\cdot B(x) &= a_3b_3x^{864} + (a_3b_2 + a_2b_3)x^{576} + (a_3b_1 + a_1b_3)x^{480} \\ &+ (a_3b_0 + a_0b_3)x^{448} + +a_2b_2x^{288} + (a_2b_1 + a_1b_2)x^{192} + (a_0b_2 + a_2b_0)x^{160} \\ &+ a_1b_1x^{96} + (a_1b_0 + a_0b_1)x^{64} + a_0b_0x^{32} = c_3x^{432} + c_2x^{144} + c_1x^{48} + c_0x^{16} \end{aligned}$$

The following reduction matrix of the product C(x) to the base $\{x^{16}, x^{48}, x^{144}, x^{432}\}$ results using the Table 11.4:

$$\begin{bmatrix} c_0 \\ c_1 \\ c_2 \\ c_3 \end{bmatrix} = \begin{bmatrix} 0\,0\,1\,2\,0\,0\,0\,2\,1\,0 \\ 0\,0\,0\,2\,0\,1\,1\,2\,0\,0 \\ 0\,1\,0\,2\,0\,0\,0\,2\,0\,1 \\ 1\,0\,0\,2\,1\,0\,0\,2\,0\,0 \end{bmatrix} \begin{bmatrix} m_0 \\ m_1 \\ \dots \\ m_8 \\ m_9 \end{bmatrix}$$

Using the multiplication matrix it results:

$$\begin{bmatrix} c_0 \\ c_1 \\ c_2 \\ c_3 \end{bmatrix} = \begin{bmatrix} 2a_2 & a_1 + 2a_3 & 2a_0 + a_3 & 2a_1 + a_2 \\ 2a_2 + a_3 & 2a_3 & 2a_0 + a_2 & a_0 + 2a_1 \\ a_1 + 2a_2 & a_0 + 2a_3 & 2a_0 & 2a_1 + a_3 \\ a_0 + 2a_2 & a_2 + 2a_3 & 2a_0 + a_1 & 2a_1 \end{bmatrix} \begin{bmatrix} b_0 \\ b_1 \\ b_2 \\ b_3 \end{bmatrix}$$

The direct calculation of each coefficient of the result involves adding seven products, four of them multiplied by 2. The same conclusion can be obtained with the cross-product table (Table B.32). In each column of this table there are three ones and four twos: a total of seven non-zero values.

Considering the coefficients with a circular structure (i.e., $a_4 = a_0$, $a_5 = a_1$, etc.), it is immediate that:

$$c_i = 2a_{i+2}b_i + (a_{i+1} + 2a_{i+3})b_{i+1} + (2a_i + a_{i+3})b_{i+2} + (2a_{i+1} + a_{i+2})b_{i+3}$$

■

11.6 Multiplication Over GF(p^m) Using the Dual Base

Let suppose a standard base $\{b_0, b_1, \ldots, b_{m-1}\} = \{1, \alpha, \alpha^2, \ldots, \alpha^{m-1}\}$ in GF(p^m) and its dual base $\{d_0, d_1, \ldots, d_{m-1}\}$. Any element A of GF(p^m), using the dual base $\{d_0, d_1, \ldots, d_{m-1}\}$, can be expressed as follows:

$$A = \sum_{i=0}^{m-1} a_i d_i$$

The coordinate a_i can be obtained from the expression $a_i = Tr(Ab_i)$. In fact:

$$Tr(Ab_i) = Tr\left(\left(\sum_{j=0}^{m-1} a_j d_j\right) b_i\right) = \sum_{j=0}^{m-1} a_j Tr\left(d_j b_i\right) = a_i$$

For a given L and N ε GF(p^m) and using the standard base for L and the dual base for N, the following development can be made:

$$L = \sum_{i=0}^{m-1} l_i b_i; \quad N = \sum_{i=0}^{m-1} n_i d_i;$$

The product $V = LN$ can be expressed in dual base as follows [Hsu88]:

$$V = \sum_{i=0}^{m-1} Tr(Vb_i) d_i = \sum_{i=0}^{m-1} Tr(LNb_i) d_i$$

Thus:

$$\begin{aligned} v_0 &= Tr(LNb_0) = Tr(LN) = Tr(l_0 N) + Tr(l_1 \alpha N) \\ &\quad + \ldots + Tr\left(l_{m-1}\alpha^{m-1}N\right) \\ &= l_0 Tr(N) + l_1 Tr(\alpha N) + \ldots + l_{m-1} Tr\left(\alpha^{m-1}N\right) \\ &= l_0 n_0 + l_1 n_1 + \ldots + l_{m-1} n_{m-1} \\ v_1 &= Tr(LNb_1) = Tr(\alpha LN) = Tr(l_0 \alpha N) + Tr\left(l_1 \alpha^2 N\right) \\ &\quad + \ldots + Tr\left(l_{m-1}\alpha^m N\right) \\ &= l_0 Tr(\alpha N) + l_1 Tr\left(\alpha^2 N\right) + \ldots + l_{m-1} Tr\left(\alpha^m N\right) \\ &= l_0 n_1 + l_1 n_2 + \ldots + l_{m-1} n_m \end{aligned}$$

Where:

$$n_m = n_0 p_0 + n_1 p_1 + \ldots + n_{m-1} p_{m-1}$$

That is, v_1 can be obtained from v_0 using the circuit for multiplying by α, given in Figure B.4, which is repeated for convenience in Fig. 11.6a. N, αN, $\alpha^2 N$, ..., $\alpha^{m-1}N$ can be obtained successively using this circuit. These terms have always to be multiplied by L to calculate the successive coefficients of the product LN, which are given in the dual base.

In general, for the coefficients v_i, $i = 0, ..., m - 1$, it results:

$$\begin{aligned} v_i &= Tr(LNbi) = Tr\left(\alpha^i LN\right) = Tr\left(l_0\alpha^i N\right) + Tr\left(l_1\alpha^{i+1}N\right) \\ &\quad + \ldots + Tr\left(l_{m-1}\alpha^{i+m-1}N\right) \\ &= l_0 Tr\left(\alpha^i N\right) + l_1 Tr\left(\alpha^{i+1}N\right) + \ldots + l_{m-1}Tr\left(\alpha^{i+m-1}N\right) \\ &= l_0 n_i + l_1 n_{i+1} + \ldots + l_{m-1} n_{i+m-1} \end{aligned} \tag{11.1}$$

where, for $k > m - 1$

$$n_k = n_{k-m}p_0 + n_{k-m+1}p_1 + \ldots + n_{k-1}p_{m-1}$$

In conclusion, using as core the circuit for multiplying by α, the circuit for the multiplication in dual base (i.e. an operand in standard base and, the other operand and the result in dual base) is the one in Fig. 11.6b. Once introduced the N coefficients into the **LFSR** (as serial or parallel data), and the coefficients of L applied as parallel data, the coefficients of the product are obtained as serial data through v in m cycles, the first v_0.

Dual base multiplication, as described above, has the disadvantage that standard and dual base are used simultaneously. Therefore, transformations between both bases should be implemented, which generally increases the hardware and prolong the computation. Last drawbacks can be overcoming using an almost self-dual base, in which case the transformation between bases can be a single permutation of the coefficients, as shown in Section B.5.3. An example using an almost self-dual base is developed at the following.

Example 11.10 **Design a circuit to multiply two polynomials of GF(3^4){x^4+ x+ 2} using the dual base.**

As obtained in Example B.16, the dual base of $\{1, \alpha, \alpha^2, \alpha^3\}$ is $\{1, \alpha^3, \alpha^2, \alpha\}$. Therefore, for going from one base to another it is only necessary to permute the coefficients. The circuit to obtain the product coefficients as serial data is shown in Fig. 11.6c.

For example, to multiply $A(x) = x^3 + 2x^2 + x + 2$ by $B(x) = x^2 + 2$ using the circuit of Fig. 11.6c, the different values involved are given in the table of Fig. 11.6d. It is assumed that $A(x)$ is given in standard base ($l_3 = 1$, $l_2 = 2$, $l_1 = 1$, $l_0 = 2$) and $B(x)$ in dual base ($n_3 = 0$, $n_2 = 1$, $n_1 = 0$, $n_0 = 0$). The result $C(x) = A(x)\bullet B(x)$ is given in dual base, resulting $C(x) = (x^3 + 2x^2 + x + 2)(x^2 + 2) = x^5 + 2x^4 + 2x + 1 = 2x^2 + x$. And indeed, this is the result provided by the circuit of Fig. 11.6c. ■

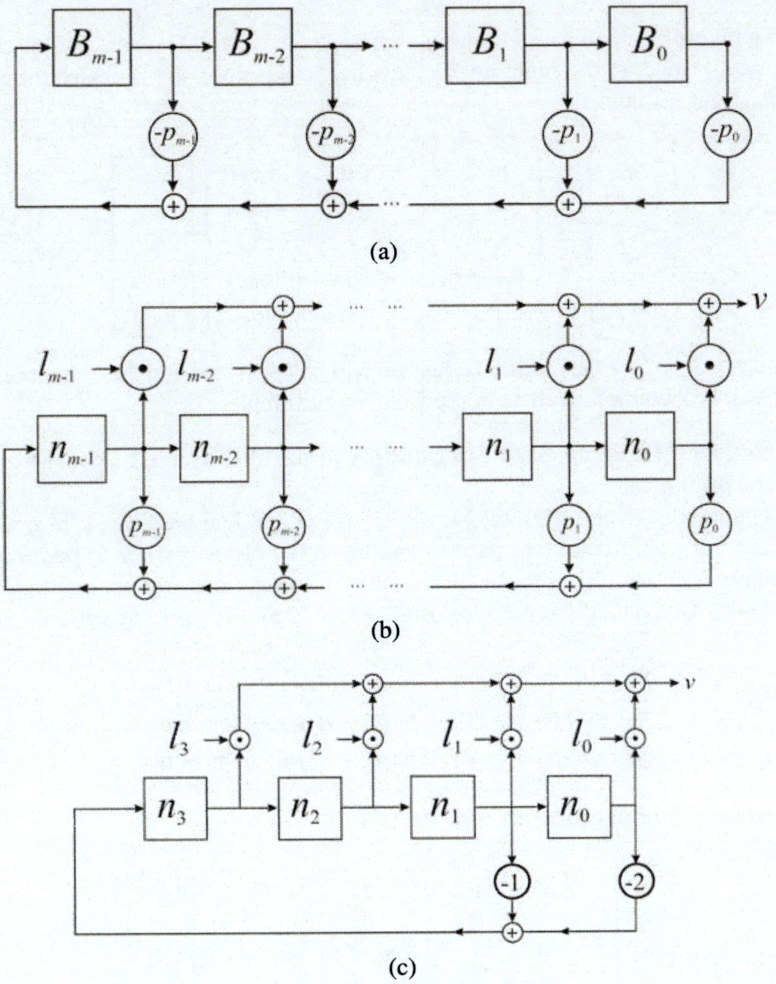

t	$n_3 \ldots n_0$	$l_3 \ldots l_0$	$\Sigma n_i l_i$
1	0102	1212	$c_0 = 0$
2	2010	1212	$c_1 = 0$
3	2201	1212	$c_2 = 2$
4	1220	1212	$c_3 = 1$

(d)

Fig. 11.6 Multiplication using dual base. **a** Circuit for multiplying by α. **b** Full circuit with serial output. **c** Serial circuit of Example 11.10. **d** Results of Example 11.10

For a multiplier with parallel output, the dependence (11.1) of the output coefficients with respect to the input coefficients has to be expressed in matrix notation. Specifically it results:

$$\begin{bmatrix} v_0 \\ v_1 \\ \cdots \\ v_{m-2} \\ v_{m-1} \end{bmatrix} = \begin{bmatrix} n_0 & n_1 & \cdots & n_{m-2} & n_{m-1} \\ n_1 & n_2 & \cdots & n_{m-1} & n_m \\ \cdots & \cdots & \cdots & \cdots & \cdots \\ n_{m-2} & n_{m-1} & \cdots & n_{2m-4} & n_{2m-3} \\ n_{m-1} & n_m & \cdots & n_{2m-3} & n_{2m-2} \end{bmatrix} \begin{bmatrix} l_0 \\ l_1 \\ \cdots \\ l_{m-2} \\ l_{m-1} \end{bmatrix}$$

For each GF(p^m){$P(x)$} the last equation will be concreted and the corresponding circuit will be obtained, such as in the following example.

Example 11.11 **Design a circuit to multiply in parallel over GF(3^4){x^4+ x+ 2} using the dual base.**

As obtained in Example B.22, $\{1, \alpha^3, \alpha^2, \alpha\}$ is the dual base of $\{1, \alpha, \alpha^2, \alpha^3\}$. Therefore, for going from one base to another it is only necessary to permute the coefficients. The matrix multiplication is used to obtain the product coefficient as parallel data. Given that, $P(x) = x^4 + x + 2$ ($p_0 = 2; p_1 = 1; p_2 = p_3 = 0$), it results:

$$n_4 = n_0 p_0 + n_1 p_1 + n_2 p_2 + n_3 p_3 = 2n_0 + n_1$$

$$n_5 = n_1 p_0 + n_2 p_1 + n_3 p_2 + n_4 p_3 = 2n_1 + n_2$$

$$n_6 = n_2 p_0 + n_3 p_1 + n_4 p_2 + n_5 p_3 = 2n_2 + n_3$$

The matrix multiplication is:

$$\begin{bmatrix} v_0 \\ v_1 \\ v_2 \\ v_3 \end{bmatrix} = \begin{bmatrix} n_0 & n_1 & n_2 & n_3 \\ n_1 & n_2 & n_3 & 2n_0 + n_1 \\ n_2 & n_3 & 2n_0 + n_1 & 2n_1 + n_2 \\ n_2 & 2n_0 + n_1 & 2n_1 + n_2 & 2n_2 + n_3 \end{bmatrix} \begin{bmatrix} l_0 \\ l_1 \\ l_2 \\ l_3 \end{bmatrix}$$

The calculation can be made (adding first $2n_0 + n_1$, $2n_1 + n_2$, $2n_2 + n_3$) using sixteen multipliers and twelve adders. ■

11.7 A^2 and A^p Over GF(p^m)

The basic operations for exponentiation are studied in 11.7.1 and 11.7.2. In 11.8 the exponentiation over GF(p^m) is aborded.

11.7.1 Square

The more convenient representation for obtaining the square is the power representation; using this representation, if a polynomial is represented by x^a, the square is:

$$\left(x^a\right)^2 = x^{2a}$$

Therefore, a is shifted one position to the left for squaring, filling with 0 the free position, and modularly reduced. Remind that if $x^a \neq 0$, a can not be all ones. The modular reduction must be done when a carry is generated in the shift, and the reduction consists on going from $n + 1$ to n bits, as seen in Sect. 3.7.1. That is, if the most significant bit of a is 0, it is required to shift and not to do modular reduction; if 1, it is require the shift and to make the modular reduction.

Any multiplier circuit can be used to obtain the square: just the two operands are the polynomial whose square was desired. But it is obvious that a circuit designed specifically for squaring will be simpler than the corresponding multiplier, as it is probed in the following example.

Example 11.12 **Design a combinational circuit over GF(3^4){x^4+ x+ 2} to obtain the square. Given $A(x) = a_3x^3 + a_2x^2 + a_1x + a_0$, it results $A^2(x) = R(x) = r_3x^3 + r_2x^2 + r_1x + r_0$.**

From Example 11.4:

$$\begin{bmatrix} r_0 \\ r_1 \\ r_2 \\ r_3 \end{bmatrix} = \begin{bmatrix} a_0 & -2a_3 & -2a_2 & -2a_1 \\ a_1 & a_0 - a_3 & -a_2 - 2a_3 & -a_1 - 2a_2 \\ a_2 & a_1 & a_0 - a_3 & -a_2 - 2a_3 \\ a_3 & a_2 & a_1 & a_0 - a_3 \end{bmatrix} \begin{bmatrix} a_0 \\ a_1 \\ a_2 \\ a_3 \end{bmatrix}$$

$$r_0 = a_0^2 + a_2^2 + 2a_1a_3$$
$$r_1 = 2a_2^2 + a_1a_3 + 2a_0a_1 + 2a_2a_3$$
$$r_2 = a_1^2 + a_3^2 + 2a_0a_2 + a_2a_3$$
$$r_3 = 2a_3^2 + 2a_0a_3 + 2a_1a_2$$

As seen in Sect. 3.8.5.1, to obtain a_i^2 it is sufficient an **OR** gate, and to multiplied by 2 consists on permuting the two bits. Thus $A^2(x)$ can be calculated using four **OR** gates, six multipliers and ten 2-input adders. ■

11.7.2 A^p

Using power representation, if a polynomial is represented by x^a, the exponentation to the power of p is:

$$(x^a)^p = x^{ap}$$

Therefore, to raise to the power of p it is necessary to multiply the exponent by p, using the corresponding modular reduction, as it is made in Sect. 3.8.2.

With polynomial representation using standard base, for raising to the power of p over GF(p^n) the following expression can be used (see Appendices A and B):

$$(a + b)^p = a^p + b^p$$

Thus, given $B(x) = b_{n-1}x^{n-1} + b_{n-2}x^{n-2} + \ldots + b_1x + b_0$, it results:

$$B^p(x) = (b_{n-1})^p x^{p(n-1)} + (b_{n-2})^p x^{p(n-2)} + \ldots + (b_1)^p x^p + (b_0)^p$$

$B^p(x)$ is a polynomial with only powers multiples of p. Remember (See Section A.39) that $(b_i)^p = b_i$ over GF(p). Therefore, it is sufficient to make modular reduction to the polynomial with coefficients $b_{n-1}\ 0 \ldots 0\ b_{n-2}\ 0 \ldots 0\ b_1\ 0 \ldots 0\ b_0$ (between the coefficients b_i and b_{i+1} there are $p - 1$ zeros). This modular reduction can be done with an **LFSR** divider, as in the following example.

Example 11.13 **Obtain $A^3(x)$ over GF(3^4){$x^4 + x + 2$)} being $A(x) = 2x^3 + x + 1$ (i.e. 2011).**

Obviously:

$$A^3(x) = 2x^9 + x^3 + 1 = 2x^2 + 2x + 1$$

This is the result that remains in the **LFSR** of Fig. 11.7a after introducing as dividend the sequence 2000001001, as it is detailed in the table of Fig. 11.7b. ■

11.8 Exponentiation Over GF(p^m)

Given a polynomial B(x) belonging to GF(p^m), the objective is to calculate $B^k(x)$, which also belongs to GF(p^m), with k integer.

First, it is immediate that k can be reduced to the range $p^m - 1$. Indeed, since $B^{p^m-1}(x) = 1$, it is clear that $B^k(x) = B^q(x)$, where $Q = k\mathbf{mod}(p^m - 1)$.

The computation of B^Q can be reduced to multiplications and raising to the power of p. For this, Q is developed as a number in base p:

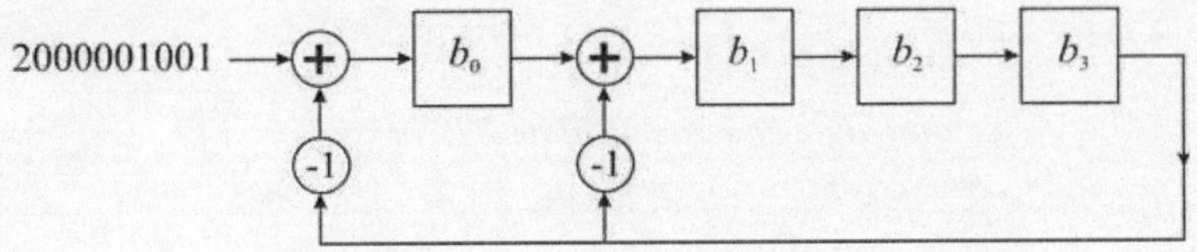

(a)

e	b_0	b_1	b_2	b_3	$b_0^+ = e + b_3$	$b_1^+ = b_0 + 2b_3$
2	0	0	0	0	2	0
0	2	0	0	0	0	2
0	0	2	0	0	0	0
0	0	0	2	0	0	0
0	0	0	0	2	2	1
0	2	1	0	0	0	2
1	0	2	1	0	1	0
0	1	0	2	1	1	0
0	1	0	0	2	2	2
1	2	2	0	0	1	2
	1	2	2	0		

(b)

Fig. 11.7 A^p using standard base. **a** Circuit. **b** Table of results

$$\begin{aligned} Q &= q_{m-1}p^{m-1} + q_{m-2}p^{m-2} + \ldots + q_1 p + q_0 \\ &= ((\ldots(q_{m-1}p + q_{m-2})p + \ldots +) + q_1)p + q_0 \end{aligned}$$

That is:

$$B^Q = \left(\ldots\left(\left(B^{q_{m-1}}\right)^p \cdot B^{q_{m-2}}\right)^p \ldots \cdot B^{q_1}\right)^p \cdot B^{q_0}$$

With this development for B^Q, the calculation involves raising to the power of p and multiplication. The core of the calculation would be:

$$\text{(1)} \quad R \leftarrow R^p$$

$$\text{(2)} \quad \text{if} \quad q_i \neq 0, \quad R \leftarrow RB^{q^i}$$

The result is obtained in register R, which initially has to be $R \leftarrow 1$. The algorithm could be as follows:

Algorithm 11.3

First algorithm to calculate B^Q

$R \leftarrow 1,\ i \leftarrow m-1;$
while $i > 0$ **do**
  begin
    1) **if** $q_i \neq 0$ **then** $R \leftarrow RB^{q_i}$;
    2) $R \leftarrow R^p$, $i \leftarrow i-1$;
  end
if $q_0 \neq 0$ **then** $R \leftarrow RB^{q_0}$;

End algorithm

A possible circuit for the exponentiation using the above algorithm is represented in Fig. 11.8a. It is supposed that B^{p-1}, …, B^2 have been previously calculated. This circuit includes a register, R, a multiplier and a circuit to rise to the power of p, and can be used with any representations for B: just implementing the multiplication and the raising to the power of p in the representation being used.

Another possible development of B^Q is:

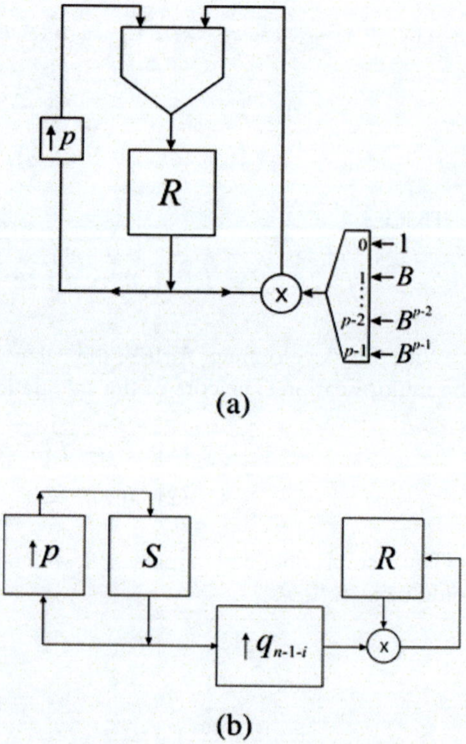

Fig. 11.8 Exponentiation: **a** First circuit. **b** Second circuit

$$B^Q = \left(B^{p^0}\right)^{q_0} \cdot \left(B^{p^1}\right)^{q_1} \cdot \ldots \cdot \left(B^{p^{m-2}}\right)^{q_{m-2}} \cdot \left(B^{p^{m-1}}\right)^{q_{m-1}}$$

Again, the calculation involves raising to the power of p and multiplication, according to the following core for the calculations:

$$\begin{aligned} &(1) \quad S \leftarrow S^p \\ &(2) \quad \text{if} \quad q_{m-1-i} \neq 0, \quad R \leftarrow RS^{q_{m-1-i}} \end{aligned}$$

Initially must be $S \leftarrow B$, $R \leftarrow 1$. The result remains in R. The algorithm in this case could be as follows:

Algorithm 11.4

Second algorithm to calculate B^Q

```
S ← B, R ← 1, i ← m−1;
while i > 0 do
    begin
        1) if q_{m−1−i} = 1 then R ← RS^{q_{m−1−i}};
        2) S ← S^p, i ← i−1;
    end
```

End algorithm

A possible circuit for the exponentiation using Algorithm 11.4 is shown in Fig. 11.8b. This circuit includes two registers, S and R, a multiplier and a circuit to raise to the power of p, and another circuit to raise to the power of q_{m-1-i}, and can be used with any representation of B: just implementing multiplication and raise to the powers that appear in the circuit, in the representation used.

11.9 Inversion and Division Over GF(p^m)

Given a polynomial $A(x) \in$ GF(p^m), to calculate the inverse $A^{-1}(x)$ it can be applied that:

$$A^{p^{2m}-2}(x) = A^{-1}(x)$$

Therefore, obtaining the inverse can be reduced to one exponentiation. As in this case the exponent is always $p^m - 2$, in the development:

$$q = q_m p^m + q_{m-1} p^{m-1} + q_{m-2} p^{m-2} + \ldots + q_1 p + q_0$$

must be $q_m = 1$; $q_i = 0$, $i = m - 1, \ldots, 1$; $q_0 = -2$. Therefore, the two exponentiation algorithms of Sect. 11.8 can be simplified considerably if $A^{-1}(x)$ has to be calculated.

An iterative calculation for a given GF(p^m) can be developed, similar to the developed for GF(2^m), as it will be studied at the following. Calling $F_n(A) = A^{p^n-1}$ and $G_n(A) = A^{p^n-2}$ it is immediate that:

$$F_{n+m}(A) = (F_m(A))^{p^n} F_n(A) = (F_n(A))^{p^m} F_m(A)$$
$$G_{n+m}(A) = (G_m(A))^{p^n} (F_n(A))^2 = (G_n(A))^{p^m} (F_m(A))^2$$

Actually:

$$F_{n+m}(A) = A^{p^{n+m}-1} = A^{p^n p^m + p^n - p^n - 1} = A^{(p^m-1)p^n} A^{p^n-1}$$
$$G_{n+m}(A) = A^{p^{n+m}-2} = A^{p^n p^m + 2p^n - 2p^n - 2} = A^{(p^m-2)p^n} A^{(p^n-1)2}$$

This facilitates the use of any development, particularly the additive chains.

Another method for calculating the inverse is based on that, if $M(x)$ and $N(x)$ are two polynomials prime to each other, then two other polynomial $R(x)$ and $S(x)$ can be found, such that:

$$1 = R(x) \cdot M(x) + S(x) \cdot N(x)$$

With $M(x) = P(x)$ and $N(x) = B(x)$, and operating over GF(p^m), it remains:

$$1 = \{R(x) \cdot P(x) + S(x) \cdot B(x)\} \mathbf{mod}\ P(x) = S(x) \cdot B(x) \mathbf{mod}\ P(x)$$

since $R(x) \cdot P(x) \mathbf{mod} P(x) = 0$. Therefore,

$$S(x) = B^{-1}(x)$$

$S(x)$ is calculated using the procedure seen in Section B.5.4, such as in the following example.

Example 11.14 **Obtain the inverse of $B(x) = 3x^3 + x^2 + 6$ over GF(7^4)$\{x^4 + x^2 + 3x + 5\}$,**

Through successive divisions becomes:

$$x^4 + x^2 + 3x + 5 = \left(3x^3 + x^2 + 6\right)(5x + 3) + \left(5x^2 + x + 1\right) \Rightarrow C_1 = 5x + 3$$
$$3x^3 + x^2 + 6 = \left(5x^2 + x + 1\right)(2x + 4) + (x + 2) \Rightarrow C_2 = 2x + 4$$
$$5x^2 + x + 1 = (x + 2)(5x + 5) + 5 \Rightarrow C_3 = 5x + 5$$

From the calculation of $S(x)$:

$$S_{-1} = 0,\ S_0 = 1,\ S_1 = S_{-1} - S_0 C_1 = -(5x + 3),$$
$$S_2 = S_0 - S_1 C_2 = 1 + (5x + 3)(2x + 4)$$
$$S_3 = S_1 - S_2 C_3 = -(5x + 3) - (1 + (5x + 3)(2x + 4))(5x + 5)$$

$$= 6x^3 + 2x^2 + 3x + 2$$

Thus, considering the last remainder no null, it results:

$$B^{-1}(x) = S_3 : 5 = 4x^3 + 6x^2 + 2x + 61$$

In fact, it is easy to probe that $\left(3x^3 + x^2 + 6\right)\left(4x^3 + 6x^2 + 2x + 6\right) = 1.$ ■

The calculation can be done by successive subtractions, as is detailed in Section B.5.4.

To divide $A(x)$ by $B(x)$ is the same as multiplying $A(x)$ by $B^{-1}(x)$. Therefore, it is sufficient to divide using any of the above algorithms to calculate $B^{-1}(x)$, and then use any of the multiplication algorithms.

11.10 Operations Over GF($(p^n)^M$)

As for GF($(2^n)^m$), all operations for GF(p^m) are transferred without any formal modification to GF($(p^n)^m$).

11.11 Conclusion

In this chapter the main circuits to implement the various operations over GF(p^n) as a generalization of the circuits corresponding to GF(2^n) have been presented.

11.12 Exercises

11.1. Calculate 19^{-1} over GF(23) using the extended Euclidean algorithm with divisions.

11.2. Calculate 19^{-1} over GF(23) using the extended Euclidean algorithm with subtractions.

11.3. Design a combinational circuit over GF(5^5)$\{x^5 + 4x + 2\}$ for the modular reduction after the product of two elements of GF(5^5).

11.4. Design a combinational circuit over GF(5^5)$\{x^5 + 4x + 2\}$ implementing the multiplication and the modular reduction of two elements of GF(5^5).

11.5. Using a divider/accumulator for GF(5^5)$\{x^5 + 4x + 2\}$, multiply $A(x) = 4x^4 + 2x^2 + 3$, provided in parallel, by $B(x) = 2x^3 + 3x^2 + x + 3$, provided in serial.

11.6. Design a circuit over GF(3^4)$\{P(x)\}$ to multiply two elements of GF(3^4), both provided as serial data, first the most significant coefficients.

11.7. Design a circuit over GF(3^4){$P(x)$} to multiply two elements of GF(3^4), both provided as serial data, first the least significant coefficients.

References

1. National Institute of Standards and Technology: FIPS 186-4: Digital Signature Standard (DSS). Gaithersburg (2013)
2. Solinas, J.A.: Generalized Mersenne Numbers. Technical Report CORR 99-39. University of Waterloo (1999)
3. Montgomery, P.L.: Modular multiplication without trial division. Math. Comput. **44**(170), 519–521 (1985)

Chapter 12
Two Galois Fields Cryptographic Applications

This last chapter is devoted to illustrating Galois Fields possibilities in cryptography. Thus, as an example, two cryptographic applications of the circuits described in previous chapters are presented.

Nowadays, cryptographic applications are becoming more and more important in communications, especially when using public channels such as Internet. The different standards available are usually software implemented, but in the following faster hardware implementations are described.

First Section introduces general concepts about cryptography, while the second one presents the discrete logarithm based cryptosystems. The third one describes elliptic curve cryptosystems.

12.1 Introduction

From old, ingenious procedures for maintaining secret and secure communications in special circumstances have been developed. Although stories about these issues are very interesting and have played decisive roles in humanity progress, they are not the objective of the present text. The interested reader can consult specialized references as [1, 6]. The science devoted to the practice and study of questions related with the encryption and the decryption of information is known as **Cryptography**.

In a first instance, two main types of cryptosystems can be identified: secret key cryptosystems, and public key cryptosystems. When using secret key cryptosystems, the transmitter and receiver of the message share a key only known by the two communicants, and allowing the encryption and decryption of a message. Anyone knowing this key can decrypt the secret message, and without knowing the key, must be very difficult performing its decryption. In this scheme, encryption and decryption are inverse operations, with similar complexity when knowing the key. Because of this, secret key cryptosystems are known also as symmetric cryptosystems.

A. Lloris Ruiz et al., *Arithmetic and Algebraic Circuits*, Intelligent Systems Reference Library 201, https://doi.org/10.1007/978-3-030-67266-9_12

When using public key cryptosystems, the receiver of a message has to establish a key known for everyone and designated as public key. This public key allows encrypting messages addressed to him. The receiver also has a secret key, only known by him and related with the public key, which allows the receiver decrypting any message encrypted using its public key. Without knowing the secret key must be impossible in practice decrypting the message. Obviously, the keys used for encrypting and decrypting a message must be different. For this reason the public key cryptosystems are also known as asymmetric cryptosystems.

Applications being described in the following are two public key cryptosystems, for which several standards have been defined. The first application is based on the discrete logarithm operation, described previously in Sect. 3.2, and the second one is based on elliptic curves, defined in Appendix C.

12.2 Discrete Logarithm Based Cryptosystems

Chapter 3 (Sect. 3.2), describes the discrete logarithm operation over finite fields, and shows the difficulty of finding the discrete logarithm of a group element with respect to a given primitive element in a reasonable amount of computing time. Using this property, public key cryptosystems can be defined, as detailed in the following.

12.2.1 Fundamentals

Ana and Carmen wish to interchange secret messages using a public channel. In order to encrypt their messages, they choose a Galois Field GF(q), being q a prime number or 2^n (with q or n large enough), and an r-order subgroup generator, p. GF(q) and p can be public.

Ana and Carmen messages will be elements of GF(q), or a correspondence between the possible messages and the elements of GF(q) will be established.

Then, Ana and Carmen select each one randomly an integer, a and b, respectively, being less than r (these are the secret keys). Ana's public key will be p^a, and Carmen's one p^b, being both of them known for the two communicants. Because the operation involved for deriving e from p^e is the discrete logarithm, it is not possible calculating e in a reasonable amount of time.

Thus, from the knowledge of the public keys corresponding to Ana and Carmen, p^a and p^b, it is not possible calculating p^{ab} without solving the discrete logarithm problem. But both Ana and Carmen can easily compute p^{ab} and p^{-ab} because each one knows its secret key.

As an standard example, the IEEE 1363-2000 [4] provides several procedures for properly selecting the components for building cryptographic systems based on the discrete logarithm. For achieving a minimal security, the standard establishes a minimum size of 1024 bits for the Galois Field, and 160 bits for the subgroup order, r.

When Ana wants to send the clear text message M to Carmen, first she multiplies M by p^{ab}, sending the encrypted message $M_c = Mp^{ab}$. Then Carmen can decrypt the message multiplying M_c by p^{-ab}, resulting the original message, $M = M_c p^{-ab} = Mp^{ab}p^{-ab}$.

The cryptosystem described before is, basically, the one proposed by El Gamal [3]. There are other cryptosystems proposals based on the discrete logarithm problem, like [2], but really the operations over Galois Fields needed in all of them are practically the same. Thus, for illustrating this type of cryptosystems, we will continue with the proposal of El Gamal, showing a simple example with a low number of subgroup elements which allows building a logarithms table (there will be no problem with computing the discrete logarithm).

Example 12.1 As an example of discrete logarithm based cryptosystem, let's consider the Galois field $GF(2^8)\{x^8 + x^4 + x^3 + x^2 + 1\}$, and the generator element x. In this situation, the subgroup order ($GF(2^8)$ itself) is 255, and the table of logarithms shown in Table 12.1 can be built.

Assuming Ana wants sending to Carmen the message $M = 10110101 = x^7 + x^5 + x^4 + x^2 + 1$, they must select two random integers less than 255, and maintain them in secret (they are the secret keys). Ana selects 143, and Carmen 98. Then, the public key corresponding to Ana is $x^{143} = x^6 + x^4 + x^2 = 01010100$, and the Carmen one is $x^{98} = x^6 + x + 1 = 01000011$.

Ana can compute $(x^{98})^{143} = (x^6 + x + 1)^{143} = x^{190} = x^7 + x^5 + x^3 + x^2 + x = 1010110$, and $M_e = M(x^{98})^{143} = (x^7 + x^5 + x^4 + x^2 + 1)\cdot(x^7 + x^5 + x^3 + x^2 + x) = x^7 + x^6 + x^5 + x^4 + x^2 + x + 1 = 11110111$. Then Ana sends the encrypted messages $M_e = 11110111$ to Carmen.

Carmen can compute $(x^{143})^{98} = (x^6 + x^4 + x^2)^{98} = x^{190} = x^7 + x^5 + x^3 + x^2 + x = 1010110$, as well as its inverse, $x^{-190} = x^7 + x^5 + x^4 + x^3 + x^2 + x = 10111110$. For decrypting the message received from Ana, she computes $M = M_e x^{-190} = (Mx^{190})x^{-190} = (x^7 + x^6 + x^5 + x^4 + x^2 + x + 1)\cdot(x^7 + x^5 + x^4 + x^3 + x^2 + x) = x^7 + x^5 + x^4 + x^2 + 1$, recovering in this way the original message. ■

Summarizing from Example 12.1, the operations to be completed by the communicants, once fixed the Galois field and the generator, are the following: random selection of an integer, exponentiation, multiplication and inversion.

Random selection of an integer is a relevant aspect for ensuring the quality of a cryptographic system. This issue is not related with the different developments presented in previous chapters, and will not be studied here. The rest of the involved operations can be completed using the procedures proposed along the text, as detailed in the next section.

Table 12.1 Logarithms table in GF(2^8){$x^8 + x^4 + x^3 + x^2 + 1$}

0:1	64:$x^6 + x^4 + x^3 + x^2 + x+1$	128:$x^7 + x^2 + 1$	192:$x^7 + x$
1:x	65:$x^7 + x^5 + x^4 + x^3 + x^2 + x$	129:$x^4 + x^2 + x+1$	193:$x^4 + x^3 + 1$
2:x^2	66:$x^6 + x^5 + 1$	130:$x^5 + x^3 + x^2 + x$	194:$x^5 + x^4 + x$
3:x^3	67:$x^7 + x^6 + x$	131:$x^6 + x^4 + x^3 + x^2$	195:$x^6 + x^5 + x^2$
4:x^4	68:$x^7 + x^4 + x^3 + 1$	132:$x^7 + x^5 + x^4 + x^3$	196:$x^7 + x^6 + x^3$
5:x^5	69:$x^5 + x^3 + x^2 + x+1$	133:$x^6 + x^5 + x^3 + x^2 + 1$	197:$x^7 + x^3 + x^2 + 1$
6:x^6	70:$x^6 + x^4 + x^3 + x^2 + x$	134:$x^7 + x^6 + x^4 + x^3 + x$	198:$x^2 + x+1$
7:x^7	71:$x^7 + x^5 + x^4 + x^3 + x^2$	135:$x^7 + x^5 + x^3 + 1$	199:$x^3 + x^2 + x$
8:$x^4 + x^3 + x^2 + 1$	72:$x^6 + x^5 + x^2 + 1$	136:$x^6 + x^3 + x^2 + x+1$	200:$x^4 + x^3 + x^2$
9:$x^5 + x^4 + x^3 + x$	73:$x^7 + x^6 + x^3 + x$	137:$x^7 + x^4 + x^3 + x^2 + x$	201:$x^5 + x^4 + x^3$
10:$x^6 + x^5 + x^4 + x^2$	74:$x^7 + x^3 + 1$	138:$x^5 + 1$	202:$x^6 + x^5 + x^4$
11:$x^7 + x^6 + x^5 + x^3$	75:$x^3 + x^2 + x+1$	139:$x^6 + x$	203:$x^7 + x^6 + x^5$
12:$x^7 + x^6 + x^3 + x^2 + 1$	76:$x^4 + x^3 + x^2 + x$	140:$x^7 + x^2$	204:$x^7 + x^6 + x^4 + x^3 + x^2 + 1$
13:$x^7 + x^2 + x+1$	77:$x^5 + x^4 + x^3 + x^2$	141:$x^4 + x^2 + 1$	205:$x^7 + x^5 + x^2 + x+1$
14:$x^4 + x+1$	78:$x^6 + x^5 + x^4 + x^3$	142:$x^5 + x^3 + x$	206:$x^6 + x^4 + x+1$
15:$x^5 + x^2 + x$	79:$x^7 + x^6 + x^5 + x^4$	143:$x^6 + x^4 + x^2$	207:$x^7 + x^5 + x^2 + x$
16:$x^6 + x^3 + x^2$	80:$x^7 + x^6 + x^5 + x^4 + x^3 + x^2 + 1$	144:$x^7 + x^5 + x^3$	208:$x^6 + x^4 + 1$
17:$x^7 + x^4 + x^3$	81:$x^7 + x^6 + x^5 + x^2 + x+1$	145:$x^6 + x^3 + x^2 + 1$	209:$x^7 + x^5 + x$
18:$x^5 + x^3 + x^2 + 1$	82:$x^7 + x^6 + x^4 + x+1$	146:$x^7 + x^4 + x^3 + x$	210:$x^6 + x^4 + x^3 + 1$
19:$x^6 + x^4 + x^3 + x$	83:$x^7 + x^5 + x^4 + x^3 + x+1$	147:$x^5 + x^3 + 1$	211:$x^7 + x^5 + x^4 + x$
20:$x^7 + x^5 + x^4 + x^2$	84:$x^6 + x^5 + x^3 + x+1$	148:$x^6 + x^4 + x$	212:$x^6 + x^5 + x^4 + x^3 + 1$
21:$x^6 + x^5 + x^4 + x^2 + 1$	85:$x^7 + x^6 + x^4 + x^2 + x$	149:$x^7 + x^5 + x^2$	213:$x^7 + x^6 + x^5 + x^4 + x$
22:$x^7 + x^6 + x^5 + x^3 + x$	86:$x^7 + x^5 + x^4 + 1$	150:$x^6 + x^4 + x^2 + 1$	214:$x^7 + x^6 + x^5 + x^4 + x^3 + 1$

(continued)

Table 12.1 (continued)

0:1	64:$x^6+x^4+x^3+x^2+x+1$	128:x^7+x^2+1	192:x^7+x
23:$x^7+x^6+x^3+1$	87:$x^6+x^5+x^4+x^3+x^2+x+1$	151:$x^7+x^5+x^3+x$	215:$x^7+x^6+x^5+x^3+x^2+x+1$
24:$x^7+x^3+x^2+x+1$	88:$x^7+x^6+x^5+x^4+x^3+x^2+x$	152:x^6+x^3+1	216:x^7+x^6+x+1
25:$x+1$	89:$x^7+x^6+x^5+1$	153:x^7+x^4+x	217:$x^7+x^4+x^3+x+1$
26:x^2+x	90:$x^7+x^6+x^4+x^3+x^2+x+1$	154:$x^5+x^4+x^3+1$	218:x^5+x^3+x+1
27:x^3+x^2	91:x^7+x^5+x+1	155:$x^6+x^5+x^4+x$	219:$x^6+x^4+x^2+x$
28:x^4+x^3	92:$x^6+x^4+x^3+x+1$	156:$x^7+x^6+x^5+x^2$	220:$x^7+x^5+x^3+x^2$
29:x^5+x^4	93:$x^7+x^5+x^4+x^2+x$	157:$x^7+x^6+x^4+x^2+1$	221:x^6+x^2+1
30:x^6+x^5	94:$x^6+x^5+x^4+1$	158:$x^7+x^5+x^4+x^2+x+1$	222:x^7+x^3+x
31:x^7+x^6	95:$x^7+x^6+x^5+x$	159:$x^6+x^5+x^4+x+1$	223:x^3+1
32:$x^7+x^4+x^3+x^2+1$	96:$x^7+x^6+x^4+x^3+1$	160:$x^7+x^6+x^5+x^2+x$	224:x^4+x
33:x^5+x^2+x+1	97:$x^7+x^5+x^3+x^2+x+1$	161:$x^7+x^6+x^4+1$	225:x^5+x^2
34:$x^6+x^3+x^2+x$	98:x^6+x+1	162:$x^7+x^5+x^4+x^3+x^2+x+1$	226:x^6+x^3
35:$x^7+x^4+x^3+x^2$	99:x^7+x^2+x	163:x^6+x^5+x+1	227:x^7+x^4
36:x^5+x^2+1	100:x^4+1	164:$x^7+x^6+x^2+x$	228:$x^5+x^4+x^3+x^2+1$
37:x^6+x^3+x	101:x^5+x	165:x^7+x^4+1	229:$x^6+x^5+x^4+x^3+x$
38:$x^7+x^4+x^2$	102:x^6+x^2	166:$x^5+x^4+x^3+x^2+x+1$	230:$x^7+x^6+x^5+x^4+x^2$
39:$x^5+x^4+x^2+1$	103:x^7+x^3	167:$x^6+x^5+x^4+x^3+x^2+x$	231:$x^7+x^6+x^5+x^4+x^2+1$
40:$x^6+x^5+x^3+x$	104:x^3+x^2+1	168:$x^7+x^6+x^5+x^4+x^3+x^2$	232:$x^7+x^6+x^5+x^4+x^2+x+1$
41:$x^7+x^6+x^4+x^2$	105:x^4+x^3+x	169:$x^7+x^6+x^5+x^2+1$	233:$x^7+x^6+x^5+x^4+x+1$
42:$x^7+x^5+x^4+x^2+1$	106:$x^5+x^4+x^2$	170:$x^7+x^6+x^4+x^2+x+1$	234:$x^7+x^6+x^5+x^4+x^3+x+1$
43:$x^6+x^5+x^4+x^2+x+1$	107:$x^6+x^5+x^3$	171:$x^7+x^5+x^4+x+1$	235:$x^7+x^6+x^5+x^3+x+1$

(continued)

Table 12.1 (continued)

0:1	64:$x^6+x^4+x^3+x^2+x+1$	128:x^7+x^2+1	192:x^7+x
44:$x^7+x^6+x^5+x^3+x^2+x$	108:$x^7+x^6+x^4$	172:$x^6+x^5+x^4+x^3+x+1$	236:$x^7+x^6+x^3+x+1$
45:x^7+x^6+1	109:$x^7+x^5+x^4+x^3+x^2+1$	173:$x^7+x^6+x^5+x^4+x^2+x$	237:x^7+x^3+x+1
46:$x^7+x^4+x^3+x^2+x+1$	110:$x^6+x^5+x^2+x+1$	174:$x^7+x^6+x^5+x^4+1$	238:x^3+x+1
47:x^5+x+1	111:$x^7+x^6+x^3+x^2+x$	175:$x^7+x^6+x^5+x^4+x^3+x^2+x+1$	239:x^4+x^2+x
48:x^6+x^2+x	112:x^7+1	176:$x^7+x^6+x^5+x+1$	240:$x^5+x^3+x^2$
49:$x^7+x^3+x^2$	113:$x^4+x^3+x^2+x+1$	177:$x^7+x^6+x^4+x^3+x+1$	241:$x^6+x^4+x^3$
50:x^2+1	114:$x^5+x^4+x^3+x^2+x$	178:$x^7+x^5+x^3+x+1$	242:$x^7+x^5+x^4$
51:x^3+x	115:$x^6+x^5+x^4+x^3+x^2$	179:x^6+x^3+x+1	243:$x^6+x^5+x^4+x^3+x^2+1$
52:x^4+x^2	116:$x^7+x^6+x^5+x^4+x^3$	180:$x^7+x^4+x^2+x$	244:$x^7+x^6+x^5+x^4+x^3+x$
53:x^5+x^3	117:$x^7+x^6+x^5+x^3+x^2+1$	181:x^5+x^4+1	245:$x^7+x^6+x^5+x^3+1$
54:x^6+x^4	118:$x^7+x^6+x^2+x+1$	182:x^6+x^5+x	246:$x^7+x^6+x^3+x^2+x+1$
55:x^7+x^5	119:x^7+x^4+x+1	183:$x^7+x^6+x^2$	247:x^7+x+1
56:$x^6+x^4+x^3+x^2+1$	120:$x^5+x^4+x^3+x+1$	184:$x^7+x^4+x^2+1$	248:x^4+x^3+x+1
57:$x^7+x^5+x^4+x^3+x$	121:$x^6+x^5+x^4+x^2+x$	185:$x^5+x^4+x^2+x+1$	249:$x^5+x^4+x^2+x$
58:$x^6+x^5+x^3+1$	122:$x^7+x^6+x^5+x^3+x^2$	186:$x^6+x^5+x^3+x^2+x$	250:$x^6+x^5+x^3+x^2$
59:$x^7+x^6+x^4+x$	123:$x^7+x^6+x^2+1$	187:$x^7+x^6+x^4+x^3+x^2$	251:$x^7+x^6+x^4+x^3$
60:$x^7+x^5+x^4+x^3+1$	124:$x^7+x^4+x^2+x+1$	188:$x^7+x^5+x^2+1$	252:$x^7+x^5+x^3+x^2+1$
61:$x^6+x^5+x^3+x^2+x+1$	125:x^5+x^4+x+1	189:$x^6+x^4+x^2+x+1$	253:x^6+x^2+x+1
62:$x^7+x^6+x^4+x^3+x^2+x$	126:$x^6+x^5+x^2+x$	190:$x^7+x^5+x^3+x^2+x$	254:$x^7+x^3+x^2+x$
63:x^7+x^5+1	127:$x^7+x^6+x^3+x^2$	191:x^6+1	255:1

12.2.2 A Real Example: GF(2^{233})

As a real example, we are going to detail the implementation of the different operations (multiplication, exponentiation and inversion) involved in a discrete logarithm based cryptosystem over the field GF(2^{233})$\{x^{233}+x^{74}+1\}$, one of the recommended in the FIPS 186-3 standard [5].

Using a standard base and assuming both operands available in parallel, multiplication can be implemented using the combinational circuit shown in Fig. 5.4, with 233 × 233 cells. Each one of the 233 cells in the first column needs one **OR** gate, the rest of the 232 cells in the first row needs one **AND** gate, and one **XOR** gate each one. There are 231 rows with $f_i=0$ needing one **AND** gate and one XOR gate, and 2 rows with $f_i=1$ needing one **AND** gate and two **XOR** gates. Summarizing, the cellular circuit will has 233 × 233 **AND** gates and 232 × 235 **XOR** gates. Multiplication could be also implemented with one or two of the operands in serial, being useful the circuits described in Sect. 5.3.

Exponentiation can be completed using any of the circuits in Fig. 5.14. In both of them, squaring is required additionally to multiplying. As shown in Example 5.15, a squarer can be built with 153 2-input **XOR** gates.

With respect to inversion, in Sect. 5.8 two different procedures are detailed: it can be reduced to an exponentiation, or it can be computed by means successive differences.

12.3 Elliptic Curve Cryptosystems

Appendix C shows how public key cryptography using elliptic curves is based on the discrete logarithm problem over elliptic curves.

12.3.1 Fundamentals

For using elliptic curves in cryptography, first a Galois Field GF(q), is chosen, with q a prime number or 2^n (being q or n large enough). Then, a public elliptic curve C defined over GF and an r-order point P in the curve are established. Each user selects randomly an integer k lower than r, that will be its secret key. For interchanging messages, the product of its secret key by the point P is made public, assuming that it is not possible computing k from the knowledge of P and kP in a reasonable amount of time.

When Ana (with secret key a) wants to send the message M, consisting of two components in GF, M = (m_1, m_2), $m_1, m_2 \in$ GF, to Carmen (with secret key b), first they must compute the products aP and bP, respectively, making them public. Note that Ana and Carmen knows aP y bP, and both of them, but only them, can compute

abP $= (x, y)$, because this calculus involves their secret key. Assuming $x, y \neq 0$ (if not, new secret keys are generated), the message being sent by Ana has two elements in GF, (xm_1, ym_2). For decrypting the message, Carmen has to divide (or multiply by the inverse) by abP $= (x, y)$, giving M $= (m_1, m_2)$.

FIPS 186-3 [5]standard recommends using five elliptic curves built over prime fields, with a number of elements given by generalized Mersenne number, as shown in Table 6.1. Also, elliptic curves over binary fields with the following primitive polynomials are recommended:

$$t^{163} + t^7 + t^6 + t^3 + 1$$
$$t^{233} + t^{74} + 1$$
$$t^{283} + t^{12} + t^7 + t^5 + 1$$
$$t^{409} + t^{87} + 1$$
$$t^{571} + t^{10} + t^5 + t^2 + 1$$

SEC2 [7] standard recommends (in addition to those given in Table 6.1) using some elliptic curves built over prime fields, with q values being:

$$q_{112} = (2^{128}-3)/76,439$$
$$q_{128} = 2^{128}-2^{97}-1$$
$$q_{160a} = 2^{160}-2^{31}-1$$
$$q_{160b} = 2^{160}-2^{32}-2^{14}-2^{12}-2^{9}-2^{8}-2^{7}-2^{3}-2^{2}-1$$
$$q_{192} = 2^{192}-2^{32}-2^{12}-2^{8}-2^{7}-2^{6}-2^{3}-1$$
$$q_{224} = 2^{224}-2^{32}-2^{12}-2^{11}-2^{9}-2^{7}-2^{4}-2-1$$
$$q_{256} = 2^{256}-2^{32}-2^{9}-2^{8}-2^{7}-2^{6}-2^{4}-1$$

and where the q subindex denotes the number of bits required for representing q. This standard also recommends using elliptic curves over binary fields defined from the following primitive polynomials (additionally to those recommended in FIPS 186-3 standard):

$$t^{113} + t^9 + 1$$
$$t^{131} + t^8 + t^3 + t^2 + 1$$
$$t^{193} + t^{15} + 1$$
$$t^{239} + t^{158} + 1$$

The security depends on the order of the subgroup generated from the point P. It is considered [5] that if the subgroup generated from the point P is given using more than 160 bits, it is practically impossible to solve the problem of the discrete logarithm in elliptic curves in an admissible time.

Example 12.2 As a simple elliptic curve cryptosystem example, the elliptic curve $Y^2 + XY = X^3 + X^2 + 1$, defined over the Galois field $GF(2^9)\{x^9 + x^4 + 1\}$, and the point of the curve P = (469, 50) will be considered. In this group, each X or Y coordinate is a polynomial of degree 8 which is represented by its decimal value: $(X, Y) = (469, 50) = (x^8 + x^7 + x^6 + x^4 + x^2 + 1, x^5 + x^4 + x)$. In this case, the order of the subgroup generated from (469, 50) is 259, and the multiples of P can be derived, as shown in Table 12.2.

Now, for Ana sending to Carmen the message $M = (m_1, m_2) = (010110101, 101101010) = (x^7 + x^5 + x^4 + x^2 + 1, x^8 + x^6 + x^5 + x^3 + x)$, the two random secret keys must be generated.

Thus, Ana and Carmen randomly and independently generate two number lower than 259, resulting 113 for Ana, and 85 for Carmen. Then, the public key of Ana is 113P, and the public key of Carmen is 85P. Using Table 12.2:

$$113P = (25, 215) = (000011001,\ 011010111)$$
$$= \left(x^4 + x^3 + 1, x^7 + x^6 + x^4 + x^2 + x + 1\right)$$

$$85P = (244, 498)$$
$$= (011110100,\ 111110010) = \left(x^7 + x^6 + x^5 + x^4 + x^2, x^8 + x^7 + x^6 + x^5 + x^4 + x\right)$$

Note that Ana can compute:

$$113(85P) = 113(244{,}498)$$
$$= (327{,}106) = \left(x^8 + x^6 + x^2 + x + 1, x^6 + x^5 + x^3 + x\right) = (q_1, q_2)$$

And

$$M_e = M(113(85P)) = (m_1 q_1, m_2 q_2)$$
$$= \left\{\left(x^7 + x^5 + x^4 + x^2 + 1\right) \cdot \left(x^8 + x^6 + x^2 + x + 1\right), \left(x^8 + x^6 + x^5 + x^3 + x\right) \cdot \left(x^6 + x^5 + x^3 + x\right)\right\}$$
$$= \left(x^8 + x^7 + x^6 + x^5 + x^4 + x^2, x^8 + x^7 + x^4 + x^3 + x\right) = (111110100,\ 110011010)$$

Resulting the encrypted message $M_e = (111110100, 110011010)$ which Ana sends to Carmen.

Carmen can compute

$$85(113P) = (327,\ 106) = \left(x^8 + x^6 + x^2 + x + 1, x^6 + x^5 + x^3 + x\right) = (q_1, q_2),$$

and its inverse:

$$\left(q_1^{-1},\ q_2^{-1}\right) = \left(x^5 + x^3 + x + 1, x^8 + x^7 + x^5 + x^3 + x^2\right)$$

Table 12.2 Multiples of (469, 50) over GF(2^9)$\{x^9 + x^4 + 1\}$

1: (469, 50)	45: (32, 219)	89: (331, 350)	133: (465, 315)	177: (335, 416)	221: (271, 439)
2: (34, 478)	46: (364, 294)	90: (422, 8)	134: (110, 335)	178: (292, 465)	222: (29, 0)
3: (160, 434)	47: (312, 330)	91: (434, 445)	135: (93, 368)	179: (347, 272)	223: (506, 476)
4: (195, 131)	48: (374, 233)	92: (38, 297)	136: (378, 218)	180: (174, 342)	224: (302, 436)
5: (438, 208)	49: (339, 336)	93: (352, 500)	137: (461, 36)	181: (508, 193)	225: (205, 324)
6: (473, 183)	50: (9, 273)	94: (98, 187)	138: (19, 467)	182: (329, 332)	226: (124, 34)
7: (193, 332)	51: (77, 135)	95: (449, 32)	139: (151, 279)	183: (42, 459)	227: (500, 107)
8: (71, 287)	52: (254, 231)	96: (310, 304)	140: (376, 409)	184: (188, 52)	228: (65, 453)
9: (482, 428)	53: (23, 135)	97: (27, 94)	141: (304, 121)	185: (333, 0)	229: (385, 183)
10: (232, 180)	54: (265, 205)	98: (380, 223)	142: (287, 294)	186: (306, 215)	230: (451, 441)
11: (486, 491)	55: (308, 402)	99: (248, 461)	143: (370, 441)	187: (147, 443)	231: (259, 336)
12: (75, 81)	56: (178, 373)	100: (246, 394)	144: (213, 507)	188: (114, 340)	232: (162, 428)
13: (498, 486)	57: (401, 242)	101: (430, 348)	145: (176, 441)	189: (393, 197)	233: (201, 91)
14: (279, 104)	58: (219, 472)	102: (283, 54)	146: (25, 206)	190: (112, 250)	234: (395, 208)
15: (217, 36)	59: (325, 476)	103: (362, 406)	147: (180, 168)	191: (457, 207)	235: (89, 238)
16: (131, 415)	60: (502, 7)	104: (91, 220)	148: (337, 337)	192: (73, 427)	236: (190, 476)
17: (316, 262)	61: (463, 142)	105: (318, 492)	149: (424, 282)	193: (426, 479)	237: (327, 301)
18: (321, 428)	62: (399, 318)	106: (133, 109)	150: (157, 268)	194: (145, 339)	238: (81, 336)
19: (490, 481)	63: (184, 28)	107: (277, 36)	151: (475, 44)	195: (275, 406)	239: (60, 236)
20: (60, 208)	64: (275, 133)	108: (475, 503)	152: (277, 305)	196: (184, 164)	240: (490, 11)

(continued)

Table 12.2 (continued)

1: (469, 50)	45: (32, 219)	89: (331, 350)	133: (465, 315)	177: (335, 416)	221: (271, 439)
21: (81, 257)	65: (145, 450)	109: (157, 401)	153: (133, 232)	197: (399, 177)	241: (321, 237)
22: (327, 106)	66: (426, 177)	110: (424, 178)	154: (318, 210)	198: (463, 321)	242: (316, 58)
23: (190, 354)	67: (73, 482)	111: (337, 0)	155: (91, 135)	199: (502, 497)	243: (131, 284)
24: (89, 183)	68: (457, 262)	112: (180, 28)	156: (362, 252)	200: (325, 153)	244: (217, 253)
25: (395, 347)	69: (112, 138)	113: (25, 215)	157: (283, 301)	201: (219, 259)	245: (279, 383)
26: (201, 146)	70: (393, 332)	114: (176, 265)	158: (430, 242)	202: (401, 355)	246: (498, 20)
27: (162, 270)	71: (114, 294)	115: (213, 302)	159: (246, 380)	203: (178, 455)	247: (75, 26)
28: (259, 83)	72: (147, 296)	116: (370, 203)	160: (248, 309)	204: (308, 166)	248: (486, 13)
29: (451, 122)	73: (306, 485)	117: (287, 57)	161: (380, 419)	205: (265, 452)	249: (232, 92)
30: (385, 310)	74: (333, 333)	118: (304, 329)	162: (27, 69)	206: (23, 144)	250: (482, 78)
31: (65, 388)	75: (188, 136)	119: (376, 225)	163: (310, 6)	207: (254, 25)	251: (71, 344)
32: (500, 415)	76: (42, 481)	120: (151, 384)	164: (449, 481)	208: (77, 202)	252: (193, 397)
33: (124,94)	77: (329, 5)	121: (19, 448)	165: (98, 217)	209: (9, 280)	253: (473, 366)
34: (205, 393)	78: (508, 317)	122: (461, 489)	166: (352, 148)	210: (339, 3)	254: (438, 358)
35: (302, 154)	79: (174, 504)	123: (378, 416)	167: (38, 271)	211: (374, 415)	255: (195, 64)
36: (506, 38)	80: (347, 75)	124: (93, 301)	168: (434, 15)	212: (312, 114)	256: (160, 274)
37: (29, 29)	81: (292, 245)	125: (110, 289)	169: (422, 430)	213: (364, 74)	257: (34, 508)
38: (271, 184)	82: (335, 239)	126: (465, 234)	170: (331, 21)	214: (32, 251)	258: (469, 487)
39: (120, 494)	83: (323, 113)	127: (261, 125)	171: (54, 145)	215: (62, 204)	259: ∞
40: (298, 509)	84: (13, 28)	128: (137, 108)	172: (504, 50)	216: (40, 101)	

(continued)

Table 12.2 (continued)

1: (469, 50)	45: (32, 219)	89: (331, 350)	133: (465, 315)	177: (335, 416)	221: (271, 439)
41: (52, 404)	85: (244, 498)	129: (102, 94)	173: (44, 30)	217: (467, 85)	
42: (467, 390)	86: (44, 50)	130: (102, 56)	174: (244, 262)	218: (52, 416)	
43: (40, 77)	87: (504, 458)	131: (137, 229)	175: (13, 17)	219: (298, 215)	
44: (62, 242)	88: (54, 167)	132: (261, 376)	176: (323, 306)	220: (120, 406)	

Thus, for decrypting the message received from Ana, can compute:

$$M = (m_1 q_1 q_1^{-1}, m_2 q_2 q_2^{-1}) = \begin{pmatrix} \left(x^8 + x^7 + x^6 + x^5 + x^4 + x^2\right) \cdot \left(x^5 + x^3 + x + 1\right), \\ \left(x^8 + x^7 + x^4 + x^3 + x\right) \cdot \left(x^8 + x^7 + x^5 + x^3 + x^2\right) \end{pmatrix}$$

$$= \left(x^7 + x^5 + x^4 + x^2 + 1, x^8 + x^6 + x^5 + x^3 + x\right)$$

recovering in this way the original message. ■

The different operations performed by the communicants must be completed on elliptic curves defined over a Galois field. Once fixed the field, the elliptic curve, and point of the curve, these operations are the same when using discrete logarithms: randomly selecting an integer, exponentiation, multiplication and inversion.

12.3.2 A Real Example: GF($2^{192} - 2^{64} - 1$)

In this section, a real example is presented, detailing the implementation of the different operations (addition/subtraction, exponentiation and inversion) involved for building a cryptosystem based on elliptic curves over the field GF($2^{192} - 2^{64} - 1$) (one of the fields recommended in the standard FIPS 186-3 [5]).

Addition can be calculated using a modification of the circuit in Fig. 3.2b. In fact, given that $M = 2^{192} - 2^{64} - 1$, $2^n - M = 2^{192} - M = 2^{64} + 1$, it results the circuit in Fig. 12.1a, containing two binary adders for 192 bits data. Other solutions can be used if the size of the circuit results excessive, at the cost of increasing the time needed for completing the calculus.

Subtraction can be computed using the circuit in Fig. 3.2c, which results in the one presented in Fig. 12.1b taking into account that M has 191 ones and one zero in position 64 in this case. Again, other solutions are possible with lower resources at the cost of performance.

Fig. 12.1 $M = 2^{192} - 2^{64} - 1$. **a** Adder. **b** Subtracter

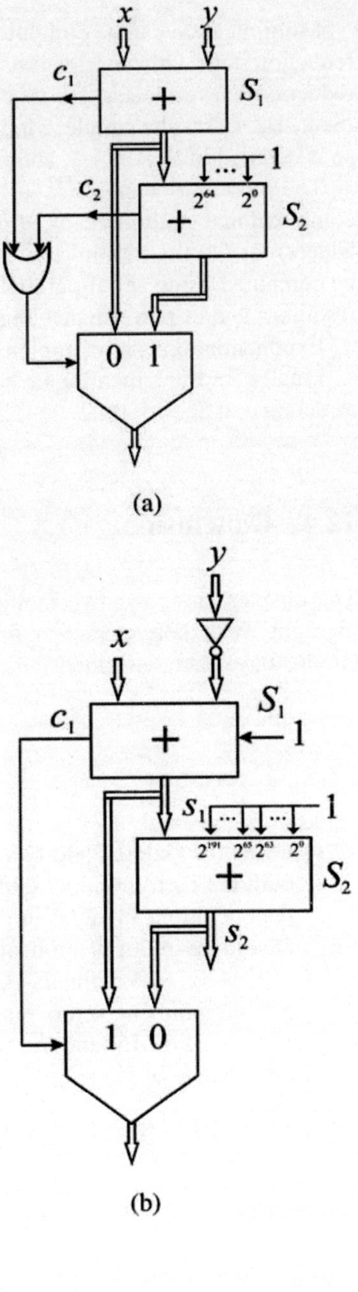

Multiplication can be computed using a binary multiplier followed by a modular reduction stage. Other option is using a Montgomery multiplier, avoiding the modular reduction. Nevertheless, for this value of M, the multiplicative modular reduction (Sect. 1.2.4), results simple. Modular reduction of the product $P = p_{382}2^{382} + \ldots + p_0$ is detailed in Sect. 6.1.1, and only an adder is required.

The product $P = p_{382}2^{382} + \ldots + p_0$, can be computed using a 192×192 bits combinational multiplier, or 96 bits multipliers (together with the corresponding adders), or 64 bits multipliers, etc., as detailed in Sect. 2.4. The product can also be computed using serial-parallel multipliers. The selected solution will depend on resources and/or performance required.

Exponentiation can be implemented using multipliers, as detailed in Sect. 2.4.4.

Finally, for implementing inversion, the modified Euclides algorithm can be used, as detailed in Sect. 1.1.1.2.

12.4 Conclusion

This chapter has shown two examples of Galois fields applications, oriented to cryptography. With these examples, finishes this book, where the most relevant algebraic circuits have been described.

12.5 Exercises

12.1 Using the Galois Field GF(29)$\{x^9 + x^4 + 1\}$ and x as the generator element, built the corresponding table of logarithms. Select two secret keys and detail the operations involves in the interchange of the massage 100011011.

12.2 Obtain the multiplication table for GF(37) and the points of the elliptic curve $y^2 = x^3 + 3x + 15$ defined in GF(37). With 37 values all alphanumeric and some orthographic characters may be codified. Using GF(37), the elliptic curve $y^2 = x^3 + 3x + 15$, and the point (5, 9), design a simple cryptosystem for the interchange of plane text.

References

1. Baldoni, M.W., Ciliberto, C., Piacentini Cattaneo, G.M.: Elementary Number Theory, Cryptography and Codes. Springer, Berlin (2009)
2. Diffie, W., Hellman, M.: New directions in cryptography. IEEE Trans. Inf. Theory **22**, 644–654 (1976)
3. ElGamal, T.: A public key cryptosystem and a signature scheme based on discrete logaritms. IEEE Trans. Inf. Theor. **31**, 472–496 (1985)

4. IEEE Standard Specifications for Public-Key Cryptography. IEEE Std. 1363–2000
5. National Institute of Standards and Technology: FIPS 186-4, Digital signature standard (DSS), Gaithersburg, MD, July 2013
6. Singh, S.: The Code Book: The Science of Secrecy from Ancient Egypt to Quantum Cryptography. Anchor Books, New York (2000)
7. Standars for Efficient Cryptography, SEC2: Recommended Elliptic Curve Domain Parameters, v1.0, 20 Sept 2000. Available at http://www.secg.org/

Appendix A
Finite or Galois Fields

Finite or Galois fields are used in the design and in the interpretation of the operation of some arithmetic and algebraic circuits. The axioms and the main properties of the finite or Galois fields are studied in this Appendix, which is oriented to have an immediate reference, excluding any demonstrations. Any of the texts listed in the Bibliography (Sect. A.5) or any other of the many that exits on these issues can be consulted in order to have a more detailed approach of the addressed issues.

A.1 General Properties

A set of axioms defining a field as well as Theorems of interest for its application to the design of circuits are enunciated bellow.

A.1.1 Axioms

Given a set of elements, C, the axioms to define a field used here are the following one:

I. **Internal laws of composition**

Two internal laws of composition, ⊕ *(operator* ***EXOR****, or addition) and* ⊗ *(operator* ***AND*** *or product) are defined in* C*, being* C *closed for the same:*

$$\forall X, y \in \mathbf{C} \quad \text{(a)} \quad x \oplus y \in \mathbf{C}$$
$$\text{(b)} \quad x \oplus y \in \mathbf{C}$$

Often the symbol ⊗ is replaced by • or deleted, writing $x \bullet y$ or just xy instead of $x \otimes y$. Also, when there is no possible confusion, the symbol ⊕ is replaced by $+$.

A. Lloris Ruiz et al., *Arithmetic and Algebraic Circuits*, Intelligent Systems Reference Library 201, https://doi.org/10.1007/978-3-030-67266-9

II. Commutativity of the internal laws of composition

Both internal laws of composition are commutative:

$$\forall x, y \in \mathbf{C}, \quad \begin{aligned} &(a) \quad x \oplus y = y \oplus x \\ &(b) \quad x \otimes y = y \oplus x \end{aligned}$$

III. Associativity of the internal laws of composition

Both internal laws of composition are associative:

$$\forall x, y, z \in \mathbf{C}, \quad \begin{aligned} &(a) \quad (x \oplus y) \oplus z = x \oplus (y \oplus z) \\ &(b) \quad (x \otimes y) \otimes z = x \otimes (y \otimes z) \end{aligned}$$

Usually parentheses are deleted and it will be written $x \oplus y \oplus z$ or $x \otimes y \otimes z$, so that, thanks to the associativity, the internal laws of composition are not restricted to be binary operators, but can be applied to any number of operands.

IV. Distributivity of the internal laws of composition

The product is distributive over addition:

$$\forall x, y, z \in \mathbf{C}, \quad x \otimes (y \oplus z) = (x \otimes y) \oplus (x \otimes z)$$

V. Neutral elements

There are neutral elements for both internal laws of composition, which are denominated 0 (neutral element for addition) and 1 (neutral element for the product):

(a) $\exists 0 \in \mathbf{C} | \forall x \in \mathbf{C}, x \oplus 0 = 0 \oplus x = x$

(b) $\exists 1 \in \mathbf{C} | \forall x \in \mathbf{C}, x \oplus 1 = 1 \oplus x = x$

VI. Opposite of each element

Every element of C has its opposite, which is represented as $-x$:

$$\forall x \in \mathbf{C}, \exists -x \in \mathbf{C} | x \oplus -x = 0$$

VII. Inverse of each nonzero element

Every element of C different from 0 has its inverse, which is represented as x^{-1}:

$$\forall x \in \mathbf{C}, x \neq 0, \exists x^{-1} \in \mathbf{C} | x \otimes x^{-1} = 1$$

VIII. 0 and 1 are different

The elements 0 and 1 are different.

Obviously, this axiom VIII also establishes that the minimum number of elements in C is two. Nothing is said in the axioms about the total number of elements or its type. There are fields with an infinite number of elements and there are others fields with a finite number of elements; these last are also known as Galois fields, in honor of the French mathematician Evariste Galois. The following sections are devoted to the Galois fields.

Under the axiom II, all the considered fields in this text are commutative.

A.1.2 Theorems

Some theorems and corollaries valids for any field are described at the following, not including its demonstration:

Theorem 1

(a) *The neutral element for addition, 0, is unique.*
(b) *The neutral element for the product, 1, is unique.*

Theorem 2 *The opposite of every element is unique.*

Theorem 3 *The inverse of each nonzero element is unique.*

Theorem 4 *The inverse of* x^{-1} *is x:*

$$\left(x^{-1}\right)^{-1} = x$$

Theorem 5 *For the opposite of the addition* $x \oplus y$ *it holds that:*

$$-(x \oplus y) = (-x) \oplus (-y)$$

Theorem 6 *There are no zero divisors. It means*:

$$xy = 0 \Rightarrow x = 0 \text{ or } y = 0$$

Theorem 7 *For the inverse of the product* $x \otimes y$ *it holds*:

$$(x \otimes y)^{-1} = x^{-1} \otimes y^{-1}$$

Theorem 8 *The inverse of 1 is 1.*

Theorem 9 *The cancellation law for multiplication is verified*:

$$x \otimes y = x \otimes z \Rightarrow y = z \text{ or } x = 0$$

Table A.1 Operation $\oplus$ in GF(2)

$\oplus$	0	1
0	0	1
1	1	0

Table A.2 Operation $\otimes$ in GF(2)

$\otimes$	0	1
0	0	0
1	0	1

Theorem 10 *The following equalities are met for the exponents*:

$$x^m \otimes x^n = x^{m+n}$$
$$X^m \otimes X^{-n} = X^{m-n}$$
$$\left(X^m\right)^n = X^{mn}$$

A.2 GF(2)

The simplest example of a Galois field is the one which only contains two elements, i.e., in this case $C = \{0, 1\}$. In this field, called the *Galois field of order* 2 or GF(2), the operations $\oplus$ and $\otimes$ are defined by Tables A.1 and A.2, and correspond to the addition and the product modulo 2. Furthermore, from the definitions of the operations $\oplus$ and $\otimes$ it is immediate that in GF(2) the inverse of x and the opposite of x are x itself:

$$x^{-1} = x;\ -x = x$$

In GF(2) the following equations are also met:

$$x \oplus x = 0$$
$$x \otimes x = x^2 = x$$

Any switching function may be expressed using the operators in GF(2), so called as Reed-Muller (**EXOR-AND** logic) expansion. On the other hand, in the processing of digital systems the Boolean algebra (logic **AND-OR**) is used, in which the used operators are **AND** (•), **OR** (+) and **NOT** ($^{-}$). Any switching function can be also developed based on these operators. It is logical to expect that these two sets of operators have a simple relationship, so it easy to move from one development to another. Specifically, to move from the development **AND-EXOR** to the development **AND-OR**, the following substitutions can be used:

Table A.3 Operation ⊕ in GF(5)

⊕	0	1	2	3	4
0	0	1	2	3	4
1	1	2	3	4	0
2	2	3	4	0	1
3	3	4	0	1	2
4	4	0	1	2	3

Table A.4 Operation ⊗ in GF(5)

⊗	1	2	3	4
1	1	2	3	4
2	2	4	1	3
3	3	1	4	2
4	4	3	2	1

$$x \otimes y = x \cdot y$$
$$x \oplus y = x \cdot \bar{y} + \bar{x} \cdot y$$

To transform an expression **AND-OR** into the **AND-EXOR** equivalent expression, the following identities can be used:

$$x \cdot y = x \otimes y$$
$$x + y = (x \cdot y) \oplus x \oplus y$$
$$\bar{x} = x \oplus 1$$

A.3 GF(p)

Let suppose p an integer number, C = {0, 1, ..., $p - 1$}, and the operations ⊕ and ⊗ are defined as the addition and the product modulo m, respectively. For example, in Tables A.3 and A.4 are the addition and product module 5. It is straightforward to check that $C_5 = \{0, 1, 2, 3, 4\}$, with the operations defined in Tables A.3 and A.4, is a Galois field, which is called as GF(5). In this case the opposite and the inverse of each element (remember that 0 has not inverse) are given in Tables A.5 and A.6.

It is easy to see that if and only if p is prime (or a power of a prime number, as shown in the next section), GF(p), defined as done in the previous paragraph, is a Galois field. It is easy to show that if there is another finite field with p elements, then it is isomorphic to GF(p). In general, it is proved that two finite fields are isomorphic if they have the same number of elements.

Table A.5 Opposites in GF(5)

	Opposite
0	0
1	4
2	3
3	2
4	1

Table A.6 Inverses in GF(5)

	Inverse
1	1
2	3
3	2
4	4

For each element $e \neq 0$, of any Galois field, its **order** is defined as the smallest integer n such that $e^n = 1$, and it is indicated as ord(e) = n. For example, in GF(5), it is immediately to check that ord(1) = 1, ord(2) = 4, ord(3) = 4, ord(4) = 2.

It is said that an element e of GF(p) is **primitive** if ord(e) = $p - 1$. A primitive element is also known as a **generator**, because the successive powers of a primitive element generate all nonzero elements of GF(p). In GF(5), 2 and 3 are primitive elements, resulting:

$$2^1 = 2, 2^2 = 4, 2^3 = 3, 2^4 = 1$$
$$3^1 = 3, 3^2 = 4, 3^3 = 2, 3^4 = 1$$

In any Galois field there is an element 0 (one and only one), an element 1 (one and only one) and primitive elements (at least one).

The **characteristic** of a Galois field is the number of different elements that can be obtained by adding the unit element with itself as many times as desired. It is straightforward to check that for a Galois field GF(p), its characteristic is p.

For any element e of G(p) the following equality, due to Fermat, is verified:

$$e^{p-1} = 1, \quad \text{or} \quad e^p = e$$

A.4 GF(p^m)

Given a Galois field GF(p), Galois fields with p^m elements GF(p^m) can be built as an extension of that field, being m any integer greater than 1, as shown in Appendix B for the case of polynomials on GF(2). GF(p^m) can be considered as a vector space of dimension m over GF(p), defining the addition in the vector space as the ordinary addition in GF(p^m), and the scalar product as the product of the elements in GF(p^m) by the elements of GF(p). As GF(p^m) is a vector space, different bases can be used for each GF(p^m).

The Galois field GF(p^m) can be defined using polynomials with coefficients belonging to GF(p), as it is detailed in Appendix B.

It is shown that the characteristic of the Galois field GF(p^m) is p. Also, if GF is a finite field with characteristic p, then the number of elements is of the form p^m, $m = 1, 2, \ldots$

For any Galois field of characteristic p, the following expression is verified for any elements e_1, …, e_n, of that field:

$$(e_1 + \cdots + e_n)_n^p = (e_1)^p + \cdots + (e_n)^p$$

For any prime number p and any integer m, it is shown that there is a Galois field GF(p^m).

For any element e of G(p^m) it is verified that $e^{p^m} = e$.

A.5 References

1. Garret, P. B.: **Abstract Algebra**, Chapman & Hall, 2008.
2. Howie, J.M.: **Field and Galois Theory**, Springer, 2006.
3. Lidl, R.; Niederreiter, H.: **Introduction to Finite Fields and Their Applications**, Cambridge University Press, 1986.
4. Stewart, I.: **Galois Theory**, Chapman and Hall, 1989.

Appendix B
Polynomial Algebra

Several applications of digital systems, such as information codification, cryptography, o digital circuit test, make use of the properties of polynomials over GF(2) and over GF(p). This Appendix summarizes the main properties of these polynomials, without showing demonstrations for most cases. The main objective is to have a close reference, as well as unifying the nomenclature. For a more detailed presentation and for inspecting the demonstrations not provided here, the list of references at the end of the Appendix may be used. The Appendix will start showing some general properties of polynomials, later particularizing to polynomials over GF(2) and over GF(p). After that, the Galois fields GF(2^m) are studied in detail, finally analyzing the Galois fields GF(p^m) and GF((p^m)^n).

B.1 General Properties

If n is a non-negative integer, a polynomial $P(x)$ in the variable x is:

$$P(x) = a_n x^n + a_{n-1} x^{n-1} + \cdots + a_1 x + a_0$$

where a_i are constants, called **coefficients**, not all null and belonging to a number field, for example, real numbers, complex numbers, or a numeric Galois field.

$P(x)$ is of **degree** n when $a_n \neq 0$; the degree of P will be represented as $g(P)$. If $a_n = 1$, the polynomial $P(x)$ is said to be **monic**.

A polynomial is defined by its coefficients. Concretely, $P(x) = \{a_i\} = (a_n, a_{n-1}, \ldots, a_1, a_0)$ will be used as definition, and the different polynomial operations may be defined depending on the coefficients, as it is done in the following. It is usual to assume $a_n \neq 0$, but it is also clear that as many coefficients $a_j = 0, j > n$ as it is desired may be added (for example, for equaling the lengths of two polynomials, if this was required for some polynomial operation).

The coefficient a_0 is called **independent term**, as it is not multiplied by x. It is obvious that when $a_0 = 0$ for $P(x)$, then $P(x) = xQ(x)$.

A. Lloris Ruiz et al., *Arithmetic and Algebraic Circuits*, Intelligent Systems Reference Library 201, https://doi.org/10.1007/978-3-030-67266-9

B.1.1 Polynomial Operations

Polynomial addition (subtraction): given the polynomials $P = \{a_i\}$ and $Q = \{b_i\}$, their addition $S = P + Q$ (subtraction $R = P - Q$) is defined as $S = \{a_i + b_i\}$ ($R = \{a_i - b_i\}$), where $a_i + b_i$, or $(a_i - b_i)$, is computed in the coefficient field.

Polynomial product: given the polynomials $P = \{a_i\}$ and $Q = \{b_i\}$, their product $M = PQ$ is defined as $M = \left\{c_i = \sum_{j+k=i} ajbk\right\}$, where $a_j b_k$ is computed in the coefficient field. It is verified that $g(M) = g(P) + g(Q)$. The cancellation law is verified for the polynomial product, i.e., if it is assumed that $P \neq 0$:

$$PQ = PR \mapsto Q = R$$

Polynomial division: given any polynomials P (dividend) and Q (divider), the division of P by Q is defined by the quotient, C, and the remainder, R:

$$P = QC + R$$

Imposing the restriction $g(R) < g(Q)$, C and R are unique. The relationship above for R can also be expressed as:

$$R = P \bmod Q, \text{ or also } R = |P|_Q$$

A polynomial P (dividend) is said to be divisible by another polynomial Q (divider) when there exists a third polynomial C (quotient) such as $P = QC$, i.e., it is an exact (null remainder) division, and Q is said to be divisor of P. Given any two polynomials, not always there exists an exact quotient. An **irreducible** polynomial is that with all its divisors being of degree zero.

Integer numbers m can be found for every polynomial $P(x)$, so $P(x)$ is a divider of the bynomial $x^m - 1$. Given e as the minimum of all of these integers m for a given $P(x)$; this value e is known as the **order of the polynomial** $P(x)$.

A **reducible** polynomial P has dividers (Q_1, ..., Q_s) whose degree is larger than zero. Given the set of all irreducible dividers of P (I_1, ..., I_t), P can always be expressed in a unique form as the product of all these irreducible polynomials:

$$P = c\Pi I_{i,j}, \quad j = 1, \ldots, t$$

Given the reducible polynomials P and Q, their **greatest common divisor** ($gcd(P, Q)$) is the monic polynomial of larger degree that divides both of them. If the greatest common divisor of two polynomials is of degree zero, **the polynomials** are said to be **relatively prime.**

The **Euclides algorithm** can be used for computing $gcd(P, Q)$ – v. [9], which is detailed below:

Algorithm B.1 *Euclides algorithm*

(1) Let $g(P) < g(Q)$. P is divided by Q:

$$P = QC_1 + R_1$$

(2) If $R_1 \neq 0$, the current divisor, Q, is divided by the current remainder, R_1:

$$Q = R_1C_2 + R_2$$

(3) Again, if $R_2 \neq 0$, the current divisor, R_1, is divided by the current remainder, R_2, and so on until a null remainder is obtained:

$$R_{k-1} = R_kC_{k+1}$$

(4) It results $gcd(P, Q) = gcd(Q, R_1) = gcd(R_1, R_2) = \cdots = gcd(R_k, 0) = R_k$.

End algorithm

It is obvious that the Euclides algorithm will find $gcd(P, Q)$ in a finite number of iterations, since the degree of the remainder is reduced in each division.

The computation of $gcd(P, Q)$ through successive divisions may have as main drawback the difficulties for implementing division. In this case, it must be taken into account that the Euclides algorithm can be also applied through successive subtractions. Given P and Q, $g(P) < g(Q)$, it is demonstrated that

$$gcd(P, Q) = gcd(P - Q, Q)$$

This equivalence is applied iteratively until it gets to a null difference, and the $gcd(P, Q)$ is the last non-null difference. $P - x^dQ$, with $d = g(P) - g(Q)$, can be used instead of $P - Q$ in order to speed up the computation process, i.e, it is applied $gcd(P, Q) = gcd(P - x^dQ, Q)$, and once more the process ends when it gest to a null difference.

It can be also easily demonstrated —*v.* [2]—that if $R = gcd(P, Q)$ there are two polynomials A and B such that:

$$R = AP + BQ$$

This decomposition, known as **extendend Euclides algorithm**, is useful for some calculations. Thus, in the following it will be shown how to obtain the polynomials A and B. In the computation of $gcd(P, Q)$ through successive divisions we have:

$$
\begin{aligned}
P &= QC_1 + R_1 \\
Q &= R_1C_2 + R_2 \\
R_1 &= R_2C_3 + R_3 \\
&\dots \\
R_{i-2} &= R_{-1}C_i + R_i \\
&\dots
\end{aligned}
$$

so the main term in this iteration is

$$R = R_{i-2} - R_{i-1}C_i$$

where $R_{-1} = P$ and $R_0 = Q$. The computation ends when $R_m = 0$, and $gcd(P, Q) = R_{m-1}$.

Assuming that R_i can be decomposed as R, i.e., that

$$R = A_iP + B_iQ$$

and substituting in the main term R_i results in:

$$
\begin{aligned}
R = R_{i-2} - R_{i-1}C_i &= (A_{i-2}P + B_{i-2}Q) - (A_{i-1}P + B_{i-1}Q)C_i \\
&= (A_{i-2} - A_{i-1}C_i)P + (B_{i-2} - B_{i-1}C_i)Q
\end{aligned}
$$

Thus, the main terms for the iterative computation of A_i and B_i are, respectively:

$$
\begin{aligned}
A_i &= A_{i-2} - A_{i-1}C_i \\
B_i &= B_{i-2} - B_{i-1}C_i
\end{aligned}
$$

where $A_{-1} = 1, A_0 = 0, B_{-1} = 0$ y $B_0 = 1$, as it is immediate to check. The computation of A and B requires as many iterations as the computation of $gcd(P, Q)$. In the way A_i y B_i have been defined, $A_iP + B_iQ$ is equal to the remainder generated in the iteration, for all iterations.

Defining the operation COC(R_1, R_2), which provides the quotient polynomial resulting from dividing the polynomial R_1 by the polynomial R_2, the following algorithm would provide the greatest common divisor of two polynomials $P(x)$ and $Q(x)$, as well as the polynomials A and B above:

Algorithm B.2

Algorithm for computing gcd(P, Q)

```
1) R1 ← P(x), R2 ← Q(x), A1 ← 1, A2 ← 0, B1 ← 0, B2 ← 1;
while R2 ≠ 0 do
    begin
        2) Q ← COC(R1, R2), TEMP ← R2;
        3) R2 ← R1 - QR2, R1 ← TEMP, TEMP ← A2;
        4) A2 ← A1 - QA2, A1 ← TEMP, TEMP ← B2;
        5) B2 ← B1 - QB2, B1 ← TEMP;
    end
```

End algorithm

After the execution of the Algorithm B.2 for computing *gcd(P, Q)*, polynomials A_1 and B_1 are stored in the registers A_1 and B_1, respectively. The greatest common divisor is stored in R_1, and can also be computed applying $gcd(P, Q) = A_1P + B_1Q$.

Example B.1 **Given $P(x) = x^8 + x^7 + x + 1$ and $Q(x) = x^5 + x + 1$, with coefficients in GF(2), compute *gcd(P, Q)*.**

Applying Euclides Algorithm B.2, through successive divisions, leads to:

$$x^8 + x^7 + x + 1 = (x^5 + x + 1)(x^3 + x^2) + x^4 + x^2 + x + 1 \quad (P = QC_1 + R_1)$$
$$x^5 + x + 1 = (x^4 + x^2 + x + 1)x + x^3 + x^2 + 1 \quad (Q = R_1C_2 + R_2)$$
$$x^4 + x^2 + x + 1 = (x^3 + x^2 + 1)(x + 1) \quad (R_1 = R_2C_3)$$

Since the remainder of the last division is null, $gcd(P, Q) = x^3 + x^2 + 1$.

The computations above yielded: $C_1 = x^3 + x^2$, $C_2 = x$, $C_3 = x + 1$. In the computation of A it holds:

$$A_1 = 1, A_0 = 0, A_1 = A_{-1} - A_2C_1 = 1, A_2 = A_2 - A_1C_2 = x$$

In the same way, in the computation of B it holds:

$$B_{-1} = 0, B_0 = 1, B_1 = B_1 - B_0C_1 = x^3 + x^2,$$
$$B_2 = B_0 - B_1C_2 = 1 + (x^3 + x^2)x = x^4 + x^3 + 1$$

It is immediate to check that $A_2P + B_2Q = R_2 = gcd(P, Q)$.

A new iteration in A and B leads to:

$$A_3 = A_1 - A_2C_3 = 1 + x(x + 1) = x^2 + x + 1$$
$$B_3 = B_1 - B_2C_3 = x^3 + x^2 + (x^4 + x^3 + 1)(x + 1)$$
$$= x^5 + x^2 + x + 1$$

It is easily checked that $A_3P + B_3Q = R_3 = 0$.

Table B.1 shows the contents of the different registers during the application of Algorithm B.1 to the polynomials above for computing *gcd(P, Q)*.

Table B.1 Algorithm for computing $gcd(P, Q)$

	Q	TEMP	R_2	R_1	A_1	A_2	B_1	B_2
1			x^5+x+1	x^8+x^7+x+1	1	0	0	1
2	x^3+x^2	x^5+x+1						
3		0	x^4+x^2+x+1	x^5+x+1				
4		1			0	1		
5							1	x^3+x^2
2	x	x^4+x^2+x+1						
3		1	x^3+x^2+1	x^4+x^2+x+1				
4		x^3+x^2			1	x		
5							x^3+x^2	x^4+x^3+1
2	$x+1$	x^3+x^2+1						
3		x	0	x^3+x^2+1				
4		x^4+x^3+1			x	x^2+x+1		
5							x^4+x^3+1	x^5+x^2+x+1

Now applying the Euclides algorithm through succesive subtractions leads to:

$$
\begin{aligned}
gcd(P, Q) &= gcd\left(x^8+x^7+x+1, x^5+x+1\right)\\
&= gcd\left(x^8+x^7+x+1-x^3\left(x^5+x+1\right), x^5+x+1\right)\\
&= gcd\left(x^7+x^4+x^3+x+1, x^5+x+1\right)\\
&= gcd\left(x^7+x^4+x^3+x+1-x^2\left(x^5+x+1\right), x^5+x+1\right)\\
&= gcd\left(x^4+x^2+x+1, x^5+x+1\right)\\
&= gcd\left(x^5+x+1, x^4+x^2+x+1\right)\\
&= gcd\left(x^5+x+1-x\left(x^4+x^2+x+1\right), x^4+x^2+x+1\right)\\
&= gcd\left(x^3+x^2+1, x^4+x^2+x+1\right)\\
&= gcd\left(x^4+x^2+x+1, x^3+x^2+1\right)\\
&= gcd\left(x^4+x^2+x+1-x\left(x^3+x^2+1\right), x^3+x^2+1\right)\\
&= gcd\left(x^3+x^2+1, x^3+x^2+1\right)\\
&= gcd\left(x^3+x^2+1-\left(x^3+x^2+1\right), x^3+x^2+1\right)
\end{aligned}
$$

$$= gcd\left(x^3 + x^2 + 1, 0\right) = x^3 + x^2 + 1.$$

Example B.2 **Given $P(x) = 5x^8 + 4x^7 + 3x^6 + x^4 + 5x^3 + x^2 + 2x + 5$ and $Q(x) = 3x^5 + x^3 + 3x + 4$, with coefficients in GF(7), compute $gcd(P, Q)$.**

Applying Euclides algorithm, through successive divisions, leads to:

$$\begin{aligned} & 5x^8 + 4x^7 + 3x^6 + x^4 + 5x^3 + x^2 + 2x + 5 \\ & = \left(3x^5 + x^3 + 3x + 4\right)\left(4x^3 + 6x^2 + 2x5\right) \\ & \quad + x^4 + x^3 + 6x^2 + 6 \quad (P = QC_1 + R_1) \\ & 3x^5 + x^3 + 3x + 4 = \left(x^4 + x^3 + 6x^2 + 6\right)(3x + 4) \\ & \quad + 4x^2 + 6x + 1 \quad (Q = R_1C_2 + R_2) \\ & x^4 + x^3 + 6x^2 + 6 = \left(4x^2 + 6x + 1\right) \\ & \quad \left(2x^2 + 6x + 6\right) \quad (R_1 = R_2C_3) \end{aligned}$$

Since the remainder of the last division is null, $gcd(P, Q) = 4x^2 + 6x + 1$.

The computations above yielded: $C_1 = 4x^3 + 6x^2 + 2x^5$, $C_2 = 3x + 4$, $C_3 = 2x^2 + 6x + 6$.

In the computation of A it holds:

$$A_{-1} = 1,\ A_0 = 0,\ A_1 = A_{-1} - A_0C_1 = 1,\ A_2 = A_0 - A_1C_2 = -(3x + 4) = 4x + 3$$

In the same way, in the computation of B it holds:

$$\begin{aligned} & B_{-1} = 0,\ B_0 = 1,\ B_1 = B_{-1} - B_0C_1 \\ & \quad = -\left(4x^3 + 6x^2 + 2x + 5\right) = 3x^3 + x^2 + 5x + 2 \\ & B_2 = B_0 - B_1C_2 = 1 + \left(4x^3 + 6x^2 + 2x + 5\right)(3x + 4) \\ & \quad = 5x^4 + 6x^3 + 2x^2 + 2x \end{aligned}$$

It is immediate to check that $A_2P + B_2Q = R_2 = gcd(P, Q)$.

A new iteration in A and B leads to:

$$\begin{aligned} & A_3 = A_1 - A_2C_3 = 1 + (3x + 4)\left(2x^2 + 6x + 6\right) \\ & \quad = 6x^3 + 5x^2 + 4 \\ & B_3 = B_1 - B_2C_3 = -\left(4x^3 + 6x^2 + 2x + 5\right) \\ & \quad - \left(5x^4 + 6x^3 + 2x^2 + 2x\right)\left(2x^2 + 6x + 6\right) \\ & \quad = 4x^6 + 5x^2 + 2 \end{aligned}$$

It is easily checked that $A_3P + B_3Q = R_3 = 0$.

Table B.2 shows the contents of the different registers during the application of Algorithm B.1 to the polynomials above for computing $gcd(P, Q)$.

Table B.2 Algorithm for computing $gcd(P, Q)$

	Q	TEMP	R_2	R_1	A_1	A_2	B_1	$B2$
1			$3x^5 + x^3 + 3x + 4$	$5x^8 + 4x^7 + 3x^6 + x^4 + 5x^3 + x^2 + 2x + 5$	1	0	0	1
2	$4x^3 + 6x^2 + 2x + 5$	$3x^5 + x^3 + 3x + 4$						
3		0	$x^4 + x^3 + 6x^2 + 6$	$3x^5 + x^3 + 3x + 4$				
4		1			0	1		
5							1	$3x^3 + x^2 + 5x + 2$
2	$3x + 4$	$x^4 + x^3 + 6x^2 + 6$						
3		1	$4x^2 + 6x + 1$	$x^4 + x^3 + 6x^2 + 6$				
4		$3x^3 + x^2 + 5x + 2$			1	$4x + 3$		
5							$3x^3 + x^2 + 5x + 2$	$5x^4 + 6x^3 + 2x^2 + 2x$
2	$2x^2 + 6x + 6$	$4x^2 + 6x + 1$						
3		$4x + 3$	0	$4x^2 + 6x + 1$				
4		$5x^4 + 6x^3 + 2x^2 + 2x$			$4x + 3$	$6x^3 + 5x^2 + 4$		
5							$5x^4 + 6x^3 + 2x^2 + 2x$	$4x^6 + 5x^2 + 2$

Applying the Euclides algorithm through successive subtractions leads to the same result above. ■

If P and Q are relatively prime, then $1 = AP + BQ$. In the case $AP = 0$, as it sometimes happens, then $BQ = 1$, i.e., $B = Q^{-1}$. Thus, it results in a procedure for computing the inverse of a polynomial $Q(x)$.

B.1.2 Congruence Relationship

The polynomials M and N are said to be congruent modulo the polynomial Q if:

$$|M|_Q = |N|_Q$$

The congruence relationship with respect to a given modulo Q is an equivalence relation and, therefore, all polynomials are classified into mutually exclusive classes, each of which may be represented by their remainder. If $g(Q) = p$, the representative of each equivalence class is the value 0 or a polynomial of degree less than p. These equivalence classes form a commutative algebra of dimension p over the field of the coefficients.

Let Q be a monic polynomial of degree n:

$$Q = x^n + \sum_{i=0}^{n-1} a_i x^i$$

It is obvious that:

$$|Q|_Q = 0$$

or also:

$$\left| x^n + \sum_{i=0}^{n-1} a_i x^i \right|_Q = 0$$

That can be written as:

$$|x^n|_Q + \left| \sum_{i=0}^{n-1} a_i x^i \right|_Q = 0$$

or:

$$|x^n|_Q = -\left| \sum_{i=0}^{n-1} a_i x^i \right|_Q \qquad \text{(B.1)}$$

So, in the modular algebra modulo $Q = x^n + \sum_{i=0}^{n-1} a_i x^i$, x^n can be substituted by the others summands of the polynomial Q, with negative sign (i.e. $-\sum_{i=0}^{n-1} a_i x^i$).

Table B.3 Polynomials over GF(2) of degree less than four

0	0	0000 (0)
1	1	0001 (1)
2	x	0010 (2)
3	$x+1$	0011 (3)
4	x^2	0100 (4)
5	x^2+1	0101 (5)
6	x^2+x	0110 (6)
7	x^2+x+1	0111 (7)
8	x^3	1000 (8)
9	x^3+1	1001 (9)
10	x^3+x	1010 (A)
11	x^3+x+1	1011 (B)
12	x^3+x^2	1100 (C)
13	x^3+x^2+1	1101 (D)
14	x^3+x^2+x	1110 (E)
15	x^3+x^2+x+1	1111 (F)

B.2 Polynomials Over GF(2)

If the coefficients of the polynomials belong to GF(2), each polynomial is given by a combination of zeros and ones. This binary information, like any other, can be given in parallel or in serial. When information is given in serial, usually the order of the coefficients is high to low (first a_n, last a_0), because of the requirements of the division.

The coefficients of polynomials over GF(2) are 0 or 1, so that there are 2^n polynomials of degree n (a_n have to be 1), and there are 2^n polynomials of degree less than n (that is, with $a_n = 0$). The 2^n polynomials of degree less than n can be represented by the 2^n possible combinations of zeros and ones, as is done in Table B.3 for the case $n = 4$. In the Table B.3, besides the binary coordinates of each polynomial, the corresponding hexadecimal value is given, in brackets, which can also be used to represent the polynomial.

Whether the information is given in serial or parallel, the operations with polynomials can be performed directly on the n-tuples of zeros and ones that represent them. For example, with parallel data, let suppose the polynomials $P(x) = (1, 0, 0, 1)$ and $Q(x) = (1, 0, 1, 0, 1)$. Equaling the lengths of $P(x)$ and $Q(x)$, it is immediate that:

$$P(x) + Q(x) = (0, 1, 0, 0, 1) + (1, 0, 1, 0, 1) = (1, 1, 1, 0, 0)$$

This addition can be made over GF(2), using binary adders as follows (without considering the carries):

$$\begin{array}{r} 01001 \\ +10101 \\ \hline 11100 \end{array}$$

The same result will be obtained for $P(x) - Q(x)$ and $P(x) + Q(x)$, since the addition and the subtraction over GF(2) is the same operation.

For the product of the polynomials given above, $P(x) \cdot Q(x)$, applying the definition it results:

$$\begin{aligned}
c_a &= a_0 b_0 \\
c_1 &= a_0 b_1 + a_1 b_0 \\
c_2 &= a_0 b_2 + a_1 b_1 + a_2 b_0 \\
c_3 &= a_0 b_3 + a_1 b_2 + a_2 b_1 + a_3 b_0 \\
c_4 &= a_0 b_4 + a_1 b_3 + a_2 b_2 + a_3 b_1 \\
c_5 &= a_1 b_4 + a_2 b_3 + a_3 b_2 \\
c_6 &= a_2 b_4 + a_3 b_3 \\
c_6 &= a_2 b_4 + a_3 b_3 \\
c_7 &= a_3 b_4
\end{aligned}$$

I.e.:

$$P(x) \cdot Q(x) = (1, 0, 0, 1) \cdot (1, 0, 1, 0, 1) = (1, 0, 1, 1, 1, 1, 0, 1)$$

It is immediate that for the product $P(x) \cdot Q(x)$, being $P(x) = a_{n\text{-}1}x^{n-1} + \cdots + a_0$ and $Q(x) = b_{n\text{-}1}x^{n-1} + \cdots + b_0$, the following matrix expression can be used:

$$\begin{bmatrix} c_0 \\ \cdots \\ c_{2n-1} \end{bmatrix} = \begin{bmatrix} a_0 & 0 & \cdots & 0 & 0 \\ a_1 & a_0 & \cdots & 0 & 0 \\ \cdots & \cdots & \cdots & \cdots & \cdots \\ a_{n-2} & a_{n-3} & \cdots & a_0 & 0 \\ a_{n-1} & a_{n-2} & \cdots & a_1 & a_0 \\ 0 & a_{n-1} & \cdots & a_2 & a_1 \\ \cdots & \cdots & \cdots & \cdots & \cdots \\ 0 & 0 & \cdots & a_{n-1} & a_{n-2} \\ 0 & 0 & \cdots & 0 & a_{n-1} \end{bmatrix} \begin{bmatrix} b_0 \\ \cdots \\ b_{n-1} \end{bmatrix}$$

The division is usually implemented by means of successive subtractions. Given $P(x)$ and $Q(x)$ defined over GF(2), the following algorithm generates the quotient $C(x)$ and the remainder $R(x)$:

Algorithm B.3

Table B.4 Operation ⊕ (Example B.3)

$\oplus$	0	1	x	$x+1$
0	0	1	x	$x+1$
1	1	0	$x+1$	x
x	x	$x+1$	0	1
$x+1$	$x+1$	x	1	0

Table B.5 Operation ⊗ (Example B.3)

$\otimes$	1	x	$x+1$
1	1	x	$x+1$
x	x	$x+1$	1
$x+1$	$x+1$	1	x

Algorithm for computing the quotient and the remainder of the division of D by d

```
    e ← g(D) − g(d), C ← 0;
    while e ≥ 0 do
        begin
            1) D ← D − x^e d, C ← C + x^e;
            2) e ← g(D)− g(d);
        end
```

End algorithm

After the execution of the Algorithm B.3, the remainder and the quotient are stored in the registers D and C, respectively.

Sometimes will be of interest to consider only irreducible polynomials over GF(2). It is easy to see that the irreducible polynomials, except $x + 1$, have an odd number of elements. Therefore, the simpler irreducible polynomials are trinomials.

For every polynomial $P(x)$, integer numbers m such that $P(x)$ is a divisor of the binomial $x^m + 1$, can be found. Let e be the minimum of all integers m for a given $P(x)$. If $P(x)$ is irreducible and $g(P) = n$, it is shown that, under these conditions, e is a divisor of $2^n - 1$ [8]. Therefore, the maximum value of e for the polynomials $P(x)$, with $g(P) = n$, is $2^n - 1$.

Example B.3 **Let be the polynomial $Q(x) = x^2 + x + 1$ with coefficients in GF(2), and consider the residues modulo $Q(x)$ of every polynomial in GF(2).**

Remember that, according to (B.1), in the modular algebra modulo $Q(x)$, x^2 can be substituted by $x + 1$. The residues modulo $Q(x)$ are all the polynomials of degree less than 2; i.e., $C = \{0, 1, x, x + 1\}$. The addition and the product of polynomials modulo $Q(x)$ are used as operations ⊕ and ⊗, respectively. Tables B.4 and B.5 correspond to these operations. It is immediate to check that it is a degree four Galois field; concretely, it is GF(2^2), that is an extension of GF(2). The elements zero an one are, obviously, 0 and 1. The inverse of each element is given in Table B.6. Moreover, observing Table B.4 it is immediate that, in this case, the opposite of each element is the element itself.

Table B.6 Inverse (Example B.3)

	Inverse
1	1
x	$x+1$
$x+1$	x

Table B.7 Operation ⊕ (Example B.4)

⊕	0	1	x	$x+1$
0	0	1	x	$x+1$
1	1	0	$x+1$	x
x	x	$x+1$	0	1
$x+1$	$x+1$	x	1	0

Table B.8 Operation ⊗ (Example B.4)

⊗	1	x	x+1
1	1	x	$x+1$
x	x	1	$x+1$
$x+1$	$x+1$	$x+1$	0

It is easy to check that x is a generator element of GF(2^2); in fact, the elements 0 and 1 are included, and multiplying by x each new element of this Galois field, all the element of C will be generated: x, $x^2 = x + 1$, $(x + 1)x = x^2 + x = x + 1 + x = 1$. In addition, $x + 1$ is also a generator element; in fact: $(x + 1)\ (x + 1) = x$, $x(x + 1) = 1$.

Given that $1 + 1 = 0$, the characteristic of this Galois field is 2. ■

Example B.4 **Let be the polynomial $Q(x) = x^2 + 1$ with coefficients in GF(2), and consider the residues modulo $Q(x)$.**

These residues, as in Example B.3, will be 0, 1, x and $x + 1$. The addition and the product of polynomials modulo $Q(x)$ are used again as operations ⊕ and ⊗, respectively. Tables B.7 and B.8 correspond to these operations. It is immediate to check that it is not a Galois field, since $x + 1$ has no inverse. It is easy to check that x is not a generator element; in fact, $1 \cdot x = x$, $x{\cdot}x = x^2 = 1$, and the element $x + 1$ is not generated. ■

From the above examples it results that not all polynomials over GF(2) can be used as modules to generate Galois fields GF(2^m). Those that can be used are known as **primitive polynomials**. It is also met that if $P(x)$ is primitive, with $g(P) = n$, the minimum value of e such that $x^e + 1$ is a multiple of $P(x)$ is exactly $2^n - 1$.

The number of primitive polynomials of a given degree n, $N(n)$, is given by the following expression [4]:

$$N(n) = \varphi\left(2^n - 1\right)/n$$

Table B.9 Number of primitive polynomials

n	$\varphi(2^n - 1)$	$N(n)$	n	$\varphi(2^n - 1)$	$N(n)$
1	1	1	10	600	60
2	2	1	11	1936	176
3	6	2	12	1728	144
4	8	2	13	8190	630
5	30	6	14	10,584	756
6	36	6	15	27,000	1800
7	126	18	16	32,768	2048
8	128	16	…	…	…
9	432	48	32	2,147,483,648	67,108,864

where $\varphi(k)$ is the Euler function (also known as totient function). The values $N(n)$ for the initial values n are given in Table B.9. The Euler function grows very rapidly, so that the number of primitive polynomials can be very large for values of n not too large; for example, there are more than 67 million primitive polynomials for $n = 32$.

Given a primitive polynomial $P(x)$, with $g(P) = n$, it is shown that its inverse $P^{-1}(x)$, defined as follows:

$$P^{-1}(x) = x^n P\left(x^{-1}\right)$$

is also primitive [13]. For example, $P(x) = x^3 + x + 1$ is a primitive polynomial, then $P^{-1}(x) = x^3P(x^{-1}) = x^3\ (x^{-3} + x^{-1} + 1) = 1 + x^2 + x^3$ is also a primitive polynomial.

Primitive polynomials of degree up to $n = 150$ are given in Table B.10. For each n a single polynomial is given: that polynomial with minimal number of summands (a trinomial, if present, or pentanomial, except for $n = 1$) and, therefore, the simpler implementation (its inverse has the same cost). Each cell of Table B.10 corresponds to a polynomial, of the exponents of x are given, except the exponent 0, which appears in all polynomials. For example, the cell 26, 6, 2, 1 corresponds to the polynomial $x^{26} + x^6 + x^2 + x + 1$. In [14] a more extended list of the primitive polynomials is shown.

As an example of the using of primitive polynomials, the FIPS 186 standard [12] proposes using the primitive polynomials $x^{163} + x^7 + x^6 + x^3 + 1$, $x^{233} + x^{74} + 1$, $x^{283} + x^{12} + x^7 + x^5 + 1$, $x^{409} + x^{87} + 1$ and $x^{571} + x^{10} + x^5 + x^2 + 1$.

Example B.5 **Let be the polynomial $Q(x) = x^4 + x^3 + 1$ over GF(2). The different remainders that can be obtained when dividing by Q(x) any other polynomial are:**

- always be the values 0 and 1;
- multiplying the successive remainders by x (so all the powers of x are generated) result x, x^2, x^3 and x^4, but x^4can be replaced by $x^3 + 1$;

Table B.10 Primitive polynomials over GF(2) up to $n = 150$

1	26, 6, 2, 1	51, 16, 15, 1	76, 36, 35, 1	101, 7, 6	126, 37, 36, 1
2, 1	27, 5, 2, 1	52, 3	77, 31, 30, 1	102, 77, 76, 1	127, 1
3, 1	28, 3	53, 16, 15, 1	78, 20, 19, 1	103, 9	128, 29, 27, 2
4, 1	29, 2	54, 37, 36, 1	79, 9	104, 11, 10, 1	129, 5
5, 2	30, 23, 2, 1	55, 24	80, 38, 37, 1	105, 16	130, 3
6, 1	31, 3	56, 22, 21, 1	81, 4	106, 15	131, 48, 47, 1
7, 1	32, 22, 2, 1	57, 7	82, 38, 35, 3	107, 65, 63, 2	132, 29
8, 4, 3, 2	33, 13	58, 19	83, 46, 45, 1	108, 31	133, 52, 51, 1
9, 4	34, 27, 2, 1	59, 22, 21, 1	84, 13	109, 7, 6, 1	134, 57
10, 3	35, 2	60, 1	85, 28, 27,1	110, 13, 12, 1	135, 11
11, 2	36, 11	61, 16, 15, 1	86, 13, 12, 1	111, 10	136, 126, 125, 1
12, 6, 4, 1	37, 12, 10, 2	62, 57, 56, 1	87, 13	112, 45, 43, 2	137, 21
13, 4, 3, 1	38, 6, 5, 1	63, 1	88, 72, 71, 1	113, 9	138, 8, 7, 1
14, 10, 6, 1	39, 4	64, 4, 3, 1	89, 38	114, 82, 81, 1	139, 8, 5, 3
15, 1	40, 21, 19, 2	65, 18	90, 19, 18, 1	115, 15, 14, 1	140, 29
16, 12, 3, 1	41, 3	66, 10, 9, 1	91, 84, 83, 1	116, 71, 70, 1	141, 32, 31, 1
17, 3	42, 23, 22,1	67, 10, 9, 1	92, 13, 12, 1	117, 20, 18, 2	142, 21
18, 7	43, 6, 5, 1	68, 9	93, 2	118, 33	143, 21, 20, 1
19, 6, 5, 1	44, 27, 26, 1	69, 29, 27, 2	94, 21	119, 8	144, 70, 69, 1
20, 3	45, 4, 3, 1	70, 16, 15, 1	95, 11	120, 118, 111, 7	145, 52
21, 2	46, 21, 20, 1	71, 6	96, 49, 2	121, 18	146, 60, 59, 1
22, 1	47, 5	72, 53, 47, 6	97, 6	122, 60, 59, 1	147, 38, 37, 1
23, 5	48, 28, 27, 1	73, 25	98, 11	123, 2	148, 27
24, 7, 2, 1	49, 9	74, 16, 15, 1	99, 47, 45, 2	124, 37	149, 110, 109, 1
25, 3	50, 27, 26, 1	75, 11, 10, 1	100, 37	125, 108, 107, 1	150, 53

- the same technique is used (by multiplying the previous remainder by x and replace x^4 by $x^3 + 1$); the remainders that occur in the second (and fifth) column of Table B.11 are obtained. Since $x^3 + x^2$ the same remainders are repeated, i.e., $x^{15} = 1$.

The power which rises x to generate the corresponding remainder with respect to $Q(x)$ is given in the first (and fourth) column of Table B.11.

The coefficients (the polynomial of the second column, in the order x^0, x^1, x^2, x^3) and the hexadecimal value (in parentheses) are given in the third (and sixth) column of Table B.11. ■

The 16 possible residues (i.e., all polynomials over GF(2)) of degree less than 4 are given in Table B.11, that form a Galois field GF(2^4) of order 16. To represent the 16 elements of GF(2^4) any of the columns in Table B.11 can be used.

Table B.11 Generation of the remainders in the Example B.5

$x^{-\infty}$	0	0000 (0)	x^7	x^2+x+1	1110 (E)
x^0	1	1000 (8)	x^8	x^3+x^2+x	0111 (7)
x^1	x	0100 (4)	x^9	x^2+1	1010 (A)
x^2	x^2	0010 (2)	x^{10}	x^3+x	0101 (5)
x^3	x^3	0001 (1)	x^{11}	x^3+x^2+1	1011 (B)
x^4	x^3+1	1001 (9)	x^{12}	$x+1$	1100 (C)
x^5	x^3+x+1	1101 (D)	x^{13}	x^2+x	0110 (6)
x^6	x^3+x^2+x+1	1111 (F)	x^{14}	x^3+x^2	0011 (3)

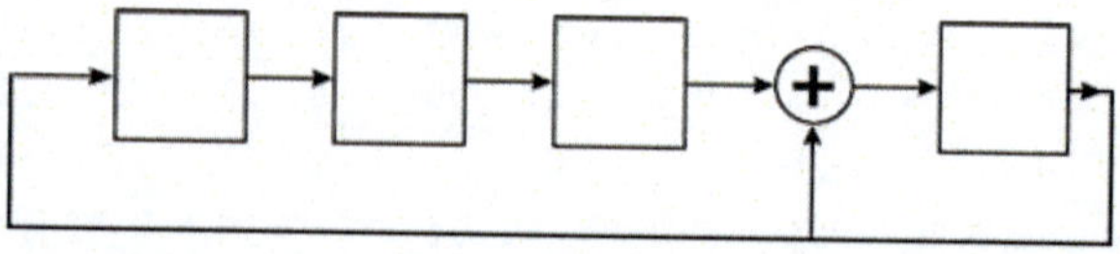

Fig. B.1 **LFSR2** of the Example B.5

In Example B.5, the successive powers of x generate all the elements of GF(2^4); that is, x is a generator element or primitive element or primitive root of GF(2^4); in this case the polynomial $Q(x)$ is a primitive polynomial. It is easily verified that x^2, x^4, x^7, x^8, x^{11}, x^{13} y x^{14} are also primitive elements, i.e., the successive powers of each of these elements generate all the polynomials of degree less than 4. Obviously, since $x^{15}=1$, any multiple of 15 can be added to any of the above exponents of x. In general, if α is a root of x^4+x^3+1, α^2, α^4, α^7, α^8, α^{11}, α^{13} and α^{14} are also roots of x^4+x^3+1.

Using a type 2 **LFSR** (see Chap. 4) whose associate polynomial is precisely $Q(x)$, as shown in Fig. B.1, it is easy to generate the residues of the Table B.11 starting from any initial content different to all zeros. For example, since 1000, the 14 rows of the third (and sixth) column of Table B.11 are generated. It can be interpreted that in each iteration, the shift to the right in the **LFSR2** of Fig. B.1 corresponds to multiplication by x of the previous content, and with the feedbacks the remainder modulo $Q(x)$ is calculated.

The elements of GF(2^5)$\{x^2+x^5+1\}$ are given in Table B.12, which will later be used.

Example B.6 **Determine the type of each one of the polynomials of degree 4.**
The possible polynomials that are strictly of degree 4 to be analyzed are: x^4+1, x^4+x^3+1, x^4+x^2+1, x^4+x+1, $x^4+x^3+x^2+1$, x^4+x^3+x+1, x^4+x^2+x+1, $x^4+x^3+x^2+x+1$. The rest of the polynomials not including the summand 1 obviously are reducible. Analyzing one by one, it results:

- x^4+1 is reducible; concretely, $x^4+1=(x^2+1)(x^2+1)$;
- x^4+x^3+1 is primitive, according to Example B.4;
- x^4+x^2+1 is irreducible and it divides to x^6+1; thus, it is not primitive;
- x^4+x+1 is the inverse of x^4+x^3+1; thus, it is primitive;

Table B.12 Generation of the remainders in GF(2^5)$\{x^5 + x^2 + 1\}$

$x^{-\infty}$	0	x^{15}	$x^4+x^3+x^2+x+1$
x^0	1	x^{16}	x^4+x^3+x+1
x^1	x	x^{17}	x^4+x+1
x^2	x^2	x^{18}	$x+1$
x^3	x^3	x^{19}	x^2+x
x^4	x^4	x^{20}	x^3+x^2
x^5	x^2+1	x^{21}	x^4+x^3
x^6	x^3+x	x^{22}	x^4+x^2+1
x^7	x^4+x^2	x^{23}	x^3+x^2+x+1
x^8	x^3+x^2+1	x^{24}	$x^4+x^3+x^2+x$
x^9	x^4+x^3+x	x^{25}	x^4+x^3+1
x^{10}	x^4+1	x^{26}	x^4+x^2+x+1
x^{11}	x^2+x+1	x^{27}	x^3+x+1
x^{12}	x^3+x^2+x	x^{28}	x^4+x^2+x
x^{13}	$x^4+x^3+x^2$	x^{29}	x^3+1
x^{14}	$x^4+x^3+x^2+1$	x^{30}	x^4+x

- $x^4 + x^3 + x^2 + 1$ is reducible: $x^4 + x^3 + x^2 + 1 = (x^3 + x + 1)(x + 1)$;
- $x^4 + x^3 + x + 1$ is reducible: $x^4 + x^3 + x + 1 = (x^3 + 1)(x + 1)$;
- $x^4 + x^2 + x + 1$ is reducible: $x^4 + x^2 + x + 1 = (x^3 + x^2 + 1)(x + 1)$;
- $x^4 + x^3 + x^2 + x + 1$ is irreducible, but it is not primitive, since it divides to $x^5 + 1$. ■

B.3 Polynomials Over GF(p)

Let suppose now that the coefficients of the polynomials belong to GF(p), being p a prime number; there will be p^n polynomials of degree n and other p^n polynomials of degree less than n. For example, for $p = 3$, all the polynomials of degree less than two are given in Table B.13. Each polynomial can be represented by the ternary coordinates that are given in the last column of Table B.13, which are the coefficients of the two possible summands.

The operations addition, subtraction, multiplication and division are performed identically or similarly to those described for GF(2), but now the operations between the coefficients are performed in GF(p). For the division, the Algorithm B.3, applicable to GF(2), has to be modified to take into account the different values of the coefficients. Specifically, for the division by means of successive subtractions, it is immediate that, given $P(x)$ and $Q(x)$ defined in GF(p), and being, in each iteration, a_n and b_m the nonzero coefficients of the highest power of x in $P(x)$ and $Q(x)$, respectively, the following algorithm generates the quotient $C(x)$ and remainder $R(x)$:

Algorithm B.4

Table B.13 Polynomial over GF(3) of degree less than two

0	0	00
1	1	01
2	2	02
3	x	10
4	$x+1$	11
5	$x+2$	12
6	$2x$	20
7	$2x+1$	21
8	$2x+2$	22

Algorithm for computing the quotient and the remainder of the division of D by d

```
e ← g(D)− g(d), C ← 0;
while e ≥ 0 do
    begin
        1) D ← D − (an:bm) xᵉd, C ← C + (an:bm) xᵉ;
        2) e ← g(D)− g(d);
    end
```

End algorithm

After applying Algorithm B.4, the remainder and the quotient are store in registers D and C, respectively.

For every polynomial $P(x)$ integer numbers m such that $P(x)$ is divisor of the binomial $x^m - 1$ can be found. Let e be the minimum of all integers m for a given $P(x)$. If $P(x)$) is irreducible and $g(P) = n$, it is shown that, under these conditions, e is a divisor of $p^n - 1$ [8]. Therefore, the maximum value of e for the polynomials $P(x)$ (i.e., the **order** of $P(x)$), with $g(P) = n$, is $p^n - 1$. The monic polynomials $P(x)$ of order $p^n - 1$ such that $P(0) \neq 0$ are called **primitive polynomials**. The primitive polynomials for $p = 3, 5, 7$ and different values of n are given in Table B.14; in this table only the primitive polynomials with minimal summands have been included; specifically, in each cell of the table the coefficients of a polynomial (first the coefficient of x^n) are given; a more complete list of primitive polynomials can be seen in [8]. Each primitive polynomial can be used as a module to generate a Galois field GF(p^n), as seen in the following example.

Example B.7 **A primitive polynomial for $p = 3$ and $n = 2$ is $x^2 + x + 2$. Construct GF(3^2)$\{x^2 + x + 2\}$.**

The elements of GF(3^2)$\{x^2 + x + 2\}$ are all the polynomials of degree lower than 2 with coefficients in GF(3): $0, 1, 2, x, x + 1, x + 2, 2x, 2x + 1, 2x + 2$. The addition is obtained immediately, such as is given in Table B.15, from what the opposite of each element is obtained, which is given in Table B.16. It is also easy to calculate the multiplication, which is given in Table B.17, from what the inverse of each element is obtained, which is given in Table B.18.

Using a Type 2 3**LFSRmod3** (see Chap. 4) which associate polynomial $x^2 + x + 2$, as shown in Fig. B.2, it is easy to generate the non-zero elements of GF(3^2)$\{x^2$

Table B.14 Primitive polynomials over GF(p) for $p = 3$, 5 and 7

	$p=3$	$p=3$	$p=5$	$p=5$	$p=5$	$p=5$	$p=7$	$p=7$	$p=7$	$p=7$	$p=7$	$p=7$
$n=1$	11		12	13			12	14				
$n=2$	112	122	112	123	133	142	113	125	145	153	155	163
							123	135				
$n=3$	1021	1201	1032	1042	1102	1403	1032	1052	1062	1504	1604	
			1033	1043	1302							
$n=4$	10012	11002	10122	11013	12042	13302	10135	10533	10663	12205	14015	15205
	10022	12002	10123	11023	12203	14012	10145	10543	11013	12303	14023	15303
			10132	11032	12302	14022	10333	10555	11063	12403	14053	15403
			10133	11042	13012	14033	10335	10565	11103	13015	14065	16013
			10412	11202	13023	14043	10343	10613	11105	13023	14103	16063
			10413	11303	13032	14202	10345	10623	11605	13053	14205	16103
			10442	12013	13043	14303	10433	10635	12025	13065	15025	16105
			10443	12022	13203		10443	10645	12055	13103	15055	16405
							10525	10653	12203	13205	15203	16605
n=5	100021	120001	100042	100102	103002	120003						
			100043	100403	103003	130002						
n=6	1000012	1200002										
	1000022											
n=7	10000201	10200001										

Table B.15 Addition table for GF(3^2){x^2+x+2}

	0	1	2	x	$x+1$	$x+2$	$2x$	$2x+1$	$2x+2$
0	0	1	2	x	$x+1$	$x+2$	$2x$	$2x+1$	$2x+2$
1	1	2	0	$x+1$	$x+2$	x	$2x+1$	$2x+2$	$2x$
2	2	0	1	$x+2$	x	$x+1$	$2x+2$	$2x$	$2x+1$
x	x	$x+1$	$x+2$	$2x$	$2x+1$	$2x+2$	0	1	2
$x+1$	$x+1$	$x+2$	x	$2x+1$	$2x+2$	$2x$	1	2	0
$x+2$	$x+2$	x	$x+1$	$2x+2$	$2x$	$2x+1$	2	0	1
$2x$	$2x$	$2x+1$	$2x+2$	0	1	2	x	$x+1$	$x+2$
$2x+1$	$2x+1$	$2x+2$	$2x$	1	2	0	$x+1$	$x+2$	x
$2x+2$	$2x+2$	$2x$	$2x+1$	2	0	1	$x+2$	x	$x+1$

Table B.16 Table of opposites for GF(3^2){x^2+x+2}

0	0
1	2
2	1
x	$2x$
$x+1$	$2x+2$
$x+2$	$2x+1$
$2x$	x
$2x+1$	$x+2$
$2x+2$	$x+1$

Table B.17 Multiplying table for GF(3^2){x^2+x+2}

	1	2	x	$x+1$	$x+2$	$2x$	$2x+1$	$2x+2$
1	1	2	x	$x+1$	$x+2$	$2x$	$2x+1$	$2x+2$
2	2	1	$2x$	$2x+2$	$2x+1$	x	$x+2$	$x+1$
x	x	$2x$	$2x+1$	1	$x+1$	$x+2$	$2x+2$	2
$x+1$	$x+1$	$2x+2$	1	$x+2$	$2x$	2	x	$2x+1$
$x+2$	$x+2$	$2x+1$	$x+1$	$2x$	2	$2x+2$	1	x
$2x$	$2x$	x	$x+2$	2	$2x+2$	$2x+1$	$x+1$	1
$2x+1$	$2x+1$	$x+2$	$2x+2$	x	1	$x+1$	2	$2x$
$2x+2$	$2x+2$	$x+1$	2	$2x+1$	x	1	$2x$	$x+2$

$+x+2$} starting with any initial content different of all zeros. For example, from 10 the remaining rows of the third (and sixth) column of Table B.19 are generated. It can be interpreted that, in each iteration, the shift to the right in the **3LFSR2** of Fig. B.2 corresponds to the multiplication by x of the previous content, and with the feedback, the remainder modulo $Q(x)$ is calculated. ■

Table B.18 Table of inverses for GF(3^2){$x^2 + x + 2$}

1	1
2	2
x	$x + 1$
$x + 1$	x
$x + 2$	$2x + 1$
$2x$	$2x + 2$
$2x + 1$	$x + 2$
$2x + 2$	$2x$

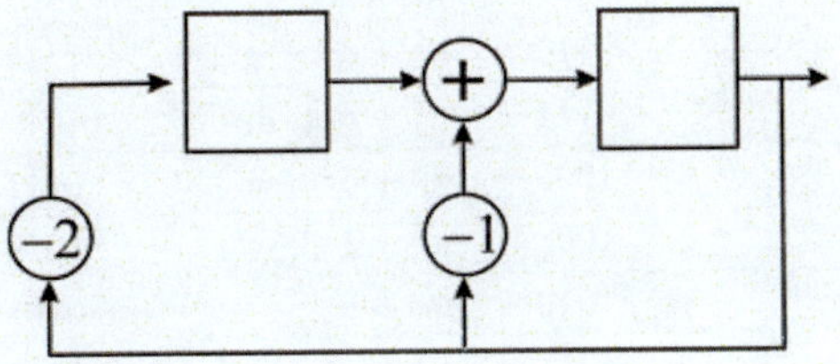

Fig. B.2 3LFSR2 of the Example B.7

Table B.19 Generation of elements of GF(3^2){$x^2 + x + 2$} with an **3LFSR2**

$x^{-\infty}$	0	00 (0)	x^4	2	20 (6)
x^0	1	10 (3)	x^5	$2x$	02 (2)
x^1	x	01 (1)	x^6	$x + 2$	21 (7)
x^2	$2x + 1$	12 (5)	x^7	$x + 1$	11 (4)
x^3	$2x + 2$	22 (8)			

B.4 Finite Fields GF(2^m)

The polynomials with coefficients over GF(2), of degree less than m, with the operation addition of polynomials and product of polynomials modulo a primitive polynomial of degree m, $P(x)$, form a finite field of characteristic two or Galois field GF(2^m); this Galois field is also represented as GF(2^m){P(x)}. GF(2^4){$x^4 + x^3 + 1$} can be used as an example of the Galois field, as described above in Example B.5.

In GF(2^m){$P(x)$} the order of any element must be a divisor of $2^m - 1$. The order of primitive elements is precisely $2^m - 1$.

As GF(2^m){$P(x)$} is a particular case of GF(p^m) (see Appendix A), for any element B of GF(2^m){$P(x)$} it is verified that $B^{2^m} = B$, that it is equivalent to $B^{2^m-1} = 1$, or $B^{2^m-2} = B^{-1}$.

As seen, the different non zero elements of GF(2^m){$P(x)$} can be represented as powers of a primitive root (potential representation) or as a polynomial of degree less than m (polynomial representation). As an example, the elements of GF(2^4){$x^4 + x^3 + 1$} are given in the first column of Table B.20 as powers of a primitive root

Table B.20 Potential representation and with standard bases $\{1, \alpha, \alpha^2, \alpha^3\}$ and $\{1, \alpha^2, \alpha^4, \alpha^6\}$ for the elements of GF(2^4)$\{x^4 + x^3 + 1\}$

		$\{1, \alpha, \alpha^2, \alpha^3\}$		$\{1, \alpha^2, \alpha^4, \alpha^6\}$	
$\alpha^{-\infty} = 0$	1111	0	0000 (0)	0	0000 (0)
$\alpha^0 = 1$	0000	1	1000 (8)	1	1000 (8)
α	0001	α	0100 (4)	$\alpha^6 + \alpha^4 + \alpha^2$	0111 (7)
α^2	0010	α^2	0010 (2)	α^2	0100 (4)
α^3	0011	α^3	0001 (1)	$\alpha^4 + 1$	1010 (A)
α^4	0100	$\alpha^3 + 1$	1001 (9)	α^4	0010 (2)
α^5	0101	$\alpha^3 + \alpha + 1$	1101 (D)	$\alpha^6 + \alpha^2$	0101 (5)
α^6	0110	$\alpha^3 + \alpha^2 + \alpha + 1$	1111 (F)	α^6	0001 (1)
α^7	0111	$\alpha^2 + \alpha + 1$	1110 (E)	$\alpha^6 + \alpha^4 + 1$	1011 (B)
α^8	1000	$\alpha^3 + \alpha^2 + \alpha$	0111 (7)	$\alpha^6 + 1$	1001 (9)
α^9	1001	$\alpha^2 + 1$	1010 (A)	$\alpha^2 + 1$	1100 (C)
α^{10}	1010	$\alpha^3 + \alpha$	0101 (5)	$\alpha^6 + \alpha^2 + 1$	1101 (D)
α^{11}	1011	$\alpha^3 + \alpha^2 + 1$	1011 (B)	$\alpha^4 + \alpha^2$	0110 (6)
α^{12}	1100	$\alpha + 1$	1100 (C)	$\alpha^6 + \alpha^4 + \alpha^2 + 1$	1111 (F)
α^{13}	1101	$\alpha^2 + \alpha$	0110 (6)	$\alpha^6 + \alpha^4$	0011 (3)
α^{14}	1110	$\alpha^3 + \alpha^2$	0011 (3)	$\alpha^4 + \alpha^2 + 1$	1110 (E)

α, and the same elements are given in the third column of this table as polynomials of the root α of degree less than 4.

With regard to the potential representation, each element can be represented by the binary value of the exponent, and to represent the zero element, the combination of all ones can be used, that does not appear in the representation of the non zero elements. This is what is done in the second column of Table B.20 for the elements of GF(2^4)$\{x^4 + x^3 + 1\}$.

The different operations with the corresponding polynomials can be more or less complex depending on the form of representation of the elements of GF(2^{m})$\{P(x)\}$. For example, the addition is more easily executed in the polynomial representation, while the potential representation is more suitable for the multiplication.

Since GF(2^{m})$\{P(x)\}$ can be considered as a vector space of dimension m over GF(2), to represent the different elements of GF(2^{m})$\{P(x)\}$ different bases may be used. It is known that $\{b_0, b_1, \ldots, b_{m-1}\}$ is a basis of GF(2^{m})$\{P(x)\}$ [8] if the following expression is verified:

$$\begin{vmatrix} b_0 & b_1 & \ldots & b_{m-2} & b_{m-1} \\ b_0^2 & b_1^1 & \ldots & b_{m-2}^2 & b_{m-1}^2 \\ \ldots & \ldots & \ldots & \ldots & \ldots \\ b_0^{2^{m-2}} & b_1^{2^{m-2}} & \ldots & b_{m-2}^{2^{m-2}} & b_{m-1}^{2^{m-2}} \\ b_0^{2^{m-1}} & b_1^{2^{m-1}} & \ldots & b_{m-2}^{2^{m-1}} & b_{m-1}^{2^{m-1}} \end{vmatrix} \neq 0$$

Different types of bases are considered at the following.

B.4.1 Standard Basis

The basis $\{1, \alpha, \alpha^2, \alpha^3, \ldots, \alpha^{m-2}, \alpha^{m-1}\}$ can be used in the representation of the elements of $GF(2^m)\{P(x)\}$, called **standard basis** (or **polynomial**, or **canonical**) being α any primitive element of $GF(2^m)\{P(x)\}$.

Selected a basis, each element of $GF(2^m)\{P(x)\}$ can be represented with an m-tuple over GF(2), which are the coefficients of the elements of the basis. The 16 elements of $GF(2^4)\{x^4 + x^3 + 1\}$ are given as powers of a primitive element, α, in the first column of Table B.20. The representation used for the exponent of the first column is given in the second column of Table B.20. Each element is expressed as a polynomial of α in the third column; in this case, the primitive element α is the initial point and the basis used is $\{1, \alpha, \alpha^2, \alpha^3\}$. The m-tuples and its hexadecimal value for the different elements of $GF(2^4)\{x^4 + x^3 + 1\}$ using the basis $\{1, \alpha, \alpha^2, \alpha^3\}$ are given in the forth column.

Previously it has been shown that α^2 is also a primitive element. From α^2, the resulting standard basis is $\{1, \alpha^2, \alpha^4, \alpha^6\}$. With this basis, the development of each element of $GF(2^4)\{x^4 + x^3 + 1\}$ is given in the fifth column of Table B.20, and the corresponding m-tuples are shown in the sixth column.

Another standard basis results when starting with any other primitive element.

B.4.2 Normal Basis

For the representation of the elements of $GF(2^m)\{P(x)\}$ other bases may be used. In fact in practice several different bases are used, depending on the operations that seek to make, because the complexity of each operation, as already indicated, may depend heavily on the basis used. One often used basis is known as **normal basis**, which is of the form $\{B, B^2, B^4, \ldots, B^{2^{m-2}}, B^{2^{m-1}}\}$, where B is a suitable element of $GF(2^m)\{P(x)\}$, not necessarily a primitive element, such that the various elements of $\{B, B^2, B^4, \ldots, B^{2^{m-2}}, B^{2^{m-1}}\}$ are linearly independent.

Normal bases always exist in a finite field. For example, $\{\alpha, \alpha^2, \alpha^4, \alpha^8\}$ is a normal basis for $GF(2^4)\{x^4 + x^3 + 1\}$, being α a primitive root. On this basis the different elements of $GF(2^4)\{x^4 + x^3 + 1\}$ are given in Table B.21 and represented vectorially as shown in the third and sixth columns of the same table. Note that, in this case, the polynomials corresponding to the different elements of $GF(2^m)\{P(x)\}$ do not have to be of lower degree than m, but will always be equal to a polynomial of degree less than m in standard basis.

It is easy to see that, when vector representation is used with a normal basis, squaring any element consists on rotating its vector representation one position to the left. This is the great advantage of normal bases. Indeed, let suppose the normal basis $\{B, B^2, B^4, \ldots, B^{2^{m-2}}, B^{2^{m-1}}\}$ and E any element:

Table B.21 Representation of the elements of GF(2^4){x^4+x^3+1} with the normal basis {α, α^2, α^4, α^8}

0	0	0000 (0)	α^7	$\alpha^8+\alpha^4$	0011 (3)
1	$\alpha^8+\alpha^4+\alpha^2+\alpha$	1111 (F)	α^8	α^8	0001 (1)
α	α	1000 (8)	α^9	$\alpha^8+\alpha^4+\alpha$	1011 (B)
α^2	α^2	0100 (4)	α^{10}	$\alpha^8+\alpha^2$	0101 (5)
α^3	$\alpha^8+\alpha^2+\alpha$	1101 (D)	α^{11}	$\alpha^4+\alpha^2$	0110 (6)
α^4	α^4	0010 (2)	α^{12}	$\alpha^8+\alpha^4+\alpha^2$	0111 (7)
α^5	$\alpha^4+\alpha$	1010 (A)	α^{13}	$\alpha^2+\alpha$	1100 (C)
α^6	$\alpha^4+\alpha^2+\alpha$	1110 (E)	α^{14}	$\alpha^8+\alpha$	1001 (9)

$$E=e_{m-1}B^{2^{m-.1}}+e_{m-2}B^{2^{m-2}}+\ldots+e_1B^2+e_0B$$

When calculating E^2, the products with coefficients e_ie_j, $i\neq j$, will appear duplicated and, therefore, will be anulated ($e_ie_j+e_je_i=0$). Moreover, $e_ie_i=e_i$ (in GF (2), $a{\cdot}a=a$). Applying this, it results:

$$E^2=e_{m-1}B^{2^m}+e_{m-2}B^{2^{m-1}}+\cdots+e_1B^4+e_0B^2$$

But $B^{2^m}=B$. Thus:

$$E^2=e_{m-2}B^{2^{m-1}}+e_{m-3}B^{2^{m-2}}+\cdots+e_1B^4+e_0B^2+e_{m-1}B$$

That is, given the vector representation of E with a normal basis, $E=(e_{m-1},e_{m-2},\ldots,e_1,e_0)$, E^2 is obtained with a rotation to the left of E; i.e,, it results $E^2=(e_{m-2},e_{m-3},\ldots,e_1,e_0,e_{m-1})$.

As a result of this, it is easy to see that the vector representation of the element 1 of GF(2^m) on a normal basis is always (1, 1, ..., 1, 1). Indeed, as $1^2=1$, the only vector different to all-zero vector (which is the representation of 0) to be reproduced after rotate one position is the one that includes only ones

Also, as a result of this, it is immediate that, with a normal basis, $\sqrt{E}$ is obtained with a rotation to the right of E, $\sqrt{E}=(e_0,e_{m-1},e_{m-2},\ldots,e_2,e_1)$.

In some cases there are normal bases that simplify the implementation of multiplication: these are the **optimal normal bases** [10, 11]. Two types of optimal normal bases can exists in GF(2^m){$P(x)$}: Type I and Type II. There is an optimal normal basis of Type I (see [11]) if $p=m+1$ is prime and 2 is a primitive element of GF(p). The elements of an optimal normal basis Type I are generated from the elements of GF(2^m){$P(x)$} of order p, as is done in Examples B.8 and B.9.

Example B.8 Given GF(2^2){x^2+x+1}. In this case $m=2$, $p=m+1=3$ is prime, and 2 is a primitive element of GF(3). All conditions are met for the existence of an optimal normal basis Type I. It is straightforward to check, using Table B.5, that the elements x and $x+1$ of GF(2^2){x^2+x+1} are of order 3. Therefore, being

α a root of $x^2 + x + 1$, $\{\alpha, \alpha^2\}$ is an optimal normal basis Type I for $GF(2^2)\{x^2 + x + 1\}$, and likewise, $\{(\alpha + 1)\ (\alpha + 1)^2\}$ is an optimal normal basis Type I for $GF(2^2)\{x^2 + x + 1\}$. ■

Example B.9 Given $GF(2^4)\{x^4 + x^3 + 1\}$. In this case $m = 4$, $p = m + 1 = 5$ is prime, and 2 is a primitive element of GF(5). All conditions are met for the existence of an optimal normal basis Type I. It is straightforward to check, using Table B.22, that the elements x^3 of $GF(2^4)\{x^4 + x^3 + 1\}$ are of order 5, since $(x^3)^5 = 1$. The elements x^6, x^9 and x^{12} are also of order 5. Therefore, being α a root of $x^4 + x^3 + 1$, $\{\alpha^3, \alpha^6, \alpha^{12}, \alpha^{24}\}$, $\{\alpha^6, \alpha^{12}, \alpha^{24}, \alpha^{48}\}$, $\{\alpha^9, \alpha^{18}, \alpha^{36}, \alpha^{72}\}$ and $\{\alpha^{12}, \alpha^{24}, \alpha^{48}, \alpha^{96}\}$ are optimal normal basis Type I for $GF(2^4)\{x^4 + x^3 + 1\}$. ■

In $GF(2^m)\{P(x)\}$ there is an optimal normal basis of Type II (see [11]) if $p = 2m + 1$ is prime and one of the following conditions is met: (a) 2 is a primitive element of GF(p), (b) $p = 3(\mathbf{mod}4)$, and in GF(p), ord (2) $= m$. Regarding the form of generation, given $A \in GF(2^{2m})$, such that $A^j \neq 1$, $1 \leq j < 2m + 1$, and $A^{2m+1} = 1$, the elements of an optimal normal basis Type II are generated from $B = A + A^{-1}$, as done in Example B.10.

Example B.10 Given $GF(2^2)\{x^2 + x + 1\}$. In this case $m = 2$, $p = 2m + 1 = 5$ is prime, and 2 is a primitive element of GF(5). All conditions are met for the existence of an optimal normal basis Type II. It is straightforward to check, using the Tables B.20 and B.22, that in $GF(2^2)\{x^2 + x + 1\}$, $(x^3)^5 = 1$ and $x^{-3} = x + 1$. Therefore, $B = x^3 + x^{-3} = x^3 + x + 1 = x$. Therefore, being α a root of $x^2 + x + 1$, $\{\alpha, \alpha^2\}$ is an optimal normal basis Type II for $GF(2^2)\{x^2 + x + 1\}$. ■

Other two normal bases that will be used lately are shown in Example B.11.

Table B.22 Multiplication table for $GF(2^4)\{x^4 + x^3 + 1\}$

	x^1	x^2	x^3	x^4	x^5	x^6	x^7	x^8	x^9	x^{10}	x^{11}	x^{12}	x^{13}	x^{14}
x^1	x^2	x^3	x^4	x^5	x^6	x^7	x^8	x^9	x^{10}	x^{11}	x^{12}	x^{13}	x^{14}	1
x^2	x^3	x^4	x^5	x^6	x^7	x^8	x^9	x^{10}	x^{11}	x^{12}	x^{13}	x^{14}	1	x^1
x^3	x^4	x^5	x^6	x^7	x^8	x^9	x^{10}	x^{11}	x^{12}	x^{13}	x^{14}	1	x^1	x^2
x^4	x^5	x^6	x^7	x^8	x^9	x^{10}	x^{11}	x^{12}	x^{13}	x^{14}	1	x^1	x^2	x^3
x^5	x^6	x^7	x^8	x^9	x^{10}	x^{11}	x^{12}	x^{13}	x^{14}	1	x^1	x^2	x^3	x^4
x^6	x^7	x^8	x^9	x^{10}	x^{11}	x^{12}	x^{13}	x^{14}	1	x^1	x^2	x^3	x^4	x^5
x^7	x^8	x^9	x^{10}	x^{11}	x^{12}	x^{13}	x^{14}	1	x^1	x^2	x^3	x^4	x^5	x^6
x^8	x^9	x^{10}	x^{11}	x^{12}	x^{13}	x^{14}	1	x^1	x^2	x^3	x^4	x^5	x^6	x^7
x^9	x^{10}	x^{11}	x^{12}	x^{13}	x^{14}	1	x^1	x^2	x^3	x^4	x^5	x^6	x^7	x^8
x^{10}	x^{11}	x^{12}	x^{13}	x^{14}	1	x^1	x^2	x^3	x^4	x^5	x^6	x^7	x^8	x^9
x^{11}	x^{12}	x^{13}	x^{14}	1	x^1	x^2	x^3	x^4	x^5	x^6	x^7	x^8	x^9	x^{10}
x^{12}	x^{13}	x^{14}	1	x^1	x^2	x^3	x^4	x^5	x^6	x^7	x^8	x^9	x^{10}	x^{11}
x^{13}	x^{14}	1	x	x^2	x^3	x^4	x^5	x^6	x^7	x^8	x^9	x^{10}	x^{11}	x^{12}
x^{14}	1	x	x^2	x^3	x^4	x^5	x^6	x^7	x^8	x^9	x^{10}	x^{11}	x^{12}	x^{13}

Example B.11 **Checking that in GF(2^5){$x^5 + x^2 + 1$}, {$\alpha^3, \alpha^6, \alpha^{12}, \alpha^{24}, \alpha^{48}$} and {$\alpha^5, \alpha^{10}, \alpha^{20}, \alpha^{40}, \alpha^{80}$}, are normal bases, where α is a root of $x^5 + x^2 + 1$.**

In fact, each element of GF(2^5){$x^5 + x^2 + 1$} is developed in the two mentioned bases in Table B.23. ■

Table B.23 Two normal bases in GF(2^5){$x^5 + x^2 + 1$}

			$\alpha^{48}\ \alpha^{24}\ \alpha^{12}\ \alpha^{6}\ \alpha^{3}$	$\alpha^{80}\ \alpha^{40}\ \alpha^{20}\ \alpha^{10}\ \alpha^{5}$
$-\infty$	$x^{-\infty}$	0	00000	00000
0	x^0	1	11111	11111
1	x^1	x	00011	01111
2	x^2	x^2	00110	11110
3	x^3	x^3	00001	11010
4	x^4	x^4	01100	11101
5	x^5	x^2+1	11001	00001
6	x^6	x^3+x	00010	10101
7	x^7	x^4+x^2	01010	00011
8	x^8	x^3+x^2+1	11000	11011
9	x^9	x^4+x^3+x	01110	01000
10	x^{10}	x^4+1	10011	00010
11	x^{11}	x^2+x+1	11010	01110
12	x^{12}	x^3+x^2+x	00100	01011
13	x^{13}	$x^4+x^3+x^2$	01011	11001
14	x^{14}	$x^4+x^3+x^2+1$	10100	00110
15	x^{15}	$x^4+x^3+x^2+x+1$	10111	01001
16	x^{16}	x^4+x^3+x+1	10001	10111
17	x^{17}	x^4+x+1	10000	01101
18	x^{18}	$x+1$	11100	10000
19	x^{19}	x^2+x	00101	10001
20	x^{20}	x^3+x^2	00111	00100
21	x^{21}	x^4+x^3	01101	00111
22	x^{22}	x^4+x^2+1	10101	11100
23	x^{23}	x^3+x^2+x+1	11011	10100
24	x^{24}	$x^4+x^3+x^2+x$	01000	10110
25	x^{25}	x^4+x^3+1	10010	11000
26	x^{26}	x^4+x^2+x+1	10110	10011
27	x^{27}	x^3+x+1	11101	01010
28	x^{28}	x^4+x^2+x	01001	01100
29	x^{29}	x^3+1	11110	00101
30	x^{30}	x^4+x	01111	10010

Optimal normal bases are interesting in the implementation of the multiplication. To display the advantage of the same, Table B.24 is the cross-product table for {α^3, α^6, α^{12}, α^{24}, α^{48}}, i.e., the result of cross-multiplying the elements of the standard basis {α^3, α^6, α^{12}, α^{24}, α^{48}} of GF(2^5){$x^5 + x^2 + 1$}. Analyzing this product table, on each of the columns of A·B there are 15 ones. It is demonstrated [11] that for the number of ones in each column of a cross-product table, called C_N, and sometimes known as **complexity of the basis**, it holds that $C_N \geq 2n - 1$. A normal basis is optimal if $C_N = 2n$ - 1. For example, Table B.25 is the cross-product table of the normal basis {α^5, α^{10}, α^{20}, α^{40}, α^{80}}, of GF(2^5){$x^5 + x^2 + 1$}, and in the same there are 9 ones in each column. In this case $n = 5$ and $C_N = 2n - 1 = 9$. Thus {α^5, α^{10}, α^{20}, α^{40}, α^{80}} is an optimal normal basis, while {α^3, α^6, α^{12}, α^{24}, α^{48}} is not.

Table B.24 Cross-product table for the normal basis {α^3, α^6, α^{12}, α^{24}, α^{48}}

		A·B				
A	B	α^{48}	α^{24}	α^{12}	α^6	α^3
α^3	α^3	0	0	0	1	0
α^3	α^6	0	1	1	1	0
α^3	α^{12}	1	0	1	1	1
α^3	α^{24}	1	1	1	0	1
α^3	α^{48}	0	0	1	1	1
α^6	α^3	0	1	1	1	0
α^6	α^6	0	0	1	0	0
α^6	α^{12}	1	1	1	0	0
α^6	α^{24}	0	1	1	1	1
α^6	α^{48}	1	1	0	1	1
α^{12}	α^3	1	0	1	1	1
α^{12}	α^6	1	1	1	0	0
α^{12}	α^{12}	0	1	0	0	0
α^{12}	α^{24}	1	1	0	0	1
α^{12}	α^{48}	1	1	1	1	0
α^{24}	α^3	1	1	1	0	1
α^{24}	α^6	0	1	1	1	1
α^{24}	α^{12}	1	1	0	0	1
α^{24}	α^{24}	1	0	0	0	0
α^{24}	α^{48}	1	0	0	1	1
α^{48}	α^3	0	0	1	1	1
α^{48}	α^6	1	1	0	1	1
α^{48}	α^{12}	1	1	1	1	0
α^{48}	α^{24}	1	0	0	1	1
α^{48}	α^{48}	0	0	0	0	1

Table B.25 Cross-product table for the normal basis $\{\alpha^5, \alpha^{10}, \alpha^{20}, \alpha^{40}, \alpha^{80}\}$

				$A{\cdot}B$		
A	B	α^{80}	α^{40}	α^{20}	α^{10}	α^5
α^5	α^5	0	0	0	1	0
α^5	α^{10}	0	1	0	0	1
α^5	α^{20}	1	1	0	0	0
α^5	α^{40}	0	0	1	1	0
α^5	α^{80}	1	0	1	0	0
α^{10}	α^5	0	1	0	0	1
α^{10}	α^{10}	0	0	1	0	0
α^{10}	α^{20}	1	0	0	1	0
α^{10}	α^{40}	1	0	0	0	1
α^{10}	α^{80}	0	1	1	0	0
α^{20}	α^5	1	1	0	0	0
α^{20}	α^{10}	1	0	0	1	0
α^{20}	α^{20}	0	1	0	0	0
α^{20}	α^{40}	0	0	1	0	1
α^{20}	α^{80}	0	0	0	1	1
α^{40}	α^5	0	0	1	1	0
α^{40}	α^{10}	1	0	0	0	1
α^{40}	α^{20}	0	0	1	0	1
α^{40}	α^{40}	1	0	0	0	0
α^{40}	α^{80}	0	1	0	1	0
α^{80}	α^5	1	0	1	0	0
α^{80}	α^{10}	0	1	1	0	0
α^{80}	α^{20}	0	0	0	1	1
α^{80}	α^{40}	0	1	0	1	0
α^{80}	α^{80}	0	0	0	0	1

Given a Galois field GF(2^{m}), there is not always optimal normal bases. So the idea is extended by defining the Gaussian normal bases [1]. In GF(2^{m}) there is a standard Gaussian basis Type T if and only if $p = Tm + 1$ is prime and $gcd\left(\frac{Tm}{k}, m\right) = 1$, where k is the order of 2 in GF(p). The optimal normal bases type I and II are Gaussian normal bases type 1 and 2, respectively.

B.4.3 Dual Basis

In GF(2^m) the dual basis are also used, similar to the dual basis as defined in any vector space. To introduce them, the Trace function, *Tr*(), is defined, for an element $e \in$ GF(2^m), as follows:

$$Tr(e) = \sum_{i=0}^{m-1} e^{2^i}$$

It holds that $Tr(e) \in$ GF(2), and $Tr(e)$ is a linear function in GF(2^m). If $A, B \in$ GF(2^m), and $a, b \in$ GF(2), it is verified:

$$Tr(aA + bB) = aTr(A) + bTr(B)$$

Given a basis $\{b_0, b_1, \ldots, b_{m-1}\}$ for GF(2^m), its dual basis is defined as $\{d_0, d_1, \ldots, d_{m-1}\}$, such that:

$$Tr(b_i, d_j) = 0 \text{ if } i \neq j$$
$$Tr(b_i d_j) = 1 \quad \text{if } i = j$$

An element A of GF(2^m) can be expressed in the dual basis as:

$$A = \beta_0 d_0 + \beta_1 d_1 + \cdots + \beta_{m-1} d_{m-1}$$

where:

$$\beta = Tr(Ad_i)$$

For any basis always exists its dual, a single basis, as shown below for standard and normal bases.

Given an standard basis $\{b_0, b_1, \ldots, b_{m-1}\} = \{1, \alpha, \alpha^2, \ldots, \alpha^{m-1}\}$ for GF(2^m){P(x)}, where α is a root of $P(x)$, it is known that its dual basis $\{d_0, d_1, \ldots, d_{m-1}\}$ can be obtained as follows (and as is done in Examples B.12 and B.13) [8]:

(a) $P(x)$ is factorized as $P(x) = (x + \alpha)\left(\sum_{i=0}^{m-1} = c_i x^i\right)$;
(b) the derivative $P'(x)$ is calculated;
(c) the elements of the dual basis are $d_i = \frac{c_i}{P'(\alpha)}$

Example B.12 **Given the standard basis, $\{1, \alpha, \alpha^2, \alpha^3\}$ in GF(2^4)$\{x^4 + x^3 + 1\}$, being α a root of $x^4 + x^3 + 1$, to calculate its dual base:**

$$P(x) = x^4 + x^3 + 1 = (x + \alpha)\left(x^3 + (\alpha + 1)x^2 + \alpha(\alpha + 1)x + \alpha^2(\alpha + 1)\right)$$

According to Table B.20:

$$\alpha + 1 = \alpha^{12}, \alpha(\alpha + 1) = \alpha^2 + \alpha = \alpha^{13}, \alpha^2(\alpha + 1) = \alpha^2 + \alpha^2 = \alpha^{14}$$

Moreover,

$$P'(\alpha) = \alpha^2$$

It results:

$$d_0 = \frac{\alpha^{14}}{\alpha^2} = \alpha^{12}; \quad d_1 = \frac{\alpha^{13}}{\alpha^2} = \alpha^{11}; \quad d_2 = \frac{\alpha^{12}}{\alpha^2} = \alpha^{10}; \quad d_3 = \frac{1}{\alpha^2} = \frac{\alpha^{15}}{\alpha^2} = \alpha^{13}$$

It is easy to check that $\{\alpha^{12}, \alpha^{11}, \alpha^{10}, \alpha^{13}\}$ is the dual basis of $\{1, \alpha, \alpha^2, \alpha^3\}$. ■

Example B.13 **Calculate the dual basis of the standard basis $\{1, \alpha, \alpha^2, \alpha^3\}$ in GF(2^4)$\{x^4 + x + 1\}$.**

In this case:

$$P(x) = x^4 + x + 1 = (x + \alpha)\left(x^3 + \alpha x^2 + \alpha^2 x + \left(\alpha^3 + 1\right)\right)$$

$$P'(\alpha) = 1$$

Thus:

$$d_0 = 1; d_1 = \alpha; d_2 = \alpha^2; d_3 = \alpha^3 + 1 = \alpha^{14}. \blacksquare$$

Given an element of GF(2^m)$\{P(x)\}$ represented in standard basis $\{1, \alpha, \alpha_2, \ldots, \alpha_{m-1}\}$, for obtaining their representation in the corresponding dual basis it is sufficient to apply the definition of dual base such as in the following example.

Example B.14 In GF(2^4)$\{x^4 + x^3 + 1\}$, given and element A represented in the standard basis $\{b_0, b_1, b_2, b_3\} = \{1, \alpha, \alpha^2, \alpha^3\}$ ($A = a_0 + a_1\alpha + a_2\alpha^2 + a_3\alpha^3$), being α a root of $x^4 + x^3 + 1$, to calculate its representation in the dual basis $\{d_0, d_1, d_2, d_3\} = \{\alpha^{12}, \alpha^{11}, \alpha^{10}, \alpha^{13}\}$ ($A = \beta_0\alpha^{12} + \beta_1\alpha^{11} + \beta_2\alpha^{10} + \beta_3\alpha^{13}$), it is applied that $\beta_i = Tr(Ab_i)$. Thus:

$$\begin{aligned}
\beta_0 &= Tr(a_0) + Tr(a_1\alpha) + \mathrm{Tr}\left(a_2\alpha^2\right) + Tr\left(a_3\alpha^3\right) \\
&= a_1 Tr(\alpha) + a_2 Tr\left(\alpha^2\right) + a_3 Tr\left(\alpha^3\right) = a_1 + + a_2 + a_3 \\
\beta_1 &= Tr(a_0\alpha) + Tr\left(a_1\alpha^2\right) + Tr\left(a_2\alpha^3\right) + Tr\left(a_3\alpha^4\right) \\
&= a_0 Tr(\alpha) + a_1 Tr\left(\alpha^2\right) + a_2 Tr\left(\alpha^3\right) \\
&\quad + a_3 Tr\left(\alpha^4\right) = a_0 + a_1 + a_2 + a_3 \\
\beta_2 &= Tr\left(a_0\alpha^2\right) + Tr\left(a_1\alpha^3\right) + Tr\left(a_2\alpha^4\right) + Tr\left(a_3\alpha^5\right) \\
&= a_0 Tr\left(\alpha^2\right) + a_1 Tr\left(\alpha^3\right) + a_2 Tr\left(\alpha^4\right) \\
&\quad + a_3 Tr\left(\alpha^5\right) = a_0 + a_1 + a_2
\end{aligned}$$

$$\begin{aligned}\beta_3 &= Tr(a_0\alpha^3) + Tr(a_1\alpha^4) + Tr(a_2\alpha^5) + Tr(a_3\alpha^6)\\ &= a_0 Tr(\alpha^3) + a_1 Tr(\alpha^4) + a_2 Tr(\alpha^5)\\ &+ a_3 Tr(\alpha^6) = a_0 + a_1 + a_3.\blacksquare\end{aligned}$$

The conversion from the representation in dual basis to the representation in standard basis, can be made as follows. Given:

$$A = \beta_0 d_0 + \beta_1 d_1 + \cdots + \beta_{m-2} d_{m-2} + \beta_{m-1} d_{m-1}$$

it consists on obtaining $a_0, a_1, \ldots, a_{m\text{-}2}, a_{m\text{-}1}$ such that:

$$A = a_0 + a_1\alpha + \cdots + a_{m-2}\alpha^{m-2} + a_{m-1}\alpha^{m-1}$$

where $\{d_0, d_1, \ldots, d_{m\text{-}1}\}$ is the dual basis of $\{1, \alpha, \alpha_2, \ldots, \alpha_{m\text{-}1}\}$.

From the definition of dual basis it is immediate that $Tr(Ad_i) = a_i$. In fact:

$$\begin{aligned}&Tr((a_0 + a_1\alpha + \cdots + a_{m-2}\alpha^{m-2} + a_{m-1}\alpha^{m-1} - 1)d_i)\\ &= a_0 Tr(d_i) + a_1 Tr(d_i\alpha) + \cdots + a_{m-2} Tr(d_i\alpha^{m-2})\\ &+ a_{m-1} Tr(d_i\alpha^{m-1}) = a_i Tr(d_1\alpha^i) = a_i\end{aligned}$$

The representation in standard basis is obtained applying this result, such as in the following example.

Example B.15 In GF(2^4)$\{x^4 + x^3 + 1\}$, given an element A represented in the dual basis $\{d_0, d_1, d_2, d_3\} = \{\alpha^{12}, \alpha^{11}, \alpha^{10}, \alpha^{13}\}$ ($A = \beta_0\alpha^{12} + \beta_1\alpha^{11} + \beta_2\alpha^{10} + \beta_3\alpha^{13}$), its representation in the standard basis $\{1, \alpha, \alpha^2, \alpha^3\}$ ($A = a_0 + a_1\alpha + a_2\alpha^2 + a_3\alpha^3$), being α a root of $x^4 + x^3 + 1$, can be obtained as follows:

$$\begin{aligned}a_0 = Tr(Ad_0) &= Tr((\beta_0\alpha^{12} + \beta_1\alpha^{11} + \beta_2\alpha^{10} + \beta_3\alpha^{13})\alpha^{12})\\ &= \beta_0 Tr(\alpha^{24}) + \beta_1 Tr(\alpha^{23}) +\\ &+ \beta_2 Tr(\alpha^{22}) + \beta_3 Tr(\alpha^{25}) = \beta_0 + \beta_1\\ a_1 = Tr(Ad_1) &= Tr((\beta_0\alpha^{12} + \beta_1\alpha^{11} + \beta_2\alpha^{10} + \beta_3\alpha^{13})\alpha^{11})\\ &= \beta_0 Tr(\alpha^{23}) + \beta_1 Tr(\alpha^{22}) +\\ &+ \beta_2 Tr(\alpha^{21}) + \beta_3 Tr(\alpha^{24}) = \beta_0 + \beta_2 + \beta_2\\ a_2 = Tr(Ad_2) &= Tr((\beta_0\alpha^{12} + \beta_1\alpha^{11} + \beta_2\alpha^{10} + \beta_3\alpha^{13})\alpha^{10})\\ &= \beta_0 Tr(\alpha^{22}) + \beta_1 Tr(\alpha^{21}) +\\ &+ \beta_2 Tr(\alpha^{20}) + \beta_3 Tr(\alpha^{23}) = \beta_1 + \beta_3\\ a_3 = Tr(Ad_3) &= Tr((\beta_0\alpha^{12} + \beta_1\alpha^{11} + \beta_2\alpha^{10} + \beta_3\alpha^{13})\alpha^{13})\\ &= \beta_0 Tr(\alpha^{25}) + \beta_1 Tr(\alpha^{24}) +\\ &+ \beta_2 Tr(\alpha^{23}) + \beta_3 Tr(\alpha^{26}) = \beta_1 + \beta_2.\blacksquare\end{aligned}$$

The transformation form a base to another in GF(2^m), in any case, is a linear application that can be represented by an $m \times m$ array, and it is implemented with relatively simple circuitry. In the case that the base is its dual (i.e. it is a self-dual basis) it is obvious that no circuit is required for this transformation. It is known that standard bases can not be self-dual [10]. However, the same situation of simplicity, with regard to the transformation circuitry is concerned, is achieved with the so-called *almost self-dual bases* that are defined below.

Given any basis $\{\alpha_0, \alpha_1, \ldots, \alpha_{m-1}\}$ of GF(2^m), it is said that this basis *is weakly self-dual* (also called *almost self-dual*) if its dual basis, $\{d_0, d_1, \ldots, d_{m-1}\}$, can be obtained from the relationship:

$$d_i = A\alpha_{\Pi(i)} \quad A \in \mathrm{GF}(2^m)$$

where $\prod$ is a permutation of the indices $\{0, 1, \ldots, m-1\}$.

I.e., the elements of the dual basis are obtained from the starting basis, making a permutation and multiplying them by a constant. In [3] it is demonstrated that in GF(2^m)$\{P(x)\}$ there is a *weakly self-dual standard basis* if $P(x)$ is a trinomial ($P(x) = x^m + x^k + 1$). In this case the permutation $\prod$ is given as follows:

$$\Pi(i) = (k - 1 - i)\,\mathbf{mod}\,m$$

The constant $A \in$ GF(2^m) is given by:

$$A = \frac{\alpha^{m-k} + 1}{P'(\alpha)}$$

For $k = 1$ and m odd ($m = 1\mathbf{mod}2$) it results $A = 1$. In this case the dual base is merely a permutation of the starting basis, as can be seen in Example B.16.

Example B.16 For $m = 7$, $x^7 + x + 1$ is a primitive polynomial. Thus, in GF(2^7)$\{x^7 + x + 1\}$, the dual basis of $\{1, \alpha, \alpha^2, \alpha^3, \alpha^4, \alpha^5, \alpha^6\}$ can be obtained using a simple permutation. Concretely this permutation, applying $\prod(i) = (k - 1 - i)$ **mod**m, is $\langle 0, 6, 5, 4, 3, 2, 1\rangle$:

$$\begin{aligned}
\Pi(0) &= 0\,\mathbf{mod}\,7 = 0\\
\Pi(1) &= (-1)\,\mathbf{mod}\,7 = 6\\
\Pi(2) &= (-2)\,\mathbf{mod}\,7 = 5\\
\Pi(3) &= (-3)\,\mathbf{mod}\,7 = 4\\
\Pi(4) &= (-4)\,\mathbf{mod}\,7 = 3\\
\Pi(5) &= (-5)\,\mathbf{mod}\,7 = 2\\
\Pi(6) &= (-6)\,\mathbf{mod}\,7 = 1
\end{aligned}$$

I.e. the dual basis of $\{1, \alpha, \alpha^2, \alpha^3, \alpha^4, \alpha^5, \alpha^6\}$ is $\{1, \alpha^6, \alpha^5, \alpha^4, \alpha^3, \alpha^2, \alpha\}$, as can easily be checked using the table of elements of GF(2^7)$\{x^7 + x + 1\}$ (Table B.26). An element is given in each cell of this table, indicating the power and the corresponding

Table B.26 Table of elements of GF(2^7){$x^7 + x + 1$}

– ∞:0000000	31:0001011	63:0001001	95:0100101
0:0000001	32:0010110	64:0010010	96:1001010
1:0000010	33:0101100	65:0100100	97:0010111
2:0000100	34:1011000	66:1001000	98:0101110
3:0001000	35:0110011	67:0010011	99:1011100
4:0010000	36:1100110	68:0100110	100:0111011
5:0100000	37:1001111	69:1001100	101:1110110
6:1000000	38:0011101	70:0011011	102:1101111
7:0000011	39:0111010	71:0110110	103:1011101
8:0000110	40:1110100	72:1101100	104:0111001
9:0001100	41:1101011	73:1011011	105:1110010
10:0011000	42:1010101	74:0110101	106:1100111
11:0110000	43:0101001	75:1101010	107:1001101
12:1100000	44:1010010	76:1010111	108:0011001
13:1000011	45:0100111	77:0101101	109:0110010
14:0000101	46:1001110	78:1011010	110:1100100
15:0001010	47:0011111	79:0110111	111:1001011
16:0010100	48:0111110	80:1101110	112:0010101
17:0101000	49:1111100	81:1011111	113:0101010
18:1010000	50:1111011	82:0111101	114:1010100
19:0100011	51:1110101	83:1111010	115:0101011
20:1000110	52:1101001	84:1110111	116:1010110
21:0001111	53:1010001	85:1101101	117:0101111
22:0011110	54:0100001	86:1001101	118:1011110
23:0111100	55:1000010	87:0110001	119:0111111
24:1111000	56:0000111	88:1100010	120:1111110
25:1110011	57:0001110	89:1000111	121:1111111
26:1100101	58:0011100	90:0001101	122:1111101
27:1001001	58:0111000	91:0011010	123:1111001
28:0010001	60:1110000	92:0110100	124:1110001
29:0100010	61:1100011	93:1101000	125:1100001
30:1000100	62:1000101	94:1010011	126:1000001

polynomial; for example, the first cell of the second column is 31:0001011, and it means $x^{31} = x^3 + x + 1$.

Using the Table B.26 it is straightforward to check that:

$$Tr(1) = 1;\ Tr(\alpha^7) = 1;\ Tr(\alpha^6) = 0;$$
$$Tr(\alpha^6) = 0;\ Tr(\alpha^4) = 0;\ Tr(\alpha^3) = 0;\ Tr(\alpha^2) = 0;$$

$$Tr(\alpha) = 0; Tr(\alpha^8) = 0; Tr(\alpha^9) = 0;$$
$$Tr(\alpha^{10}) = 0; Tr(\alpha^{11}) = 0; Tr(\alpha^{12}) = 0. \blacksquare$$

Let suppose an standard basis $\{b_0, b_1, \ldots, b_{m\text{-}1}\} = \{1, \alpha, \alpha^2, \ldots, \alpha^{m-1}\}$ over GF(2^m)$\{P(x)\}$, being α a root of $P(x)$, and its dual basis $\{d_0, d_1, \ldots, d_{m-1}\}$. An element $A \in$ GF(2^m)$\{P(x)\}$ is given in the dual basis as:

$$A = \beta_0 d_0 + \beta_1 d_1 + \cdots + \beta_{m-2} d_{m-2} + \beta_{m-1} d_{m-1}$$

If A is multiplied by α it results:

$$Q = \alpha A = q_0 d_0 + q_1 d_1 + \ldots + q_{m-2} d_{m-2} + q_{m-1} d_{m-1}$$

Applying the definition of dual basis, for the different coefficients of αA it results:

$$\begin{gathered} q_0 = Tr(Q) = \beta_1; \\ q_1 = Tr(\alpha\, Q) = \beta_2; \\ \ldots; \\ q_i = Tr(\alpha^i Q) = \beta_{i+1}; \\ \ldots; \\ q_{m-1} = Tr(\alpha^{m-1} Q) = Tr(\alpha^m A) \end{gathered}$$

I.e., αA can be obtained from A by a simple rotation of the coefficients, except for the most significant, that can be calculated using $P(x)$. If:

$$P(x) = x^m + p_{m-1} x^{m-1} + \cdots + p_1 x + p_0$$

then:

$$\alpha^n = p_{m-1} \alpha^{m-1} + \cdots + p_1 \alpha + p_0$$

Thus:

$$\begin{aligned} q_{m-1} = Tr(\alpha^m A) &= Tr\big((p_{m-1})\alpha^{m-1} + \cdots + p_1 \alpha + p_0\big) \\ &\quad (\beta_0 d_0 + \beta_1 d_1 + \cdots + \beta_{m-1} d_{m-1}) \\ &= \beta_0 p_0 + \beta_1 p_1 + \cdots + \beta_{m-1} p_{m-1} \end{aligned}$$

Using a circuit whose kernell is an **LFSR1**, αA, $\alpha^2 A$ $\alpha^3 A$, …, can be obtained from A as it is shown in Fig. B.3.

Let suppose $\{b_0, b_1, \ldots, b_{m-1}\} = \{B, B^2, B^4, \ldots, B^{2^{m-2}}, B^{2^{m-1}}\}$, a normal basis whose generator is B. The steps to calculate its dual basis, that is also a normal basis, are the following [10, 16] (as it is made in Example B.17):

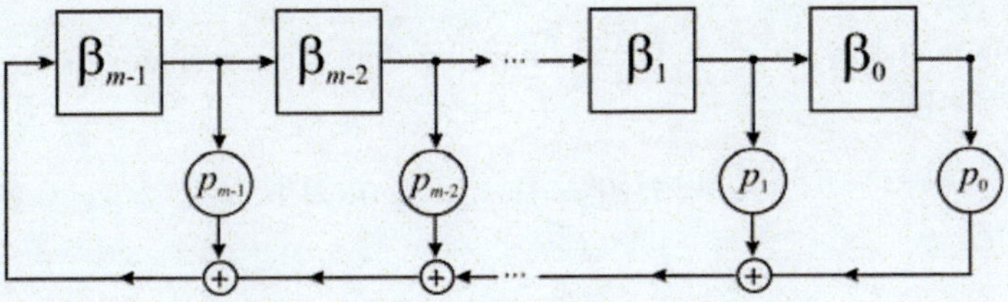

Fig. B.3 **LFSR1** for calculating αA

(a) the coefficients $c_i = Tr(Bb_i)$ of the polynomial $p(x) = \sum_{i=0}^{m-1} = c_i x^i$ are calculated;
(b) the polynomial $q(x) = \sum_{i=0}^{m-1} d_i x^i$ such that $p(x)q(x) = 1\mathbf{mod}(x^n + 1)$ is calculated;
(c) the dual basis of $\{B, B^2, B^4, \ldots, B^{2^{m-2}}, B^{2^{m-1}}\}$ is:

$$\{D, D^2, D^4, \ldots, D^{2^{m-2}}, D^{2^{m-1}}\}$$

being $D = \sum_{i=0}^{m-1} b_i d_i$.

Example B.17 In Example B.9 it is concluded that $\{b_0, b_1, b_2, b_3\} = \{\alpha^3, \alpha^6, \alpha^{12}, \alpha^{24}\}$ is a Type I optimal normal basis for GF(2^4)$\{x^4 + x^3 + 1\}$, being α a root of $x^4 + x^3 + 1$. The generator element is B = α^3. For calculating its dual basis using the above procedure, the following coefficients have to be calculated:

$$\begin{aligned} c_0 &= Tr\left(\alpha^3 \cdot \alpha^3\right) = 1; \\ c_1 &= Tr\left(\alpha^3 \cdot \alpha^6\right) = 1; \\ c_2 &= Tr\left(\alpha^3 \cdot \alpha^{12}\right) = 0; \\ c_3 &= Tr\left(\alpha^3 \cdot \alpha^{24}\right) = 1 \end{aligned}$$

It results:

$$p(x) = x^3 + x + 1$$

Given:

$$q(x) = d_3x^3 + d_2x^2 + d_1x + d_0$$

for the product $p(x)\cdot q(x)$ it results:

$$\begin{aligned} p(x) \cdot q(x) = d_3x^6 + d_2x^5 + (d_1 + d_3)x^4 \\ + (d_0 + d_2 + d_3)x^3 + (d_1 + d_2)x^2 \\ + (d_0 + d_1)x + d_0 \end{aligned}$$

Dividing $p(x) \cdot q(x)$ by $x^4 + 1$, and imposing than the remainder of the division equals 1, it results:

$$d_0 = 1;\ d_1 = 1;\ d_2 = 0;\ d_3 = 1$$

Thus:

$$D = \sum_{i=0}^{m-1} b_i d_i = \alpha^3 + \alpha^6 + \alpha^{24} = \alpha$$

and $\{\alpha, \alpha^2, \alpha^4, \alpha^8\}$ is the dual basis of $\{\alpha^3, \alpha^6, \alpha^{12}, \alpha^{24}\}$, as can be easily checked. ■

In [7] a method for obtaining self-dual normal bases, if exists, is presented. Specifically, irreducible polynomials whose roots can be used as sources of self-dual normal bases are obtained, as can be seen in the following example.

Example B.18 **In [7] it is concluded that for $n = 5$, using the primitive polynomial $x^5 + x^4 + x^2 + x + 1$, and being α a root of this polynomial, $\{\alpha, \alpha^2, \alpha^4, \alpha^8, \alpha^{16}\}$ is a self-dual normal basis of GF(2^5)$\{x^5 + x^4 + x^2 + x + 1\}$. Check this statement.**

Using Table B.27, in which the different elements of GF(2^5)$\{x^5 + x^4 + x^2 + x + 1\}$ are given, it is immediate to check that:

$$Tr(\alpha^2) = 1;\ Tr(\alpha^4) = 1;\ Tr(\alpha^8) = 1;\ Tr(\alpha^{16}) = 1;\ Tr(\alpha^{32}) = 1$$
$$Tr(\alpha^3) = 0;\ Tr(\alpha^5) = 0;\ Tr(\alpha^9) = 0;\ Tr(\alpha^{17}) = 0;\ Tr(\alpha^6) = 0;$$
$$Tr(\alpha^{10}) = 0;\ Tr(\alpha^{18}) = 0;\ Tr(\alpha^{12}) = 0;\ Tr(\alpha^{20}) = 0;\ Tr(\alpha^{24}) = 0.$$

It is therefore a self-dual basis.

Calculating cross-products for this self-dual basis, as given in Table B.28, it can be seen that the number of ones in each column is 9 (that is, $9 = 2n - 1$). It is therefore an optimal basis. ■

Table B.27 Elements of GF(2^5)$\{x^5 + x^4 + x^2 + x + 1\}$

$-\infty$	0	7	x^2+1	15	x^4+x^2+1	23	x^2+x+1
0	1	8	x^3+x	16	$x^4+x^3+x^2+1$	24	x^3+x^2+x
1	x	9	x^4+x^2	17	x^3+x^2+1	25	$x^4+x^3+x^2$
2	x^2	10	$x^4+x^3+x^2+x+1$	18	x^4+x^3+x	26	x^3+x^2+x+1
3	x^3	11	x^3+1	19	$x+1$	27	$x^4+x^3+x^2+x$
4	x^4	12	x^4+x	20	x^2+x	28	x^3+x+1
5	x^4+x^2+x+1	13	x^4+x+1	21	x^3+x^2	29	x^4+x^2+x
6	x^4+x^3+1	14	x^4+1	22	x^4+x^3	30	x^4+x^3+x+1

Table B.28 Cross-product table for the normal basis $\{\alpha, \alpha^2, \alpha^4, \alpha^8, \alpha^{16}\}$

				$A \cdot B$		
A	B	α^{16}	α^8	α^4	α^2	α
α	α	0	0	0	1	0
α	α^2	0	1	0	0	1
α	α^4	1	1	0	0	0
α	α^8	0	0	1	1	0
α	α^{16}	1	0	1	0	0
α^2	α	0	1	0	0	1
α^2	α^2	0	0	1	0	0
α^2	α^4	1	0	0	1	0
α^2	α^8	1	0	0	0	1
α^2	α^{16}	0	1	1	0	0
α^4	α	1	1	0	0	0
α^4	α^2	1	0	0	1	0
α^4	α^4	0	1	0	0	0
α^4	α^8	0	0	1	0	1
α^4	α^{16}	0	0	0	1	1
α^8	α	0	0	1	1	0
α^8	α^2	1	0	0	0	1
α^8	α^4	0	0	1	0	1
α^8	α^8	1	0	0	0	0
α^8	α^{16}	0	1	0	1	0
α^{16}	α	1	0	1	0	0
α^{16}	α^2	0	1	1	0	0
α^{16}	α^4	0	0	0	1	1
α^{16}	α^8	0	1	0	1	0
α^{16}	α^{16}	0	0	0	0	1

Other irreducible polynomials over GF(2) whose roots can be used as sources of self-dual normal bases and are also given in [7]; these are $x^3 + x^2 + 1$; $x^6 + x^5 + x^4 + x + 1$.

Besides the above, other bases have been proposed for different applications; as example, the Refs. [5, 17] can be consulted.

B.4.4 Inverse

The algorithm for calculating the greatest common divisor of two polynomials can be used to obtain the inverse in GF(2^m){$P(x)$} of a polynomial $Q(x) \in$GF(2^m){$P(x)$}. Indeed, as $Q(x)$ is always prime to $P(x)$, $gcd\{P(x), Q(x)\}=1$, and as indicated in Sect. B.1.1, given $Q(x)$, another polynomial $B(x)$ such that $B(x) \cdot Q(x) = 1$, $B(x) = Q^{-1}$ (x) can be found. The Algorithm B.1 is readily adapted to calculate $B(x)$, being as follows:

Algorithm B.5

First algorithm for the inverse of a polynomial

```
1) R1 ← P(x), R2 ← Q(x), B1 ← 0, B2 ← 1;
while R2 ≠ 0 do
    begin
    2) Q ← COC(R1, R2), TEMP ← R2;
    3) R2 ← R1 - QR2, R1 ← TEMP, TEMP ← B2;
    4) B2 ← B1 - QB2, B1 ← TEMP;
    end
```

End algorithm

After applying Algorithm B.5, the register $B1$ contains $Q^{-1}(x)$. An example of application is developed at the following.

Example B.19 **Calculate the inverse of $Q(x) = x^3 + x + 1$ over GF(2^4){$x^4 + x^3 + 1$}.**

Table B.29 shows the content of the registers in the different steps of the application of Algorithm B.5. It results $Q^{-1}(x) = x^3 + x$. In fact, it is immediate to check that $(x^3 + x + 1)(x^3 + x) = 1$ over GF(2^4){$x^4 + x^3 + 1$}. ■

In the Algorithm B.5 the process of division by successive subtraction can be incorporated, given in the Algorithm B.3. Moreover, in this case really only the remainder of the division is interesting. The modified algorithm is the following:

Algorithm B.6

Table B.29 Example B.19

	$R1$	$R2$	$B1$	$B2$	Q	TEMP
1	x^4+x^3+1	x^3+x+1	0	1		
2					$x+1$	x^3+x+1
3	x^3+x+1	x^2				1
4			1	$x+1$		
2					x	x^2
3	x^2	$x+1$				$x+1$
4			$x+1$	x^2+x+1		
2					x	$x+1$
3	x	$x+1$				x^2+x+1
4			x^2+x+1	x^3+x^2+1		
2					1	$x+1$
3	$x+1$	1				x^3+x^2+1
4			x^3+x^2+1	x^3+x		
2					$x+1$	1
3	1	0				x^3+x
4			x^3+x	x^3+x		

Modified algorithm for the inverse of a polynomial

```
1) R1 ← P(x), R2 ← Q(x), B1 ← 0, B2 ← 1;
while R2 ≠ 0 do
    begin
      2) e ← g(R1) − g(R2);
     while e ≥ 0 do
         begin
           3) R1 ← R1 - x^e R2, B1 ← B1- x^e B2;
           4) e ← g(R1) − g(R2);
          end
      5) R1 ← R2, B1 ← B2;
    end
```

End algorithm

Again, $B1$ contains $Q^{-1}(x)$.

B.5 Finite Fields GF(p^m)

Polynomials with coefficients over GF(p) and of degree less than m, with the operations addition of polynomials and product of polynomials modulo a primitive polynomial of degree m, $P(x)$, form a finite field of characteristic p or Galois field GF(p^m);

Table B.30 Power base and standard base (using $\{1, \alpha, \alpha^2\}$) representation of the elements of $GF(3^3)\{x^3 + 2x + 1\}$

		$\{1, \alpha, \alpha^2\}$				$\{1, \alpha, \alpha^2\}$	
$\alpha^{-\infty} = 0$	11111	0	000	α^{13}	01101	2	200
$\alpha^0 = 1$	00000	1	100	α^{14}	01110	2α	020
α	00001	α	010	α^{15}	01111	$2\alpha^2$	002
α^2	00010	α^2	001	α^{16}	10000	$2\alpha + 1$	120
α^3	00011	$\alpha + 2$	210	α^{17}	10001	$2\alpha^2 + \alpha$	012
α^4	00100	$\alpha^2 + 2\alpha$	021	α^{18}	10010	$\alpha^2 + 2\alpha + 1$	121
α^5	00101	$2\alpha^2 + \alpha + 2$	212	α^{19}	10011	$2\alpha^2 + 2\alpha + 2$	222
α^6	00110	$\alpha^2 + \alpha + 1$	111	α^{20}	10100	$2\alpha^2 + \alpha + 1$	112
α^7	00111	$\alpha^2 + 2\alpha + 2$	221	α^{21}	10101	$\alpha^2 + 1$	101
α^8	01000	$2\alpha^2 + 2$	202	α^{22}	10110	$2\alpha + 2$	220
α^9	01001	$\alpha + 1$	110	α^{23}	10111	$2\alpha^2 + 2\alpha$	022
α^{10}	01010	$\alpha^2 + \alpha$	011	α^{24}	11000	$2\alpha^2 + 2\alpha + 1$	122
α^{11}	01011	$\alpha^2 + \alpha + 2$	211	α^{25}	11001	$2\alpha^2 + 1$	102
α^{12}	01100	$\alpha^2 + 2$	201				

this Galois field can also be represented as $GF(p^m)\{P(x)\}$. As an example of Galois field, $GF(3^2)\{x^2 + x + 2\}$ can be used, described above in Example B.7.

In $GF(p^m)\{P(x)\}$ the order of any element must be a divisor of $p^m - 1$. The order of the primitive elements is precisely $p^m - 1$.

As it is known (see Appendix A), for any element B of $GF(p^m)\{P(x)\}$ it is verified that:

$$B^{p^m} = B$$

This is equivalent to $B^{p^m-1} = 1$, or also that $B^{p^m-2} = B^{-1}$.

As seen, the different non-zero elements of $GF(p^m)\{P(x)\}$ can be represented as powers of a primitive root **(potential representation)** or as a polynomial of degree less than m **(polynomial representation)**. As an example, in the first and fifth columns of Table B.30, the different elements of $GF(3^3)\{x^3 + 2x + 1\}$ as powers of a primitive root α are given. In the second and sixth columns the binary representation of these powers is shown. In the third and seventh columns, the same elements expressed as polynomials of the root α, of degree less than 3 are given. Finally, in the fourth and eighth columns the coefficients of the polynomial development are given.

With regard to the potential representation, each element can be represented by the binary value of the exponent, and to represent the zero element the combination of all ones can be used, that does not appear in the representation of the non-zero elements.

Depending on the form of representation of the elements of $GF(p^m)\{P(x)\}$, the different operations with the corresponding polynomials can be more or less complex.

For example, the addition is more easily executed in the polynomial representation, while the potential representation is more suitable for the multiplication. Next different types of bases are considered.

B.5.1 Standard Basis

In the representation of the elements of GF(p^{m}){$P(x)$} can be used the basis $\{1, \alpha, \alpha^2, \alpha^3, \ldots, \alpha^{m-2}, \alpha^{m-1}\}$, called as standard basis (or polynomial), being α any primitive element of GF(p^{m}){$P(x)$}.

Selected a basis, each element of GF(p^{m}){$P(x)$}can be represented with an m-tuple over GF(p), which are the coefficients of the elements of the basis. This is what is done in Table B.30 using the basis $\{1, \alpha, \alpha^2\}$, where α is a root of $x^3 + 2x + 1$. Starting with any other primitive element, other standard bases will result.

B.5.2 Normal Basis

A commonly used basis is known as the **normal basis**, which is of the form $\{B, B^p, B^{p^2}, \ldots, B^{p^{m-2}}, B^{p^{m-1}}\}$, where B is a suitable element of GF(p^{m}){$P(x)$}, non necessarily a primitive element, such that the various elements of $\{B, B^p, B^{p^2}, \ldots, B^{p^{m-2}}, B^{p^{m-1}}\}$ are linearly independent.

Normal basis always exit for each finite field. For example, in GF(3^3){$x^3 + 2x + 1$}, a normal basis is $\{\alpha^2, \alpha^6, \alpha^{18}\}$, α being a primitive root of $x^3 + 2x + 1$. On this basis the different elements of GF(3^3){$x^3 + 2x + 1$} are given in Table B.31 and they are represented vectorially as shown in the third and sixth columns of the same table. Note that, in this case, the polynomials corresponding to the different elements of GF(2^{m}){$P(x)$} do not have to be of lower degree than m, but will always be equal to a polynomial of degree less than m in standard base.

It is easy to see that, when the vector representation is used with a normal basis, raising any element to the power p consists on rotating one position to the left its vector representation. This is the great advantage of normal bases. Recall that, by Theorem 3.3, *if p is a prime number and a is a positive integer, where p and a are relatively primes, then R(a^p, p) = R(a, p), i.e., operating modulo p, a^p=a.*

Let suppose the normal basis $\{B, B^p, B^{p^2}, \ldots, B^{p^{m-2}}, B^{p^{m-1}}\}$, and E any element represented in this basis:

$$E = e_{m-1}B^{p^{m-1}} + e_{m-2}B^{p^{m-2}} + \cdots + e_1 B^p + e_0 B$$

When calculating E^P, by applying that $(e_1 + \cdots + e_n)^p = (e_1)^p + \cdots + (e_n)^p$ (see Apendix A), it results:

Table B.31 Representation of the elements of GF(3^3){$x^3 + 2x + 1$} with the normal basis {α^2, α^6, α^{18}}

0	0	000	α^{13}	$\alpha^{18} + \alpha^6 + \alpha^2$	111
1	$2\alpha^{18} + 2\alpha^6 + 2\alpha^2$	222	α^{14}	$2\alpha^{18} + \alpha^6$	012
α	$\alpha^{18} + 2\alpha^6$	021	α^{15}	$2\alpha^2$	200
α^2	α^2	100	α^{16}	$\alpha^{18} + 2\alpha^2$	201
α^3	$2\alpha^{18} + \alpha^2$	102	α^{17}	$\alpha^{18} + 2\alpha^6 + 2\alpha^2$	221
α^4	$2\alpha^{18} + \alpha^6 + \alpha^2$	112	α^{18}	α^{18}	001
α^5	$2\alpha^{18}$	002	α^{19}	$2\alpha^6$	020
α^6	α^6	010	α^{20}	$\alpha^6 + \alpha^2$	110
α^7	$2\alpha^6 + 2\alpha^2$	220	α^{21}	$2\alpha^{18} + 2\alpha^6$	022
α^8	$\alpha^{18} + \alpha^6$	011	α^{22}	$2\alpha^6 + \alpha^2$	120
α^9	$\alpha^6 + 2\alpha^2$	210	α^{23}	$2\alpha^{18} + \alpha^6 + 2\alpha^2$	212
α^{10}	$\alpha^{18} + 2\alpha^6 + \alpha^2$	121	α^{24}	$\alpha^{18} + \alpha^2$	101
α^{11}	$2\alpha^{18} + 2\alpha^2$	202	α^{25}	$2\alpha^{18} + 2\alpha^6 + \alpha^2$	122
α^{12}	$\alpha^{18} + \alpha^6 + 2\alpha^2$	211			

$$
\begin{aligned}
E^P &= (e_{m-1}B^{p^{m-1}})^p + (e_{m-2}B^{p^{m-2}})^p + \cdots + (e_1B^p)^p + (e_0B)^p \\
&= (e_{m-1})^p B^{p^m} + (e_{m-2})^p B^{p^{m-1}} + \cdots + (e_1)^p B^{p^2} + (e_0)^p B^p \\
&= e_{m-1}B^{p^m} + e_{m-2}B^{p^{m-1}} + \cdots + e_1 B^{p^2} + e_0 B^p
\end{aligned}
$$

But $B^{p^m} = B$. Thus:

$$E^P = e_{m-2}B^{p^{m-1}} + \cdots + e_1 B^{p2} + e_0 B^p + e_{m-1}B$$

That is, given the vector representation of E with a normal base, $E = (e_{m-1}, e_{m-2}, \ldots, e_1, e_0)$, E^p is obtained with a rotation to the left of E, i.e., it results $E^p = (e_{m-2}, e_{m-3}, \ldots, e_1, e_0, e_{m-1})$.

As a result of this, it is easy to see that the vector representation of any constant of GF(p^m) on a normal basis is always of the form (c, c, ..., c). Indeed, as $a^p = a$, the single vector that is reproduced after rotating one position is what only includes the same constant.

Also, as a result of this, it is immediate that, with a normal basis, $\sqrt[p]{E}$is obtained with a rotation to the right of E, $\sqrt[p]{E} = (e_0, e_{m-1}, e_{m-2}, \ldots, e_2, e_1)$.

In some cases there are normal bases that simplify the implementation of the multiplication: these are the **optimal normal bases** [10, 11]. For the **complexity of a normal basis**, C_N, defined as the number of non-zero values in each column of their cross-products table (as in GF(2^m)), it holds that $C_N \geq 2m - 1$. If the basis is optimal then $C_N = 2m - 1$.

In GF(p^m){$P(x)$} there is an optimal normal basis (see [11]) if $m + 1$ is prime and p is a primitive element of GF(m +1); this optimal normal basis is generated from

Table B.32 Cross-product table for the normal basis $\{x^{16}, x^{48}, x^{144}, x^{432}\}$

A	*B*	x^{432}	x^{144}	x^{48}	x^{16}
x^{16}	x^{16}	1	0	0	0
x^{16}	x^{48}	0	1	0	0
x^{16}	x^{144}	2	2	2	2
x^{16}	x^{432}	0	0	1	0
x^{48}	x^{16}	0	1	0	0
x^{48}	x^{48}	0	0	0	1
x^{48}	x^{144}	1	0	0	0
x^{48}	x^{432}	2	2	2	2
x^{144}	x^{16}	2	2	2	2
x^{144}	x^{48}	1	0	0	0
x^{144}	x^{144}	0	0	1	0
x^{144}	x^{432}	0	0	0	1
x^{432}	x^{16}	0	0	1	0
x^{432}	x^{48}	2	2	2	2
x^{432}	x^{144}	0	0	0	1
x^{432}	x^{432}	0	1	0	0

a root $m + 1$ of the unit that is different from 1. An example of construction of an optimal normal basis is developed at the following.

Example B.20 **Let suppose *p*=3 and *m*+1=5. These are prime numbers.**

It is verified that 3 is a primitive element of GF(5). All the conditions for the existence of an optimal normal basis are met.

The optimal normal basis for GF(3^4) is constructed from an element H such that H^5=1. Since $x^{80} = 1$, it is possible to make $H = x^{16}$ ($H^5 = x^{80} = 1$). Thus, an optimal normal basis is $\{x^{16}, x^{48}, x^{144}, x^{432}\}$.

Using Table 11.4, in which the different elements of GF(3^4)$\{x^4 + x + 2\}$ are included, cross-product table for this basis is obtained, that is Table B.32, and indeed it is found that is optimal, as its complexity is $C_N = 2 \times 4 - 1 = 7$. ■

B.5.3 *Dual Basis*

In GF(p^m) the dual bases can be also used, that are defined in a similar way that for GF(2^m). For introducing them, the *Trace* function, *Tr*(), is defined, for an element $e \in$ GF(p^m):

$$Tr(e) = \sum_{i=0}^{m-1} e^{p^i}$$

It holds that $Tr(e) \in$ GF(p), and that $Tr(e)$ is a lineal function over GF(p^m). If A, $B \in$ GF(p^m), and $a, b \in$ GF(p):

$$Tr(aA + bB) = aTr(A) + bTr(B)$$

Given a basis $\{b_0, b_1, \ldots, b_{m-1}\}$ for GF(p^m), its dual basis $\{d_0, d_1, \ldots, d_{m-1}\}$ is defined, such that:

$$Tr(b_i d_j) = 0 \text{ if } i \neq j$$
$$Tr(b_i d_j) = 1 \text{ if } i = j$$

An element A of GF(p^m) can be expressed in the dual basis as:

$$A = \beta_0 d_0 + \beta_1 d_1 + \cdots + \beta_{m-1} d_{m-1}$$

where:

$$\beta = Tr(Ad_i)$$

For any basis its unique dual basis always exists, such as it is shown at the following for standard and normal bases.

Given an standard basis $\{b_0, b_1, \ldots, b_{m-1}\} = \{1, \alpha, \alpha^2, \ldots, \alpha^{m-1}\}$ for GF(p^m)$\{P(x)\}$, being α a root of $P(x)$, it is known, [6, 8] , that its dual basis $\{d_0, d_1, \ldots, d_{m-1}\}$ can be obtained as follows (and such as it is made in the Examples B.21 and B.22):

(a) $P(x)$ is factorized as $P(x) = (x - \alpha)\left(\sum_{i=0}^{m-1} c_i x^i\right)$;
(b) the derivate $P'(x)$ is calculated;
(c) the elements of the dual basis are $d_i = \frac{c_i}{P'(\alpha)}$.

Example B.21 **Given the standard basis $\{1, \alpha\}$ over GF(3^2)$\{x^2 + x + 2\}$, being α a root of $x^2 + x + 2$, calculate it dual basis.**

$$P(x) = x^2 + x + 2 = (x - \alpha)(x + (\alpha + 1))$$

According to Table B.33, in which the elements of GF(3^2)$\{x^2 + x + 2\}$ are shown:

Table B.33 Elements of GF(3^2)$\{x^2 + x + 2\}$

$x^{-\infty}$	x^1	x^2	x^3	x^4	x^5	x^6	x^7
0	x	$2x + 1$	$2x + 2$	2	$2x$	$x + 2$	$x + 1$

$$(\alpha + 1) = \alpha^7$$

On the other hand,

$$P'(\alpha) = 2\alpha + 1 = \alpha^2$$

It results:

$$b_0 = \frac{\alpha^7}{\alpha^2} = \alpha^5; \quad b_1 = \frac{1}{\alpha^2} = \alpha^6$$

It is easy to check that $\{\alpha^5, \alpha^6\}$ is the dual basis of $\{1, \alpha\}$. ■

Example B.22 **Given the standard basis $\{1, \alpha, \alpha^2, \alpha^3\}$ over $GF(3^4)\{x^4 + x + 2\}$, being α a root of $x^4 + x + 2$, calculate it dual basis.**

$$P(x) = x^4 + x + 2 = (x - \alpha)\left(x^3 + \alpha x^2 + \alpha^2 x + (\alpha^3 + 1)\right)$$

According to Table 11.4, $(\alpha^3 + 1) = \alpha^{79}$.
On the other hand, $P'(\alpha) = \alpha^3 + 1 = \alpha^{79}$.
It results:

$$b_0 = \frac{\alpha^{79}}{\alpha^{79}} = 1; \quad b_1 = \frac{\alpha^2}{\alpha^{79}} = \alpha^{-77} = \alpha^3;$$

$$b_2 = \frac{\alpha}{\alpha^{79}} = \alpha^{-78} = \alpha^2; \quad b_3 = \frac{1}{\alpha^{79}} = \alpha$$

It can be easily checked that, $\{1, \alpha^3, \alpha^2, \alpha\}$ is the dual basis of $\{1, \alpha, \alpha^2, \alpha^3\}$. ■

Given an element of $GF(p^m)\{P(x)\}$ represented in an standard basis $\{1, \alpha, \alpha^2, \ldots, \alpha^{m-1}\}$, its representation in the dual basis can be obtained just applying the definition of dual basis, such as it is made in the following example.

Example B.23 Given an element A of $GF(3^2)\{x^2 + x + 2\}$ represented in the standard basis $\{b_0, b_1\} = \{1, \alpha\}$ ($A = a_0 + a_1\alpha$), being α a root of $x^2 + x + 2$, in order to calculate its representation in the dual basis $\{d_0, d_1\} = \{\alpha^5, \alpha^6\}$ ($A = \beta_0\alpha^5 + \beta_1\alpha^6$), it is applied that $\beta_i = Tr(Ab_i)$. Thus, taking into account that, according to Table B.33:

$$Tr(1) = 2; \; Tr(\alpha) = 2; \; Tr\left(\alpha^2\right) = 0$$

It results:

$$\beta_0 = Tr(a_0) + Tr(a_1\alpha) = a_0 Tr(1) + a_1 Tr(\alpha) = 2a_0 + 2a_1$$

$$\beta_1 = Tr(a_0\alpha) + Tr\left(a_1\alpha^2\right) = a_0 Tr(\alpha) + a_1 Tr\left(\alpha^2\right) = 2a_0. \; ■$$

The conversion from the dual basis representation to the standard basis representation can be made as follows. Given:

$$A = \beta_0 d_0 + \beta_1 d_1 + \cdots + \beta_{m-2} d_{m-2} + \beta_{m-1} d_{m-1}$$

The values $a_0, a_1, \ldots, a_{m-2}, a_{m-1}$ have to be obtaining, such that:

$$A = a_0 + a_1\alpha + \cdots + a_{m-2}\alpha^{m-2} + a_{m-1}\alpha^{m-1}$$

being $\{d_0, d_1, \ldots, d_{m\text{-}1}\}$ the dual basis of $\{1, \alpha, \alpha^2, \ldots, \alpha^{m-1}\}$.

From the definition of dual basis it is immediate that $Tr(Ad_i) = a_i$. In fact:

$$\begin{aligned} &Tr\left(\left(a_0 + a_1\alpha + \cdots + a_{m-2}\alpha^{m-2} + a_{m-1} + \alpha^{m-1}\right)d_i\right) \\ &= a_0 Tr(d_i) + a_1 Tr(d_i\alpha) + \cdots + a_{m-2} Tr\left(d_i\alpha^{m-2}\right) \\ &+ a_{m-1} Tr\left(d_i\alpha^{m-1}\right) = a_i Tr\left(d_1\alpha^i\right) = a_i \end{aligned}$$

The representation on standard basis can be obtained applying the above equality, such as it is made in the following example:

Example B.24 Given an element A of GF(3^2)$\{x^2 + x + 2\}$ represented in the dual basis $\{d_0, d_1\} = \{\alpha^5, \alpha^6\}$ ($A = \beta_0\alpha^5 + \beta_1\alpha^6$), its representation in the standard basis $\{1, \alpha\}$ ($A = a_0 + a_1\alpha$), being α a root of $x^2 + x + 2$, can be obtained as follows:

$$\begin{aligned} a_0 &= Tr(Ad_0) = Tr\left(\left(\beta_0\alpha^5 + \beta_1\alpha^6\right)\alpha^5\right) \\ &= \beta_0 Tr\left(\alpha^{10}\right) + \beta_1 Tr\left(\alpha^{11}\right) = 2\beta_1 \\ a_1 &= Tr(Ad_1) = Tr\left(\left(\beta_0\alpha^5 + \beta_1\alpha^6\right)\alpha^6\right) \\ &= \beta_0 Tr\left(\alpha^{11}\right) + \beta_1 Tr\left(\alpha^{12}\right) = 2\beta_0 + \beta_1. \blacksquare \end{aligned}$$

The transformation of a basis to another in GF(p^m), in any case, is a linear application that can be represented by an $m \times m$ matrix, and it is implemented with relatively simple circuitry. In case that the basis itself is its dual basis—i.e. it is a self-dual basis—it is obvious that no circuit is required for the transformation. It is known that the standard bases (or polynomial bases) can not be self-dual [10]. However, the same situation of simplicity, with regard to the transformation circuitry is concerned, is achieved with the so-called *almost self-dual bases,* that are defined below.

Given a basis $\{\alpha_0, \alpha_1, \ldots, \alpha_{m-1}\}$ of GF(p^m), this basis is *weakly self-dual* (also called *almost self-dual*) if its dual basis, $\{d_0, d_1, \ldots, d_{m-1}\}$, can be obtained from the relationship:

$$d_i = A\alpha_{\Pi(i)} \quad A \in \text{GF}(\text{p}^\text{m})$$

where $\prod$ is a permutation of the indices $\{0, 1, \ldots, m - 1\}$.

That is, the elements of the dual basis are obtained from the elements of the starting basis, permuting and multiplying by a constant. In [3] it is shown that in GF(p^m){$P(x)$} there is a weakly self-dual standard basis if $P(x)$ is a trinomial ($P(x) = x^m - cx^k - 1$). In this case the permutation $\prod$ is given as follows: $\prod(i) = (k - 1 - i)$ **mod**m. The constant $A \in$ GF(p^m) is given by:

$$A = \frac{\alpha^{m-k} - c}{P'(\alpha)}$$

For $k = 1$ and $m = 1$**mod**p it results $A = 1$. In this case the dual base is merely a permutation of the starting basis, as in the Example B.22.

Let suppose a standard basis $\{b_0, b_1, \ldots, b_{m-1}\} = \{1, \alpha, \alpha^2, \ldots, \alpha^{m-1}\}$ for GF(p^m){$P(x)$}, being α a root of $P(x)$, and its dual basis $\{d_0, d_1, \ldots, d_{m-1}\}$. An element $A \in$ GF(p^m){$P(x)$} is given in the dual basis as:

$$A = \beta_0 d_0 + \beta_1 d_1 + \ldots + \beta_{m-2} d_{m-2} + \beta_{m-1} d_{m-1}$$

If A is multiplied by α it results:

$$Q = \alpha A = q_0 d_0 + q_1 d_1 + \cdots + q_{m-2} d_{m-2} + q_{m-1} d_{m-1}$$

Applying the definition of dual basis, the different coefficients of αA are:

$$\begin{gathered} Q_0 = Tr(N) = \beta_1; \\ Q_1 = Tr(\alpha N) = \beta_2; \\ \ldots; \\ Q_i = Tr(\alpha^i N) = \beta_{i+1}; \\ \ldots; \\ Q_{m-1} = Tr\left(\alpha^{m-1} N\right) = Tr(\alpha^m A) \end{gathered}$$

This is αA can be obtained from A by a simple rotation of the coefficients, except the most significant, that can be obtained using $P(x)$. If

$$P(x) = x^m + p_{m-1} x^{m-1} + \cdots + p_1 x + p_0$$

then:

$$\alpha^m = -\left(p_{m-1}\alpha^{m-1} + \cdots + p_1\alpha + p_0\right)$$

Thus:

$$\begin{aligned} q_{m-1} &= Tr\left(\alpha^m A\right) \\ &= Tr\left(-\left(p_{m-1}\alpha^{m-1} + \cdots + p_1\alpha + p_0\right)(\beta_0 d_0 + \beta_1 d_1 + \cdots + \beta_{m-1} d_{m-1})\right) \\ &= -(\beta_0 p_0 + \beta_1 p_1 + \cdots + \beta_{m-1} p_{m-1}) \end{aligned}$$

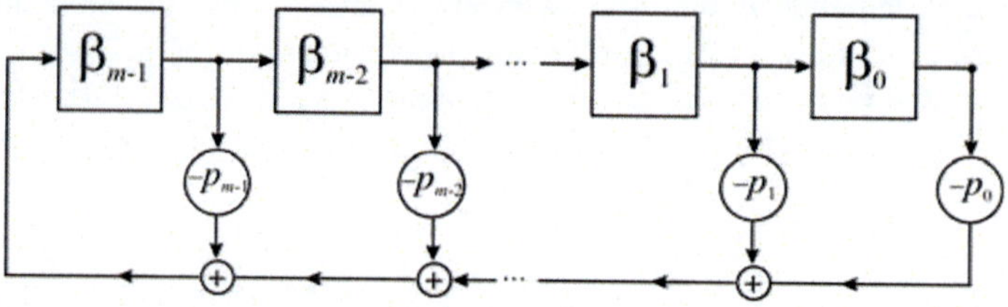

Fig. B.4 **LFSR1** for calculating αA

Using a circuit whose kernel is an **LFSR1,** αA, $\alpha^2 A$ $\alpha^3 A$, etc. can be obtained from A, such as it is represented in Fig. B.4.

Let suppose $\{b_0, b_1, \ldots, b_{m-1}\} = \{B, B^p, B^{p^2}, \ldots, B^{p^{m-2}}, B^{p^{m-1}}\}$ a normal basis whose generator is B. The procedure to calculate its dual basis, that it is also a normal basis, is the following [10, 16] (such as it is made in Example B.25):

(a) the coefficients $c_i = Tr(Bb_i)$ of the polynomial $P(x) = \sum_{i=0}^{m-1} c_i x^i$ are calculated;

(b) the polynomial $q(x) = \sum_{i=0}^{m-1} d_i x^i$ is obtained, such that $p(x)q(x)$=1**mod**$(x^n - 1)$;

(c) the dual basis $\{B, B^p, B^{p^2}, \ldots, B^{p^{m-2}}, B^{p^{m-1}}\}$ is:

$$\{D, D^p, D^{p^2}, \ldots, D^{p^{m-2}}, D^{p^{m-1}}\}$$

being $D = \sum_{i=0}^{m-1} b_i d_i$.

Example B.25 In Example B.20 it is concluded that $\{b_0, b_1, b_2, b_3\} = \{\alpha^{16}, \alpha^{48}, \alpha^{144}, \alpha^{432}\}$ is an optimal normal basis for GF(3^4)$\{x^4 + x + 2\}$, being α a root of $x^4 + x + 2$. The generator element is $B = \alpha^{16}$. The following values are calculated in order to obtain its dual basis according to the above procedure:

$$c_0 = Tr\left(\alpha^{16} \cdot \alpha^{16}\right) = 2;\ c_1 = Tr\left(\alpha^{16} \cdot \alpha^{48}\right) = 2;$$
$$c_2 = Tr\left(\alpha^{16} \cdot \alpha^{144}\right) = 1;\ c_3 = Tr\left(\alpha^{16} \cdot \alpha^{432}\right) = 2$$

It results:

$$p(x) = 2x^3 + x^2 + 2x + 2$$

Given:

$$q(x) = d_3x^3 + d_2x^2 + d_1x + d_0$$

for the product $p(x) \cdot q(x)$ it results:

$$p(x)\cdot q(x)=2d_3x^6+(d_3+2d_2)x^5+(2d_1+d_2+2d_3)x^4$$
$$+(2d_0+d_1+2d_2+2d_3)x^3+(d_0+2d_1+2d_2)x^2$$
$$+(2d_0+2d_1)x+2d_0$$

Dividing $p(x)\cdot q(x)$ by $x^4 - 1$, and imposing than the rest of the division equals 1, it results:

$$d_a=2;\ d_1=2;\ d_2=1;\ d_3=2$$

Thus, using Table 11.4, it results:

$$D=\sum_{i=0}^{m-1}b_id_i=2\alpha^{16}+2\alpha^{48}+\alpha^{144}+\alpha^{432}$$
$$=2(2\alpha^3+\alpha+2)+2(2\alpha^2+2\alpha+2)$$
$$+2\alpha^2+2+2(\alpha^3+2\alpha^2+2)=\alpha^2+2=\alpha^{17}$$

and $\{\alpha^{17}, \alpha^{51}, \alpha^{153}, \alpha^{459}\}$ is the dual basis of $\{\alpha^{16}, \alpha^{48}, \alpha^{144}, \alpha^{432}\}$, as can be easily checked. ■

In [7, 10] procedures, if exists, for obtaining self-dual normal basis are presented.

B.5.4 Inverse

The algorithm for calculating the greatest common divisor of two polynomials can be used to obtain the inverse over $GF(p^m)\{P(x)\}$ of a polynomial $Q(x) \in GF(p^m)\{P(x)\}$. Indeed, as $Q(x)$ is always prime to $P(x)$, $gcd\{P(x), Q(x)\} = 1$, and as indicated in Sect. B.1.1, given $Q(x)$, another polynomial $B(x)$ such that $B(x)\cdot Q(x)=1$, $B(x)=Q^{-1}(x)$ can be found. The Algorithm B.1 is easily adapted to compute $B(x)$. Before detailing this adaptation for $GF(p^m)$, it has to be commented that the adaptation done for $GF(2^m)$ in the Sect. B.4.4, produces a polynomial $B(x)$ such that $B(x)\cdot Q(x) = c \neq 0$, where c is the last nonzero remainder. Therefore, to obtain the inverse, $B(x)$ must be divided by c (over $GF(2^m)$ is always $c = 1$). After entering this final correction, the algorithm is as follows:

Algorithm B.7

Algorithm for the inverse of a polynomial

```
1) R1 ← P(x), R2 ← Q(x), B1 ← 0, B2 ← 1;
while R2 ≠ 0 do
    begin
        2) Q ← COC(R1, R2), TEMP ← R2;
        3) R2 ← R1 - QR2, R1 ← TEMP, TEMP ← B2;
        4) B2 ← B1 - QB2, B1 ← TEMP;
    end
    5) B1 ← B1:R1;
```

End algorithm

After applying the Algorithm B.7, the register $B1$ contains $Q^{-1}(x)$. An example of application is developed at the following.

Example B.26 **Calculate the inverse of $Q(x) = x^2 + 2x + 1$ over $GF(3^3)\{x^3 + 2x + 1\}$.**

The contents of the registers in the different steps of the application of the Algorithm B.7 are shown in Table B.34. It results $Q^{-1}(x) = 2x^2 + 2$. Indeed, it is straightforward to check that $(x^2 + 2x + 1)(2x^2 + 2) = 1$ over $GF(3^3)\{x^3 + 2x + 1\}$. ■

In the Algorithm B.7 the process of division by successive subtractions given in the Algorithm B.4 can be introduced. Moreover, in this case only the remainder of the division is of interest. The modified algorithm, being, in each iteration, a_n and b_m the nonzero coefficients of the highest power of x in $R1$ and $R2$, respectively, would be:

Algorithm B.8

Table B.34 Corresponding to the Example B.26

	R1	*R2*	*B1*	*B2*	*Q*	TEMP
1	x^3+2x+1	x^2+2x+1	0	1		
2					$x+1$	x^2+2x+1
3	x^2+2x+1	$2x$				1
4			1	$-(x+1)$		
2					$2x+1$	$2x$
3	$2x$	1				$-(x+1)$
4			$-(x+1)$	$2x^2+2$		
2					$2x$	1
3	1	0				$2x^2+2$
4			$2x^2+2$	$2x^3+x+1$		
5			$2x^2+2$			

Table B.35 Addition and product over GF(2^2){$x^2 + x + 1$}

⊕	A	B	C	D
A	A	B	C	D
B	B	A	D	C
C	C	D	A	B
D	D	C	B	A
⊗	A	B	C	D
A	A	A	A	A
B	A	B	C	D
C	A	C	D	B
D	A	D	B	C

Modified algorithm for the inverse of a polynomial

```
1) R1 ← P(x), R2 ← Q(x), B1 ← 0, B2 ← 1;
while R2 ≠ 0 do
   begin
        2) e ← g(R1)-g(R2);
        while e ≥ 0 do
           begin
              3) R1 ← R1 - (an:bm) x^e R2, B1 ← B1 -(an:bm) x^e B2;
              4) e ← g(R1)-g(R2);
           end
        5) R1 ← R2, B1 ← B2;
   end
```

End algorithm

Again, $B1$ contains $Q^{-1}(x)$.

B.6 Finite Fields GF((p^m)^n)

The coefficients of the polynomials can be elements of a finite field GF(p^m), originating in this way the composite Galois fields, GF((p^m)^n), as it is built in the following example.

Example B.27 Let suppose GF(2^2){$x^2 + x + 1$}; its elements are {0, 1, x, x+1}, that can be also represented, using binary coordinates, as {00, 01, 10, 11}. The operations addition and product for GF(2^2){$x^2 + x + 1$} are given in Tables B.4 and B.5. For convenience, in what follows we call these elements as $A(0)$, $B(1)$, $C(x)$ and $D(x + 1)$, and the addition and product operations are repeated in Table B.35 with the new names.

The coefficients of polynomials over GF(2^2){$x^2 + x + 1$) are A, B, C and D. It is easy to check that the polynomial $Y^2 + Y + C$ is primitive over GF(2^2){$x^2 + x + 1$).

Table B.36 Generation of the elements of GF($(2^2)^2$)$\{\{x^2+x+1\}Y^2+Y+C\}$

$Y^{-\infty}$	0	AA	Y^7	$CY+D$	CD
Y^0	1	AB	Y^8	$BY+D$	BD
Y^1	$BY+A$	BA	Y^9	$CY+C$	CC
Y^2	$BY+C$	BC	Y^{10}	$AY+D$	AD
Y^3	$DY+C$	DC	Y^{11}	$DY+A$	DA
Y^4	$BY+B$	BB	Y^{12}	$DY+B$	DB
Y^5	$AY+C$	AC	Y^{13}	$CY+B$	CB
Y^6	$CY+A$	CA	Y^{14}	$DY+D$	DD

Thus, GF($(2^2)^2$)$\{\{x^2+x+1\}Y^2+Y+C\}$ can be defined; specifically in Table B.36, the 16 elements of this Galois field are given; generated as successive powers of a generating element; in the first and fourth columns of the table the successive powers of the generator element are given; in the second and fifth columns, the corresponding polynomials are given; each element can be given with the two binomial coefficients: is what is done in the third and sixth columns .

Obviously GF($(2^2)^2$)$\{\{P_1(x)\}P_2(Y)\}$ is isomorphic to GF(2^4)$\{P_3(z)\}$. Once defined $P_1(x)$, $P_2(Y)$ and $P_3(z)$, it is easy to obtain the transformation matrix which leads from one field to another. ■

The composite Galois fields GF($(p^m)^n$) can be represented using the same procedures that for GF(p^m) and the same operations can be made.

B.7 Conclusion

This Appendix shows the required concepts for understanding the algebraic circuits presented in Chaps. 9, 10 and 11. Readers interested in more profound details or in demonstrations of the properties here shown can make use of some of the references listed in the following section.

B.8 References

1. Ash, D. W.; Blake, I. F.; Vanstone, S. A.: *Low Complexity Normal Bases*, Discrete Applied Mathematics, vol. 25, pp. 191–210, 1989.
2. Garret, P. B.: **Abstract Algebra**, Chapman & Hall, 2008.
3. Geiselman, W.; Gollmann, D.: *Sel-Dual Bases in* F_q^n, Design, Codes and Cryptography, vol. 3, pp. 333–345, 1993.

4. Green, D. H.; Taylor, I. S.: *Irreducible polynomials over composite Galois fields and their applications in coding techniques*. Proc. IEE, vol. 121, n° 9, pp. 935–939, September 1974.
5. Hsu, I. S.; Truong, T. K.; Deutsch, L. J.; Reed, I. S.: *A Comparison of VLSI Architecture of Finite Field Multipliers Using Dual, Normal, or Standard Bases*, IEEE Trans. On Compt., Vol. 37, (6), Jun. 1988, pp. 735–739.
6. Lee, CH.; Lim, JL.: *A New Aspect of Dual Basis for Efficient Field Arithmetic*, Lecture Notes in Computer Science 1560, pp. 12–28, 1999.
7. Lempel, A.: *Characterization and Synthesis of Self-Complementary Normal Basis in Finite Fields*, Linear Algebra and its Applications, 98, pp. 331–346, 1988.
8. Lidl, R.; Niederreiter, H.: **Introduction to Finite Fields and Their Applications**, Cambridge University Press, 1986.
9. McCoy, N. H.; Janusz, G. J.: **Introduction to Abstract Algebra**, Academic Press, 2001.
10. Menezes, A., Ed., **Applications of Finite Fields**, Kluwer Academic Publisher, 1993.
11. Mullin, R. C.; Onyszchuk, I. M.; Vanstone, S. A.: *Optimal normal bases in* $GF(p^n)$, Discrete Applied Mathematics 22 (1988/89), pp. 149–161, 1989.
12. National Institute of Standards and Technology: **FIPS 186-3, Digital signature standard (DSS)**, Gaithersburg, MD, June 2009.
13. Peterson, W. W.; Weldon, E. J.: **Error-Correcting Codes**. MIT Press, 1972.
14. Rajski, J.; Tyszer, J.: *Primitive Polynomials Over GF(2) of Degree up to 660 with Uniformly Distributed Coefficients*, Journal of Electronic Testing: Theory and Applications 19, 645–657, 2003.
15. Stahnke, W.: *Primitive Binary Polynomials*, Math. Comput., vol. 27, N° 124, 1973, pp. 977–980
16. Wan, Z-X; Zhou, K.: *On the complexity of the dual basis of a type I optimal normal basis*, Finite Fields and Their Applications 13 (2007), pp. 411–417.
17. Wu, H.; Hasan, M.A.; Blake, I.F.: *New Low-Complexity Bit-Parallel Finite Field Multiplier Using Weakly Dual Bases*, IEEE Trans. On Compt., Vol. 47, (11), Nov. 1998, pp. 1223–1234.

Appendix C
Elliptic Curves

This Appendix is devoted to the essential elliptic curves properties when used for public key encryption. The objective will be providing the tools for the understanding of the examples presented in Chap. 12 without the needing of additional texts about cryptography. Thus, like in the other appendices, the exposition is oriented to provide an immediate reference, without the inclusion of demonstrations and without approaching other aspects. Moreover, only elliptic curves defined over Galois fields and recommended for cryptographic applications are considered.

For detailed information about elliptic curves, the references provided at the end of the Appendix (Sect. C.8), can be read.

C.1 General Properties

Elliptic curves can be defined over different mathematic structures (real numbers, complex numbers, rational numbers, etc). Specifically, they can be defined over finite fields, being interesting because of their cryptographic applications. In general, being K a field, an elliptic curve is defined by the Weirstrass equation:

$$y^2 + a_1xy + a_3y = x^3 + a_2x^2 + a_4x + a_6, \text{ with } a_i \in K \tag{C.1}$$

where K is a finite field, and depending on the Galois field being GF(2^n), GF(3^n), or GF(p^n), $p > 3$, the Weirstrass equation can be simplified as detailed in the following.

When K is a binary Galois field, GF(2^n), the known as non supersingular equation can be used:

$$y^2 + xy = x^3 + ax^2 + b, \text{ with } a, b \in \mathrm{GF}\left(2^n\right), b \neq 0 \tag{C.2}$$

Other simplification when using GF(2^n) is the known as supersingular Weierstrass equation:

A. Lloris Ruiz et al., *Arithmetic and Algebraic Circuits*, Intelligent Systems Reference Library 201, https://doi.org/10.1007/978-3-030-67266-9

$$y^2 + cy = x^3 + ax^2 + b, \text{ with } a, b, c \in \mathrm{GF}(2^n), c \neq 0 \tag{C.3}$$

When using GF(3^n), the non supersingular Weierstrass equation is:

$$y^2 = x^3 + ax^2 + b, \text{ with } a, b \in \mathrm{GF}(3^n), a \neq 0, b \neq 0 \tag{C.4}$$

Other Weirstrass equation simplification for GF(3^n) is the known as supersingular, used in some applications [3]:

$$y^2 = x^3 + ax + b, \text{ with } a, b \in \mathrm{GF}(3^n), a \neq 0 \tag{C.5}$$

For GF(p^n), $p > 3$, the simplified Weierstrass equation is:

$$y^2 = x^3 + ax + b, \text{ with } a, b \in \mathrm{GF}(p^n) \tag{C.6}$$

with the condition that $x^3 + ax + b$ not having multiple radix.

Given a Galois field GF, an elliptic curve over GF is the set of all points (x, y) in GF^2 satisfying the corresponding simplified Weierstrass equation, with the singular point of infinite, ∞.

The points in an elliptic curve establish an abelian group with the point addition operation (which will be defined later). The number of points in an elliptic curve E, including the point at infinite, is named the curve order and noted as #E(GF).

The number of points #E(GF) over a finite field GF with q elements is limited, as established in the Hasse theorem [4]:

$$q + 1 - 2\sqrt{q} \leq \#\mathrm{E(GF)} \leq q + 1 + 2\sqrt{q} \tag{C.7}$$

For large values of q, the order of an elliptic curve is approximately q.

Example C.1 Elliptic curves over GF(5) are those corresponding to all possible values of a and b in equation $y^2 = x^3 + ax + b$ such as $x^3 + ax + b$ not having multiple radix. Thus, there are 25 curves least those including multiple radixes, as detailed in Table C.1. Taking into account that $x^3 + 2x + 3 = (x + 1)^2(x + 3)$, $x^3 + 3x + 4 = (x + 2)^2(x + 2)$, $x^3 + 3x + 1 = (x + 3)^2(x + 4)$, $x^3 + 2x + 2 = (x + 4)^2(x + 2)$, the four corresponding curves together with $y^2 = x^3$, are discarded.

In Table C.1 the number of points, including the point at the infinite, is detailed for each curve. As an example, the curve $y^2 = x^3 + 4x$ has 8 points, which are {(0,0), (1,0), (2,1), (2,4), (3,2), (3,3), (4,0), ∞}. ■

The Hasse theorem for $q = 5$ establishes that $2 \leq$ #E(GF) ≤ 10, as shown in Table C.1.

Example C.2 **Obtaining the points in the elliptic curve $y^2 + xy = x^3 + x^2 + 1$ over GF(2^3){$t^3 + t + 1$}.**

Table C.2 shows the multiplication table for GF(2^3){$t^3 + t + 1$}, and Table C.3 presents the value of $x^3 + x^2 + 1$ for each of the elements A $\in$ GF(2^3){$t^3 + t + 1$}.

Table C.1 Elliptic curves over GF(5)

Curve	Number of points	Curve	Number of points
$y^2 = x^3$	No (triple radix)	$y^2 = x^3 + 2x + 3$	No (double radix)
$y^2 = x^3 + 1$	6	$y^2 = x^3 + 2x + 4$	7
$y^2 = x^3 + 2$	6	$y^2 = x^3 + 3x$	10
$y^2 = x^3 + 3$	6	$y^2 = x^3 + 3x + 1$	No (double radix)
$y^2 = x^3 + 4$	6	$y^2 = x^3 + 3x + 2$	5
$y^2 = x^3 + x$	4	$y^2 = x^3 + 3x + 3$	5
$y^2 = x^3 + x + 1$	9	$y^2 = x^3 + 3x + 4$	No (double radix)
$y^2 = x^3 + x + 2$	4	$y^2 = x^3 + 4x$	8
$y^2 = x^3 + x + 3$	3	$y^2 = x^3 + 4x + 1$	8
$y^2 = x^3 + x + 4$	9	$y^2 = x^3 + 4x + 2$	3
$y^2 = x^3 + 2x$	2	$y^2 = x^3 + 4x + 3$	3
$y^2 = x^3 + 2x + 1$	7	$y^2 = x^3 + 4x + 4$	8
$y^2 = x^3 + 2x + 2$	No (double radix)		

Table C.2 Multiplication table for $GF(2^3)\{t^3 + t + 1\}$

	1	t	$t+1$	t^2	t^2+1	t^2+t	t^2+t+1
1	1	t	$t+1$	t^2	t^2+1	t^2+t	t^2+t+1
t	t	t^2	t^2+t	$t+1$	1	t^2+t+1	t^2+1
$t+1$	$t+1$	t^2+t	t^2+1	t^2+t+1	t^2	1	t
t^2	t^2	$t+1$	t^2+t+1	t^2+t	t	t^2+1	1
t^2+1	t^2+1	1	t^2	t	t^2+t+1	$t+1$	t^2+t
t^2+t	t^2+t	t^2+t+1	1	t^2+1	$t+1$	t	t^2
t^2+t+1	t^2+t+1	t^2+1	t	1	t^2+t	t^2	$t+1$

Table C.3 $A^3 + A^2 + 1$ calculation

A	A^2	A^3	$A^3 + A^2 + 1$
0	0	0	1
1	1	1	1
t	t^2	$t+1$	t^2+t
$t+1$	t^2+1	t^2	0
t^2	t^2+t	t^2+1	t
t^2+1	t^2+t+1	t^2+t	0
t^2+t	t	t^2+t+1	t^2
t^2+t+1	$t+1$	t	0

Table C.4 $y^2 + xy = x^3 + x^2 + 1$ solutions

x	$x^3 + x^2 + 1$	$y^2 + xy$	y
000 (0)	001	y^2	001
001 (1)	001	$y^2 + y$	No
010 (t)	110	$y^2 + ty$	101, 111
011 (t+1)	000	$y^2 + (t + 1)y$	000, 011
100 (t^2)	010	$y^2 + t^2y$	010, 011
101 ($t^2 + 1$)	000	$y^2 + (t^2 + 1)y$	000, 101
110 ($t^2 + t$)	100	$y^2 + (t^2+t)y$	011, 101
111 ($t^2 + t + 1$)	000	$y^2 + (t^2+t+1)y$	000, 111

In Table C.4, the different solutions of equation $y^2 + xy = x^3 + x^2 + 1$ are shown for all x values. The first column in Table C.4 corresponds to the different elements in GF(2^3), in the second column, the values of $x^3 + x^2 + 1$ (Table C.3) are shown, and in the third column, $y^2 + xy$ for each value of x is presented. Finally, in the fourth column the values of y making true the equation $y^2 + xy = x^3 + x^2 + 1$ are detailed. The following points result: {(000, 001), (010, 101), (010, 111), (011, 000), (011, 011), (100, 010), (100, 011), (101, 000), (101, 101), (110, 011), (110, 101), (111, 000), (111, 111), ∞}. Thus, the order of this curve is 14. ■

C.2 Points Addition

Given two point of an elliptic curve, their sum can be defined. In the following, this operations is introduced for GF(2^n) non supersingular curves and for GF(p), $p > 3$ (the two types or curves used in Chap. 12). First, the point doubling and the addition of two different points will be defined separately. In any case, the ∞ point is the neutral element for addition:

$$P + \infty = \infty + P = P, \infty + \infty = \infty \tag{C.8}$$

Given the point P ($P \neq \infty$) in curve $y^2 + xy = x^3 + ax^2 + b$ over GF(2^n), the point can be doubled, $P_3 = P + P$, as follows:

$$\begin{aligned}
&\text{being } P = (x_1, y_1), P_3 = (x_3, y_3).\\
&\quad x_3 = \lambda^2 + \lambda + a = x_1^2 + b/x_1^2,\\
&\quad y_3 = \lambda(x_1 + x_3) + x_3 + y_1 = x_3(\lambda + 1) + x_1^2\\
&\text{where } \lambda = x_1 + y_1/x_1
\end{aligned} \tag{C.9}$$

Given the points P_1 and P_2 ($P_1, P_2 \neq \infty$) in the curve $y^2 + xy = x^3 + ax^2 + b$ over GF(2^n), the sum $P_3 = P_1 + P_2$ is computed as follows:

$$
\begin{aligned}
&\text{being } P_1 = (x_1, y_1),\ P_2 = (x_2, y_2),\ P_3 = (x_3, y_3).\\
&\quad \text{if } x_2 = x_1 \text{ and } y_2 = x_1 + y_1, \text{ then}\\
&\quad P_3 = \infty(\text{i.e.}, -P_1 = (x_1, x_1 + y_1)).\\
&\text{else, } x_3 = \lambda^2 + \lambda + a + x_1 + x_2,\ y_3 = \lambda(x_2 + x_3) + x_3 + y_2\\
&\text{where } \lambda = (y_2 + y_1)/(x_2 + x_1)
\end{aligned} \tag{C.10}
$$

Given the point P ($P \neq \infty$) in the curve $y^2 = x^3 + ax + b$ over GF(p), $p > 3$, it can be doubled, $P_3 = P + P$, as follows:

$$
\begin{aligned}
&\text{being } P = (x_1, y_1),\ P_3 = (x_3, y_3).\\
&\quad x_3 = \lambda^2 - 2x_1,\ y_3 = \lambda(x_1 - x_3) - y_1\\
&\text{where } \lambda = \left(3x_1^2 + a\right)/2y_1
\end{aligned} \tag{C.11}
$$

Given the points P_1 and P_2 (P_1, $P_2 \neq \infty$) in the curve $y^2 = x^3 + ax + b$ over GF(p), with $p > 3$, their sum $P_3 = P_1 + P_2$ can be computed as:

$$
\begin{aligned}
&\text{being } P_1 = (x_1, y_1),\ P_2 = (x_2, y_2),\ P_3 = (x_3, y_3).\\
&\quad \text{if } x_2 = x_1 \text{ and } y_2 = -y_1, \text{ then}\\
&\quad P_3 = \infty(\text{i.e.}, -P_1 = (x_1, -y_1)).\\
&\text{else, } x_3 = \lambda^2 - x_1 - x_2,\ y_3 = \lambda(x_1 - x_3) - y_1\\
&\text{where } \lambda = (y_2 - y_1)/(x_2 - x_1)
\end{aligned} \tag{C.12}
$$

For adding two points in any of the cases before, addition, subtraction, multiplication and division of elements over GF must be performed. Note that the subtraction of two points ($P_1 - P_2 = P_1 + (-P_2)$) has the same complexity than addition.

C.3 Scalar Multiplication

A point P can be doubled as: $2P = P + P$. In general, a point P can be multiplied by an integer N as: $NP = P + \cdots + P$ (N times). This operation is known as scalar multiplication. Usually, N is assumed to be positive because $-NP = N(-P)$.

The lowest integer k satisfying that $kP = \infty$ is known as the order of the point P. It can be probed that the order of a point P divides the order of the elliptic curve where the point is included, as shown in the following example.

Example C.3 The elliptic curve $y^2 = x^3 + 3x$ over GF(5) has ten points: {(0, 0), (1, 2), (1, 3), (2, 2), (2, 3), (3, 1), (3, 4), (4, 1), (4, 4), ∞}.

For $P = (0, 0)$ we have:

$$2P = (0, 0) + (0, 0) = [\lambda = \infty] = \infty$$

Table C.5 Addition table for the subgroup generated from $P = (1, 2)$

	∞	(1, 2)	(1, 3)	(4, 1)	(4, 4)
∞	∞	(1, 2)	(1, 3)	(4, 1)	(4, 4)
(1, 2)	(1, 2)	(4, 1)	∞	(4, 4)	(1, 3)
(1, 3)	(1, 3)	∞	(4, 4)	(1, 2)	(4, 1)
(4, 1)	(4, 1)	(4, 4)	(1, 2)	(1, 3)	∞
(4, 4)	(4, 4)	(1, 3)	(4, 1)	∞	(1, 2)

For $P = (1, 2)$ we have:

$$\begin{aligned} 2P &= (1,2) + (1,2) = [\lambda = 4] = (4,1) \\ 3P &= (1,2) + (4,1) = [\lambda = 3] = (4,4) \\ 4P &= (1,2) + (4,4) = [\lambda = 4] = (1,3) \\ 5P &= (1,2) + (1,3) = [\lambda = \infty] = \infty \end{aligned}$$

For $P = (2, 2)$ we have:

$$\begin{aligned} 2P &= (2,2) + (2,2) = [\lambda = 0] = (1,3) \\ 3P &= (2,2) + (1,3) = [\lambda = 4] = (3,4) \\ 4P &= (2,2) + (3,4) = [\lambda = 2] = (4,4) \\ 5P &= (2,2) + (4,4) = [\lambda = 1] = (0,0) \\ 6P &= (2,2) + (0,0) = [\lambda = 1] = (4,1) \\ 7P &= (2,2) + (4,1) = [\lambda = 2] = (3,1) \\ 8P &= (2,2) + (3,1) = [\lambda = 4] = (1,2) \\ 9P &= (2,2) + (1,2) = [\lambda = 0] = (2,3) \\ 10P &= (2,2) + (2,3) = [\lambda = \infty] = \infty \end{aligned}$$

Thus, the point (0, 0) order is 2, the point (1, 2) has order 5 and the point (2, 2) has order 10, which are the dividers of the order of the elliptic curve including these points. ■

Given a point P in an elliptic curve, E, the successive points obtained from P multiplying it for the successive integers until its order, P, $2P$, $3P$, etc., ∞, establish a cyclic group with the point addition operation. This E(GF) subgroup derived from P will be named as EP(GF), and its order will be designed as #EP(GF).

Example C.4 In the Example C.3 has been shown that for $P = (1,2)$, $2P = (4,1)$, $3P = (4,4)$, $4P = (1,3)$, $5P = \infty$. Thus, the subgroup {(1,2), (4,1), (4,4), (1,3), ∞} results. The addition table for this subgroup is detailed in Table C.5.

Note that – (1,2) = (1,3), and that – (4,1) = (4,4). ■

Multiplication of a point P by an integer N can be methodized using the binary expansion of N ($N = n_{r-1}2^{r-1} + n_{r-2}2^{r-2} + \cdots + n_1 2 + n_0$), as follows:

$$NP = 2(2(\ldots.(2(n_{r-1}P) + n_{r-2}P) + \cdots) + n_1P) + n_0P \tag{C.13}$$

With this expansion for *NP*, the computation can be completed doubling and adding. The core of the computation will be:

$$\begin{aligned} &(1)\ \ \mathsf{R} \leftarrow 2\mathsf{R} \\ &(2)\ \ \textbf{if}\ \mathsf{n_i} = 1, \mathsf{R} \leftarrow \mathsf{R} + \mathsf{P} \end{aligned} \tag{C.14}$$

The result remains in the *R* register, initialized to R $\leftarrow \infty$. The algorithm could be the following:

Algorithm C.1

First algorithm for the multiplication of a point P by an integer

```
R ← ∞, i ← r − 1;
while i > 0 do
    begin
    1) if ni = 1 then R ← R + P;
    2) R ← 2R, i ← i − 1;
    end
if n0 = 1 then R ← R + P;
```

End algorithm

The core of the Algorithm C.1 consists of two stages. If it is desired executing the algorithm in only one stage (thus reducing the execution time to the half), in each iteration, additionally to the computing of the partial result of the present iteration, the contribution to the partial result over the next iteration must be computed. Using A as the partial result and B as a support for the calculation, the previous algorithm can be written as follows, where the final result remains in A:

Algorithm C.2

Second algorithm for the multiplication of a point P by an integer

```
A ← ∞, B ← P, i ← r − 1;
while i > 0 do
    begin
      if ni = 0 then A ← 2A, B ← A + B,
      else A ← A + B, B ← 2B;
    end
```

End algorithm

Other NP expansion is:

$$NP = n_0P + n_1(2P) + n_2\left(2^2P\right) + \cdots + n_{r-2}\left(2^{r-2}P\right) + n_{r-1}\left(2^{r-1}P\right) \tag{C.15}$$

Note that each term P, $2P$, 2^2P, …, $2^{r-2}P$, $2^{r-1}P$, is the double of the previous one. Thus, the new calculation consists on doubling and adding, as carried out in the following core for the computation:

$$\begin{aligned} &(1)\quad \mathsf{S} \leftarrow 2\mathsf{S} \\ &(2)\quad \textbf{if } \mathsf{n}_{r-1-i} = 1, \mathsf{R} \leftarrow \mathsf{R} + \mathsf{S} \end{aligned} \tag{C.16}$$

Initially, $S \leftarrow P$, $R \leftarrow \infty$. The result remains in R. Then algorithm can be written as:

Algorithm C.3

Third algorithm for the multiplication of a point P by an integer

```
S ← P, R ← ∞, i ← r - 1;
while i > 0 do
    begin
      1) S ← 2S, i ← i - 1,
         if n_{r-1-i} = 1 then R ← R + S;
    end
```

End algorithm

Given the binary expansion of N, the Algorithm C.2 starts with the most significant coefficient, i.e., it is executed from the left to the right. The Algorithm C.3 starts with the less significant coefficient, being executed from the right to the left. Both of them have the same computational cost, and both of them can be used if the binary expansion of N is available. When the binary expansion is not available, the computing of the expansion can be incorporated to the Algorithm C.3, because the coefficients are used in the same order than they are computed. Thus, this will be the recommended algorithm for this situation.

Example C.5 In order to compare the Algorithms C.2 and C.3, 19P is going to be computed using both of them:

$$19_{10} = 10011_2 = n_4n_3n_2n_1n_0$$

Table C.6 shows the application of Algorithm C.2. Each of iterations requires an addition and a doubling, resulting a total of 5 additions and 5 doublings.

Table C.7 details the application of Algorithm C.3. Each of iterations requires an addition and a doubling. ■

Because subtraction have the same cost than addition, for reducing the number of operations, a canonical codification can be used (see Sect. 1.8.3.2) for both of the previous expansions, thus providing the fewer number of non-zero summands. When using negative digits, the Algorithm C.3 can be rewritten as follows:

Table C.6 Computing of 19P using Algorithm C.2

i	A	B	n_i
4	∞	P	1
3	$A \leftarrow A + B\ (P)$	$B \leftarrow 2B\ (2P)$	0
2	$A \leftarrow 2A\ (2P)$	$B \leftarrow A + B\ (3P)$	0
1	$A \leftarrow 2A\ (4P)$	$B \leftarrow A + B\ (5P)$	1
0	$A \leftarrow A + B\ (9P)$	$B \leftarrow 2B\ (10P)$	1
	$A \leftarrow A + B\ (19P)$	$B \leftarrow 2B\ (20P)$	

Table C.7 Computation of 19P using Algorithm C.3

i	S	R	$n_{r\text{-}1\text{-}i}$
4	P	∞	1
3	$S \leftarrow 2P$	$R \leftarrow P$	1
2	$S \leftarrow 4P$	$R \leftarrow 3P$	0
1	$S \leftarrow 8P$	$R \leftarrow 3P$	0
0	$S \leftarrow 16P$	$R \leftarrow 3P$	1
	$S \leftarrow 32P$	$R \leftarrow 19P$	

Algorithm C.4

Fourth algorithm for the multiplication of a point P by an integer

```
S ← P, R ← ∞, i ← r - 1;
while i > 0 do
    begin
      1) S ← 2S, i ← i - 1, if n_{r-1-i} ≠ 0 then R ← R + n_{r-1-i}S;
    end
```

End algorithm

As an example, if $N = 119 = 1110111 = 1000\bar{1}00\bar{1}$, only 3 additions-subtractions are required instead of 6 additions.

C.4 Discrete Logarithm in Elliptic Curves

In Chap. 3, when studying multiplicative groups, the discrete logarithm problem was introduced, related with the exponentiation operation (as corresponds with the logarithm definition). In a similar way, when working with the additive group formed by the elliptic curve points generated from one of them, the discrete logarithm problem can be defined, but now related with the scalar multiplication. Thus, really the operation implied is not the logarithm.

Given two points, P and Q, in an elliptic curve, the discrete logarithm problem in elliptic curves consists on finding, if it exists, an integer k such $P = kQ$.

As in the discrete logarithm case, if the group order is large enough, and the group is appropriately chosen, the discrete logarithm problem over elliptic curves cannot be solved in a reasonable computing time. Cryptography over elliptic curves is based on this problem [6].

C.5 Koblitz Curves

Cryptographic procedures using elliptic curves are based on the scalar multiplication, which requires doubling and addition of points. There is a particular type of elliptic curves over GF(2^n), non supersingular, for which doubling is not required for scalar multiplication. These curves are known as binary anomalous curves or *Koblitz curves* [7], and the corresponding expression is:

$$y^2 + xy = x^3 + ax^2 + 1, \text{ with } a \in \{0, 1\} \tag{C.17}$$

Depending of the a value, the Koblitz curves are named as E_0 or E_1. The points subgroup used for cryptographic applications, $EP_a(GF(2^n))$, must be chosen for making difficult solving the discrete logarithm problem over elliptic curves. This requires the order $\#EP_a(GF(2^n))$ being divisible by a large prime number [4], and thus, n must be a prime number.

Because in GF(2^n), $(A + B)^2 = A^2 + B^2$, the squaring of the equation defining Koblitz curves results in:

$$y^4 + x^2y^2 = x^6 + ax^4 + 1, \text{ with } a \in \{0, 1\} \tag{C.18}$$

If the point $P = (x, y)$ is in the Koblitz curve, also the point (x^2, y^2) is in the curve, and also (x^4, y^4). Thus, it can be probed that for the Koblitz curves:

$$(x^4, y^4) + 2(x, y) = s(x^2, y^2), s = -1 \text{ for } E_0, s = 1 \text{ for } E_1 \tag{C.19}$$

and the point doubling can be computed as an addition. Note that squaring only requires a rotation when using a normal base. From the expression before, a procedure for computing the scalar multiplication using only points additions can be developed [10].

Naming F to the application of GF^2 over GF^2, knows as Frobenius application and defined as:

$$F(x, y) = (x^2, y^2) \tag{C.20}$$

The expression (C.19) can be written as:

$$F^2P + 2P = sFP \tag{C.21}$$

or:

$$F^2 + 2 = sF \tag{C.22}$$

Each of the $2^n P$ terms, with $n \le 4$, can be computed by means of point additions-subtractions [7]. In fact, for $s = 1$ it results:

$$2 = F - F^2 \tag{C.23}$$

$$4 = 2F - 2F^2 = (F - F^2)F - 2F^2 = -F^3 - F^2 \tag{C.24}$$

$$8 = 2 \times 4 = (F - F^2)(-F^3 - F^2) = F^5 - F^3 \tag{C.25}$$

$$\begin{aligned} 16 = 4^2 &= (-E^3 - F^2)^2 = F^6 + 2F^5 + F^4 = F^6 + (F - F^2)F^5 + F^4 \\ &= -F^7 + 2F^6 + F^4 = -F^7 + (F - F^2)F^6 + F^4 = -F^8 + F^4 \end{aligned} \tag{C.26}$$

For $s = -1$ we have:

$$2 = -F - F^2 \tag{C.27}$$

$$4 = -2F - 2F^2 = (F + F^2)F - 2F^2 = F^3 - F^2 \tag{C.28}$$

$$8 = 2 \times 4 = -(F + F^2)(F^3 - F^2) = F^3 - F^5 \tag{C.29}$$

$$\begin{aligned} 16 = 4^2 &= (F^3 - F^2)^2 = F^6 - 2F^5 + F^4 \\ &= F^6 + (F + F^2)F^5 + F^4 \\ &= F^7 + 2F^6 + F^4 = F^7 + (-F - F^2)F^6 + F^4 \\ &= -F^8 + F^4 \end{aligned} \tag{C.30}$$

In general, for any k, kP can be computed by using point additions-subtractions. In [10] an algorithm for expressing any k as an expansion in terms of the F operator is detailed.

C.6 Projective Coordinates

The addition of two points requires division (or inversion) over GF, which is the more complex field operation. In order to avoid this issue, a coordinate conversion can be introduced, allowing point addition without field division. Specifically, each point, represented by two coordinates (x, y) known as affine coordinates, will be converted to a new representation, with three coordinates (X, Y, Z), known as projective coordinates. There are several ways of relating (x, y) with (X, Y, Z), being the one described in the standard IEEE 1363-2000 [5], the best in terms of performance [1]. This standard establishes the following expressions for obtaining the affine coordinates from the projective ones:

$$x = X/Z^2, \quad y = Y/Z^3 \tag{C.31}$$

Note that the projective coordinates corresponding to a point (x, y) are not unique, because it is possible to use any set (c^2X, c^3Y, cZ) with $c \neq 0$. For the point at the infinite, the coordinates $(c^2, c^3, 0)$ with $c \neq 0$, can be used.

The projective coordinates of a point from the affine ones can be obtained as:

$$X = x, \quad Y = y, \quad Z = 1 \tag{C.32}$$

Projective coordinates allows avoiding division when performing point operations. As an example, for the (x_1, y_1) point doubling in the curve $y^2 = x^3 + ax + b$ over GF(p), $p > 3$, we have:

$$\lambda = \left(3x_1^2 + a\right)/2y_1 = \left(3X_1^2/Z_1^4 + a\right)/\left(2Y_1/Z_1^3\right) = \left(3X_1^2 + aZ_1^4\right)/(2Y_1Z_1) \tag{C.33}$$

Defining:

$$M = 3X_1^2 + aZ_1^4, Z_3 = 2Y_1Z_1 \tag{C.34}$$

From $x_3 = \lambda^2 - 2x_1$, it results:

$$\begin{aligned} X_3/Z_3^2 &= (M/Z_3)^2 - 2X_1/Z_1^2, \rightarrow \\ X_3 &= M^2 - 8X_1Y_1^2 \end{aligned} \tag{C.35}$$

From $y_3 = \lambda(x_1 - x_3) - y_1$, it results:

$$\begin{aligned} Y_3/Z_3^3 &= (M/Z_3)\left(X_1/Z_1^2 - X_3/Z_3^2\right) - Y_1/Z_1^3, \rightarrow \\ Y_3 &= M\left(4X_1Y_1^2 - X_3\right) - 8Y_1^4 \end{aligned} \tag{C.36}$$

For doubling a point (x_1, y_1) in the curve $y^2 + xy = x^3 + ax^2 + b$ over GF(2^n), we have:

$$\lambda = x_1 + y_1/x_1 = X_1/Z_1^2 + Y_1/X_1Z_1 = \left(X_1^2 + Y_1Z_1\right)/Z_1^2X_1 \tag{C.37}$$

Defining:

$$L = Y_1 Z_1, M = X_1^2 + L, Z_3 = Z_1^2 X_1, T = X_1^2 Z_3 \quad (\lambda = M/Z_3) \tag{C.38}$$

From $x_3 = \lambda^2 + \lambda + a$, it results:

$$X_3/Z_3^2 = (M/Z_3)^2 + M/Z_3 + a, \rightarrow X_3 = M^2 + MZ_3 + aZ_3^2 \tag{C.39}$$

From $y_3 = \lambda(x_1 + x_3) + x_3 + y_1$, it results:

$$\begin{aligned} Y_3/Z_3^3 &= (MIZ_3)\left(X_1/Z_1^2 + X_3/Z_3^2\right) + X_3/Z_3^2 + Y_1/Z_1^3, \rightarrow \\ Y_3 &= MT + MX_3 + X_3 Z_3 + LT = (M + L)T + (M + Z_3)X_3 \end{aligned} \tag{C.40}$$

Thus, no division is required in any of the calculations before for point doubling using projective coordinates. In a similar way, expansions for point addition without division can be obtained using projective coordinates (see IEEE Std. 1363-2000 [5]). As an example, for adding the points (x_1, y_1) and (x_2, y_2) in curve $y^2 = x^3 + ax + b$ over GF(p), $p > 3$, we have:

$$\lambda = (y_2 - y_1)/(x_2 - x_1) = \left(Y_2/Z_2^3 - Y_1/Z_1^3\right)/\left(X_2/Z_2^2 - X_1/Z_1^2\right) = N/Z_1 Z_2 D \tag{C.41}$$

being:

$$N = Y_2 Z_1^3 - Y_1 Z_2^3, D = X_2 Z_1^2 - X_1 Z_2^2 \tag{C.42}$$

From $x_3 = \lambda^2 - x_1 - x_2$, we have:

$$X_3/Z_3^2 = N^2/(Z_1 Z_2 D)^2 - X_1/Z_1^2 - X_2/Z_2^2 \tag{C.43}$$

Making:

$$Z_3 = Z_1 Z_2 D \tag{C.44}$$

It results:

$$X_3 = N^2 - D^2\left(X_1 Z_2^2 + X_2 Z_1^2\right) \tag{C.45}$$

From $y_3 = \lambda(x_1 - x_3) - y_1$, it results:

$$\begin{aligned} Y_3/Z_3^3 &= (N/Z_1 Z_2 D)\left(X_1/Z_1^2 - X_3/Z_3^2\right) - Y_1/Z_1^3 \\ Y_3 &= N\left(X_1 (DZ_2)^2 - X_3\right) - Y_1 (DZ_2)^3 \end{aligned} \tag{C.46}$$

For adding points (x_1, y_1) and (x_2, y_2) of the curve $y^2 + xy = x^3 + ax^2 + b$ over GF(2^n), we have:

$$\lambda = (y_2 + y_1)/(x_2 + x_1) = \left(Y_2/Z_2^3 + Y_1/Z_1^3\right)/\left(X_2/Z_2^2 + X_1/Z_1^2\right) = N/Z_1Z_2D \tag{C.47}$$

being:

$$N = Y_2Z_1^3 + Y_1Z_2^3,\ D = X_2Z_1^2 + X_1Z_2^2 \tag{C.48}$$

From $x_3 = \lambda^2 + \lambda + a + x_1 + x_2$, it results:

$$X_3/Z_3^2 = N^2 I(Z_1Z_2D)^2 + N/(Z_1Z_2D) + a + X_1/Z_1^2 + X_2/Z_2^2 \tag{C.49}$$

Making:

$$Z_3 = Z_1Z_2D \tag{C.50}$$

we have:

$$X_3 = N^2 + NZ_3 + aZ_3^2 + D^2\left(X_1Z_2^2 + X_2Z_1^2\right) \tag{C.51}$$

From $y_3 = \lambda(x_2 + x_3) + x_3 + y_2$, it results:

$$\begin{aligned} Y_3/Z_3^3 &= (N/Z_1Z_2D)\left(X_2/Z_2^2 + X_3/Z_3^2\right) + X_3/Z_3^2 + Y_2/Z_2^3 \\ Y_3 &= N\left(X_2(DZ_1)^2 + X_3\right) + X_3Z_3 + Y_2(DZ_1)^3 \end{aligned} \tag{C.52}$$

Usually, projective coordinates are used only for computing operations, converting from affine coordinates to projective ones before the computations, and performing the inverse conversion once completed them.

C.7 Conclusion

This Appendix is intended as a support for the cryptographic applications of the algebraic circuits presented in Chap. 12, based on the discrete logarithm problem in elliptic curves. Only the most relevant aspects have been presented, being essential for the following of the mentioned examples without the need of resorting to other references.

C.8 References

1. Chudnovsky, D. V., Chudnovsky, G. V.: *Sequences of Numbers Generated by Addition in Formal Groups and New Primality and Factorizations Tests*. Advances in Applied Mathematics 7, pp. 385–434, 1987.
2. Cohen, H., Frey, G., Avanzi, R., Doche, C., Lange, T., Nguyen, K.; Vercauteren, F.: **Elliptic and Hyperelliptic Curve Cryptography**. CRC Press, Boca Raton, Florida, 2006.
3. Granger, R., Page, D., Stam, M.: *Hardware and Software Normal Basis Arithmetic for Pairing Based Cryptography in Characteristic Three*. Cryptology ePrint Archive, report 2004/157. Available from http://eprint.iacr.org/2004/157.pdf.
4. Hankerson, D., Menezes, A., Vanstone, S.: **Guide to Elliptic Curve Cryptography**, Springer, 2004.
5. **IEEE Standard Specifications for Public-Key Cryptography**. IEEE Std. 1363-2000.
6. Koblitz, N.: *Elliptic Curve Cryptosystems*. Mathematics of Computation, vol. 48, n° 77, January 1987, pp. 203–209.
7. Koblitz, N.: *CM-Curves with Good Cryptographic Properties*. Pro. Crypto'91, Springer-Verlag, 1992, pp. 279–287.
8. Menezes, A.; van Oorschot, P.; Vanstone, S.: **Handbook of Applied Cryptography**. CRC Press, Boca Raton, Florida, 1996.
9. Silverman, J.: **The Arithmetic of Elliptic Curves**, Springer-Verlag, 1986.
10. Solinas, J. A.: *Efficient Arithmetic on Koblitz Curves*. Design, Codes and Cryptography, 19, 195–249, 2000.

Appendix D
Errors

An important question to take into account when calculating is the amount of error associated to the results. This error may be produced by inexact data (for instance, the result of a measurement may introduce some uncertainty), or by inexact operations. It is known that the floating point representation of real numbers is inexact by construction. So, it is fundamental to establish the precision of the different calculations when using floating point representation.

This Appendix is devoted to describe the different types of errors in numerical data, and to the propagations of errors through the different operations.

D.1 Types of Errors

Errors may have many origins. To understand the nature of errors, this section is devoted to describe the different types of errors. Firstly avoidable and unavoidable errors are deal with.

D.1.1 Avoidable and Unavoidable Errors

Avoidable errors are produced by mistakes in the process of measurement or during calculation. For instance, when calculating manually it may happen to introduce an erroneous value; or it may happen to use an erroneous program. In the following it is supposed that this kind of avoidable errors has been depurated; also it is supposed that programs are adequately implemented, and human errors are not the object of this Appendix.

Unavoidable errors are produced by the measurement process or by the representation of the involved quantities or by the calculation procedure. So they may be classified as follows:

A. Lloris Ruiz et al., *Arithmetic and Algebraic Circuits*, Intelligent Systems Reference Library 201, https://doi.org/10.1007/978-3-030-67266-9

A. **Measurement errors**, which appear when quantifying a physical value because measurement instruments have a finite precision. So the result of a measurement must include its error, and in subsequent calculations have to be managed.

B. **Representation errors**, owing to the nature of the represented quantities or to the representation system used. Some cases of representation errors are detailed at the following.
Irrational numbers have infinite decimal figures. But, independently of the representation system used, always a finite number of decimal figures will be used, and an error is introduced.
Rational numbers, depending of the representation system used, also may have infinite decimal figures. In this case its representation includes an error.
The floating point representation allows to represents exactly only a finite numbers of values; the others will include an error.

C. **Calculation errors**, produced when the operations are based on approximations or it is impossible to use infinite precision. Some cases are:
When a function is expressed as a power series, with infinite terms, unavoidably an error is introduced, because in practice always a finite number of terms will be used.
In a calculation based on an iterative process of approximations, as the Newton-Raphson method for instance, always a finite number of iterations will be executed, but the exact solution is only guarantee with infinite iterations.

In the following will be assumed that errors are randomly distributed.

D.1.2 Absolute and Relative Errors

An exact quantity A is represented in its approximate value as A', so the absolute error ΔA associated to this representation is:

$$\Delta A = \left|A' - A\right|$$

Or $A = A' \pm \Delta A$, or $A' = A \pm \Delta A$.
The relative error $\varepsilon(A)$, define as:

$$\varepsilon(A) = \frac{\Delta A}{A}$$

is a more meaningful measure of the precision in the calculations under process.

D.2 Generated and Propagated Errors in Arithmetic Operations

The calculation with approximate operands gives a result also approximate, and the question now is to obtain the error of the result as a function of the error of the operands. But, as previously stated, the own calculation may produce an error. Let it be an arithmetic operation (sum, difference, multiplication, division, etc), represented by the Θ operator when its implementation do not introduce any error, and by the θ operator when its implementation introduce some error and so the result is approximate even if the operands are exacts. Let it be two operands, A y B, whose approximate representations are A' and B'. If $A'\theta\, B'$ is calculated, the error introduced compared with the exact value AΘB is:

$$\begin{aligned}|A'\theta\mathrm{B}' - A\Theta B| = |A'\theta\, B' - A'\Theta B' + A'\Theta B' - A\Theta B| \le |A'\theta\, B' - A'\Theta B'| \\ + |A'\Theta B' - A\Theta B|\end{aligned}$$

The first term of this final expression, $|A'\theta\, B' - A'\Theta B'|$, gives the difference between the results generated by both operators (the operands are repeated). So this term gives the error generated by the approximate operator, and it is known as **generated error** by the operator. In the second term of this final expression, $|A'\Theta B' - A\Theta B|$, we have the same operator and different operands; so it gives the error due to the operands, and propagated by the operator, so it is known as **propagated error**. In the following expressions for the propagated error in different arithmetic operations are obtained.

D.2.1 Sum and Difference

The propagated error in these two arithmetic operations is $|A' \pm B' - A \pm B|$. But, $A = A' \pm \Delta A$ and $B = B' \pm \Delta B$. So,

$$|A' \pm B' - A \pm B| = |A' \pm B' - A' \pm \Delta A \pm B' \pm \Delta B| = |\pm\Delta A \pm \Delta B|$$

This expression may also be written as:

$$\Delta(A \pm B) \le \Delta(A) + \Delta(B)$$

That is, the absolute error of a sum (or of a difference) is lower or equal than the sum of the absolute errors of the operands. In general, for many addends:

$$\Delta(A \pm B \pm C \pm \cdots \pm Z) \le \Delta(A) + \Delta(B) + \Delta(C) + \cdots + \Delta(Z)$$

When many addends, errors will be as positives as negatives and tend to cancel each other; for normally-distributed errors the Statistic [2] gives that:

$$\Delta(A \pm B \pm C \pm \cdots \pm Z) \approx \sqrt{\Delta^2(A) + \Delta^2(B) + \Delta^2(C) + \cdots + \Delta^2(Z)}$$

Example D.1 **Being x_i, $i = 0, \ldots, 2047$, normal real floating point numbers (Standard IEEE 754, $n = 32$) with the same exponent and assuming that all errors are positive, obtain the worst total error when calculating**

$$S = \sum_{i=0}^{i=2047} x_i$$

As all numbers have the same exponent, no shifting of the mantissas is necessary. Then the maximum absolute error value for each number (see Chap. 4) is ½*ulp*, so the worst total error (WTE) is:

$$\text{WTE} = 2048 \times 1/2ulp = 1024ulp$$

This (1024*ulp*) means to loose ten bits of precision. The 2048 initial data have 24 bits of precision; the result has only 14. ■

The relative error in a sum may have a relevant value, as it is shown in the following example:

Example D.2 **Being $a = 0.43584201 \times 10^8$, $b = -0.43584044 \times 10^8$ two exact numbers, obtain the relative error when calculating $a + b$, assuming that all arithmetic operations are implemented in floating point, using six decimal ciphers and rounding half to even.**

The exact result is

$$S = a + b = 0.00000157 \times 10^8$$

The approximate result using six decimal ciphers is

$$S' = a + b = 0.000002 \times 10^8$$

So

$$\Delta S = |S' - S| = 0.000002 \times 10^8 - 0.00000157 \times 10^8 = 0.00000043 \times 10^8$$

It results:

$$\varepsilon(S) = \frac{\Delta S}{S} = \frac{0.00000043}{0.0000157} \approx 0.27$$

That is, the relative error is 27%. ■

D.2.2 Multiplication

The propagated error in multiplication is:

$$\begin{aligned}\Delta(A \times B) &= \left|A' \times B' - A \times B\right| = |(A \pm \Delta A) \times (B \pm \Delta B) - A \times B| \\ &= |\pm A\Delta B \pm B\Delta A \pm \Delta A \times \Delta B|\end{aligned}$$

Dividing by $A \times B$ it results:

$$\begin{aligned}\varepsilon(A \times B) &= \frac{\Delta(A \times B)}{A \times B} \leq \frac{A\Delta B}{A \times B} + \frac{B\Delta A}{A \times B} + \frac{\Delta A\Delta B}{A \times B} \\ &= \varepsilon(B) + \varepsilon(A) + \varepsilon(A) \times \varepsilon(B)\end{aligned}$$

Normally the relative errors $\varepsilon(B)$ y $\varepsilon(A)$ are sufficiently small so $\varepsilon(A) \times \varepsilon(B) \ll \varepsilon(B)$ or $\varepsilon(A)$. If this is the case, the last inequality means that the relative error of the product is lower or equal than the sum of the relative errors of the operands:

$$\varepsilon(A \times B) \approx \varepsilon(A) + \varepsilon(B)$$

It is clear from this that the relative error of a quantity is unaltered when multiply by an exact number.

For normally-distributed errors the Statistic [2] gives that:

$$\varepsilon(A \pm B \pm C \pm \cdots \pm Z) \approx \sqrt{\varepsilon^2(A) + \varepsilon^2(B) + \varepsilon^2(B) + \cdots + \varepsilon^2(Z)}$$

D.2.3 Division

The propagated error in division is

$$\begin{aligned}\Delta(A/B) &= \left|\frac{A'}{B'} - \frac{A}{B}\right| = |\{(A \pm \Delta A)/(B \pm \Delta B)\} - A/B| \\ &= |(B\Delta A \pm A\Delta B)/B(B \pm \Delta B)|\end{aligned}$$

Dividing by A/B it results:

$$\varepsilon(A/B) = \left|\frac{\Delta(A/B)}{A/B}\right| = \frac{(\Delta A/A) \pm (\Delta B/B)}{1 \pm \Delta B/B} \leq \frac{\varepsilon(A) + \varepsilon(B)}{1 - \varepsilon(B)}$$

If the relative error $\varepsilon(B)$ is sufficiently small, $\varepsilon(B) \ll 1$, the last inequality may be written as

$$\varepsilon(A/B) \leq \varepsilon(A) + \varepsilon(B)$$

In this case the relative error of the quotient is lower or equal than the sum of the relative errors of the operands, as in the product.

It is clear from this that the relative error of a quantity is unaltered when divided by an exact number.

D.2.4 Square Root

The relative error propagated when calculating the square root of a positive number A whose approximate representation is $A' \pm \Delta A$ amounts:

$$\varepsilon\left(\sqrt{A}\right) = \frac{\sqrt{A \pm \Delta A} - \sqrt{A}}{\sqrt{A}} = \left|\sqrt{1 \pm \varepsilon(A)} - 1\right|$$
$$= \left|\frac{\pm\varepsilon(A)}{1 + \sqrt{1 \pm \varepsilon(A)}}\right| \leq \frac{\varepsilon(A)}{1 + \sqrt{1 - \varepsilon(A)}}$$

If $\varepsilon(A) \ll 1$,

$$\sqrt{1 - \varepsilon(A)} \cong 1$$

In this case:

$$\varepsilon\left(\sqrt{A}\right) \cong 1/2\varepsilon(A)$$

D.3 Errors and Laws of Algebra

If the results of the arithmetic operations may be affected by errors, some laws of Algebra do not always hold, as may be tested with the following examples.

Example D.3 **Being $a = 0.435842 \times 10^8$, $b = -0.435725 \times 10^8$, $c = 0.145862 \times 10^5$, test if the associative law for sum holds:**

$$(a + b) + c = a + (b + c)$$

It is assumed that all arithmetic operations are implemented in floating point, using six decimal ciphers and rounding half to even. It results:

$$(a + b) + c = \left(0.435842 \times 10^8 - 0.435725 \times 10^8\right) + 0.145862 \times 10^5$$
$$= 0.117000 \times 10^5 + 0.145862 \times 10^5 = 0.262862 \times 10^5$$

But:

$$a + (b + c) = 0.435842 \times 10^8 + \left(-0.435725 \times 10^8 + 0.145862 \times 10^5\right)$$
$$= 0.435842 \times 10^8 - 0.435579 \times 10^8 = 0.263000 \times 10^5$$

Evidently, the equality $(a + b) + c = a + (b + c)$ not holds. ■

Example D.4 **Being $a = 0.435842 \times 10^8$, $b = -0.435725 \times 10^8$, $c = 0.145862 \times 10^5$, test if the following distributive law holds:**

$$(c \times a) + (c \times b) = c \times (a + b)$$

It is assumed that all arithmetic operations are implemented in floating point, using six decimal ciphers and rounding half to even. It results:

$$\begin{aligned}
c \times a &= 0.145862 \times 10^5 \times 0.435842 \times 10^8 \\
&= 0.635728 \times 10^{12} \\
c \times b &= 0.145862 \times 10^5 \times \left(-0.435725 \times 10^8\right) \\
&= -0.635557 \times 10^{12} \\
(c \times a) + (c \times b) &= 0.635728 \times 10^{12} - 0.635557 \times 10^{12} \\
&= 0.171000 \times 10^9 \\
a + b &= 0.435842 \times 10^8 - 0.435725 \times 10^8 \\
&= 0.117000 \times 10^5 \\
c \times (a + b) &= 0.145862 \times 10^5 \times 0.117000 \times 10^5 \\
&= 0.170659 \times 10^9
\end{aligned}$$

Evidently, the equality $(c \times a) + (c \times b) = c \times (a + b)$ not holds. ■

Example D.5 **Being $a = 0.435842 \times 10^8$, $b = -0.145862 \times 10^8$, test if the following equality holds:**

$$a^2 - b^2 = (a + b) \times (a - b)$$

It is assumed that all arithmetic operations are implemented in floating point, using six decimal ciphers and rounding half to even. It results:

$$\begin{aligned}
a^2 &= 0.189958 \times 10^{16}; b^2 = 0.212757 \times 10^{15}; \\
a^2 - b^2 &= 0.168682 \times 10^{16} \\
a + b &= 0.581704 \times 10^8; a - b = 0.289980 \times 10^8; \\
(a + b) \times (a - b) &= 0.168683 \times 10^{16}
\end{aligned}$$

Evidently, rounding produces different results calculating $a^2 - b^2$ than calculating $(a + b) \times (a - b)$. ■

The obvious conclusion from these examples is that the order in the implementation of the arithmetic operations may affect the result.

D.4 Interval Arithmetic

When the control of errors is important, an option to control the limits of the error is the use of the Interval Arithmetic (also known as Range Arithmetic), introduced in 1966 [1].

Using Interval Arithmetic a quantity Q is given by two real numbers, (q_{inf}, q_{sup}) $q_{inf} \leq q_{sup}$, corresponding to the inferior and superior limits of the quantity. Of course, when $q_{inf} = q_{sup}$, Q is an exact quantity. For instance, (3,141; 3,142) gives the quantity π using three decimals.

Given two quantities $A = (a_{inf}, a_{sup})$ and $B = (b_{inf}, b_{sup})$, arithmetic operations in Interval Arithmetic may be defined as follows:

$$\begin{aligned} A + B &= \left(a_{inf} + b_{inf}, a_{sup} + b_{sup}\right) \\ -B &= \left(-b_{sup}, -b_{inf}\right) \\ A - B &= \left(a_{inf} - b_{sup}, a_{sup} - b_{inf}\right) \\ A \times B &= (\min P, \max P), \text{ being} \\ P &= \left\{a_{inf} \times b_{inf}, a_{inf} \times b_{sup}, a_{sup} \times b_{inf}, a_{sup} \times b_{sup}\right\} \\ A^{-1} &= \left(a_{sup}^{-1}, a_{inf}^{-1}\right), \text{ provided } 0 \notin \left[a_{inf}, a_{sup}\right] \\ A/B &= A \times B^{-1} \end{aligned}$$

Example D.6 **Given $A = (3.141; 3.142)$ and $B = (0.6983; 0.6984)$, calculate $A + B$, $A - B$, $A \times B$ and A/B, using Interval Arithmetic.**

$$\begin{aligned} A + B &= (3, 141 + 0.6983; 3, 142 + 0, 6984) \\ &= (3, 8393; 3, 8404) A - B \\ &= (3.141 - 0.6984; 3.142 - 0.6983) \\ &= (2.4426; 2.4437) \\ P &= \{3.141 \times 0.6983; 3.141 \times 0.6984; 3.142 \\ &\quad \times 0.6983; 3.142 \times 0.6984\} \\ &= \{2.1933603; 2.1936744; 2.1940586; 2.1943728\} \\ A \times B &= (minP, maxP) = (2.1933603; 2.1943728) B^{-1} \\ &= \left(0.6984^{-1}; 0.6983^{-1}\right) \end{aligned}$$

$$= (1.43184422\ldots; 1.43204926\ldots)$$
$$= (1.4318; 1.4320)$$
$$P' = \{3.141 \times 1.4318; 3.141 \times 1.4320; 3.142 \times 1.4318; 3.142 \times 1.4320\}$$
$$= \{4.4972838; 4.497912; 4.4987156; 4.499344\}$$
$$A/B = A \times B^{-1} = \left(min P', max P'\right)$$
$$= (4.4972838; 4.499344). \blacksquare$$

With Interval Arithmetic also happens that some algebraic identities do not always holds, as may be tested in the following example.

Example D.7 **Given $A = (1; 2)$, $B = (3; 4)$, and $C = (-2, -1)$, test if the following equality holds in Interval Arithmetic**:

$$A \times (B + C) = A \times B + A \times C$$
$$A \times (B + C) = (1; 2) \times (1; 3) = (1; 6)$$
$$A \times B + A \times C = (1; 2) \times (3; 4) + (1; 2) \times (-2, -1)$$
$$= (3; 8) + (-4, -1) = (-1, 7)$$

The proposed equality not holds. ■

D.5 Conclusion

This appendix gathers some general ideas about errors and its impact on the confidence of calculations. In each case it is advisable to introduce this question and select the adequate strategy in our calculations.

D.6 References

1. Moore, R. E. Interval Analysis. Prentice Hall, Englewood Cliffs, 1966.
2. Box, E. P., Hunter, J. S., Hunter, W. G., 'Statistics for Experimenters", John Wiley & Sons, New York, 2005.

Appendix E
Algorithms for Function Approximation

In this Appendix some important algorithms for function approximation are described in detail. Chebyshev and Legendre sequences of orthogonal polynomials are also presented. These algorithms and the orthogonal polynomials are used in Chap. 8 for the evaluation of hardware implementation of special functions.

E.1 Newton-Raphson Approximation

The Newton-Raphson method [5] is a powerful method for solving equations numerically and it is based on the idea of linear approximation. Newton-Raphson method is based on the simple idea of linear approximation and requires one initial value x_0, which we will refer to as the initial guess for the root. Let $f(x)$ be a function, and let r be a root of the equation $f(x) = 0$. The Newton-Raphson iterative formula to approximate the root of this function is:

$$x_{i+1} = x_i - \frac{f(x_i)}{f'(x_i)}$$

where $f'(x)$ denotes the first derivate of $f(x)$ with respect to x, x_i is an approximation to the root, and x_{i+1} is a better approximation. Basically, the method finds the tangent to the function $f(x)$ at x_0 and extrapolates it to intersect the X axis to get x_1. This point of intersection is taken as the new approximation to the root and the procedure is repeated until convergence is obtained whenever possible. This process is shown in Fig. E.1, where it can be observed that for each iteration, the value $(f(x_i) - 0)/(x_i - x_{i+1})$ is used to approximate $f'(x_i)$. In order to guarantee the convergence this initial value should be selected as close as possible to the root sought.

The Newton-Raphson iteration can be also obtained from the Taylor-series expansion of the function in $(x - x_i)$. Thus:

A. Lloris Ruiz et al., *Arithmetic and Algebraic Circuits*, Intelligent Systems Reference Library 201, https://doi.org/10.1007/978-3-030-67266-9

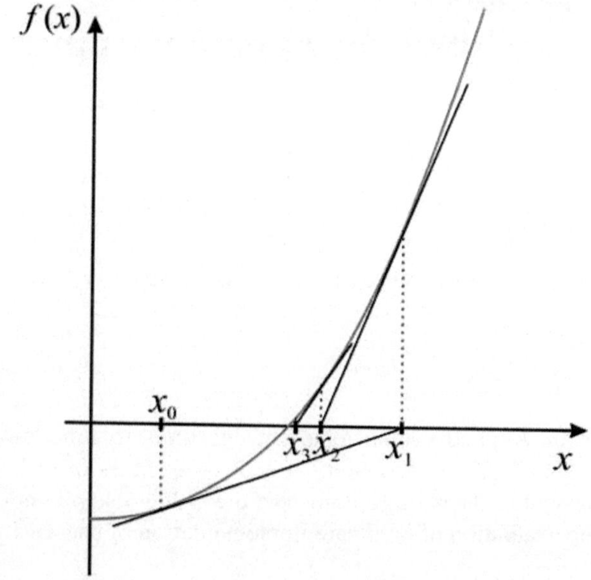

Fig. E.1 Newton-Raphson iteration

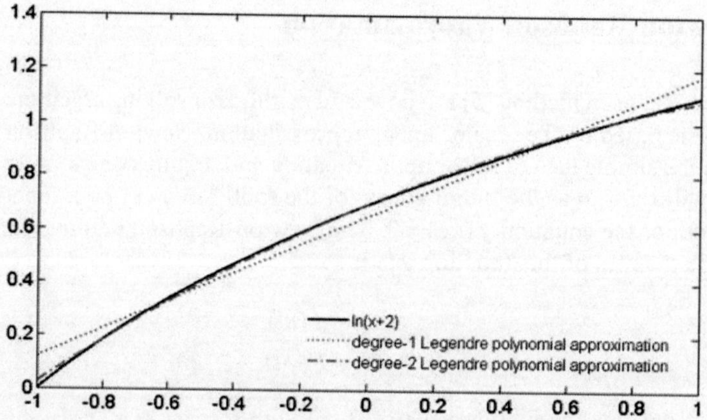

Fig. E.2 Function ln(*x*+2), degree-1 and degree-2 Legendre polynomial approximations

$$f(x) = f(x_i) + f'(x_i)(x - x_i) + (x - x_i)^2 \frac{f''(x_i)}{2!} + \cdots$$

Applying the truncation at the second term and evaluating at x_{i+1}:

$$f(x_{i+1}) = f(x_i) + f'(x_i)(x_{i+1} - x_i)$$

If it is accepted that x_{i+1} is an approximation to the root, $f(x_{i+1}) = 0$, and it is obtained the iteration formula. It is important to mention that this truncation suppose the initial value x_0 is pretty close to the real root.

E.2 Polynomial Approximation

This section is devoted to polynomial approximation methods. Concretely, Least Squares and Least Maximum polynomial methods are detailed.

E.2.1 Least Squares Polynomial Methods

Least squares polynomial methods approximate a function to a polynomial minimizing the "average error" or the worst-case. The criterion depends on a continuous, nonnegative, weight function, $w(x)$, that can be used to select parts where it is desirable the approximation to be more accurate. For this approximations the distance d is:

$$d = \|p^* - f\|_2 = \sqrt{\int_a^b w(x)(f(x) - p^*(x))^2 dx}$$

and the polynomial p^* satisfies:

$$\int_a^b w(x)\left(f(x) - p^*(x)\right)^2 dx = \min_{p \in P^n} \int_a^b w(x)(f(x) - p(x))^2 dx$$

Given the pair of polynomials, f and g, its inner product $\langle f, g\rangle$ is given by:

$$\langle f, g\rangle = \int_a^b w(x) f(x) g(x) dx$$

Orthogonal polynomial sequence [2] is a family of polynomials such that any two different polynomials in the sequence are orthogonal to each other under some inner product. This is, given a sequence of polynomials (T_m) of degree m, these polynomials are orthogonal if:

$$\langle T_i, T_j\rangle = 0 \text{ for } i \neq j$$

Thus, the approximation function $p*$ can be calculated as [1]:

$$p^* = \sum_{i=0}^{n} a_i T_i$$

where (T_m) is a sequence of orthogonal polynomials of degree $m (m \leq n)$ and a_i are the coefficients obtained from the following expression:

$$a_i = \frac{\langle f, T_i \rangle}{\langle T_i, T_i \rangle}$$

Legendre, Chebyshev, Jacobi and Laguerre Polynomials [2] are examples of sequences of orthogonal polynomials. Chebyshev and Legendre sequences of orthogonal polynomials are detailed at the following.

E.2.1.1 Chebyshev Orthogonal Polynomials

The Chebyshev polynomials are a set of orthogonal polynomials defined as the solutions to the Chebyshev differential equation:

$$(1 - x^2)\frac{d^2y}{dx^2} - x\frac{dy}{dx} + n^2 y = 0, \quad n > 0, \quad |x| < 1$$

Chebyshev polynomial of degree $n \geq 0$ is defined as:

$$T_n(x) = \cos(n \arccos(x)), x \in [-1, 1]$$

or, in a more instructive form,

$$T_n(x) = \cos(n\theta), \theta \in [0, \pi]$$

The recursive relation of Chebyshev polynomials is the following:

$$T_0(x) = 1$$
$$T_1(x) = x$$
$$\vdots$$
$$T_{n+1}(x) = 2xT_n(x) - T_{n-1}(x), n \geq 1$$

The Chebyshev polynomials are orthogonal polynomials in the interval $[-1, 1]$ with respect to the weighting function $\omega(x) = 1/\sqrt{1 - x^2}$:

$$\int_{-1}^{1} \frac{T_n(x)T_m(x)}{\sqrt{1-x^2}} dx = \begin{cases} 0 \text{ if } m \neq n \\ \frac{\pi}{2} \text{if } m = n \text{ for each } n \geq 1 \end{cases}$$

E.2.1.2 Legendre Orthogonal Polynomials

Legendre's differential equations is:

$$\left(1 - x^2\right)\frac{d^2y}{dx^2} - 2x\left(1 - x^2\right)\frac{dy}{dx} + n(n+1)y = 0, \quad n > 0, \quad |x| < 1$$

The solutions of the Legendre's differential equations are called Legendre functions of order n, being the Legendre Polynomials functions of the first kind. Legendre Polynomials or order n are given by Rodrigue's formula:

$$T_n(x) = \frac{1}{2^n n!}\frac{d^n}{dx^n}\left(x^2 - 1\right)^n$$

The recursive relation of Legendre polynomials is the following:

$$T_0(x) = 1$$
$$T_1(x) = x$$
$$\vdots$$
$$T_{n+1}(x) = \tfrac{2n+1}{n+1}T_n(x) - \tfrac{n}{n+1}T_{n-1}(x), n \geq 1$$

The Legendre polynomials are said to be orthogonal in the interval $-1 \leq \mathrm{x} \leq 1$ with respect to the weighting function $\omega(x) = 1$:

$$\int_{-1}^{1} T_n(x)T_m(x)dx = \begin{cases} 0 \text{ if } m \neq n \\ \frac{2}{2n+1} \text{if } m = n \text{ for each } n \geq 1 \end{cases}$$

Let us present now an example of approximations using Legendre orthogonal polynomials.

Example E.1 **Using least square method, approximate the function $\ln(x+2)$ by degree-1 and degree-2 Legendre polynomials.**

We are going to calculate the degree-1 and degree-2 polynomial approximations to the function $\ln(x+2)$ on the interval $[-1, 1]$.

For the degree-1 approximation:

$$p^* = a_0 T_0 + a_1 T_1$$

Thus, we need the first two Legendre polynomials:

$$T_0(x) = 1$$

$$T_1(x) = x$$

The coefficients a_0 and a_1 can be computed as follows:

$$a_0 = \frac{\langle f, T_0 \rangle}{\langle T_0, T_0 \rangle}$$
$$a_1 = \frac{\langle f, T_1 \rangle}{\langle T_1, T_1 \rangle}$$

Let first compute the inner products:

$$\langle f, T_0 \rangle = \int_{-1}^{1} ln(x+2)dx$$
$$= (x+2)(ln(x+2)-1)]_{-1}^{1} = 3\ln(3) - 2$$

$$\langle f, T_1 \rangle = \int_{-1}^{1} ln(x+2) \cdot x \cdot dx$$

$$= \left. \frac{(2x^2-8)\big(ln(x+2) - x^2 + 4x\big)}{4} \right]_{-1}^{1} = 2 - \frac{3\ln(3)}{2}$$

$$\langle T_0, T_0 \rangle = 2$$
$$\langle T_1, T_1 \rangle = \frac{2}{3}$$

Thus:

$$a_0 = \frac{3}{2}\ln(3) - 1$$
$$a_1 = 3 - \frac{9}{4}\ln(3)$$

And the degree-1 polynomial approximation is:

$$p^* = \left(3 - \frac{9}{4}\ln(3)\right)x + \left(\frac{3}{2}\ln(3) - 1\right)$$
$$\cong 0.528122350496753x + 0.647918433002164$$

For the degree-2 approximation:

$$p^* = a_0 T_0 + a_1 T_1 + a_2 T_2$$

The third Legendre polynomial is:

$$T_3(x) = \frac{3}{2}x^2 - \frac{1}{2}$$

On the other hand, the values of a_0, a_1 are the same that we have calculated for degree-1 approximation, and we have to calculate the coefficient a_2:

$$\begin{aligned}
a_2 &= \frac{\langle f, T_2 \rangle}{\langle T_2, T_2 \rangle} \\
\langle f, T_2 \rangle &= \int_{-1}^{1} \ln(x+2)\left(\frac{3}{2}x^2 - \frac{1}{2}\right)dx \\
&= \left.\frac{(3x^3 - 3x + 18)\ln(x+2) - x^3 + 3x^2 - 9x}{6}\right]_{-1}^{1} \\
&= 3\ln(3) - \frac{10}{3} \\
\langle T_2, T_2 \rangle &= \frac{2}{5}
\end{aligned}$$

And thus:

$$a_2 = \frac{15}{2}\ln(3) - \frac{50}{6}$$

The degree-1 polynomial approximation is:

$$\begin{aligned}
p^* &= \left(\frac{15}{2}\ln(3) - \frac{50}{6}\right)\left(\frac{3}{2}x^2 - \frac{1}{2}\right) \\
&\quad + \left(3 - \frac{9}{4}\ln(3)\right)x + \left(\frac{3}{2}\ln(3) - 1\right) \\
&\cong -0.140611752483767x^2 \\
&\quad + 0.528122350496753x + 0.694789017163419
\end{aligned}$$

The function $\ln(x+2)$ and its degree-1 and degree-2 Legendre polynomial approximations are plotted in Fig. E.2. ■

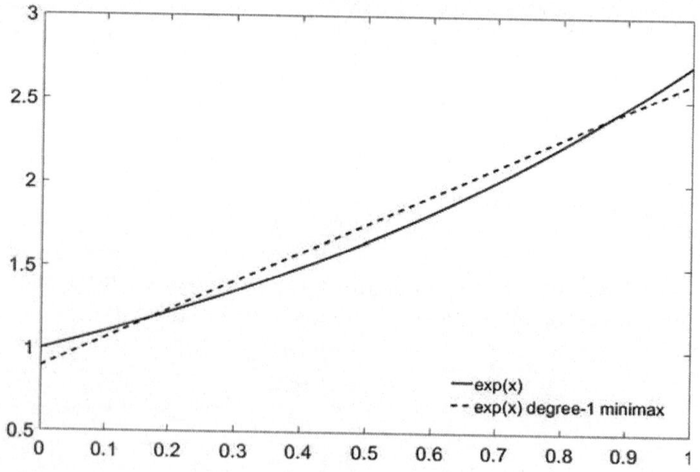

Fig. E.3 Function e^x and and degree-1 minimax polynomial approximation

E.2.2 Least Maximum Polynomial Methods

Least maximum polynomial approximation is a method to find an approximation of a function that minimizes the worst-case error, called least maximum or minimax approximation. For this approximation the distance is:

$$\left\| p^* - f \right\|_\infty = \max_{a \le x \le b} w(x) \left| p^*(x) - f(x) \right|$$

Let us assume the weight function $w(x)$ equals 1, thus, in the following $\|f - p\|_\infty$ denotes de distance:

$$\|f - p\|_\infty = \max_{a \le x \le b} |f(x) - p(x)|$$

The minimax degree-n polynomial approximations to f on $[a, b]$ is the polynomial p^* such that:

$$\left\| f - p^* \right\|_\infty = \max_{a \le x \le b} \left| f(x) - p^*(x) \right| = \min_{p \in P_n} |f(x) - p(x)|$$

The *Weierstrass approximation theorem* [3] states that every continuous function defined on a closed interval can be uniformly approximated as closely as desired by a polynomial function. In order to reduce the computational expense of repeated evaluation, it is possible to minimize the maximum absolute or relative error of a polynomial fit for any given number of terms.

The *Chebyshev theorem* [6] gives also a characterization of the minimax approximation to a function. This theorem states that for any $f(x)$ in closed interval $[a, b]$

a polynomial function p^* is the minimax degree-n approximation to this function if and only if there exist at least $n+2$ points in $[a, b]$, denoted as x_i with i from 0 to $n+1$, $(a \leq x_0 \leq x_1 \leq \cdots \leq x_{n+1} \leq b)$ at which $f(x_i) - p^*(x_i)$ attains its maximum absolute value, namely $\|f - p^*\|_\infty$ with alternating signs, this is:

$$p^*(x_i) - f(x_i) = (-1)^i \left[p^*(x_i) - f(x_0)\right] = \pm\|f - p^*\|_\infty$$

Example E.2 **Using minimax approximation method, find a degree-1 polynomial which best approximates the function $f(x) = e^x$ in terms of minimax approximation in [0, 1].**

Considering the function $f(x) = e^x$ we have to find the best linear approximation$p^*(x) = a_0 + a_1 x$. The maximum error must be attained at exactly $n+2$ points according to the *Chebyshev theorem*. For this case, the maximum error must be attained at three points, x_0, x_1 and x_2 in [0, 1]. The minimax theory tell us that we must have:

$$\epsilon = \max_{0 \leq x \leq 1} \left(e^x - (a_0 + a_1 x)\right)$$

Two of these three points where the maximum error must be attained are $x_0 = 0$ and $x_2 = 1$, due to the convexity of the exponential function, and the third point x_1 is in (0, 1), and the sign of the error alternates. This is:

$$\epsilon = e^0 - (a_0 + a_1 \cdot 0) = 1 - a_0 \tag{E.1}$$

$$-\epsilon = e^{x_1} - (a_0 + a_1 \cdot x_1) \tag{E.2}$$

$$\epsilon = e^1 - (a_0 + a_1 \cdot 1) = e - (a_0 + a_1) \tag{E.3}$$

We have four variables and there are only three equations, so we add one more equation. Since $q(x) = f(x) - p^*(x)$ has a local maximum at x_1, we have that the derivate of $q(x)$, $q'(x) = e^x - a_1$, is equal to zero for $x = x_1$. This gives:

$$q'(x_1) = e^{x_1} - a_1 = 0 \tag{E.4}$$

Equations (E.1), (E.2), (E.3) and (E.4) allow to get the solution for the required approximation:

- Equaling (E.1) and (E.3):

$$a_1 = e - 1$$

- Replacing the value of a_1 in (E.4):

$$e^{x_1} = a_1$$

$$x_1 = \ln(e - 1)$$

- Adding the Eqs. (E.2) and (E.3):

$$e - a_1 - 2a_0 + e^{x_1} - a_1 \cdot x_1 = 0$$

Replacing the values of a_1, x_1 and e^{x_1}, it is obtained:

$$a_0 = \frac{\mathrm{e} - (\mathrm{e} - 1)\ln(\mathrm{e} - 1)}{2}$$

- From Eq. (E.1):

$$\epsilon = 1 - a_0$$

Thus, the solution of this system of equations is:

$$a_1 \approx 1.718282$$

$$a_0 \approx 0.894067$$

$$x_1 \approx 0.541325$$

$$\epsilon \approx 0.105933$$

The obtained linear approximation is $p^*(x) = 1.718282x + 0.894067$ and the largest approximation error is 0.105933. Figure E.3 shows the graph the exponential function and its degree-1 minimax polynomial approximation $p^*(x)$. ■

E.3 Tang's Algorithm for the Exponential Function

The algorithm proposed by Tang [4] follows the guideline described at Sect. 8.6.2 considering the evaluation of the exponential function in IEEE floating-point arithmetic. First, for an integer $L \geq 1$ chosen beforehand, the input argument x is reduced to r in the interval:

$$\left[-\frac{\ln 2}{2^{L+1}}, \frac{\ln 2}{2^{L+1}}\right]$$

Thus, the input can be expressed as:

$$x = \left(m2^L + j\right)\frac{ln\,2}{2^L} + r, \quad j = 0, 1, 2, \ldots, 2^L - 1$$

where m and j are integers, with $j = 0, 1, 2, \ldots, 2^L - 1$. For the evaluation of the exponential function in IEEE double-precision floating-point arithmetic, Tang selected $L = 5$, and thus, the input argument x is reduced to the interval $\left[-\frac{\ln 2}{64}, \frac{\ln 2}{64}\right]$. Then, it is necessary to obtain the integers m and j, with $0 \leq j \leq 31$, such that the input x is expressed as:

$$x = (32m + j)\frac{\ln 2}{32} + r$$

That can also be expressed as:

$$x = (m \ln 2) + \frac{j \ln 2}{32} + r$$

In order to get a more accurate computation, the reduced argument r is represented as the sum of two floating-point numbers, R_1 and R_2, thus, being $r = R_1 + R_2$. Then,

$$x = (m \ \ln 2) + \frac{j \ln 2}{32} + (R_1 + R_2)$$

From the previous expression, the exponential function can be developed as follows:

$$\begin{aligned} e^x &= e^{\left((m\ \ln 2) + \frac{j \ln 2}{32} + (R_1 + R_2)\right)} \\ &= e^{\left((\ln 2^m) + \left(\ln 2^{\frac{j}{32}}\right) + (R_1 + R_2)\right)} \\ &= 2^m \cdot 2^{\frac{j}{32}} \cdot e^{(R_1 + R_2)} \end{aligned}$$

For the implementation of this algorithm, it is important to consider that the values of $\cdot 2^{\frac{j}{32}}$ for $0 \leq \mathrm{j} \leq 31$ are prestored constants, and multiplication by 2^m is fast. Hence, we just need to calculate $e^{(R_1+R_2)}$ for $(R_1 + R_2) \in \left[-\frac{\ln 2}{64}, \frac{\ln 2}{64}\right]$. This is done by a polynomial approximation $p(r) \approx e^{(R_1+R_2)} - 1$, where:

$$p(r) = r + a_1 r^2 + a_2 r^3 + \ldots + a_n r^{n+1}$$

Table E.1 Exceptional cases for input arguments for double precision implementation

Input argument x	Output
NaN	NaN
$+\infty$	$+\infty$
$-\infty$	$-\infty$
$>2610 \cdot \ln^2$	+inf overflow signal or +0 underflow signal
$<2^{-54}$	$1+x$

Finally, the exponential function is reconstructed as follows:

$$e^x = 2^m \cdot 2^{\frac{j}{32}} \cdot (p(r) + 1)$$

The main objective of this algorithm is isolating m and j, and evaluating the approximating polynomial. For this algorithm, large-degree polynomials with rather small lookup-tables are used. This paradigm is useful for high precisions (for example double precision) where the required size of lookup-tables can be incredibly large for low-degree piecewise polynomials to be implemented.

On the other hand, an accurate implementation of a function under this algorithm depends on the errors on it and how they can be controlled. We can distinguish three types of errors: error in reduction, error in approximation and rounding error. Error in reduction can be easily controlled by means of the arithmetic and the maximum magnitude for a legal input argument. Error in approximation can be also controlled and thus reduced by using high order polynomials. Besides, the rounding errors can be minimized considering a small price in storage. Thus, Tang demonstrated that the implementation will have an overall error below 0.54 ulp over practically the entire input domain, without using extra-piece arithmetic.

The implementation of this algorithm requires the following steps.

Step 1 **To consider the exceptional cases**

First, it is necessary to detect some cases for the input argument x and to force a specific output. Table E.1 resume all the cases studied by Tang [4].

Step 2 **To perform the argument reduction accurately**

As it was mentioned above, r is represented as $r = R_1 + R_2$. To make the computation more accurate, $R_1 \ll R_2$, an thus, $R_1 + R_2$ approximates r to a precision higher than the working precision. For the computation of the numbers R_1 and R_2 it is necessary to calculate first some values:

- $N = \text{round}\left(x \cdot \frac{32}{\ln 2}\right)$
  - $N_2 = N \mod 32$

Table E.2 Constants for double precision Tang's algorithm (step 1 and step 2) [4]

Constant	Decimal representation	IEEE double-precision format (hexadecimal representation)
L_1	0.2166084939017309800	3F962E42 FEF00000
L_2	$0.2325192846878874000 \times 10^{-11}$	3D8473DE 6AF278ED
a_1	0.5000000000000000000	3FE00000 00000000
a_2	0.1666666666652608700	3FC55555 55548F7C
a_3	$0.4166666666622608000 \times 10^{-1}$	3FA55555 55545D4E
a_4	$0.8333367984342196000 \times 10^{-2}$	3F811115 B7AA905E
a_5	$0.1388894908637772000 \times 10^{-2}$	3F56C172 8D739765

- $N_1 = N - N_2$

- $\frac{\ln 2}{32}$ is approximated to $L_1 + L_2$, in order to get a precision much higher than the working one:
  - L_1 has a few trailing of zeros,
  - $L_2 \ll L_1$

Four double precision, these constants L_1 and L_2 are shown in Table E.2 [4]. The numbers R_1 and R_2 are computed as follows:

$$R_2 = -N \cdot L_2$$

- If $N < 2^9$:

$$R_1 = x - N \cdot L_1$$

- Else:

$$R_1 = (x - N \cdot L_1) - N_2 \cdot L_1$$

The values m and j are computed as:

$$m = N_1/32$$

$$j = N_2$$

Step 3 **To compute the approximation polynomial**

The polynomial $p(r)$ is computed as follows:

$$r = R_1 + R_2$$

$$q(r) = r \cdot r(a_1 + r \cdot (a_2 + r \cdot (\cdots + r \cdot a_n)))$$

$$p(r) = R_1 + (R_2 + q(r))$$

The coefficients a_i were obtained from a Remez algorithm [4], and are shown in Table E.2.

Step 4 **To compute the reconstruction**

The reconstruction expression required one of the values $s_j = 2^{\frac{j}{32}}$, with $j = 0, 1, 2, \ldots, 31$. These values are precomputed and represented by the sum of two working-precision numbers s_{j1} and s_{j2}, that can be stored in look-up tables.This sum approximates $2^{\frac{j}{32}}$ to roughly double the working precision. The constants s_{j1} and s_{j2} for $j = 0, 1, 2, \ldots, 31$ are shown in Table E.3 [4]. The computation of the reconstruction is:

$$s_j = s_{j1} + s_{j2}$$
$$e^x = 2^m \cdot \left(s_{j1} + \left(s_{j2} + s_j \cdot p(r)\right)\right)$$

Example E.3 **Calculate exp(0.56998) using the Tang's algorithm for double precision.**

The representation of 0.56998 in IEEE single-precision format using hexadecimal representation is 3FE23D46 B26BF87.

We have to follow the steps detailed above, using the decimal value of the data, but operating using IEEE single-precision format.

Step 1

The value 0.56998 is out of the exceptional cases.

Step 2

$$N = \text{round}\left(0.56998 \cdot \frac{32}{\ln 2}\right) = 26$$
$$N_2 = N \mod 32 = 26$$
$$N_1 = N - N_2 = 0$$

Table E.3 Constants for double precision Tang's algorithm (step 3), IEEE double-precision format (hexadecimal representation) [4]

j	s_{j1}	s_{j2}
0	3FF00000 00000000	00000000 00000000
1	3FF059B0 D3158540	3D0A1D73 E2A475B4
2	3FF0B558 6CF98900	3CEEC531 7256E308
3	3FF11301 D0125B40	3CF0A4EB BF1AED93
4	3FF172B8 3C7D5140	3D0D6E6F BE462876
5	3FF1D487 3168B980	3D053C02 DC0144C8
6	3FF2387A 6E756200	3D0C3360 FDGD8E0B
7	3FF29E9D F51FDEC0	3D009612 E8AFAD12
8	3FF306FE 0A31B700	3CF52DE8 D5A46306
9	3FF371A7 373AA9C0	3CE54E28 AA05E8A9
10	3FF3DEA6 4C123400	3D011ADA 0911F09F
11	3FF44E08 60618900	3D068189 B7A04EF8
12	3FF4BFDA D5362A00	3D038EA1 CBD7F621
13	3FF5342B 569D4F80	3CBDF0A8 3C49D86A
14	3FF5AB07 DD485400	3D04AC64 980A8C8F
15	3FF6247E B03A5580	3CD2C7C3 E81BF4B7
16	3FF6A09E 667F3BC0	3CE92116 5F626CDD
17	3FF71F75 E8EC5F40	3D09EE91 B8797785
18	3FF7A114 73EB0180	3CDB5F54 408FDB37
19	3FF82589 994CCE00	3CF28ACF 88AFAB35
20	3FF8ACE5 422AA0C0	3CFB5BA7 C55A192D
21	3FF93737 B0CDC5C0	3D027A28 0E1F92A0
22	3FF9C491 82A3F080	3CF01C7C 46B071F3
23	FFA5503 B23E2540	3CFC8B42 4491CAF8
24	3FFAE89F 995AD380	3D06AF43 9A68BB99
25	3FFB7F76 F2FB5E40	3CDBAA9E C206AD4F
26	3FFC199B DD855280	3CFC2220 CB12A092
27	3FFCB720 DCEF9040	3D048A81 E5E8F4A5
28	3FFD5818 DCFBA480	3CDC9768 16BAD9B
29	3FFDFC97 337B9B40	3CFEB968 CAC39ED3
30	3FFEA4AF A2A490C0	3CF9858F 73A18F5E
31	3FFF5076 5B6E4540	3C99D3E1 2DD8A18B

Then we use the value of the constants L_1 and L_2 shown in Table 8.15 to calculate R_1 and R_2:

$$R_1 = x - N \cdot L_1 = 0.006797915855499$$
$$R_2 = -N \cdot L_2 = -6.045501401885073 \times 10^{-11}$$

The values m and j are computed:

$$m = \frac{N_1}{32} = 0$$
$$j = N_2 = 26$$

Step 3

We first calculate the value r:

$$r = R_1 + R_2 = 0.006797915795044$$

Then, we calculate the polynomial $q(r)$ for this r value and using the coefficients a_i of Table E.2:

$$q(0.006797915795044) = 2.315827584054626 \times 10^{-5}$$

Finally, we calculate the polynomial $p(r)$:

$$p(0.006797915795044) = 0.006821074070885$$

Step 4

From step 1 we obtained $j = 26$. Thus, from Table E.3:
$s_{26,1} = 1.7562521603732932362618157640\,1$ (3FFC199B DD855280)
$s_{26,2} = 6.2468503448553655751467998306\,1 \times 10^{-15}$(3CFC2220 B12A092)
Using the reconstruction formula:

$$e^x = 2^m \cdot \left(s_{26,1} + \left(s_{26,2} + s_{26} \cdot p(r)\right)\right)$$

and computing using the calculated valued for m, $s_{26,1}$, $s_{26,1}$, s_{26} and $p(r)$, we obtain:

$$e^{0.56998} = 1.768231686446358$$

The real value for $e^{0.56998}$ using octave is 1.768231686446358, which gives a difference of 0 between the real value and the approximation. ■

E.4 Conclusion

This Appendix has presented some particular algorithms and methods for the approximation of special function, as well as the description of Chebyshev and Legendre sequences of orthogonal polynomials required for the development of method based on these sequences. Thus, it is an important support for the structures of the special functions presented in Chap. 8.

E.5 References

1. Gautschi W., *Numerical Analysis : an Inroduction,* Birkhäuser, Boston, MA, 1997.
2. Gautschi W., Bachelis G.H., "An accurate elementary mathematical library for the IEEE floating point standard", *ACM Transactions on Mathematical Software*, vol. 17, no. 1, pp. 26–45, March, 1991.
3. Stone, Marshall H. "The Generalized Weierstrass Approximation Theorem." *Mathematics Magazine*, vol. 21, no. 5, pp. 237–254, 1948.
4. P.T.P. Tang. "Table driven implementation of the exponential function in IEEE float-ing point arithmetic," *ACM Transactions on Mathematical Software,* vol. 15, no. 2, pp. 144–157, 1989.
5. Tjalling J. Ypma, "Historical development of the Newton-Raphson method", *SIAM Review*, vol. 37, no. 4, pp. 531–551, 1995.
6. Touhey P. "Chebyshev's Theorem: A Geometric Approach", *The College Mathematics Journal*, 26:2, 139–141, 1995.

Index

A. Lloris Ruiz et al., *Arithmetic and Algebraic Circuits*, Intelligent Systems Reference Library 201, https://doi.org/10.1007/978-3-030-67266-9

D

E

F

N

O